3rd International Meeting for Researchers in Materials and Plasma Technology (IMRMPT 2015) & 1st Symposium on Nanoscience and Nanotechnology 2015

AA002698

Journal of Physics: Conference Series Volume 687

Bucaramanga, Colombia
4 – 9 May 2015

Editor:

Ely Dannier V. Nino

ISBN: 978-1-5108-2095-1
ISSN: 1742-6588

Printed from e-media with permission by:

Curran Associates, Inc.
57 Morehouse Lane
Red Hook, NY 12571

Some format issues inherent in the e-media version may also appear in this print version.

Printed by Curran Associates, Inc. (2016)

For permission requests, please contact the Institute of Physics
at the address below.

Institute of Physics
Dirac House, Temple Back
Bristol BS1 6BE UK

Phone: 44 1 17 929 7481
Fax: 44 1 17 920 0979

techtracking@iop.org

Additional copies of this publication are available from:

Curran Associates, Inc.
57 Morehouse Lane
Red Hook, NY 12571 USA
Phone: 845-758-0400
Fax: 845-758-2634
Email: curran@proceedings.com
Web: www.proceedings.com

Table of contents

Volume 687

3rd International Meeting for Researchers in Materials and Plasma Technology (IMRMPT) & 1st Symposium on Nanoscience and Nanotechnology 4–9 May 2015, Bucaramanga, Colombia

Accepted papers received: 8 January 2016
Published online: 12 February 2016

Preface

011001

3rd International Meeting for Researchers in Materials and Plasma Technology (IMRMPT) & 1st Symposium on Nanoscience and Nanotechnology OPEN ACCESS

011002

Peer review statement OPEN ACCESS

Papers

012001

Operational management of offshore energy assets OPEN ACCESS A J Kolios and M Martinez Luengo pg. 1

012002

Development of an experimental methodology for assessing the growth of scale ($CaCO_3$) in pipelines OPEN ACCESS M García, J Vega, Y Pineda and E Vera pg. 8

012003

Mathematical-statistical model for predicting the tendency of the mechanical integrity of a buried pipeline OPEN ACCESS H Aponte Betancur, E vera Lopez and Y Pineda Triana pg. 12

012004

Electrochemical characterization of the steel wire used as reinforcement in the conductors transmission networks electricity nitride by ion implantation OPEN ACCESS J J Castro Maldonado, H J Dulcé Moreno and W Aperador pg. 16

012005

Artificially modulated hard coatings produced with a vacuum arc evaporator OPEN ACCESS P A Tsygankov, F F Parada Becerra, V D Dugar-Zhabon, A Plata and E D V-Niño pg. 20

012006

Obtaining Au thin films in atmosphere of reactive nitrogen through magnetron sputtering OPEN ACCESS J H Quintero, R Ospina and A Mello pg. 24

012007

Comparison of three systems of solar water heating by thermosiphon OPEN ACCESS E Hernández and R E Guzmán pg. 29

012008

Effects of texturization due to chemical etching and laser on the optical properties of multicrystalline silicon for applications in solar cells OPEN ACCESS D Vera, J Mass, M Manotas, R Cabanzo and E Mejia pg. 33

012009

Estimation of incident solar radiation on the roof of the cultural and sports university centre of the Foundation University Los Libertadores OPEN ACCESS S A Jiménez, V M Carrillo and L C Rátiva pg. 38

012010

Objects tracking using a periodic array of volume: numerical evaluation OPEN ACCESS J A Sanchez Daza, M L Molina Prado and N A Arias Hernández pg. 42

012011

Estimate of the diameter of surface defects on a pattern specimen made of Plexiglas through active infrared thermography OPEN ACCESS M H Medina-Barreto, J M Giraldo, A L Ramírez-Hurtado, L Llamosa and B Cruz pg. 46

012012

HOS cell adhesion on Ti_6Al_4V surfaces texturized by laser engraving OPEN ACCESS A Sandoval Amador, H Carreño Garcia, P Escobar Rivero, D Y Peña Ballesteros and H A Estupiñán Duran pg. 50

012013

Design of TEOS-GPTMS sol-gel coatings on rare-earth magnesium alloys employed in the manufacture of orthopaedic implants OPEN ACCESS L M Rueda, C Nieves, C A Hernández Barrios, A E Coy and F Viejo pg. 54

012014

Corrosion resistance of multilayer hybrid sol-gel coatings deposited on the AISI 316L austenitic stainless steel OPEN ACCESS Y T Caballero, E A Rondón, L Rueda, C A Hernández Barrios, A Coy and F Viejo pg. 58

012015

Influence of alumina and titanium dioxide coatings on abrasive wear resistance of AISI 1045 steel OPEN ACCESS A Santos, A Remolina and J Marulanda pg. 62

012016

Formation of fouling deposits on a carbon steel surface from Colombian heavy crude oil under preheating conditions OPEN ACCESS D A Muñoz Pinto, S M Cuervo Camargo, M Orozco Parra, D Laverde, S García Vergara and C Blanco Pinzon pg. 66

012017

Corrosive effect of the type of soil in the systems of grounding more used (copper and stainless steel) for local soil samples from the city of Tunja (Colombia), by means of electrochemical techniques OPEN ACCESS L Guerrero, Y Salas and J Blanco pg. 70

012018

Electrochemical evaluation of the corrosion behaviour for structural steel SAE 1005 exposed to two different atmospheres (urban and industrial) and comparison with atmospheric corrosion gravimetric measurements OPEN ACCESS Y Salas, L Guerrero, H Castiblanco, C Torres and A Palacios pg. 74

012019

Presence of nitrate NO_3 a ects animal production, photocalysis is a possible solution OPEN ACCESS Heli Barba-Molina, J Barba-Ortega and M R Joya pg. 78

012020

Alumina and quartz as dielectrics in a dielectric barrier discharges DBD system for CO_2 hydrogenation OPEN ACCESS E Y Mora, A Sarmiento and E Vera pg. 81

012021

Evaluation of the stiffness and friction of Ti6Al4V ELI treated by glow discharge nitriding OPEN ACCESS J R Tavera, D Y Peña Ballesteros and H A Estupiñán Duran pg. 86

012022

Simulation of the magnetic field generated by wires with stationary current and magnets with constant magnetization applied to the mirror trap, minimum-B and zero-B OPEN ACCESS M T Murillo and O Otero pg. 90

012023

Analytical study of the magnetic field generated by multipolar magnetic configuration OPEN ACCESS M T Murillo Acevedo, V D Dugar-Zhabon and O Otero pg. 96

012024

Development of a sintering methodology through abnormal glow discharge for manufacturing metal matrix composites OPEN ACCESS S Pérez, Y Pineda, A Sarmiento and A López pg. 102

012025

Effect of glow discharge sintering in the properties of a composite material fabricated by powder metallurgy OPEN ACCESS A Cardenas, Y Pineda, A Sarmiento Santos and E Vera pg. 106

012026

Diffraction pattern by nanometric thin films under illumination of an orbital angular momentum beam with integer topological charge OPEN ACCESS J H Mendoza, C F Díaz, C H Acevedo and Y Torres pg. 110

012027

Gas-phase supersaturation effects on morphology properties of ZnO nano and microstructures grown by PVT OPEN ACCESS D N Montenegro, M C Martínez Tomas, V Muñoz Sanjosé and V Sallet pg. 115

012028

Effects of post-growth thermal annealing on room temperature pulsed laser deposited ZnO thin films OPEN ACCESS D Padilla Rueda, J M Vadillo and J J Laserna pg. 119

012029

Automated and inexpensive method to manufacture solid- state nanopores and micropores in robust silicon wafers OPEN ACCESS M Vega, P Granell, C Lasorsa, B Lerner and M Perez pg. 123

012030

Effect of forming technique $Bi_xSi_yO_z$ coatings obtained by sol- gel and supported on 316L stainless steel OPEN ACCESS J Bautista Ruiz, J Olaya Flórez and W Aperador pg. 127

012031

Evaluation of micro-abrasion-corrosion on SiO_2-TiO_2-ZrO_2 coatings synthesized by the sol-gel method OPEN ACCESS J Bautista Ruiz, W Aperador and J Caballero Gómez pg. 131

012032

Tribo-electrochemical characterization of hafnium multilayer systems deposited on nitride/vanadium nitride AISI 4140 steel OPEN ACCESS M Mora, E Vera and W Aperador pg. 135

012033

Chromatographic analysis of phytochemicals components present in mangifera indica leaves for the synthesis of silver nanoparticles by $AgNO_3$ reduction OPEN ACCESS D Martínez-Bernett, A Silva-Granados, S N Correa-Torres and A Herrera pg. 139

012034

Evaluation of the photocatalytic activity of iron oxide nanoparticles functionalized with titanium dioxide OPEN ACCESS A Herrera, A Reyes and J Colina-Márquez pg. 143

012035

Ultrasonic cavitation for obtainment of nanometric sized particles OPEN ACCESS A Santos, R Guzmán, J Espinosa and J Estrada pg. 147

012036

Raman spectroscopy of ZnMnO thin films grown by pulsed laser deposition OPEN ACCESS S Orozco, H Riascos and S Duque pg. 151

012037

Growth of PEO ceramic coatings on AA 2024-T3 aluminium alloy OPEN ACCESS S Forero Sotomonte, C Blanco Pinzon and S García Vergara pg. 155

012038

Electrochemical preparation and structural characterization of $CuGaSe_2$ thin films OPEN ACCESS L Manfredy, O P Marquez, S A Lopez-Rivera, J Marquez, Y Martínez and D A Miranda pg. 159

012039

Analysis of the influence of the applied voltage and the scan speed in the atomic force microscopy local oxidation technique OPEN ACCESS C Mendoza, A Plata, Z Lizarazo and C A Chacón pg. 164

012040

Structural and impedance analysis of Co-doped SrTiO3 perovskite OPEN ACCESS E Echeverri and O Arnache pg. 167

012041

Study of optical and structural properties of CZTS thin films grown by co-evaporation and spray pyrolysis OPEN ACCESS R Moreno, E A Ramirez and G Gordillo Guzmán pg. 171

012042

Synthesis of an air-working trilayer artificial muscle using a conductive cassava starch biofilm (manihot esculenta, cranz) and polypyrrole (PPy) OPEN ACCESS Y E Núñez D, Á A Arrieta A, J A Segura B and S D Bertel H pg. 176

012043

First principles study of structural, electronic and magnetic properties of magnesium
OPEN ACCESS G P Abdel Rahim, J A Rodríguez M and M G Moreno-Armenta pg. 179

012044

Quicklime as an alternative in the photodegradation of contaminants OPEN ACCESS M R Joya, J H Bautista Ruiz and A M Raba pg. 184

012045

Coal tar, material used in soil improvement for use in road engineering OPEN ACCESS R Ochoa Díaz, A Montañez and J Cuentas pg. 188

012046

Synthesis of praseodymium doped cerium oxides by the polymerization-combustion method for application as anodic component in SOFC devices OPEN ACCESS A F Cruz Pacheco, J A Gómez Cuaspud and E Vera López pg. 191

012047

Structural properties, electric response and magnetic behaviour of $La_2SrFe_2CoO_9$ triple complex perovskite OPEN ACCESS F Casallas, E Vera, D Landínez, C Parra and J Roa pg. 195

012048

Theoretical investigation of GaN carbon doped OPEN ACCESS M J Espitia Rico, M G Moreno Armenta, J A Rodríguez and N Takeuchi pg. 200

012049

Synthesis of (azelaic-co-dodecanedioic) polyanhydride by microwave technique OPEN ACCESS M Gutiérrez, C Sierra, M Acevedo Morantes and A P Herrera pg. 204

012050

Mercury removal in wastewater by iron oxide nanoparticles OPEN ACCESS E Vélez, G E Campillo, G Morales, C Hincapié, J Osorio, O Arnache, J I Uribe and F Jaramillo pg. 208

012051

Structural and magnetic properties $CuAl_{1-x}Cr_xS_2$ alloys OPEN ACCESS C Ortega López, G Casiano Jiménez and M J Espitia pg. 212

012052

Electronic and magnetic properties GaN/MnN/GaN and MnN/GaN/MnN interlayers OPEN ACCESS C Ortega López, G Casiano Jímenez and M J Espitia pg. 216

012053

Structural and electronic properties of $Er_{0.5}Lu_{0.5}B_2$ and LuB_2 compounds by using the PBE0 hybrid functional OPEN ACCESS C Ortega, D Rasero and L C Sánchez pg. 220

012054

A study ab-initio of relative stability and electronic properties of substitutions of C by B and N atoms in the w-BN compound OPEN ACCESS G Casiano Jimenéz, C Ortega López and M J Espitia pg. 224

012055

Feasibility for the use of coal tar as a new material for road surfaces (pavement) construction OPEN ACCESS M A Romero Farfán, H E Murillo Vega and F A Trujillo Pinto pg. 228

012056

Effect of sulfonation and diethanolamine addition on the mechanical and physicochemical properties of SEPS copolymer OPEN ACCESS D Patiño, E Correa and M Acevedo-Morantes pg. 232

012057

A structural study of the hydrogen absorption properties by replacing vanadium with zirconium in metal alloys OPEN ACCESS D Bellon, A Martinez, D Barreneche and D S dos Santos pg. 236

012058

Development of a web application for estimate the total charge in an electric discharge OPEN ACCESS F H Vera-Rivera, V D Dugar-Zhabon, H J Dulce-Moreno and E D V-Niño pg. 240

012059

PIC numerical study of ECR plasmas confinement in a minimum-B and zero-B magnetic traps with GPU OPEN ACCESS M T Murillo Acevedo and V D Dugar-Zhabon pg. 244

012060

Influence of thermal residual stress on behaviour of metal matrix composites reinforced with particles OPEN ACCESS R E Guzmán and E Hernández Arroyo pg. 250

012061

Numerical modeling of high temperature fracture of metallic composites OPEN ACCESS E D Cendales and A García pg. 254

012062

Phase diagram of the Kondo lattice model with a superlattice potential OPEN ACCESS J Silva-Valencia, R Franco and M S Figueira pg. 258

012063

Spiral to ferromagnetic transition in a Kondo lattice model with a double-well potential OPEN ACCESS R C Caro, R Franco and J Silva-Valencia pg. 262

012064

Density matrix renormalization group study of the Anyon-Hubbard model OPEN ACCESS J Arcila-Forero, R Franco and J Silva-Valencia pg. 266

012065

Mott insulator and superfluid phases in bosonic superlattices OPEN ACCESS G J Cruz, R Franco and J Silva-Valencia pg. 270

012066

Quantum phases of AB_2 fermionic chains OPEN ACCESS L S Murcia-Correa, R Franco and J Silva-Valencia pg. 274

012067

Computational modeling of the mechanism of hydrogen embrittlement (HE) and stress corrosion cracking (SCC) in metals OPEN ACCESS E D Cendales, F A Orjuela and O Chamarraví pg. 278

012068

Mathematical modeling and validation of the carburizing of low carbon steels OPEN ACCESS A García Mariaca, E D Cendales and O Chamarraví pg. 282

012069

Ab initio calculations of half-metallic ferromagnetism in $(VN)_1/(AlN)_1$, $(VN)_1/(GaN)_1$ and $(VN)_1/(InN)_1$ superlattices OPEN ACCESS M J Espitia Rico, J H Díaz F and C Ortega López pg. 286

012070

Computational two-dimensional modeling of the stress intensity factor in a cracked metallic material OPEN ACCESS J E Rolón, E D Cendales and I M Cruz pg. 290

012071

Mott lobes evolution of the spin-1 Bose-Hubbard model OPEN ACCESS A F Hincapie-F, R Franco and J Silva-Valencia pg. 293

012072

2D electrostatic PIC algorithm for laser induced studying plasma in vacuum OPEN ACCESS C A. Álvarez, H Riascos and C Gonzalez pg. 297

012073

Control of atom-atom entanglement by cavity detuning OPEN ACCESS O Calderón, M R Joya and K M Fonseca Romero pg. 302

012074

From sequential to correlated tunneling of two bosons OPEN ACCESS R Avella, R R Rey González and K M Fonseca Romero pg. 305

012075

Morphological analysis of carbon steels using fractal geometry OPEN ACCESS D A Prada, C P González, P E Vera and M A Álvarez pg. 309

012076

Self-consistent simulation of an electron beam for a new autoresonant x-ray generator based on TE_{102} rectangular mode OPEN ACCESS V D Dugar-Zhabon, E A Orozco and A M Herrera pg. 313

012077

3D electromagnetic simulation of spatial autoresonance acceleration of electron beams OPEN ACCESS V D Dugar-Zhabon, J D González and E A Orozco pg. 319

012078

Virtual experimentation in electromagnetism, mechanics and optics: web-based learning OPEN ACCESS J H Escobar, H Sánchez, J R Beltrán, J De la Hoz and J D González pg. 325

012079

Donor-assisted resonant electron tunneling in double-barrier heterostructures under tilted magnetic fields: A theoretical study OPEN ACCESS H Paredes, C L Beltrán Ríos and J C Arce pg. 329

012080

Effect of a high intensity laser beam on impurity binding energy in a nanocone OPEN ACCESS H Paredes, C L Beltrán Ríos and W Gutíerrez pg. 333

012081

Current density and state density in diluted magnetic semiconductor nanostructures OPEN ACCESS S T Pérez Merchancano, H Paredes Gutiérrez and J A Zuñiga pg. 338

012082

Biosynthesis and characterization of gold nanoparticles using extracts of tamarindus indica L leaves OPEN ACCESS S N Correa, A M Naranjo and A P Herrera pg. 343

012083

Borax as flux on sintering of iron Ancor Steel 1000® under glow discharge OPEN ACCESS H G Ariza Suarez, A Sarmiento Santos and C A Ortiz Otálora pg. 347

012084

Synthesis and characterization of the oxide nanoparticles obtained by the polymeric precursor method OPEN ACCESS A M Raba, J Bautista and E Murillo pg. 351

012085

Energy transfer in two-wave mixing quasi-degenerated in photorefractive crystals OPEN ACCESS B Barrera, N A Arias Hernández, M L Molina Prado and M Tebaldi pg. 355

012086

Structure, mechanical and thermal behaviour of mixtures of polyester resin and dental ceramic waste OPEN ACCESS G Peña Rodríguez, L Martínez Maldonado and H J Dulce Moreno pg. 360

012087

Structural and magnetic characterization of the new $GdMn_{1-x}Fe_xO_3$ perovskite material OPEN ACCESS J A Cardona Vasquez, D A Landínez Téllez, C A Collazos and J Roa Rojas pg. 364

012088

Hardness and compression resistance of natural rubber and synthetic rubber mixtures OPEN ACCESS J M Arguello and A Santos pg. 368

012089

Production a characterization of new system superconductor $TR_3Ba_8Cu_{11}O_\delta$ OPEN ACCESS S Sandoval Gutiérrez, I Supelano G, J Roa-Rojas and C A Parra Vargas pg. 372

012090

Experimental set up of a magnetoelectric measuring system operating at different temperatures OPEN ACCESS K Gil, J Gil, B Cruz, A Ramirez, M Medina and J Torres pg. 376

012091

Structural study of yttrium substituted $BiFeO_3$ OPEN ACCESS J A Mejía Gómez, C Canaria, R Ochoa Burgos, C A Ortiz, G I Supelano and C A Parra Vargas pg. 380

012092

Prediction of the quality of coke obtained from vacuum residues by using spectroscopy infrared FTIR-ART OPEN ACCESS A Y León, N A Rodríguez, E Mejía and R Cabanzo pg. 384

012093

Prediction and extension of curves of distillation of vacuum residue using probability functions OPEN ACCESS A Y León, P A Riaño and D Laverde pg. 388

012094

Reactivity study on thermal cracking of vacuum residues OPEN ACCESS A Y León, S D Díaz, R C Rodríguez and D Laverde pg. 392

012095

Treatment of vacuum residues in hydroconversion conditions OPEN ACCESS A Y León, D L Mendoza, J O Espinosa and D Laverde pg. 396

012096

Structural magnetic and Mossbauer study of ^{57}Fe doped $La_{2/3}Ca_{1/3}Mn_{1-x}Fe_xO_3$ ($0 \geq x \geq 0.1$) manganites OPEN ACCESS L C Sánchez, D Durango and O Arnache pg. 400

012097

Spectroscopy characterization and quantum yield determination of quantum dots OPEN ACCESS S N Contreras Ortiz, E Mejía Ospino and R Cabanzo pg. 404

012098

The relationship between ferroelectric domain pattern and properties of PLT21 ceramics OPEN ACCESS F A Londoño, J A Eiras and D Garcia pg. 408

012099

Influence of the pH of the synthesis using sol-gel method on the structural and optical properties of TiO_2 OPEN ACCESS J Jaramillo, B A Garzón and L Tirado Mejía pg. 411

012100

Measurement of nonlinear refractive index of organic materials by z-scan OPEN ACCESS M Fontalvo, A Garcia, S Valbuena and F Racedo pg. 415

012101

Considerations on electrical impedance measurements of electrolyte solutions in a four-electrode cell OPEN ACCESS C V Chaparro, L V Herrera, A M Meléndez and D A Miranda pg. 419

012102

Magnetic effect in viscosity of magnetorheological fluids OPEN ACCESS H A Fonseca, E Gonzalez, J Restrepo, C A Parra and C Ortiz pg. 423

012103

Measurement of the thermal expansion coefficient of Guadua angustifolia-Kunth using the photoacoustic technique OPEN ACCESS A Bedoya, F Gordillo Delgado and E Marin pg. 427

012104

Synthesis, structural and morphological characterization of the perovskite $LaYbO_3$ OPEN ACCESS A Cabrera Ramírez, F E Charry Pastrana, J Roa Rojas, D A Landinez Tellez and F Fajardo pg. 431

012105

Influence of thermal spraying parameters on the corrosion resistance of aluminium oxide coatings deposited on steel 1020 OPEN ACCESS Y Salas, E Vera, M Moreno and Y Pineda pg. 435

012106

Evaluation of the structural steel corrosion behaviour, covered with epoxy-type paints, by means of electrochemical DC techniques OPEN ACCESS Y Salas, L Guerrero, R Martinez, T Chicino and C Devia pg. 439

012107

Processing of spent Ni-MH batteries for the recovery of cobalt, nickel and rare earth elements bearing materials by means of a chemical and electrochemical sequential process OPEN ACCESS P Delvasto, R Orta Rodríguez and S Blanco pg. 443

012108

Geometric, electronic and intrinsic chemical reactivity properties of mono- and bi-substituted quinoline derivatives for the ground state in gas phase OPEN ACCESS O L Neira Bueno, L Hincapié H and C García Madrid pg. 447

012109

AC magnetic susceptibility and influence of heat treatment on obtaining the nanocrystalline structure for the amorphous alloy of $Fe_{37}Co_{35}Nb_6B_{11}Si_{10}Cu_1$ composition OPEN ACCESS A Orozco, Z Caamaño and A Rosales pg. 452

012110

Changes on electrical and structural properties of polyaniline and polypyrrol by mullite doping OPEN ACCESS C P González, A M Montaño, S E Estrada and C A Ortiz pg. 456

012111

Surface characterization of stainless HP-40 steel using laser induced μ-breakdown spectroscopy (μ -LIBS) OPEN ACCESS M Pinto, X Calderón, E Mejía Ospino, R Cabanzo and Juan C Poveda pg. 460

012112

Adsorption of vapreotide on gold colloids studied by surface enhanced Raman spectroscopy OPEN ACCESS J A Gómez, R Cabanzo and E Mejia Ospino pg. 464

012113

Physico-chemical characterization of products from vacuum oil under delayed coking process by infrared spectroscopy and chemometrics methods OPEN ACCESS L V Meléndez, R Cabanzo, E Mejía-Ospino and A Guzmán pg. 468

012114

Effects of substitutions of C atoms by Al and N in the w-AlN compound OPEN ACCESS J F Murillo, C Ortega and M J Espitia pg. 473

Preface

These proceedings present the written contributions of the participants of the 3rd International Meeting for Researchers in Materials and Plasma Technology (3rd IMRMPT) and the 1st Symposium on Nanoscience and Nanotechnology which was held from 4 to May 9, 2015 at the Dann Carlton Hotel Bucaramanga, Colombia, organized by the faculty of science of the Universidad Industrial de Santander (UIS) and the basic science department of the Universidad Pontificia Bolivariana. This was the third version of biennial meetings that began in 2011.

The five-day scientific program of the 3rd IMRMPT consisted of 24 Magisterial Conferences, 70 Oral Presentations, 185 Poster Presentations, 3 Courses and 1 Discussion Panel with the participation of undergraduate and graduate students, professors, researchers and entrepreneurs from Colombia, Russia, Germany, France, Spain, England, United States, Mexico, Argentina, Uruguay, Brazil, Venezuela, among others. Moreover, the objective of IMRMPT was to bring together national and international researchers in order to establish a network of scientific cooperation with a global impact in the area of the science and the technology of materials; to promote the exchange of creative ideas and the effective transfer of scientific knowledge, from fundamental research to innovation applied to industrial solutions and to advances in the development of new research allowing to increase the lifetime of the materials used in the industry by means of efficient transference of the knowledge between sectors academia and industry.

The topics covered in the 3rd IMRMPT include New Materials, Surface Physics, Structural Integrity, Renewable Energy, Online Process Control, Non Destructive Evaluation, Characterization of Materials, Laser and Hybrid Processes, Thin Films and Nanomaterials, Surface Hardening Processes, Wear and Corrosion/Oxidation, Plasma Applications and Technologies, Modelling, Simulation and Diagnostics, Biomedical Coatings, Surface Treatments and Surface Modification (Ion Implantation, Ion Nitriding, PVD, CVD). The editor hopes that those interested in the area of the science and the technology of materials and plasma enjoy the reading that reflect a wide range of topics.

It is a pleasure to thank the sponsors, partners and all the participants and contributors for making possible this international meeting of researchers.

It should be noted that the event organized by UIS, through of the research groups FITEK and CMN, was a very significant contribution to the national and international scientific community, achieving the interaction of different research groups from academia and business sector.

On behalf of the research groups FITEK and CMN of the UIS, we greatly appreciate the support provided by the Sponsors and Partners, who allowed to continue with the dream of research which contributing to Innovation and development in the present and future of Colombia.

The Editor

Ely Dannier V. Niño

Journal of Physics: Conference Series **687** (2016) 011001 doi:10.1088/1742-6596/687/1/011001

Organizing Committee
Eduardo Alberto Orozco Ospino (Chairman)
David Alejandro Miranda Mercado
Germán Moreno Arenas
Fernando Durán Flórez
Claudia Paulina González Cuervo
José José Barba Ortega
Ely Dannier V. Niño

National Scientific Committee
Dr. Héctor Jaime Dulce Moreno
Dr. Henry Riascos Landázuri
Dr. Juan José Pavón Palacio
Dra. Miryam Rincón Joya
Dr. Valeriy D. Dugar-Zhabon
Dr. Eduardo Alberto Orozco Ospino
Dr. David Alejandro Miranda Mercado
Dr. Jorge Humberto Martínez Téllez
Dr. Rafael Cabanzo Hernández
Dr. William Gutierrez Niño
Dr. Arturo Plata Gómez
Dr. Germán Moreno Arenas
Dr. Darío Yesid Peña Ballesteros
Dr. Alejandro David Martínez Amariz
Dr. Aduljay Remolina Millan
Dra. Claudia Paulina González Cuervo
Dr. (c) Cesar Augusto Rodríguez Adaime
Dr. (c) Jesús David González Acosta
MSc. Hermes Barrera Ávila
MSc. Fernando Durán Flórez
MSc. Carlos Javier Jaimes Ochoa
MSc. Luis Carlos Caicedo Barrera
MSc. Fredy Fabián Parada Becerra
MSc. Hernán Alfonso Garnica

International Scientific Committee
Dr. Petr Tsygankov - Russia
Dr. Jean Paul Allain – Unite State of America
Dr. Edmanuel Eduardo Torres Amarís – Unite State of America
Dr. Wilson Vesga Rivera – Unite Kingdom
Dr. Eduardo Méndez Morales - Uruguay
Dr. Rogelio Ospina Ospina - Brazil
Dr. Elvis Oswaldo López Meza - Brazil
Dr. Humberto José Jiménez - Venezuela
Dr. (c) Sergio Eduardo Rios Contesse – United Kingdom
Dr. (c) Jaime Andrés Pérez Taborda - España
Dr. (c) Manuel José Andrade Romero - Brasil
Dra. (c) Ana María Herrera Rodríguez - Germany
Dra. (c) Angela Lorena Cardenas Manosalva - Brazil
Dra. (c) Monica Arenas - Brazil
Dra. (c) Sugheis Liscano Durán - Venezuela

Journal of Physics: Conference Series **687** (2016) 011001 doi:10.1088/1742-6596/687/1/011001

Sponsors

- ➢ Universidad Industrial de Santander
- ➢ Universidad Pontificia Bolivariana
- ➢ Centro Colombiano de Tecnología
- ➢ Universidad del Magdalena
- ➢ Electrificadora de Santander

Partners

- ✓ Grupo de Investigación en Física y Tecnología del Plasma y Corrosión (FITEK)
- ✓ Grupo de Investigación en Tecnología del Plasma
- ✓ Centro de Materiales y Nanociencias (CMN)
- ✓ Universitaria de Investigación y Desarrollo
- ✓ Sociedad Colombiana de Ingeniería Física
- ✓ Asociación Colombiana de Microscopía
- ✓ Centro de Microscopía y Microanálisis
- ✓ Banco de la República de Colombia
- ✓ Hotel Dann Carlton Bucaramanga
- ✓ J & G Soluciones Industriales
- ✓ Canaán Multiservicios
- ✓ S & S Ingnieria S.A.S
- ✓ Vortex Company
- ✓ Coca-Cola
- ✓ Avianca

doi:10.1088/1742-6596/687/1/011001

Peer review statement

All papers published in this volume of *Journal of Physics: Conference Series* have been peer reviewed through processes administered by the proceedings Editors. Reviews were conducted by expert referees to the professional and scientific standards expected of a proceedings journal published by IOP Publishing.

Operational management of offshore energy assets

A J Kolios[1] and M Martinez Luengo[1]
[1] Cranfield Energy, Cranfield University, Bedfordshire, UK.

E-mail: a.kolios@cranfield.ac.uk

Abstract. Energy assets and especially those deployed offshore are subject to a variety of harsh operational and environmental conditions which lead to deterioration of their performance and structural capacity over time. The aim of reduction of CAPEX in new installations shifts focus to operational management to monitor and assess performance of critical assets ensuring their fitness for service throughout their service life and also to provide appropriate and effective information towards requalification or other end of life scenarios, optimizing the OPEX. Over the last decades, the offshore oil & gas industry has developed and applied various approaches in operational management of assets through Structural Health and Condition Monitoring (SHM/CM) systems which can be, at a certain level, transferable to offshore renewable installations. This paper aims to highlight the key differences between offshore oil & gas and renewable energy assets from a structural integrity and reliability perspective, provide a comprehensive overview of different approaches that are available and applicable, and distinguish the benefits of such systems in the efficient operation of offshore energy assets.

1. Introduction

A generic definition of Operations Management refers to the process of maximizing the creation of goods and services in the most efficient and effective way on a business [1] by planning, organizing, coordinating, and controlling all the necessary assets and resources. Therefore, people, equipment, technology, IT and all other assets involved in production are managed always looking for an increase in revenues (or value), a reduction in costs through an improvement of internal effectiveness [2] and at the same time, complying with strict HSE regulations. Operational management (OM) of assets extends this scope to also consider the efficient planning and execution of activities relating to delivery of a service or a process. For this reason, OM can be considered as a critical management function of every business, being its performance determinant for success.

Energy assets, and especially those deployed offshore, are designed for long periods of operation, which often exceed 30 years and during which they are subject to extensive environmental and operational loads. Performance of structures as they age is expected to deteriorate hence decisions need to be taken on how their operational regime should be modified in order to comply with the supply and demand requirements. Especially for renewable energy assets, which contrary to Oil & Gas once have tighter profit constraints and minimization of CAPEX is a fundamental requirement, efficient operational management strategies should be employed towards overall optimization of the return on investment (ROI).

This paper aims to present a review of the current status of operational management for energy assets, particularly distinguishing similarities and differences in approach between Oil & Gas and Renewable Energy assets, identifying common practices, key standards and fundamental characteristics. A discussion and generic presentation of an operational management strategy will also be presented for a generic offshore structure, highlighting the need for availability of appropriate data and procedures for maximising the information through appropriate post processing frameworks.

2. Comparison between oil and gas and renewable energy assets

Providing a formal definition for the term "asset" it can be defined as "a resource controlled by the enterprise as a result of past events and from which future economic benefits are expected to flow to the enterprise" [3], while [4] introduces to the definition the possibility of having intangible assets by "assets are resources or rights" and [5] goes a bit further introducing the term "probable future economic benefits" allowing consideration of those known but unproduced resources.

Within this paper, two different applications are considered, the offshore Oil & Gas and the offshore Renewables Energy industries and for the latter with a focus to offshore wind energy. Within the Offshore Oil & Gas industry the following assets can be considered: oil and gas platforms, access equipment, onshore facilities, stored gas or fuel, the gas and fuel expected to be extracted, technicians and personnel training programs, warehouses, among others. In the Offshore Wind industry assets can be considered to be: the wind farm structures, taking into account the turbines, submarine cables, offshore and onshore substations, onshore facilities (operations base, warehouses, spares, equipment, technicians and personnel training programs, access equipment, etc. There is an important difference between Oil & Gas and renewables assets, which is that oil and gas can be considered assets when they are in the reservoir, as they are "probable future economic benefits"; and also once they have been extracted and stored or processed. However, wind resource cannot be considered an asset as firstly, it is a renewable source, which means it cannot be depleted and secondly, the electricity produced by it, cannot be stored.

This two applications are governed by different types of risks. In the Offshore Wind industry the most critical risks are the ones related to the integrity of the units, its downtime and the loss of production due to poor availability of the turbines and high maintenance requirement, while in the offshore Oil & Gas industry it is crucial to ensure that the resource can be extracted and stored in the same quality and required quantity; it is also critical to ensure the structural integrity of the platforms as there are manned structures and the loss of lives is a fatality not comparable to the collapse of a wind turbine, which only result to economic losses.

For these risks some mitigation options can be identified: designing platforms complying with up-to-date Health and Safety (H&S) protocols and design standards, continuous investment in personnel's trainings, development and installation of Structural Health Monitoring systems or a combination of Condition-based with Risk-based maintenance of structural components; the latter constituting proactive operational management approaches. The risk-based methodology presents a cost-effective way of minimising life cycle costs in the management of assets whilst maintaining reliability or availability targets, and operating within safety and environmental regulations [6].

Moreover, there is a significant difference in the supply chain of these structures. Taking into account that an offshore wind farm can easily have more than 50 turbines and specially nowadays that the trend is to deploy in more remote locations, it can be concluded that Renewable Energy assets are designed to be fabricated in volumes, while each Oil & Gas structure can be considered as unique. Therefore it is reasonable to consider that maintenance replacements and spares need to be planned efficiently for Oil & Gas assets, as the sudden necessity of a non-predicted spare could take a significant amount of time to be manufactured, with the subsequent loss of revenue.

3. Operational Management of offshore Oil and Gas assets

3.1. Overview

Offshore Oil & Gas assets may have a significant impact on the profits they generate and their efficient operation should constitute a strategic aspect for operators. For that reason, operational management plays an important role not only ensuring the performance and the fitness for service of those assets, but also aiming for Operational Excellence required for maximising profits. Assets performance can be influenced by numerous factors such as inefficient maintenance planning, communication issues, increased downtime due to extreme weather conditions and unexpected failures, human factors, constant change in operating and environmental conditions, etc. Thus, asset damage, business interruption, pollution, injuries to people, and damage to components are key potential consequences in conventional activities [7].

Whether some of these factors occur due to uncertainties and cannot be controlled, as for example extreme weather or frequent changes in operating conditions, the majority of the rest can be addressed by using appropriate Operational Strategies (OS). This term was defined by [8] as the pattern of decisions that shape the long term capabilities of an operation and their contribution to the overall strategy of the company. These strategies revolve around patterns of choices that usually are from medium to long term nature and to reflect both the core capabilities and competencies of the company [9]. For example, the mature Oil & Gas industry uses Structural Health Monitoring and Condition Monitoring (SHM and CM) Systems combined with the experience developed during the years of operations, and this is recognised as an appropriate and efficient operational management approach.

SHM and CM Systems are important for collecting the data required for carrying out the appropriate risk assessment based on the structures current (and historically recorded) operating conditions, which may vary from the conditions they were initially designed for. Moreover, based on the data collected, several assessments related to the Structural Integrity and Reliability (SI/SR) of the unit can be determinant for both evaluating if the structure's performance degrades as well as establishing a bespoke inspections and maintenance plan. This Operational Risk Assessment is also necessary in case the assets are intended to operate outside normal operating conditions [7]. Furthermore, poor data quality heavily impacts the decision-making process, increasing the risks of operational mistakes. Oil & Gas operators need to carefully tackle this issue to avoid reducing effectiveness of operations. Additionally inconsistent data across systems increase the risk non-compliance to regulators. One of the most significant challenges for the Oil & Gas sector is to manage and post process the huge amount of data and SHM and CM Systems data, and at the same time maximise the information that can be derived from combination of monitoring systems.

3.2. Legislation and standards

The main UK legislation related to risk assessment, H&S Management and Installation of offshore Oil and Gas Assets are: Health and Safety at Work etc. Act 1974 [10], Management of Health & Safety at Work Regulations 1999 [11], Offshore Installations (Safety Case) Regulations 2005 [12], Offshore Installations and Wells (Design & Construction etc.) Regulations 1996 [13], Offshore Installations (Prevention of Fire & Explosion, and Emergency Response) Regulations 1995 [14], Pipeline Safety Regulations 1996 [15], Offshore Installations and Pipeline Works (Management and Administration) Regulations 1995 [16], Provision and Use of Work Equipment Regulations 1998 [17], among others.

Further, the most widely used Standards for Offshore Operations are:

- API RP2MET/ISO 19901-1:2006 - Deprivation of Metocean Design and Operating Conditions.
- API RP2MOP/ISO 19901-6:2009 - Marine Operations.
- API RP2SIM: Recommended Practice for Structural Integrity Management of Fixed Offshore Platforms.
- API RP75, Recommended Practice for Development of a Safety and Environmental Management Program (SEMP) for Offshore Operations and Facilities.

Relevant standards developed by British Standards (BSI) are: BS EN ISO 13628 parts 1-11: 1999-2007 [18] and BS 6349-1-1:2013 Maritime works. General. Code of practice for planning and design for operations [19].

Moreover, due to the fact that the presence of ice, snow, extreme cold temperatures, and extremes in daylight hours requires operational procedures for ice management, some Standards for Artic Oil & Gas Operations are currently being developed [20]. These standards aim to regulate the use of specialized equipment and materials, for working in both cold temperatures and long periods of darkness, as well as how to handle emergencies in these extreme conditions. It is expected that these standards will help to reduce operational costs [21].

3.3. Operational excellence

Oil & Gas operators are constantly working towards achieving Operational Excellence (OE) required for maximising the profits of the business. This aspiration assures that their assets are operated safely, reliably, sustainably and cost effectively [22]. Many of these companies have high quality management programs, codified manuals and active measurement and training. Top management support, employee training, and employee involvement are some of the key variables that have to be enhanced in order to improve operational performance [23].

The convergence theory [24] states that learning will allow managers to adopt different cultures (and therefore, different ways of doing things) to carry out the same efficient management practices. Moreover [25] agree in the importance of personnel training in increasing productivity, while [26] was able to show empirically how significant is the improvement that quality information and analysis have on process management.

4. Operational Management of offshore Renewable Energy assets

Offshore Renewables Energy assets Operation Strategies (OS) are in principle similar to those employed in the Oil & Gas industry. However, due to the differences already mentioned (manned vs unmanned; single vs multiple structures, etc), and the multidisciplinary risks of offshore renewables [27,28], these need to be adapted to the application and optimised to make this industry as efficient as possible, due to its tight economic constraints and related financial risks. This aim will encourage optimisation not only of the design, but also the operation, maintenance and inspection strategies [23]. Design and operation of offshore wind farms is influenced by several factors as have been studied in [29,30,31].

Due to the continuous expansion and development that is currently taking place within the Offshore Wind industry, it had to rapidly adapt OS to the allow deployment further offshore and in larger Wind Farms (WF), without proportionally optimising these strategies, collecting and processing data regarding the structural integrity of units. However, as previously happened with the Oil & Gas industry, it is expected that as soon as the sector gets some experience, OS will tend to become harmonised, following the convergence theory, which mixes both, the Culture Free [32] and the Culture Specific [33] theories in the evolution of management practices [23]. Another important difference between the two applications is the associated level of risks associated with their O&M activities. Therefore, Offshore Wind OS should address this factor adapting, among other things, relevant Health and Safety (H&S) protocols. Some examples of such risks are:

- The permanent personnel on the structure: while Oil and Gas platforms have in average 80 people and possibility of fatalities is completely unacceptable, there is no personnel inside offshore wind turbines during operation; therefore H&S protocols regarding this issue do not need to be as strict as for Oil and Gas platforms.
- The level of exposure the structure has during maintenance activities and H&S protocols required. In case of wind turbines, even though the structures are unmanned and during operations no life is risked, safety during maintenance has to be carefully assessed as a wind turbine is much more exposed to weather conditions than an Oil & Gas platform.
- The level of risk associated with the spares and equipment transport, storage and access to the turbines is significantly more complicated and costly in the Offshore Wind Industry, as inside a nacelle there is not much storage and working space compared to Oil & Gas platforms. This factor makes maintenance costs proportionally higher and an economical risk for the operator that has to be carefully considered and optimised.

According to [34], the cost of offshore O&M activities currently varies from 2 to 5 times the cost onshore. One of the reasons is the increased loading WTs have due to the offshore environment and the variation from the loadings they were initially designed for [35]. Moreover, the additional effort of building wind turbines that resist the environmental harshness of the offshore environment and avoid hazards such as corrosion, would drive not only CAPEX, but also O&M costs offshore even higher [36,37]. This is the reason why operators are currently looking for strategies that enhance the efficient operation of turbines reducing the O&M costs.

One of the best ways of enhancing efficient operation is by employing SHM/CM systems, as part of a Condition Based Maintenance (CBM) paradigm with smart loads management [35]. CBM consists of preventive maintenance carried out based on the variation of different parameters that are continuously monitored [38]. CBM strategies allow planning of maintenance activities and inspections based on the data of SHM/CM Systems [39].

By constantly monitoring the condition of structural components of every turbine, required maintenance activities can be planned in advance and carried out when necessary offshore rather than in predetermined time intervals [40,41]. Currently, these systems are being adapted to Offshore Wind. SHM/CM systems are already widely used in some subsystems, such as generators and gearboxes, however in others are still waiting for the development of relevant frameworks that will assesses collected data and draw conclusion for the integrity/performance of the subsystem (ie support structures). Therefore, the goal for Operational Management of Renewable Energy assets is to work towards integrating and developing SHM and Structural Health and Prognosis (SHP) Systems for carrying out smart loads management.

Operation based on their integrity status can potentially allow the control variation of a wind turbine when a certain level of damage is detected in an early stage by the SHM/SHP System. This may lead to reduction in the power production but at the same time it will delay damage propagation, increase their service life, and allow for multiple turbines to be serviced during the same visit to the plant in order to maximize the overall profit of the WF, decreasing maintenance costs and increasing operational efficiency. Some simulations carried out by [34] showed that according to this practice reducing the power production by 5%, fatigue life extension in blades would increase by a factor of 3 due to a decrease of a 10% in the corresponding loading.

5. Discussion

As mentioned earlier, the transition from Oil & Gas to Renewable Energy applications (unmanned structures and hence lower risk profile, volume production, and significant operational loads) bring a great motivation for further studies in Operational Management. The fact that renewables are forced to reduce CAPEX as much as possible, considering that their payoff period is longer, introduce the challenge of optimization of the CAPEX/OPEX ratio, transferring expenditure in later periods of operation of the units and more specifically when they will be generating profit. Further with the introduction of new materials (of higher compositional consistency and lower variation of mechanical properties), consideration of new operational conditions and hence new prevailing failure modes and mechanisms constitute efficient operational management as pertinent towards increased cost competitiveness of the industry. Further, information gathered throughout the asset's service life will provide valuable input towards requalification and more well informed selection of the most appropriate End of Life (EOL) scenario.

Ability to reassess the performance and integrity of the structure is driven by its structural deterioration, accidental damage due to impact, changing demands and regulations, life extension, deployment conditions etc. More specifically, aging mechanisms in offshore assets can be considered fatigue, corrosion, accumulated damage, scour etc. Based on the above a justified requirement becomes the development and application of a widely accepted framework for the adequate evaluation of the condition of a structure throughout its service life, considering updated data from real operations.

Design standards have evolved over the last decades including general information regarding the design, maintenance and operation of structures. Based on the provisions of API RP-2A Section 17, the key aspects of reassessment should include: Inspections, surveys and data management, Environmental conditions and forces, Structural elements, systems and analysis, Foundation elements, systems and analysis, Operational considerations, Policy considerations and consequences. The aforementioned provisions yield for a strong requirement for sufficient data and appropriate processing methodologies, however generic methodologies for the latter tasks are not available.

Data collected should adequately inform Structural Integrity Management (SIM) or Structural Reliability Assessment (SRA), in order to allow assessment of the current status and contribute to a cost effective inspection and maintenance strategy. Concepts of SRA have been successfully applied to other engineering systems of increased complexity such as manufacturing, power generation and offshore environmental loading [42, 43, 44]. Figure 1, illustrates a framework for Integrity/Reliability Re-Assessment of Offshore Support Structures. An important element that should be considered, is the calibration of the initial simulation models (stresses, reliability estimations) after installation in order to take into consideration the actual deployment conditions (ie geotechnical conditions, and particularly for wind turbines operational loads). This activity will allow for more realistic assessment for critical parameters, such as fatigue life (deterministically calculated or through reliability analysis), to be determined, allowing for a more efficient operational management and planning. Further, collecting information during the structure's service life, will allow for tailored interventions (with effective scheduling), rather than in regular intervals, utilising the residual resistance of the structure.

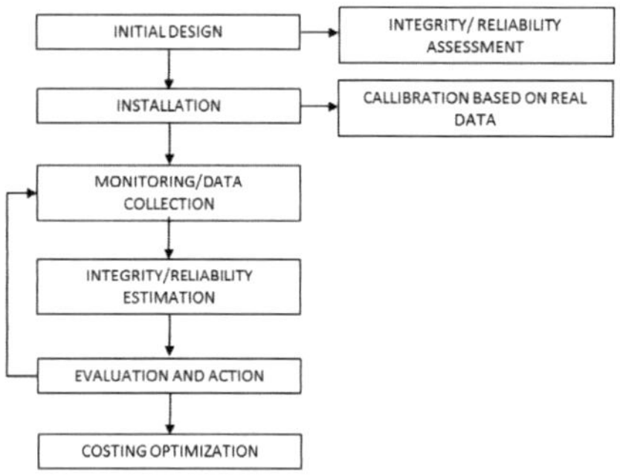

Figure 1. Integrity/reliability assessment of offshore support structures.

Collection of data in the most efficient way, is a challenge of great research priority, in order to ensure that sufficient information is collected potentially from multiple monitoring systems (such as accelerometers, strain gauges) etc, focusing more on the quality of data rather than in the (often un-necessarily) increased resolution. Another challenge, specifically applicable to offshore wind farms, is that of the increased cost of installation of SHM systems, as the number of units increases. The ability to extrapolate information from a limited number of instrumented units to the entirety of the wind farm, will allow customised inspection and maintenance planning and well informed EOL scenario selection for each unit or group of units.

6. Conclusions

This paper has presented a review of asset operational management approaches for the offshore Oil & Gas and Renewable Energy sectors, in order to highlight similarities, differences and influential parameters. Moving towards a broader deployment of offshore wind farms, with thousands of units already installed or planned to be installed over the next few years, provides a great motivation for further research in the efficient collection and processing of data in order to optimise OPEX through optimised planning of inspection and maintenance activities as well as well informed decisions for the most profitable End of Life scenario.

References

[1] N R Sanders 2014 *The definitive guide to manufacturing and services operations* (USA: Pearson FT Press)
[2] N Snack, S Chambers, R Johnston 2010 *Operations Management* (UK: Pearson Education)
[3] IASC 1989 *Framework for the preparation and presentation of financial accounting statements, international accounting standards committee*
[4] D Solomons 1989 *Guidelines for financial reporting standards* (UK: Institute of Chartered Accountants in England and Wales)

[5] FASB 1985 *Statement of financial accounting concepts No 6 elements of financial statements* (USA: Financial Accounting Standards Board)
[6] U R Bharadwaj, V V Silberschmidt, J B Wintle 2012 *J of Quality in Maintenance Engineering* **18-4** 417
[7] R Bigliani 2013 *Reducing Risks in Oil and Gas Operations*, (USA: IDC Energy Insights)
[8] N Slack, M Lewis 2002 *Operations Strategy* (UK: Financial Times and Prentice-Hall)
[9] R H Lowson 2002 *International Journal of Operations & Production Management* **22-10** 1112
[10] The Health and Safety at Work Act 1974 (UK: Health and Safety Executive) chapter 37
[11] HSE 2000 *Management of health and safety at work regulations 1999 approved code of practice and guidance ACOP L21* (UK: Health and Safety Executive)
[12] HSE 2006 *Offshore installations (safety case) regulations 2005* (UK: Health and Safety Executive)
[13] HSE 1996 *Offshore installations and wells (design and construction, etc) regulations 1996 (DCR)* (UK: Health and Safety Executive)
[14] HSE 1995 *Offshore installations (prevention of fire & explosion, and emergency response) regulations 1995* (UK: Health and Safety Executive)
[15] HSE 1996 *A guide to the pipelines safety regulations 1996* (UK: Health and Safety Executive)
[16] HSE 1995 *Offshore installations and pipeline works (management and administration) regulations 1995* (UK: Health and Safety Executive)
[17] HSE 1998 *Provision and use of work equipment regulations 1998* (UK: Health and Safety Executive)
[18] BS EN ISO 13628 2005 *Petroleum and natural gas industries design and operation of Subsea production systems* (UK: British Standards)
[19] BS 6349-1-1 2013 *Maritime works general code of practice for planning and design for operations* (UK: British Standards)
[20] S Green, P Steenhof, B Walsh 2014 *Oceans St John's* 1 (DOI: 10.1109/OCEANS.2014.7003107)
[21] S Fournier and M Caron-Vuotari 2013 *Economic development in Canada's northern marine waters* (Ottawa: Conference Board of Canada)
[22] J McCreery, E Phillips and F Cigala 2013 *Operational excellence the imperative for oil and gas companies* (http://www.bain.com/publications/articles/operational-excellence-the-imperative-for-oil-and-gas-companies.aspx)
[23] M M Parast, S G Adams, E C Jones 2011 *Inter Journal of Quality & Reliability Management* **28-4** 426
[24] W Form 1979 *Annual Review of Sociology* **5** 1
[25] S L Ahire, P Dreyfus 2000 *Journal of Operations Management* **18-5** 549
[26] S M Lee, B H Rho, S G Lee 2003 *International Journal of Production Research* **41-9** 2003
[27] A J Kolios, G Read, A Ioannou 2014 *International Journal of Sustainable Energy* **35-1** 59
[28] A J Kolios, G Read 2013 *Energies* **6-10** 5023
[29] A J Kolios, A Rodriguez-Tsouroukdissian, K Salonitis 2014 *Ships and Offshore Structures* **12** (DOI: 10.1080/17445302.2014.961295)
[30] H Martin, G Spano, J F Küster, M Collu, A J Kolios 2013 *Ships and Offshore Structures* **8-5** 477
[31] A J Kolios, M Collu, A Chahardehi, F P Brennan, M H Patel 2010 *Proceedings of the European Wind Energy Conference and Exhibition* **6** 4778
[32] S S Rao, L E Solis, T S Raghunathan 1999 *Total Quality Management* **10-7** 1047
[33] D A Ralston, D H Holt, R H Terpstra, Y Kai-Cheng 1997 *J of International Business Studies* **28-1** 177
[34] D T Griffith, N C Yoder, et al 2014 *Wind Energy* **17** 1737
[35] D T Griffith, N C Yoder, et al 2012 *Proceedi of the European Wind Energy Conference* (Denmark: EWEA)
[36] A C Levitt, W Kempton, A P Smith, W Musial and J Firestone 2011 *Energy Policy* **39-10** 6408
[37] W Musial and B Ram 2010 *Large-scale offshore wind energy for the United State assessment of opportunities and barriers* (USA: NREL Report No TP-500-49229)
[38] SIS SS-EN 13306 2001 *Maintenance terminology* (Sweden: Swedish Standard Institute)
[39] F Besnard, K Fischer, L Bertling 2010 *Reliability centred asset maintenance a step towards enhanced reliability, availability, and profitability of wind power plants* (Sweden: ISGT Europe and IEEE PES)
[40] Y Amirat, M E H Benbouzid, B Bensaker, R Wamkeue 2007 *Proceedings IEEE International Electric Machines and Drives Conference* **2** 1434
[41] C C Ciang, J R Lee, H J Bang 2008 *Measurement Science and Technology* **19** 1
[42] D P Hanak, A J Kolios, C Biliyok, V Manovic 2015 *Applied Energy* **139** 350
[43] B J Snow, I Moulitsas, A J Kolios, M De Dominicis 2014 *Geoscientific Model Development* **7-4** 1507
[44] K Salonitis, A J Kolios 2014 *International Journal of Advanced Manufacturing Technology* **71-5-8** 1197

Development of an experimental methodology for assessing the growth of scale ($CaCO_3$) in pipelines

M García[1], J Vega[1], Y Pineda[1] and E Vera[1]
[1] Universidad Pedagógica y Tecnológica de Colombia, Tunja, Colombia.

E-mail: marceladelpilar.garcia@uptc.edu.co

Abstract. One of the problems that affect the mechanical integrity of pipelines transporting aqueous fluids is the formation of fouling deposits as $CaCO_3$. Today different theoretical models that establish criteria trend but are not decisive in establishing the true rate of growth of scale in a pipeline there. This paper presents experimental development which was conducted to deposit scale of calcium carbonate by using a loop in which the deposits formed into a capillary of stainless steel AISI 316 and thus study the influence of various physical - chemical factors such as the composition of the solution, flow rate, and surface temperature and the presence of CO_2 generated on amount of scale. The evolution of the thickness and the amount of fouling obtained by tests was subsequently determined and by X-ray diffraction (XRD) and scanning electron microscopy (SEM) were characterized the deposits. With the results of the study show a strong influence of carbon dioxide in the amount of fouling obtained and according to the composition of the solution the flow rate fouling can generate greater or lesser scale. Under the test conditions is not possible to obtain deposits at room temperature.

1. Introduction

Pipes that transporting aqueous fluids are affected their mechanical integrity due to the formation of scales, which occur when the fluid is oversaturated by compounds having in solution and this starts at the time when variations occur in the operational conditions of system (pressure, temperature, etc.) [1]. The literature indicate that the operating conditions have an important effect on the growth of the scales in the industries [2].

Deposits occur in different types of industries in which a solution aqueous is in contact with a surface [3], said layers may be formed in pumps, pipes, heaters, heat exchangers, boilers, evaporators and various accessories and equipment [4]. Among the industries most affected are those using equipment heat transfer and extraction and production of oil [5], causing costs due to production losses, removal treatments (chemical or mechanical) and repair or replacement of equipment. In industrial processes can present different types of fouling as strontium sulphate, barium sulphate, calcium sulphate and calcium carbonate [6], the latter is one of the most common in the petroleum industry.

Due to problems caused by scales, posits have gained a lot of research have get deposits under laboratory conditions in order to understand the causes, variables and trend criteria involved in the formation, growth and removal of the layer formed. Methods of studying the deposition of layers in the laboratory are: electrochemical techniques, rotating ring electrode, micro quartz crystal balance (QCM) [7] and loops, these last time-consuming for the test performance [8], but considered as a subject of study time of blockage of the pipes can establish good predictions. Zhang et al. [1], showed the performance of a kinetic model to predict the rate of fouling of calcium carbonate, which is based on measurements of growth of single crystals of calcite, the results for the model were compared to those obtained with experiments and obtained a good correlation between the blocking time predicted by the model and obtained from laboratory tests also evaluated the potential application of the model to compared the results with the profile of scale that had an oilfield.

In this paper we present experimental development used to study the formation of calcium carbonate scale on steel AISI 316 with a loop and correlated different physicochemical factors such as the composition of the solution, flow rate, temperature of the surface and presence of CO_2 on the amount generated of scale.

2. Experimental equipment and procedure

The substrate on which were deposited of the scales was a capillary of 15cm length and 4.1cm internal diameter, made of stainless steel AISI 316. The test tube was attached to a pump to impulse the solution to inside of capillary and back to the vessel which initially contained. To increase the temperature of capillary was used a heating bath, a schematic representation is shown in Figure 1.

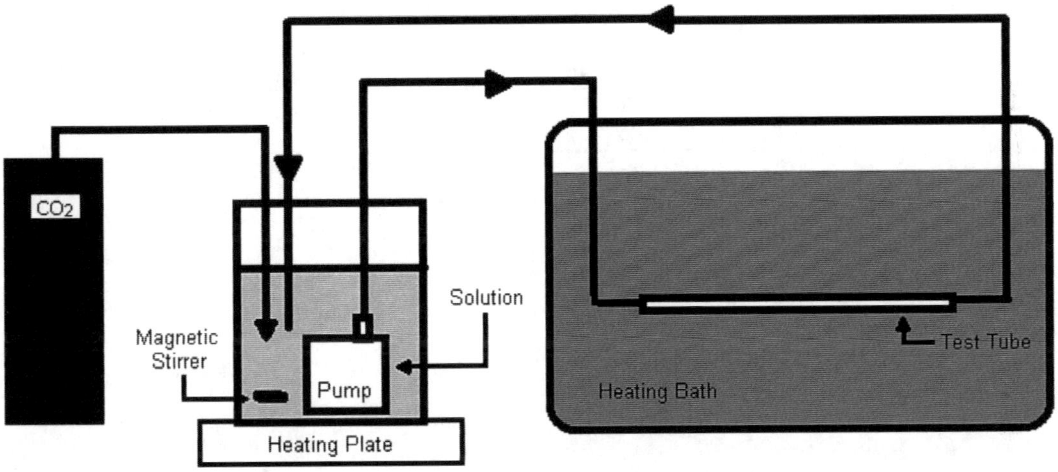

Figure 1. Experimental assembly for the precipitation of calcium.

In the performing test two different solutions were used: first consisting of H_2O-NaCl-$CaCO_3$ and second: $CaCl_2.2H_2O$-$MgCl_2.6H_2O$-$NaHCO_3$-NaCl-H_2O [9]. For solutions analytical grade reagents which were weighed on a sartorius balance AC121S with a precision of four tenths were used.

The experiments were conducted for 101 hours, during this time the heating bath temperature was controlled by an immersion thermometer, test at room temperature (18-21), 60 and 80 *°C* and velocities of 0.0063 and 0.035*m/s* which cause laminar flow which creates favourable conditions for scaling are executed. Besides was injected bubbling controlled of CO_2 a container containing the fouling solution when required it. After each test performed the evolution of scale thickness obtained was measured, taking data every centimetre of capillary length, this was done by means of a microscope stereographic SMZ800 NIKON. For the characterization of the deposits obtained X-ray diffraction was used by a team PANalytical X'Pert PRO with copper anode and a sweep between 5 and 90° driving a step time of 0.5 seconds.

3. Results

The evolution of scale obtained reveals growth as presented in Figure 2 and Figure 3 shows measurements of the layer thickness.

The results obtained through this study are presented in Table 1 for each correlation variables used, the obtained maximum thickness in micrometres and the amount of fouling micrograms is shown.

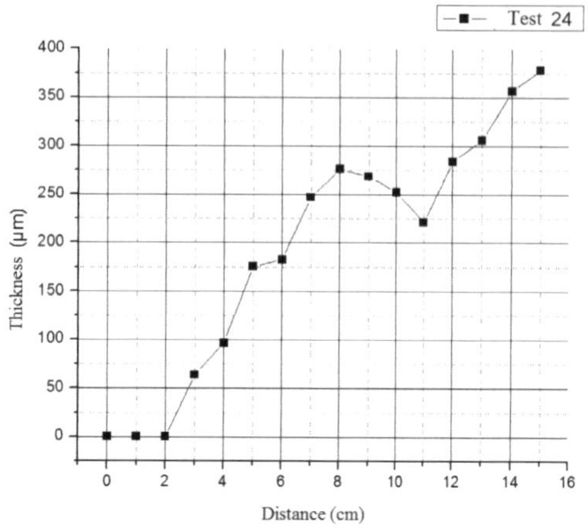

Figure 2. Thickness vs distance for the test capillary 24.

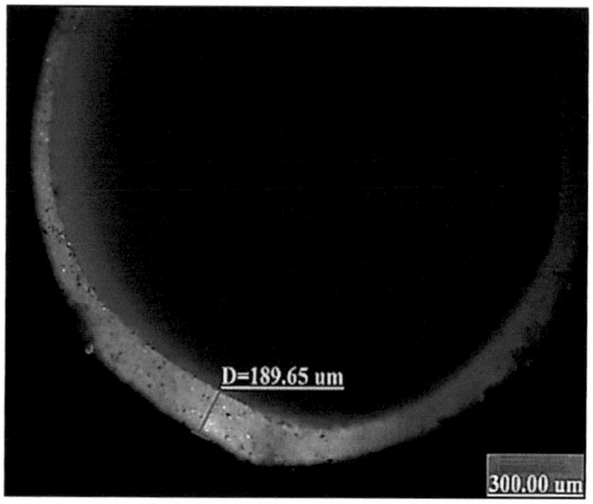

Figure 3. Thickness 14 centimetre length of capillary for test 17.

Table 1. Maximum thickness and amount of scale for each of the tests.

P.No.	Conditions				Thickness	Amount
	T SP [a]	VF (m/s) [b]	CS [c]	CO_2	(μm)	(μg)
9	60°C	0.0063	1	Yes	193,04	1896,03
10	60°C	0.0063	1	No	nm [d]	nm [d]
11	60°C	0.0063	2	Yes	426,20	2707,98
12	60°C	0.0063	2	No	256,12	2217,53
13	60°C	0.035	1	Yes	31,24	117,06
14	60°C	0.035	1	No	nm [d]	nm [d]
15	60°C	0.035	2	Yes	329,39	2818,99
16	60°C	0.035	2	No	147,08	1331,16
17	80°C	0.0063	1	Yes	267,37	2152,23
18	80°C	0.0063	1	No	nm [d]	nm [d]
19	80°C	0.0063	2	Yes	111,24	1024,08
20	80°C	0.0063	2	No	173,70	1345,36
21	80°C	0.035	1	Yes	146,70	1028,79
22	80°C	0.035	1	No	nm [d]	nm [d]
23	80°C	0.035	2	Yes	222,28	2011,53
24	80°C	0.035	2	No	377,92	2918,92

For test between 1 to 8 is not get scale measurable.
a Temperature of the surface
b Flow rate
c Composition of the solution
d Scale obtained unmeasurable

SEM photograph in the Figure 4 indicated that the deposit structure in uniform and that in few places is possible observed crystals of $CaCO_3$ completely build. Too by XRD, it was determined that for tests with solution 1 calcite precipitated it in Figure 5 is observed and solution 2 identified compounds as: calcium carbonate as aragonite, double carbonate of calcium and magnesium, magnesium carbonate as magnesite and calcium. We conclude that aragonite precipitated because that the presence of Mg^{+2} inhibits the growth of the calcite [10], and dolomite and magnesite exists by partial or complete replacement of calcium by magnesium in the calcium carbonate.

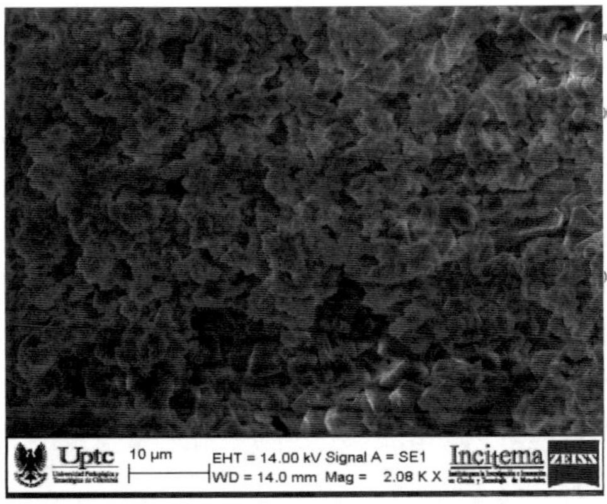

Figure 4. Scanning electron microscopy picture of $CaCO_3$ crystals.

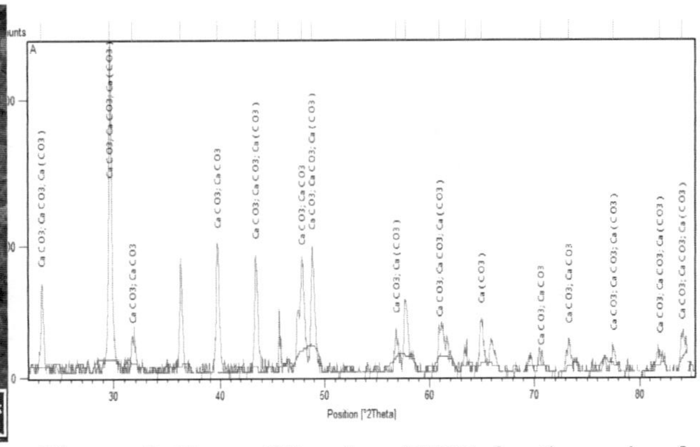

Figure 5. X-ray diffraction (XRD) for the scale of $CaCO_3$.

4. Conclusions

Due to the different nature of the solutions different thicknesses with solution 1 and 2 are presented and also as fluids were not filtered, the flow rate generates different cutting forces due to the particle size of each solution because differences grades solubility of the reagents used in each.

Solution 1 generates no measurable deposits without the presence of CO_2 this occurs because the low solubility their reagents, which is enhanced when carbon dioxide is present.

The higher the temperature the energy of the system increases causing higher speed chemical reactions. It also generates a decrease of solubility causing increased degree of supersaturation, so it requires use temperature above room ambient so that the deposits are formed.

With solution 2 is obtained aragonite which its crystallographic system is orthorhombic generates a deposit more porous that obtained with solution 1 because this is entirely formed by calcite.

References

[1] Zhang Y, Shaw H, Farquhar R and Dawe R and 2001 *Journal of Petroleum and Engineering* **29** 85
[2] Hasan B, et al 2012 *Applied Thermal Engineering* **36** 210
[3] Al Nasser W, et al 2008 *Chemical Engineering Science* **63** 1381
[4] Helalizadeh A, Müller-Steinhagen H and Jamialahmadi M 2005 *Chemical Enginee and Process* **61** 2069
[5] Martinod A, et al 2009 *Chemical Engineering Science* **64** 2413-21
[6] Bin Merdhah A and Yassin A 2008 *Emirates Journal for Engineering Research* **13** 69
[7] Abdel-Aal N, Satoh K and Sawada K 2002 *Journal of Crystal Growth* **245** 87
[8] Quddus A 2002 Desalination **142** 57
[9] NACE TM0374 2001 *Laboratory screening tests to determine the ability of scale inhibitors to prevent the precipitation of calcium sulfate and calcium carbonate from solution (for oil and gas production systems)* (USA: NACE)
[10] Helalizadeh A, Müller–Steinhagen H and Jamialahmadi M 200 *Chemical Enginee and Process* **39** 29

Mathematical-statistical model for predicting the tendency of the mechanical integrity of a buried pipeline

H Aponte Betancur[1], E vera Lopez[1] and Y Pineda Triana[1]

[1] Universidad Pedagógica y Tecnológica de Colombia, Tunja, Colombia.

E-mail: hector.apontebetancur@uptc.edu.co

Abstract. This paper is intended to use a process model stochastic renewal to analyze the disintegration of the pipeline due to the time of use and service conditions, classifying the pipeline into segments according to the conditions to which it is exposed (temperatures, pressures, speed flow, environmental conditions, etc.). The model will determine the material wear over time, periods of renewal of the material and life chances at any time.

1. Introduction

The transport of dangerous pipeline fluids are exposed to dependent and no dependent threats of time. The deterioration of pipeline with time is inevitable despite different coatings and protections that help extend the life of the material used, but eventually require renovation or change [1]. The disintegration pipeline thickness depends on many variables (temperature, pressure, flow rate, environmental conditions, etc.) that are not easy to relate all at once [2]. This model is intended to apply to predict the wear of the material, refresh rates and probabilities of life at any time t.

2. Wear equation

There are several factors that determine the pipeline desintegration, then it is assumed that the rate of change of the thickness of the pipe wall is proportional to $E(t)$ the remaining material in time t [3]. Mathematically it can be expressed as:

$$\frac{dE}{dt} = -\lambda E \tag{1}$$

Where $E(t) = 1 - d(t)/h$ represents the remaining percentage of the thickness of the pipe wall in a time t, $\lambda > 0$, a proportionality constant having wear duct over time and conditions thereof, the minus sign represents the disintegration suffered reducing material thickness with time. The Figure 1 represents: Pit length (Lp), maximum pit depth at a time (d), nominal wall thickness of the pipe (h), and nominal outside diameter of the pipe (D).

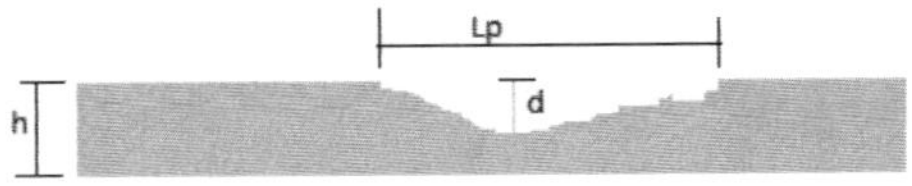

Figure 1. Corrosion parameters.

ASME B31 [4] sets a maximum length (L) of permissible sting, calculated as:

$$L = 1.12 * \sqrt{\left((\frac{\frac{d}{h}}{1.1\frac{d}{h} - 0.15})^2 - 1 \right) * D * h} \tag{2}$$

If $Lp > L$ is considered the critical pitting and suggests renew the material, then the solution of equation (Equation 1) as a function of driveability and maximum lengths to predict areas as required by the ASME B31 [4] is taken and possibly the probability of life with respect to time t in a process of renewal.

Solving the equation (1) is $f(t) = E(t) = C_0 \exp(-\lambda t), t \geq 0$, where $C0 = 1$, representing the 100% the thickness of the pipe wall in a time $t = 0$, when $d(t = 0) = 0$. Therefore

$$\frac{d(t)}{h} = 1 - E(t) = 1 - \exp(-\lambda t) \tag{3}$$

The corrosive maximum length allowed in a time t, according to the above equation and the standard ASME B31 [4] is given by $L_{max} = 1.12 * \sqrt{\left[(\frac{1-\exp(-\lambda t)}{1.1(1-\exp(-\lambda t))-0.15})^2 - 1 \right] * D * h}$, the maximum permitted corrosive area at a time t is $A_{max}(t) = 0.893\frac{L_{max}(t)}{\sqrt{Dh}}$ and the pressure maximum (P') is $P'(t) = 1.1P\exp(-\lambda t)$, where P is the value of the relationship established by maximum allowable pressure eleoducto.

3. Renewal processes

The pipeline has a lifespan that could be modeled by a random variable T. Once the pipeline fails or is in poor conditions of service is removed or restructured. This process generates a collection of random variables $T_1, T_2, T_3...$, positive and independent representing the succession of pipeline lifetimes put into operation. It is clear that whenever revive restructuring process probabilistically.

be $Q_n = T_1 + T_2 + T_3...T_n, Q_0 = 0$ representing the time of the n-th renewal and N_t as the number of renewals in time t [5, 6].

The density function of the form is considered $f(t) = \exp(-\lambda t)$ y $F(t) = \int_0^t f(s)ds$ [6, 7] representing a distribution function of the lifetimes of the pipeline.

The distribution function of N_t depends on $F(t)$, as indicated by the equation $P(N_t = n) = \exp(-\lambda t)\frac{(\lambda t)^n}{n!}$ [5].

The number of renewals that are made to the pipeline during a given time t, then defined $\Lambda(t) = E(N_t)$ (Expected value of the number of renewals at a time t) renewal and function that satisfies the following proposition.

Proposition 1 *Renewal function $\Lambda(t)$ satisfies the Equation 1.*

$$\Lambda(t) = F(t) + \int_0^t \Lambda(t-s)dF(s) \tag{4}$$

Where $\Lambda(t) = \lambda t$ and $F(t) = 1 - \exp(-\lambda t)$ satisfies the Equation 4.

3.1. Lifetimes

For this process three non-negative random variables defined: γ_t Remaining lifetime of the tube, δ_t Time elapsed tube life and β_t The total lifetime, as it is shown in Figure 2. Time remaining life is: $\gamma_t = Q_{N_{t+1}} - t$ (see Figure 2).

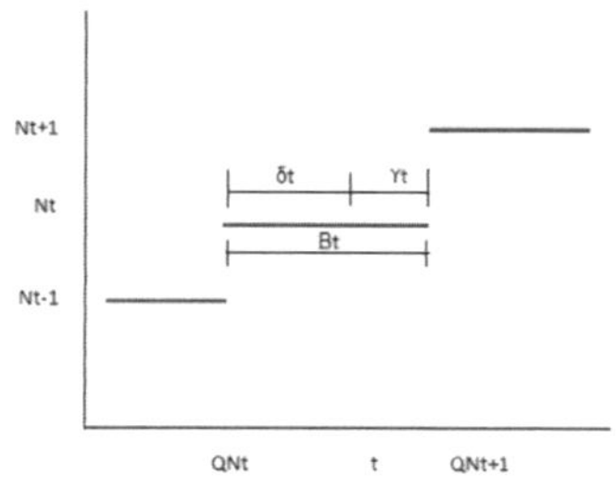

Figure 2. Graphic of useful life [5].

Proposition 2 *For $x > 0$, the function $h(t) = P(\gamma_t > x) = 1 - P(\gamma_t < x)$ satisfies the renewal Equation 1.*

$$h(t) = 1 - F(t + x) + \int_0^t h(t - s)ds \quad for \quad x > 0 \tag{5}$$

Where $h(t) = P(\gamma_t > x) = \exp(-\lambda x)$, satisfies the Equation 4 of the previous proposition. Lifetime elapsed is: $\delta_t = t - Q_{N_t}$ [5].

Proposition 3 *For $t \in (x, \infty)$, the function $g(t) = P(\delta_t > x) = 1 - P(\delta_t < x)$ satisfies the Equation 1.*

$$g(t) = 1 - F(t) + \int_0^{t-x} g(t - s)dF(s) \tag{6}$$

Where $P(\delta_t < x) = 1 - \exp(-\lambda x)$ for $0 < x < t$; if $x > t$ then $P(\delta_t < x) = 1$ and the total lifetime is (see Figure 2): $\beta_t = \gamma_t + \delta_t$ [5].

Proposition 4 *The function $g(t) = P(\beta_t > x) = 1 - P(\beta_t < x)$ renewal satisfies the Equation 1.*

$$g(t) = 1 - F(max\{t, x\}) + \int_0^t g(t - s)dF(s) \quad for \quad x > 0 \tag{7}$$

Where the total lifetime has the following distribution function; $P(\beta_t < x) = 1 - \lambda(1 + min\{t, x\})\exp(-\lambda x)$ [5].

4. Result

The data of Table 1 is used to apply the equations discussed above (see Equations 1-7).

Table 1. To apply the model data.

Specifications	Measures	Specifications	Measures
Renewal Time	4 años	Pit depth	0.0405 in
Diameter	16 in	Pipe wall thickness	0.312 in
Length sting	0.3386 in	Design pressure	1684 in

Table 2 shows the predicted depth, length and pressure versus time t allowed by ASME.

Table 2. Prediction depth, length and maximum pressure allowed by ASME at time t.

Time (years)	Depth %	Length sting (in)	Pressure (psi)
4	0.130000011	46.40565441	1611.58798
5	0.159768307	15.32631443	1556.445188
7	0.216284056	5.624921409	1451.755416
9	0.268998436	3.876184499	1354.107297
12	0.341497025	2.84263713	1219.810911
15	0.406805417	2.334069374	1098.833646
20	0.501579111	1.87063979	923.274855
25	0.581210972	1.603842059	775.7647947

Table 3 shows the reliability of service life remaining, elapsed time and total for a t.

Table 3. Prediction of reliability (probability) of life.

Time (years)	$\Lambda(t)$	$P(\gamma_t > x)$	$P(\delta_t < x)$	$P(\beta_t < x)$
4	0.139262068	0.869999999	0.130000001	0.977369912
5	0.174077585	0.840231706	0.159768294	0.966625503
7	0.243708619	0.783715961	0.216284039	0.93975967
9	0.313339653	0.731001584	0.268998416	0.906346803
12	0.417786204	0.658502999	0.341497001	0.845437856
15	0.522232755	0.59319461	0.40680539	0.773389741
20	0.69631034	0.498420919	0.501579081	0.633282533

5. Conclusions

This model predicts the percentage of pit depth, lengths, pressures allowed by ASME and likely lifespan of the pipeline at a time t.

These equations can generate a permanent monitoring with respect to the likelihood of failure of transmission lines mitigating risk in the pipeline.

References

[1] W Kent Muhlbauer 2008 *Pipeline risk assessment* (USA: Copyright and WKM Consultancy)
[2] W Kent Muhlbauer 2004 *Pipeline risk management manual 3rd* (USA: Elsevier)
[3] Dennis G Zill 1997 *Ecuaciones diferenciales con aplicacin de modelado 6th ed* (Mexico: Thomson)
[4] ASME B31G 1991 *Manual for determining the remaining strength of corroded pipelines* (New York: The American Society of Mechanical Enaineers)
[5] Rincon L 2011 *Introducción a los procesos estocasticos* (Mexico: http://www.matematicas.unam.mx/lars)
[6] Rober B ASH 1970 *Basic probability theory* (New York: Wiley -Sons)
[7] Martínez C 2006 *Estadística básica aplicada* (Bogota: Ecoe Ediciones)

Electrochemical characterization of the steel wire used as reinforcement in the conductors transmission networks electricity nitride by ion implantation

J J Castro Maldonado[1,2], H J Dulcé Moreno[2] and W Aperador[3]
[1] Servicio Nacional de Aprendizaje (SENA), San José de Cúcuta, Colombia.
[2] Universidad Francisco de Paula Santander, San José de Cúcuta, Colombia.
3 Universidad Militar Nueva Granada, Bogotá, Colombia.

E-mail: johnjacama@gmail.com , hjdulce@ufps.edu.co

Abstract. The power company feature infrastructure, which are generally shaped so the transmission and distribution lines, here is why it is necessary to characterize the process of electrochemical corrosion of these components. In this case the steel wire coated with zinc or aluminium, as it is undergoes the rigor of corrosive environments. Given the geographical diversity and different climatic environments, atmospheric corrosion carried affecting service life of structures. For example in very humid environments such as coasts and high altitudes, wetting time (TOW), parameter that meets the conditions of temperature and relative humidity, it affects large proportion, accelerating the corrosion of ferrous materials. Given the importance of establishing mechanisms that lessen the impact on degradation in transmission and distribution lines of both the reliability and the availability of the same. This paper presents the implementation in nitride steels as an alternative or complement to zinc coating.

1. Introduction

The power companies have made infrastructure, in general, for the transmission and distribution lines [1]. The stability of the supply system and electricity distribution is affected by the deterioration of basic infrastructure components; in this case the most critical is the steel wire, [2] which works not only as a support but is exposed to highly corrosive environments. Traditionally to increase corrosion resistance wire that is coated with aluminium or zinc (galvanized)[3]. Considering potential advantages (such as lightening, reduced size, increased resistance to hydrogen permeation, adhesion and surface hardness greater) [4,5] it was considered as an alternative treatment ion implantation.

2. Experimental procedure

Steel core samples of a commercial wire ACSR 2/0 (according to the NTC 461 has the following chemical composition: Carbon 0.50 to 0.88%, manganese 0.50 to 1.10%, 0.035% maximum phosphorus, 0.045% maximum sulphur and 0.10 to 0.35% silicon). The length of the specimens was 70mm long, this in order that upon immersing the samples shielded saline in 3.5% NaCl is exactly a contact area of $300mm^2$, for testing electrochemical impedance spectroscopy (EIS). According to NTC 461 standard wire corresponds to class A. In Figure 1, the zinc layer is shown generated by galvanizing a thickness of 20 microns is equivalent to about $210g/m^2$, equivalent to approximately 240g per 100m length of wiring.

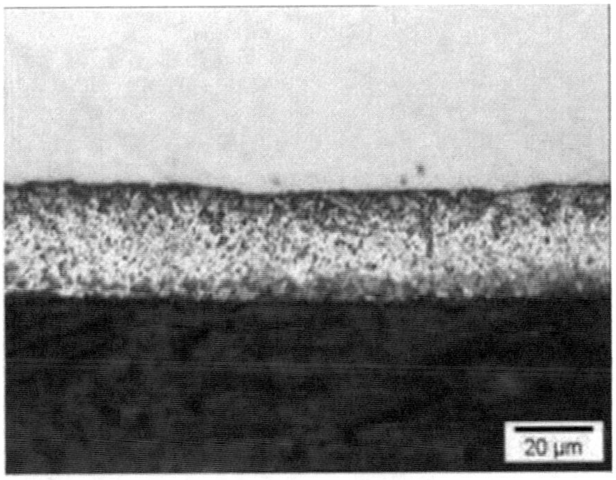

Figure 1. Zinc coating microscopy.

Using the software, it was determined as the optimal amount of 15 specimens, for experimental design, we proceeded to divide into three groups as follows: uncoated wire, zinc coated steel (galvanized) and nitrided steel cable. The nitriding was performed by high voltage electrical discharges at low pressures JUPITER device, using nitrogen as a working gas, a range of pressures comprised between 0.83 and 0.85Pa. The discharge pulses were 20kV, 30Hz, 0.25ms pulse width, and a total treatment time of 30 minutes [3,6].

The electrochemical characterization was performed on a Gamry PCI-4 model, using techniques Electrochemical Impedance Spectroscopy (EIS) at room temperature using a cell composed of the working electrode with an exposed area of 300mm^2, a reference electrode of Ag/AgCl and a wire as a counter electrode graphite, all immersed in a solution of sodium chloride (NaCl 3.5%w/v) prepared with distilled water, this solution was chosen because it simulates a marine solution also forming active metals corrode metal chlorides on. Bode plots were obtained and Nyquist performing frequency sweeps in the range of 0.001Hz to 100kHz using an amplitude of the sinusoidal signal of 10mV [7,8].

3. Results and discussion

Interpreting data obtained by analogous modeling requires electrical circuits to physical system studied, also called equivalent circuit. As can be seen in Figures 2, 3 and 4, total impedance in Tables 1, 2 and 3 [9].

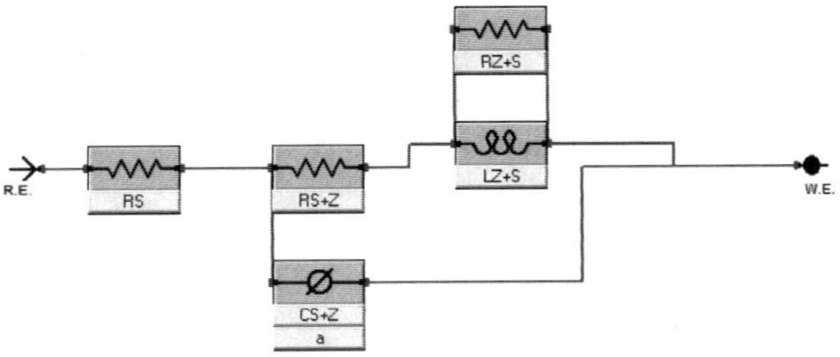

Figure 2. Bare steel wire equivalent circuit.

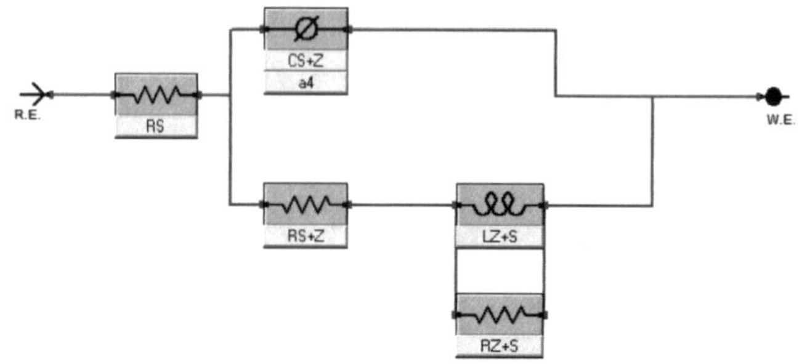

Figure 3. Nitrided specimens equivalent circuit.

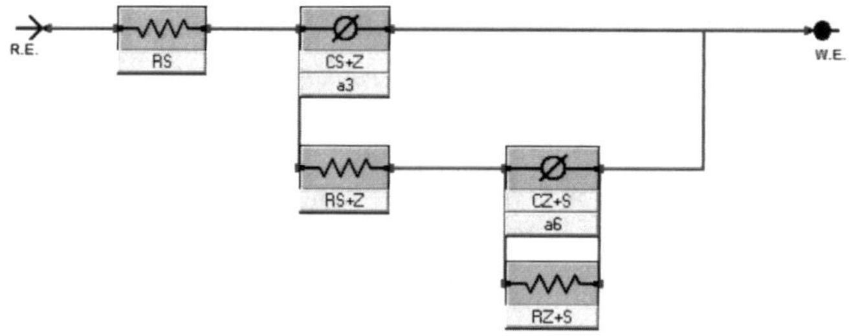

Figure 4. Galvanized specimens equivalent circuit.

Based on these equivalent circuits may be tabulated values of total impedance as seen in Tables 1, 2 and 3.

Table 1. Values of the circuit elements of bare steel wire specimens.

	Bare steel wire					
Specimen	1	2	3	4	5	
RS	3,75E+00	4,39E+00	3,53E+00	3,86E+00	3,07E+00	
RS+Z	3,60E+01	5,08E+01	9,11E+00	2,31E-04	3,50E+01	
CS+Z	1,92E-04	1,33E-04	4,22E-05	1,04E-04	1,47E-04	
a	7,85E-01	8,57E-01	9,84E-01	8,41E-01	8,43E-01	
RZ+S	1,41E+01	2,86E+01	1,07E+01	4,74E+01	1,50E+01	
LZ+S	1,74E+01	2,18E+01	1,81E+00	1,68E+02	1,41E+01	
R TOTAL	5,38E+01	8,38E+01	2,33E+01	5,13E+01	5,31E+01	5,31E+01

Table 2. Values of the circuit elements of the samples nitride.

	Nitrided steel					
Specimen	N1	N2	N3	N4	N5	
RS	1,85E+00	1,43E+00	2,54E+00	1,81E+00	2,16E+00	
RS+Z	4,61E+01	3,68E+01	4,00E+01	1,08E+01	2,34E+01	
CS+Z	2,54E-05	2,87E-05	5,89E-05	1,77E-05	1,94E-05	
a4	9,25E-01	9,09E-01	8,90E-01	9,77E-01	9,30E-01	
RZ+S	4,49E+01	3,15E+01	2,40E+01	1,42E+01	2,10E+01	
LZ+S	2,73E+01	1,62E+01	1,62E+01	2,11E+00	9,51E+00	
R TOTAL	9,28E+01	6,97E+01	6,65E+01	2,68E+01	4,65E+01	7,08E+01

Table 3. Values of the circuit elements of the specimens galvanized.

	Galvanized steel					
Specimen	A	B	C	D	E	
RS	1,35E+00	1,77E+00	3,78E+00	4,53E+00	3,42E-01	
CS+Z	2,46E-04	1,38E-03	2,03E-03	1,82E-03	1,35E-03	
a3	6,83E-01	3,27E-01	7,18E-02	2,79E-01	2,98E-01	
RS+Z	1,57E+02	6,95E-01	2,69E-04	2,62E-06	3,13E+00	
CZ+S	7,30E+03	7,20E-05	8,85E-04	6,06E-05	5,84E-05	
a6	9,15E-01	8,34E-01	6,62E-01	8,53E-01	8,39E-01	
RZ+S	3,65E-09	1,86E+02	3,06E+02	6,99E+02	2,96E+02	
R TOTAL	1,59E+02	1,89E+02	3,10E+02	7,04E+02	2,99E+02	3,32E+02

As shown in the equivalent circuits, impedance equivalent circuit of the test pieces of bare steel was lower in relation to the impedance manifested circuit the nitrided samples, this is due to the implantation of nitrogen ions it has a great influence on the behaviour of the material. Implantation nitriding decreases the hydrogen in steel, so it is important to consider the possibility of using implantation as a tool to prevent corrosion and hydrogen permeation of the wire.

4. Conclusions

It was observed that the impedance of the equivalent circuit increase in the specimens that had undergone the nitriding process with respect to the bare specimens. However it was less than that of galvanized steel, this mainly due to the changes undergone by the steel surface during the implantation process. Therefore it is more advisable to significantly extend the average wiring life, would perform a nitriding and then galvanized. To implement this combined technique is recommended further study.

Acknowledgements

W Aperador, thanks the financial support from the Universidad Militar Nueva Granada, contract number ING-1572.

References

[1] Castaño J G, Botero C, Restrepo A, Correa E and Echeverría F 2010 *Corros Sci* **52** 216
[2] Molina J D, Villada F and Mesa N 2011 *Rev Fac Ing Univ Antioquia* **59** 257
[3] V Niño E D, Peña D and Dugar Zhabon V 2011 *Rev Colomb Física* **43** 712
[4] Mesa D H and Tschiptschin A P 2007 *Sci Tech* **7** 279
[5] Dulce Moreno H J and Contreras de la Ossa G 2009 *Bistua Rev de la Fac Cien Bás* **7-2** 10
[6] V Niño E D, Peña B D, Reyes M and Dugar-Zhabon V 2012 *Rev Latin Metal y Mater* **33** 138
[7] Peña D, Fontalvo P, Estupiñan H, V Niño E D and Vesga W 2009 *Dyna* **76** 43
[8] Castro-Maldonado J J, Dulcé-Moreno H J and V-Niño E D 2013 *J Phys Conf Ser* **466** 8
[9] Castro-Maldonado J J 2014 *Efectos de la nitruración por implantación en alambre de acero recubierto con zinc empleado como refuerzo en los conductores de las redes de transmisión de energia eléctrica* (Cúcuta: Universidad Francisco de Paula Santander)

Artificially modulated hard coatings produced with a vacuum arc evaporator

P A Tsygankov[1], F F Parada Becerra[2], V D Dugar-Zhabon[2], A Plata[2] and E D V-Niño[2,3]

[1] Bauman Moscow State Technical University, Moscow, Russia.
[2] Universidad Industrial de Santander, Bucaramanga, Colombia.
[3] Universidad Politécnica de Madrid, Madrid, España.

E-mail: tsyg@bmstu.ru , deydannv@gmail.com

Abstract. The experimental set for artificially modulated structures production through an advanced vacuum arc evaporator with a magnetically-driven cathode spots on the cathode surface is described. The main features of vacuum arc as a vapor source with time-modulated compositions are discussed. The characteristics of the obtained multilayer coatings with artificially modulated Ti/TiB structures are presented.

1. Introduction

Modern advanced coatings for wide area applications represent the multilayer periodic structures with a bilayer thickness of 15-50nm and the total thickness of up to 10μm. A comprehensive guide to a microstructural design of hard coating is presented in [1]. The multilayers can be manufactured as a structure consisting of different materials, such as TiCN/AlN bilayers, or can be artificially modulated. The latter compositions consist of the layers, each possessing individual distribution of the same chemical element, i.e. the concentration of one or more chemical components is varied periodically, for example, titanium in Ti/TiN structure. Certain multilayer coatings are characterized by a good combination of high hardness and shock-stress resistance, crack and heat resistance properties; that is why these coating are now widely applied in industry. As a rule, the composites of sintered powder materials are widely used in PVD technologies to fabricate the targets. The sintered composite targets based on titanium carbonitride, boride, silicide and aluminide compounds [2-4] are assumed as promising materials for production of the super hard and multifunctional coatings.

One of the most productive sources of metal vapour is a vacuum arc evaporator with an integrally cold cathode [5]. However, a specific property of this type of discharge consists in significant thermal stress in the cathode spots [6] which restricts the use of sintered powder cathodes. The paper presents the experimental results on the artificially modulated Ti-B and Ti-B-N systems through the advanced vacuum arc evaporator with magnetically-driven arc spots.

2. Experimental

The schematic view of the experimental set is shown in Figure 1. A stainless vacuum chamber is pumped out up to the residual pressure of 0.1mPa by 1500l/s turbomolecular pump. The vacuum arc evaporator, which works with the disk-type cathode of 178mm in diameter, is mounted on the chamber top wall opposite the substrates on the bottom flange of the chamber. The arc current can be varied in the range of 100-300A at the voltage of 20-30V [7-9]. The cathode is water cooled to maintain it in integrally cold conditions close to the ambient temperature. It is well known [5,6] that vacuum arc evaporators are highly productive; our experimental set provides thin films growth with a rate up to 10nm/s.

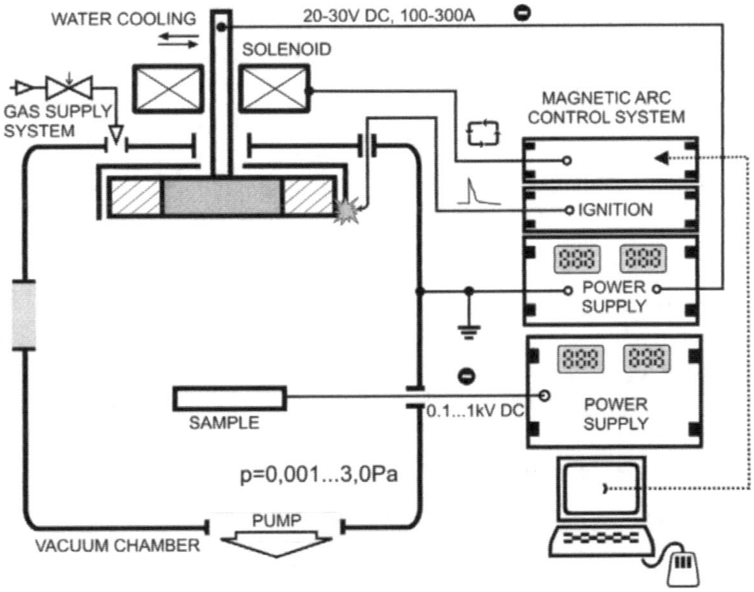

Figure 1. Experimental setup scheme.

To allow the magnetic control over the arc discharge, a solenoid system with magnetic circuit is housed on the back side of the cathode disc to predetermine the localization and motion of the arc spots on the cathode surface. The used solenoid is supplied by DC-modulated current which is specified by a programmed source. Figures 2(a) and 2(b) demonstrate the cathode spot trajectories and their location on the cathode surface according to the solenoid magnetic field amplitude: in the first case (Figure 2(a)) the amplitude is low and the spot trajectories are located near the central area of the cathode in the current feed area where the spots velocity is very high. In the second case (Figure 2(b)), the superposition of the external magnetic field and the field produced by the arc forces the spots to move along the ring trajectories located on the cathode outer area. The duration of the spot stay in different cathode areas is also programmed by the magnetic control system.

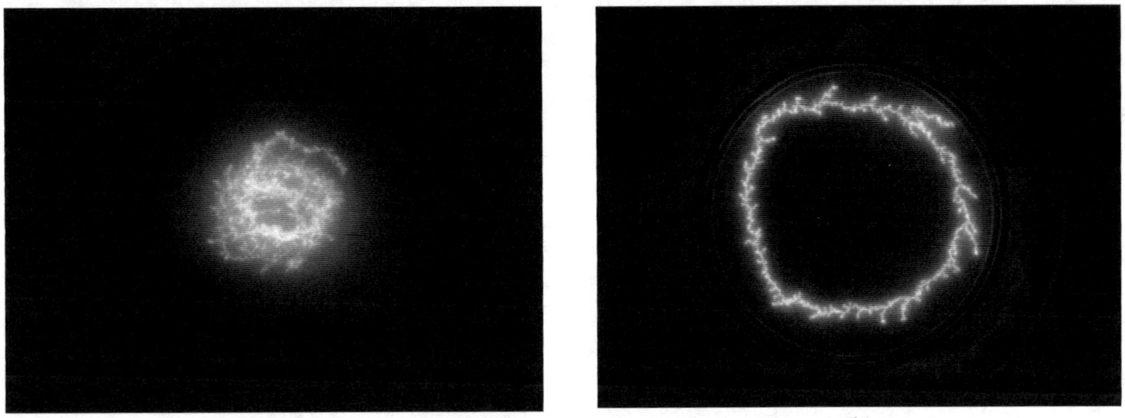

(a) (b)

Figure 2. Cathode spot trajectories for different levels of external magnetic field.

The spot velocity determined experimentally by using a digital registration camera is found in the range of 15-20m/s, which is three times higher than the value obtained when the external magnetic field is turned off. These data correlate well with the concepts of vacuum arc movement on the cathode surface [10,11]. Furthermore, the external magnetic field does not only control the spot trajectories but it produces a higher number of spots in comparison with the case of this field absence. Of importance is the fact that the spot velocity increases and the increase of the total number of spots results in a reduction of the thermal tension in the area where the spots vaporize the cathode material.

The fact that the cathode thermal load can be stepped significantly makes it possible to apply the powder cathodes in the arc coating technologies. For the regular arc evaporators, it is impossible to employ the powder cathode owing to its fissuring at high thermal tensions. It is significant that the use of the arc cathode fabricated of the composite powder stimulates the production of new surface coatings with predicted properties.

To create the multilayer systems, we use a composite cathode which consists of two parts (see Figure 3). The inner part is a disc made of titanium and boron powders mixed in equal atomic proportions (TiB) and later sintered by self-propagated reaction [4]. The external part, which is a plane ring, is made of titanium of 99.9% purity through the traditional metallurgical technology. The external magnetic field which is generated by the solenoid current in the pre-established regime localizes the cathode spot zones in the central or outer cathode areas, thus producing a stoichiometrically modulated film.

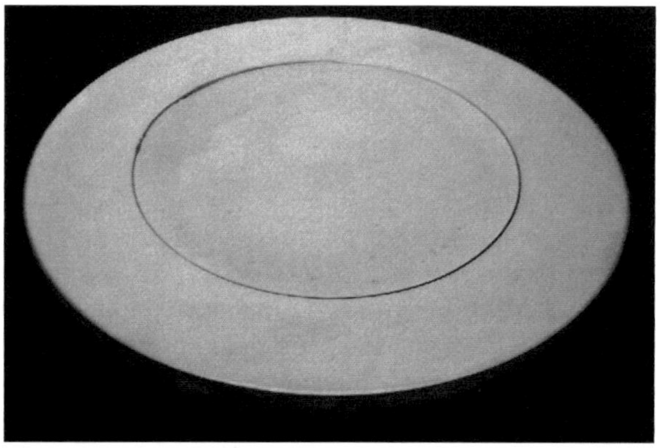

Figure 3. Composite cathode: the inner disc part is formed of mixed titanium and boron powders and the external ring part made of 99.9% titanium.

It is important to note that the vacuum arc discharge generates a highly ionized vapor of the cathode material that makes it possible to control the particle flux onto the substrate and to program the properties of a growing film and its crystallite structure [12]. The detailed mechanism of the thin film growth will be presented in the upcoming papers. This unique feature provides a powerful tool for producing the high quality and industry applicable coatings.

3. Results

Scanning electron microscope (SEM) pattern of cross-section of the multilayers structure obtained through the above described vacuum arc technique is presented in Figure 4. The space period of this artificially modulated structure is of 40.75nm, whose bilayers are composed of deposited titanium and boron titanium layers, Ti/TiB. The arc current of 180A at an operating voltage of 25V provides a growth rate of the multilayer of approximately 3nm/s. The magnetic field is modulated by a rectangular control signal ensuring a periodical localization of the arc in the central part of the disc cathode and its periphery. This technology can easily be extended for processing in the reactive nitrogen atmosphere to guarantee the superhard TiN/TiN-BN systems formation.

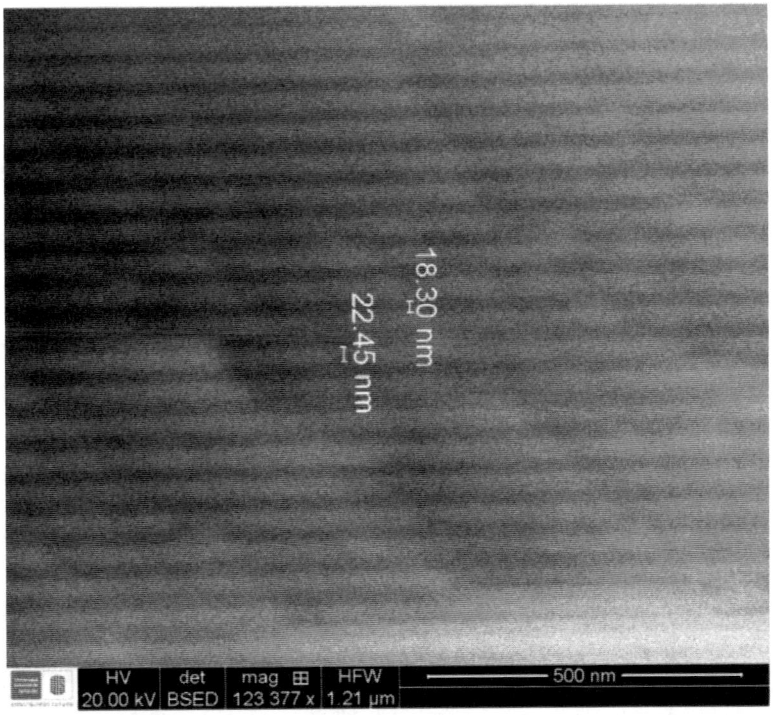

Figure 4. Cross-section of the SEM image of the obtained Ti/TiB multilayer structure.

4. Conclusions

In our experiments, the multilayers with the bilayer structure basic element composed of deposited titanium and boron titanium nanolayers Ti/TiB are obtained. The process of this structure shaping is controlled by a magnetic field in such a manner that the discharge is ignited sequentially on the central disc and the outer ring parts of the composite cathode. Furthermore, the possibility to evaporate composite cathodes made from sintered powder compositions due to decreasing of the local power release level is shown.

References

[1] Mayrhofer P H, Mitterer C, Hultman L, Clemens H 2006 *Progress in Material Science* **51** 1032
[2] Park I W, Kim K H, Kunrath A O, Zhong D, Moore J J, Voevodin A A, Levashov E A 2005 *Journal of Vac Sci Technol B* **23-2** 588
[3] Shtansky D V. Sheveiko A N, Petrzhik M I, Kiryukhantsev Korneev F V, Levashov E A, Leyland A, Yerokhin A L, Matthews A 2005 *Surface and Coatings Technology* **200** 208
[4] Levashov E A, Pogozhev Yu S, Rogachev A S, Kochetov N A, Shtanskiy D V 2012 *Russian Journal of Non-Ferrous Metals* **53-1** 77
[5] Anders A 2008 *Cathodic arcs* (New York: Springer)
[6] Raizer Y P 1991 *Gas discharge physics* (New York: Springer)
[7] F F Parada Becerra, R Cabanzo Hernandez, V Dugar-Zhabon, P Tsygankov, E Mejia-Ospino and E D V-Nino 2012 *Journal of Physics Conference Series* **370** 012054
[8] Dugar-Zhabon V D, Dulcé-Moreno H, Garnica-Villamizar H A and V-Niño E D 2012 *Materials Research* **15-6** 969
[9] Valbuena Niño E D, Tsygankov P, Dugar-Zhabon V, Jaimes Ochoa C J, Parada Becerra F F and Plata A 2011 *Revista Colombiana de Física* **43-2** 458
[10] Sethuraman S K and Barrault M R 1980 *Journal of Nuclear Materials* **93-94** 791
[11] P D Swift, D R McKenzie, I S Falconer 1989 *Journal of Appl Phys* **66-2** 505
[12] W D Munz 1991 *Surface and Coating Technology* **48** 81

Obtaining Au thin films in atmosphere of reactive nitrogen through magnetron sputtering

J H Quintero[1], R Ospina[2] and A Mello[2]
[1] Universidad de Medellín, Medellín, Colombia.
[2] Centro Brasileiro de Pesquisas Físicas (CBPF), Rio de Janeiro, Brasil.

E-mail: jhquintero@udem.edu.co

Abstract. 4d and 5d series of the transition metals are used to the obtaining nitrides metallic, due to the synthesis of PtN, AgN and AuN in the last years. Different nitrides are obtained in the Plasma Assisted Physics Vapour Deposition system, due to its ionization energy which is necessary for their formation. In this paper a Magnetron Sputtering system was used to obtain Au thin films on Si wafers in Nitrogen atmosphere. The substrate temperature was varied between 500 to 950°C. The samples obtained at high temperatures (>500°C) show Au, Si and N elements, as it is corroborated in the narrow spectrum obtained for X-Ray Photoelectron Spectroscopy; besides the competition of orientation crystallographic texture between (111) and (311) directions was present in the X-Ray Diffraction analysis to the sample heated at 950°C.

1. Introduction

Actually, the developments in the binaries nitrides are in the inclusion of element as: Al or Si, the which are calls ternary nitrides of the Me-X-N form (where Me: Transition Metal, N: N and X: B, C, Al or Si), due to that improve the hardness and thermodynamic stability when are compared with binary metal correspondingly [1,2]. Particularly, TiN/Si_3N_4 present high elastic modules, high hardness and high resistance to the oxidized, because, the Si_3N_4 amorphous phase is found as nanocrystals into TiN phase [3]. Besides, different studies has shown the relation between Si content with the hardness in the coatings [4]. On the other side, CrN/Si_3N_4 has been reported with 26 GPa [5], and 34GPa [6] hardness and low friction coefficients due to the presence of Si (9%) [7]. AuN thin films has been produced for: Arc Pulsed and Reactive Ion Sputtering [8-10] and only it has been characterized for N1s narrow spectra, while in the structural part it has shown small widenings in the diffraction pattern to Au-fcc phase, therefore, the research are heading to the obtaining of some change in the Difraction Patterns [11-13]. In this work we present the growth of Au films with N and Si in the surface through a sputtering reactive system and we show the competition of orientation crystallographic texture between the directions (111) and (311) in the Au-fcc phase and through the narrow spectres we show the Au, N and Si in the surface of the samples.

2. Experimental setup

To produce the films a Magnetron Sputtering Reactive (AJA International) was used. The films were deposited on Si wafers of $10mm^2$ in Argon and Nitrogen environment. The substrate temperature was varied from 500 to 950°C with a step size of 100°C, only is shown the samples of 500°C to 950°C, because they were the samples that presented Au, N and Si. The total discharge conditions are shown in Table 1. A X-Ray Photoelectron Spectometer - XPS (Specs) and a X-Ray Diffractometer (XRD, PANalytical) were used to obtaining narrow spectra and diffraction pattern respectively.

Tabla1. Growth conditions for the films.

S	Current (mA)	Pressure (mTorr)	Time (S)	Temperature (°C)	% Nitrogen/Argon	Power supply (Watts)
M1	50	30	180	500	20/80	400
M2	50	30	180	600	20/80	400
M3	50	30	180	700	20/80	400
M4	50	30	180	800	20/80	400
M5	50	30	180	950	20/80	400

3. Results and discussion

In the Figure 1 is shown N1s, Si2p and Au4f narrows spectrum of the sample M5 (950°C), it is not shown the spectrums of the all samples because they present the same behaviour. The presence of N and Si compounds in the surface of the samples can be due to heating of the substrates (N and Si only are present to temperatures above 500°C), because is known that substrate temperature during the deposition process increase the particles mobility and the nucleation sites, the which lead to the densification and the appearing of new stable thermally species [14,15].

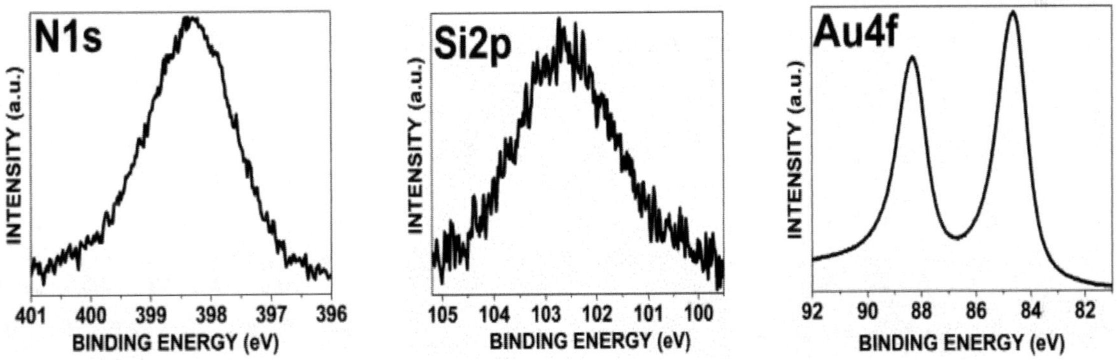

Figure 1. Narrow spectra of N1s, Si2p and Au4f to samples 2A6 (950°C).

It is observed that the Si atoms act as precursor in the hybridization between N and Au, because they always appear with the presence of Si atoms. The low temperatures of the different compounds that are produced in a film depend strongly of the first nucleation stage, which is present in the interface. In the literature is found that the diffusivity and solubility of the Au atoms on Si substrate is increasing with high temperatures (temperatures near to eutectic temperature) and it improving the presence of bonds [16]. When the surface layer has Si-Au eutectic compounds (how would expect in the PAPVD systems), an alloy of liquid phase is produced, which accelerating the processes of diffusivity and mix to reach the composition saturation [17], which generate small Au-Si polycrystalline islands where is possible that the N atoms arrive to the surface [18] and they can find the best stabilization state to the film growth (called crystallization) [19]. Other authors has found that the Au-Si eutectic compound is based on Si diffusion [20], besides that thermal treatment and structural in the interface can be controlled through a-Si or c-Si compounds [21].

In the Figure 2 is shown the superposition of X-Ray Diffraction Patterns, it is observed the characteristic pattern of the Au-fcc phase for all samples. A small change on the range of 54 to 56° were observed for the samples grown to 800 and 950°C (new peaks appear), see Figure 3. In the sample M5 (950°C) a fighting on the preferential orientation between (111) and (311) direction was observed, because the texture coefficient to the M5 sample (950°C) was of 0.62, while to the other samples were approximately of 0.80 (it was found through of the crystallographic texture coefficient between directions (111) and (311)) [11-13]. This change is a big advance to this nitrides class due to that in the literature is found only small widening of the Diffraction Pattern, we want to leave it to discussion to future papers.

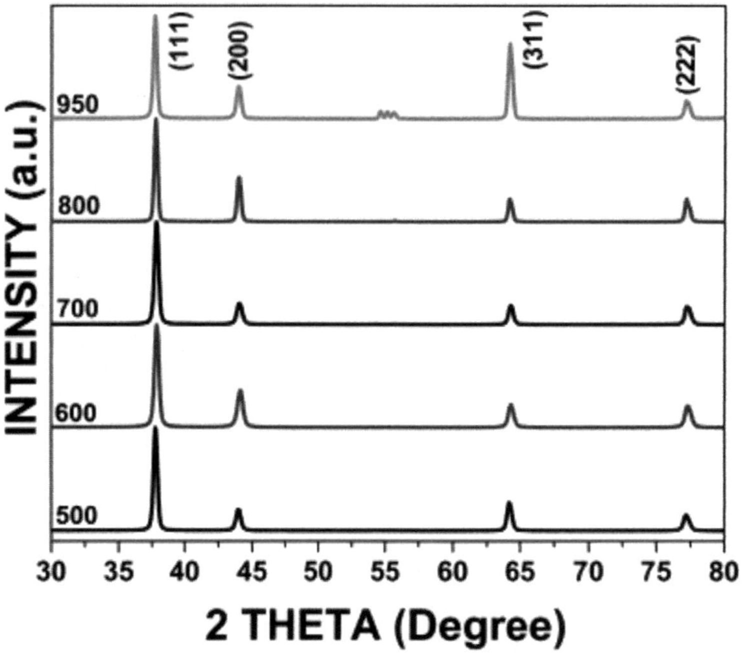

Figure 2. Superposition of X-ray Diffraction Pattern.

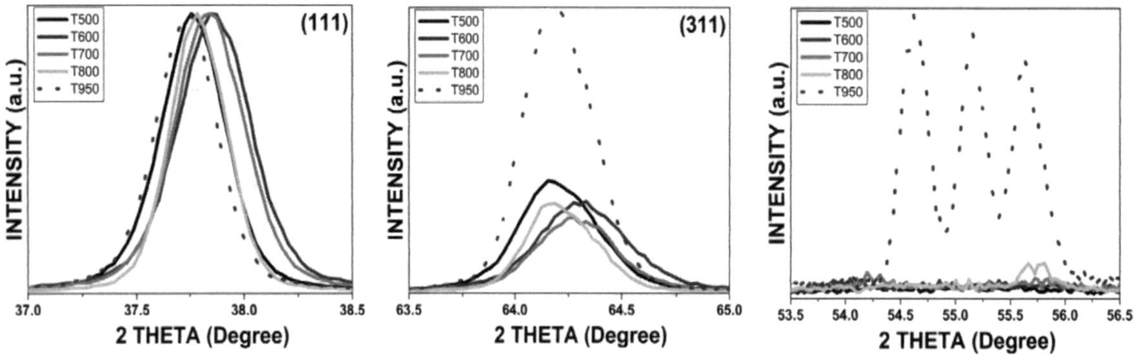

Figure 3. Superposition of direction (111) and (311) and appearing of a new peak in 53 to 56° range.

4. Conclusions

The Au-Si eutectic compounds are highly involved with the N atoms, because the nitrogen appears in the surface when the Si atoms appear. N, Si and Au atoms are favoured for the increase of substrate temperature, because to that, to higher temperatures, the diffraction patterns presented changes in its peaks for example the fighting on the preferential orientation between (111) and (311) peak.

Acknowledgments

The authors gratefully acknowledge the financial support of the vicerrectoría de Investigaciones of the Universidad de Medellín under grant 748.

References

[1] S Zhang, X L Bui, J Jiang, X Li 2005 *Surf Coat Technol* **198** 206
[2] J Musil, H Hruby 2000 *Thin Solid Films* **365** 104
[3] F Vaz, L Rebouta, S Ramos, M F da Silva, J C Soares 1998 *Surf Coat Technol* **236** 108
[4] C Barshilia, B Deepthi, A S Arun Prabhu, K S Rajam 2006 *Surf Coat Technol* **201** 329
[5] E Martinez, R Sanjines, A Karimi, J Esteve, F Levy 2004 *Surf Coat Technol* **570** 180
[6] J H Park, W S Chung, Y R Cho, K H Kim 2004 *Surf Coat Technol* **425** 188
[7] I W Park, D S Kang, J J Moore, S C Kwon, J J Rha, K H Kim 2007 *Surf Coat Technol* **201** 5223
[8] J H Quintero, A Mariño, P J Arango 2013 J Phys Conf Series **466** 012002
[9] J H Quintero, R Ospina, O O Cárdenas, G I Alzate, A Devia 2008 *Phys Scr* **T131** 014013
[10] Yu V Butenko, L Alves, A C Brieva, J Yang, S Krishnamurthy, L Siller 2006 *Chem Phys Lett* **430** 89
[11] A Devia, V Benavides, H Castillo, J Quintero 2006 *AIP Conf Proc* **875** 258
[12] J H Quintero, P J Arango, R Ospina, A Mello and A Mariño 2015 *Surf Inter Anal* **47** 701
[13] L Alves, T P A Hase, M R C Hunt, A C Brieva, L Siller 2008 *J Appl Phys* **104** 113527
[14] A Anders 2002 *Surf & Coat Techn* **156** 3
[15] R Ospina, D Escobar, E Restrepo-Parra, P J Arango, J F Jurado 2013 *Tribol Inter* **62** 124
[16] Y T Cheng, L Lin and K Najafi 2000 *Journal of Microelectromechanical System* **9** 3
[17] R F Wolffenbuttel and K D Wise 1994 *Sensors and Actuators A Physical* **43** 223
[18] Y C Lin, M Baum, M Haubold, J Fromel, M Wiemer, T Gessner and M Esashi 2009 *Proceedings of International Solid-State Sensors Actuators and Microsystems Conference (TRANSDUCERS 2009)* 244 (DOI: 10.1109/SENSOR.2009.5285519)
[19] R M Walser and R W Bene 1976 *Appl Phys Lett* **28** 624
[20] R F Wolffenbuttel 1997 *Sensors and Actuators A Physical* **62** 680
[21] M Novakovica, M Popovic, K Zhang, K P Lieb, N Bibic 2014 *Appl Surf Sci* **295** 158

Obtaining Au thin films in atmosphere of reactive nitrogen through magnetron sputtering

J H Quintero[1] , R Ospina[2] and A Mello[2]

[1] Universidad de Medellín, Medellín, Colombia.
[2] Centro Brasileiro de Pesquisas Físicas (CBPF), Rio de Janeiro, Brasil.

E-mail: jhquintero@udem.edu.co

CORRIGENDUM TO: J H Quintero, R Ospina and A Mello 2016 *J. Phys.: Conf. Ser.* **687** 012006

The authors would like to add an affiliation to this work. The correct list of authors and affiliations for the paper 'Obtaining Au thin films in atmosphere of reactive nitrogen through magnetron sputtering' is:

Obtaining Au thin films in atmosphere of reactive nitrogen through magnetron sputtering

J H Quintero[1], R Ospina[2,3] and A Mello[2]

[1] Universidad de Medellín, Medellín, Colombia.
[2] Centro Brasileiro de Pesquisas Físicas (CBPF), Rio de Janeiro, Brasil.
[3] Universidad Industrial de Santander, Bucaramanga, Colombia.

E-mail: jhquintero@udem.edu.co

Comparison of three systems of solar water heating by thermosiphon

E Hernández[1] and R E Guzmán[1]
[1] Universidad Pontificia Bolivariana, Floridablanca, Colombia.

E-mail: emil.hernandez@upb.edu.co

Abstract. The main purpose of this project was to elaborate a comparison between three water heating systems; using two plane water heating solar collector and another using a vacuum tube heater, all of them are on top of the cafeteria´s roof on building of the "Universidad Pontificia Bolivariana" in Bucaramanga, Colombia. Through testing was determined each type of water heating systems´ performance, where the Stainless Steel tube collector reached a maximum efficiency of 71.58%, the Copper Tubing Collector a maximum value of 76.31% and for the Vacuum Tube Heater Collector a maximum efficiency of 72.33%. The collector with copper coil was the system more efficient. So, taking into account the Performance and Temperature Curves, along with the weather conditions at the time of the testing we determined that the most efficient Solar Heating System is the one using a Vacuum Tube Heater Collector. Reaching a maximum efficiency of 72.33% and a maximum temperature of 62.6°C.

1. Introduction

Colombia emits 62 million tons of CO_2 per year and the United States about 5286, the latter, is one of the countries with the highest carbon footprint. A solution to these problems is the use of solar energy [1].

Use of the sun energy by the human being it is lost in the time. However, applying the knowledge gained from ancient times and the development of new materials, it has created applications for obtain welfare at home. The water tank exposed to sun in the past, has led to modern solar collectors to have domestic hot water (DHW), heating and similar services, with the same purpose as photovoltaic and wind systems [2].

This investigation is about study of testing three different types of solar water heaters. Two flat plate collectors with different material of the coil, one with stainless steel coil, the other with a copper coil and vacuum tubes are analysed. This in order to do a comparison of their performance curves and determine the optimal system. This project seeks to find a balance between environment and economy. Solar domestic hot water system by thermosyphon, where the liquid flows from the solar collector to the storage tank by buoyancy forces, has spread widely in the field of practice and research [3]. Numerous studies have investigated the performance of solar water heaters operating by thermosiphon [4-7]. Burbano et al. [4] they built a thermosyphon solar water heater with an interesting design process. Zerrouki A. et al. [5] found experimentally that the rate of mass flow increases in the hour of greatest radiation (11*am* - 2*pm*). Sakhrieh and Al-Ghandoor [6] presented an experimental study of the performance of different types of solar collectors and conclude that the vacuum tube collector is the top performer with respect to efficiency, with a maximum value of 72%. Gupta and Garg [7] experimentally proved that the flow rate of a thermosyphon water heater can be increased if the relative height between the collector and the storage tank is increased, but the efficiency is not increased.

2. Theory

Energetic balance of a flat collector is:

$$Q_l=Q+Q_2 \tag{1}$$

Where:
Q_l: is the incident energy (direct+diffuse+albedo) in the unit time.
Q: is useful energy, by collecting the heat transfer fluid.
Q_2: is the energy lost by dissipation to the environment.

Given the definition of intensity of radiant energy, Q_l is the product of the intensity and the surface. Not all the incident energy Q_l will be absorbed in the absorber. First, if there is deck, you have to have the transmittance in a count, which will leave to flow part of this energy (τSI). Moreover, the absorption coefficient α or absorptance of the absorber plate never becomes equal to unity, so that the fraction of energy actually absorbed is:

$$Q_l=\tau\alpha SI \tag{2}$$

As for the energy lost Q_2, detailed calculation is very complex because, must be taken into account simultaneously and in different proportions losses by radiation, convection and conduction. However, in order to be able to use a simple formulation, it has been agreed to include these influences in the overall loss coefficient called U, which is measured experimentally and is data supplied by the manufacturer. Experience has shown that supposing losses per unit of surface proportional to the difference between the average temperatures t_C of the absorber plate and the environment t_a, being the proportionality factor the coefficient U, is a good approximation. So that:

$$Q_2=SU(t_c-t_a) \tag{3}$$

Substituting (2) and (3) in (1) is obtained:

$$Q=S[I(\tau*\alpha)-U(t_c-t_a)] \tag{4}$$

Where:
S: collector surface.
I: radiant intensity (*W/m²*).
τ: transmittance of the transparent cover.
α: absorptance of the absorber plate.
U: overall loss coefficient ($W/(m^2 °C)$).
t_c: average temperature of the absorbing plate (*°C*).
t_a: environment temperature (*°C*).

The efficiency ratio of the collector, η, is defined by the relationship between energy captured and received at a given instant according to the Equation (5).

$$\eta=Q/(SI) \tag{5}$$

Substituting (4) in (5) is obtained (6)

$$\eta=F_R[(\tau\alpha)-U(t_m-t_a)/I] \quad (6)$$

Where t_m is the mean temperature of the collector according to the Equation (7).

$$t_m=(t_e+t_s)/2 \quad (7)$$

For a flat plate collector the Equation (6) is approximate to the Equation (8).

$$\eta=0.83-6.8(t_m-t_a)/I \quad (8)$$

3. Methodology

The solar collector has the dimensions 1.6*m**0.8*m**0.13*m* with stainless steel plates and copper according to Figure 1. Internally the collector has seven copper tubes 1.5*m* long and 1.5 inches in diameter. The collector has a tempered glass dimensions 1.6*m**0.8*m*, which determines the catchment area of solar radiation.

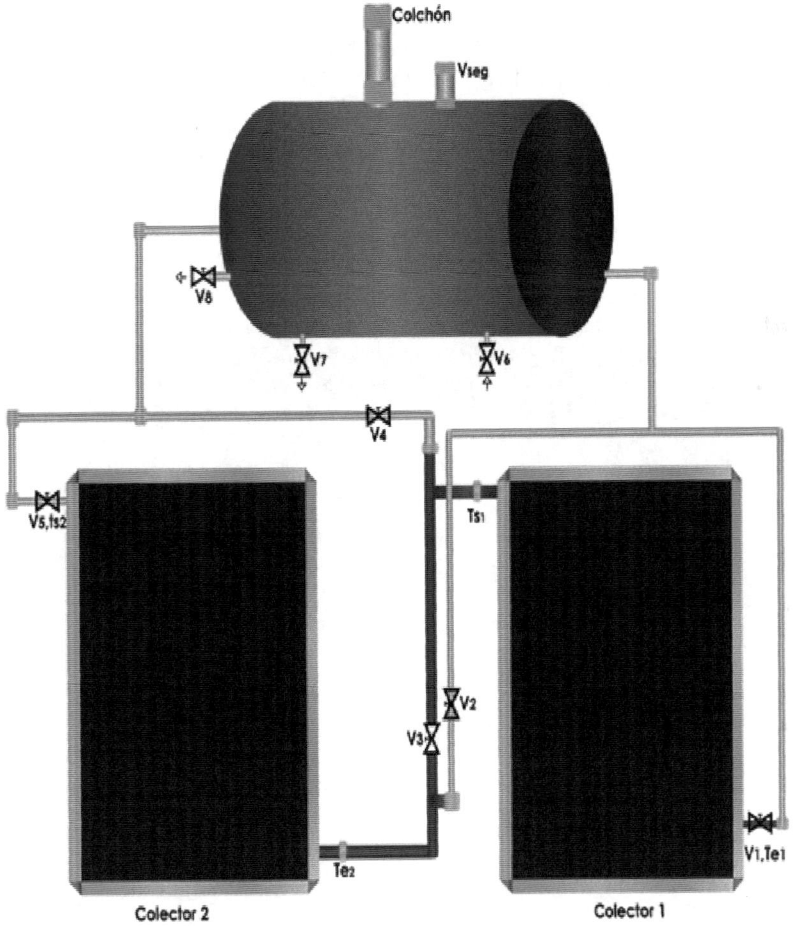

Figure 1. Thermosiphon system.

Radiation I(W/m^2) was measured with a pyranometer Kipp & Zonen CM3 and temperatures with type K thermocouples located at the ends of the manifold, the results shown in Figure 2, are obtained.

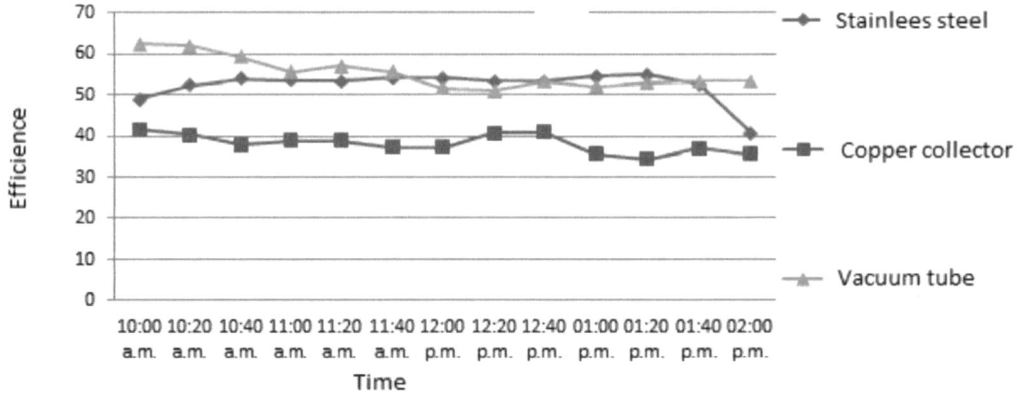

Figure 2. Efficience vs time for three systems.

4. Conclusions

The system of solar water heating was implemented, two flat plate collectors, giving as a result of flat plate collectors with different material coil, one coil of copper and the other with stainless steel coil.

The maximum efficiency for the collector with stainless steel coil was 71.58% (±0.0099%) of 76.31% (±0.0003%) for the collector with copper coil and 72.33% (±0.0063%) for the vacuum tube collector.

References

[1] T Boden, G Marland, and R Andres 2013 *Global, regional and national fossil-fuel CO_2 emissions carbon dioxide information analysis center* (USA: Department of Energy) p 1751
[2] T Perales 2006 *Guía del instalador de energías renovables* (Mexico: Ed Limusa) p 145
[3] B Norton, S D Probert 1986 *Thermosiphon Solar Energy Wáter Heater* **3** 125
[4] C Burbano, H Restrepo y J Sabogal 2006 *Sciencia et Technica* **31** 85
[5] A Zerrouki, A Boumédien y K Bouhadef 2002 *Renewable Energy* **26** 549
[6] A Sakhrieh, A Al-Ghandoor 2013 *Energy Conversion and Management* **65** 715
[7] Cl Gupta, HP Garg 1968 *Solar Energy* **12** 163
[8] Censolar 2008 *Curso de proyectista instalador de energía solar tomo 3* (España: Censolar)

Effects of texturization due to chemical etching and laser on the optical properties of multicrystalline silicon for applications in solar cells

D Vera[1], J Mass[2], M Manotas[2], R Cabanzo[3] and E Mejia[3]
[1] Universidad de la Costa (CUC), Barranquilla, Colombia.
[2] Universidad del Atlántico, Barranquilla, Colombia.
[3] Universidad Industrial de Santander, Bucaramanga, Colombia.

E-mail: davem1804@gmail.com

Abstract. In this work we carried out the texturization of surfaces of multicrystalline silicon type-p in order to decrease the reflection of light on the surface, using the chemical etching method and then a treatment with laser. In the first method, it was immersed in solutions of $HF:HNO_3:H_2O$, $HF:HNO_3:CH_3COOH$, $HF:HNO_3:H_3PO_4$, in the proportion 14:01:05, during 30 seconds, 1, 2 and 3 minutes. Subsequently with a laser (ND:YAG) grids were generated beginning with parallel lines separated $50\mu m$. The samples were analyzed by means of diffuse spectroscopy (UV-VIS) and scanning electron micrograph (SEM) before and after the laser treatment. The lowest result of reflectance obtained by $HF:HNO_3:H_2O$ during 30 seconds, was of 15.5%. However, after applying the treatment with laser the reflectance increased to 17.27%. On the other hand, the samples treated (30 seconds) with acetic acid and phosphoric acid as diluents gives as a result a decrease in the reflectance values after applying the laser treatment from 21.97% to 17.79% and from 27.73% to 20.03% respectively. The above indicates that in some cases it is possible to decrease the reflectance using jointly the method of chemical etching and then a laser treatment.

1. Introduction

Silicon is the main element in the construction of solar cells, besides its extensive applications in the electronics industry for the construction of semiconductor devices. The multicrystalline silicon is the one that is used the most in the manufacture of terrestrial solar cells [1] due to its low cost compared to monocrystalline silicon. Since that the first solar cells were manufactured with multicrystalline silicon, was tried to implement methods to improvement of the efficiency. One methods is texturization through a chemical etching on surface of the cell. [2-4]. The acid solutions are based on a combination of flourhydric acid with nitric acid, this composition has been studied very much, as well as its effects on the optical and electrical properties [5], and the alkaline solutions are based on a solution of potassium hydroxide or sodium hydroxide, and it is the method that is currently used for the texturization of monocrystalline silicon that is implemented in the manufacture of the solar cells. Another method in order to texturize is by means of laser beam generating grooves on the surfaces in the shape of parallel lines and in the shape of grids [6,7]. The objective of this work is to see the effects caused on the optical properties of multicrystalline silicon in the process of texturization by means of chemical etching and laser applying the two methods simultaneously.

2. Experimental methods

The samples used in this work were of multicrystalline silicon type p, cut in 2*cm*x2*cm* pieces, with a thickness of 200μm and a resistivity of 0.5$\Omega .m$. First we carried out texturizations of multicrystalline silicon samples by means of chemical etching of HF: HNO_3, with three different diluents: HF:HNO_3:H_2O, HF:HNO_3: CH_3COOH, HF:HNO_3:H_3PO_4, in the proportion 14:01:05, during 30 seconds, 1, 2 and 3 minutes. Subsequently on the surface of these texturized samples, grooves of 20μm of width were generated in the form of parallel lines separated 50μm from center to center, using a laser pulsed of Nd:YAG with a wave longitude of 1064*nm*, a pulse power of 265*mW* and a pulse energy of 94.3*mJ*. For the generation of the groove (line) in a straight line on the sample, it was passed through a cylindric lens that was 2.2*cm* of diameter, separated from the sample 12.8*cm*, 3 pulses of laser beam. Then, beginning with these parallel grooves grids with grooves perpendicular to the first ones were generated and also at 50μm of separation. The results obtained were compared with a sample of raw multicrystalline silicon. Each one of the samples was treated before and after the laser application, they were analyzed by means of diffuse spectroscopy using a Shimadzu 2600 UV–VIS spectrophotometer and scanning electron microscopy (SEM) using a Joel JSM 5600 microscope.

3. Results and discussion

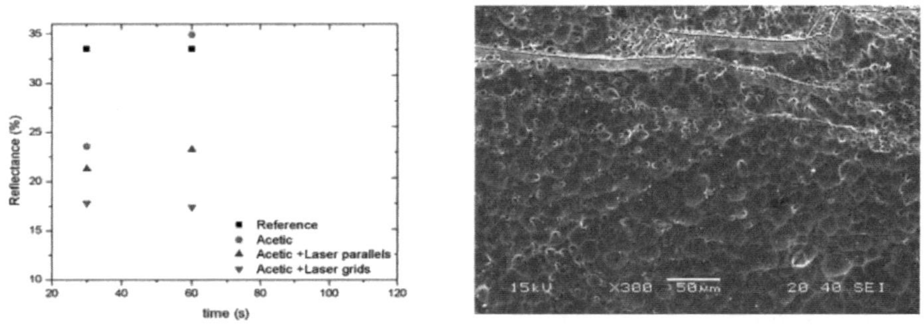

(a) (b)

Figure 1. (a) Average Reflectance for samples that were etched chemically with acetic acid during 30 seconds and 1 minute and then with laser in the form of parallel lines and grids. (b) Images obtained by SEM micrograph (300X)for a sample chemically etched using acetic acid as diluent during 1 minute.

3.1. Chemical texturization with acetic acid as diluent and laser over-texturization

In Figure 1(a), shown average reflectance of the raw samples, those texturized with acetic acid as diluent, and those that have the laser over-texturized in parallel grooves and in grids for 30 and 60 seconds of chemical texturizing. These values show a single trend for the two times indicated, in the decrease of the average reflectance, and we could observe that beginning with the raw sample with an average reflectance of 33%, and up to the grooves in grid with laser on the chemically texturized samples, we obtained a decrease of this coefficient up to 17.41% of average reflectance. However, an anomalous behavior occurs for the chemical texturization with the acetic acid as diluent for a period of 60 seconds, overcome the average reflectance of the raw sample. It is possible that when we increase the time with the acetic acid as diluent this produces on the surface something like bowls whose diameter - depth ratio increases, which could produce an increase in the average reflectance. In Figure 1(b), we show the SEM micrograph of this chemical texturization at 60 seconds, and we could observe a morphology that is mostly uniform with the semblance of depressions of an almost circular section with the diameter – depth (d/h) ratio greater than the unit and some sectors that suggest formations of defects. However, in the SEM micrograph of the sample over – texturized with laser, we observed a greater homogenization of the surface, but now the width - depth (W/H) ratio, still continues being greater than the unit. In the SEM micrograph show the laser grid, and we could observe a much rougher surface that includes some defects of the sample. These grids have the shape of trunks and they generate a greater trapping of the light in accordance with the smaller average reflectance that was obtained.

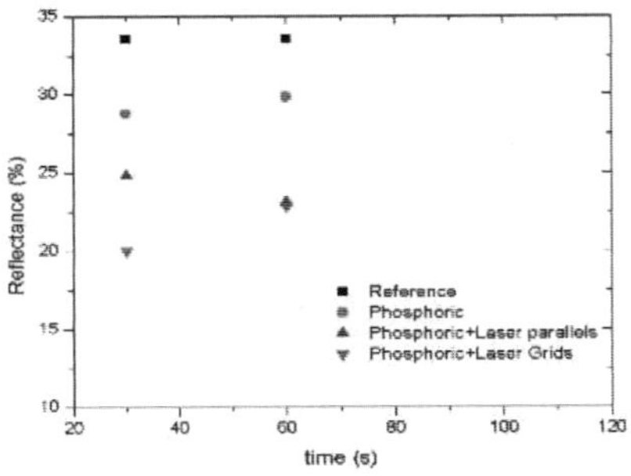

(a)

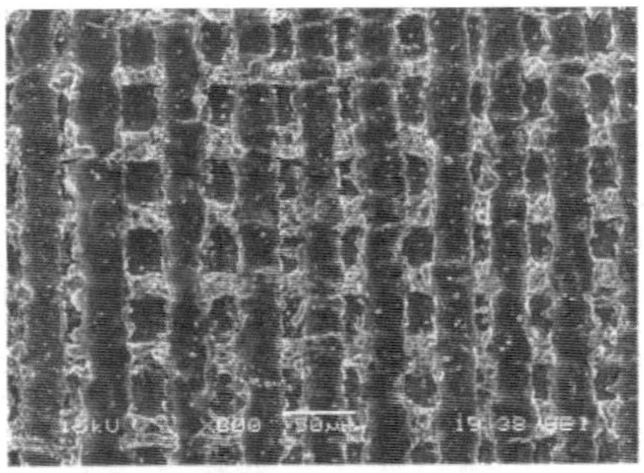

(b)

Figure 2. (a) Average reflectance for samples that were etched chemically with phosphoric acid during 30 seconds and 1 minute and then with laser in the form of parallel lines and grids. (b) SEM micrograph (300X) Sample initially etched with acetic acid as diluent during 1 minute and then with laser in grids.

3.2. Chemical texturization with phosphoric acid as diluent and laser over-texturization

In Figure 2(a), we could observe that the samples that were chemically texturized with phosphoric acid as diluent and then over-texturized with laser during 30 and 60 seconds decrease their average reflectance with a very similar trend, and we were able to reduce this coefficient by tracing the grooves in grids with the laser up to a minimum of 20.03% of average reflectance. In this figure we can also observe that the minimum average reflectance, is very close for the parallel grooves and of the grooves gridded, which suggests that the effective area of reflection of the light is not modified significantly for the two samples over-texturized with the laser. Also, we could observe in SEM micrographs that texturization with phosphoric acid generates deeper depressions towards the inside of the sample but with a distribution that is more spaced out. This produce an average reflectance a smaller than the one obtained with the acetic acid as diluent. In the Figures 2(b) we can see the surfaces in grids. In these micrographs we can see that the draw of the laser in order to form grids produced the shaping of trunks in an alternated way, which makes it possible that the average reflectance does not differentiate itself significantly with respect to the average reflectance obtained with the parallel grooves.

3.3. Texturization with deionized water as diluent and laser over-texturization

The samples that were chemically texturized with deionized water as diluent decreased their average reflectance by 15.55% and increased after applying the over texturization with laser in the form of parallel lines and in grids, as we can see in Figure 3(a). In the micrograph of the Figure 3(b), we can observe a very uniform morphology for the chemically texturized sample, we do not see defects such as boundaries that the etching is mostly isotropic, with a roughness the like undulations of which contribute to a better trapping of the incident light. With the application of the laser we obtain a flat effective surface with a width depth ratio that is much greater than the unit, as we can observe in the Figures 3(c) and 3(d).

The draw of grids does not have a great influence on the decrease of the average reflectance. This result suggests that the over-texturization with laser for samples that were chemically texturized with deionized water as diluent, do not introduces any improved.

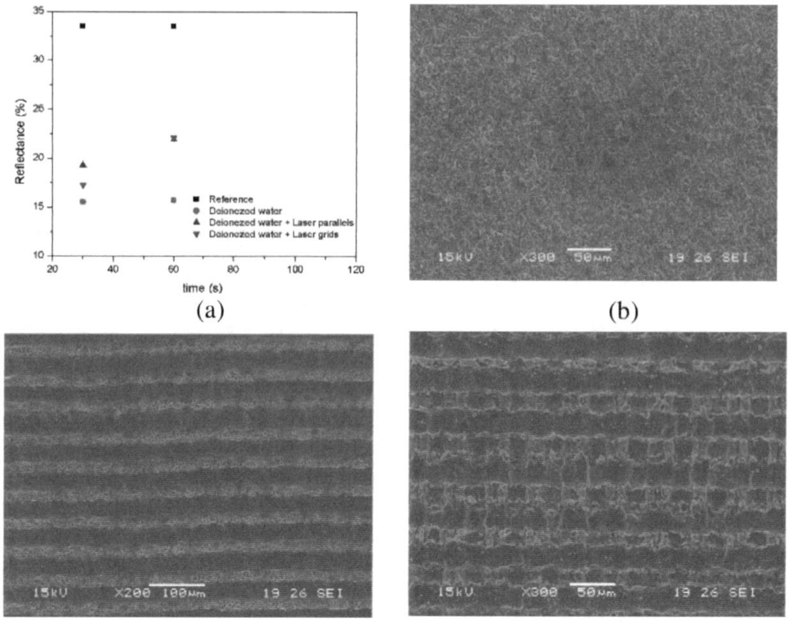

(a) (b)

(c) (d)

Figure 3. (a) Average reflectance for samples than were chemically etched with deionized water as diluent, and then with laser in the form of parallel lines and grids. (b) Images SEM (300X) for a sample that was chemically etched using deionized water as diluent during 1 minute, (c) Images SEM (200X) of sample that was etched initially with deionized water as diluent during 1 minute and then textured with laser in the form of parallel lines, (d) Images SEM (300X) of sample that was etched initially with deionized water as diluent during 1 minute and then with laser in grids.

4. Conclusions

We applied chemical texturization and laser methods simultaneously in order to reduce the reflectance on the surface of the multicrystalline silicon. The lowest reflectance results were obtained with the solution $HF:NHO_3:H_2O$ in the proportion 14:01:05 during 30 seconds with a value of 15.55%. Besides, this solution shown that the etching is totally isotropic and a very uniform morphology is obtained. For the samples that were treated with this solution it is not convenient to treat them with laser, because increases the reflectance applying the laser in the form of parallel lines or in grids. On the other hand, the samples that were etched initially with acetic acid and phosphoric acid decrease the average reflectance with respect to the reference sample, and they decrease even more when we apply over those surface etched the laser texturization in the form of parallel lines as also in the form of grid.

References

[1] J L Balenzategui Manzanares 2007 *Modúlo: Gestión eficiente de la energía, tecnologías de celdas solares de silicio cristalino* (Madrid: EOI Escuela de Negocios)
[2] Chapin D M, Fuller C S and Pearson 1954 *J Appl Phys* **25** 676
[3] U Gangopadhyay and S K Dhungel 2007 *Solar Energy Materials & Solar Cells* **91** 285
[4] P Panek, M Lipinski and J Dutkiewcz 2005 *Journal of materials science* **40** 1459
[5] G Kulesza, P Panek and P Zięba 2014 *Archives of Civil and Mechanical Engineerin* **14-4** 595
[6] L A Dobrzanski and A Drygała 2008 *J of Achieves in Mater and Manufac Enginer* **29-1** 7
[7] L A Dobrzanski and A Drygała 2007 *Journal of Materials Processing Technology* **191** 228

Effects of texturization due to chemical etching and laser on the optical properties of multicrystalline silicon for applications in solar cells

D Vera[1], J Mass[2], M Manotas[2], R Cabanzo[3] and E Mejia[3]

[1] Universidad de la Costa (CUC), Barranquilla, Colombia
[2] Universidad del Atlántico, Barranquilla, Colombia.
[3] Universidad Industrial de Santander, Bucaramanga, Colombia

E-mail: davem1804@gmail.com

CORRIGENDUM TO: D Vera, J Mass, M Manotas, R Cabanzo and E Mejia 2016 *J. Phys.: Conf. Ser.* 687 012008

The authors would like to correct an affiliation in this work. The correct list of affiliations for the paper 'Effects of texturization due to chemical etching and laser on the optical properties of multicrystalline silicon for applications in solar cells' is:

Effects of texturization due to chemical etching and laser on the optical properties of multicrystalline silicon for applications in solar cells

D Vera[1], J Mass[2], M Manotas[2], R Cabanzo[3] and E Mejia[3]

[1] Universidad de la Costa (CUC), Barranquilla, Colombia
[2] Universidad del Norte, Barranquilla, Colombia.
[3] Universidad Industrial de Santander, Bucaramanga, Colombia

E-mail: davem1804@gmail.com

Estimation of incident solar radiation on the roof of the cultural and sports university centre of the Foundation University Los Libertadores

S A Jiménez[1], V M Carrillo[1] and L C Rátiva[1]
[1] Fundación Universitaria Los Libertadores, Bogotá, Colombia.

E-mail: sajimenezc@libertadores.edu.co , lcrativap@libertadores.edu.co

Abstract. This document shows the estimate of the total solar irradiance incident for the set of solar collectors to be located on the roof of cultural and sports university centre (CSUC) of the Foundation University Los Libertadores (FULL) in Bogotá, Colombia, and they will be part of the climate control system of the pool built inside. The calculation was based on experimental data of global solar radiation on the horizontal surface on March, July, October, November and December, through the three most commonly models used to determine the total solar radiation on tilted surfaces: isotropic sky, HDKR and Perez. The results show differences of less than 5% between the values calculated by the three models for December, the month with lower irradiance. For this month, reductions up to 15% and 19% were observed in the estimated irradiance, relative to those obtained on a horizontal surface on a surface under ideal orientation and inclination, respectively.

1. Introduction

Pool heating for swimming is a process that consists on raising and maintain temperature of water on values generally ranging between 22°C and 27°C approx., depending on use, to ensure the comfort and enjoyment of users. When this temperature is below 18°C, the sensation in the pool is unpleasant. This is one of the most profitable ways to use solar energy because the temperatures are not high, and the energy demand matches with periods of increased solar radiation [1].

When any solar thermal system is designed, the solution must ensure the highest energy efficiency with maximum architectural integration [2]. To ensure the first requirement, it should be determined as accurately as possible, both energy demand of the system to be conditioned as solar radiation available at the site where the solar collectors will be located.

Most of the radiation data are available for horizontal surfaces and simultaneously include the Diffuse and the Direct. However, there are a few solar thermal systems in which collectors are set horizontally and in many applications, for aesthetic or architectural reasons, the emplacement of the collector does not allow optimal positions of azimuth and inclination. These situations cause losses of solar radiation and therefore it is important to calculate its magnitude [2].

Some related studies about the use of models for estimating the solar radiation on solar collecting surfaces are: Skeiker [3], Mendoza [4] and Chang [5].

Moreover, among the studies that explicitly report the calculations of radiation data for the design of climate control system using solar energy for pool heating are: Szeicz and McMonagl [6], Rakopoulos and Vazeos [7], Singh et al. [8] and Woolley et al. [9].

In this paper the application of equations of isotropic sky, HDRK and Perez mathematical models are presented for estimating total solar irradiance incident on the roof of CSUC and thus to know if this amount differs significantly from the data on a horizontal surface.

The CSUC of FULL is a three-story building, located in Bogota, Colombia, approximately on 4°39'6" North Latitude and West Longitude 74°03'56.3"and azimuth +102°. Its roof is tilted with two flaps ("two waters"), both with 9° of slope. The enclosures where the water does not fall, end up in a triangular shape.

2. Methodology

Based on 180748 measurements made every minute by a precision spectral pyranometer EPPLEY PSP, from July 9 to 27, 2011, October 29 to December 13, 2011, March 2 to 3, 2012 and March 7 to 10, 2012, from 6:00 to 18:00, horizontal solar irradiance monthly average schedules for five typical days of the months March, July, October, November and December were obtained. These averages and three sky models were used to calculate the monthly hourly averages of the total solar irradiance for tilted surface G_T (CSUC roof) and for the same surface in optimum orientation (inclination=4° and surface azimuth=0°). Relating the averages obtained, inclination and optimum orientation factors K_1 y K_2 were calculated.

3. Models for calculation of solar radiation on tilted surfaces

The data of solar radiation on tilted surfaces can be measured experimentally or can be estimated by mathematical models. This second approach is the one used in this work.

Equations 1, 2 and 3 show the most used models to calculate the total irradiance on tilted surface G_T: isotropic sky, HDKR and Perez [10, 12].

$$G_T = G_b R_b + G_d\left(\frac{1+\cos\beta}{2}\right) + G\rho_g\left(\frac{1-\cos\beta}{2}\right) \quad (1)$$

$$G_T = (G_b + G_d A_i)R_b + G_d(1-A_i)\left(\frac{1+\cos\beta}{2}\right)\left[1 + f\,\mathrm{sen}^3\left(\frac{\beta}{2}\right)\right] + G\rho_g\left(\frac{1-\cos\beta}{2}\right) \quad (2)$$

$$G_T = G_b R_b + G_d(1-F_1)\left(\frac{1+\cos\beta}{2}\right) + G_d F_1 \frac{a}{b} + G_d F_2 \mathrm{sen}\beta + G\rho_g\left(\frac{1-\cos\beta}{2}\right) \quad (3)$$

These three models, with some variations, include direct, diffuse (isotropic, circumsolar and horizon) and reflected contributions (see Figure 1). The latter is not considered here because the CSUC roof is the uppermost surface of those found in the vicinity of the building.

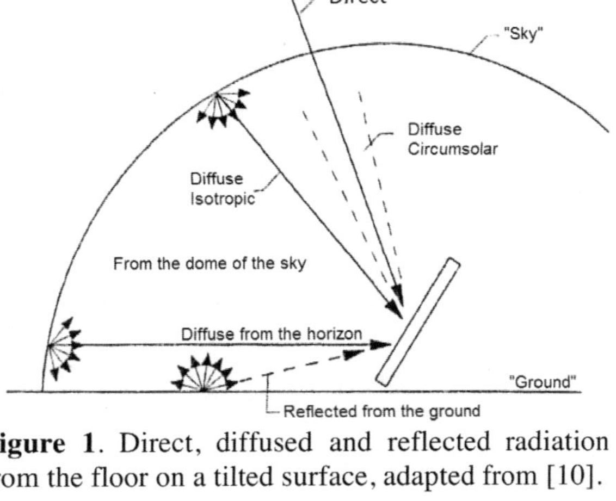

Figure 1. Direct, diffused and reflected radiation from the floor on a tilted surface, adapted from [10].

4. Results and discussion

4.1. Month identification with less favourable irradiance

Figures 2, 3 and 4 show the hourly averages of the total solar irradiance on the tilted surface, estimated from isotropic Sky, HDKR and Perez models, for the months for which experimental data were available.

From the Figures 2, 3 and 4, we can make the following general observations: a) except at 6:00 and 18:00, the figures are very similar, which means that the three models are suitable for predicting total irradiance on tilted surface b) December is having a larger number points of time with lower irradiance.

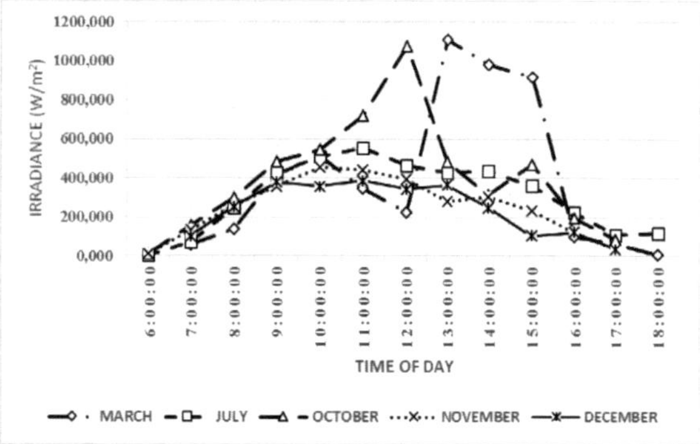

Figure 2. Solar radiation available, tilted surface–HDKR model.

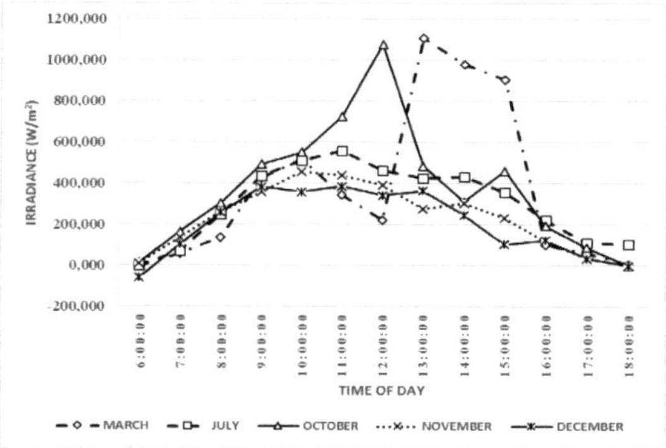

Figure 3. Solar radiation available, tilted surface–Isotropic Sky model.

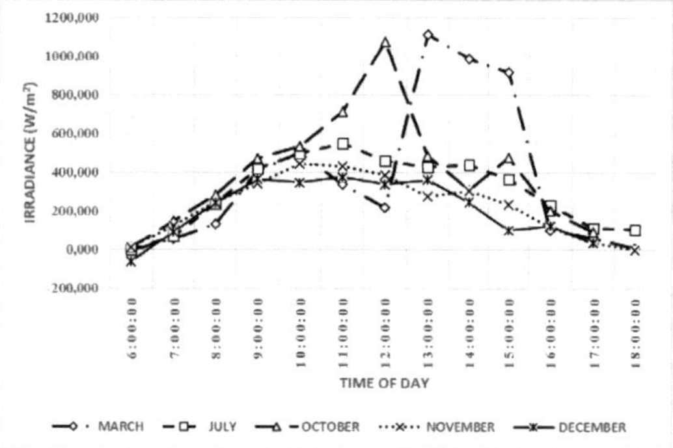

Figure 4. Solar radiation available, tilted surface–Perez model.

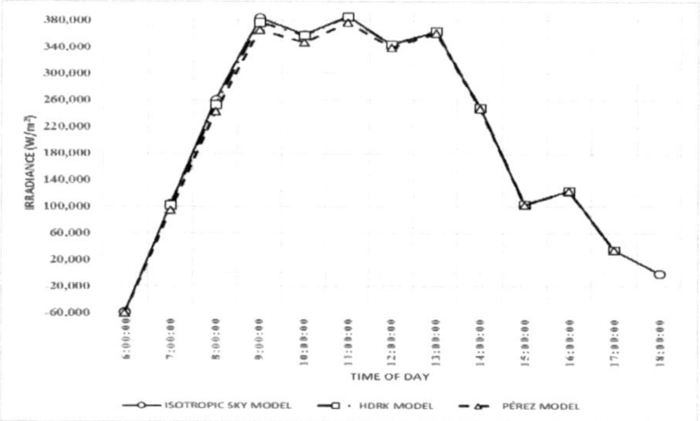

Figure 5. Solar radiation available, tilted surface–December.

4.2. Selection of the prediction model

Figure 5 show the hourly averages of the global solar irradiance on the tilted surface, for December, from 6:00 to 18:00, obtained by isotropic sky, HDKR and Perez models. The Perez's model shows some slightly lower irradiance levels from 7:00 to 9:00, with even lower values from 9:00 to 13:00 and with coinciding values with those of the other models, from 13:00. Following the criteria of the worst case condition, Perez model was selected to predict the values of global solar irradiance on the tilted surface (CSUC's roof).

4.3. Tilt factors and optimum orientation estimation

Figure 6 shows that on December and according to Perez model, from 7:00 to 14:00, the irradiance for tilted surface with real orientation (9° tilt, 102° azimuth) are below of the corresponding to the horizontal surface and to the tilted surface with optimum orientation (4° tilt, 0° azimuth). From 14:00

to 18:00 the values match.

Figure 7 shows that lower values, for tilt factors K_1 and optimum orientation K_2, are presented at morning hours and higher values at afternoon hours. Both coefficients matches approx after 14:00 hours. K_1 value ranges between 0.85 y 0.99, with an average value of 0.95; K_2 value ranges between 0.81 y 0.99, with an average value of 0.93.

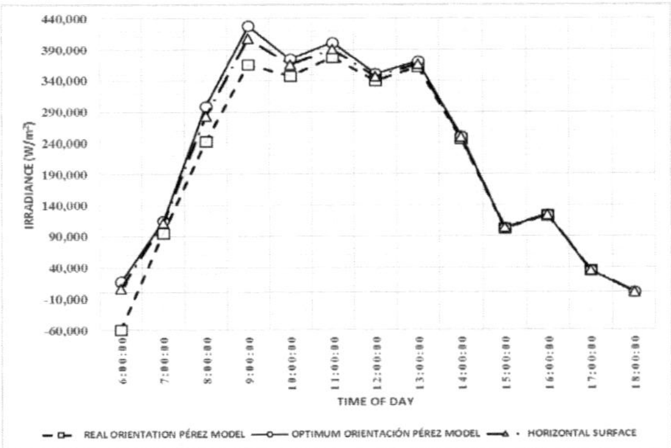

Figure 6. Solar radiation available on the CSUC's roof.

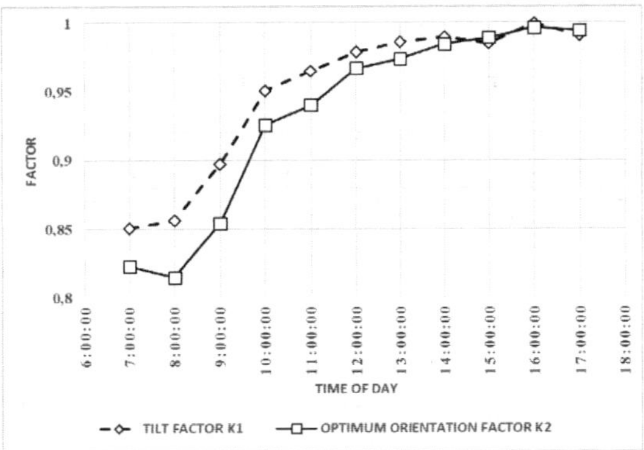

Figure 7. Correction factors, solar radiation on the CSUC's roof.

5. Conclusions

Total Solar irradiance values to be taken to design the climate control system of the pool inside the CCDU shall correspond to December.

Isotropic Sky, HDKR and Perez models predict similar values of solar irradiance on the tilted surface with real orientation (CCDU roof). For the calculation of the catchment area, predicted values for December will be used by Perez model.

To ensure maximum architectural integration of solar collectors on the CCDU roof, total solar horizontal irradiance available will be multiplied by a factor K_1 equal 0.85.

In the most unfavourable conditions, 19% of the energy will not be catched by the collect surface because it has no an optimal orientation.

References

[1] Méndez J M y Cuervo G R 2008 *Energía térmica solar* (Madrid: FC Editorial) p 183
[2] Perales T 2008 *Instalación de paneles solares térmicos* (México: Alfaomega) p 83
[3] Skeiker K 2009 *Energy Conversion and Management* **50** 2439
[4] Mendoza D 2006 *Revista Colombiana de Física* **38-4** 1435
[5] Chang T P 2009 *Solar Energy* **83** 1274
[6] Szeicz G and McMonagl R C 1983 *Solar Energy* **30** 247
[7] Rakopoulos C D and Vazeos E 1987 *Energy Conversion and Management* **27** 189
[8] Singh M et al 1989 *Energy Conversion and Management* **29** 239
[9] Woolley J, Harrington C and Modera M 2011 *Building and Enviroment* **46** 187
[10] Duffie J A y Beckman W A 2006 *Solar engineering of thermal process* (New York: Wiley) p 9
[11] Manrique J A 1984 *Energía solar: Fundamentos y aplicaciones fototérmicas* (México: Harla) p 19
[12] Kalogirou S 2014 *Solar energy engineering: Processes and systems* (UK: Elsevier) p 91

Objects tracking using a periodic array of volume: numerical evaluation

J A Sanchez Daza[1], M L Molina Prado[1] and N A Arias Hernández[1]
[1] Universidad de Pamplona, Pamplona, Colombia.

E-mail: nesariher@unipamplona.edu.co

Abstract. A method for tracking 3D objects using sensitivity phase of its Fourier transform to determine the position, displacement and orientation of a moving object is presented. The orientation and 3D position of the object from the 3D imaging system with subvoxel accuracy are calculated. A reference pattern should be fixed on the surface of the object and should be registered by the system of 3D images. This method is based on a prior knowledge of the association between the phase of the reference pattern and object motion, inherent property of the Fourier Transform. The reference pattern consists of a regular distribution of intensity that can be represented mathematically by three fringes of orthogonal patterns of volume whose 3D spectrum it allows to generate a 3D distribution of continued phase and absolute for each strip of the pattern orthogonal. A local maximum intensity value in the reference pattern represents a phase value $2\pi N$ where N corresponds to the fringe where is encountered the desired value. Reconstruction of the absolute phase associated with each set of fringes of volume in each direction orthogonal allows obtaining an exact point of the pattern, leading to the determination of the position of the object's location with subvoxel accuracy. The displacements from two consecutive positions are calculated. The accuracy of this method considering the parameters such as sampling, the number of holes, noise and contrast is evaluated numerically. Also its efficiency measuring the orientation is evaluated.

1. Introduction

Position and displacement measurement with high accuracy and resolution have been widely studied and applied in various fields of science and engineering. They have been proposed several devices that measure the position and rotation of an object with high precision through techniques optical, of radiofrequency, digital image processing for localization of marks between other [1].

A method in the field of the image processing which aims to locate a mark is computationally developed and simulated. The high sensitivity of the phase with respect to displacement allow measure the position and displacement of an object in a 3D scene, with high accuracy and resolution. This method is based on priori knowledge of the association between the phase of the reference pattern and the displacement of the object, inherent property of the Fourier transform [2]. The reference pattern or target consists of a regular intensity distribution which can be represented mathematically by three patterns of volume fringes orthogonal whose 3D spectrum can generate a three-dimensional distribution of continuous phase and absolute to each orthogonal fringes pattern allowing efficient analysis in the frequency space which is performed with the Fourier transform [3]. The spectral content of the reference pattern shows information of this periodic arrangement of volume in the frequency space. The reconstructions of the absolute phase associated at orthogonal fringes system allows the precise location of the centre of the target that determine the positioning of the object with precision subvoxel. The displacements are calculated from two consecutive positions.

2. Measurement principle

A combination of strategies of Takeda and Sandoz is used [4,5]. In the setup, a cubic target is used as shown in the Figure 1, with 10x10x10 hols dark of four pixels wide, this target is introduced into an empty matrix 256x256x256 which is called workspace, distribution in the spatial frequencies is observed, which is obtained using Fourier Transform (TF),

$$F(u,v,w) = \int_{x,y,z} f(x,y,z)e^{2\pi i(xu+yv+zw)}dxdydz \tag{1}$$

Where the spatial function $f(x,y,z)$ is the distribution of intensity of the image of volume acquired (target) and $F(u,v,w)$ is its Fourier Transform, which can be seen in Figure 2. Then, three of the first harmonics are filtered independently, one for each direction orthogonal of the fringes (see Figure 2) contained in the input information. The location of the corresponding harmonics can be found because of that the period of the target is known. In these positions the filters must be located. This can be seen in Figure 2. The filtering described mathematically as:

$$F_i(u,v,w) = F(u,v,w)\cdot W(u-u_i,v-v_i,w-w_i) \qquad i=1,2,3 \tag{2}$$

Where $W(u-u_i,v-v_i,w-w_i)$ is the filter type Hanning centred in (u_i,v_i,w_i) with a width of 2/3 the cut-off frequency. Then, each harmonic are relocated in the origin of coordinates independently and three-dimensional inverse Fourier transform is applied to each one

$$f_i(x,y,z) = \int_{x,y,z} F_i(u,v,w)e^{-2\pi i(xu+yv+zw)}du\,dv\,dw \tag{3}$$

Once recovered the inverse Fourier transform, the modulus $M_i(x,y,z)$ and the phase $p_i(x,y,z)$ of each function is calculated:

$$\begin{aligned} M_i(x,y,z) &= f_i(x,y,z)\cdot f_i^*(x,y,z) \\ p_i(x,y,z) &= \arctan\left[\frac{\Im m(f_i(x,y,z))}{\Re e(f_i(x,y,z))}\right] \qquad i= 1,2,3 \end{aligned} \tag{4}$$

Where $f_i^*(x,y,z)$ is the complex conjugate of the function $f_i(x,y,z)$. In the Figure 3, the module and the phase is obtained from the function $f_1(x,y,z)$ corresponding to lobe u_1,v_1,w_1. The module is used order to build a binary mask which when multiplied with the phase $p_1(x,y,z)$ allows to define the phase region to working.

In the Figure 3 is shown the phase of the work in each of the orthogonal directions. Similarly, the function $f_2(x,y,z)$ and $f_3(x,y,z)$ the same treatment are applied.

Once phase region is obtained and as the phase $p_i(x,y,z)$ recovered has jumps of phase of 2π and as they is wrapped between $-\pi$ and π, a three-dimensional algorithm the phase unwrapping is applied [6]. As observed in the Figure 4, the phase obtained is linear (hyperplane), the direction of growth for each of the harmonics filtered is orthogonal. From each phase obtained, the coefficients are calculated by polynomial approximation a_i,b_i,c_i and d_i [7,8], which help describe the hyperplane of phase $P_i(x,y,z)$ best fit of the calculated phase. The Figure 4 shows the hyperplanes of phase and the associated phase, which can be described by the following equation:

Journal of Physics: Conference Series **687** (2016) 012010 doi:10.1088/1742-6596/687/1/012010

$$P_i(x,y,z) = a_i x + b_i y + c_i z + d_i, \qquad i = 1,2,3 \quad y \quad \begin{bmatrix} a_1 & b_1 & c_1 \\ a_2 & b_2 & c_2 \\ a_3 & b_3 & c_3 \end{bmatrix} \begin{bmatrix} x \\ y \\ z \end{bmatrix} = \begin{bmatrix} P_1 - d_1 \\ P_2 - d_2 \\ P_3 - d_3 \end{bmatrix} \tag{5}$$

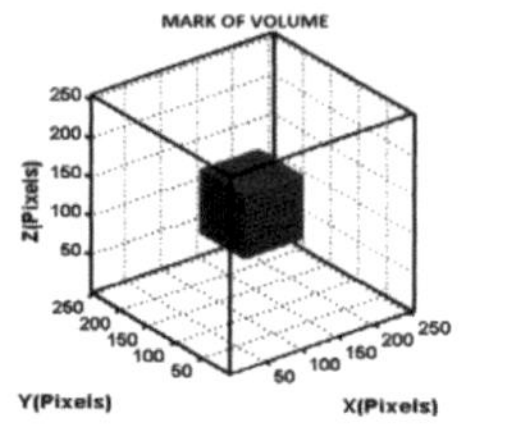

Figure 1. Target of volume inside of the workspace of 256x256x256 pixels.

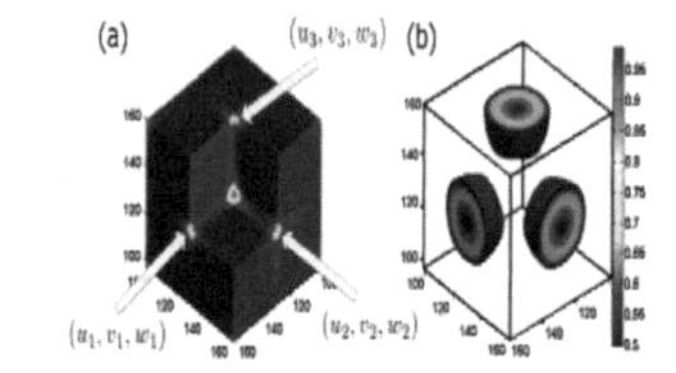

Figure 2. (a) Location of the harmonics at filter and (b) filter hanning for to extract the harmonics.

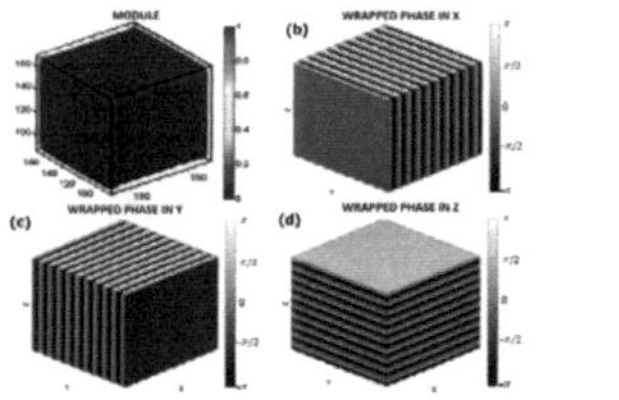

Figure 3. (a) Module of the inverse Fourier transform (b), (c) y (d) Phase wrapped in x, y, z.

For to find the coordinates of the centre of the target or any other point on this, a phase value for each hyperplane is fixed, counting $2\pi N$ from the origin. The value of x,y,z is found solving the system of linear Equations (5), for the values of phase P_1, P_2, P_3, corresponding to the desired location. The value of phase associated to the centre of the target is $P_1, P_2, P_3 = 0$, because of target symmetry. Similarly, for values of phase corresponding the faces and corners are calculated, Figure 8.

3. Results

For evaluating the positioning method are performed simulations in order to obtain the accuracy of the system with respect at position without displacement versus to sampling (pixels by each hole), the number of holes used in the target, the noise, the contrast and the measurement of rotation. The target is centred on a workspace comprised of 2563 voxels, therefore theoretically the centre coordinate should be (128.5, 128.5, 128.5). For each case a target with the desired characteristics is taken, are generated 100 targets with standard deviation of 2 grey levels and it proceeds to calculate the centre of each one with the positioning method implemented, then the parameter varies evaluation and the procedure is repeated. The results obtained are described in the Figure 5 and 6.

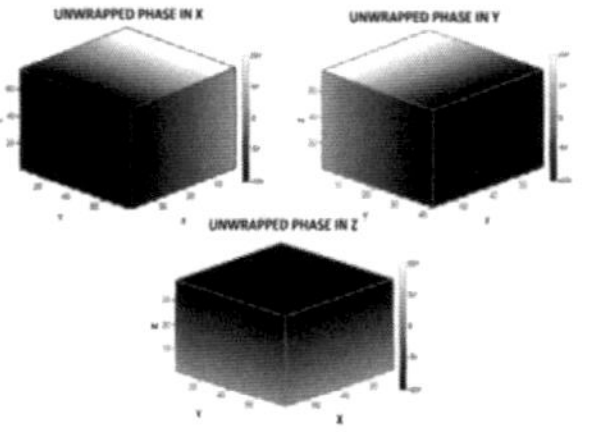

Figure 4. Hyperplanes of the unwrapped phase in each orthogonal direction.

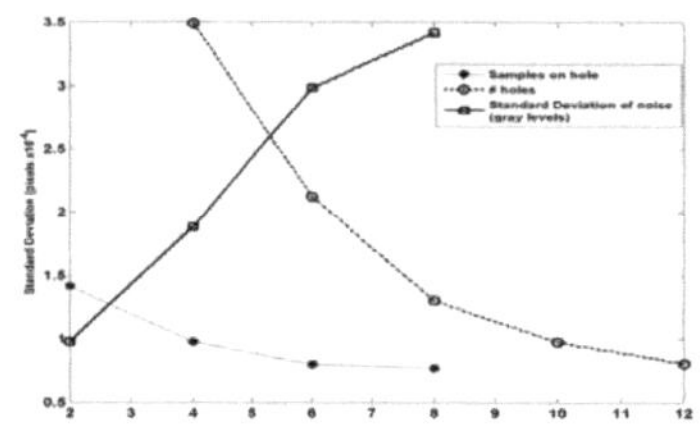

Figure 5. Analysis of the dependence of the accuracy in calculating the position with the sampling, number of holes and noise in the grey levels from 100 measurements.

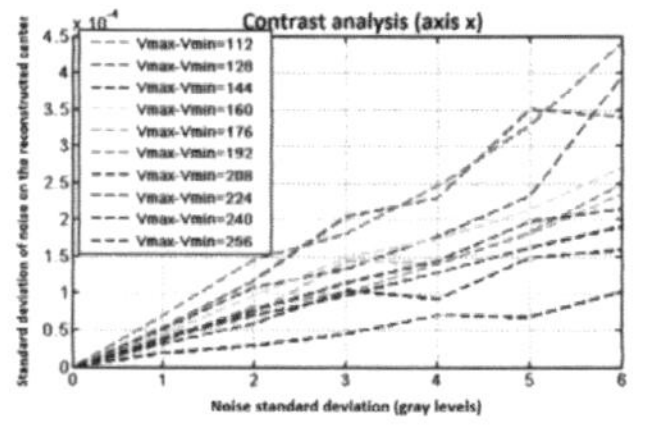

Figure 6. Standard deviation in the reconstruction of the centre vs. standard deviation of the greys levels for different contrasts of the target (date of axis x).

For measuring the rotation of the target, the reference points are calculated the centre (padding), a face (hole) and a corner (padding), Figure 7.

 doi:10.1088/1742-6596/687/1/012010

The Figure 9 shows the target of volume rotated 45° parallel to the axis z. This is located in the centre of the target of volume.

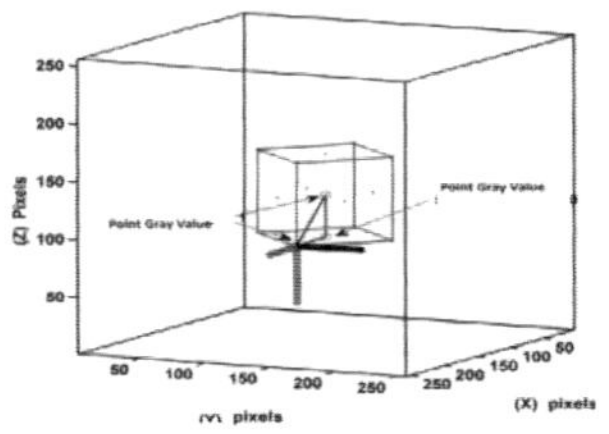

Figure 7. Location of points of reference in the target.

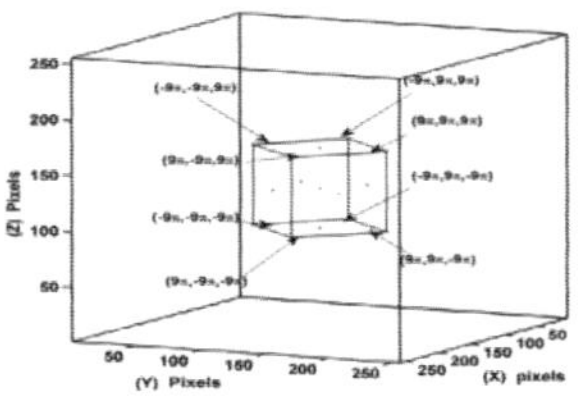

Figure 8. Phase values of the corners.

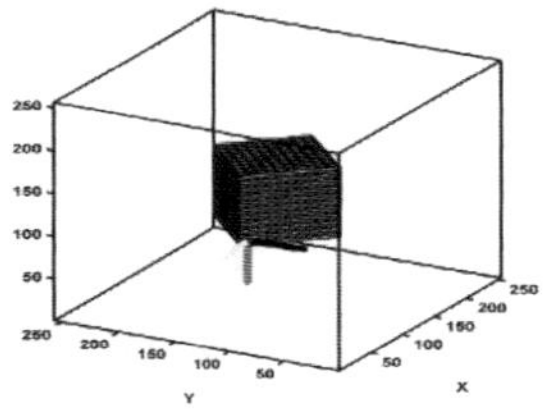

Figure 9. Target of volume rotated an angle of 45 degree.

4. Conclusions

In this paper, a method for measuring the positioning and rotation of an object are evaluated numerically using a periodic array of volume. This array is composed of periodic patterns of stripes in each orthogonal direction. The simulation shows that is possible to determine the position and angular displacement of an object, with subvoxel precision. The accuracy depends on the specific characteristics of the periodic array of volume used (sampling, number of hole, noise and contrast). Similarly, the effect of to add a defect on the volume target (padding a hole) allows for a rotation of 360° without degrading system accuracy in the positioning calculation of the target and without significant changes in the position calculation algorithm. Is observed that minimum accuracy of the system for when the target is of 10 holes, with a contrast of 240 grey levels and standard deviation of the noise of 2 grey levels, is the order of 4×10^{-4} voxels, that is, is possible to measure 2500 positions between voxels in any direction only when the target is sampled at 2 pixels per hole. When increasing the number of pixels per hole, increases accuracy, similarly the computational cost, because the area of phase interest is equivalent to the size of the target, this weighs in the calculation algorithm applied. When the target is sampled at 8 pixels per hole then the accuracy it is approximately doubled, that is, will could measure 5000 positions between voxels.

Acknowledgments

This research was performed under the grants: Universidad de Pamplona, Project: PR130-00-012 (GA160-BP-II-2013-2.1.2.2.1), and Vicerrectoría de Investigaciones Universidad de Pamplona (Pamplona-Colombia).

References

[1] Guido B, Giancarlo F, Roberto O and Antonio P 2000 *Radiotherapy and Oncology* **54** 21
[2] Gasvik Kjell J 2002 *Optical metrology* (Chichester: Wiley) chapter 11 pp 269-295
[3] Hussein A R 2007 *Three-dimensional Fourier fringe analysis and phase unwrapping* (Liverpool: Liverpool University)
[4] Sandoz P, Ravassard J C, Dembele S and Janex A 2000 *IEEE Trans Instrum Meas* **44** 867
[5] Takeda M 1982 *J Opt Soc Am* **72** 156
[6] Itoh K 1982 *Applied Optics* **21** 2470
[7] Arias N A, Meneses J, Suarez M A and Gharbi T 2009 *Revista BISTUA* **7** 70
[8] Arias N A, Sandoz P, Meneses J, Suarez M A and Gharbi T 2010 *Optics Express* **18** 1094

Estimate of the diameter of surface defects on a pattern specimen made of Plexiglas through active infrared thermography

M H Medina-Barreto[1], J M Giraldo[1], A L Ramírez-Hurtado[1], L Llamosa[1] and B Cruz[1]

[1] Universidad Tecnológica de Pereira, Pereira, Colombia.

E-mail: mmedina@utp.edu.co

Abstract. In recent years the use of Infrared Thermography (IRT) has reached a special importance in the industrial sector as a method of Non-Destructive Testing (NDT), for inspection of surface and subsurface defects of different types of materials. In order to determine the size of such defects by implementing the technique of Infrared Active Thermography, a Plexiglass sample (PMMA) with circular defects of different diameters and depths was designed and built. The sample was thermally stimulated by a continuous radiation from a halogen lamp, and subsequently the thermographic images were acquired by using the time-resolved infrared radiometry technique (TRIR). Images were obtained using Fluke TI-30 and TI-32 cameras, and a program was designed in Matlab for its further processing. Through a user interface, the program can display and filter the image; then it chooses a particular defect and calculates the diameter with their respective uncertainties. The best results for the calculated diameter were obtained with Ti-32 camera. The estimated uncertainties were below of 1.2mm, regard of defect diameter. Importantly, uncertainties grew when the diameter/depth ratio was 1 for depths of 4.0mm.

1. Introduction

Quantitative evaluation and periodical monitoring of discontinuities in newly manufactured parts, or currently in use, are key aspects to establish protocols to provide reliability and high quality. The active infrared thermography (IRT) is a non-destructive evaluation technique (NDE) that has great impact on the analysis of delamination, impact damage and inclusions in metallic and composite materials [1-3]; hence the IRT technique has been widely accepted by the scientific community [4]. This technique requires an external excitation source of heat which stimulates the test sample and an infrared camera that scans the surface temperature, yielding an image that is further processed [5].

For this paper was implemented a time-resolved infrared radiometry technique (TRIR) for the quantification of artificial defects on a standard Plexiglas sample designed with 16 circular holes of different diameters and depths. Subsequently, the IRT images were processed in Matlab in order to characterize such defects and to calculate their respective uncertainty.

2. Materials and methods

A PolyMethylMethAcrylate (PMMA) reference sample was designed and built with 16 circular defects, whose diameters are from 4.0mm (4th column) to 10.0mm (1st column), as shown in Figure 1(a); and whose depths are from 1.0mm (4th row) to 4.0mm (1st row), as shown in Figure 1(b). The sample type and geometric characteristics defined in this study are similar to the standard sample proposed by Montanini [6]. Along the same lines, we performed a supplementary Fourier Transform Infrared spectroscopy (FTIR) that revealed an emissivity coefficient of 0.86 for the PMMA sample.

The diameters of the artificial defects were corroborated using a scanning by Coordinate Machine Measurement (CMM) Brown & Sharpe, Model Gage 2000-R454 with a tolerance of 0.001mm. The measurements from CMM scanning were taken as reference values for the ensuing comparative analysis with respect to IRT results.

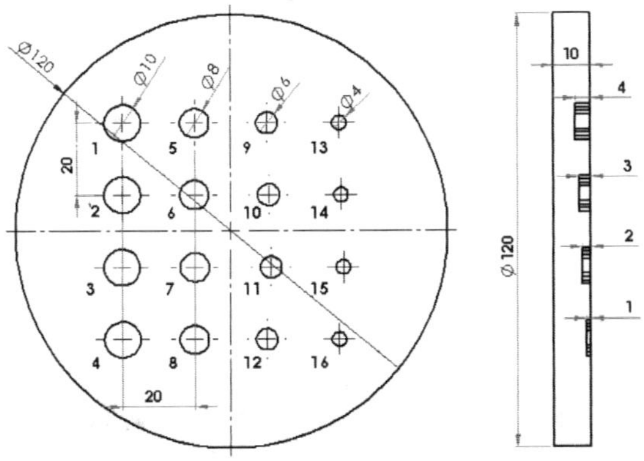

(a) (b)

Figure 1. Drawing showing nominal dimensions (in mm) and relative position of defects.

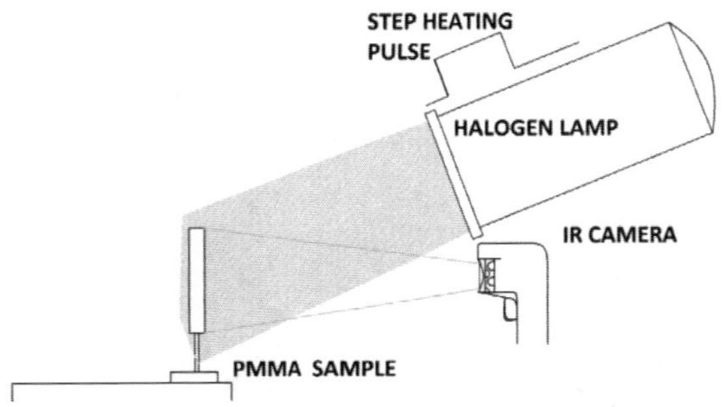

Figure 2. TRIR technique Schematic representation applies to a Plexiglas sample.

Time-Resolved Infrared Radiometry (TRIR) was implemented so that the infrared images from the pattern sample could be registered during and after the heating pulse generated with a halogen lamp. The experimental setup consists of a 1000W halogen lamp, infrared IR cameras and the PMMA sample placed on a holder (Figure 2). The halogen lamp was always positioned above the IR camera and the latter was always in frontal position with respect to the sample. Two different Fluke cameras were used: Ti-30 and Ti-32, the distance between the sample and the Ti-30 camera was 590mm, and the distance between the sample and the Ti-32 camera was 426mm. The heating pulse was applied to the PMMA sample during 300 seconds. The specifications of the cameras are shown in Table 1.

Table 1. IR cameras technical characteristics specifications.

	Ti-30	Ti-32
Thermic Sensitivity	200 mK	45 mK
Resolution	120 x 160	240 x 320
Precision	± 2°C	± 2°C
Visual Field	17° x 12.6°	23° x 17°

The sequence of TRIR images capture had three stages: 180 seconds prior to the heating pulse, 300 seconds during heating pulse (10 images each 30 seconds) and 300 seconds of cooling (10 images each 30 seconds). The IR images obtained were exported as bitmaps (.BMP). A user interface designed in Matlab allows to visualize and process the IR images.

In order to assess the accuracy of the TRIR technique, both type A and B uncertainty sources have to be taken into account when using CMM scanning as calibration method; then, the combined standard uncertainty of the defects size measurements can be estimated according to Guide to the Expression of Uncertainty in Measurement (GUM) [7].

3. Results and discussions

Figure 3 shows the thermal infrared images acquired by means of TRIR technique, using Fluke Ti-30 and Ti-32 infrared cameras. As it can be seen, when impinging the sample surface with an external radiation, the defects are noticeable during the stages of heating and cooling for both cameras. However, the number of defects that can be discriminated for the analysis of images processing with the TI-30 camera is limited because of its low resolution. Nevertheless, due its better resolution, Ti-32 camera is capable to show the thermal contrast of deeper defects at the start of the heating stage. The thermal waves start to propagate through the material surface, when they find discontinuities a not uniform local heating takes place in a short period of time. Although images were taken each 30 seconds, actually relevant changes in the thermal contrast only appeared each 60 seconds.

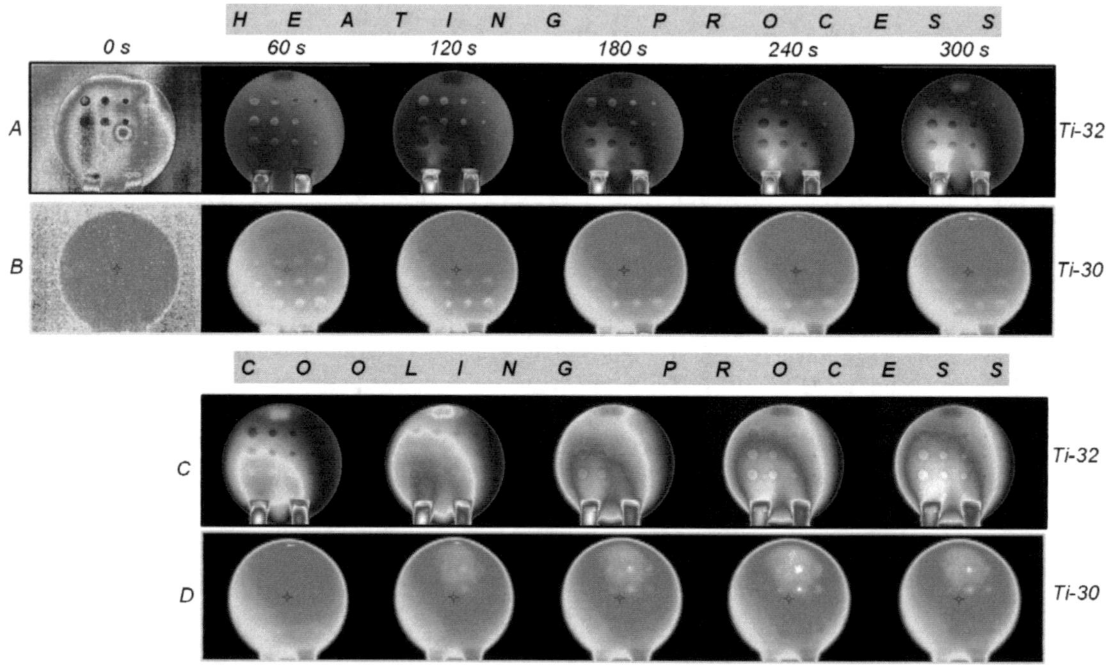

Figure 3. TRIR thermographic images acquired with a step width of 300 seconds.

For further processing of the images, it was designed a user interface which allows to filter (smoothing, remove noise, enhance edges and detect edges) each image through a sequence of instructions. From the processed image, a defect was chosen to calculate the diameter value with its respective uncertainty. These values are presented in Table 2, where uncertainties with the CMM scanning and the TRIR technique, are related. In all cases the measured diameter values were above the nominal value, considering that are allied to the inherent process of counting pixels in the filtered infrared image. The relative error calculated for the defect #6 is 14%, whereas for all other defects are lower values. In particular, when the diameter/depth ratio is about 1, the data suggest that their relative errors tend to be lower; by contrast, when the diameter/depth ratio reaches values close to 2, the relative errors increase. The high values found for the relative errors could be related to the limited number of experimental data, since in the images were not possible to discriminate all the defects. Furthermore, the incidence radiation from the heat source on the PMMA specimen generates radial heat diffusion, making it difficult to measure the actual size of the border. However, it is necessary to clarify that the here reported error for the defect #1, 11.7% is significantly lower than the relative error of -27.5% found by Montanini [6] for a defect of 10 mm in diameter and 3.6mm in depth.

Table 2. Comparison between nominal (MMC) and estimated (TRIR) diameters during the heating stage.

Defect	Depth (mm)	MMC		TRIR		Error (%)
		Diameter $_{nominal}$ (mm)	U_E (mm)	Diameter $_{measured}$ (mm)	U_E (mm)	
# 1	4	10.023	0.02	11.20	1.07	11.7
# 5	4	8.006	0.01	9.06	0.88	13.2
# 6	3	8.006	0.01	9.19	0.89	14.8
# 9	4	6.010	0.01	6.78	0.69	12.8
# 11	2	6.010	0.01	6.72	0.69	11.8
# 13	4	4.021	0.05	4.31	0.55	7.1
# 14	3	4.021	0.05	4.35	0.46	8.1
# 15	2	4.021	0.05	4.51	0.48	12.3

4. Conclusions

The TRIR technique was implemented enabling circular surface defects on a sample of PMMA were located. The infrared images obtained allowed to calculate the uncertainties of the measurements of the diameters of such defects through post-processing of the thermal images obtained

Although the errors obtained were relatively high, they are lower than those reported for similar samples analysed in this study. The calculations of the diameters derived from thermal images indicate a correlation between the values of the expanded uncertainty with the diameter / depth ratio, which can be examined in more detail using pulsed active thermography (PIRT).

References

[1] D Wu, G Busse 1998 *Gén Therm* **37** 693
[2] C Meola, G M Carlomagno, A Squillace, G Giorleo 2002 *Meas Sci Technol* **13** 1583
[3] C Meola, G M Carlomagno, G Giorleo 2004 *J Mater Process Technol* **155-156** 1132
[4] X P V Maldague 2010 *Proceedings of the 10th edition of the Quantitative Infrared Thermography (QIRT)* (Quebec: Université Laval) p 993 (http://qirt.gel.ulaval.ca/dynamique/index.php?idD=50&Lang=0)
[5] Benítez Restrepo H D, H Loaiza Correa y E Caicedo Bravo 2011 *Termografía activa pulsada en inspección de materiales. Técnicas avanzadas de procesado* (Colombia: Editorial Universidad del Valle y Universidad Javeriana) chapter 18 p 154
[6] R Montanini 2010 *Infrared Physics & Technology* **53** 363
[7] JCGM/WG 1 2008 *Evaluation of measurement data. Guide to the expression of uncertainty in measurement vol 1* (Paris: JCGM BIPM)

HOS cell adhesion on Ti_6Al_4V surfaces texturized by laser engraving

A Sandoval Amador[1], H Carreño Garcia[1], P Escobar Rivero[1], D Y Peña Ballesteros[1] and H A Estupiñán Duran[2]
[1] Universidad Industrial de Santander (UIS), Bucaramanga, Colombia.
[2] Universidad Nacional de Colombia, Medellín, Colombia.

E-mail: anderson84f@outlook.com

Abstract. The cell adhesion of the implant is determinate by the chemical composition, topography, wettability, surface energy and biocompatibility of the biomaterial. In this work the interaction between human osteosarcoma HOS cells and textured Ti_6Al_4V surfaces were evaluated. Ti_6Al_4V surfaces were textured using a CO_2 laser in order to obtain circular spots on the surfaces. Test surfaces were uncoated (C1) used as a control surface, and surfaces with points obtained by laser engraving, with 1mm spacing (C2) and 0.5mm (C3). The HOS cells were cultured in RPMI-1640 medium with 10% fetal bovine serum and 1% antibiotics. No cells toxicity after one month incubation time occurred. The increased cell adhesion and cell spreading was observed after 1, 3 and 5 days without significant differences between the sample surfaces (C2 and C3) and control (uncoated) at the end of the experiment.

1. Introduction

Some papers have shown that the adhesion, spreading and proliferation process depend on the physicochemical surface properties. Composition, electric charge, material conductivity, roughness, and micro or nanopatterning play an important role in the cellular response of the material when it is implanted [1-3]. For example, the chemical composition of the surface determines the surface energy, polarity, wettability, Zeta potential and consequently the characteristics of the cell-material interaction [4-6]. Therefore, a suitable surface treatment can promote osteosynthesis ability of the implant material, so producing cells to adhere and proliferate on the major surface and likewise lead to an increase in mineralization of the surface, which forms the basis for the formation of new bone [7].

In this work the effect of the geometry of a pattern obtained by laser engraving on the adhesion of human osteosarcoma cells on surfaces of Ti_6Al_4V was evaluated.

2. Experimental

2.1. Samples

A mill annealed Ti_6Al_4V cylindrical disc with 14mm diameter and 3mm thick was used to study the patterning by laser engraving effect on cells/surface interactions. Three different types of surfaces were used in this study. They include: smooth surface (C1) and laser engraving with two different circular patterns (C2 and C3). The smooth surface was obtained by grinding with silicon carbide paper in successive grades from 220 to 1200 grit and subsequently polished with slurry containing $3\mu m$ and $0.5\mu m$ Al_2O_3 powder. The patterning surfaces were obtained by laser engraving using a *laserpro X380-RX* equipment; this has a sealed CO_2 laser with maximum power of 100W. The circular patterns consist of points with 100μm diameter, 1mm vertical spacing between points, 1mm horizontal distance for pattern 1 and 0.5mm for pattern 2. All samples were sterilized by UV radiation for 30 minutes on each side and then autoclaved at 120°C for 20 minutes.

2.2. Biological test

The human osteosarcoma cell line (HOS, ATTC, CRL-1543) was cultured in 25cm^2 Falcon Culture plates and maintained in an incubator at a temperature of 37°C regulated with 5% CO_2, 95% air, and saturated humidity. A Roswell Park Memorial Institute medium (RPMI-1640) supplemented with 10% fetal bovine serum and 1% penicillin/streptomycin was used as the cell cultured medium.

For cell toxicity assay, each sample (C2 and C3) and controls (sample without coating, C1) were placed in 1.5ml of RPMI-1640 culture medium and incubated at 37°C. After a month, the culture medium was recollected and HOS cells were incubated with serial dilutions of each supernatant or medium alone for 72 hours at 37°C. Cell toxicity was evaluated using a colorimetric MTT assay as described by J. E. Gomez et. al [8]. The percentage of cytotoxicity was calculated with the equation: cytotoxicity (%)=(OD control group-OD treatment)/OD control group x100.

For adhesion assay, 5.0x10^{4}cells/ml in cultured medium were placed on each material and incubated at 37°C. After 24, 72 and 120 hours the cells were detached using 0.5ml Trypsin-EDTA for 5 minutes and counted microscopically in a Neubauer chamber. In addition a fluorescence staining were made directly in each sample using Hoechst 33342 fluorescent stain and epifluorescence microscopy (Nikon Eclipse E4000), B-2A filter (Ex=450-490, DM=500, Ba=515). Images were obtained using a camera Nikon Coolpix 5000. UV2A. Each time experiments were performed in duplicate and the results were calculated as number of cells per surface area.

3. Results and discussion

The surface morphology of Ti_6Al_4V smooth C1 and patterned C2 and C3 shown in the Figure 1. C1 sample has a smooth and homogeneous surface like a mirror finish surface. Also on surface of C2 and C3 it can be observed that the patterns obtained, have a quasi-circularity shape attributed to sweep the laser beam on the surface to generate this kind of geometry. The diameter of the spots is about 150μm, the separation action on center between the spots on the vertical axes is about 1370μm and is the horizontal separation of 1475μm for C2 and 730μm for C2 and C3.

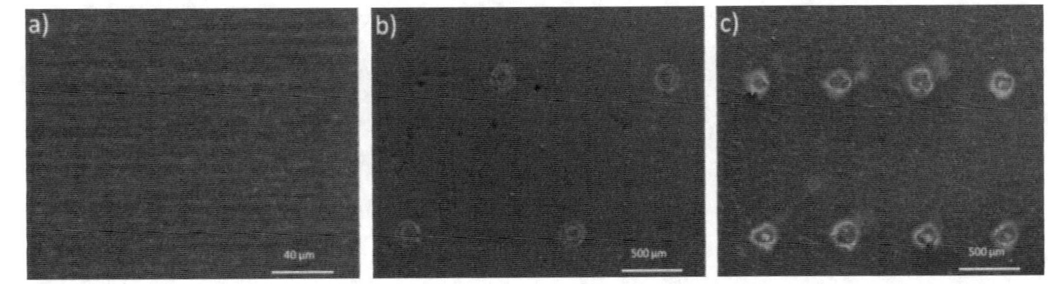

Figure 1. Ti_6Al_4V surface morphology (a) smooth C1, (b) Pattern 1 C2, and (c) Pattern 2 C3.

As shown in the Figure 2, the spots have a dendritic structure from the center to the edge. This maybe is due to rapid heating cooling cycle during the laser engraving process. Furthermore, the high temperature can be caused by absorption of a large number of photons in the small area of the spot, producing then combustion of the material and moving this to the vicinity of the spots, this has been previously reported by M. E. Khososhahi et. al [9].

The contact angle results showed that the surfaces of C1, C2 and C3 samples have a hydrophilic behaviour, since the angles are less than 90°. As shown in the Figure 3, the laser engraving generated a change in the wettability of the surface, making the C3 surface slightly more hydrophobic. There were no statistically significant differences between the contact angle of C1 and C2 but respect C3 surface the difference is around 43% calculated.

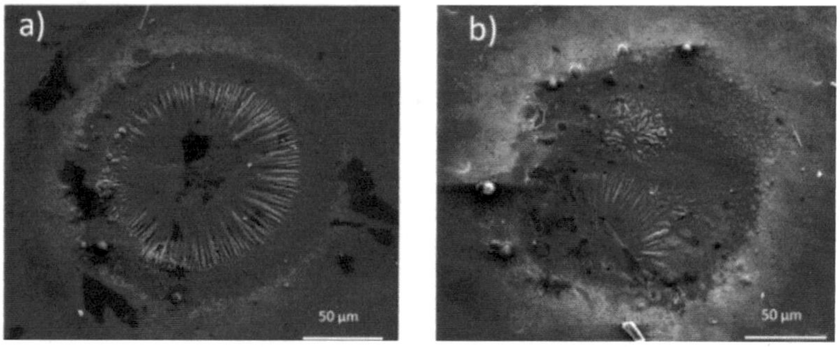

Figure 2. Pattern surface morphology, (a) Pattern 1 C2 and (b) Pattern 2 C3.

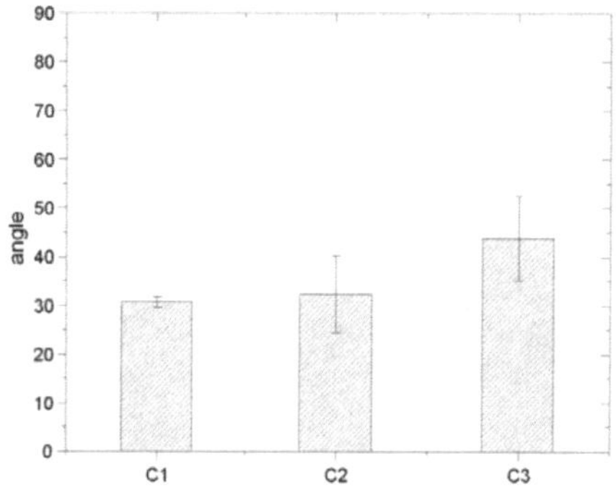

Figure 3. Contact angle.

Surface of Ti_6Al_4V with and without laser engraving were not toxic to the HOS cells any evaluated concentration, which is consistent with that reported by P. Sevilla et. all in [10], as shown in Figure 4, the percentage of cytotoxicity are below 15%.

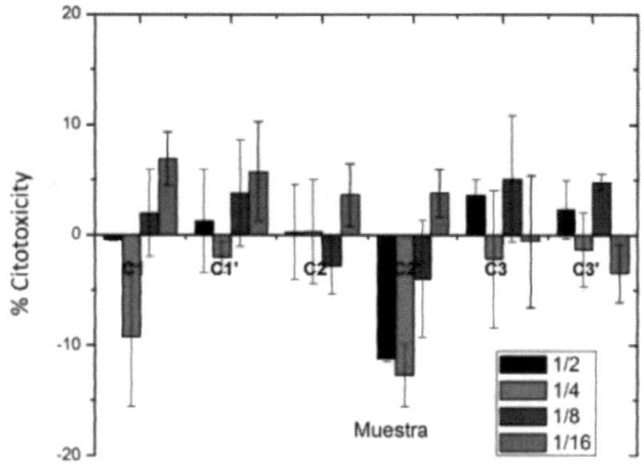

Figure 4. MMT Assay.

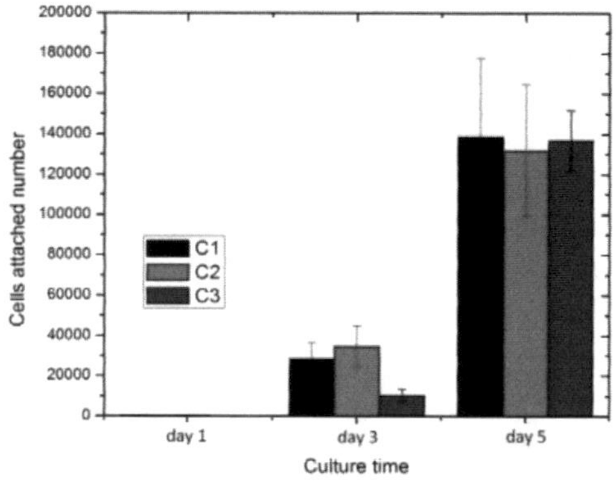

Figure 5. Cells adhesion.

The results of cell adhesion are shown in the Figure 5. At third day, C3 surface have a low adherence respect at the other surfaces. The number of attached cells increased exponentially on the fifth day but there were no significant statistically differences in the cell adhesion on the studied surfaces. Although the number of adherent cells on the surfaces at the first day was very low, these few cells adhered with a good cell affinity for the three surfaces. Cells showed oval shape at first day and become each day more elongated which indicates the spread of the cells on the surfaces and it can be observed in Figure 6, these results are similar to those reported in [9].

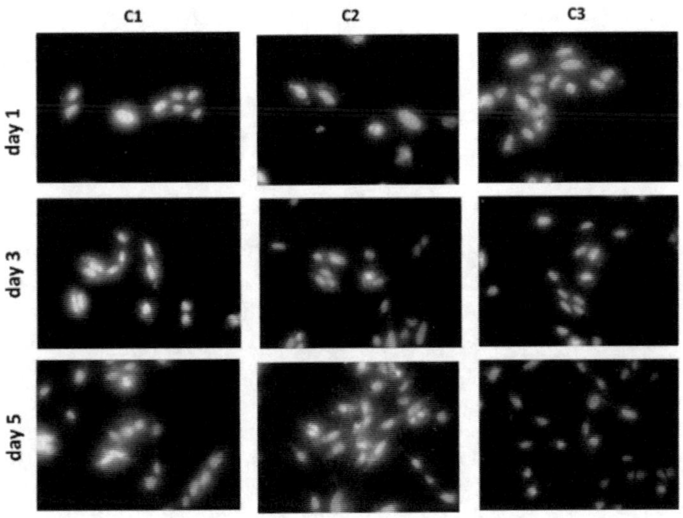

Figure 6. Fluorescence micrographs of HOS cells attached on different studied surfaces.

4. Conclusions

The patterns obtained in C3 achieved modify the wettability of the surface alloy by 43% which is associated to the number of spots on the surface. Thus the patterns obtained by laser engraving, not produce a cytotoxic response; therefore the biocompatible nature of alloy is maintained.

The cell proliferation in three surfaces exhibit good cell adhesion with random orientations, however, no significant differences in cell adhesion of HOS cells by comparing the samples with and without pattern were observed, which concludes that requires extending the study variables to evaluate this method of surface engineering for osteosynthesis process.

Acknowledgment

Our sincere thanks to COLCIENCIAS, for funding this project under contract No. 780-2011.

References

[1] Xuanyong Lui, Paul k Chu and Chuanxian Ding 2010 *Materials Science and Engineering Reports* **70** 275
[2] Koon Gee Neoh, Xuefeng Hu, Dong Zheng and En Tan Kang 2012 *Biomaterials* **33** 2813
[3] K Anselme 200 *Biomaterials* **21** 667
[4] L Bacakova, E Filova, M Parizek, T Rumi and V Svorcik 2011 *Biotechnology Advances* **29** 739
[5] Y Hibata et al 2010 *Biomaterials* **31** 8546
[6] Paulo Bartolo et al 2012 *CIRP Annals Manufacturing Technology* **61** 635
[7] Juan Carlos Vanegas Acosta, Diego Garzón Alvarado, Martin Casale 2010 *Revista Cubana de Investigaciones Biomedicas* **29-1** 51
[8] Javier Enrrique Gomez R, Luis Emilio Foreror Gomez, Patricia Escobar Rivero y Wilfredo Valdivieso Q 2007 *Scientia Et Technica* **13-36** 85
[9] M E Khosroshahi, M Mahmoodi, J Tavakoli 2007 *Applied Surface Science* **253** 8772
[10] P Sevilla, M Godoy, E Salvagni, D Rodriguez, F J Gil 2010 *Journal of Physics Conferences Series* **252-1** (doi:10.1088/1742-6596/252/1/012009)

Design of TEOS-GPTMS sol-gel coatings on rare-earth magnesium alloys employed in the manufacture of orthopaedic implants

L M Rueda[1], C Nieves[1], C A Hernández Barrios[1], A E Coy[1] and F Viejo[1]
[1] Universidad Industrial de Santander, Bucaramanga, Santander, Colombia.

E-mail: anaecoy@uis.edu.co , fviejo@uis.edu.co

Abstract. In the present work hybrid sol-gel coatings were synthesized on different magnesium alloys with potential interest in the fabrication of orthopaedic implants. Hybrid sols were obtained from a mixture of the inorganic precursor tetraethoxysilane (TEOS) and the organic precursor 3-glycidoxypropyltrimethoxysilane (GPTMS), employing ethanol as solvent and acetic acid as catalyst. The characterization of the sols was performed using pH measurements, rheological tests and infrared spectroscopy (FTIR) for different aging times. On the other hand, the coatings were characterized by scanning electron microscopy (SEM), while the corrosion resistance was evaluated using anodic polarization in SBF solution at 37±2°*C*. The results confirmed that, under specific conditions, uniform and homogeneous sol-gel coatings were obtained, which enhanced the corrosion resistance so that the corrosion current density was reduced in about two orders of magnitude with regard to the parent alloy.

1. Introduction

In biomedicine, metallic materials perform an essential role as biomaterials for the fabrication of temporary orthopaedic implants, assisting during regeneration of bone tissue damaged. For this purpose, AISI 316L stainless steel and Co-Cr and Ti alloys has been conventionally employed due to their excellent combination of mechanical properties and corrosion resistance. Nevertheless, one of the main disadvantages of the use of these alloys is their high mechanical properties (density, elastic modulus, and compressive yield strength), which increases the tendency to generate a stress shielding phenomenon between the implant and bone tissue. In this regard, rare-earth magnesium alloys are recently considered potential candidates to replace those alloys since their mechanical properties are close to those of the human bone so that, in theory, might reduce the stress shielding difficulty [1,2]. Moreover, magnesium alloys exhibit excellent biocompatibility, in which rare earth alloying elements provide anticarcinogenic properties to the alloy [3].

However, the main limitation of magnesium alloys for their implementation as biomaterial is the high degradation rate of magnesium as element in physiological environments, which may bring negative consequences such as loss of the mechanical integrity of the implant before the bone has healed. Further, according to its corrosion mechanism, magnesium degradation also generates hydrogen gas bubbles, which, in case of accumulation, can lead to undesired subcutaneous swelling and necrosis of the adjacent tissue [4].

Aiming to improve the corrosion resistance of magnesium alloys, synthesis of hybrid sol-gel coatings (mixture of organic and inorganic precursors) have arisen as one of the most promising alternatives due to its convenient adhesion to the metallic substrate, low cracking susceptibility, high density and flexibility [5]. The latter has been associated with the presence of organic precursors, which promote crosslinking of the sol-gel network and reduces both temperature and time during the curing process, this being an important factor for alloys with a low melting point like magnesium ones [6].

Based on these statements, the aim of the present work was to design hybrid sol-gel barrier coatings to enhance the corrosion resistance of the ZE41 rare-earth magnesium alloy.

2. Experimental procedure

The material used in the present investigation was the ZE41 magnesium alloy. The nominal composition (%wt) of the alloy is: 3,5-5,0 Zn; 0,4-1,0 Zr; 0,75-1,75 RE (La,Ce,Ga); 0,15 Mn and balance Mg. Hybrid sols were prepared from a mixture of the inorganic precursor tetraethoxysilane, TEOS, and the organic precursor 3-glycidoxypropyltrimethoxysilane, GPTMS, in 3:1 molar proportion, employing ethanol as a solvent and acetic acid as catalyst. Sols were characterized by pH measurements, rheological tests and Fourier Transform Infarred (FTIR) spectroscopy for different ageing times in order to study the hydrolysis and condensation process. Later, sol-gel coatings were obtained by deposition of the sol following the dip-coating route and then were cured at 120*°C* for 2 hours. Coatings were characterized using scanning electron microscopy (SEM), while their corrosion resistance was evaluated by anodic polarization test in simulated body fluid (SBF) solution at 37±2*°C*.

3. Results

3.1. Characterization of hybrid sols

Figure 1 comprises the results of the hybrid sols characterization. The evolution curves of viscosity and pH for different aging times are given in Figure 1(a). Hybrid sols showed a moderated increment in viscosity until a critical point that was reached at the fifth day, when the viscosity suffered an abrupt slope change associated with the definitive progress of the condensation reactions (beginning of the gelation state). On the other hand, pH values of sol remain constant in 5, from the first day, due to the employment of acetic acid (weak acid) as catalyst, which stabilizes the acidity of the sol.

Regarding the infrared spectroscopy results, Figure 1(b) shows the FTIR spectra of the sol for different aging times. Vibration bands related to the O-H and C-H (stretching and bending) can be observed at wavenumber of 3400-3200cm^{-1} (υ(O-H)), 3000-2900cm^{-1} (υ(C-H)), 1650cm^{-1} (δ(O-H)) and 1500-1300cm^{-1} (δ(C-H)). On the other hand, Figures 1(c-e) comprises the deconvolution analysis of the region 1250-1000cm^{-1}, where the bands related to Si-O-Si, Si—OH and Si-C bonding are masked. It can be noticed the presence of bands associated with: a) the symmetric and asymmetric vibrations of Si-OH bonding centred at 1200 and 1170cm^{-1}, which increases in intensity with the ageing time indicating that the hydrolysis of the precursors is occurring as part of the gelation process; the vibrations of the Si-O-C bonding (1100 and 1070cm^{-1}) related to the presence of the no-hydrolyzed precursor that decrease in intensity, as expected; finally, the vibration band centred at 1040cm^{-1} was related to the formation of Si-O-Si bonding, which increases in intensity associated to the advance of the condensation reactions.

3.2. Characterization of the hybrid sol-gel coatings

Electron micrographs of the coatings synthesized on the ZE41 magnesium alloy at different aging times are shown in Figure 2. For the shortest aging time (2 hours), the coating revealed low uniformity being the microstructure of the alloy still visible (Figure 2(a)). Conversely, after 3 days it was possible to obtain homogeneous coating with a good wettability and virtually free of cracks (Figure 2(b)). Further, longer aging time (5 days) resulted in high viscosity sols, which deposition originates thicker, cracked coatings, attributed to the generation of internal stress during the curing treatment (Figure 2(c)).

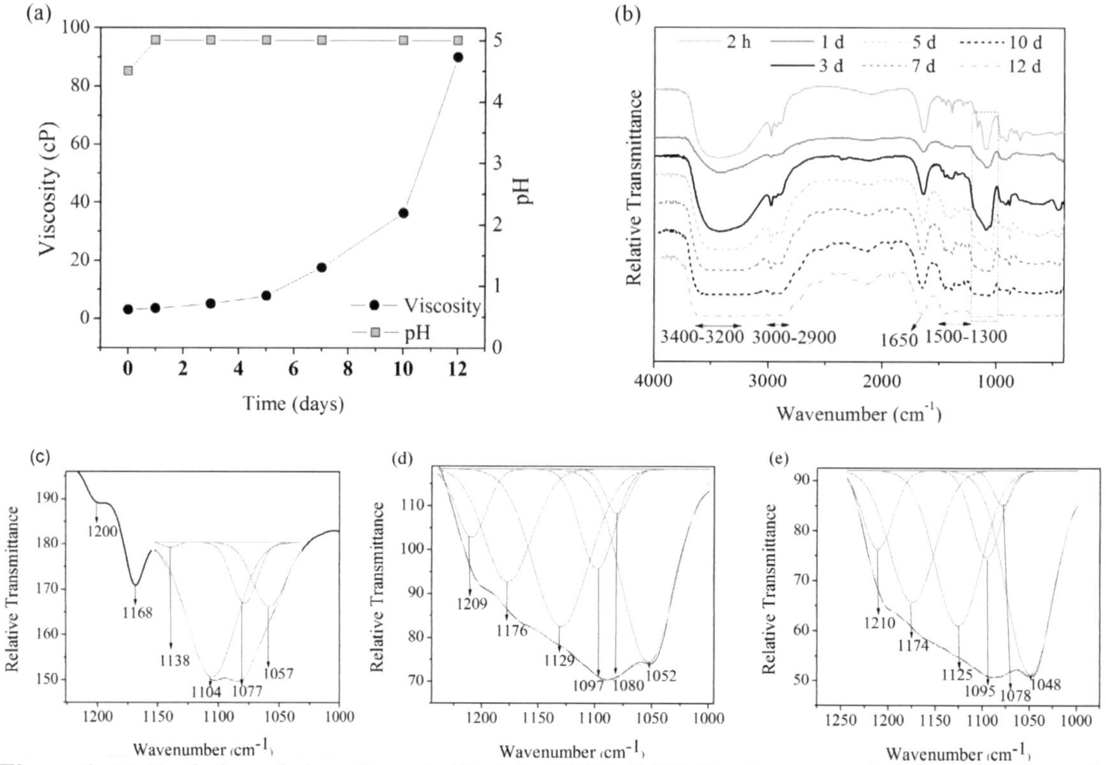

Figure 1. (a) Evolution of viscosity and pH of the TEOS-GPTMS sol versus aging time. (b) General FTIR spectrum. (c-e) Deconvolution analysis of the region 1250-1000cm^{-1}.

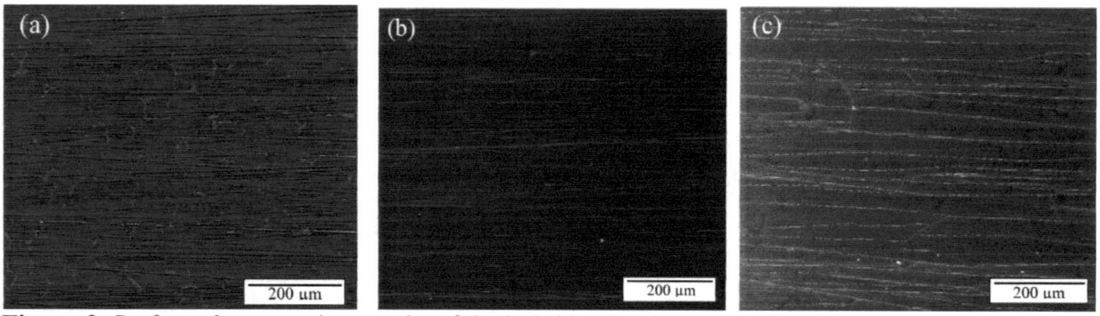

Figure 2. Surface electron micrographs of the hybrid sol-gel coatings for different aging times: (a) 2 hours, (b) 3 days and (c) 5 days.

3.3. Evaluation of the corrosion resistance of the coatings

Figure 3 shows the anodic polarization curves of the ZE41 alloy and the hybrid sol-gel coatings synthesized at different aging times. In general, the hybrid sol-gel coatings enhanced the corrosion resistance of the alloy, except for the shortest aging time (2 hours). In this case, the coating did not offer a convenient protection, showing similar values of corrosion current density with the alloy (Table 1). On the other hand, the best corrosion behaviour was exhibited by the coating synthesized at 3 days of aging, which corrosion current density was reduced by about two orders of magnitude. However, for longer times, the cracks favoured the diffusion of the electrolyte through the coating, diminishing its barrier properties.

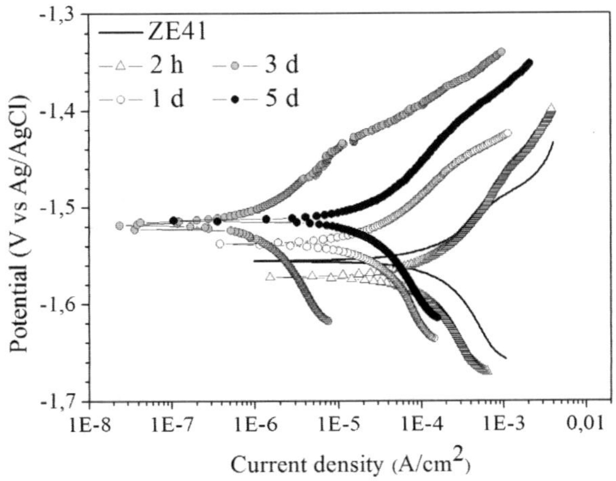

Figure 3. Anodic polarization curves of the ZE41 magnesium alloy and the hybrid sol-gel coatings obtained at different aging times.

Table 1. Electrochemical parameters obtained from anodic polarization curves.

	Corrosion potential (V vs Ag/AgCl)	Corrosion current density (A/cm^2)
ZE41	-1,55	1,4E-4
2h	-1,57	8,8E-5
1d	-1,54	1,7E-5
3d	-1,51	1,0E-6
5d	-1,51	2,4E-5

4. Conclusions

The results confirmed that the employment of TEOS-GPTMS systems and under specific ageing conditions allows obtaining uniform and homogeneous sol-gel coatings, with no evidence of cracking, which enhanced the corrosion resistance properties so that the corrosion current density was reduced up to two orders of magnitude with regard to the ZE41 alloy.

Ageing time played a key role in the morphology and barrier properties of the hybrid sol-gel coating. In this regard, coatings synthesized at shorter aging times did not improve the corrosion resistance, due to the hydrolysis-condensation reactions were still not significant. By contrast, for extended aging times, the high viscosity of the sol resulted in coatings with presence of cracks and diminished corrosion properties.

Acknowledgments

The authors wish to acknowledge to the Vicerrectoría de Investigación y Extensión of the Universidad Industrial de Santander, Co (Project 1332) and Colciencias for the financial support of this work.

References

[1] Dorozhkin S 2014 *Acta Biomaterialia* **10** 1919
[2] Shadanbaz S and Dias G 2012 *Acta Biomaterialia* **8** 202
[3] Xin Y et al 2011 *Acta Biomaterialia* **7** 1452
[4] Carboneras M, García M C and Escudero M L 2011 *Corrosion Science* **53** 1433
[5] Murillo-Gutiérrez N et al 2014 *Applied Surface Science* **309** 62
[6] Zomorodian A 2012 *Surface & Coating Technology* **206** 4368

Corrosion resistance of multilayer hybrid sol-gel coatings deposited on the AISI 316L austenitic stainless steel

Y T Caballero[1], E A Rondón[1], L Rueda[1], C A Hernández Barrios[1], A Coy[1] and F Viejo[1]
[1] Universidad Industrial de Santander, Bucaramanga, Santander, Colombia.

E-mail: fviejo@uis.edu.co ; anaecoy@uis.edu.co

Abstract. In the present work multilayer hybrid sol-gel coatings were synthesized on the AISI 316L austenitic stainless steel employed in the fabrication of orthopaedic implants. Hybrid sols were obtained from a mixture of inorganic precursor, TEOS, and organic, GPTMS, using ethanol as solvent, and acetic acid as catalyst. The characterization of the sols was performed using pH measurements, rheological tests and infrared spectroscopy (FTIR) for different ageing times. On the other hand, the coatings were characterized by scanning electron microscopy (SEM), while the corrosion resistance was evaluated using anodic potentiodynamic polarization in SBF solution at 37±2°C. The results confirmed that sol-gel synthesis employing TEOS-GPTMS systems produces uniform and homogeneous coatings, which enhanced the corrosion resistance with regard to the parent alloy. Moreover, corrosion performance was retained after applying more than one layer (multilayer coatings).

1. Introduction

The AISI 316L austenitic stainless steel has been widely used in fabrication of orthopaedic implants and biomedical internal fixation devices such as bone plates, spinal fixations, screws, etc., due to its good mechanical properties, acceptable biocompatibility, and low cost compared with those of other metallic materials used for implants, allowing it to be affordable to everybody in developing countries [1]. Nevertheless, this material is prone to suffer pitting corrosion in physiological environments, inducing the release of metallic ions, i.e. as Ni^{2+}, Fe^{2+}, Mo^{3+} and Cr^{3+}, which can develop different diseases such as allergy, arthritis, cancer, or genotoxic and mutagenic activities, if these exceed the limit concentration in the human body [2,3].

One way to provide corrosion protection is to coat the surface to prevent contact with the aggressive environment. In this sense, hybrid sol-gel coatings (mixture of organic and inorganic precursors) have arisen as one of the most promising alternatives due to their convenient adhesion to the metallic substrate, low cracking susceptibility and flexibility. The latter has been associated with the presence of organic precursors, which promote crosslinking of the sol-gel network and reduces both temperature and time during the curing process. Furthermore, it is possible to obtain greater layer thicknesses, even design multilayer coatings [4-6].

Therefore, the aim of this work was to perform hybrid sol-gel multilayer coatings for improving the corrosion resistance of the AISI 316L stainless steel employed in fabrication of orthopaedic implants. It is should mentioned that the present work takes part of a large-scale research project, which final purpose is the design of multilayer sol-gel coatings with dual (barrier-bioactivity) properties on different metallic materials used for biomaterial devices.

2. Experimental procedure

Hybrid sols were prepared from a mixture of the inorganic precursor tetraethoxysilane, TEOS, and the organic precursor 3-glycidoxypropyltrimethoxysilane, GPTMS, in 3:1 molar proportion, employing ethanol as a solvent and acetic acid as catalyst. Sols were characterized by pH measurements, rheological tests and Fourier Transform Infarred (FTIR) spectroscopy for different ageing times in order to study the hydrolysis and condensation process. Later, sol-gel monolayer and multilayer (two layer) coatings were deposited the sol on the AISI 316L stainless steel (SS) following the dip-coating route and then were cured in two steps where temperature was increased progressively to prevent cracking: 60ºC (4 hours) and 120ºC (2 hours). Hybrid sol-gel coatings were characterized by scanning electron microscopy (SEM). Finally, their corrosion resistance was evaluated by anodic polarization test in simulated body fluid (SBF) solution at 37±2°C.

3. Results

3.1. Characterization of hybrid sols

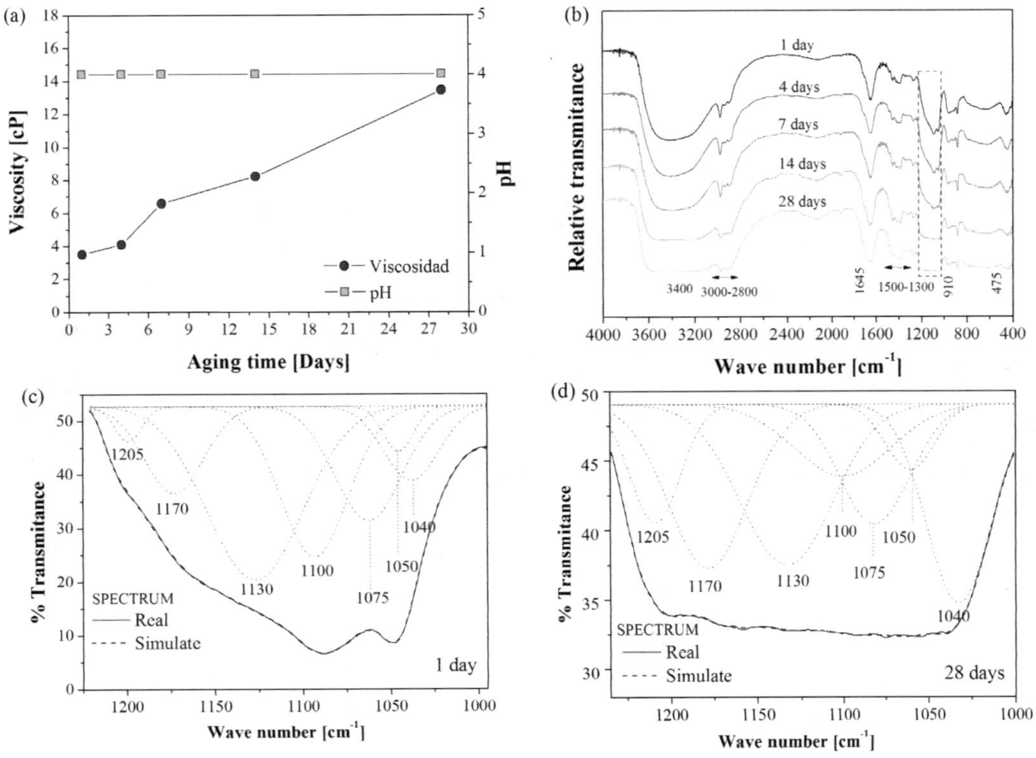

Figure 1. (a) Evolution of viscosity and pH of the sol synthetized. Evolution of the FTIR spectra: (b) General spectra. Deconvolution analysis of the region 1250-1000cm^{-1}: (c) 1 day and (d) 28 days.

Evolution of viscosity and pH versus the aging time, up to 28 days, are presented in Figure 1(a). Hybrid sols showed a progressive increment of viscosity ranging from 3,50cP (initial) to 13,43cP (final) (the gelation state was reached at 60 days of aging). Also it is observed that the pH of the sol remained invariant in a value of 4, during the whole gelation process, probably due to the presence of acetic acid (weak organic acid) as a catalyst. Complementary, Figure 1(b) shows FTIR spectra of sols where the O-H absorption bands, associated with the presence of solvent/water, can be observed at 3400 and 1645cm^{-1}. Similarly, precursors distinctive bands are located at 3000-2800cm^{-1} and 1500-

1300cm^{-1} related to C-H bands by methyl (-CH_3) and methylene (-CH_2-) groups, which lose intensity as the ageing time increases, fact attributed to the progress of the hydrolysis reactions of the precursors [7]. In order to study hydrolysis and condensation reactions that occurred in the sol, deconvolution analysis was carried out for the region of the spectra around 1250-1000cm^{-1} (this region is related to the silicon chemistry) (Figures 1(c)-(d)). It can be noticed that the intensity of the stretching vibrations bands concerned to the Si-OH groups (1200 and 1170cm^{-1}) increased, while that for the Si-O-C groups (1100 y 1080cm^{-1}) declined, both as the ageing time progressed. This result is indicative of hydrolysis reaction of the precursors during the gelation process. Likewise, characteristic Si-O-Si band at 1040cm^{-1} attributed to the condensation reaction was also observed [8].

3.2. Characterization of hybrid coating

Electron micrographs of the coatings synthesized on the SS at different aging times are shown in Figure 2. As general result, the longer ageing time the more uniform coatings were obtained. In this regard, lines resulting of the surface preparation of substrate, which were initially evident for short ageing time, were completely covered when the coating was deposited under ageing longer than 7 days. This behaviour can be explained as a result of evolution of the hydrolysis and condensation reaction, which allow: i) the release of silanol groups necessary to ensure wettability and coating adhesion, ii) and the increase of the viscosity of sol required to generate continuous and uniform coatings. Further, cross-sectional analysis revealed that both monolayer and multilayer coatings were continuous and virtually free of cracks, with a thickness about 1.9 and 4.2μm respectively (Figure 3). In addition, multilayer coatings exhibited an apparent good adhesion between layers.

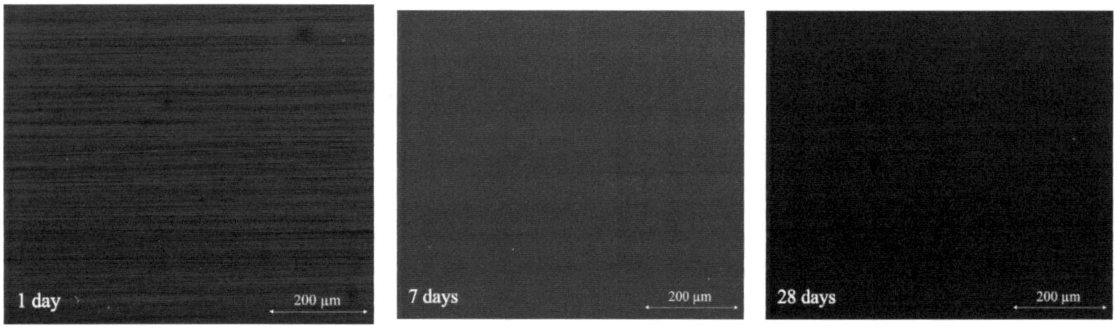

(a) (b) (c)

Figure 2. Surface electron micrographs of the hybrid sol-gel coatings for different aging times: (a) 1 day, (b) 7 days and (c) 28 days.

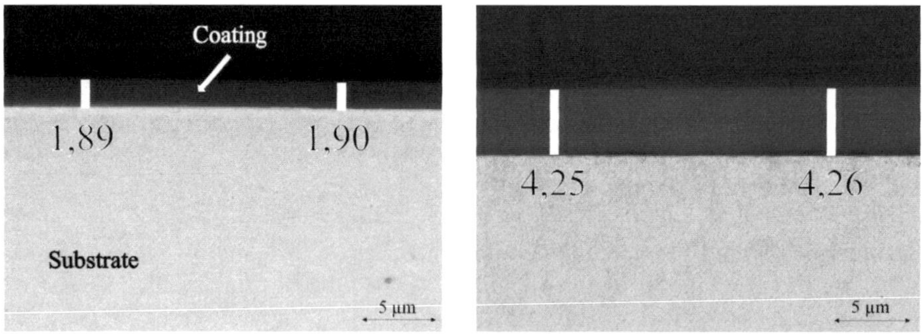

(a) (b)

Figure 3. Cross-sectional SEM micrographs of coatings obtained at 7 days of aging time: (a) monolayer and (b) multilayer.

3.3. Evaluation of the corrosion resistance of the coatings

Figure 4 shows the anodic polarization curves of the SS and monolayer coatings at different aging times and multilayer coating (obtained under ageing of 7 days). Polarization curves are presented with relative potential units to facilitate their visual analysis. In general, an increase of the aging time resulted in coatings with enhanced corrosion resistance. Thus, the corrosion current density of the coatings gradually decreased until reach a value around 2,0E-8A/cm^2 (about one order of magnitude lower with regard to the base material (1,6E-7A/cm^2)) at seven days of aging, and remained virtually constant for longer times (Figure 4(a)). The latter can be related with the progressive formation of SiO_2, as a result of the advance of the condensation reaction, which confers a hydrophobic character and therefore good protection properties to the coating. Moreover, multilayer deposition did not show significant influence on the anodic polarization response when compared to the monolayer coating, indicating that the corrosion performance of the coating was retained (Figure 4(b)).

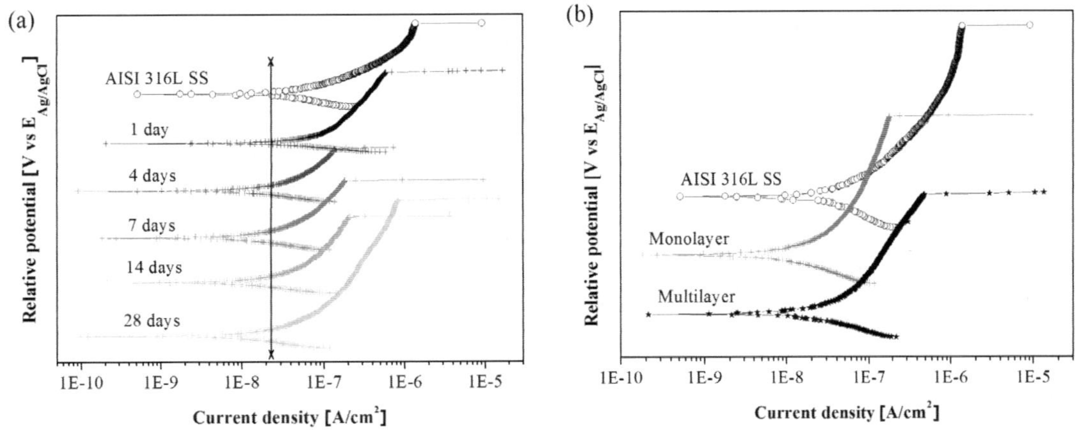

Figure 4. Anodic polarization curves of the SS and the hybrid sol-gel coatings obtained at different conditions: (a) by variation of aging time and (b) by variation of the number of layers.

4. Conclusions

The results confirmed that TEOS-GPTMS sol-gel systems allow obtaining uniform coatings, with no evidence of cracking. Longer ageing times resulted in coatings with better barrier properties, so that the corrosion current density was reduced up to one order of magnitude with regard to the SS for ageing of seven days. Moreover, multilayer coatings showed good adhesion between films, increased thickness, without affecting the corrosion performance of the monolayer coatings.

Acknowledgments

The authors wish to acknowledge to the Vicerrectoría de Investigación y Extensión of the Universidad Industrial de Santander, (Project 1798) for the financial support of this work.

References

[1] Schmidt C, Ignatius A, Claes L E 2001 *J Biomed Mater Res* **54** 209
[2] Wataha J C 2000 *J Prosthet Dent* **83** 223
[3] Y Okazaki, E Gotoh 2008 *Corros Sci* **50** 3429
[4] Murillo-Gutiérrez N, et al 2014 *Appl Surf Sci* **309** 62
[5] López D A, et al 2008 *Surf& Coat Tech* **202-10** 2194
[6] Catauro M, Bollino F, Papale F, Pacifico S 2015 *J Drug Deliv Sci and Tech* **26** 10
[7] Limpo J, Rubio J, Oteo L 1993 *Bol Soc Esp Ceram Vidr* **32** 31
[8] Macan J, Ivankovi H, Ivankovi M, Mencer H J 2004 *Thermochim Acta* **414** 219

Influence of alumina and titanium dioxide coatings on abrasive wear resistance of AISI 1045 steel

A Santos[1], A Remolina[1] and J Marulanda[2]
[1] Universidad Pontificia Bolivariana Seccional Bucaramanga, Floridablanca, Colombia.
[2] Universidad Tecnológica de Pereira, Pereira, Risaralda, Colombia.

E-mail: alfonso.santos@upb.edu.co

Abstract: This project aims to compare the behaviour of an AISI 1045 steel´s abrasive wear resistance when is covered with aluminium oxide (Al_2O_3) or Titanium dioxide (TiO_2), of nanometric size, using the technique of thermal hot spray, which allows to directly project the suspension particles on the used substrate. The tests are performed based on the ASTM G65-04 standard (Standard Test Method for Measuring Abrasion Using the Dry Sand/Rubber Apparatus). The results show that the amount of, lost material increases linearly with the travelled distance; also determined that the thermal treatment of hardening-tempering and the alumina and titanium dioxide coatings decrease in average a 12.9, 39.6 and 29.3% respectively the volume of released material during abrasive wear test.

1. Introduction
The surface of a component is susceptible to mechanical (friction and wear), chemicals, electrical, optical and thermal interactions. As a result in any design process should be considered the quality and the protection of the surface of a component; a possible method to protect it, consists of applying coatings, which can be metallic, polymeric and ceramic. Ceramic coatings are widely used because they can transform a substrate surface making it chemically inert, abrasion resistant, low-friction and easy to clean. The project seeks to determine how are affected the mechanical properties of wear resistance on a standard AISI 1045 steel, when it is thermally treated with a hardening-tempering process or is coated with particles of nanometric size of Al_2O_3 or TiO_2, through a process of thermal hot spray.

2. Conceptual framework

2.1. Materials nanostructured
Ultrafine microstructures with an average phase or grain size on the order of nanometers ($1nm=10^{-9}m$) are classified as nanostructured materials [1]. Currently, in a broader sense of the term, any material containing grains or clusters below 100 nm, or layers or strands of that dimension, can be considered nanostructured [2]. The interest in these materials has been stimulated by the fact that, due to the small size of the particles, grains, or phases, and the high ratio of volume-surface is expected that these can prove unique mechanical properties [3]. In nanophase materials, a variety of size-related effects can be incorporated by controlling sizes of constituent elements [4]. One of the most critical features of nanoparticles is a very high relationship surface-volume, with a large proportion of surface atoms. The large specific surface area provides exceptional properties for industrial applications [5].

2.2. Coatings
A coating is a layer of a material that is applied to the surface of an object, called the substrate, which is usually of other material. The purpose of the coating is the improvement of the surface properties of the substrate: appearance, adhesion, corrosion, dissolution, wear or scratch resistance (hardness), among others. The coatings with improved properties are very demanded in various industrial sectors, either to optimize the characteristics of the structural elements to protect or create functionalized surfaces [1,6].

2.3. Thermal spray by flame
The thermal spraying by conventional flame was the first process of thermal spray developed and is still in common use. Processes by flame use chemical energy of the combustion gases to generate heat, being the oxy-acetylene torches the most common. These systems use one gun for projection of gases containing materials used for the coating and in some designs; additional air flow restricts the flame. The restriction of the flame affects expanding gases, creating in this way an area of combustion with very high temperatures. The nozzles also give shape to the pattern of the flame [7].

3. Methodology

3.1. Morphological and chemical characterization of Al_2O_3 and TiO_2
Morphological and chemical characterization of Al_2O_3 and TiO_2, was performed through the technique of Scanning Electron Microscope; for which was used a microscopy equipment FEI brand QUANTA FEG 650 model. The images obtained from microscope allow to identify the shape and calculate the size of the particles that compose the material. Additionally, the microscope has integrated a system of Energy Dispersive X-ray Spectroscopy (EDS).

3.2. Deposit and characterization of alumina and titanium dioxide coatings
The coatings of Al_2O_3 and TiO_2 of nanometric size were carried out through the process of thermal spraying by flame, using an Eutalloy gun 85BX, which is adapted to a conventional gas burning equipment. The distance used between the torch and the base material was approximately 30cm, with an inclination from the horizontal of 60° [7]. The oxygen pressure used was 25psi and acetylene was 5psi, which defines a pressure relation of 1:5.

3.3. Testing
The tests were conducted in three different types of specimens: the first were produced in AISI 1045 standard steel; the second were of the same steel, but thermally treated with a process of hardening (warming 850°C and cooling in water) and tempering (warming 350°C and cooling with air); and the last were AISI 1045 steel with Al_2O_3 and TiO_2 coating of nanometric size, deposited through thermal spray by flame process. The tests are performed based on the ASTM G65-04 standard (Standard Test Method for Measuring Abrasion Using the Dry Sand/Rubber Wheel Apparatus) [8].

4. Results

4.1. Morphological and chemical characterization of Al_2O_3 and TiO_2
Based on images obtained, of each of the materials used as coating, is analysed a pre-sample of the dimensions of the smaller side of 10 different particles, and with them is defined a sample size of 48 data to achieve a reliability of 95% and a maximum error of 5%. The average of the particle size of Al_2O_3 was 92.9nm; while the TiO_2 was 36.7nm. The chemical analysis of the powders used was carried out through the values of microanalysis by EDS, which are listed in Figure 1. The error percentage obtained between the calculated theoretical and the experimental atomic weight retrieved by analysis EDS, for Alumina and titanium dioxide was 3.3% and 9% respectively.

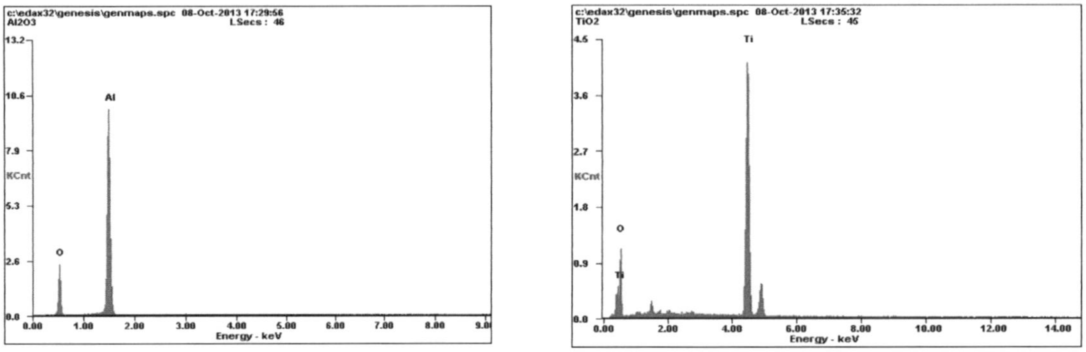

(a) (b)

Figure 1. Spectrum obtained in EDS analysis. (a) Alumina powder; (b) titanium dioxide powder.

4.2. Deposit and characterization of alumina and titanium dioxide coatings

The coatings obtained through the process of thermal spray by flame, Figure 2(a), over the specimens of AISI 1045 steel, can be seen in Figure 2(b); it can be concluded that coatings are uniform on the surface. Also can be said that the nanometric size materials used in deposits is preserved during the process, however, there is a tendency to agglomerate to form larger structures.

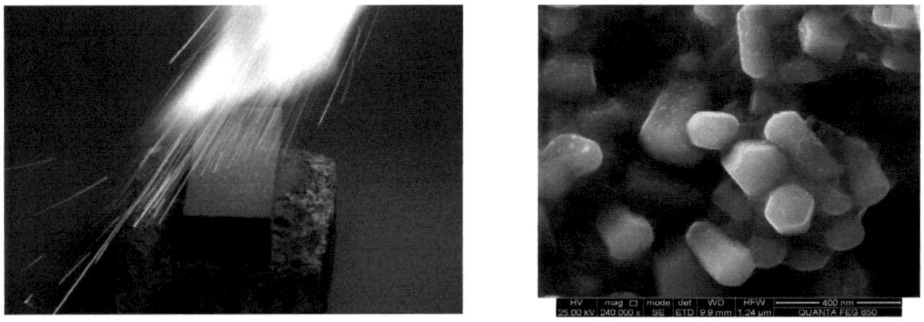

(a) (b)

Figure 2. Coatings deposition. (a) Process of thermal spray by flame; (b) coating of TiO_2.

4.3. Test results of abrasive wear resistance

Abrasive wear resistance tests result in a straight mark on the specimen, and therefore a removal of material. The results are listed in Figure 3.

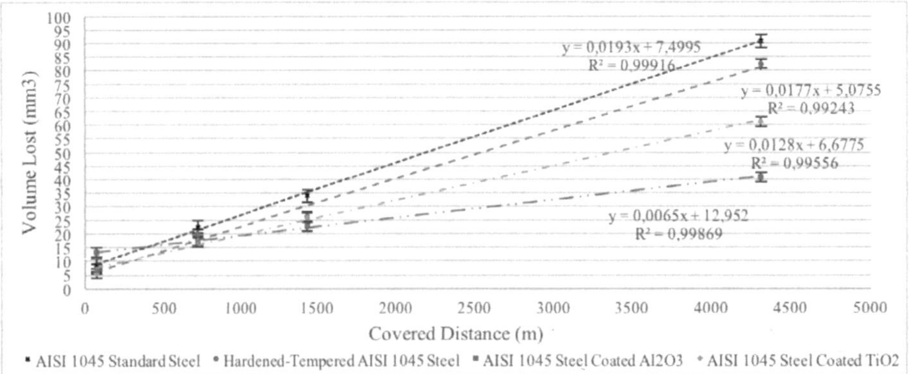

Figure 3. Comparison of the abrasive wear resistance in specimens of AISI 1045 steel.

Figure 3 allows concluding that the thermally treated specimens with a process of hardening-tempering decrease on average a 12.9% the loss of material during the test of abrasive wear. Also can be seen that TiO_2 or Al_2O_3 coatings deposited on AISI 1045 Steel presents a more significant improvement in this property, reaching average reduction of 29.3 and 39.6%, respectively, with respect to standard steel.

To compare the obtained results of abrasive wear tests is calculated the loss of material per unit of linear distance covered, these values are graphed in Figure 4, which shows that for short distances the loss of material of the specimens is more significant, this mainly due to surface roughness caused by the machining of the specimens of normalized and thermally treated steel; and to the irregularity of the coating of alumina made through the process of thermal spray by flame; at greater distances the loss of material is less and tends to stabilize.

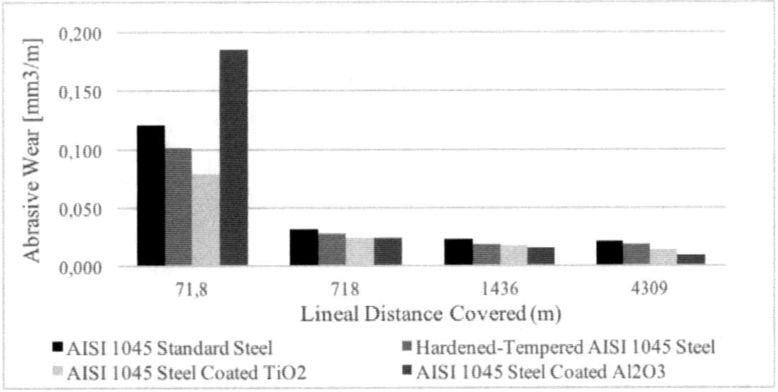

Figure 4. Abrasive wear resistance per unit of distance covered in specimens of AISI 1045 Steel.

5. Conclusions

The process of thermal spray by flame allows the deposition of alumina and titanium dioxide, of nanometric size, on a base material of AISI 1045 Steel, but the use of an anchor material is necessary. However obtained coatings, although they retain their nanoscale dimensions, have a high porosity and discontinuity, which affects the hardness and resistance to wear of them.

The average loss of material, per unit of length travelled, during abrasive wear tests of the standard AISI 1045 steel, thermally treated with a process of hardening and tempering, and coated with Al_2O_3 or TiO_2, nanosized, is 0,0493mm^3/m; 0,0416mm^3/m; 0,0586mm^3/m and 0,0337mm^3/m respectively.

The thermally treated specimens with a process of hardening-tempering decrease on average a 12.9% the loss of material during the abrasive wear test. However titanium dioxide or alumina nanosized coatings deposited on AISI 1045 steel presented a more significant improvement in this property, reaching an average reduction of 29.3 and 39.6%, respectively, with respect to standard steel, this is due to the large capacity of the oxides to resist abrasion.

References

[1] H Gleiter 1989 *Progress in Materials Science* **33** 223
[2] R W Siegel 1993 *Nanostructured Materials* **3** 1
[3] R W Siegel 1993 *Materials Science and Engineering: B* **189** 37
[4] R W Chan 1990 *Nature* **348** 389
[5] S C Tjong and H Chen 2004 *Materials Science and Engineering R Reports* **45** 1
[6] J He and J M Schoenung 2002 *Materials Science and Engineering A* **336** 274
[7] M G Itzel and R P David 2009 *Caracterización del proceso de rociado térmico de acero inoxidable sobre aceros de bajo contenido de carbono* (México: Universidad Nacional Autónoma de México) p 120
[8] ASTM G65–04 2010 *Standard test method for measuring abrasion using the dry sand/rubber wheel apparatus* (United States: American Society for Testing Materials-ASTM)

Formation of fouling deposits on a carbon steel surface from Colombian heavy crude oil under preheating conditions

D A Muñoz Pinto[1], S M Cuervo Camargo[1], M Orozco Parra[1], D Laverde[1], S García Vergara[1] and C Blanco Pinzon[2]
[1] Universidad Industrial de Santander, Bucaramanga, Colombia.
[2] Instituto Colombiano del Petróleo (ICP), Piedecuesta, Colombia.

E-mail: diegoarmandomunozpinto@hotmail.com

Abstract. Fouling in heat exchangers is produced by the deposition of undesired materials on metal surfaces. As fouling progresses, pressure drop and heat transfer resistance is observed and therefore the overall thermal efficiency of the equipment diminishes. Fouling is mainly caused by the deposition of suspended particles, such as those from chemical reactions, crystallization of certain salts, and some corrosion processes. In order to understand the formation of fouling deposits from Colombian heavy oil (API≈12.3) on carbon steel SA 516 Gr 70, a batch stirred tank reactor was used. The reactor was operated at a constant pressure of 340psi while varying the temperature and reaction times. To evaluate the formation of deposits on the metal surfaces, the steel samples were characterized by gravimetric analysis and Scanning Electron Microscopy (SEM). On the exposed surfaces, the results revealed an increase in the total mass derived from the deposition of salt compounds, iron oxides and alkaline metals. In general, fouling was modulated by both the temperature and the reaction time, but under the experimental conditions, the temperature seems to be the predominant variable that controls and accelerates fouling.

1. Introduction

The fouling is defined as the deposition of unwanted materials on heat transfer equipment, which results in an increase in thermal resistance to heat transfer and loss of the equipment thermal efficiency. Fouling in crude oil pre-heat trains is a complex phenomenon with several mechanisms interacting with each other [1]. The phenomenon is mainly caused by deposition of particles, chemical reactions, crystallization and some corrosion processes [2-4].

The causes of fouling include:

- Impurities from the crude, gum or polymeric species formed through oxidation of reactive species in the oils.
- Insoluble asphaltene.
- Iron sulphide formation.
- Coke formation.

A lot of research has focused on the study of fouling generated in units of heat transfer in the petrochemical industry. These studies were performed at high temperatures (T>350°C), where the fouling is generated mainly by coke formation [5-8]. The present work aims to study the fouling phenomenon at temperatures below 300°C using Colombian heavy crude oil (CHCO) like fouling agent.

2. Experimental

Fouling experiments were performed using a 1L Parr instrument Model 4848 stirred batch reactor. The reactor is shown schematically in Figure 1.

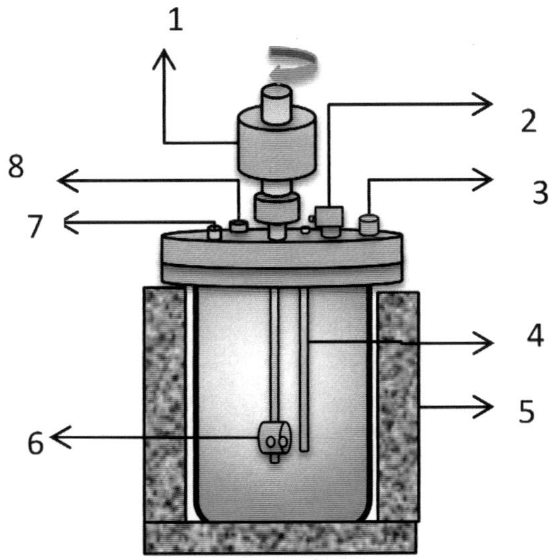

1. Magnetic stirred
2. Pressure controller
3. Gas injection valve
4. Thermocouple
5. Heater
6. Flat coupon holder
7. Pressure controller valve
8. Safety valve

Figure 1. Diagram of stirred batch reactor.

The carbon SA 516 Gr 70 steel (CS) coupons were prepared according to the NACE SP0775 2013 norm [9]. They were processed by 120 grit sand paper, rinsed with acetone and dried in hot air.

CHCO used for the tests contains about 1.6%wt sulphur; API gravity approx. 12.2 and total salt content about 9 Pound/1000Bls.

Tests were carried out at temperatures of 220, 260 and 300°C, using for each experiment 750mL of CHCO in the reactor. Pressure and the rotation speed were maintained constant at 340psi and 15rpm, respectively. For each selected temperature, three different exposure times were used: 25, 30 and 35 hours. After each experiment, the coupons were washed with acetone to remove oil and dried in warm air. To evaluate the formation of deposits on the metal surfaces, the steel samples were characterized by gravimetric analysis and Scanning Electron Microscopy (SEM). After each run, spent oil was analysed to determine salt contents [10].

3. Results and discussion

The Figure 2(a) shows the relationship between final mass and initial mass of the coupons (mass gain) at different experimental conditions. The major mass gain of coupons occurs at temperatures of 220°C. This behaviour is proportional to the increase of experimental time. However, as the temperature increases the behaviour of the curve is different. Both intermediate and high experimental temperature (260 and 300°C), present a major mass gain at 25 hours and decrease at 30 hours. At 35 hours, the mass gains increase significantly at temperature of 260°C while at 300°C the mass gain presents a similar value that at 30 hours. The lowest point occurs at 30 hours for the two higher temperatures.

Figure 2(b) shows the salt intake for each temperature after 35 hours. Salt intake is expressed as the difference between total salt content of crude before the test and total salt content of crude after the test. The major intake was at 220°C. This result indicates that the fouling mechanism may be related to the crystallization of salts.

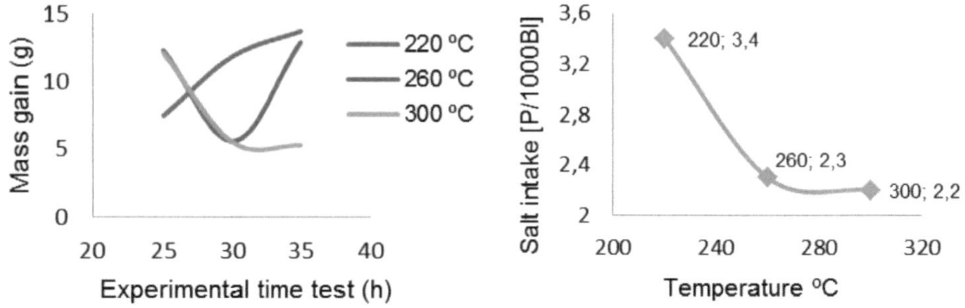

(a) (b)

Figure 2. (a) Mass gain of coupons at experimental conditions and (b) Salt intake after 35 hours experiment at 220, 260 and 300°C.

The results of coupon surface characterization are shown in Figure 3(a) and (b). Crystal deposits on coupon surface are observed in Figure 3(a). Energy Dispersive X-ray Spectroscopy (EDS) analysis (Figure 3(b)) shows the presence of sodium and chlorine on carbon steel surface probably by deposition and crystallization of crude salts. This result explains the decrease of the total percentage of salt present in the crude after each experiment (Figure 2(b)).

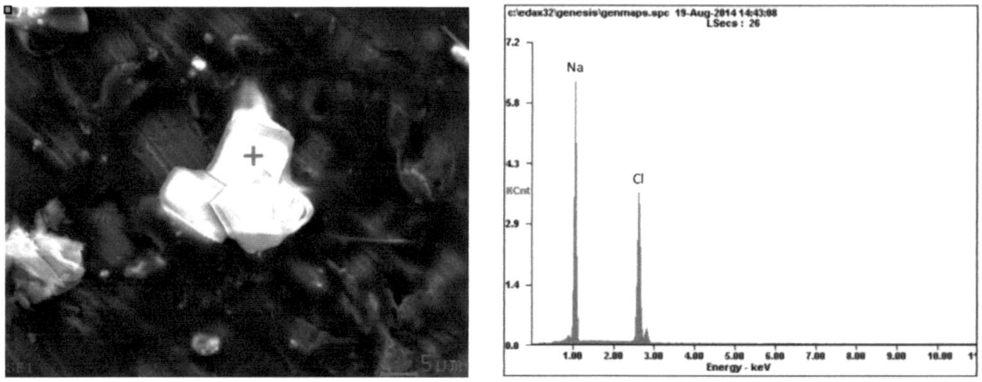

(a) (b)

Figure 3. SEM image of crystal deposit on CS after 35 hours experiment at 300°C. (a) Crystal deposit and (b) EDS of crystal deposit.

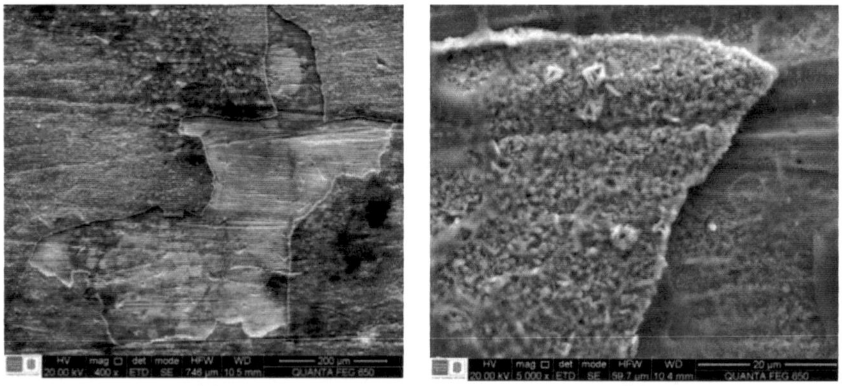

(a) (b)

Figure 4. SEM images of deposits on CS after 35 hours experiment at 300°C. (a) Mag 500x and (b) Mag 5000x.

Qualitative assessments of coupon surfaces by SEM (Figure 4) showed formation of a layer, which could have been produce by the reaction between the sulphur from the oil and the iron atoms in the alloy [11-12]. Layer formation occurred at 300°C. No layer formations on metal surface at the other two temperatures occurred. This result confirms the temperature effect on fouling processes.

4. Conclusions

At 220°C, the fouling is related with decreasing of total salt content in the crude. Crystal deposits on metal surface are evidence of the results.

According to the results, both temperature and time have influence on fouling, however under these experimental conditions, the temperature seems to be the predominant variable that controls and accelerates the fouling.

References

[1] Ratel M, Kapoor Y, Anxionnaz Minvielle Z, Seminel L and Vinet B *2013 Investigation of fouling rates in a heat exchanger using an innovative fouling rig* (Budapest: Proceedings of 10th International conference on Heat Exchanger Fouling and Cleaning) p 36

[2] Srinivasan M and Watkinson A P 2003 *Fouling of some Canadian crude oils* (New Mexico: Heat Exchanger Fouling and Cleaning: Fundamentals and Applications)

[3] Bott T R 1995 *Fouling of heat exchanger* (Amsterdam: Elsevier science) p 185

[4] Watkinson A P Wilson D 1997 *Chemical Reaction Fouling: A Review. Exp Therm Fluid Sci* **14** 361

[5] Wang W and Wattkinson A P 2015 *Heat Transfer Engineering* **36-7** 623

[6] Petkovic B and Watkinson A P 2014 *Heat Transfer Engineering*. **35-3** 302

[7] Watkinson A P 2007 *Heat Transfer Engineering*. **28-3** 177

[8] Wang W and Watkinson A P 2011 *Iron sulphide and coke fouling from sour oils: review and initial experiments* (Crete Island: Proceedings of 9th International conference on Heat Exchanger Fouling and Cleaning) p 23

[9] NACE SP0775 2013 *Preparation, installation, analysis and interpretation of corrosion coupons in oilfield operations* (Houston: NACE international)

[10] ASTM D3230 2013 *Standard test method for salts in crude oil (Electrometric method)* (DOI: 10.1520/D3230-13)

[11] Dettman D. Dettman, Nana Li, Dhanuka Wickramasinghe, Zhengrong, Xi Ning Chen, Glen R.D. Elliott, Jingli Luo 2012 *NACE international* C2012-0001326

[12] Tebbal S and Kane R 1996 *NACE International* **607** 96607

Corrosive effect of the type of soil in the systems of grounding more used (copper and stainless steel) for local soil samples from the city of Tunja (Colombia), by means of electrochemical techniques

L Guerrero[2,3], Y Salas[1,2,3] and J Blanco[2]
[1] Universidad Antonio Nariño, Bogotá, Colombia.
[2] Universidad Antonio Nariño, Tunja, Colombia.
[3] Universidad Pedagógica y Tecnológica de Colombia, Tunja, Colombia.

E-mail: jose.guerrero@uan.edu.co

Abstract. In this work electrochemical techniques were used to determine the corrosion behaviour of copper and stainless steel electrodes, used in grounding varying soil type with which they react. A slight but significant change in the corrosion rate, linear polarization resistance and equivalent parameters in the technique of electrochemical impedance spectroscopy circuit was observed. Electrolytes in soils are slightly different depending on laboratory study, but the influence was noted in the retention capacity of water, mainly due to clays, affecting ion mobility and therefore measures such as the corrosion rate. Behaviour was noted in lower potential for copper corrosion, though the corrosion rate regardless of the type of soil, was much higher for electrodes based on copper, by several orders of magnitude.

1. Introduction

According to the technical regulation of electrical installations RETIE, all electrical systems must have a grounding system, to prevent people in contact with it, remain subject to dangerous tensions. The objectives of a grounding system are the safety of people, the protection of installations and electromagnetic compatibility [1,2].

The electrodes in a grounding system must ensure conductivity and the resistance to corrosion, which must be minimum 15 years from the date of installation. The method of immersion in salt spray for 1000 hours, is the most used to test the corrosion resistance of electrodes for grounding systems [3].

The purpose of this work is to analyse the influence of soil characteristics on morphology and properties defined for the grounding electrodes [4]. Decay rates were established by electrochemical techniques using potentiodynamic curves and linear polarization resistance measures.

2. Materials and methods

2.1. Experimental soils samples

For the analysis, two types of soil characteristic of the city of Tunja, Colombia were referenced [5]. The experimental design matrix was then defined with two soil types and two types of electrodes.

Soil samples are taken to the laboratory (Universidad Pedagógica y Tecnológica de Colombia (UPTC) to obtain data such as texture, pH, percentage of organic matter, phosphorus, aluminium, potassium, calcium, magnesium, cation exchange capacity and electrical conductivity characterization throws us two types of soil as well: sandy loam and clay loam. The electrolyte used in the cell was obtained from soil samples prepared with a relative humidity of 20 percent.

2.2. Electrodes
According to table 15.2 of technical regulation of electrical installations RETIE [1], test electrodes copper and stainless steel were taken. These electrodes are the most used in grounding systems.

2.3. Experimental Equipment
2.3.1. Potentiosta-galvanostat. The Gamry G-750 Potentiostan-Galvanostats, an electrode Ag/AgCl and an electrochemical cell was used for generate the potentiodynamics and LPR curves and is located in the Materials Lab of the Universidad Antonio Nariño, Tunja, Colombia.

2.3.2. Tellurometer measurement of soil resistivity. The Wenner four–pin method, is the most commonly used technique. In brief, four probes are driven into the earth along a straight line, at equal distances an apart, driven to a depth b. The voltage between the two outer (current) electrodes to give a value of resistance R [6].

The equipment used, located in the laboratories of the Universidad Antonio Nariño, Tunja, Colombia, consists of a eurotest™ 61557 MI 2086 and his accessories.

2.4. Experimental design
The experimental matrix considered two factors (Table 1) type of soil, with two levels: sandy loam and clay loam; and electrodes type, with two levels: copper and stainless steel.

Table 1. Experimental matrix.

Soil	Electrode	Variables Response
Sandy Loam	Copper	Ecorr (mV)
	Stainless Steel	Jcorr (μA/cm^2)
Clay Loam	Copper	CR (mm/yr)
	Stainless Steel	LPR (Ω*cm^2)

2.5. Electrochemical tests
2.5.1. Tafel. Tafel polarization curves are used to obtain an accurate estimate of the corrosion rate of a metal in a solution. The current in the cell is measured over a sweep potential and is presented in logarithmic form. A scan is made from -250 to 250mV with respect to the open circuit potential.

2.5.2 Linear polarization resistance. The linear polarization resistance following the ASTM G5 standard, standard reference method of anodic polarization, potentiostatic and potentio-dynamic [7].

3. Results
The obtained results was resume in the next potentio-dynamics graphics. In the potentio-dynamics graphics, the x axis is the base 10 logarithm of current (Amperes) and the y axis is the Potential E in Volts (see Figure 1) [8].

The main response variables for each experiment were the corrosion potential, the corrosion current, anodic and cathodic pending by means of Tafel approximation and linear polarization resistance, obtained by linear regression (see Table 2).

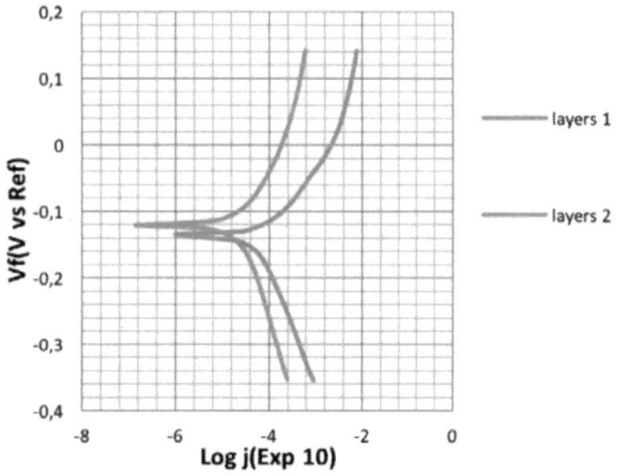

(a)

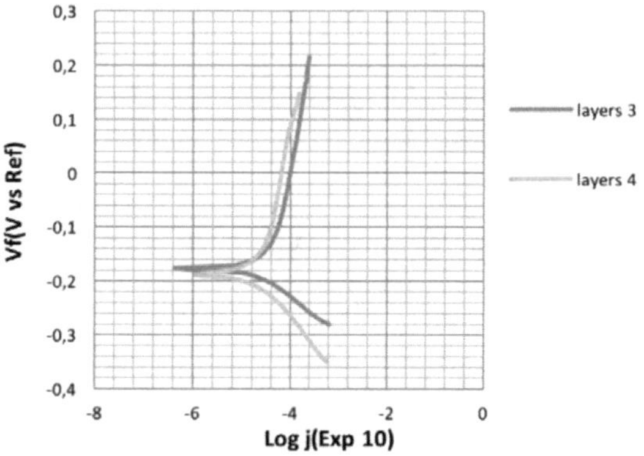

(b)

Figure 1. Curves for the different soils (a) copper (layers 1 is for sandy loam and layers 2 is for clay loam) and (b) stainless steel (layers 3 is for sandy loam and layers 4 is for clay loam).

Table 2. Experimental data from the electrochemical test.

Soil	Material	Ecorr (mV)	Icorr (μA/cm^2)	βa (mV/dec)	βc (mV/dec)	CR (mm/yr)	LRP (Ω-cm^2)
Sandy Loam	Copper	-1,10E+02	1,13E+00	8,67E+01	-2,51E+02	1,99E+00	6,01E+03
	Inox	-2,19E+02	6,35E-02	3,74E+02	-7,75E+01	3,35E-01	1,54E+05
Clay Loam	Copper	-1,29E+02	1,76E-01	1,55E+02	-2,42E+02	3,08E-01	8,20E+04
	Inox	-1,96E+02	2,96E-02	2,87E+02	-1,01E+02	1,56E-01	2,18E+05

4. Discussion

Comparisons for each factor are made in the Figure 2, 3, and 4. The scale for each plot is particular, but it indicates the change in the response variable for each of the considered factors, where Figure 2 shows the lowest corrosion potentials of copper in the two soil types, Figure 3 shows the lowest corrosion currents for stainless steel and Figure 4 shows the lowest corrosion rate for stainless steel.

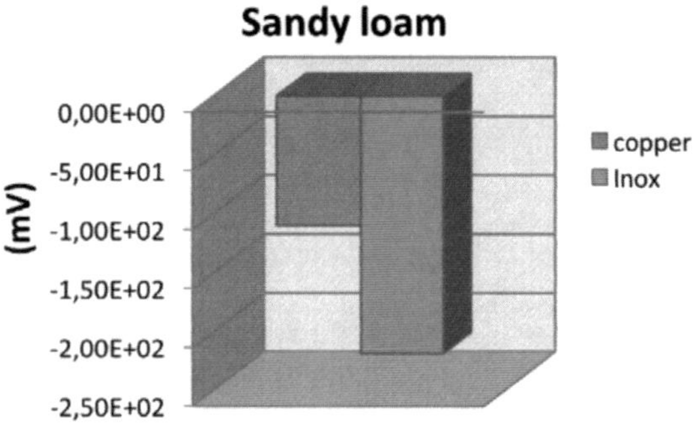

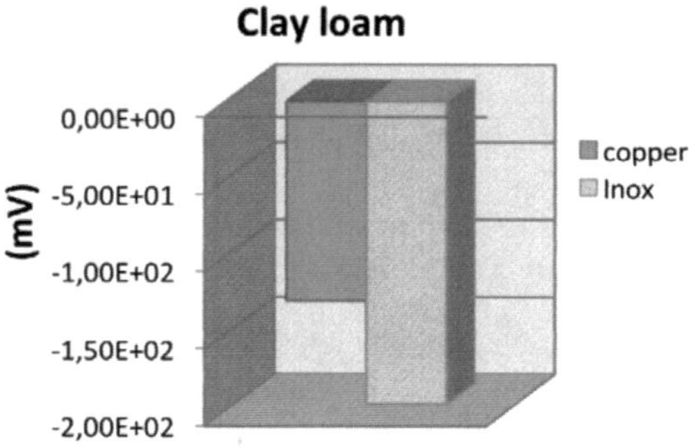

Figure 2. Ecorr (mV) Corrosion potentials for the two types of soils.

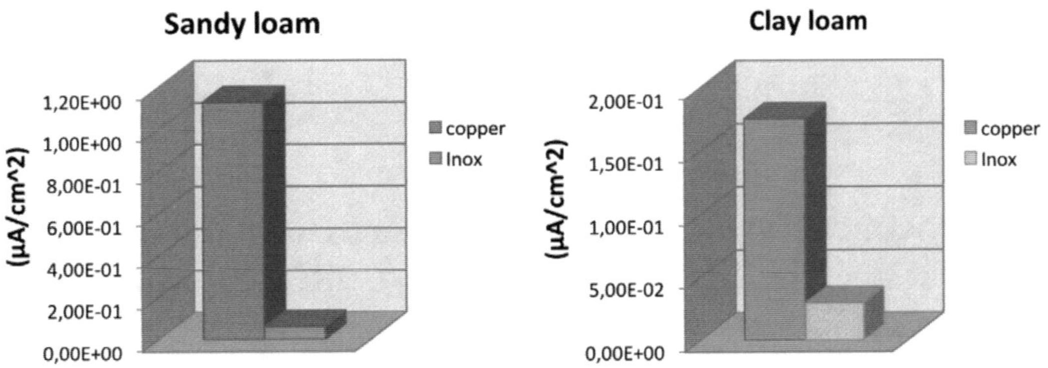

Figure 3. j_{corr} Corrosion current for the two types of soils.

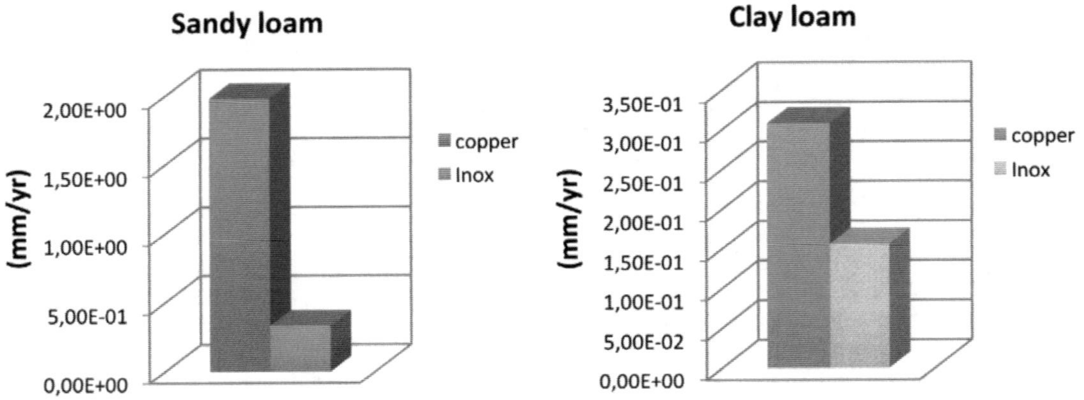

Figure 4. Corrosion rate.

5. Conclusions

Corrosion rates for different types of electrodes and soils were established. Appreciable differences in the currents and corrosion potentials are observed.

The copper electrodes are more corrosives than the stainless steel electrodes, in both types of soils under study. Corrosion rates are determined by electrochemical techniques that serve as reference for future studies.

The resistivity of the two types of soil was modelled by the wenner method. Cannot establish a direct relationship between the soil resistivity and corrosion rates. Should be consider other control variables as moisture, ph, redox potential and others that will be the subject of further studies.

References

[1] Ministerio de Minas y Energía 2013 *RETIE* (Bogotá: Ministerio de Minas y Energía)
[2] ICONTEC 1998 *Codigo eléctrico colombiano NTC 2050* (Bogotá: ICONTEC)
[3] ICONTEC 1998 *Procedimiento para el ensayo en cámara salina NTC 1156* (Bogotá: ICONTEC)
[4] ICONTEC 2001 *Equipo de conexión y puesta a tierra NTC 2206* (Bogotá: ICONTEC)
[5] Instituto Geográfico Agustin Codazzi et al 2005 *Estudio general de suelos y zonificación de tierras del departamento de Boyacá* (Bogotá: Instituto Geográfico Agustin Codazzi)
[6] IEEE 2012 *Guide for measuring earth resistivity, ground impedance, and earth surface potentials of a grounding system* (Portland: IEEE)
[7] ASTM-G5 1994 *Standard reference test method for making potentiostatic and potentiodynamic anodic polarization measurements* (West Conshohocken, PA: ASTM International)
[8] Gamry Instruments 2009 *GAMRY echem analyst help* (Warminster: Gamry Inc)

Electrochemical evaluation of the corrosion behaviour for structural steel SAE 1005 exposed to two different atmospheres (urban and industrial) and comparison with atmospheric corrosion gravimetric measurements

Y Salas[1], L Guerrero[2], H Castiblanco[3], C Torres[3] and A Palacios[3]
[1] Universidad Antonio Nariño, Bogotá, Colombia.
[2] Universidad Pedagógica y Tecnológica de Colombia, Tunja, Colombia.
[3] Universidad Antonio Nariño, Tunja, Colombia.

E-mail: yecsalas@uan.edu.co

Abstract. The Atmospheric corrosion is a phenomenon we see every day in our environment that arises due to environmental pollution we generate, there is currently very little information on atmospheric corrosion in the department of Boyacá and in general, in Colombia. The aim of this paper is to analyse which of these two environments is more aggressive and wherein the steel corrodes faster. To analyse these phenomenon specimens made in steel SAE 105 exposed for five months to the atmosphere in the municipalities of Tunja and Nobsa (an urban atmosphere and other industrial atmosphere) were installed, a control was carried to verify the amount of time that will be exposed each of these samples to the atmosphere, of Thus it may determine the lifetime of a structural steel. For the analysis of these samples electrochemical tests were carried out to calculate the rate of corrosion and resistance to polarization, also the gravimetric method be conducted to compare what was the amount of mass lost during the time of exposure to each of the samples.

1. Introduction

The atmospheric corrosion of steel passes through a mechanism of electrochemical nature. The electrolyte consists of an extremely thin film of moisture or an aqueous film, the metal appears noticeably when wet [1]. Electrolyte film under most exposed metals corrode in the atmosphere by means of cathodic oxygen reduction process. This phenomenon of corrosion is present everywhere causing considerable losses in terms of structures because of this electrochemical analysis being made so far have taken great importance due to the destruction of the materials [2,3].

This work aims at comparing two environments: one unpolluted (Tunja) and other industrial (Nobsa). SAE 1005 reference used in the structural steel industry was taken, and thus to determine their differences of atmospheric corrosion behaviour in these two municipalities, considering simultaneous and during the first five months of exposure posed a year's exposition, to study regarding the deterioration manifested by gravimetric measurements and compared to the respective electrochemical measurements in the laboratory.

2. Methodology

Steel is characterized by spectroscopy arc. Plates 10cm wide, 15cm long and 1mm thickness are subjected to a first pickling with hydrochloric acid. Immediately the plates are weighed on an analytical balance, and installed in test stations in Tunja and Nobsa. After a month of exhibition, the first sample is taken and this is again weighed. The sample is taken to the laboratory where electrochemical tests will be done to find their corrosion rate and polarization resistance. Finally the sample is analysed and is subjected to a second pickling with Hexamethylene tetramine and finally is weighed again on an analytical balance. For the electrochemical test was used Na_2SO_4 electrolyte of 3.5%, a scan rate of 2mV/s, an exposed area of 0,95cm^2 with a sweep from -250mV to +250mV with respect to Ecorr (Corrosion Potential).

2.1. Experimental design
The experimental matrix has two factors: sampling site, with two levels that are Tunja and Nobsa; and exposure time, which corresponds to the first five months of this study. The response variables arise from electrochemical tests and gravimetric method [4].

2.2. Experimental equipment
2.2.1 Analytical balance. This balance allows monitoring changes in mass with a definition of thousandths of a gram and is essential for gravimetric measurements by weight little difference formed oxide.

2.2.2. Electrochemical equipment. The Gamry G-750 Potentiostan-Galvanostats, an electrode Ag/AgCl and a flat electrochemical cell was used for generate the potentio-dynamics and LPR curves and is located in the Materials Lab of the Universidad Antonio Nariño, Tunja, Colombia.

2.3. Methods
The plates are located in two stations belonging to the University of Nariño in Tunja and Nobsa, assembled according to ASTM G50 standard [5].

2.3.2 Gravimetric. It is performed according to ASTM standard setting G1L cleaning for different chemicals for this etching an acid is used in a solution of 3.5g hexamethylene tetramine serving as inhibitor and 1000ml of distilled water. Additionally calculates the Pilling-Bedworth ratio that relates the volume of oxide formed with the volume of metal consumed [6].

2.3.2. Electrochemical. The response variables for electrochemical tests are back extracted from the Tafel approximation, in the case of potentio-dynamic curves and LPR test shows a value of the linear polarization resistance [7].

3. Results
Mass measurements exhibit the behaviour of oxidation of the samples at each station as a function of exposure time.

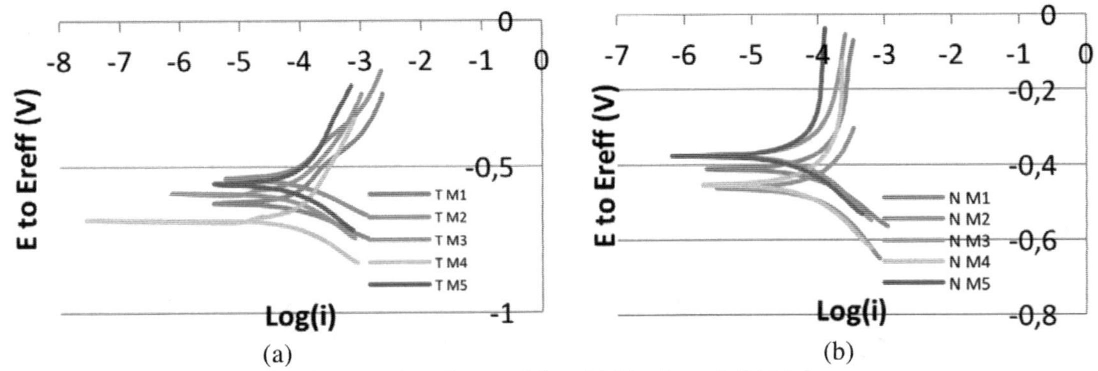

Figure 1. Curves Tafel for the months of exposition (a) Tunja and (b) Nobsa.

The data obtained are compared with Bedworth-Pilling factor that determines the behaviour of the mass loss of the substrate and the oxide layer generated (see Table 1).

The obtained results are reported in the Figure 1 and in the Table 2. In the potentio-dynamics graphics (Figure 1), the x axis is the base 10 logarithm of current (Ampere) and the y axis is the Potential E in Volts.

Table 1. Resume of the gravimetric measurements.

City and month	Initial weight (g)	Net weight after exposure (g)	Weight Without Rust (g)	Rust weight (g)	Weight loss of steel (g)	Percentage loss (%)	Exposure time (hours)	C.R. (mm/y)	Bedworth Pilling Ratio
Nobsa 01	134,09	135,46	133,36	2,10	-0,732	0,0055	720	0,0761197	4,2978038
Nobsa 02	133,75	135,91	130,64	5,27	-3,116	0,0233	1440	0,1620142	2,5336787
Nobsa 03	134,71	135,17	130,63	4,54	-4,078	0,0303	2160	0,1413552	1,6678116
Nobsa 04	134,67	136,98	133,36	3,62	-1,315	0,0098	2880	0,0341863	4,1240240
Nobsa 05	134,48	135,18	130,63	4,55	-3,845	0,0286	36000	0,0079967	1,7727742
Tunja 01	132,1	132,07	131,97	0,10	-0,120	0,0110	720	0,0124786	1,2484097
Tunja 02	133,6	133,96	133,36	0,60	-0,280	0,0021	1440	0,0145584	3,2101963
Tunja 03	134,5	135,17	133,35	1,82	-1,170	0,0087	2160	0,0405556	2,3303647
Tunja 04	133,3	134,26	132,26	2,00	-1,030	0,0077	2880	0,0267771	2,9089157
Tunja 05	133	133,36	130,63	2,73	-2,320	0,0174	36000	0,0048251	1,7628406

Table 2. Resume of the electrochemical variables response with the B-P ratio.

Nobsa (Industrial Atmosphere)					
Coupon	βa (mV/dec)	βc (mV/dec)	CR 10^{-3}(mm/y)	LPR (Ohm)	B. P. ratio
N1	391,9	-180,9	1,34	108,7	4,2978
N2	351,9	-228,8	1,38	286,6	2,5336
N3	318,2	-178,5	0,71	410,08	1,6678
N4	362,7	-160,5	0,91	375,3	4,1240
N5	295,1	-120	0,48	503,3	1,7727
Tunja (Urban Atmosphere)					
Coupon	βa (mV/dec)	βc (mV/dec)	CR 10^{-3}(mm/y)	LPR (Ohm)	B. P. ratio
T1	260,1	-97,2	0,42	148,8	1,2484
T2	191,1	-93,6	0,34	143,7	3,2102
T3	268,8	-155,8	0,99	250,4	2,3303
T4	281,9	-124,6	0,84	245,9	2,9089
T5	351	-153,6	0,59	273,5	1,7628

4. Discussion

The gravimetric method yields real values of the loss of material. It is shown in Figure 2 that the attack Nobsa is much more aggressive than in Tunja for any month.

In Figure 3 it shows that the corrosion rates are very similar, unlike what happens with the gravimetric method. This may be due to oxide layers which are applied potentiodynamic curves have in common the same Na2SO4 electrolyte at the same concentration. However we note that the attack Nobsa is more pronounced, especially in the first two months.

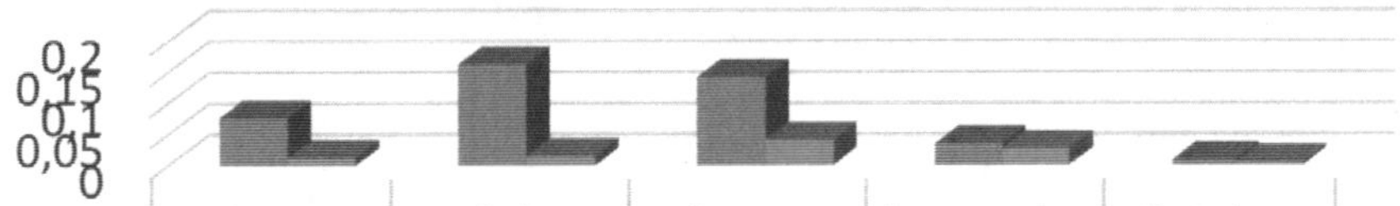

	June	July	August	September	October
Nobsa	0,0761197	0,1620142	0,1413552	0,0341863	0,0079967
Tunja	0,0124786	0,0145584	0,0405556	0,0267771	0,0048251

Figure 2. Corrosion Rate Gravimetric Method (mm/y) comparison for the two stations in function of exposition time.

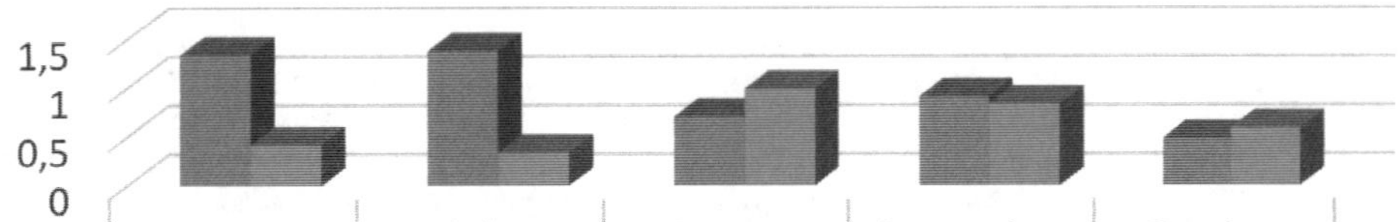

	June	July	August	September	October
Nobsa	1,34	1,38	0,708	0,914	0,481
Tunja	0,421	0,339	0,992	0,836	0,591

Figure 3. Corrosion Rate Electrochemical Method (mm/y) comparison for The two stations in function of exposition time.

5. Conclusions

As noted in the study of air quality for Sugamuxi [8] this industrial environment is high presence of SO_2 unlike urban environments such as the city of Tunja. Faster attack Nobsa (about the first and second month) compared to Tunja (3 and 4 months) was observed.

The oxide layer presents a series of etched ment that is related to rainfall. The rain acts as a cleaning mechanism of the plates. This was evident when calculating the Pilling-Bedworth ratio for each month and a more pronounced effect was noted in Nobsa.

The corrosion rates are not as high after the third month due to the amount of oxide that accumulated on the surface layer causes which serve as protective preventing more contaminants from seeping inside the steel on forming.

References

[1] Leygraf C and Graedel T 2000 *Atmospheric corrosion* (New York: Wiley)
[2] Yamashita M et al 1994 *Corrosion Science* **36** 283
[3] Morcillo M et al 2011 *Revista de Metalurgia* **47** 5
[4] Fernandez J and Vidales J 2007 *Corrosión en las estructuras de hormigón armado fundamentos medida diagnosis y prevención* (Madrid: CSIC Press)
[5] ASTM G50 2003 *Standard atmospheric corrosion tests* (West Conshohocken: ASTM International)
[6] Gonzalez M 2010 *Determinación de los índices de corrosividad atmosférica y parámetros agresivos en coatza coalcos* (Veracruz: Universidad Veracruzana)
[7] Mansfeld F 1981 *Electrochemical corrosion testing a symposium* (Bal Harbour: ASTM International)
[8] Díaz V 2014 *Revista Ambiental Agua Aire y Suelo* **5** 1

Presence of nitrate NO_3 affects animal production, photocalysis is a possible solution

Heli Barba-Molina[1], **J Barba-Ortega**[2] **and M R Joya**[2]

[1] Universidad Cooperativa de Colombia, Bucaramanga, Colombia.
[2] Universidad Nacional de Colombia, Bogotá, Colombia.

E-mail: mrinconj@unal.edu.co , jjbarbao@unal.edu.co

Abstract. Farmers and ranchers depend on the successful combination of livestock and crops. However, they have lost in the production by nitrate pollution. Nitrate poisoning in cattle is caused by the consumption of an excessive amount of nitrate or nitrite from grazing or water. Both humans and livestock can be affected. It would appear that well fertilised pasture seems to take up nitrogen from the soil and store it as nitrate in the leaf. Climatic conditions, favour the uptake of nitrate. Nitrate poisoning is a noninfectious disease condition that affects domestic ruminants. It is a serious problem, often resulting in the death of many animals. When nitrogen fertilizers are used to enrich soils, nitrates may be carried by rain, irrigation and other surface waters through the soil into ground water. Human and animal wastes can also contribute to nitrate contamination of ground water. A possible method to decontaminate polluted water by nitrates is with methods of fabrication of zero valent iron nanoparticles ($FeNps$) are found to affect their efficiency in nitrate removal from water.

1. Introduction

The water resources in Colombia have always suffered indifference by government, due to the absence of attention to development plans and the environmental policies of government in power, the same happens in the scientific field. Water is the essential element of the life, all living beings need it for survival; therefore, its quality has a major influence on the development of plants, animals and humans, any alteration in the composition of this molecule effects on different levels to each of these entities. Water should also be considered as dynamic element that interacts to different levels with millions of organisms, in the case of livestock production, the water takes a special importance because it impacts directly on two levels, first as a source of life and nutrition of forage plants that are consumed by livestock, and the second as an essential element of nutrition and maintenance of animals. Thus, if the water reached farms, is not the top quality, the affection generated on animals is twofold due to consumption of contaminated plants and water directly. This problem of water pollution has been highly studied during recent years, as scientists around the world have realized the strong impact on the human health when poor quality of water used in agricultural work [1–3].

Nitrate pollution on groundwater and surface water has become a serious environmental problem over the world, due to indiscriminate use of urea ($CO(NH_2)_2$) as a fertilizer of the soil in the production of fooder to feed animals. Nitrate is an inorganic compound having a nitrogen atom and three oxygen atoms, is one of the chemical with the highest concentration in the atmosphere, however animals and plants are unable to absorb directly from the air. The plants must absorb from the soil and animals meanwhile they get their dietary intake and drinking water [2,3].

The most common source of poisoning by nitrate and nitrite in animals are plants. The toxic principle existing in plants is always nitrate. Nitrates are absorbed by the roots and normally incorporated in tissues as amino acids, proteins and other nitrogen compounds. These nitrates may also be present in high concentrations in some foods, especially in broad leaf plants, fertilized with large quantities of nitrogen [1,3]. In a published paper the authors write: the photocatalytic degradation method is fast, effective, ecofriendly, economically viable and efficient method in the waste water treatment. Figure 1 shows the schematic representation of the advantages of photocatalysis in the treatment of organic pollutants in waste water [4].

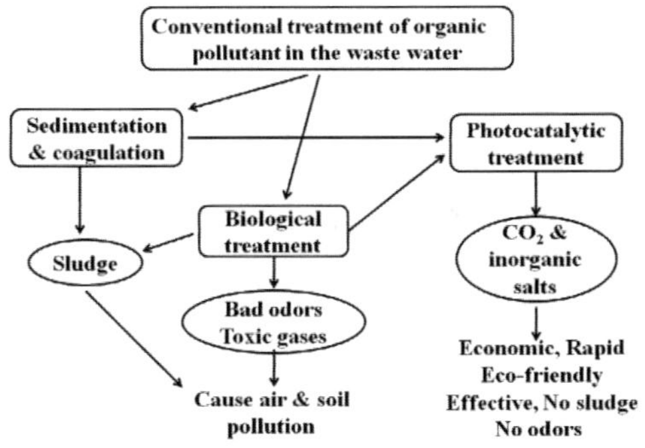

Figure 1. Schematic diagram showing the advantages of photocatalysis in the treatment of organic pollutants in waste water [4].

The objective of this work is expound about the effect of nitrites and nitrates in animal nutrition and toxic effects of these compounds in animals, as a result of the indiscriminate use of these elements in production systems. This article reviews advances in the field of nitrate removal processes.

2. Theoretical formalism

Research into semiconductor photocatalysis has largely focused on the removal of organic pollutants from water, since a wide range of organic substrates can be completely mineralized in the presence of disolved oxygen photosensitized. One of the very attractive features of semiconductor photocatalysis is its effectiveness against a broad range of pollotutants, including those that are recalcitrant to other water treatment processes [5,6].

Nitrate is the main form of nitrogen associated to water pollution, it is also the most abundant form of nitrogen in the soil, this ion has a high solubility being easily dragged by rainwater and often causing pollution water. Mineral nitrogen fertilizers and animal manures handled inappropriately are the mayor contaminant sources for existing nitrates [7].

Heterogeneous photocatalysis is a process of great potential for pollutant abatement and waste treatment. In order to improve the overall performance of the photo process, heterogeneous photocatalysis is being combined with physical or chemical operations [8]. Thus semiconductor photocatalysis can be used to oxideze nitrite, sulphite and cyanide anions to form relatively harmless products. Due to the non-selective action of photocatalysis, it finds application beyond the degradation of organic species in water. Semiconductor photocatalysis has also found application in the degradation of toxic inorganic compounds to harmless or less toxic compounds [9].

3. Results and discussion

The photocatalytic degradation method is fast, effective, ecofriendly, economically viable and efficient method in the waste water treatment. Figure 2 shows the schematic representation of the advantages of photocatalysis in the treatment of organic pollutants in waste water [4].

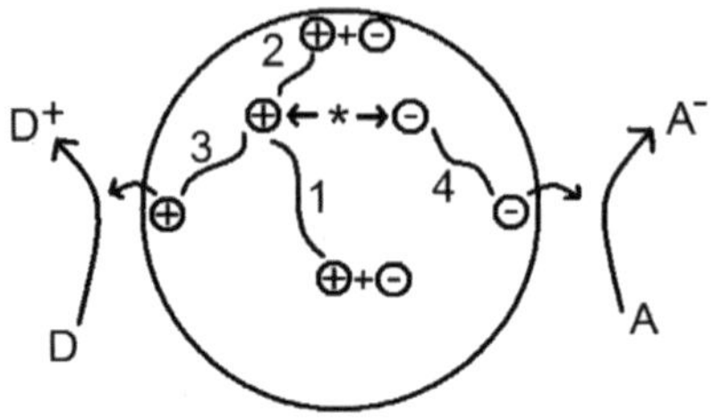

Figure 2. Schematic image showing different pathways for an electron-hole pair created at. (1) Recombination inside the bulk of the particle. (2) Recombination at the surface of the particle. (3) A hole migrates to the surface and reacts with a donor molecule. (4) An electron migrates to the surface and reacts with acceptor molecule, [4].

Semiconductor materials possess a region that is void of energy bands, thereby creating an energy space between the VB and the CB. This energy difference between the VB and the CB is called the bandgap energy. Reactions of photo generated electron hole pairs on the surface the photocatalyst occur in the presence of electron donors or electron acceptors adsorbed on the photocatalyst surface. Thus, if there is an electron acceptor, A, then the photo generated electrons can react with A generating a reduced product A^-. Similarly the photogenerated holes that migrate to the surface can react with an electron donor, D; to form an oxidized product D^+ [10,11].

We hope that in animal production, the organic pollutant degradation efficiency was over a wide range of pH values, irradiation time and optimum catalyst load. In additional effects like aeration and stirring increase the photodegradation efficiency of organic. Easy recovery and reuse of supported photocatalyst, along with higher degradation efficiency is the major advantages and it reduces the over treatment cost [4].

We know that the problem of nitrate pollution in Colombia, must be improved and animal production will be of better quality. Pesticides and herbicides, intentionally released into the environment, are ubiquitous in aquatic systems; they are often detected at low levels and commonly occur in the form of complex mixtures [4]. Leaching of chemical fertilizers and pesticides, applied to agricultural and forest land, is one of the main reasons for organic pollution in several water streams. Pesticides and herbicides are harmful to life because of their toxicity, carcinogenicity.

Acknowledgements
Authors acknowledge financial support from the Colciencias-Colombia and CNPq-Brazil agencies.

References
[1] M Palaniappan, P H Gleick, L Allen, M J Cohen, J Christian-Smith, C Smith and N Ross 2010 *Clearing the waters a focus on water quality solutions* (Nairobi: Publishing Services Section)
[2] M Z Kassaee, E Motamedi, A Mikhak, R Rahnemaie 2011 *Chem Eng J* **166** 490
[3] W Wang, Z Jin, T Li, H Zhang, S Gao 2006 *Chemosphere* **65** 1396
[4] H P Shivaraju 2011 *Int J Envir Sci Sci Technol* **1** 05
[5] A O Ibhadon, P Fitzpatrick 2013 *Catalysts* **3** 189
[6] D Beydoun, R Ammall, G Low, S McEvoy 1999 *J Nanop Resech* **1** 439
[7] S J A Moniz, S A Shevlin, D J Martin, Z X Guob 2015 *J Tang Energy Environ Sci* **8** 731
[8] M I Litter 1999 *Appl Cataly B Envir* **23** 89
[9] S K Lee, A Mills 2004 *J Ind Engi Chem* **10** 173
[10] E E Bamuza-Pemu 2014 *Photocatalytic degradation of phenolic compounds and algal metabolites in water* (South Africa: University of Pretoria)
[11] R M Nageeb 2013 *Adsorption technique for the removal of organic pollutants from water and wastewater* (Croatia: INTECH Open Access Publisher)
[12] W Aktar, D Sengupta, Chowdhury 2009 *Inter Toxic* **2** 1

Alumina and quartz as dielectrics in a dielectric barrier discharges DBD system for CO_2 hydrogenation

E Y Mora[1], A Sarmiento[1] and E Vera[1]
[1] Universidad Pedagógica y Tecnológica de Colombia (UPTC), Tunja, Colombia.

E-mail: eduin.mora@uptc.edu.co

Abstract. In this work was studied the CO_2 carbon dioxide treatment, which is a pollutant gas and the main cause of global warming. For this aim, plasma was generated, through dielectric barrier discharges DBD, using hydrogen H_2 together with the CO_2 as reaction gases. There were used as dielectrics, alumina and quartz tubes of identical geometry. It was studied the CO_2 conversion in function of mixture composition CO_2+H_2, of the electrical power and the operation frequency, for three different gas flows. In all cases it was achieved better conversion levels with the alumina; this is because the alumina has a relative dielectric permittivity coefficient higher than the quartz. As products of CO_2 conversion in the chemical reactions, water H_2O and methane gas CH_4 were identified. The CO_2 conversion percentage to fixed work conditions was higher with the decrease the quantity of this gas in the mixture, with increase the active electrical power, and with decrease the operation electrical frequency.

1. Introduction

In the manufacture of products at industrial level, emissions of CO_2, SO_2 and NOx among others are generated; for example, in cement production about 60% of CO_2 emissions come from the heat treatment to which it is subjected limestone, and 40% of emissions come from the fuels used in production [1]. During this process gases SO_2 and NOx are also generated [1], forcing the industry to find alternative ways to avoid or reduce emissions, mainly CO_2 given its impact on global warming [2]; then appear possible solutions such as exchange from coal to renewable energy sources, improvements in energy efficiency in the combustion of fossil fuel, replacement of raw material for waste materials rich in CaO and the use of combustible waste materials, etc [3]. An alternative is the use of plasma in the conversion of gases such as NOx, SOx and CO_2 [3-6]. Thus the implementation of the dielectric barrier discharge DBD, allowed in some studies, the transformation of other hydrocarbons CH_4 [7], the reaction between CO_2 and H_2 to produce CH_4 and CO [8], the NO_2 gas treatment catalytic processes attending, reducing nitrogen oxides [9], the production of H_2 from CH_4 using zinc and chromium oxide as catalysts [10] or from hydrocarbons [11], the use of cold plasma for dissociation of NO_2 into NO radicals, N and O [12] and the removal of SO_2 of the atmosphere [13].

The DBD have advantages over traditional techniques when pollution concentrations are low, typically in the range of 10 to 1000ppm. In the DBD, the energetic electrons are generated in the microdischarges without substantially increasing initial gas flow enthalpy; also have the advantage over conventional techniques when different pollutants are treated simultaneously [14]. The transformations and use of CO_2 have become an important issue. One proposal is to recycle CO_2 as an energy carrier, perhaps combined with hydrogen in the form of liquid fuel or as a feedstock in the chemical industry. New catalysts are being developed for methanol synthesis from CO_2 and H_2 [14]. Eliasson [15] found that methanol may be used as a fuel in an engine of a conventional vehicle, directly or together with a small reformer to generate hydrogen in a fuel cell for an electric motor.

Different researchers work in the laboratory on the CO_2 hydrogenation and partial oxidation of CH_4 in the DBD. Quartz has been used mainly as dielectric material in these investigations. Methanol formation has been observed in CH_4/Oxygen and CH_4/air mixture [15-18]. Zhou [19] and other researchers obtained high rates of change for conversion in mixtures of CO_2/CH_4 become Syngas (H_2+CO), which is a feedstock for many chemical processes including methanol synthesis. Jan Sentek and others [20] experimentally studied the plasma conversion of methane and CO_2 in DBD, using catalysts of silver-alumina and palladium-alumina.

2. Experimental

In Figure 1 is shown the schematic diagram of the experimental setup. The reactor camera was a tube of quartz (alumina was also used); the inside diameter of 26mm, thickness of 2.0mm and length of 215mm. A thin stainless steel sheet, serving as the outer electrode, adapted to the outer wall of the reactor. The inner electrode was one dipstick of stainless steel located in the center of the reactor of 15.87mm diameter, giving a discharge volume of approximately 36.67mL. For plasma generation was used a high-voltage alternating current (AC) power supply (Didriv10, PWM500) that has maximum voltage of 40kVp-p and frequency varying of 20kHz to 60kHz. The power was measured by a watt-meter digital (Intertek, GS cat II). The frequency was measured in an oscilloscope (Tektronix, TBS 1062) by mean a high voltage probe. The gas flows of CO_2 and H_2 were controlled by flow meters.

To measure the CO_2 percentage in the mixture was used the digital meter (CO_2 meter, CM-0006) which allows readings of 0 to 100% CO_2, using the no-dispersive infrared technology; this instrument was connected in the output of the reactor. Were analyzed two samples in order to identify the species generated in the plasma. To gas chromatography analysis was used the equipment AGILENT 7890 B and for the study of the liquid sample by mean FTIS, was used an equipment SHIMADZU Prestige 21. The amount of CO_2 introduced was measured at the output of the reactor without electric discharge. The amount of CO_2 consumed is calculated by subtracting the measurement taken at the reactor outlet after discharge of the measurement without discharge. CO_2 conversion was studied for three total flows 52, 74 and 96Nml/min, changing the percentage of CO_2 in the CO_2+H_2 mixture from 10 to 100%, the frequency from 20 to 25kHz and the applied power from 50 to 114W.

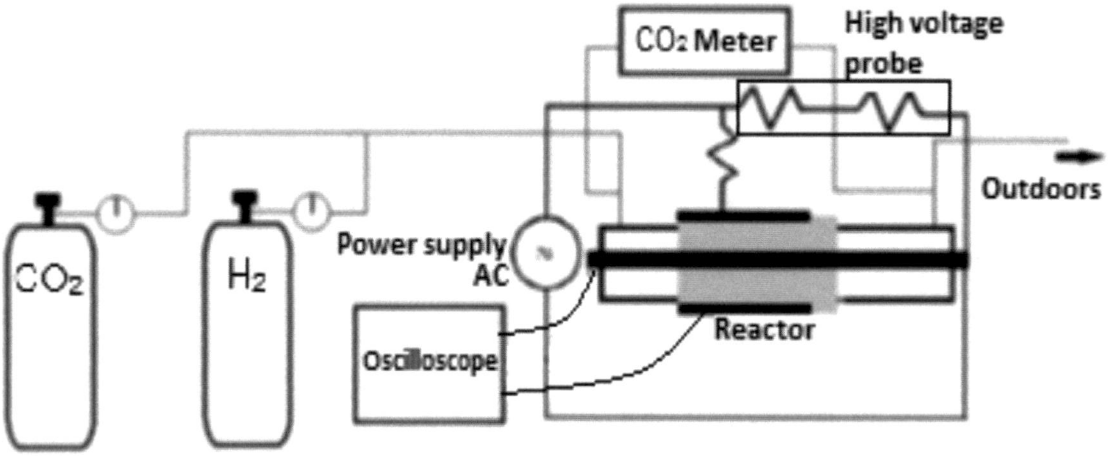

Figure 1. Schematic diagram for experimental set-up.

The CO_2 conversion was calculated as follows:

$$(\%)CO_2 = \frac{amount\ of\ CO_2\ consumed}{amount\ of\ CO_2\ introduced} * 100\ \ (\%) \quad (1)$$

3. Results

3.1. *Effects of the CO_2+H_2 mixture on the reaction*

In this case the conversion of CO_2 was studied varying the percentage of CO_2 in the feed gas mixture of 10 to 100% in flows of 52, 74 and 96Nml/min, for operation frequency and electric power delivered of 19.72kHz and 105W respectively, using the two dielectric materials. The behaviour can be seen in Figure 2. As shown in the three flows, both the alumina and quartz, with increasing CO_2 content in the mixture, the transformation decreases. The highest conversion is 43.71%; this value was obtained with alumina using the lowest flow, and remains the mixture the possible time highest within the reactor. For all proportions of CO_2 in the mixture, the transformation was higher in alumina than quartz. The reason is due alumina has a coefficient of relative dielectric permittivity higher than that of quartz, 9.8 and 3.75 at 1MHz, respectively. Methane was detected according with chromatography test (see Figure 3(a)) and water according FTIS test due to the vibration bands of 1635 and 3298cm^{-1} are characteristic of water (see Figure 3(b)). Regardless of the used dielectric, in intermediate ratios in the mixture of CO_2 (30 to 60% approximately), the transformation is favoured due to the increase in the water formation.

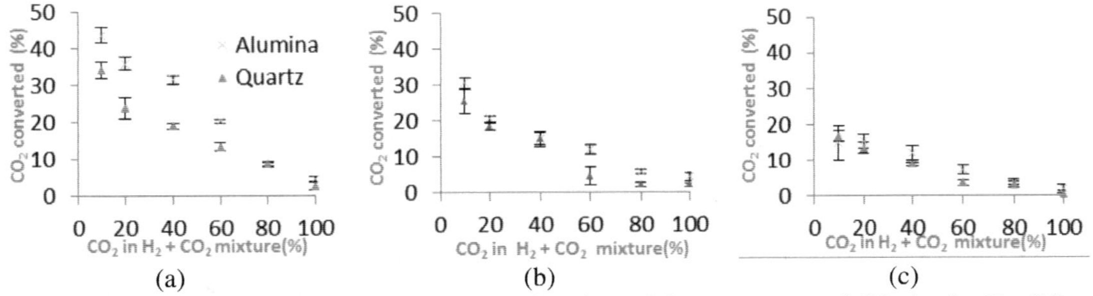

Figure 2. Percentage of CO_2 transformed as a function of the percentage of CO_2 in the H_2+CO_2 mixture at frequency of 19.72kHz and power of 105W for (a) 52Nml/min (b) 74Nml/min (c) 96Nml/ min, using alumina and quartz as dielectrics.

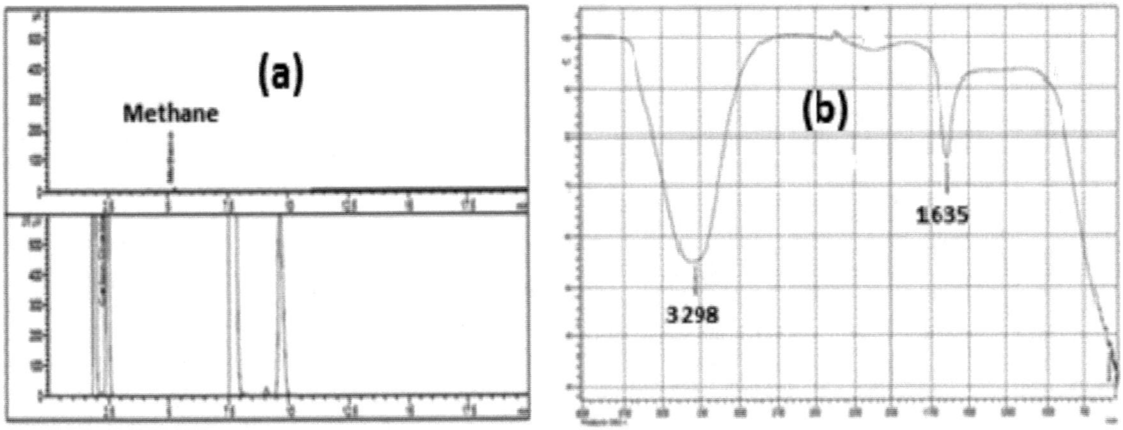

Figure 3. Tests for samples (a) Chromatography, (b) FTIS.

3.2. *Effects of delivered power on the reaction*

The degree of conversion of CO_2 was studied by varying the power supplied between 50 and 114W, in flows 52, 74 and 96Nml / min, for operating frequency of 19.72kHz and composition of the mixture of 20% CO_2 and 80% H_2. For both alumina as quartz and for the three adjusted flows, the trend is the increase in the transformation of CO_2 when the power supplied to the system is increased, as shown in Figure 4. Again best transformations are confirmed when the flow decreases independently of the dielectric. For the same power applied to the system in the three flows, the conversion is higher in the alumina.

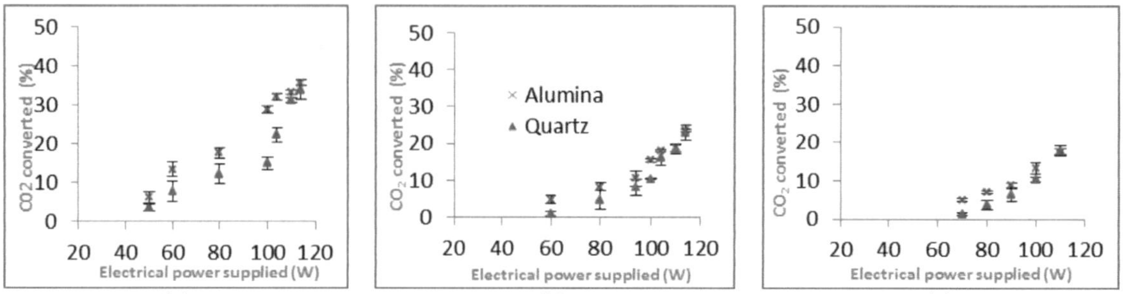

(a) (b) (c)

Figure 4. Percentage of CO_2 transformed as a function of the electric power supplied to frequency of 19.72kHz and composition of the mixture of 20% CO_2 and 80% H_2 for (a) 52Nml/min, (b) 74Nml/min, (c) 96Nml/min, using alumina and quartz as dielectrics.

3.3. Effects of the frequency on the reaction

In this case, the degree of conversion of CO_2 was studied by varying the electrical frequency of 20 to 25kHz, in streams of 52, 74 and 96Nml/min to 13.15kV voltage and composition of the mixture of 20% CO_2 and 80% H_2. The behaviour can be seen in Figure 5. For the three flows and the two dielectric materials, the transformation of CO_2 decreases as the operating frequency increases. The highest percentages of transformation are obtained with alumina in the three streams studied, getting as maximum value 31.04%, for flow of 52Nml/min and 20kHz operation frequency. As the frequency is increased, the transformation decreases in spite of small increases in near 22kHz, associated them to system complexity; this behaviour is due to source inductive predominance given for coupling impedance. Furthermore, the minimum frequency fixed in the source is higher than the resonance frequency of the circuit, resulting in lower levels of transformation with increasing frequency.

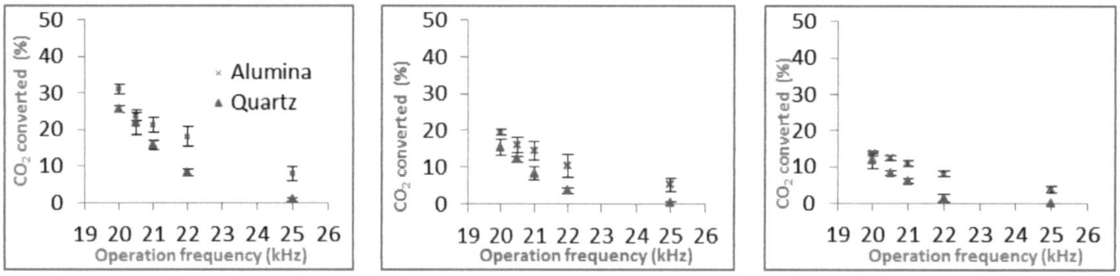

(a) (b) (c)

Figure 5. Percentage of CO_2 transformation as a function of electrical frequency in voltage of 13.15kV and composition of the mixture of 20% CO_2 and 80% H_2 for (a) 52 Nml/min, (b) 74 Nml/min, (c) 96Nml /min using alumina and quartz as dielectrics.

4. Conclusions

CO_2 transformation was studied as a function of the mixture composition, of the electrical power and the operation frequency using alumina and quartz as dielectrics in DBD. Reduction in conversion of CO_2 with increasing CO_2 in the mixture regardless of flow and dielectric material used, is associated to CH_4 methane gas formation, which decreases with increasing presence of CO_2 in the mixture; increases in the amount of water to ratios of 30 to 60% of CO_2, increases the conversion. As the power supply increases, so does the conversion, due to the formation of new species such as methane and water into the reactor, is associated directly with the energy supplied. CO_2 conversion decreases with increasing frequency regardless of dielectric material. According to the characteristics of the electrical circuit of the system, occurs inductive predominance due to the source coupling impedance. Higher levels of conversion to alumina were presented in all cases. This is because the alumina has a relative dielectric permittivity coefficient higher than the quartz. As the capacitance is higher in the alumina, the capacitive reactance is lower, therefore is easier to deliver power to gaseous mixture.

References

[1] Márquez S 2005 *Estabilización de suelos* (Argentina: Universidad Nacional de la Patagonia San Juan Bosco)
[2] Francisco Blanco Álvarez 2008 *Materiales de construcción tema 8 cementos* (España: Universidad de Oviedo) (http://www6.uniovi.es/usr/fblanco/Tema7.CEMENTOS.R.pdf)
[3] Hoyos A, Jimenez E, Ortíz M and Montes C 2008 *Revista Ingeniería e Investigación* **28-3** 41
[4] Leray A, Khacef A and Makerov M 2011 *Diesel oxidation catalyst Ccombined to non-hermal plasma effect on activation catalyst temperature and by-products formation 20th internatinal symposium on plasma chemistry*. (United States: HAL Id: hal-00680790) pp 1-4
[5] Young Sun M 2006 *Plasma Science & technology* **8-2** 207
[6] Nunnally T, Gutsol K, Rabinovich A and Fridman V 2009 *Inter Journal of Hydrogen Energy* **34-18** 7618
[7] Sazal K, Eric M and Kennedy V 2011 *Chemical Engineering Journal* **180** 178
[8] Kyung T and Won K 2012 *Journal of Industrial and Engineering Chemistry* **18** 1710
[9] Hoon D and Taegyu K 2009 *IPCS Proceedings International Plasma Chemestry* (http://www.ispc-conference.org/ispcproc/ispc20/488.pdf)
[10] Indarto A 2012 *Journal of the Chinese Institute of Chemical Engineers* **39** 23
[11] Sarmiento B and Brey J 2007 *Journal of Power Sources* **169** 140
[12] Welzel S, Ponduri S and Brehmer 2009 *IPCS Proceedings International Plasma Chemestry* (http://www.ispc-conference.org/ispcproc/ispc20/243.pdf)
[13] S Yanzhou, Q Yuchang, Y Fashan and Y Xingcheng 2004 *Plasma Science and Technology* **6-6** 2589
[14] Hippler R, Kersten H and Scmidt M 2008 *Low temperature plasmas* (Berlin: Wiley-vch)
[15] Jan Paul and Claire-Marie Pradier 1994 *Carbon dioxide chemistry environmental issues* (England: Royal Society of Chemistry)
[16] Okazaki K and Nozaki T 1995 *12th Simposy on Plasma Chemistry Proceedings* (Minneapolis: University of Minnesota)
[17] Kogelschatz U and Eliasson B 1996 *Phys Blatter* **52** 360
[18] Eliasson B, Kogelschatz , Xue B and Li-Ming Zhou 1998 *Ind Eng Chem* **37-8** 3350
[19] Zhou L, Xue B, Kogelschatz K and Eliasson B 1998 *Energy Fuels* **12-6** 1191
[20] Sentek J, Krzsztof K, Mlotek M, Kalczewska M, Kroker T, Kolb T, Schenk A, Karl-Heinz G and Schmidt-Szałowski K 2010 A*pplied Catalysis B Environmental* **94** 19

Evaluation of the stiffness and friction of Ti6Al4V ELI treated by glow discharge nitriding

J R Tavera[1], D Y Peña Ballesteros[1] and H A Estupiñán Duran[2]
[1] Universidad Industrial de Santander (UIS), Bucaramanga, Colombia.
[2] Universidad Nacional de Colombia, Medellín, Colombia.

E-mail: haestupinand@unal.edu.co

Abstract. In this study, an evaluation of the elastic-plastic surface hardening on Ti6Al4V ELI titanium nitride films obtained by glow discharge method was carried out by nanoindentation tests according to the standard ISO 14577. The nanotribological properties (metal-metal) were also evaluated using the pin-on-disc system Ti6Al4V surface deposition ELI with nitrogen, obtaining a correlation between the coefficient of friction of Ti6Al4V ELI treated by PVD and the Young's modulus of the respective substrate modified by PVD. To characterize the substrate for the characterization tests, scanning electron microscopy, atomic force microscopy and X-ray diffraction and contact angle were carried out. The results demonstrated that the substrates nitrided improved mechanical and tribological properties, hardness, Young's modulus and coefficient of friction, making the alloy Ti6Al4V ELI support axial loads in tension and compression.

1. Introduction

Nitriding of metallic materials is well known to enhance resistance to wear and corrosion, thereby reducing the coefficient of friction, and getting better tribological properties [1] technique. It is obtained by using different methods gaseous. For this work, the need for a suitable substrate, with acceptable hardness, in which there are no phase changes, the need to nitriding is suggested by the method of intense plasma discharge, which is a gas was raised (rich in nitrogen) in which an electric glow discharge generates nitrogen ions onto the desired substrate, favouring the diffusion in the substrate [2], in our case Ti6Al4V ELI alloy. Nitriding by physical vapour deposition (PVD for short) is a proprietary process engineering materials which ions of a material can be implanted into another solid, where Implanted ions are distributed in depth ranging between 200 and 2000 atomic layers ($0.05\mu m$ $0.5\mu m$) depending on the type of ion, the base material and the sputtering power. The maximum concentration, which for light ions such as carbon or nitrogen can exceed 50%, is located at a certain depth and then decreases smoothly [3]. The aim of this study was to provide the tools to analyse nanomechanical behaviour of nitriding by the method of intense plasma discharge and propose recommendations that lead to improved standards substrate.

2. Experimental

The material used was ELI alloy Ti6Al4V, its chemical composition is presented in Table 1. The dimensions of the specimens used were 12mm diameter and 2mm thick.

The samples were polished with silicon carbide paper, from 240 to 1200 and then from coarse and fine cloth polishing, of $0.3\mu m$ and $0.05\mu m$ with alumina slurry. Thereafter ultrasonic cleaning with alcohol is performed for 15 minutes. The surface modification of Ti6Al4V ELI, nitriding was performed by intense plasma discharge method, this process for a NitrEos T700x900x1300 Tandem System, which has a working volume of 700mmx900mm and 700mmx1300mm and having a heating technology used hot wall and a system operating range 350 to 600°C, with 25 kwatt power source. 800V up to 10kHz, and pulse power of 70kWatt. Specimens worked at a temperature of 450, 480 and 520°C, and residence time of 4 hours. The characterization by x-ray diffraction (XRD) was performed in a brand diffractometer Bruker D8 Discover. Nanoindentation test was performed using a penetrator that this investigation was a nanoindenter Berkovich tip. For tribological tests, the contact pair used was metal-metal (Ti6Al4V ELI pin - disc treated Ti6Al4V ELI) for the respective calculation of the friction coefficient.

Table 1. Chemical composition of the titanium alloy Ti6Al4V obtained by fluorescence spectrometry energy dispersive X-ray.

Alloy element	Al	V	Fe	N	Ti
wt (%)	6.102	4.119	0.183	0.05	88.66

3. Results and discussion

XRD results showed possible changes generated by the nitriding treatment on the microstructure that manifest at very small scales. Figure 1 shows the appearance of nitrides at low temperatures, between 450 and 520°C. These directly affect the hardness obtained by the substrate and is confirmed by other authors obtained nitrides below 773K, ϵ-Ti2N y δ-TiN and which are clearly present to an atmosphere N/H=3/2, after treatment of more than 240min [4].

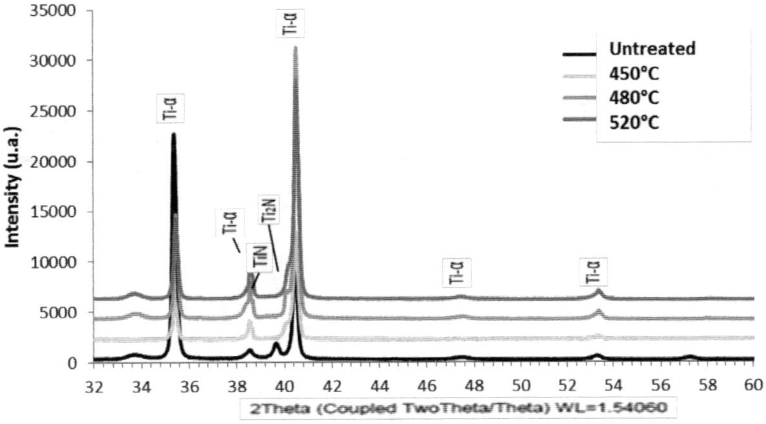

Figure 1. XRD for samples treated at different temperatures.

The effect of the intense plasma discharge in the composition on the surface of Ti-6Al-4V ELI, is a very complex function of the process parameters: treatment time, the proportion of nitrogen, sample temperature and pressure. Figure 2 shows the graphs nanoindentation with temperature variation. These curves are nonlinear; the download data are used to determine the mechanical properties based on the theory of indentation, in which the portion of the initial discharge depth-load curve represents the purely elastic recovery.

The theory developed to determine the hardness and rigidity of a material by testing nanoindentation is based on the material is ideally elastoplastic and the material deforms around the indenter [5]. From the depth-load curve, the elastic modulus and hardness are calculated by the following equations:

$$E_r = \frac{\sqrt{\pi}S}{2\sqrt{A}} \quad (1)$$

Where A is the contact area of the penetrator, and Er is the reduced modulus, this is defined by:

$$\frac{1}{E_r} = \frac{(1-v_m^2)}{E_m} + \frac{(1-v_i^2)}{E_i} \quad (2)$$

Where Em and Vm are the Young's modulus and Poisson's ratio of indented material respectively, and Ei and saw the indenter. The hardness of the alloy H is defined as the maximum load Pmax, divided by the projected area of the indentation area under this load, that is, H=Pmax/Amax. Table 2 shows the results obtained in nanoindentation tests.

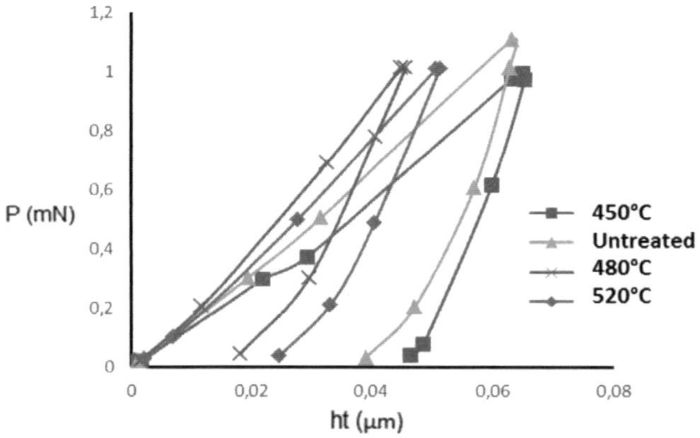

Figure 2. Nanoindentation curves for samples treated at different temperatures.

Table 2. Hardness and Young's modulus obtained by nanoindentation.

Samples	Nanoindentation hardness (GPa)	Young's modulus (GPa)
Untreated	3.9±0.2	271.8±2.4
450° C	6.4±0.4	197.7±8.1
480° C	9.8±0.4	161.7±6.1
520° C	7.9±0.3	145.5±2.4

Regarding the sample without nitriding, hardness increased until the treatment temperature of 480°C and a slight decrease in this tendency to 520°C is observed. However, the Young's modulus decreases with temperature nitriding, which relate it to some authors the accommodation of the nitrogen in the structure assuming a decrease in the interatomic distances of Ti6Al4V ELI, decreasing its rigidity and preventing it when subjected high stress, decrease the delamination and debris [6,7].

Glow discharge nitriding, reducing wear and friction in the alloy Ti6Al4V ELI, leading to low mass loss. In Figure 3, the comparison of the friction coefficient variation ELI deTi6Al4V untreated with nitrided samples at different temperatures is observed.

versus time is observed in the sample of Ti6Al4V ELI, not treated by nitriding, with coefficient values about 0.8. Coefficient of friction values of about less than half, compared to the value given in the untreated sample (between 0.29 and 0.32) were obtained in all samples nitrided, being less friction in the treated sample at 520°C. In Figure 4(a) and 4(b), the morphology of friction damage is observed from the pin-on-disc tests.

In these SEM images observed in the nitrided samples, no wear debris, contrary to the presence of such wear on the untreated sample nitriding.

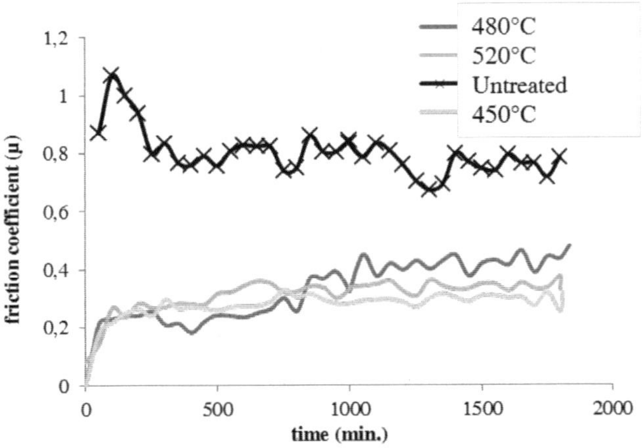

Figure 2. 3. Friction behaviour over time for different samples of Ti6Al4V ELI nitrided at different temperatures.

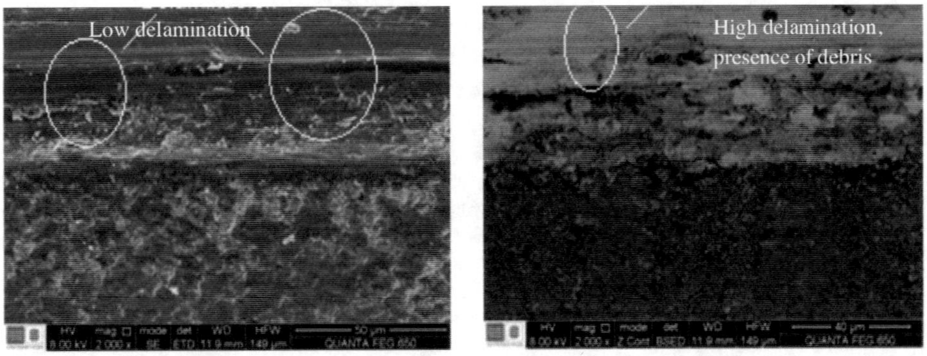

Figure 4. Images by scanning electron microscopy: (a) sample nitrided at 520°C and (b) not nitrided.

4. Conclusions

Nitriding by glow discharge in Ti6Al4V ELI allows decreasing the rigidity of the alloy to improve the tribological behaviour of mass loss by wear and friction, avoiding the presence of debris.

Acknowledgment

Our sincere thanks to COLCIENCIAS, for funding this project under contract no. 780-2011.

References

[1] T Fraczek, M Olejnik, J Jasiñski, Z Skuza 2011 *Metabk* **50-3** 151

[2] N Kashaev, H R Stock and P Mayr 2004 *Metal Science and Heat Treatment* **46-7–8** 294

[3] Mohammad K Khalaf, Bahaa T Chiad, Thair L Al-Zubaydi and Zenab T Alani 2011 *Iraqi Journal of Physics* **9-16** 34

[4] S L R da Silva, L O Kerber, L Amaral, C A dos Santos 1999 *Surface and Coatings Technology* **116–119** 342

[5] C Y Tang, C P Tsui, D j Janackovic, P S Uskokovic 2006 *Journal of Optoelectronics and Advanced Materials* **8-3** 1194

[6] E Bemporad, M Sebastiani, M H Staia, E Puchi Cabrera 2008 *Surface & Coatings Technology* **203** 566

[7] L Ceschini, E Lanzoni, C Martini, D Prandstraller, G Sambogna 2008 *Science Direct Wear* **264** 86

Simulation of the magnetic field generated by wires with stationary current and magnets with constant magnetization applied to the mirror trap, minimum-B and zero-B

M T Murillo[1,2] **and O Otero**[2]

[1] Universidad Santo Tomás, Bucaramanga, Colombia.
[2] Universidad Industrial de Santander, Bucaramanga, Colombia.

E-mail: `oswaldoterolarte@gmail.com`

Abstract. As a contribution to the computational simulation of magnetic confinement and heating of plasmas ECR (Electron Cyclotron Resonance), this work is dedicated to the calculation and subsequent analysis of the magnetic fields generated by permanent magnets and coils required in magnetic traps between which we can mention the mirror trap, minimum-B and zero-B. To do this, we solved numerically the Biot-Savart law in the case of the coils with stationary current and the Ampere law in the case of the permanent magnets. The study includes the characterization of the ECR areas as well as the display of the vector field all of this applied to the magnetic traps mentioned above. Additionally, in the case of the mirror type trap and minimum-B trap, it is determined the ratio of the mirror, because it is important in the description of confinement.

1. Introduction

The magnetic confinement searches to capture and retain charged particles using two properties of the magnetic field; first one is that every particle tends to move in a spiral way along the field lines, so that their displacement in parallel direction to the field lines is not affected. Second one, according to the magnitude of the field, the particle is ejected or reflected to areas where the field is weak, this phenomenon is known as magnetic mirror [1]. Taking the relevance of these premises have been proposed and built many devices trying to find an appropriate configuration of the geometry and magnitude of the field and prolong the lifetime of the particles into the trap. This problem has been studied in many works from the middle of last century [2]. The research group of plasma physics and technology and corrosion (FITEK), ascribed to the school of physics at the University Industrial of Santander is working on a theoretical study of plasmas ECR confined in magnetic mirror type trap, minimum-B trap and and zero-B trap [3]. To generate these confinement fields usually is used elements such as coils and permanent magnets [2, 4, 5]. For this reason, it is necessary to develop a tool to reproduce the field generated by those elements. Using these results and physical schemes presented in [2, 4, 5], we determined the confinement fields for each one of the traps, the ECR surfaces and the mirror radio to the mirror trap and minimum-B trap [2], because it is important in the description of confinement.

Journal of Physics: Conference Series **687** (2016) 012022 doi:10.1088/1742-6596/687/1/012022

2. Coils and magnets modelling

The magnetic field generated by a coil is obtained from of the superposition of contributions produced by circulars loops with steady current, according to that the Biot-Savart's law [1] in this case is used as:

$$\mathbf{B} = \frac{I\, r_0}{c}\int_0^{2\pi} \frac{\left[(z-z_s)\cos\varphi\;\hat{\mathbf{i}} + (z-z_s)\sin\varphi\;\hat{\mathbf{j}} + (r_0 - y\sin\varphi - x\cos\varphi)\;\hat{\mathbf{k}}\right] d\varphi}{\left[(x-r_0\cos\varphi)^2 + (y-r_0\sin\varphi)^2 + (z-z_s)^2\right]^{\frac{3}{2}}} \tag{1}$$

where r_0 is the radius of the loop. We have determined the field produced by a coil in the following conditions: the current flowing through the loops is $I = 9.6 \times 10^{12}$ stat amperes, 8 longitudinal loops and 5 transversal loops, the radius of the first loop is 5cm and the wire diameter is 0.5cm. The obtained field is presented in Figure 1, which is stronger in the central region due to its proximity to the loops constituting the coil.

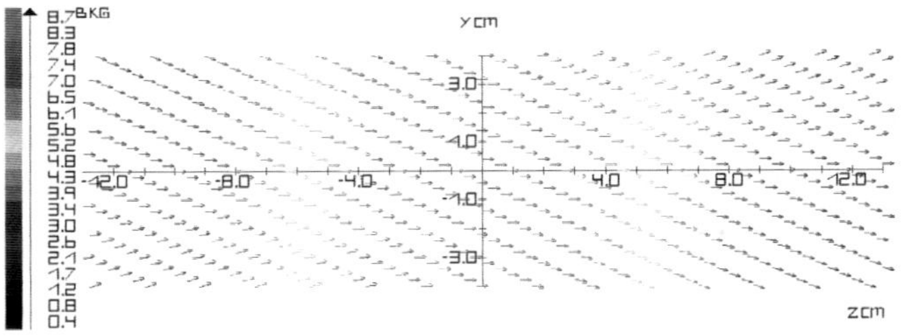

Figure 1. YZ view (plane $x = 0$cm) of magnetic field.

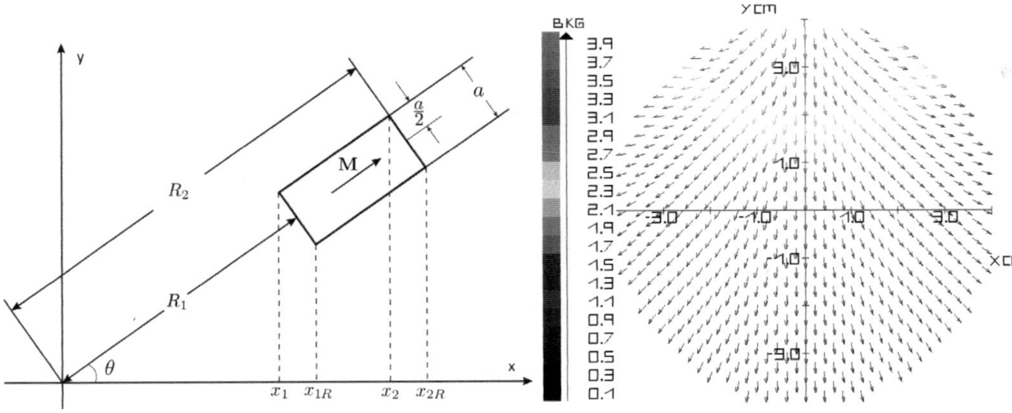

(a) (b)

Figure 2. Reference system (magnet bar cross section) and XY view (plane $z = 0$cm) of the magnetic field.

In the case of magnets, we consider a constant magnetization into the volume of the material, this vector is oriented radially from one pole to the opposite pole, as is shown in Figure 2. The region of interest is located outside of the magnet, the field can be calculated through the gradient of the magnetic scalar potential [6], because there is no current density in that region:

$$\phi_m(\mathbf{r}) = \int_{s_{(v)}} \frac{\mathbf{M}(\mathbf{r}')}{|\mathbf{r}-\mathbf{r}'|}\cdot d\vec{s'} - \int_v \frac{\nabla'\cdot\mathbf{M}(\mathbf{r}')}{|\mathbf{r}-\mathbf{r}'|}dv' \tag{2}$$

Where ϕ_m is the magnetic scalar potential, $s_{(v)}$ is the boundary surface of the magnet, v es the volume of the magnet. The contribution made by the second term is zero because the vector $\mathbf{M}$ is constant.Therefore, we can write the expression for the potential as follows:

$$\phi_m(\mathbf{r}) = \int_{z_1}^{z_2}\int_{x_2}^{x_{2R}} \frac{dx'dz'}{|\sin\theta|}\frac{M_0}{|\mathbf{r}-\mathbf{r}'|} - \int_{z_1}^{z_2}\int_{x_1}^{x_{1R}} \frac{dx'dz'}{|\sin\theta|}\frac{M_0}{|\mathbf{r}-\mathbf{r}'|} \quad (3)$$

Where x_{2R}, x_2, x_{1R} y x_1 are the integration limits on the axis X and z_1 and z_2 are the limits on the axis Z, as presented in the Figure 2(a). An analytical result was published in [7] for the modelling of magnets with various polarization directions, however, we solved the Equation 3 in the case when the polarization vector is constant in all magnet and can turn around an axis. Now, we study a particular case, according to Figure 2 we define $R_1 = 4.7$cm, $R_2 = 7.1$cm, $\theta = 90°$, $a = 2.5$cm, $b = 21$cm and $M_0 = 2.21$KG. The magnet defined here is outside of the simulation area. For a better visualization of the vector field, we present the Figure 2(b), as we can see, this figure shows that in the area near to the pole of the magnet (north pole), the field is intense, but as we move away from it, the magnitude decreases. The field lines are curved at $x = 1.5$cm because it are looking for the south pole, which it corresponds to presented in [1].

3. Mirror trap

The mirror trap is obtained by the superposition of the field produced by two coils [12] and [10]. Such elements are placed at the end of a cylindrical cavity with which they share the same axis of symmetry; steady currents flowing through them with equal magnitudes and directions, see Figure 3. The radius of the cylindrical cavity is 6.4cm and its length is 16.1cm, the plasma is heated by microwave whose frequency is 2.45GHz. The current flowing through the coils is 3.3×10^{12} stat amperes, we used 10 longitudinal loops and 3 transversals, the radius of the first loop is 8.43 cm and the loop separation is 1cm.

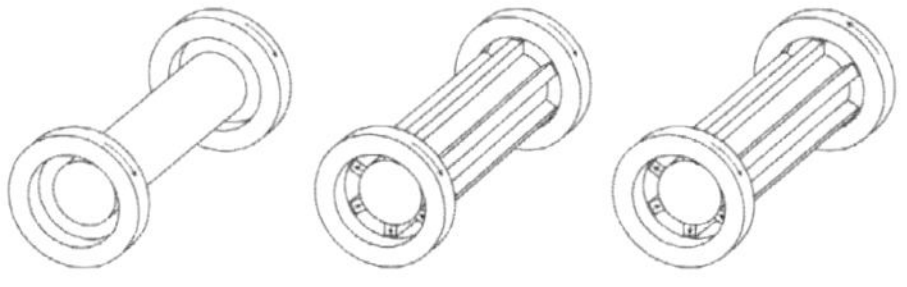

Figure 3. Physical schemes: mirror trap, minimum-B and zero-B.

The magnetic field of the mirror trap has an axial symmetry, we present a view of the field in the YZ plane (see Figure 4). In which, the magnitude of the field decreases as you move on the Z axis, approaching to the transversal center plane of the cavity, placed at $z = 0$cm. Additionally, in this region the field decreases with distance from the axis of the cavity, it means, the field is weak near to the walls. The magnetic field near to the coils is 1584.3G, while in the center of the cavity is 906.63G, with a mirror ratio equal to 1.75.

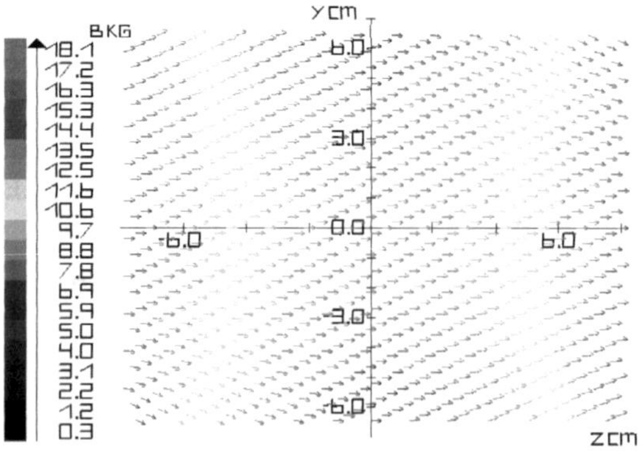

Figure 4. YZ view (plane $x = 0$cm) of the magnetic field.

The ECR surface is obtained when $B = mc\omega/e$, where m and e is the mass and charge of the electron, c is the speed of light and ω is the frequency of microwaves with which the plasma is heated [10, 12].In the case of 2.45GHz, the resonant field is 875.89G, the ECR surface takes the form of a one-sheet hyperboloid. The cross section of the hyperboloid in the central plane is a circle of radius 2.3cm. The ECR surface is closed and is not a regular one-sheet hyperboloid, which tends to the infinity, as we can see in the Figure 5.

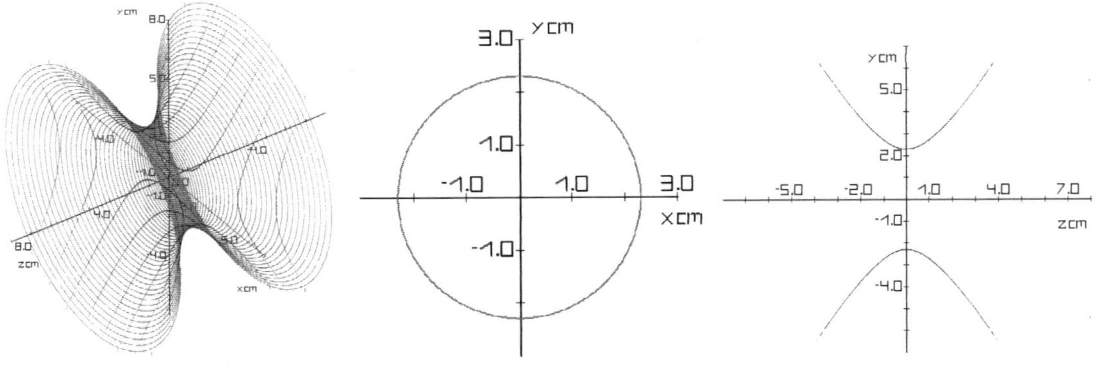

(a) (b) (c)

Figure 5. XYZ, XY (plane $z = 0$cm) and YZ (plane $x = 0$cm) view of ECR surface.

4. Minimum-B trap

The minimum-B configuration is obtained adding a transversal magnetic field to the mirror trap, the complementary field increases in the radial direction, which is placed throughout the cavity [4], see Figure 3(c). The longitudinal field is generated by two coils, the current flowing through the coils is 9.6×10^{12} stat amperes, the coil has 8 longitudinal loops and 5 transversals, the radius of the first loop is 7.29cm and the loop separation is 1cm. The transversal field is produced by permanent magnets (hexapole system), according to the Figure 2 we define the following parameters: the radius of the inner cap is 4.7cm and the outer 7.1cm, the dimensions bars are 2.5cm and 21cm, with a magnetization of 2.21KG. For hexapole field simulation we take the following angles $\theta = 30°$, $90°$, $150°$, $210°$, $270°$ and $330°$. To get a better idea about the minimum-B field, we present several figures. The longitudinal field has a mirror ratio equal to 1.82. The transversal field produced by a hexapole system corrects magnetohydrodynamic plasma instabilities confined on a mirror trap because it strengthens the radial component of the field in the central region of the cavity, given that in this region the contribution generated by the coils is zero. The magnitude of the field produced by the Minimum-B configuration increases in the longitudinal direction as $|z|$ increases, which can be seen in the color scale shown in Figure 6. Additionally, the field is minimized at the center of the cavity (see Figure 6).

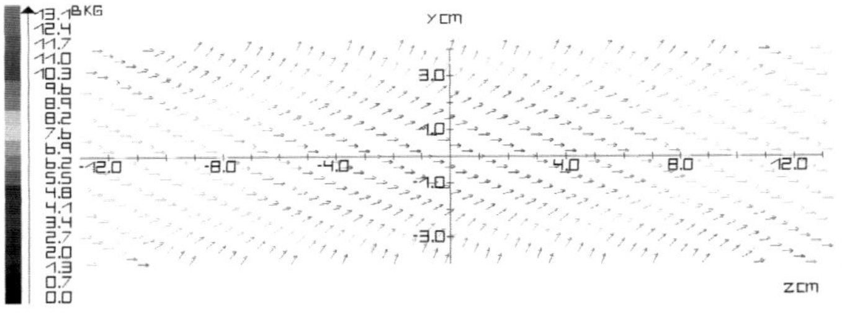

Figure 6. YZ view (plane $x = 0$cm) of magnetic field.

The value that take in this case the ECR magnetic field is 5005.102G and the ECR surface takes the form of an ellipsoid flattened at the ends. To determine the dimensions of the surface obtained, we present a cross-section on $z = 0$cm which allows us to visualize a circle of radius 1.9cm and a longitudinal-section on $x = 0$cm obtaining a value of 2.85cm, as we can see on the Figure 7.

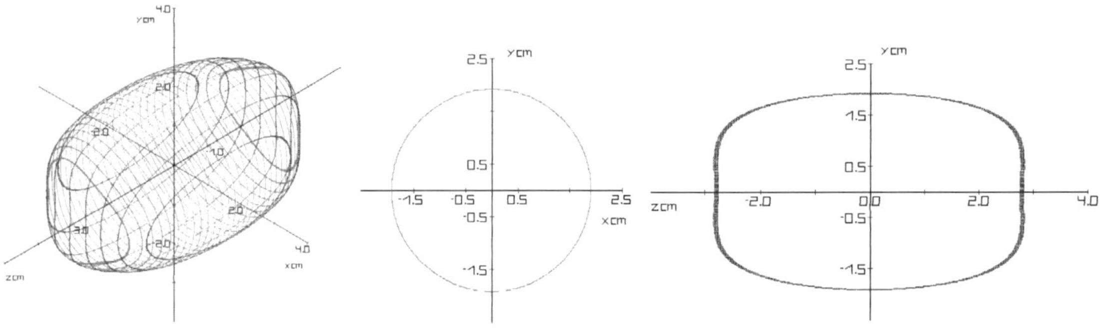

(a) (b) (c)

Figure 7. XYZ, XY (plane $z = 0$ cm) and YZ (plane $x = 0$cm) view of ECR Surface.

5. Zero-B trap

In the year 2000, Valeriy Dougar proposed a new trap to confine plasmas called zero-B trap. The main feature of this new system is the absence of magnetic field in the center of the cavity and the increase in the value of the same in all directions of the space [5]. The magnetic trap zero-B is obtained by the superposititon of a cusp longitudinal field and a transversal multicusp field, the first one is generated by two coils while the latter is produced by a hexapole system, see Figure 3(right). The values used for the construction of the zero-B trap are the same that we use at the previous section for the minimum-B configuration. The radial component of the longitudinal field is positive and grows when are close to the central transversal plane of the cavity. The field generated by magnets strengthens the field in the region adjacent to the cavity walls, the magnitude of the field in the region close to south poles of the magnets is strengthened, because they have the same direction of the field produced by coils, while that near to the north poles is weakened, because they have opposite directions. The Figure 8 shown how the magnitude of the zero-B trap increases as $|z|$ grows. This view also allows us to see how the field becomes zero at the center of the cavity. The magnitude of the magnetic field at the ECR surface is 5001.07G and its shape looks like an ellipsoid flattened at the ends but noncircular cross-section like that obtained in the minimum-B trap. In this case, the dimensions of the ellipsoid on the X axis is 2.58cm, Y axis is 2.95 cm and Z axis is 7.55cm (see Figure 9).

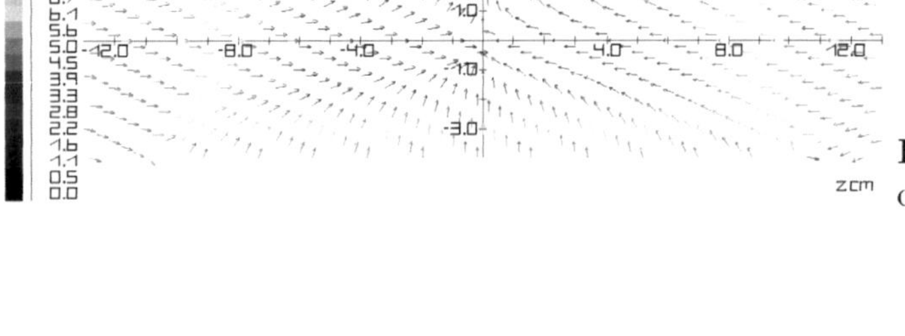

Figure 8. YZ view (plane $x = 0$cm) of the magnetic field.

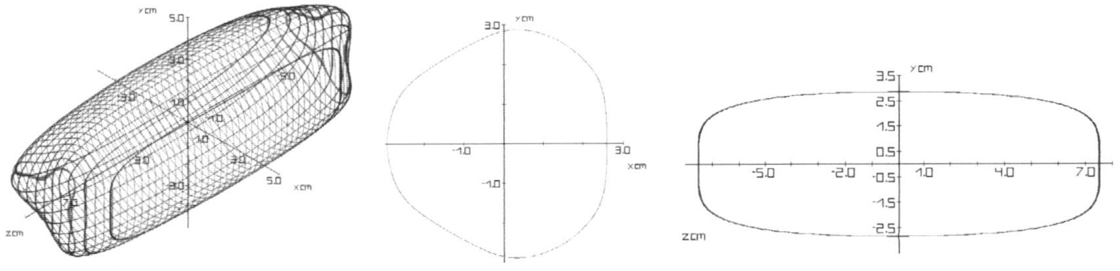

(a) (b) (c)

Figure 9. XYZ, XY (plane $z = 0$cm) and YZ (plane $x = 0$cm) view of ECR surface.

6. Conclusion

The graphs obtained for the magnetic traps allow the researcher to know the direction of field and its magnitude gradient, which is important to determine the direction of the diamagnetic force, essential in the study of magnetic confinement. The study of the ECR surfaces is significant for plasma heating by microwaves, allowing understand the fields involved in each of the magnetic traps.

References

[1] W Greyner 1998 *Classical electrodynamics 1st ed* (United States of America: Springer-Verlag) pp 186-195
[2] R Geller 1996 *Electron cyclotron resonance ion sources and RCE plasmas* (Bristol: Institute of Physics Publishing) pp 117-127
[3] V Dougar-Jabon, A Umnov and D Suescun 2002 *Rev Sci Instrum* **73** 629
[4] Y Baiborodov, M Ioffe, V Petrov and R Sobolev 1963 *J Nucl Energy Part C Plasma Phys* **5** 409-410
[5] V Dugar-Jabon, F Vivas and A Umnov 2000 *Phys Scr* **62** 183-185
[6] J Jackson 1999 *Classical electrodynamics 3rd ed* (United States of America: John Wiley&Sons Inc) pp 195-197
[7] R Ravaud and G Lemarquand 2009 *Progress In Electromagnetics Research* **98** 207-219
[8] V Dougar-Jabon 2001 *Phys Scr* **63** 322
[9] V Dugar-Zhabon 2004 *Phys Scr* **69** 313
[10] V Dugar-Zhabon and M Murillo 2010 *IEEE Transaction on Plasma Science* **38** 3449-3454
[11] M Murillo 2008 *Simulación de un plasma RCE en una trampa tipo espejo mediante el método aárticula en celda* (Colombia: Universidad Industrial de Santander)
[12] M Murillo, J Guerrero and V Dougar Jabon 2008 *Revista Colombiana de Física* **40** 180-182

Analytical study of the magnetic field generated by multipolar magnetic configuration

M T Murillo Acevedo[1,2], V D Dugar-Zhabon[2] and O Otero[2]

[1] Universidad Santo Tomás, Bucaramanga, Colombia.
[2] Universidad Industrial de Santander, Bucaramanga, Colombia.

E-mail: oswaldoterolarte@gmail.com

Abstract. The magneto-statics field from a parallelepiped magnet which can turn around an axis, is the first step to find the whole magnetic field in a multipolar configuration. This configuration is present in the ion sources, which are heated by electron cyclotron resonance. We present the analytic formulas to calculate this magnetic field outside the volume of the magnet. To model the magnet, we considered a constant magnetization vector inside of magnet volume. Therefore, the magnetic scalar potential method can be used. We present the results by a hexapolar system. Their magnetic field components are calculated on confinement region, several graphics are shown with directions and magnitude's gradients of the magnetic field to help understand better the confinement system. Our results are confronted with experimental ones. These formulas are very useful in research of plasma magnetic confinement in ion sources through computational simulations.

1. Introduction

To confine plasma in an electron cyclotron resonance (ECR) ion source, a transversal multicusp magnetic field is used. This field helps to remove magnetohydrodynamics instabilities due to the convex curvature of mirror's magnetic field [1]. It is common to use six or eight parallelepiped magnets around cylinder discharge camera to create a cusp geometric form of magnetic field, which changes the magnetic field curvature. The analytic calculation of magnetic field is useful in plasma dynamics computational simulations. Although several models use multipolar approximation, the interaction between the magneto-statics field of the trap and the microwaves field exerts the biggest influence over plasma behaviour; therefore is very important to get a better model to calculate magnetic field. An analytical result was published in [4] for the modelling of parallelepipedic magnets of various polarisation directions. However, we present the case, when the polarisation vector stays constant, but the whole magnet can turn around an axis and we solve each indetermination present in the formulas. Therefore, our formulas are more useful for calculate multipolar magnetic field. The multipolar magnetic field has been important in plasma studying from microelectronics fabrication to the fabrication flat panel display device [2]. This system of magnetic field is used in some configurations of magnetic confinement [3]. It starts from a magnetic scalar potential [5], but its equations are not solved, instead the magnetic fields components are calculated by using gradients, transforming those equations until they transform into a integrable form. The final equations are using to calculate the magnetic field into a cubic mesh. The magnet is considered as a material with constant magnetisation vector inside it, and zero outside. The equations are solved only in confinement volume, which is found outside magnets inner region. Several pictures show the curvature and magnitudes gradients.

2. Magnet modelling

The magnet is modelling by considering a constant magnetisation vector inside their volume, it is oriented radially from a pole toward the opposite pole, as is showed in Figure 1. Since the interest region is found outside magnet, the magnetic field can be calculated through the method of gradient of scalar magnetic potential [5] since, not conduction density current exists in this place: $\phi_m(\mathbf{r}) = \int_{s_{(v)}} \frac{\mathbf{M}(\mathbf{r}')}{|\mathbf{r}-\mathbf{r}'|} \cdot d\mathbf{s}' - \int_v \frac{\nabla' \cdot \mathbf{M}(\mathbf{r}')}{|\mathbf{r}-\mathbf{r}'|} dv'$.Where, ϕ_m is the scalar magnetic potential, $s_{(v)}$ is the magnet boundary surface, v is the magnet volume. However, the second term is zero since the vector $\mathbf{M}$ is constant inside. Then the potential equation can be changed by:

$$\phi_m(\mathbf{r}) = \int_{z_1}^{z_2} \int_{x_2}^{x_{2R}} \frac{dx'dz'}{|\sin\theta|} \frac{M_0}{|\mathbf{r}-\mathbf{r}'|} - \int_{z_1}^{z_2} \int_{x_1}^{x_{1R}} \frac{dx'dz'}{|\sin\theta|} \frac{M_0}{|\mathbf{r}-\mathbf{r}'|} \tag{1}$$

Where x_{2R}, x_2, x_{1R} and x_1 are the limits on X axis, as is showed in Figure 1. The z_1 and z_2 are the limits on z axis. The limits to x axis can be found in this way: $x_1 = R_1\cos\theta - \frac{a}{2}|\sin\theta|$ and $x_{1R} = R_1\cos\theta + \frac{a}{2}|\sin\theta|$ for inner pole's top. $x_2 = R_2\cos\theta - \frac{a}{2}|\sin\theta|$ and $x_{2R} = R_2\cos\theta + \frac{a}{2}|\sin\theta|$ for outer pole's top.

We use the follow notation $\phi(x_l, x_r, \theta, R)(\mathbf{r}) = \int_{z_1}^{z_2} \int_{x_l}^{x_r} \frac{dx'dz'}{|\sin\theta|} \frac{M_0}{|\mathbf{r}-\mathbf{r}'|}$, therefore: $\phi_m(\mathbf{r}) = \phi(x_2, x_{2R}, \theta, R_2)(\mathbf{r}) - \phi(x_1, x_{1R}, \theta, R_1)(\mathbf{r})$. The integral respect to z can be solved, so we get:

$\phi(x_l, x_r, \theta, R)(\mathbf{r}) = f(x_l, x_r, \theta, R, z_2)(\mathbf{r}) - f(x_l, x_r, \theta, R, z_1)(\mathbf{r})$, where: $f(x_l, x_r, \theta, R, z_b)(\mathbf{r}) = -\frac{M_0}{\sin\theta} \int_{x_l}^{x_r} \ln(\sqrt{(z-z_b)^2 + \rho^2(\theta, R, x')} + z - z_b)\, dx'$ and $\rho^2(\theta, R, x') = (x - x')^2 + \left(y - \frac{R}{\sin\theta} + x'\cot\theta\right)^2$.

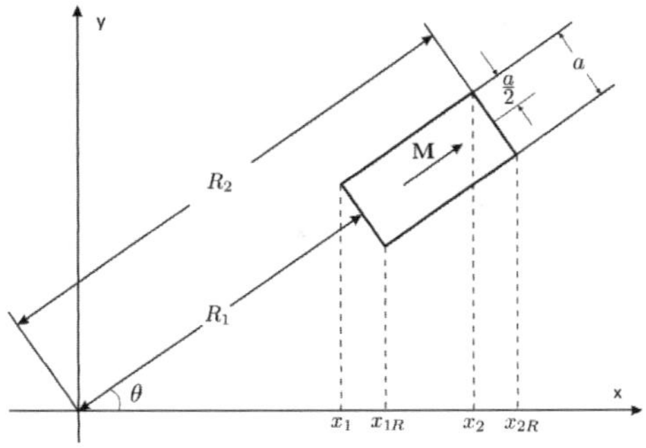

Figure 1. Reference system (Magnet bar cross-section).

B_x component

Since $B_x = -\frac{\partial\phi_m}{\partial x}$ we need to solve:

$$\frac{\partial f(x_l, x_r, \theta, R, z_b)(\mathbf{r})}{\partial x} = -\frac{M_0}{|\sin\theta|} \int_{x_l}^{x_r} \frac{(x-x')}{\sqrt{(z-z_b)^2 + \rho^2(\theta, R, x')} + z - z_b} \frac{dx'}{\sqrt{(z-z_b)^2 + \rho^2(\theta, R, x')}}$$

We can solve this equation through two substitutions and using an integral table for one of integrals, [6]. Then we get the following solutions for B_x component:

$$
\begin{aligned}
B_x = &- p(R_2,\theta)\Big[S(\theta,R_2,z_2,x_{2R})(\mathbf{r}) - S(\theta,R_2,z_2,x_2)(\mathbf{r}) - S(\theta,R_2,z_1,x_{2R})(\mathbf{r}) + S(\theta,R_2,z_1,x_2)(\mathbf{r})\Big] \\
&+ p(R_1,\theta)\Big[S(\theta,R_1,z_2,x_{1R})(\mathbf{r}) - S(\theta,R_1,z_2,x_1)(\mathbf{r}) - S(\theta,R_1,z_1,x_{1R})(\mathbf{r}) + S(\theta,R_1,z_1,x_1)(\mathbf{r})\Big] \\
&+ |\sin\theta|\Big[D(x_{2R},\theta,R_2,z_1,z_2) + D(x_2,\theta,R_2,z_2,z_1) + D(x_{1R},\theta,R_1,z_2,z_1) + D(x_1,\theta,R_1,z_1,z_2)\Big]
\end{aligned}
$$

Where:

$$
p(R,\theta) = \begin{cases} \dfrac{|\sin\theta|(x+\cos\theta(y\sin\theta - R) - x\sin^2\theta)}{\sqrt{x^2\sin^2\theta + (y\sin\theta - R)^2 - (\cos\theta(y\sin\theta - R) - x\sin^2\theta)^2}} & \text{If } y\sin\theta - R \neq 0, \\ |\cos\theta| & \text{If } y\sin\theta - R = 0 \end{cases}
$$

$$
S(\theta,R,z_b,x')(\mathbf{r}) = -2M_0\arctan\left(\frac{u(\theta,R,x') + \sqrt{q^2(\theta,R,z_b)+u^2(\theta,R,x')} + (z-z_b)|\sin\theta|}{l(\theta,R)}\right)
$$

$$
D(x_b,\theta,R,z_r,z_l) = \begin{cases} M_0\ln\left(\dfrac{z-z_l}{z-z_r}\right) & \text{If } \sin^2\theta = 1 \wedge y\sin\theta - R = 0 \wedge x_b - x = 0, \\ M_0\ln\left(\dfrac{D_t(x_b,\theta,R,z_r)}{D_t(x_b,\theta,R,z_l)}\right) & \text{Elsewhere} \end{cases}
$$

$$
\begin{aligned}
D_t(x_b,\theta,R,z_b) &= \sqrt{q^2(\theta,R,z_b)+u^2(\theta,R,x_b)} + (z-z_b)|\sin\theta| \\
u(\theta,R,x') &= x' + \cos\theta(y\sin\theta - R) - x\sin^2\theta \\
h^2(\theta,R) &= x^2 + \left(y - \frac{R}{\sin\theta}\right)^2 - \left(\cos\theta\left(y-\frac{R}{\sin\theta}\right) - x\sin\theta\right)^2
\end{aligned}
$$

Then:

$$
\begin{aligned}
q^2(\theta,R,z_b) &= \left[(z-z_b)^2 + x^2\right]\sin^2\theta + (y\sin\theta - R)^2 - \left[\cos\theta(y\sin\theta - R) - x\sin^2\theta\right]^2 \\
l(\theta,R) &= |\sin\theta| h(\theta,R) = \sqrt{x^2\sin^2\theta + (y\sin\theta - R)^2 - \left[\cos\theta(y\sin\theta - R) - x\sin^2\theta\right]^2}
\end{aligned}
$$

B_y component

For B_y component, we have:

$$
\begin{aligned}
B_y = &\, j(R_2,\theta)\Big[-S(\theta,R_2,z_2,x_{2R})(\mathbf{r}) + S(\theta,R_2,z_2,x_2)(\mathbf{r}) + S(\theta,R_2,z_1,x_{2R})(\mathbf{r}) - S(\theta,R_2,z_1,x_2)(\mathbf{r})\Big] + \\
&j(R_1,\theta)\Big[S(\theta,R_1,z_2,x_{1R})(\mathbf{r}) - S(\theta,R_1,z_2,x_1)(\mathbf{r}) - S(\theta,R_1,z_1,x_{1R})(\mathbf{r}) + S(\theta,R_1,z_1,x_1)(\mathbf{r})\Big] + \\
&\frac{|\sin\theta|}{\sin\theta}\cos\theta\Big[D(x_1,\theta,R_1,z_2,z_1) + D(x_{1R},\theta,R_1,z_1,z_2) + D(x_{2R},\theta,R_2,z_2,z_1) + D(x_2,\theta,R_2,z_1,z_2)\Big]
\end{aligned}
$$

Where:

$$
j(R,\theta) = \begin{cases} \dfrac{y|\sin\theta| - R\dfrac{|\sin\theta|}{\sin\theta} + \Big(x\sin^2\theta - \cos\theta(y\sin\theta - R)\Big)\cos\theta\dfrac{|\sin\theta|}{\sin\theta}}{\sqrt{x^2\sin^2\theta + (y\sin\theta - R)^2 - (\cos\theta(y\sin\theta - R) - x\sin^2\theta)^2}} & \text{If } y\sin\theta - R \neq 0, \\ \dfrac{\sin\theta\cos\theta}{|\cos\theta|} & \text{Otherwise} \end{cases}
$$

B_z component

For B_z component, the integral is more simple; first we integrate respect to x' variable doing the previous substitutions and then integrate with z':

$$B_z = M_0 \ln\left(\frac{k(\theta,R_2,z_2,x_{R2})k(\theta,R_2,z_1,x_2)k(\theta,R_1,z_1,x_{R1})k(\theta,R_1,z_2,x_1)}{k(\theta,R_2,z_1,x_{R2})k(\theta,R_2,z_2,x_2)k(\theta,R_1,z_2,x_{R1})k(\theta,R_1,z_1,x_1)}\right)$$

Where, $k(\theta,R,z_b,x_b) = \sqrt{q^2(\theta,R,z_b)+u^2(\theta,R,x')}+u(\theta,R,x')$.

Results for $\theta = 0$ or $\theta = 180°$

Since these formulas do not work when $\theta = 0$ or $\theta = 180°$, because in this cases we have: $|\mathbf{r}-\mathbf{r}'| = \sqrt{(z-z')^2+(x-R\cos\theta)^2+(y-y')^2}$.

We use the follow notation, $\phi_0(y_l,y_r,\theta,R)(\mathbf{r}) = \int_{z_1}^{z_2}\int_{y_l}^{y_r} \frac{M_0\,dy'dz'}{\sqrt{(z-z')^2+(x-R\cos\theta)^2+(y-y')^2}}$, therefore, we have:

$$\phi_0(y_l,y_r,\theta,R)(\mathbf{r}) = f_0(y_l,y_r,\theta,R,z_2)(\mathbf{r}) - f_0(y_l,y_r,\theta,R,z_1)(\mathbf{r})$$

Where:

$$f_0(y_l,y_r,\theta,R,z_b)(\mathbf{r}) = -M_0\int_{y_l}^{y_r}\ln\left(\sqrt{(z-z_b)^2+\rho_0^2(\theta,R,y')}+z-z_b\right)dy'$$

B_{x0} component (with $\theta = 0$)

$$\begin{aligned}B_{x0} = &-S_0\left(\theta,R_2,z_2,\frac{a}{2}\right)(\mathbf{r}) + S_0\left(\theta,R_2,z_2,-\frac{a}{2}\right)(\mathbf{r}) + S_0\left(\theta,R_2,z_1,\frac{a}{2}\right)(\mathbf{r}) - S_0\left(\theta,R_2,z_1,-\frac{a}{2}\right)(\mathbf{r})\\ &+ S_0\left(\theta,R_1,z_2,\frac{a}{2}\right)(\mathbf{r}) - S_0\left(\theta,R_1,z_2,-\frac{a}{2}\right)(\mathbf{r}) - S_0\left(\theta,R_1,z_1,\frac{a}{2}\right)(\mathbf{r}) + S_0\left(\theta,R_1,z_1,-\frac{a}{2}\right)(\mathbf{r})\end{aligned}$$

Where:

$$S_0(\theta,R,z_b,y_b)(\mathbf{r}) = \begin{cases} 0 & \text{If } x-R\cos\theta = 0,\\ 2M_0\frac{x-R\cos\theta}{|x-R\cos\theta|}\arctan\left(\frac{y-y_b+\sqrt{q_0^2(\theta,R,z_b)+(y-y_b)^2}+(z-z_b)}{|x-R\cos\theta|}\right) & \text{Otherwise}\end{cases}$$

$$q_0^2(\theta,R,z_b) = (z-z_b)^2+(x-R\cos\theta)^2$$

B_{y0} component (with $\theta = 0$)

$$\begin{aligned}B_{y0} = &D_0\left(\theta,R_2,\frac{a}{2},z_1,z_2\right)(\mathbf{r}) + D_0\left(\theta,R_2,-\frac{a}{2},z_2,z_1\right)(\mathbf{r}) + D_0\left(\theta,R_1,\frac{a}{2},z_2,z_1\right)(\mathbf{r})\\ &+ D_0\left(\theta,R_1,-\frac{a}{2},z_1,z_2\right)(\mathbf{r})\end{aligned}$$

Where:

$$D_0(\theta,R,y_b,z_n,z_d)(\mathbf{r}) = \begin{cases} M_0\ln\left(\frac{|z-z_d|}{|z-z_n|}\right) & \text{If } x-R\cos\theta = 0 \wedge y-y_b = 0,\\ M_0\ln\left(\frac{D_{t0}(\theta,R,y_b,z_n)}{D_{t0}(\theta,R,y_b,z_d)}\right) & \text{otherwise}\end{cases}$$

B_{z0} **component (with $\theta = 0$)**

$$B_{z0} = M_0 \ln \left(\frac{k_0\left(\theta, R_2, -\frac{a}{2}, z_2\right) k_0\left(\theta, R_2, \frac{a}{2}, z_1\right) k_0\left(\theta, R_1, \frac{a}{2}, z_2\right) k_0\left(\theta, R_1, -\frac{a}{2}, z_1\right)}{k_0\left(\theta, R_2, \frac{a}{2}, z_2\right) k_0\left(\theta, R_2, -\frac{a}{2}, z_1\right) k_0\left(\theta, R_1, -\frac{a}{2}, z_2\right) k_0\left(\theta, R_1, \frac{a}{2}, z_1\right)} \right)$$

Where:

$$k_0(\theta, R, y_b, z_b) = \sqrt{q_0^2(\theta, R, z_b) + (y - y_b)^2} + y - y_b$$

We present a picture of magnetic field from a cubic magnet of 4 cm length, with a magnetization vector of 518.67G of magnitude (see Figure 2). To confront this results, we take data from Dexter Magnetic Tecnologies [7]. We take a cubic magnet with $\theta = 270°$, $R_1 = 0$ in, $R_2 = 2$ in, $z_1 = -1$ in, $z_2 = 1$ in, and $a = 2$ in; Residual induction (Br) G= 500, Material type, Nd-Fe-B. We use a point (x_p, y_p, z_p) to calibrate the magnetic field magnitude, for example we take $M_0 = 1$ and calculate the magnetic field magnitude in this point, through our formulas, called $B_f(x_p, y_p, z_p)$. Then we get the experimental result called $B_e(x_p, y_p, z_p)$, then $M_0 = B_e(x_p, y_p, z_p)/B_f(x_p, y_p, z_p)$, this way we can get the correct magnetization vector, in this case is -37.894G. We put both experimental and simulated results in a Table 1. Therefore these results help us to show the validity of our formulas. Now we present the results for a hexapole system.

Table 1. Confront of data.

Coordinates (in)			B (G) Experimental			B (G) Simulated		
x	y	z	B_x	B_y	B_z	B_x	B_y	B_z
0	1	0	0	64.182	0	0	64.18203	0
0	2	0	0	21.6	0	0	21.59964	0
0	3	0	0	9.352	0	0	9.35169	0
0	4	0	0	4.824	0	0	4.82424	0
0	10	0	0	0.455	0	0	0.45542	0
0	1	1	0	38.530	29.885	0	38.53027	29.88534
0	1	2	0	6.546	20.439	0	6.54579	20.43920
0	1	3	0	-0.621	8.948	0	-0.62051	8.94804
0	0	3	0	-6.803	8.258	0	-6.80321	8.25755
0	-1	2	0	-32.091	0	0	-32.09102	0
1	1	1	20.59	23.287	20.59	20.59032	23.28694	20.59032
2	1	1	15.491	3.856	7.452	15.49089	3.85553	7.45216

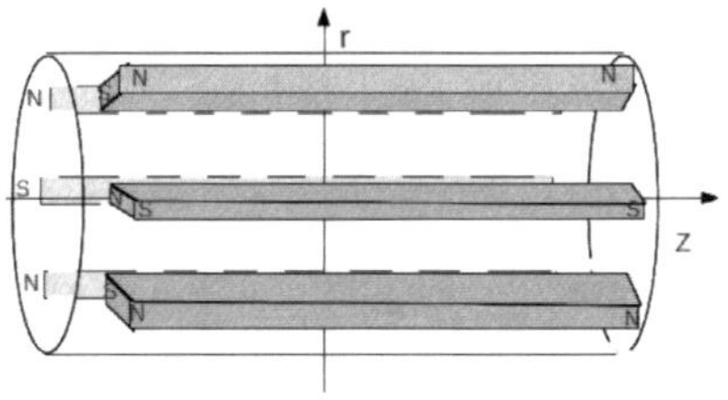

Table 2. Physical scheme (hexapolar system).

3. Hexapolar system

The physical system consists in a six parallelepiped bar magnets placed on the outside of a cylinder, in alternating North/South polarizations, whose poles are targeted radially; this system is showed in Figure 2.

The cylinder chamber radius is 3.9956cm, with length of 25.9720cm. Longitudinal size magnet bar is 21.0cm, the inner pole radius is $R_1 = 4.1$cm in Figure 2; the external pole radius is $R_2 = 7.1$cm, width $a = 2.5$cm. For hexapole field simulation we take the following angles $\theta = 30°$, $90°$, $150°$, $210°$, $270°$ and $330°$. The field graphic on XY section (plane $z = 0$cm) is showed in Figure 3; where the cusp curvature is seen near to magnetic pole zone. In this figure we can see a gradient radial from the colour scale of field magnitude toward Z axis; this property is necessary to press the plasma back inside camera discharge.

The field on $z = 10.0$cm plane (See Figure 4) is like one central configuration since it is inside region covered by hexapole system. The field on $z = 12.5$cm plane (Figure 5) is more longitudinal, specially in regions near magnet poles; since this zone is outside of hexapole region. The longitudinal field view on $x = 0$cm plane, is given in Figure 6, the magnetic field becomes more longitudinal, outside of region covered by magnets, in the zone near to the magnets.

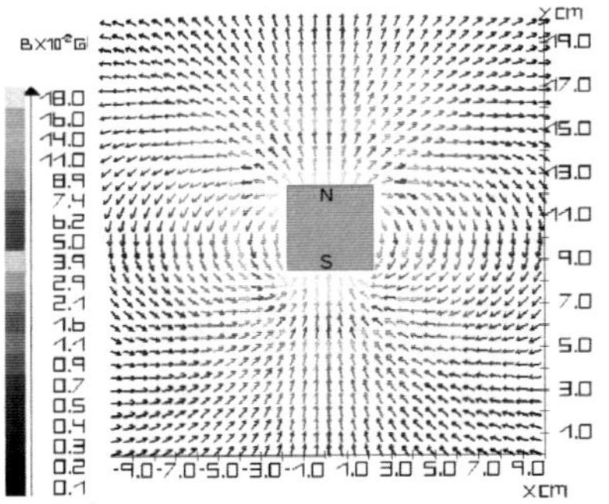

Figure 2. Physical system.

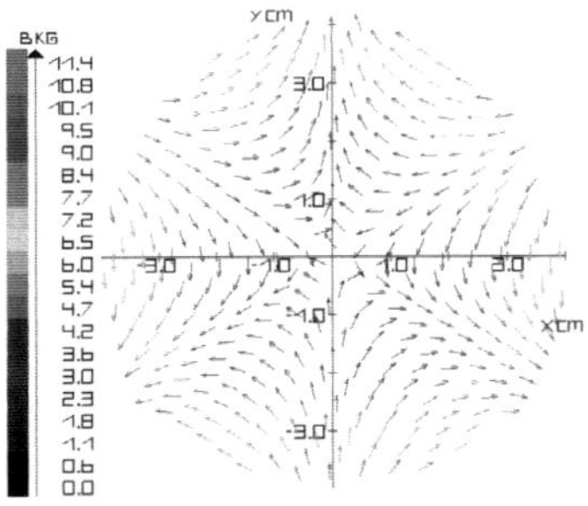

Figure 3. XY view (plane $z = 0.0$cm) of the magnetic field.

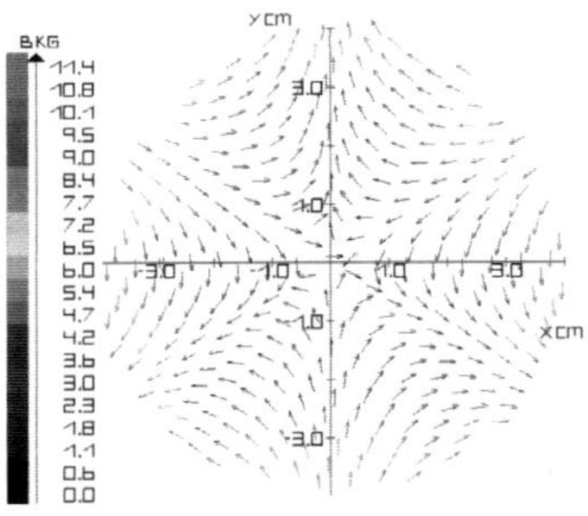

Figure 4. XY view (plane $z = 10.0$cm) of the magnetic field.

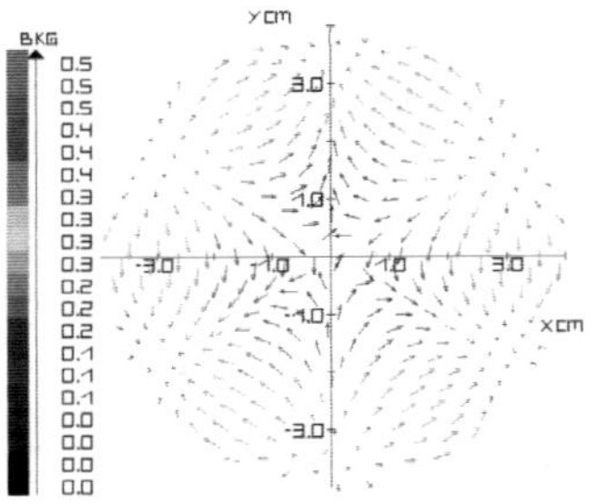

Figure 5. XY view (plane $z = 12.5$cm) of the magnetic field.

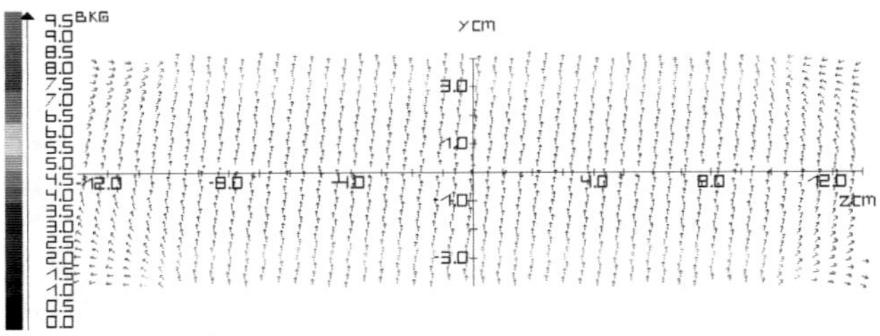

Figure 6. YZ view (plane $x = 0.0$cm) of the magnetic field.

4. Conclusion

The analytic calculation from a parallelepiped magnet, which can turn around an axis is very important to researchers on plasma physics under multipolar magnetic field, for example, in plasma magnetic confinement in ECR sources, since hexapole magnetic field configurations are an important term to improve the plasma dynamics simulations. Using an approximate value of magnetic field can cause mistakes. Although these errors can be small, the plasma simulations cycles need to be performed many times to reach the stability of the system. Therefore the errors can increase when the number cycles grow. These calculations serve to help reduce computational effects over the plasma behaviour simulated.

References

[1] A Dinklage T Klinger G Marx and L Schweikhard 2005 *Plasma physics confinement, transport and collective effects 1st ed* (Berlin: Springer-Verlag) p 148

[2] K Kim, Y Lee, S Kyong and G Yeom 2004 *Surface and Coatings Technology* **177-178** 752

[3] H Takekida and K Nanbu 2004 *Phys D Appl Phys* **37** 1800

[4] R Ravaud and G Lemarquand 2009 *Progress in Electromagnetics Research* **98** 207

[5] J Jackson 1999 *Classical Electrodynamics 3rd ed* (United States of America: John Wiley & Sons Inc) pp 195-197

[6] H Bristol 1957 *Tables of integrals and other mathematical data 3rd ed* (New York: The macmillan company) p 93

[7] Dexter Magnetic Tecnologies 2015 *Field calculations for a rectangle* (http://www.dextermag.com/resource-center/magnetic-field-calculators/field-calculations-for-rectangle)

Development of a sintering methodology through abnormal glow discharge for manufacturing metal matrix composites

S Pérez[1], Y Pineda[1], A Sarmiento[1] and A López[1]
[1] Universidad Pedagógica y Tecnológica de Colombia (UPTC), Tunja, Colombia.

E-mail: patricia.perez@uptc.edu.co

Abstract. In this study, a sintering methodology is presented by using abnormal glow discharge to metal matrix composites (MMC), consisting of 316 steel, reinforced with titanium carbide (TiC). The wear behaviour of these compounds was evaluated according to the standard ASTM G 99 in a tribometer pin-on-disk. The effect of the percentage of reinforcement (3, 6, and 9%), with 40 minutes of mixing in the planetary mill is analysed, using compaction pressure of 700MPa and sintering temperature of 1,100°C±5°C, gaseous atmosphere of H_2 - N_2, and sintering time of 30 minutes. As a result of the research, it shows that the best behaviour against wear is obtained when the MMC contains 6% TiC. Under this parameter the lowest percentage of pores and the lowest coefficient of friction are achieved, ensuring that the incorporation of ceramic particles (TiC) in 316 austenitic steel matrix significantly improves the wear resistance. Also, it is shown that it is possible to sinter such materials using the abnormal glow discharge, being a novel and effective method in which the working temperature is reached in a short time.

1. Introduction

In search of new materials with improved properties, an inquiry is made about the behaviour of metal matrix composite (MMC) materials that can provide properties that are not possible to obtain with only one of the original materials (matrix or reinforcement). One of the most used techniques for the manufacture of composite materials is the powder metallurgy techniques (PM). In this study, type 316 stainless steel is used as the matrix, given that it is one of the most used to present excellent mechanical properties, corrosion resistance and thermal conductivity [1]. Although these steels exhibit low wear resistance, superior properties are achieved if particles as reinforcing material are added and if the sintering method of the parts [2] is varied. Reinforcement particles commonly used in MMC are oxides (Al_2O_3, ZrO_3), carbides (TiC, WC) and borides (TIB_2, CrB_2). Among these reinforcements, the TiC is considered a good alternative to be used to present high melting point and being chemically stable at high temperatures. Its pure structure allows the combination of ionic, covalent and metallic links; for that reason, it combines the physical properties of the ceramic and the electrical properties of metals, such as high hardness and high thermal conductivity [3]. Some studies [4] suggest that the reason for that studied steel (316) is sintered in inert atmospheres, at high temperatures is to thereby achieve a better performance. For the development of this research, the MMC are sintered by generating an abnormal glow discharge in H_2 – N_2 atmosphere, which is a recent technique that represents some decrease in the time required for the sintering of samples and lower power consumption, because heating is performed directly by the bombardment of ions and neutral atoms on the surface of the cathode where the sample [5] is usually located. Considering that the material wear is a major problem in industry, which largely affects the production sectors [6], the objective of this work is to study the wear behaviour of MMC manufactured with 3, 6 and 9% (vol.) of TiC. Thus useful information to ensure that the incorporation of ceramic particles (TiC) in austenitic steel matrix (316) may lead to an improvement in wear resistance.

2. Experimental

The metal matrix composite is obtained through conventional PM using powders of austenitic 316 stainless steel and metal matrix, with the following chemical composition: 17% Cr, 12% Ni, 2.5% Mo, 2% Mn, 1% Si, 0.08% C and Fe (balance), using as reinforcement titanium carbide TiC (99.95% purity), with average particle sizes for the steel 106.140 μm and for the TiC 4.056μm.

The mixture is performed in a planetary mill (RESTCH), consisting of two stainless steel vessels with balls of the same material of 10mm in diameter with a balls\loading ratio of 4:1, for 40 minutes at 180rpm; the mixtures are prepared with three different amounts of reinforcement 3, 6 and 9% (vol.). The homogeneity of the mixtures was checked by scanning electron microscopy. The samples are uniaxially compacted at 700MPa in a hydraulic press (ELE International) of 1000KN capacity, using a matrix to which was applied zinc stearate on the wall as a lubricant to obtain a uniform density, resulting in green preforms with approximate dimensions of 12mm diameter and 4mm thick. These preforms were sintered by means of plasma abnormal glow discharge of direct current in an atmosphere of H_2–N_2, under the following operating parameters: pressure 3torr, flow rate of 2cm^3/min, voltage 534V, 160mA current and heating rate of 100°C/min, conditions that reach the temperature of (1,100°C±5°C), with a stay of 30 minutes, ensuring the presence of a stable plasma. Cooling is performed in the same sintering atmosphere under the same flow conditions [4,6].

Characterization of the interface between TiC particles and the matrix is performed by chemical microanalysis, to thereby images are taken in the scanning electron microscope at 5000X, in the mode of backscattered electrons with an acceleration voltage of 20kV and spot size of 40.

To analyse the properties achieved in the MMC the following are determined: The density of the samples by the Archimedes method; porosity, with the software IQMaterials, version 2, which quantifies the amount of pores in the MMC with the images obtained in the scanning electron microscope, under the guidelines of ASTM E112 standards, E1382 and E562; wear, with the pin-on-disk test according to ASTM G 99 standard in a tribometer (NANOVEA TRB), using as contramaterial a steel pin of 6mm diameter and hardness of RC 71.4. The tests are carried out at room temperature without lubrication. The applied load is of 5N, with sliding speed of 0.1m/s, footprint diameter of 3.5 mm and sliding distance of 377m. The friction coefficient is measured continuously during the test and the wear is determined by the wear coefficient k: k(m^2/N)= loss of material volume (m^3)/[applied load (N)*Sliding distance (m)].

With the obtained results, the effect of the percentage of reinforcement against wear produced in the MMC when sintered through abnormal glow discharge is evaluated.

3. Results and discussion

In Figure 1, the chemical EDX microanalysis is shown, performed with the samples by Line Scan, in order to characterize the interface between TiC particles and the matrix, indicating the start (P_o) and final (P_f) points of the line scan.

For all MMC made with different percentages of reinforcing the behaviour is similar. Observing that chromium oxide appears as a continuous layer defining TiC particles; a situation that is evidenced by the presence of peaks of O and Cr located on one or both sides of the region comprising the reinforcing particle. Likewise, the elements comprised by the matrix are shown: Fe, Cr, Ni, Mo and Si, which are located around the particle and no direct chemistry interaction with the elements that constitute the reinforcement is evidenced. Considering that titanium carbide is a chemically stable reinforcement that does not react with the matrix, ZiFei Ni, et. al [7], hence the addition of TiC aids to the grain refinement of the austenite matrix without forming a new phase in the microstructure, and improves the mechanical properties and wear resistance [8]. The sintered samples exhibit the following densities: steel PM (6.07g/cm^3), MMC 3% TiC (5.58g/cm^3), MMC 6% TiC (5.45g/cm^3) and MMC 9% TiC (5.31g/cm^3). It is evident that when reinforcing the stainless steel with titanium carbide a decrease in density occurs due to the lower specific weight of the reinforcing material. The relative density is maintained around 70% in all compounds.

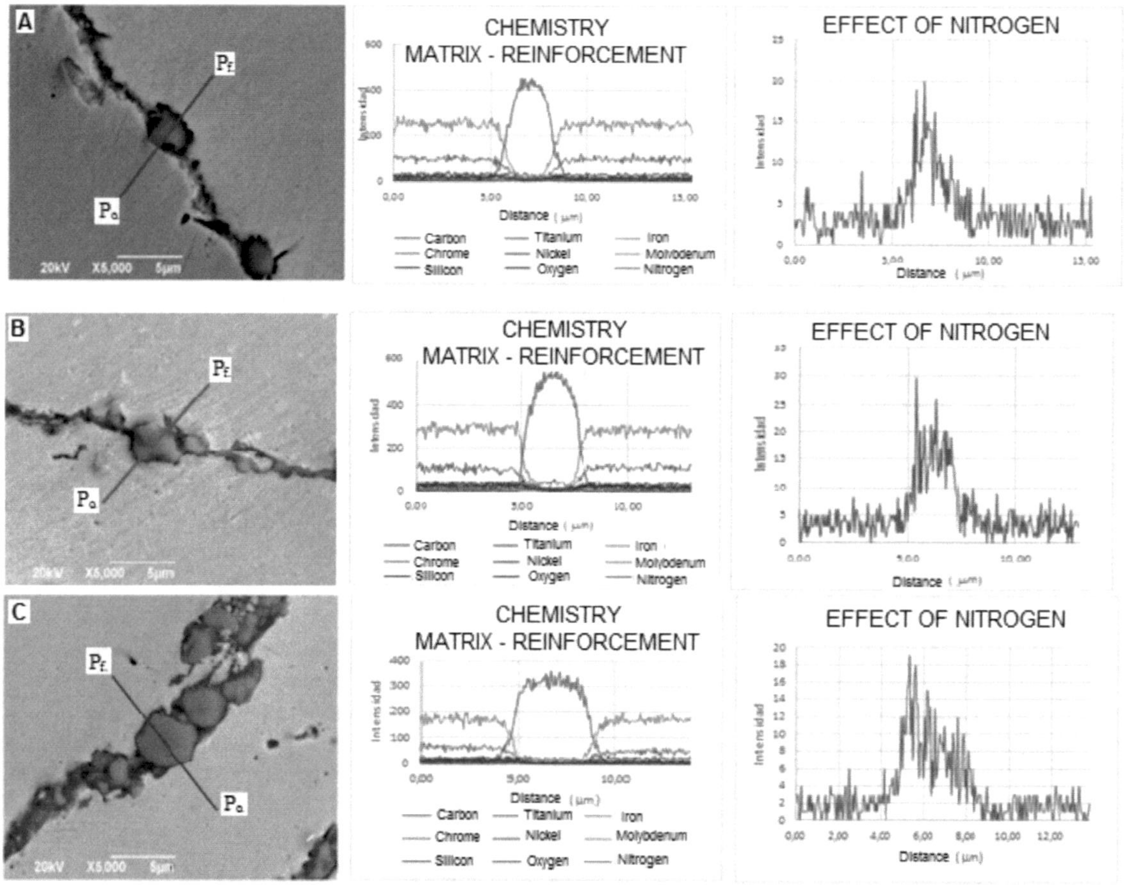

Figure 1. Line Scan performed at MMC reinforced: (A) (3% TiC), (B) (6% TiC), (C) (9% TiC).

In Figure 2, the analysis of images obtained in a scanning electron microscope of the sintered samples by using the software IQMaterials, Version 2, is presented, showing that the MMC exhibiting lower percentage of pores is the composite reinforced with 6% TiC, which is possibly due to the fact that with this amount of reinforcement there is better interaction between the matrix and reinforcement.

The effect of sintering on the mechanical properties is evaluated, emphasizing on the friction coefficient and wear in the manufactured compounds with different percentages of reinforcement, see Table 1. It is noted that the lower coefficient of friction is presented in the MMC made with 6% of TiC and, as expected, it exhibits less wear. Also, it shows that by increasing the percentage of reinforcing the compound undergoes greater wear, which can be caused by the TiC tendency to agglomerate around itself, preventing any proper interaction between the matrix and the reinforcing; an effect that is observed in Figure 1(C), causing that the reinforcing be removed from the material and help generate three-body abrasive wear.

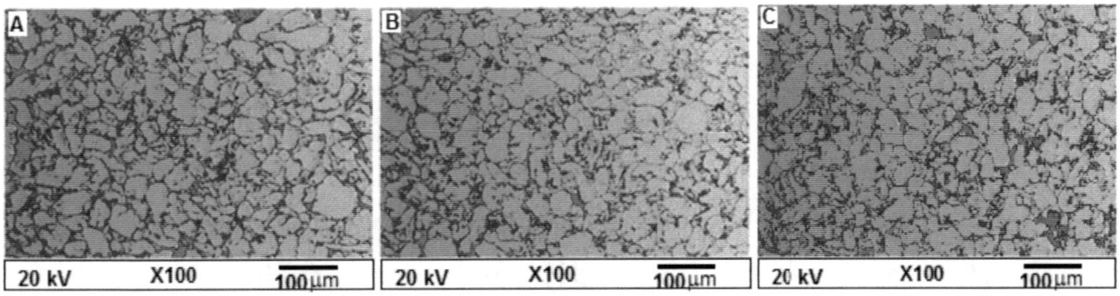

VARIABLE	A (3% TiC)	B (6% TiC)	C (9% TiC)
% of Poros (red)	4.6203	2.2205	4.934

Figure 2. Percentage of porosity in the sintered MMC, obtained with software IQMaterials.

Table 1. Effect of the percentage of reinforcement in the friction coefficient and wear coefficient.

% TiC	Coefficient of friction	Wear coefficient (m^3/N.m)
0	0,7441	8,29 X 10^{-9}
3	0,5956	6,43 X 10^{-9}
6	0,4471	3,68 X 10^{-9}
9	0,6469	7,73 X 10^{-9}

4. Conclusions

The best condition for manufacturing the MMC in the percentage of reinforcement is obtained when the compound contains 6% of TiC. Under this parameter the smaller pore size and lower coefficient of friction was obtained by ensuring the incorporation of ceramic particles (TiC) in austenitic steel matrix (316) representing improvements in wear resistance. In materials produced with the powder metallurgy technique it is important to analyse different manufacturing conditions that lead to obtain better MMC behaviour against wear; for this reason it is recommended to compact the preforms in green at pressures greater than 700MPa and sinter at temperatures above 1,100°C.

References

[1] J Meng, N H. Loh, B Y Tay, G Fu and S B Tor 2010 *Wear* **268-7-8** 1013
[2] F Akhtar, S J Askari, K A Shah, X Du and S Guo 2009 *Mater Charact* **60-4** 327
[3] M Sheikhzadeh and S Sanjabi 2012 *Mater Des* **39**366
[4] J Abenojar, F Velasco and A Bautista 2003 *Sci Technol* **63** 69
[5] U Guerrero, C Parra and A Sarmiento 2011 *Dyna* **78-169** 201
[6] L Ferreiro, A García, A Varela, C Camba, J L Mier and F Barbadillo 2010 *Correlación entre los resultados obtenidos mediante diversas técnicas de ensayo de desgaste* (Spaña: XVIII Congreso Nacional de Ingeniería Mecánica)
[7] Z Ni, Y Sun, F Xue, J Bai and Y Lu2011 *Mater Des* **32-3** 1462
[8] Wang Zhen-ting, Zhou Xiao-hui and Zhao Guo-gang 2008 Transactions of Nonferrous Metals Society of China **18** 831

Effect of glow discharge sintering in the properties of a composite material fabricated by powder metallurgy

A Cardenas[1], Y Pineda[2], A Sarmiento Santos[2] and E Vera[2]
[1] Universidade Federal do Rio de Janeiro, Rio de Janeiro, Brasil.
[2] Universidad Pedagógica y Tecnológica de Colombia, Tunja, Colombia.

E-mail: asarmiento.santos@uptc.edu.co

Abstract. Composite samples of 316 stainless steel and SiC were produced by powder metallurgy. Starting materials were mixed in different proportions and compacted to 700MPa. Sintering stage was performed by abnormal glow discharge plasma with direct current in an inert atmosphere of argon. The process was conducted at a temperature of 1200°C±5°C with a heating rate of 100°C/min. This work shows, the effectiveness of plasma sintering process to generate the first contacts between particles and to reduce vacancies. This fact is confirmed by comparing green and sintered density of the material. The results of porosity show a decrease after plasma sintering. Wear tests showed the wear mechanisms, noting that at higher SiC contents, the wear resistance decreases due to poor matrix-reinforcement interaction and by the porosity presence which causes matrix-reinforcement sliding.

1. Introduction

The use of sintered powder metallurgical of stainless steels is represented mainly in the automotive industry, on items such as exhaust pipes, flanges, valves high temperature control, among others, where it requires specific characteristics such as resistance to oxidation at high temperatures, corrosion resistance, rupture strength and controlled porosity, [1,2]. Porosity is one of the most important physical properties in materials, for metalworking applications, more specifically on self-lubricating bushings, [3] and metal filters [4], where the type of porosity, inter connected or isolated plays an important role in the final properties of the material. The primary objective of sintering in powder metallurgy processes is the union of particles where the resulting component acquires mechanical strength [5]. Adding particulate ceramic reinforcements such as carbides in metal matrix, is obtained a composite material with improved mechanical properties, SiC is an example of particulate ceramic reinforcement, which has good microstructural stability, high temperature resistance and hardness, among other properties. Reinforced powder metallurgy materials, sintered by standard methods in atmospheres as H_2, N_2 or vacuum, provide good characteristics [6]. However, processes such as SPS (Spark plasma sintering) are a good choice for quickly powder metallurgy materials processing, due to its similarity to the hot pressing where the compaction pressure is applied to the sintering process [7,8]. Sintering by unconventional methods such as the use of abnormal glow discharge plasma, which is evidenced by electrical conduction through the ionized gases that becomes electrically conductive [9,10], is a candidate for optimization of sintering processes of powder metallurgy materials. The aim of this study is to show the influence of glow discharge DC (Direct current) sintered with a protective atmosphere of Argon, in 316 steel samples previously reinforced with silicon carbide (SiC) in proportions of 2, 4, 6 and 8% in volume, manufactured following a conventional powder metallurgy route. The efficiency of the sintering process and the effect of the addition of SiC particles were evaluated according to the porosity of green and sintered samples, hardness, density of the material and wear test after sintering processes.

2. Experimental

The morphologically characterized the materials used in this research :316 stainless steel powder matrix and silicon carbide reinforcement, was carried out by SEM (scanning electron microscopy). For producing green samples, mixtures with four different amounts of reinforcement were used: 2.0, 4.0, 6.0% and 8.0 vol.%. The mixture was prepared to generate the first contacts between the steel particles and the SiC reinforcement and to obtain a homogeneous composite. This procedure was performed using a planetary mill for 30 minutes at 90rpm with a ball charge ratio 2:1. The mixture powders were pressed to obtain cylindrical samples 12mm in diameter and high 5mm, this procedure was made through uniaxial pressed in a double effect matrix at 700MPa [11]. The green samples were sintered in a plasma reactor using the abnormal glow discharge with electrodical configuration shown in Figure 1.

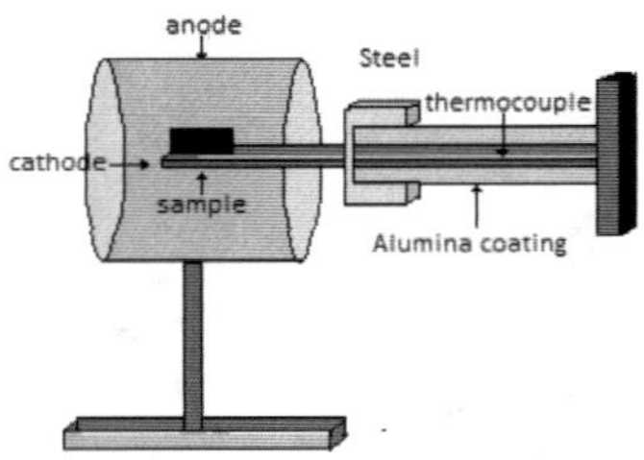

Figure 1. Electrode configuration for plasma sintering.

The green samples were sintered for abnormal glow discharge in argon atmosphere with flow of 50cm^3/min at 4.0torr of pressure. To turn the glow discharge, a pulsed current source was used. The samples were heated at a rate of 100°C/min, until reaching the sintering temperature (1200°C±5°C) [12]. The specimens remained at the sintering temperature for a period of 35 minutes after which suspends the discharge energy and cooling was performed within the vacuum chamber. Green and sintering porosity, of the samples, was determined in order to observe the sintering effect. The quantification and comparison of the results of open porosity was made to the procedures described in ISO 2738. Material hardness was determined in the scale Rocwell B. Finally the ceramic particles reinforced steel and without reinforcing material steel, were evaluated according to ASTM G99-05 standard for wear test in pin-on-disk machine, using the following test conditions: pin of 8mm diameter, sliding distance 400m, equipment revolutions 60rpm and applied load of 5N.

3. Results and discussion

Were determined the theoretical density 3.22g/cm^3 and Angular morphology for SiC powders, typical of ceramic powders and particle size <75µm, observed by SEM and confirmed by standard mesh sieve. The 316 stainless steel powders showed typical morphology of water atomization process which confers it features as irregularly and makes them suitable for compaction processes, providing good sinterability and high green and sintered density, with particle size were <150µm and theoretical density of 8.0g/cm^3. BET results, represented by absorption-desorption isotherms, showed the macro porous characteristic of green and sintered samples, being a reduction in the pore volume of approximately 60% in the sintered samples. In addition it is determined that as the reinforcing content increases, the pore volume also increases, primarily due to the interaction matrix-reinforcement, which prevents that steel completely surrounds the SiC particles, behaviour observed in studies by J. Abenojar et al [2]. Figure 2(a) shown the density of the green samples (black colour) and in sintered samples (red colour) for each amounts of reinforcement, is possible to observe a decreasing trend related to the density value with increases the percentage of reinforcement, this behaviour was for the green and the sinterized samples. It was determined that this variation is mainly due to the low density of SiC that makes that if its content increases, the green density of the samples decreases, showing little contribution, of the ceramic particulate content, to composite material densification as observed by J. Abenojar et al studies [2]. It can also be an increased density (red columns) in sintered samples, with respect to green samples, which shows a good sample processing and sintering in glow discharge, showing the consolidation of individual powder particles through the bonding there between due to high temperature exposure.

Results of porosity (Figure 2(b)), shown increased necking between particles, closing gaps or vacancies, after sintering. The reinforcement volume percentage having higher porosity in green and sintered is the corresponding, consequently, to the composition with a higher content of reinforcement (8% SiC), which should have lower matrix-reinforcement interaction and therefore greater difficulty to make bonding between the particles. Figure 3(a) shows the microstructure corresponding to green samples obtained by SEM (200x). Microstructure consists of the fully austenitic matrix with poorly defined grain boundaries and presence of the reinforcing material (dark areas) with porosity in its boundaries.

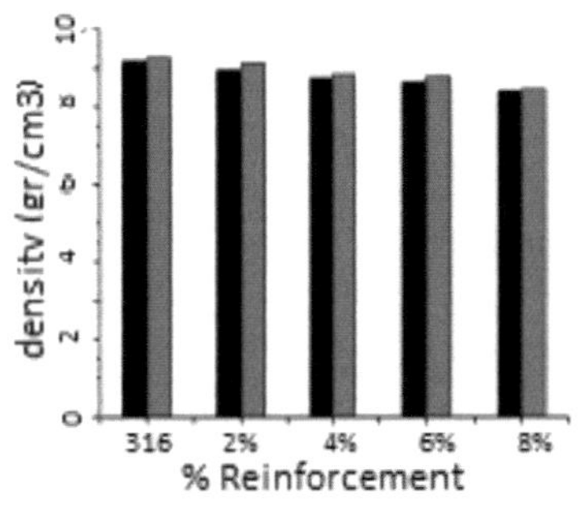

(a)

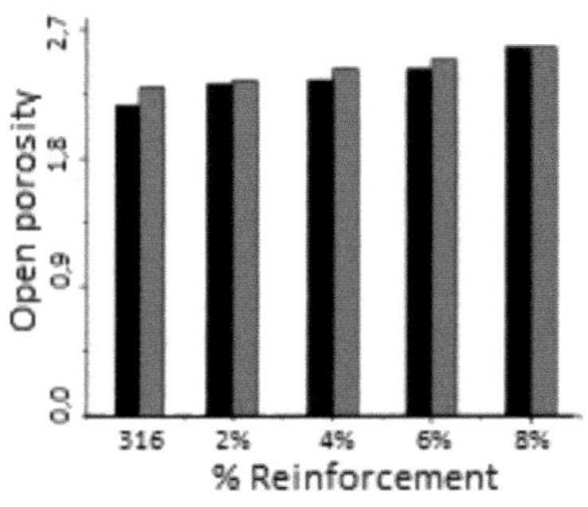

(b)

Figure 2. Density and porosity of green and sintered samples for various SiC compositions.

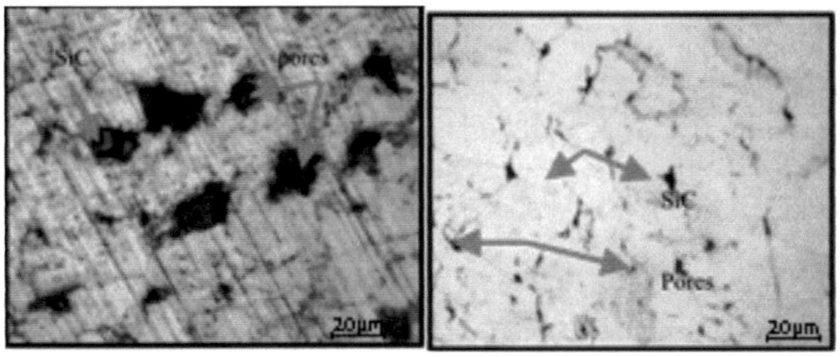

(a) (b)

Figure 3. Samples microstructure at 200x: (a) green samples and (b) sintered samples

Figure 3(b) shows the microstructure of the sintered samples in the glow discharge plasma (1200°C±5°C for 35 minutes). Clear defining of the grain boundaries and the reinforcement and rounding of pores and decreasing in its size were observed. Thus, the densification achieved during plasma sintering is shown. Figure 4 shows micrographs of samples with different proportions of reinforcement obtained by SEM (5000X) after wear tests on a pin-on-disk tribometer.

Figure 4. SEM (5000X) of samples with all reinforcement proportions after wear test.

Were observed the morphology of worn after sliding distance of 400m and the abrasive wear mechanism, due to generation of scratches and material removal. Were observed that the lowest percentages of reinforcement 2.0 and 4.0%, show the best features in this property. On the other hand, high percentages of reinforcing the matrix can not maintain the immersed reinforcements, causing detachment thereof which, in turn, contributes to the attrition process the material.

4. Conclusions

In this paper, samples of 316 stainless steel reinforced with silicon carbide particles in different proportions were produced by powder metallurgy. The sintering stage of the samples was made by an unconventional method: plasma of the glow discharge. This type sintering provides densification to the samples under study, in addition to having reduced the cost of production, as the supply direct current required to generate the plasma in the vacuum chamber is very small, found it that this method allows reaching the sintering temperature (1200°C±50°C) in a short time (12 minutes) generating processes faster and more accurate compared with the conventional sintering processes for ferritic or austenitic stainless steel, can take approximately 1h, as shown the study of Martin et al [13]. The porosity content for the compositions of SiC prepared, increases with the content of reinforcement (SiC), mainly due to the low interaction matrix-reinforcement which prevents material densification and promotes increased porosity, positively affecting related applications to Self-lubricating bushings to filters and where low weight and mainly controlled porosity are required. The samples after the sintering process in glow discharge exhibit good densification, decreased porosity and rounding of pores, showing a new application of plasma in consolidating the green compacts.

Acknowledges

The authors thank the School of Materials Engineering of the Universidad del Valle, the INCITEMA, GESEC and, catalysis research group, of the Universidad Pedagógica y Tecnológica de Colombia, for their support in the development of this research.

References

[1] Causton Robert, Mumau Claudia and M Tina 1998 *Elsevier Science* **22** 22
[2] J Abenojar, F Velasco, J M Torralba, J A Bas, J A Calero and R Marce 2002 *Materials Science and Engineering* **A335-1** 1
[3] Surapol Raadnui, Sithipong Mahathanabode and Ruangdaj Tongsri 2008 *Wear* **265** 546
[4] R A Tolosa 2004 *Revista Ciencia e Ingenieria* **23-1** 67
[5] Morral F R and Gimeno Molera P 2004 *Metalurgia general vol 2* (Barcelona: Editorial Reverte) p 977
[6] J Abenojar, F Velasco, A Bautista, M Campos, J A Bas and J M Torralba 2003 *Composite Science and Technology* **63** 69-79
[7] Binghong et al 2009 *Journal of Materials Processing Technology* **210** 91
[8] Dustin M Hulbert et al 2009 *Scripta Materialia* **60** 835
[9] Brian Chapman 1980 *Glow discharge process sputtering and plasma etching* (New York: Jhon willey)
[10] Michael A Lieberman and Allan J Lichtenberg 2005 *Principles of plasma discharges and materials processing 2nd ed* (United states of America: Jhon willey & sons)
[11] A Bautista et al 2006 *Revista de Metalurgia* **42** 175
[12] F Velasco et al 2003 *Tribology International* **36** 547
[13] F Martín, C García et al 2015 *Materials science and Engineering* **A642** 360

Diffraction pattern by nanometric thin films under illumination of an orbital angular momentum beam with integer topological charge

J H Mendoza[1], C F Díaz[1], C H Acevedo[1] and Y Torres[2]
[1] Universidad Industrial de Santander, Bucaramanga, Colombia.

E-mail: nachomendozac@hotmail.com

Abstract. The orbital angular momentum of light has a big contribution in many engineering applications like optical communications, because this physical property allows eigenstates characteristic of the wavefront rotation when the beam is propagated. The nature of these eigenstates allows that information can be encoded and gives immunity to electromagnetic interference, allowing an increase of bandwidth, cadence and capacity of the communication channel. This work shown the methodology using nanometric thin films like Titanium based (TiO_2) grown over strontium titanate ($SrTiO_3$) support, to distinguish and discriminate a well-defined integer value of the topological charge of an OAM beam.

1. Introduction

Optical beams with orbital angular momentum OAM, have a spatial phase structure $e^{jm\theta}$ where m is the topological charge TC, which can take integer values [1]. These beams have great potential for generating and improving existing engineering applications such as optical communications [2], optical tweezers and particle manipulation [3], encoding and multiplexing [4]; since no need to interrupt the light beam for the transfer of data, provides immunity to electromagnetic interference, increases bandwidth, the rate of transmission and the ability of information transmitted by the channel, making the communication process more efficient [5]. Because of these properties, it is necessary to deepen the study of the TC detection. The current used techniques are interferometry [6] and digital image processing [7]; however, it can be found in a simple way from the associated pattern when the AOM beam is diffracted by a triangular opening [8]. Numerical and experimental far-field diffraction pattern by a border for OAM beams with integer topological charge, are analyzed by a triangular equilateral aperture. Results are presented below and we show that thanks to use of those patterns, we can discriminate the spatial distribution information on the front-wave of OAM beams.

2. Results and discussion

For a Bessel-Gauss beam with topological charge m, we use the description of the propagating electrical field amplitude denoted by [9]:

$$U(\rho,\theta,z_0) = c_0 e^{-\frac{\rho^2}{2w^2(z_0)}} e^{im\theta} \rho \left[I_{\left(\frac{m-1}{2}\right)}\left(\frac{\rho^2}{2w^2(z_0)}\right) - I_{\left(\frac{m+1}{2}\right)}\left(\frac{\rho^2}{2w^2(z_0)}\right) \right] \quad (1)$$

Where $c_0 = t\frac{\sqrt{\pi}}{2w_0} e^{i\frac{z}{z_r}} e^{i\frac{3\pi}{2}m} e^{ikz_0}$ is a complex number, $w(z_0)$ the radius of the beam on the opening plane, $R(z) = z\left[1 + \left({}^{z_r}/_{z}\right)^2\right]$ the wavefront radius, $z_r = \frac{kw_0^2}{2}$ the Rayleigh range, w_0 the beam-waist, k the wave number of the monochromatic beam and, $I\big((m \pm 1)/2\big)$ the modified Bessel function. The factor $e^{im\theta}$ expresses OAM, so that the value of m indicates the number of wavefront rotations around the propagation axis, is the topological charge (See Figure 1).

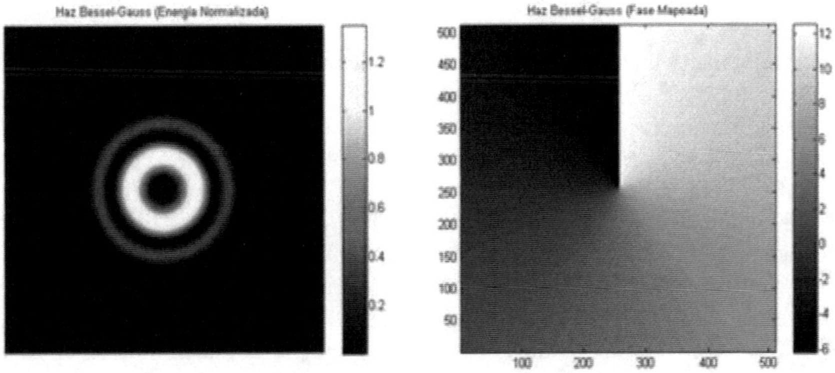

Figure 1. Magnitude and phase of a Bessel-Gaussian beam with topological charge *m*=3.

This equation describes the field distribution at $z = z_0$, where is located the nanometric thick edge, centered on the optical axis of the system. After, an equilateral triangular aperture, AT, is used. This equilateral triangular aperture provides a simple and easy method for determining the magnitude and sign of the topological charge of the optical vortex, by counting the number of light spots along one side of the diffraction pattern [6, 7]. When an OAM beam is diffracted by a nanometric thick edge and after far field analyzed through AT, a characteristic pattern of lines that offers an alternative way to find the magnitude and sign of TC occurs (see Figure 2). Experimentally, we used nanometric thin films with transmittance very close to zero (TiO_2@ 532nm) and thickness comparable to the wavelength.

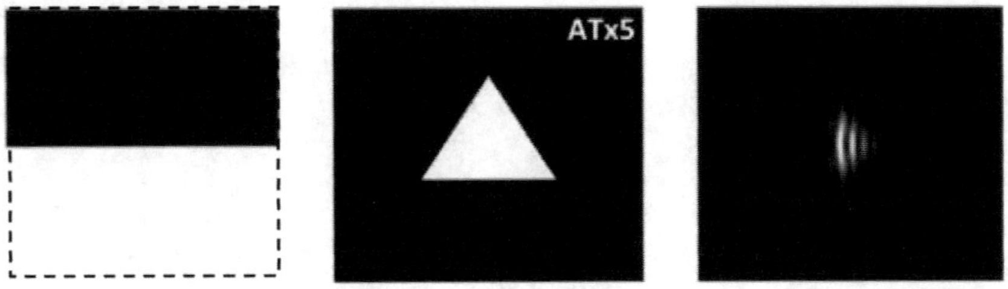

(a) **(b)** **(c)**

Figure 2. Diffraction pattern of an OAM beam with *m*=3, after analysis by a triangular equilateral aperture AT.

For openings, generally the far field diffraction is mainly due to the interference of the diffracted beam by each edge, where each one has finite length with infinitely steepening. By combining the edge with the equilateral triangular aperture (see Figure 4), used here for analysis, we obtain the results shown in Figure 3. Linking already intensity distributions from the Equation (2).

$$I_{Bordes\,ABC} = In_A + In_B + In_C - In_{AB} - In_{BC} - In_{CA} \qquad (2)$$

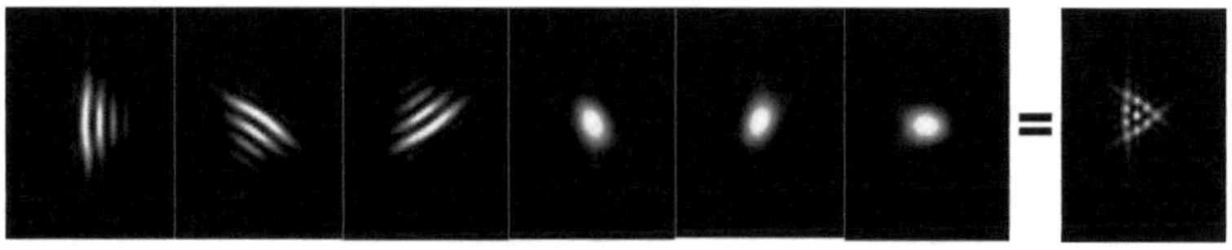

Figure 3. $I_{Bordes\,ABC}$ for an OAM beam with $m=+3$ (Numerical simulation results).

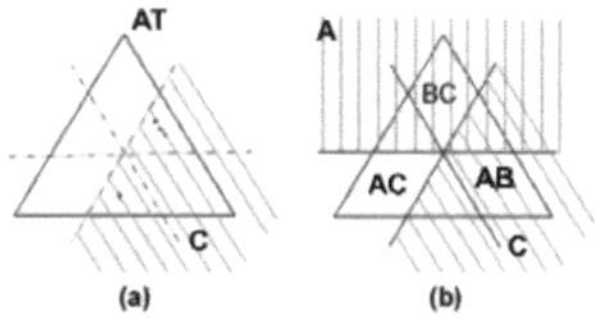

Figure 4. Edge-interaction geometry for a Triangular Aperture.

Where In_i is the intensity diffraction pattern due to the opening i (The addition is not the Born rule, it's just a simple way to determine the topological charge).

Suppose the edge thickness d, with nanoscale dimensions higher than wavelength ($d \gg \lambda$), and transmittance $T_i = T_F \cdot T_S$ (See Figure 5). The film transmission, in a fixed position relative to the axis of the donut modifies its far field diffraction pattern.

A simple model which describes through their transmittance parameters, like thickness and refractive index, the influence produced by the nanometer thin film edge over a vortex, is proposed. As shown in Figure 5, the geometry of the proposed model, allows to introduce a complex term in mathematical Equation 1. The refractive index of a material (n_i), generally is associated with a propagation delay of the electromagnetic wave, we relates directly to a phase term. The phase must keep a proportionality between the total revolution (2πrad) and wavelength (λ) according to the thickness (d) of the film. From the standpoint of strength, it is considered a term of proportionality (T_i). When scanning transmittance with the model of Figure 5, is obtained the results of Figure 6.

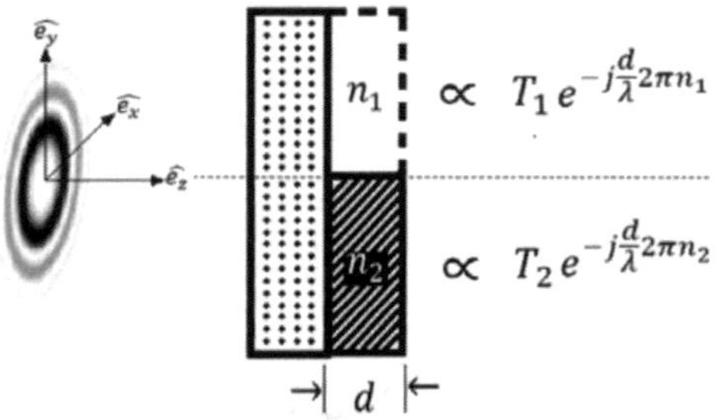

Figure 5. Geometry for interaction between a OAM diffracted beam by a straight edge of comparable thickness to the incident wavelength.

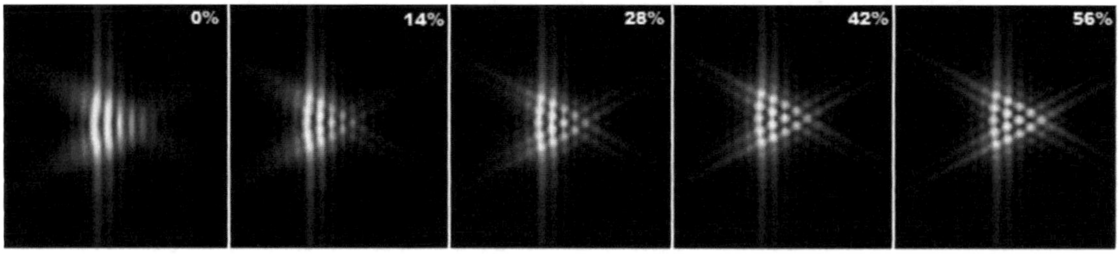

Figure 6. Far-field diffraction of an OAM beam with TC positive and integer, $m=+3$, simulated for different values of transmittance of the edge, seen through the triangular equilateral aperture.

Numerically, diffracted beams are calculated with TC for m=±1, ±5, diffraction patterns obtained are shown in Figure 7. The experimental implementation was performed with the setup shown schematically in Figure 8. Patterns obtained for far-field diffraction as shown in Figure 9.

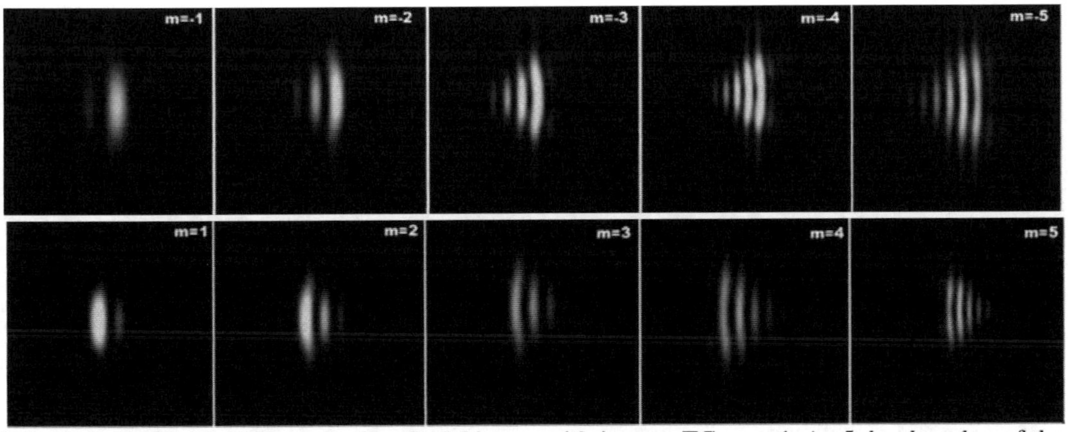

Figure 7. Far-field diffraction of an OAM beam with integer TC, m=±1 →±5, by the edge of the thin film, seen through AT (Numerical simulation results).

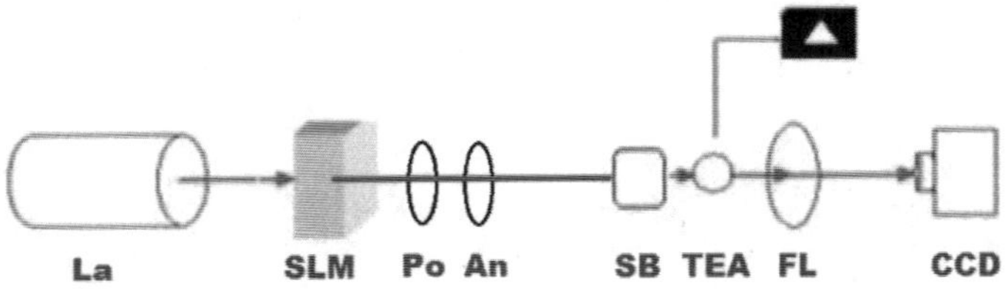

Figure 8. Schematic experimental setup: Laser (La), spatial light modulator (SLM) Polarizer (Po), Analyzer (An), Straight Borde (SB), Triangular Equilateral Aperture (TEA), Fourier lens (FL) and CCD camera.

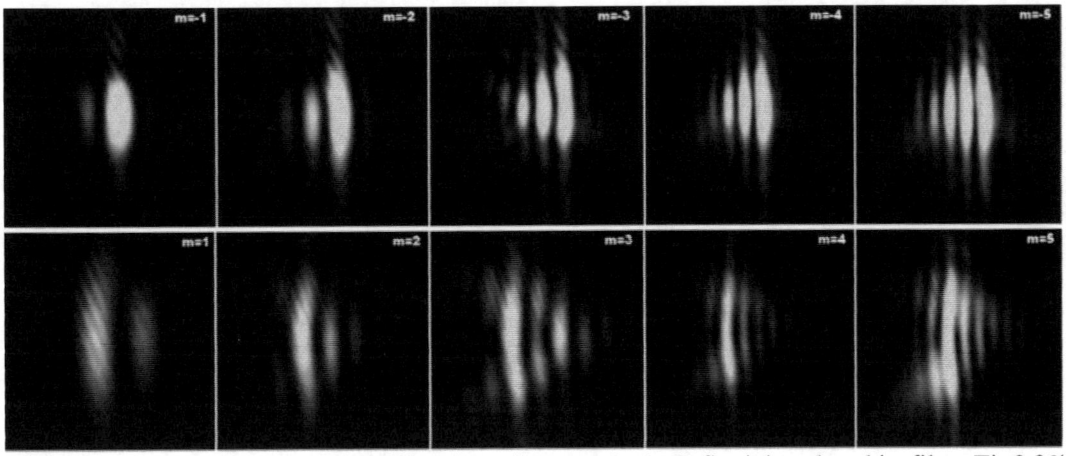

Figure 9. Experimental far-field diffraction, for $m=\pm 1$ →± 5. Straight edge thin film: Ti=0.2%, e≈660nm.

3. Conclusions

The Fraunhofer diffraction patterns obtained by simulation and experimentally for light beams with integer orbital angular momentum, by the edge of different thin film thickness and transmittance are reported for first time, to our knowledge. The proposed mathematical model is in accord with numerical simulations and experimental results. The results obtained, experimentally and numerically, are in agreement. The proposed method is useful especially for high topological charges, bigger than m=20.

From far-field diffraction pattern of an OAM beam by the edge of a thin film and seen through an equilateral triangular aperture, we can discern the value and sign of its topological charge, by a quick inspection. It is also possible to infer the value and sign of the topological charge, adding the diffraction patterns of rotated borders, individually and by couples, without using the Born rule.

We propose a new technique for determining the transmittance in thin films, using a beam with very high topological charge and analysis of their far-field diffraction pattern.

4. Acknowledgements

To Universidad Industrial de Santander UIS, by support this research from Research and Services Vice-presidency VIE, with funding projects 5708 and 1788 both, of the institutional program for consolidate research groups, years 2013 and 2014 respectively. To Colciencias, by support this research with the project 110256934957, "Optics devices for quantum key distribution, high dimensionality systems based in orbital angular momentum of light", funds obtained from the National call for the Bank of Projects in Science, Technology and Innovation 2012.

References

[1] Allen L, Beijersbergen M, Spreeuw R, Woerdman J 1992 *Physical Review* A **45-11** 8185
[2] Díaz C F 2013 *Revista Colombiana de Física* **45-1** 733
[3] Grier D and Curtis E 2003 *Physics review letters* **90** 133901
[4] Marinelli P et al 2011 *Electronics Letters* **47** 17
[5] Wang J et al 2012 *Nature Photonics* **6** 488
[6] Acevedo C H, Díaz C F and Torres Y 2014 *Ingeniería y Desarrollo* **32-2** 161
[7] Díaz C F, Acevedo C H, Torres Y and Barrero J 2013. *Revista UIS Ingenierías*, **11-1** 3193
[8] Hickmann J M, Fonseca E J S, Soares W C and Chávez-Cerda S 2010 *Phys Rev Lett* **105-5** #053904
[9] Acevedo C H Diaz C F Torres Moreno Y 2015 *Dyna* **82-190** 214

Gas-phase supersaturation effects on morphology properties of ZnO nano and microstructures grown by PVT

D N Montenegro[1], M C Martínez Tomas[2], V Muñoz Sanjosé[2] and V Sallet[3]
[1] Universidad del Atlántico, Barranquilla, Colombia.
[2] Universitat de València, Valencia, España.
[3] Université de Versailles Saint-Quentin-en-Yvelines, Versailles, France.

E-mail: dianamontenegro@uniatlantico.edu.co

Abstract. A systematic study of the morphology evolution of ZnO nanostructures grown by physical vapour transport was carried out. The evolution of the shape with the growth time is shown to depend on the different gas-phase supersaturation and temperature conditions encountered in the crystallization zone of the tube furnace. The observed morphology transitions are discussed, and a growth model for ZnO nanostructures is given.

1. Introduction

The synthesis of ZnO at nanoscale has paid a renewed interest during the last decade due to the potential applications of this material. Indeed, ZnO nanostructures are viewed as potential candidates for application in biomedical nanodevices thanks to the non-toxic and biocompatible character of this semiconductor [1,2]. In addition, ZnO nanostructures can be applied in numerous technology devices due to the unique combination of physical properties exhibited by this material [1,2]. Nanostructured ZnO has been achieved to grow by using several methods involving in most cases the transport of precursors in the vapour phase [3,4] and wet chemistry processes [5,6]. Despite the efforts made during the last years, the growth mechanisms of ZnO nanostructures and the control of their physical properties are no totally understood. Low-cost growth methods have attracted intensive interest due to the requirements of the crystal growth technology. Physical vapour transport (PVT) has shown to be well-adapted for the growth of nanostructures with good quality and physical properties [4]. Unlike other vapour phase methods with industrial character, PVT does not offer an accurate and direct control of the richness of gas-phase precursors during the growth process. This aspect makes difficult the control of nanostructure properties. Considering that studies on PVT growth mechanisms are still needed, we have investigated in this work the role of gas-phase supersaturation. We have also emphasized the influence of growth time on the behaviour of the gas-phase supersaturation along the crystallization region and their effect on morphological characteristics of ZnO nanostructures.

2. Experimental details

Catalyst-free ZnO nano and microstructures were grown on c-sapphire substrates by physical vapour transport method under atmospheric pressure. The experiments were carried out into a horizontal tube furnace with two temperature control zones. An argon flow (100l/min) and a mixture of ZnO and graphite powder [100:1] were used as carrier gas and material source, respectively. The temperature gradient was obtained by setting the temperature of the evaporation zone (material source) at 1015^oC while the temperature of the crystallization zone (substrates) was set at 400^oC. The morphology of

ZnO nanorods was assessed by using scanning electron microscopy (SEM Jeol 840) and the structural properties of the samples were determined by powder X-ray diffraction (BrukerD8Advance).

3. Results and discussion

Several sets of samples were grown by employing different growth times (5, 15, 30 and 60 minutes) and using the same temperature gradient as illustrated in Figure 1(a). The analysis of each set of samples evidenced substantial changes such as the grown material quantity on the substrates and the nanostructures morphology, depending on the position in the crystallization zone. The Figure 1(b) shows the typical behaviour of the grown material quantity of our sets of samples as a function of the distance along the crystallization zone, measured from the material source. The corresponding graph exhibits an asymmetric bell shape with a maximum centred in the zone corresponding to 420-450*°C*.

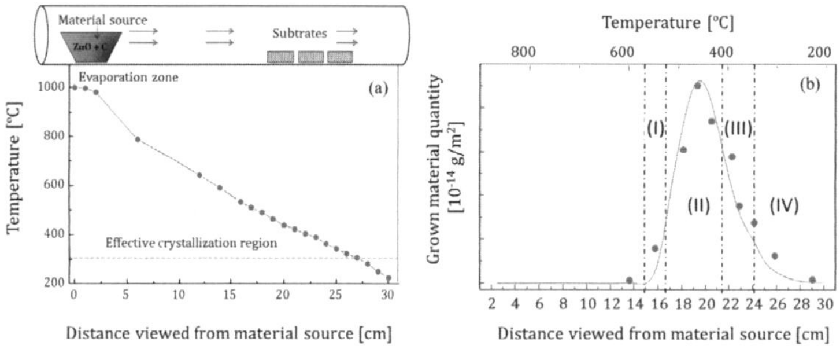

Figure 1. (a) Temperature gradient (temperatures of controllers: 1015*°C* (evaporation zone) and 400*°C* (crystallization region)). (b) Estimation of grown material quantity along crystallization region.

Crystal growth fundamentals emphasize the primordial importance of parameters such as supersaturation of precursors and surface energies of crystal facets because they tend to govern the morphological development of crystals [7,8]. Regarding nanostructures it was previously reported that the final morphology is governed by the competition between the processes of adsorption and incorporation of atoms on the different crystal surfaces under a particular supersaturation conditions [9]. It is commonly admitted that the anisotropic growth is favourable under low supersaturation conditions [10] while some studies carried out on whiskers growth have shown that rich supersaturation conditions favour the lateral growth of the structures [11]. Indeed Ye et al. [9] have particularly studied the behaviour of gas-phase supersaturation of precursors along the crystallization region during a growth process using a CVD system. These authors showed that gas-phase supersaturation exhibits a decreasing behaviour along the crystallization region when the growth is made without carrier-gas flow, while it exhibits a curving behaviour with an asymmetric bell shape when a carrier-gas flow is used. According to our results, the behaviour of the grown material quantity on our samples seems to be well-correlated with the above described behaviour of gas-phase supersaturation using a carrier-gas flow. Therefore, and similarly to the procedure done by these authors [9], we have divided the crystallization region into 4 different zones regarding their gas-supersaturation level associated to a grown material quantity: zone I, low supersaturation (480-500°C); zone II, high supersaturation (420-480*°C*); zone III, intermediate gas-phase supersaturation (380-400°C) and zone IV, low gas-phase supersaturation (300-370*°C*), as shown in Figure 1(b).

The typical morphology characteristics of the samples grown into zones II, III and IV, with different growth times (5, 15, 30 and 60 minutes) are shown in Figure 2. Samples grown into zone I (polyhedrons structures) are not shown here due to the very low material quantity found on these samples. The SEM images reveal three principal features: (i) nanopolyhedrons with well-defined hexagonal facets are synthetized into all crystallization zones during the early growth stages (5 minutes); (ii) depending on the crystallization zone, these nanopolyhedrons exhibit different morphological evolution as the growth time is increased: zone II, nanopolyhedrons evolve to a rough layer composed by coalesced polyhedrons (Figures 2(c), (f), (i), (l)); zone III, nanopolyhedrons to isolated micropolyhedrons (figures 2(b), (e), (h), (k)) and zone IV, nanopolyhedrons to needles (Figures 2(a), (d), (g), (j)); (iii) the number of structures/area seems to be constant as the growth time is increased. The last observation suggests that, under our conditions, the nucleation process is not continuous during the growth, i.e. once a number of seeds have been nucleated during the early growth stage, the subsequent atoms will tend to be incorporated into different crystal planes of the nucleated seeds instead of creating new ones on the substrate surface. The morphology evolution will depend on the temperature and supersaturation conditions. In our case, the morphological transitions agree with what is usually admitted, i.e. isolated polyhedrons with well-defined hexagonal crystal facets are obtained under intermediate gas-phase supersaturation conditions, while under low gas-phase supersaturation conditions our structures develop into needles (IV zone).

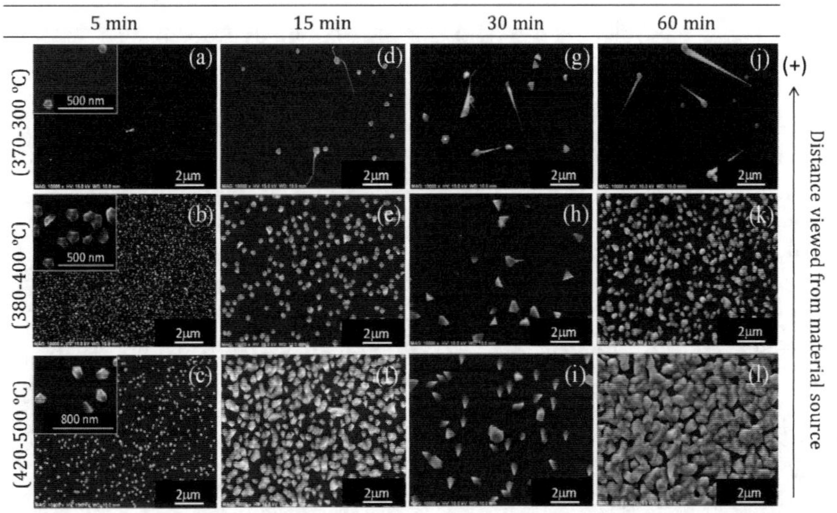

Figure 2. SEM images of the samples grown into the three zones, II (420-500 C), III (380-400°C) and IV (300-370°C) with different growth times: (a-c) 5*min*, (d-f) 15 min, (g-i) 30*min* and (j-l) 60*min*.

Otherwise, SEM images of zone II samples show that polyhedrons exhibit a faster lateral growth as time growth is increased. As it was mentioned, previous studies have shown that high supersaturation conditions favour the lateral growth of the structures [11]. This aspect can explain the morphology transition exhibited by our samples in this zone, i.e. polyhedrons to a rough layer. Under high gas-phase supersaturation, the lateral growth ("m" and "a" planes) of our ZnO polyhedrons is enhanced, thus favouring the coalescence of the structures and the subsequent development of a rough layer. Moreover, higher magnification SEM images of needles structures (zone IV) show that they tend to develop from some pyramidal or basal plane of a polyhedron structure previously synthetized, Figure 3(a). Into the BCF (Burton, Cabrera and Frank) theory [7] and the studies developed by Sears [8], a screw dislocation is proposed to be at the origin of the whisker growth. However, reported studies have associated the anisotropic growth of ZnO nanostructures to another kind of crystal defect, such as stacking faults [12] or twin planes [13]. Recently, Perillat et al. have focused on the growth of ZnO nanorods grow by MOCVD [14]. These authors suggest an inversion domain boundary (IDB) to be at the origin of the nanorod growth. The IDBs could come from the segregation of aluminium impurities; nevertheless the occurrence of such a polarity inversion is still to be fully understood. In our study, the same kind of mechanism could explain the development of needles, emerging from the polyhedrons. According to our experimental results, a growth model which summarizes the morphology evolutions observed for our ZnO nanostructures, can be drawn when the growth time is increased, depending on the gas-phase supersaturation conditions in each crystallization zone, Figure 3.

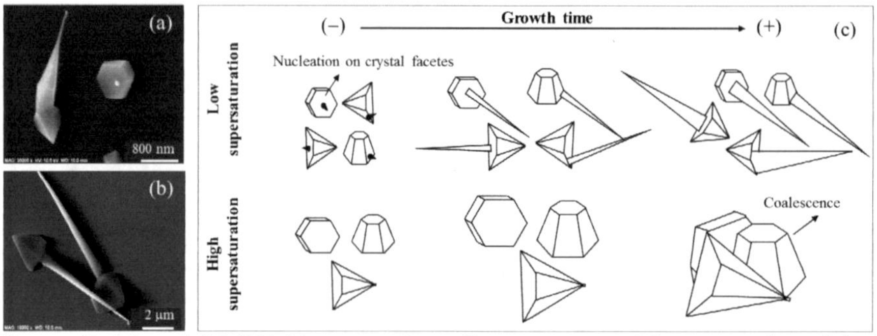

Figure 3. (a) and (b) SEM images of needles structures; (c) proposed model of the morphology development of ZnO structures, as the growth time increases under different gas-phase supersaturation of precursors.

4. Summary and conclusions

Several ZnO structures were grown at low temperature range by varying the growth time using a well selected value of the carrier gas flow and temperature gradient. The study carried out has allowed correlating the morphology transitions observed in our ZnO structures with the level of gas-phase supersaturation along the crystallization zone. The results show that: (i) under low supersaturation conditions, the anisotropic growth is more favourable. The needle shape structures tend to develop from the basal o pyramidal facets of a nanopolyhedron synthesized during the early growth stages; (ii) under high supersaturation conditions, rough layers tend to be formed, due to the coalescence of polyhedrons. This latter case can be explained by the enhanced lateral growth of the polyhedrons.

Acknowledgments

The authors acknowledge the SCSIE–University Valencia for providing the FESEM facilities and the financial support from the MINECO under project TEC2011-28076-C02-01/02, and the Generalitat Valenciana under projects Prometeo/2011-035 and ISIC/2012/008.

References

[1] Wang Z L 2009 *Materials Science and Engineering R* **64** 33
[2] Özgür Ü, Alivov Ya I, Liu C, Teke A, Reshchikov M A, Doğan S, Avrutin V, Cho S-J and Morkoç H 2005 *J Appl Phys* **98** 041301
[3] Wang L, Zhanga X, Zhao S, Zhou G, Zhou T Y and Qi J 2005 *Appl Phys Lett* **86** 024108
[4] Montenegro D N, Agouram S, Martínez-Tomás M C, Llorens C, Reig C and Muñoz-Sanjosé V 2010 *Physics Procedia* **8** 121
[5] Ling B, Wang Y, Sun X W, Dong Z L and Yang N X 2011 *IEEE J Selected Topics in Quantum Electronics* **17** 801
[6] Xu F, Lu Y, Xie Y and Liu Y 2010 *J Solid State Electrochem* **14** 63
[7] Burton W K, Cabrera N and Frank F C 1951 *Philosophical Transactions of the Royal Society of London Series A Mathematical and Physical Sciences* **243** 299
[8] Sears G W 1955 *Acta Metall* **3** 361
[9] Ye C, Fang X, Hao Y, Teng X and Zhang L 2005 *J Phys Chem B* **109** 19758
[10] Cao G Z and Wang Y 2011 *Nanostructures and nanomaterials synthesis, properties and applications 2nd ed* (Singapore: World scientific publishing)
[11] Ruth V and Hirth J P 1964 *J Chem Phys* **41** 3139
[12] Ming N B J 1992 *Synth Cryst* **21** 217
[13] Takeuchi S, Iwanaga H and Fujii M 1994 *Philos Mag A* **69** 1125
[14] Perillat-Merceroz G, Thierry R, Jouneau P H, Ferret P and Feuillet G 2012 *Nanotechnology* **23** 125702

Effects of post-growth thermal annealing on room temperature pulsed laser deposited ZnO thin films

D Padilla Rueda[1], J M Vadillo[2] and J J Laserna[2]
[1] Universidad del Atlántico, Barranquilla, Colombia.
[2] Universidad de Málaga, Málaga, España.

E-mail: dianapadilla@mail.uniatlantico.edu.co , laserna@uma.es

Abstract. ZnO thin films have been elaborated by Pulsed Laser Deposition (PLD) onto glass substrates, at room temperature. Post-growth annealing treatment was applied to the films in air background. Morphology, chemical composition and optical characteristics of ZnO films were evaluated as a function of annealing temperature. With the increase in annealing temperature, the 002 axis peak intensity is improved and optical band gap is shifted to lower values. The PL peak positions shift are associated at stoichiometry change, it is confirmed with the experimental results. The films elaborated with this experimental setup could be used to applications in short wavelength optoelectronic devices, chemical and optical sensors.

1. Introduction

In the last years, ZnO thin films have become more important because of their special properties compared with the bulk material. ZnO is a semiconductor with a direct wide band gap of 3.37eV at room temperature [1-5], with unique properties and wide range of applications [1-6]. Pulsed Laser Deposition allows tailoring the deposition conditions [3-5] to obtain films with good stoichiometry, multilayered samples and the possibility of using substrates with low thermal tolerance like polymers or glass [7,8]. In this paper, we present the influence of post-growth annealing developed in air atmosphere on the morphological, chemical and optical properties of PLD ZnO thin films prepared at room temperature over glass substrate.

2. Experimental setup

The PLD deposition system and as-growth film growth parameters have been described previously [7]. Substrates are hydrolytic class D263M1 glass slides (10mm in diameter, 0.1mm thick). As-growth films were treated with a post-growth thermal annealing process (100-600°C), for 1 hour, in a conventional furnace in air atmosphere over on flat surface; the samples were allowed to cool down to room temperature in air atmosphere. Photoluminescence data were acquired at room temperature using a Cary Eclipse fluorometer (λ_{exc}=310nm; λ_{emi}=330–750nm, spectral resolution 2nm). Before test measurements, films were exposed for 1h to dry air. UV-Vis absorption spectroscopy was performed on a Cary Eclipse 50 UV-Vis Spectrophotometer. SEM analysis was made with SEM microscope JEOL JSM-840. AFM data were recorded in a Veeco Nanoscope V in contact mode over an area of 2μm×2μm. Diffractograms were recorded with a Philips X'Pert PRO-MPD with a Cu $K_{\alpha 1}$ (1.5406A°) source. X-ray photon spectroscopy (XPS) was performed on a Physical Electronics PHI model 5700 equipped with Mg K_{α} radiation (1253.6eV) in combination with 4keV Ar+ gun sputtering.

Journal of Physics: Conference Series **687** (2016) 012028 doi:10.1088/1742-6596/687/1/012028

3. Results and discussion

As-growth films are homogeneous, transparent and colourless, smooth surface with clusters, the average roughness is 10nm. With annealing process is observed reduction of cluster on film surface; increment in grain size and roughness, in agree with other authors (Figure 1 and 2) [2-5]. Coalescence effects in film structure is not observed.

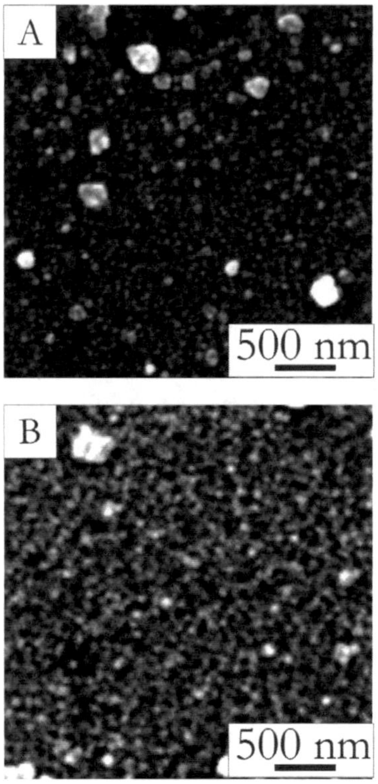

Figure 1. SEM images (A) As-growth (B) annealed film at 400°C.

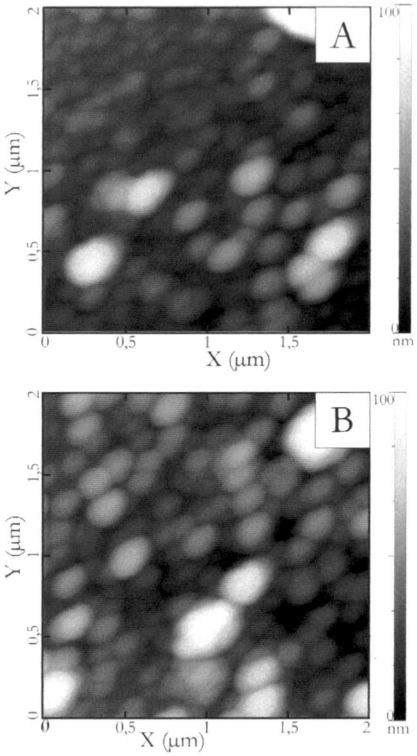

Figure 2. AFM images (A) as-growth (B) annealed film at 400°C.

As-growth films are polycrystalline; the peaks (100), (002) and (101) axes are detected; (002) peaks showed the highest intensity, at regardless of previous reports of amorphous films (Figure 3) [2]. The (002) peak shifting indicates an increase of interplanar spacing; there is a transformation from press stress to strain stress causing by the annealing as shown the calculated values (Table 1) [1,4,5,9]. The press stress may be derived from lattice mismatch between glass substrate and ZnO film from differences in thermal expansion coefficients [4,5,9] or owing to observed increasing crystalline grain size [1,9]; the strain stress is attributed at surface contraction owing to the elimination of defects (interstitial or vacancy) [4,5,9,10] or the elimination of chemisorbed O [3]. The increment in annealing temperature stimulates the diffusion of the atoms at more advantageous positions in the structure, it improve the film orientation in c-axis direction and reduce the crystalline defects [4-6,10].

XPS photoelectron peaks (O 1s, Zn 2p3/2 and Zn LMM) exhibited a shifting in peak position (spectra do not show), it has been associated at films defects as interstitial or vacancies of Zn and O [3,9,11]. Film stoichiometry is modified with the increment in annealing temperature. This indicates atomic redistribution and variations in film defects, it can be associated at changes in the film stress observed with XRD results.

Films present a good transmittance in the visible range, with an increment as function of annealing temperature (Figure 4). The band gap energy of ZnO films can be determined by Tauc's plot

($\ln(T)\propto\alpha$) (inset Figure 4) [1,4,5,7,9,10]. The calculated band gap values (Table 1) are similar to those already reported [4,5]. Some reports indicate that the optical properties are not modified by annealing process [1], although shifts in optical band gap energy is generally observed in the annealed direct-transition-type semiconductor films [1,6]. The effect of the annealing temperature in optical properties of films can be attributed at grain size variations [4-5], film stress alterations [1,4,5,9] or changes in the film stoichiometry [9,10]. In agree with our analysis; the incrementing in annealing temperature causes that press stress changes at tensile stress, indicating stoichiometry degradation, it explain the band gap shift observed as effect of O variations.

Table 1. Structural and optical parameters of ZnO films as function of annealing temperature.

Annealing temperature (°C)	002 Peak position (°)	FWHM (°)	d-spacing (Å)	Grain size (nm)	Stress (GPa)	Band gap (eV)
As-growth	34,24	0,409	2,61669	23	-1,19753	3,27
100	34,30	0,305	2,61211	32	-0,78761	3,31
200	34,35	0,246	2,60855	42	-0,46899	3,27
300	34,36	0,191	2,60780	59	-0,40186	3,26
400	34,29	0,225	2,61307	41	-0,87353	3,25
500	34,43	0,216	2,60308	50	0,02059	3,24
600	34,43	0,222	2,60309	48	0,01969	(a)

(a) Not evaluated

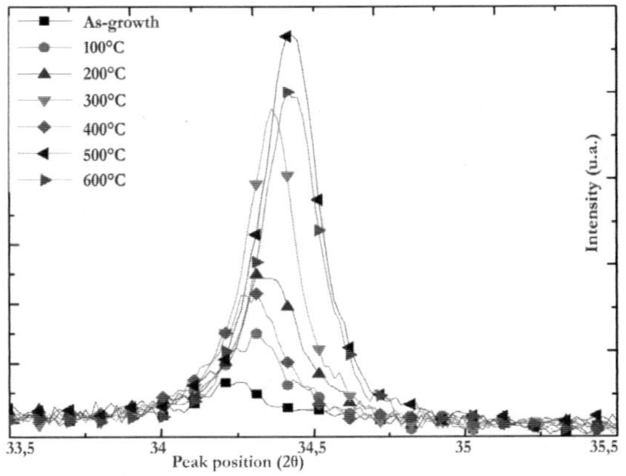

Figure 3. 002 XRD peak intensity and position as function of annealing temperature.

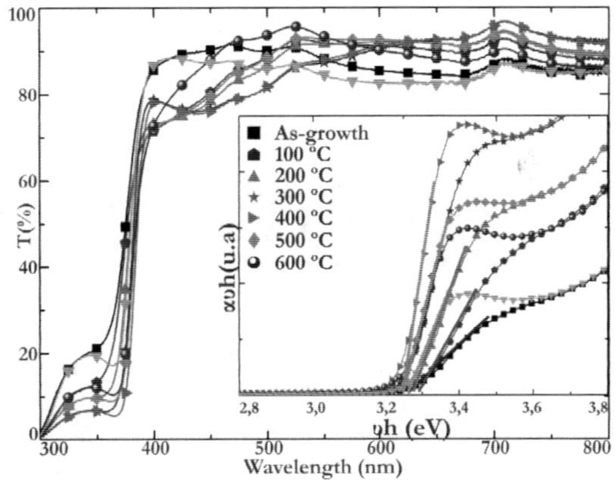

Figure 4. Spectra transmittance of ZnO films as function of annealing temperature. Inset: Tauc plot.

In Photoluminescence data, the peak position and intensity in PL spectra are function of the structure and stoichiometry: DLE band (500-650nm) is associated at defects like vacancies or interstitials atoms [3,5,6,11,12] and UV band is attributed to the near band edge emission (NBE), is originated from the recombination of free excitons [1-3,6,7,11]. Figure 5 shows peak position of UV and Visible PL bands as function of annealing temperature. The blue shift of DLE band as function of the rise in annealing temperature is associated at changes in crystalline defects ($O_i \rightarrow V_O, O_{Zn}$) [12]; the red shift of UV peak is related with the increase in grain size [6], changes in O vacancies (film stoichiometry) [2,3,7] or type of film strain [2]. In agree with the experimental data, the stoichiometry degradation and crystallographic changes causing the peak shift observed.

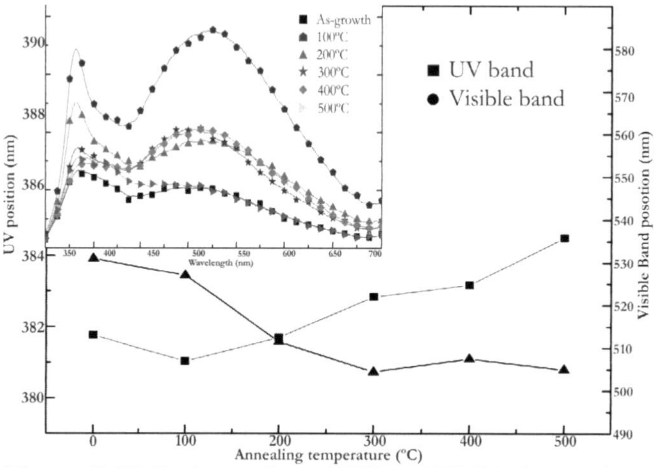

Figure 5. PL Peak position of UV and Visible band of ZnO films, as function of annealing temperature. Insert. PL spectra.

4. Conclusions

In the present work, we showed that is possible to elaborate ZnO thin films using the PLD technique with IR laser over glass substrates at room temperature. Analysis showed defects in as-growth ZnO thin films. The effect of post-growth annealing treatment on film properties has been studied. Film morphology is strongly modified by annealing, film thickness is reduced and roughness is incremented; film cracking can be observed at highest temperature, by differences in thermal expansion coefficients between glass substrate and ZnO film. With the annealing the crystalline properties are improved in c-axis; 002 peak exhibited reduction of FWHM and increment in the intensity and grain size; all this is associated at correction of defects in the structure. The film stoichiometry study confirmed the increment in O vacancies with the annealing in agree with the shifting in the band gap and PL peak position of the UV and visible band. The post-growth annealing modifies the film properties and allows the preparation of films with tailored properties.

References

[1] Mathew J P, Varghese G and Mathew J 2015 *IOP Conf Series Materials Science and Engine* **73** 012065
[2] Shan F K, Liu Z F, Liu G X, Lee W J, Lee G H, Kim I S, Shin B C and Yu Y S 2004 *J Electrocer* **13** 189
[3] Wei X Q, Man B Y, Liu M, Xue C S, Zhuang H Z and Yang C 2007 *Physica B* **388** 145
[4] Zhu B L, Sun X H, Zhao X Z, Su F H, Li G H, Wu X G, Wu J, Wu R and Liu J 2008 *Vacuum* **82** 495
[5] Zhu B L, Zhao X Z, Su F H, Li G H, Wu X G, Wu J and Wu R 2010 *Vacuum* **84** 1280
[6] Yadav A B, Periasamy C and Jit S 2015 *IOP Conf Series Materials Science and Engineering* **73** 012060
[7] Padilla-Rueda D, Vadillo J M and Laserna JJ 2012 *Applied Surface Science* **259** 806
[8] Han J, Gonga H, Yanga X, Qiua Z, Zia M, Qiua X, Wang H and Cao B 2015 *Applied Surface Science* **332** 549
[9] Chen T, Liu S Y, Xie Q, Detavernier C, Van Meirhaeghe R L and Qu X P 2010 *J Mater Sci Mater Electron* **21** 88
[10] Rusop M, Uma K, Soga T and Jimbo T 2006 *Mater Sci Eng B Adv Funct Solid State Mater* **127** 150
[11] Hsieh P T, Chen Y C, Kao K S and Wang C M 2008 *Appl Phys A* **90** 317
[12] Wang C, Chen Z, Hua H and Zhang D 2009 *Physica B* **404** 4075

Automated and inexpensive method to manufacture solid-state nanopores and micropores in robust silicon wafers

M Vega[1], P Granell[2], C Lasorsa[1], B Lerner[1] and M Perez[1]
[1] Universidad Tecnológica Nacional, Buenos Aires, Argentina.
[2] INTI-CMNB, San Martín, Buenos Aires, Argentina.

E-mail: mvega@fullgen.com

Abstract. In this work an easy, reproducible and inexpensive technique for the production of solid state nanopores and micropores using silicon wafer substrate is proposed. The technique is based on control of pore formation, by neutralization etchant (KOH) with a strong acid (HCl). Thus, a local neutralization is produced around the nanopore, which stops the silicon etching. The etching process was performed with 7M KOH at 80°C, where 1.23μm/min etching speed was obtained, similar to those published in literature. The control of the pore formation with the braking acid method was done using 12M HCl and different extreme conditions: i) at 25ºC, ii) at 80°C and iii) at 80°C applying an electric potential. In these studies, it was found that nanopores and micropores can be obtained automatically and at a low cost. Additionally, the process was optimized to obtain clean silicon wafers after the pore fabrication process. This method opens the possibility for an efficient scale-up from laboratory production.

1. Introduction

Nanopore-based technologies are emerging as a powerful tool for single molecule analysis. They can be used in different applications such as detection of ssDNA [1-3], dsDNA segment [4-6] and proteins [7,8]. Micropores have been used as well to detect cancer cells [9]. The production of solid state nanopores has aroused great interest because silicon wafers are inexpensive, resistant and durable. In particular, solid state nanopores included inside the wafer are easy to handle and show great resistance to washing (including sonication) when compared to nanopores made on silicon nitride/dioxide membranes, which are extremely fragile and must be handled carefully. Normally, the methods to produce solid-state nanopores involve the use of advanced microscopy tools to complete the manufacturing process, namely Transmission Electron Microscopy (TEM) [10], Scanning Transmission Electron Microscopy (STEM) or Focused Ion Beam (FIB) [11]. However, the most traditional methods for making nanopores in silicon wafers involve standard microfabrication tools such as photolithography for pattern definition and Reactive Ion Etching (RIE) to remove the silicon nitride layer on the exposed area. This work describes a method to manufactured pores up to 20nm diameter which relies on automatic braking pore formation, without the intervention of an external agent. Nanopores and micropores were manufactured using KOH etching and subsequently chemical braking was performed with HCl, in order to automatically control the pore opening. Chemical breaking is accomplished as the neutralizing acid HCl becomes in contact the basic etchant KOH through the nanopore.

2. Materials and methods

2.1. Pattern formation
A schematic picture of the pattern forming process is shown in Figure 1. The substrates used were 700μm thick double-side polished <100> oriented silicon wafers (Virginia Semiconductor In.). A silicon nitride layer 50nm thick was deposited on a wafer using Plasma Enhanced chemical vapour deposition (PECVD) at 600°C. The gas used for plasma deposition was hexamethyldisilazane $(CH_3)_3SiNHSi(CH_3)_3$ and nitrogen as reactive gas.

2.2. Chemical etching
The chemical anisotropic etching of silicon wafers was carried out using7M KOH at 80°C.First, the etching was performed simultaneously on both sides of the silicon wafer. The reaction was stopped before the inverted pyramids joined. Subsequently, the silicon wafer was removed from the container in which the first etching was performed and the wafer was washed with deionized water and introduced into a PDMS device, where a second etching step was performed (chemical braking). The pore size was adjusted by controlling the reaction temperature and / or applying an electric potential. Fifteen samples were studied for each condition. Chemical braking was performed at two temperatures: 25°C and 80°C. These two temperatures were chosen to study, according to previous studies [12] where the greater rate of nanopores formation was obtained. A potential of 12V was also applied to control the reaction rate. At this stage KOH 4M and HCl 12M was used, being a simple neutralization reaction, where the HCl is used in excess to ensure the OH^- neutralization.

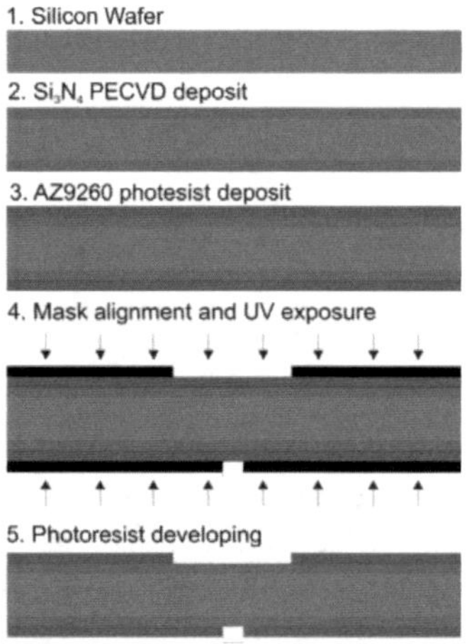

Figure 1. Steps for the manufacturing process of nanopores on a silicon wafer.

3. Results
Figure 2 shows the results of the etch depth versus time for the three studied processes. In the three cases, the joining of the two cavities produces an abrupt chemical braking of the etching process. However, only in cases of chemical braking at 25°C and 80°C with applied voltage it was possible to

obtain nanopores. Additionally, in all cases the braking reaction occurred when a window depth of approximately 510±2μm was achieved.

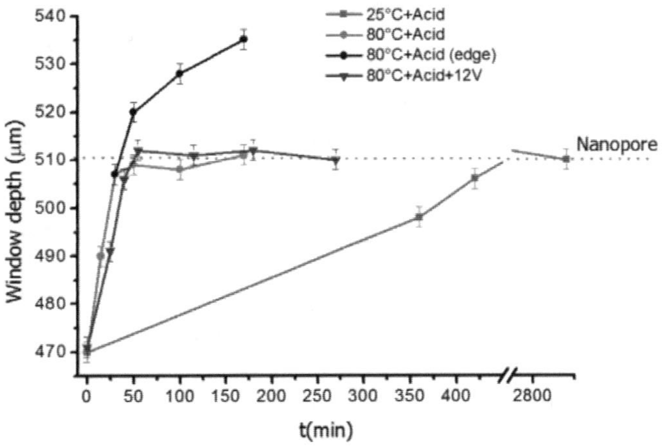

Figure 2. Comparison of the window depth of 2.1mm, according to the attack time for the three procedures studied.

Figure 3(a) and 3(b) shows a nanopore and micropore respectively, obtained by this technique. Thus, HCl protons have more time to traverse the newly formed nanopore and neutralize the KOH. So, the etching stops and allows nanopore formation. Micropores can be produced if the second etching step is allowed to continue further, because as the pore opening increases, the acid braking process cannot be sustained. The silicon nitride layer is degraded, exposing the underlying silicon. It was found that after about 3300 minutes etching micropores are formed, which will be larger with longer etching times.

Figure 3. (a) Nanopore SEM image obtained by etching combined with braking acid, 4M KOH at room temperature and (b) Micropore SEM obtained by etching with 4M KOH and 12M HCl, at room temperature after 4000 minutes.

The application of electric potential also has the advantage of acting as a cleaning process of the wafer surface. In cases in which the etching agent and the braking agent which can react to form salts crystallize or precipitate, applying a potential attracts ions to the electrode, leaving the substrate surface free of ions or crystals. This process is useful for industrial application, because it avoids the need for extensive subsequent washes. Figure 3 shows two wafers treated with the braking acid at 80°C temperature. In the wafer of Figure 4(a), a 12V potential was applied and in the wafer of Figure 4(b), potential was not applied. On the wafer on which the potential was applied, the surface was completely clean and free from reaction products. The other wafer (no applied voltage) shows a white precipitate on the surface, corresponding to KCl that formed upon neutralization.

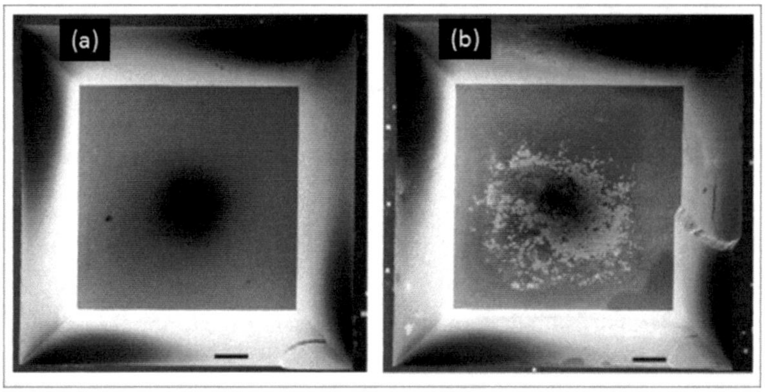

Figure 4. Windows formed in silicon wafers etched with acid braking method. (a) With electric potential. (b) No electrical potential. The scale bar in both images represents 200μm.

4. Conclusions

The results presented here for the manufacture of nanopores with chemical braking method have shown several advantages over other manufacturing methods. First, the method allows to stop the nanopore manufacturing process automatically, without any feedback system or the intervention of an external agent. It is also possible to obtain micropores, if the second etching process has longer exposure time. Over 200 minutes for 80°C experiments, 250min for 80°C with applied voltage experiments and finally 2800 minutes at room temperature experiments. Second, this process has a low cost compared to other manufacturing methods wich require the use of FIB and e-beam. For this reason this method can be very useful for the development of new technologies for biomolecules sensing using nanopores, providing an easily adaptable methodology to mass production processes at low cost. This paper introduces a new concept for the manufacture of nanopores and micropores, demonstrating that the proposed method works and defines a direction for future work. An extensive characterization of the method as well as the study of nanopore arrays production will be the subject of future publications.

Acknowledgments

The authors thank the financial support from CONICET-YPF (PIO-2014), ANPCyT (PICT2012-575/PICT2013- 0840), and FAN 2014, Argentina.

References

[1] Iqbal S M, Akin D, Bashir R 2007 *Nat Nanotechnol* **2** 243
[2] Chang H, Kosari F, Andreadakis G, Alam M A, Vasmatzis G, Bashir R 2004 *Nano Lett* **4** 1551
[3] Zwolak M, Di Ventra M 2005 *Transport Nano Lett* **5** 421
[4] Rodríguez-Manzo J, Puster M, Nicolaï A, Meunier V, Drndic M 2015 *DNA AcsNano* **9** 6555
[5] Kwok H, Briggs K, Tabard-Cossa V 2014 *PLoS One* **9** 1
[6] Yanagi I, Akahori R, Hatano T, Takeda K I 2014 *Scientific Reports* **4** 1
[7] Yu J S, Lim M C, Huynh D T N, Kim H J, Kim H M, Kim Y R, Kim K B 2015 *AcsNano* **9** 5289
[8] Yusko C, Johnson J M, Majd S, Prangkio P, Rollings R C, Li J, Yang J 2011 *Nature Nanotech* **6** 253
[9] Asghar W, Wan Y, Ilyas A, Bachoo R, Kim Yt, Iqbal S M 2012 *Lab Chip* **12** 2345
[10] Rollings R C, McNabb D S, Li J 2012 *Methods Mol Biol* **870** 79
[11] Kuan T, Golovchenko J A 2012 *Appl Phys Lett* **100** 213104
[12] Lerner B, Perez M, Toro C, Lasorsa C, Rinaldi C, Boselli A 2012 *Appl Surf Sci* **258** 2914

Effect of forming technique $Bi_xSi_yO_z$ coatings obtained by sol-gel and supported on 316L stainless steel

J Bautista Ruiz[1,2], J Olaya Flórez[1] and W Aperador[3]
[1] Universidad Nacional de Colombia, Bogotá, Colombia.
[2] Universidad Francisco de Paula Santander, San José de Cúcuta, Colombia.
[3] Universidad Militar Nueva Granada, Bogotá, Colombia.

E-mail: jorgebautista@ufps.edu.co

Abstract. $Bi_xSi_yO_z$ type coatings via sol-gel synthesized from bismuth nitrate pentahydrate, and tetraethyl orthosilicate as precursors; glacial acetic acid and 2-ethoxyethanol as solvents, and ethanolamine as complexing. The coatings were supported on AISI 316L stainless steel substrate through dip-coating and spin-coating techniques. The study showed that the spin-coating technique is efficient than dip-coating because it allows more dense and homogeneous films.

1. Introduction

Bismuth silicon $Bi_{12}SiO_{20}$ (BSO), belong to the sillenite family, with the general formula $Bi_{12}MO_{20}$ (M= Si, Ge, Ti, Pb, Mn, $B_{1/2}$, $P_{1/2}$). BSO is a stoichiometric sillenite with a fully occupied oxygen sublattice that meets all the major criteria for application as a low sintering temperature (Ts=850°C) [1].

Sol-gel synthesis is a very promising, since it offers potential advantages over traditional solid-state synthesis methods. The sol-gel method allows a precise control over the composition, the homogeneity and deposition over a large area. The most important advantage of sol-gel synthesis in comparison with other coating methods is the ability to tailor the microstructure of the deposited film by varying the concentration, the pH and the viscosity of the sol. [1art]. The key aspect of any sol-gel thin film process is the chemical stability of solution that directly determines properties of the films [2].

Spin coating has been shown to be an effective approach for the fabrication of inorganic films or membranes with controlled structure and crystal orientation [3]. Spin coating is a well-established technology which in the past has been used in the manufacture of photoresists and oxide coatings for screens [4,5]. There is a substantial scientific literature on the spin-coating process and useful reviews on the subject are given by [6,7].

The dip-coating process was first commercially used to produce thin films in a sol–gel technology in 1939 [8]. It has also been used to produce thin films in other technologies, such as photoresists films [9] and lubricant layers for magnetic hard disks [10]. Dip coating is a simple process for depositing a thin film of solution onto a plate, cylinder, or irregular-shaped object. The fact that the geometry of the substrates can vary widely is a distinguishing feature of the dip-coating technique. The process involves immersing a substrate into a reservoir of solution for some time thereby ensuring that the substrate is completely wetted, and then withdrawing the substrate from the solution bath. The liquid film formation is achieved by two main mechanisms, i.e. gravity draining of liquid solution and evaporation of solvent. An early dip-coating analysis was presented by Landau and Levich [11]. This model is a 1-D model derived purely from hydrodynamics of a Newtonian fluid flow, ignoring solvent evaporation [12].

2. Experimental

For the deposition of the BSO thin films, the sols were prepared using bismuth nitrate $Bi(NO_3)_3 \cdot 5H_2O$ (Alfa Aeser, 98%) and $Si(OC_2H_5)_4$ TEOS (Aldrich, 98%) as the precursors, 2-ethoxyethanol (Aldrich, 99%) and glacial acetic acid (Aldrich, 99%) as the solvents, and ethanolamine (Aldrich, 99%) as the complex agent [1]. First, the bismuth nitrate was drying at a temperature of 65°C/96h. Next, bismuth nitrate (10g) was dissolved in the acetic acid (25ml) under constant stirring for 3h at room temperature. Second, the TEOS (0.38ml) was then diluted with 2-ethoxyethanol to the concentration from 0.9M under constant stirring for 0.5h at room temperature. Finally, was mixing together solutions forming the sol. Ethanolamine was added to control pH=4. Sol with a concentration from 0.9M, was deposited onto 316L stainless steel substrate using the spin-coating method at 2500 and 4000rpm for 10s and dip-coating method at 13.2cm/min.

Films sintering were carried out at heating rate of 1°C/min. permitting removal in a controlled manner, the organic component of the films. This process seeks to avoid pore formation and the difference in thermal expansion present in the metal-coating interface, does not favour the formation of cracks in the coatings. Then the heating program employed for this study relates: Initial temperature of 20°C to 300°C and stabilized at this temperature for one hour. Heating was restarts to 400°C and equilibrated for half an hour.

The microstructure development was investigated with Scaning Electron Microscopy (SEM) (Tescan vega3 Bruker probang EDS with 15keV power) and Confocal Laser Microscopy (CLM) (LSM 700 Carl Zeiss).

3. Results and discussion

Figures 1 and 2 shows the BSO coatings obtained by dip-coating. Very porous, cracked and low adhesion coatings are observed. According to these results, the dip-coating technique is not recommended for films BSO system. This behaviour is most likely due to high viscosity of the sol. When removing the stainless steel substrate of the sol, abrupt dehydration of solvents that make it up is not presented, this makes the film drains and does not adhere to the substrate. After the sintering process, the coatings obtained are very thick, because the difference between the thermal expansion coefficients of the film and the substrate of poor quality coatings are obtained.

Figure 1. BSO films by dip–coating technique.

Figure 2. 2M2 film detail by dip-coating technique.

The spin-coating technique allows obtaining dense coatings with good adhesion. The results obtained with the spin-coating technique are shown in Figure 3 and Figure 4. Figure 3 shows the

Journal of Physics: Conference Series **687** (2016) 012030 doi:10.1088/1742-6596/687/1/012030

results of SEM for coatings BSO. Figure labelled 2M2, corresponding to a monolayer coating obtained at a speed of 2500rpm and corresponds to a monolayer 2M3 obtained speed of 4000rpm. Figures 3 and 4 show the EDS analysis which verifies the existence of the film BSO. In Figure 3(a) coating failure detailed with the appearance of pores.

In Figure 3(b) shows the coating 2M3, corresponds to a monolayer obtained at a speed of 4000rpm by spin coating. A denser coating without pores and trails that are characteristic of the forming process by spin-coating at high speed (4000rpm) is observed.

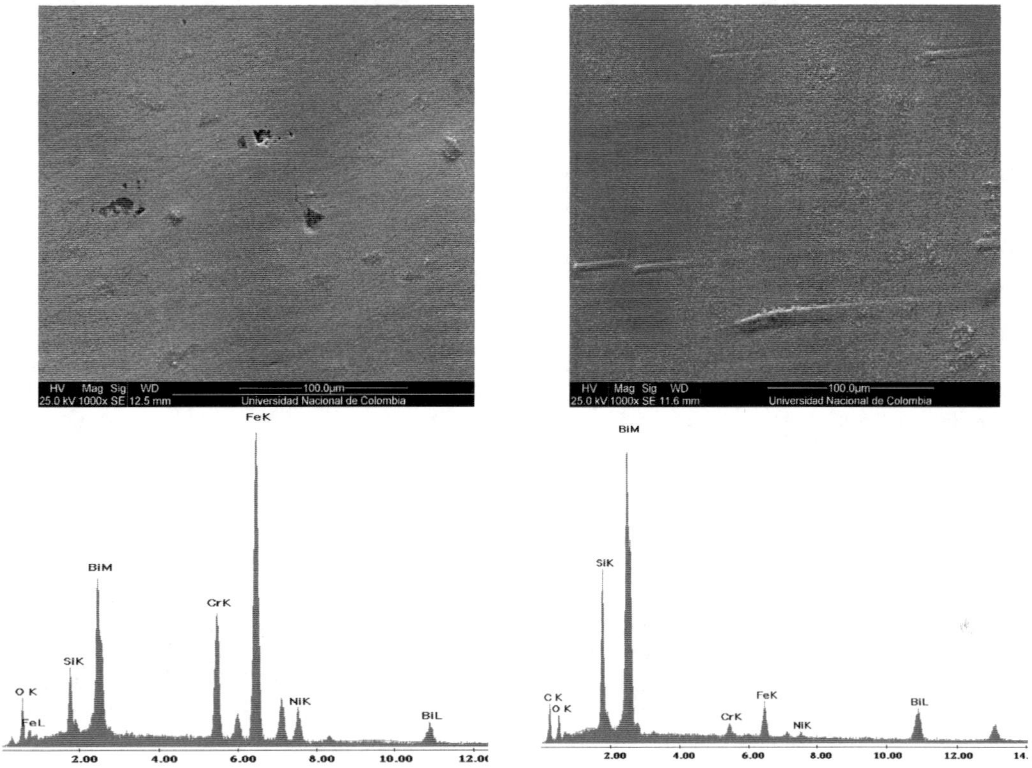

(a) (b)

Figure 3. Representative SEM photomicrograf and EDS analysis of samples. (a) 2M2 film and (b) 2M3 film.

In Figure 4, the micrographs for determining the roughness values shown. Regarding the roughness values of the films 2M2 and 2M3 0.029 m difference. The films are rougher than the substrate, the 2M3 samples are a little rough due to the formation of contrails in the coating and that emerges from the SEM micrograph of Figure 3(b).

Table 1 registers the roughness values obtained for the substrate and the films obtained by spin-coating.

Table 1. Rugosity Values.

Sample	Rugosity (m)
Substrate	0.072
2M2	0.228
2M3	0.257

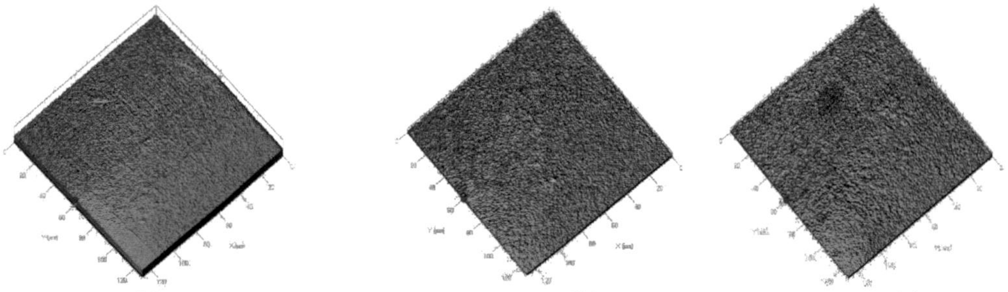

(a) (b) (c)

Figure 4. Representative CLM photomicrograf analysis of samples. (a) substrate, (b) 2M2 film and (c) 2M3 film.

4. Conclusions

A sol–gel dip/spin coating method was proposed to prepare BSO films, and this was demonstrated for fabricating films. According to SEM the best results were obtained coatings speeds 4000rpm. According to these results, the dip-coating technique is not recommended for films BSO system

Acknowledgement

Thanks the financial support from the Universidad Francisco de Paula Santander FINU, contract number FINU 014-2013. W Aperador, thanks the financial support from the Universidad Militar Nueva Granada, contract number ING-1775-2015.

References

[1] Veber A, Kunej S, Korosec R C and Suvorov D 2010 *Journal of the European Ceramic Society* **30** 2475
[2] Haoshuang G, Chao D, Peizhi C, Dinghua B, Anxiang K and Xingjiao L 1998 *Journal of Crystal Growth* **186** 403
[3] Zhang F, Sun M, Xu S, Zhao L and Zhang B 2008 *Chemical Engineering Journal* **141** 362
[4] Emslie A G, Bonner F T and Peck L G 1958 *Journal of Applied Physics* **29** 858
[5] Hall B, Underhill P and Torkelson 1998 *polymer Engineering and Science* **38** 2039
[6] Scriven L E 1988 *Materials Research Society Symposium Proceedings* **121** 717
[7] Larson R G and Rehg T J 1997 *Liquid film coating scientific principles and their technological implications* (London: Chapman & Hall)
[8] Brinker C L and Hurd A J 1994 *Journal de Physique III* **4** 1231
[9] Gibson M, Frejlich J and Machorro R 1985 *Thin Solid Films* **128** 167
[10] Gao C, Lee Y C, Chao J and Russak M 1995 *IEEE Transactions on Magentics* **31** 2982
[11] Landau L and Levich B 1942 *Acta Physicochimica U.R.S.S* **17** 42
[12] Yimsiri P and Mackley M R 2006 *Chemical Engineering Science* **61** 3496

Evaluation of micro-abrasion-corrosion on SiO_2-TiO_2-ZrO_2 coatings synthesized by the sol-gel method

J Bautista Ruiz [1], W Aperador[2] and J Caballero Gómez[2]
[1] Universidad Francisco de Paula Santander, San José de Cúcuta, Colombia.
[2] Universidad Militar Nueva Granada, Bogotá, Colombia

E-mail: jorgebautista@ufps.edu.co

Abstract. The medical science and the engineering, work to improve the materials used in the manufacture of joint implants, since they have a direct impact on the quality of people life. The surgical interventions are increasing worldwide with a high probability of a second or even a third intervention. Around these circumstances, it was evaluated the behaviour against micro-abrasion-corrosion phenomena on SiO_2 TiO_2 ZrO_2 coatings, synthesized by the sol-gel method with concentration of the Si/Ti/Zr precursors: 10/70/20 and 10/20/70. The coatings were deposited on AISI 316 LVM stainless steel substrates. The morphological characterization of the wear was made by AFM techniques. It was observed that the coatings with higher levels of titanium have a good response to the phenomena of microabrasion-corrosion.

1. Introduction

The main characteristic of stainless steel is its high resistance to corrosion, but when getting in contact with Cl– and Br– they undergo of pitting. The sol-gel is one of the methods of synthesis of materials that presents great interest at the present time; by means of this method it is possible to obtain ceramic coatings that act as an anticorrosive barrier. Research projects carried out on this topic have indicated that sol-gel method has allowed applying binary systems (SiO_2-TiO_2 and SiO_2-ZrO_2) on stainless steel substrates AISI 304 and has allowed to diminish the attack of corrosive agents [1].

The sol-gel method consists of three main parts: a) the preparation of the sol, b) gelation of the same, and c) removing the solvents [2-5]. This technology has been widely applied in different fields of materials science, [6-11].

In the present study, system SiO_2-TiO_2-ZrO_2 was synthesized with the sol-gel method at room temperature. These stable soles were formed in concentrations Si/Ti/Zr: 10/70/20 and Si/Ti/Zr: 10/20/70 using as precursors tetraethyl orthosilicate $Si(OC_2H_5)_4$, titanium tetrabutoxide $Ti(OBu)_4$ and zirconium tetrabutoxide $Zr(OC_3H_7)_4$. The morphological characterization of the wear was made by Atomic Force Microscopy (AFM) technique. It was observed that the coatings with higher levels of titanium have a good response to the phenomena of microabrasion-corrosion.

2. Experimental procedure

The process developed for the formation of the sols was as follows [1,12]. The substrates used for depositing the multi-component system, were steel sheets AISI 316LVM with dimensions of 3.5cmx2.5cmx0.2cm, which were polished to metallographic brightness. The next step was to form films by dip-coating, at a velocity of 13.2cm/min. The films sintering were carried out at heating rate of

1°C/min. The initial temperature was 20°C to 300°C and stabilized at this temperature for one hour. Heating was restarted to 400°C and equilibrated for 30 minutes.

The micro-abrasion-corrosion tests were carried out using an equipment which has a sample holder attached to a lever arm, which rotates on its pivot that controls the load on the specimen upon contact with the ball by the movement of a dead weight and a counter weight. The sphere is fixed between two coaxial shafts supported on bearings, when one of them is driven by a DC motor with encoder to ensure the speed and number of revolutions of test (see Figure 1). In order to observe the micro-abrasive process was provided abrasive particles to the system by a peristaltic pump controlled manually, these particles were suspended in Ringer's solution with a magnetic stirrer. For the study of the corrosive phenomena it was adapted to the equipment a potentiostat that contains an electrochemical cell.

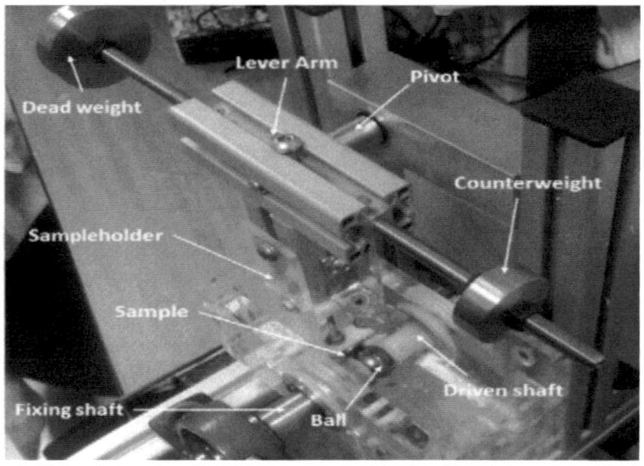

Figure 1. Micro-abrasion equipment.

Two types of tests were performed, the first tests were performed with only micro-abrasive wear, due to it was not used the electrochemical cell. Followed by micro-abrasion-corrosion tests where the wear tests were repeated with the same parameters but using the electrochemical cell, performing the electrochemical impedance spectroscopy (EIS) technique. The used parameters in tests are in the Table 1.

Table 1. Tests parameters.

Specifications micro-abrasion test (wear only)	
Sample material	316 LVM steel, SiO_2-TiO_2- ZrO_2 10/70/20 and 10/20/70 coatings
Ball material	AISI 52100 steel (25mm of diameter)
Particles size	Alumina 0.3μm
Solution	Ringer's solution
Abrasive concentration	0.02%
Rotation speed	40rpm
Load	2N
Number of revolutions	100 revolutions
Sliding distance	15.7m
Slurry feed rate	30ml/min
Specifications micro-abrasion-corrosion test (wear with corrosion)	
Reference electrode	Ag/AgCl
Auxiliary electrode	Platinum wire
Working electrode	Samples (316 LVM steel, SiO_2-TiO_2- ZrO_2 10/70/20 and 10/20/70 coatings)
EIS	Frequency: 10kHz to 0.1Hz

3. Results and discussion

3.1. Wear rate estimation

For the wear estimations a mathematical model was used, this model relates the dimensions of the wear track with a loss volume rate and a wear constant. Therefore, after each test were taken parallel and perpendicular measurements of the generated wear scar according to the rotation of the ball and then determine their average, see Figure 2.

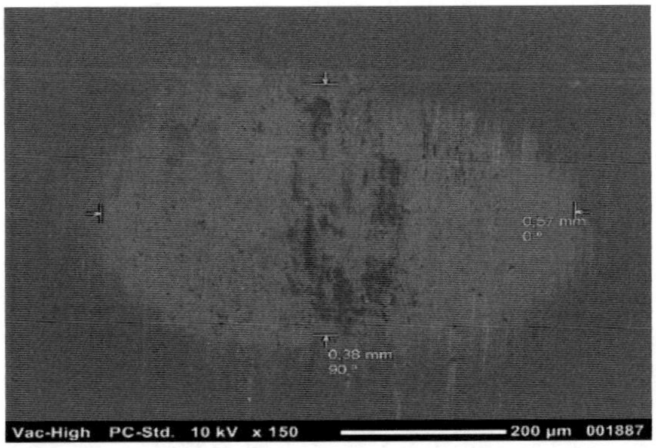

Figure 2. Measurement on wear scar of the 10/70/20 coating using SEM.

With the average lengths, the parameters of normal load (2N) and the sliding distance (15.7m) were used to calculate the volumes and constants of wear for the different performed tests as is shown in the Table 2; where is appreciated a lower wear volume for the substrate, followed by the 10/20/70 coating. The higher wear volume was exhibited by the 10/70/20 coating, however when is subjected to the micro-abrasion-corrosion presents a better behaviour in contrast with the micro-abrasion tests and the behaviour of the substrate and the 10/20/70 coating.

Table 2. Volumes and constants of wear.

Samples	Wear volume (mm^3)	Wear constant (m^2/N)
(316 L) Micro-abrasion	$1.70x10^{-05}$	$5.41x10^{-16}$
(316 L) Micro-abrasion-corrosion EIS	$4.54x10^{-05}$	$1.45x10^{-15}$
10/70/20 Micro-abrasion	$1.93x10^{-04}$	$6.15x10^{-15}$
10/70/20 Micro-abrasion-corrosion EIS	$1.00x10^{-04}$	$3.18x10^{-15}$
10/20/70 Micro-abrasion	$3.48x10^{-05}$	$1.11x10^{-15}$
10/20/70 Micro-abrasion-corrosion EIS	$6.41x10^{-05}$	$2.04x10^{-15}$

3.2. Morphological study

In the Figure 3 is observed the surfaces before and after the micro-abrasion-corrosion tests using the AFM technique, where the 316 LVM substrate and the 10/20/70 coating presents similar rugosity in comparison with the 10/70/20 coating that exhibit a lower value due to its high Ti content (see Table 3).

Table 3. Rugosity of the substrate and the coatings before and after micro-abrasion-corrosion

Sample	Rugosity Ra (nm)	
	Before test	After test
316 LVM	9.5712	169.81
10/70/20	9.0644	67.746
10/20/70	8.9521	135.39

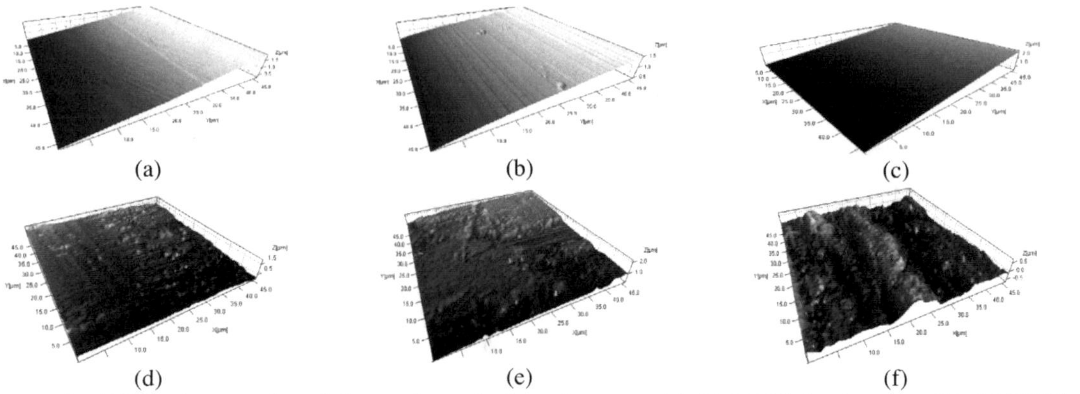

Figure 3. AFM micrographs of the surfaces before the tests (a) 316 LVM substrate, (b) 10/70/20 coating, (c) 10/20/70 coating; and the worn surfaces (d) 316 LVM substrate, (e) 10/70/20 coating, (f) 10/20/70 coating.

4. Conclusion

Despite of the higher values of wear volume exhibited by the 10/70/20 coating, it had a good performance due to the reduction of the wear volume when is subject to corrosive phenomena in contrast with the presented by substrate and the 10/20/70 coating which increase. Furthermore, the 10/70/20 coating presents the lowest rugosity at the micro-abrasion-corrosion mechanism and a similar morphology with the 10/20/70, where it was observed small grooves characteristic of the micro-abrasive wear.

Acknowledgments

This research was supported by the Universidad Militar Nueva Granada, Bogota, Colombia project IMP-ING-1767 (2015).

References

[1] Bautista-Ruiz J, Sandoval J and Ortiz C 2009 *Revista Mexicana de Física* **55** 144
[2] Brinker C and Scherer 1990 *Sol-gel science the physics and chemistry of sol-gel processing* (New York: Academic Press)
[3] Hech L and Ulrich D 1986 *Science of ceramic chemical processing* (New York: John Wiley & Sons)
[4] Klein L 1988 *Sol-gel technology for thin films, fibers, performs, electronics and speciality shapes* (New Yersey: Noyes Publications)
[5] Bautista-Ruiz J, Ortiz C, Vera E and Rodríguez–Páez J 2007 *Rev Acad Colomb Cienc* **31** 235
[6] Yao T and Takashima K 1998 *Biosens* **13** 67
[7] Li H, Jin D, Yu Q and Tu H 2011 *J Power Sources* **196** 3836
[8] Zalga A, Reklaitis J, Norkus E, Beganskiene A and Kareiva A 2006 *Chem Phys* **327** 220
[9] Tao S, Wu Q, Zhan Z and Meng G 1999 *Solid State Ionics* **124** 53
[10] Olding T, Sayer M and Barrow D 2001 *Thin Solid Films* **398** 581
[11] Castro Y 2003 *Recubrimientos protectores por deposición electroforética (EPD) a partir de suspensiones sol-gel* (Madrid: Instituto de Cerámica y Vidrio)
[12] Bautista-Ruiz J, Aperador W, Díaz M and Delgado E 2014 *Int J Electrochem Sci* **9** 4144
[13] Stack M M, Rodling J, Mathew M T, Jawan H, Huang W, Park G and Hodge C 2010 *Wear* **269** 376
[14] Gee M G, Gant A, Hutchintgs I, Schiffmann K, Van Acker K, Poulat S, Gachon Y and Von Stebut J 2002 *Cratering or micro-abrasion wear testing of coating, measurement good practice guide No 57* (Teddington: National Physical Laboratory)
[15] Li Y S, Wang K, He P, Huang B X and Kovacs P 1999 *J Raman Spectrosc* **30** 97

Tribo-electrochemical characterization of hafnium nitride/vanadium nitride multilayer systems deposited on AISI 4140 steel

M Mora[1], E Vera[2] and W Aperador[3]
[1] Fundación Universitaria Juan de Castellanos, Tunja, Colombia.
[2] Universidad Pedagógica y Tecnológica de Colombia Tunja, Colombia.
[3] Universidad Militar Nueva Granada, Bogotá, Colombia.

E-mail: g.ing.materiales@gmail.com

Abstract. In this work is presented the synergistic behaviour among corrosion/wear (tribocorrosion) of the multilayer coatings hafnium nitride/vanadium nitride [HfN/VN]n. The multilayers were deposited on AISI 4140 steel using the technique of physical vapor deposition PVD magnetron sputtering, the tests were performed using a pin-on-disk tribometer, which has an adapted potentiostat galvanostat with three-electrode electrochemical cell. Tribocorrosive parameters such as: Friction coefficient between the coating and the counter body (100 Cr6 steel ball); Polarization resistance by means of electrochemical impedance spectroscopy technique and corrosion rate by polarization curves were determined. It was observed an increase in the polarization resistance, a decrease in the corrosion rate and a low coefficient of friction in comparison with the substrate, due to an increase on the number of bilayers.

1. Introduction

Various investigations have been made about wear of different materials with or without lubrication, similarly it has analysed the behaviour of corrosion when materials are exposed to aggressive environments [1-2]; however these analyses were carried out separately, and according to experimental observations, the most materials are subjected to the combined effect of corrosion and wear (tribocorrosion), therefore it is necessary to study the synergy, because it allows at real time the characterization of the electrochemical surface condition of a material and its evolution during friction tests [3-9]. Considering; these statements, the development of this work contributes to the assessment of the tribocorrosion properties of [HfN/VN]n multilayer coatings, which exhibit similar properties to titanium nitride under static conditions of corrosion and wear, such as high hardness, wear and corrosion resistance, among others [10-14]. Likewise, it will serve as a reference for application of these coatings, in improving the tribological and electrochemical properties of different steels for industrial use.

2. Experimental procedure

The sample preparation of AISI 4140 steel was performed using metallographic techniques, which consisted of passing the samples on an abrasive silicon carbide paper, and observe the process of polishing in the optical microscope. This process was repeated with different grain papers (80-1500) and a cloth using alumina as an abrasive solution. Then, the multilayer coatings of [HfN/VN]n were obtained by the PVD magnetron sputtering technique multi-target rf (13.56 MHz). The coatings were deposited on AISI 4140 steel substrates; (2cm diameter, thickness 4mm); with a total coating thickness 0.04 microns 3.22 ± multilayer for each system. The substrates were subjected for 15 minutes to a bias voltage of - 400V (rf), with a power of 60W (rf) and a plasma of argon (Ar) to remove the oxide layer. Similarly, an intermediate layer of V and Hf films were deposited for the VN and HfN respectively for the four deposited multilayer systems. In order to increase adhesion between the coatings and the substrate, it was used a low power rf 350W and 400W respectively, in an argon atmosphere for 5 minutes. Thereafter, the samples were subjected to tribocorrosion tests using the system shown in Figure 1, which consisted of a pin-on-disk tribometer with a galvanostat potentiostat and a three electrode electrochemical cell, that has as reference electrode a Silver/Silver Chloride (Ag/AgCl), a platinum as counter electrode; and as working electrode the samples AISI 4140 steel with and without coating, immersed in a solution of NaCl 3.5% at room temperature (25°C).

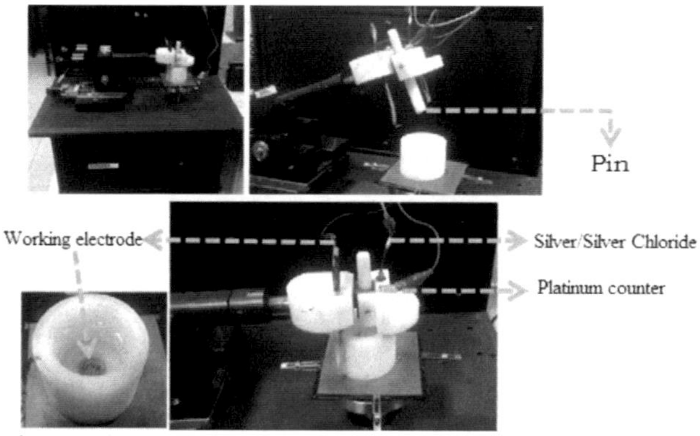

Figure 1. Experimental system used for evaluation of corrosion-wear (fretting corrosion) of the [HfN/VN]n multilayer coatings synergy deposited on AISI 4140 steel.

3. Results and discussion

In each of the graphs in Figure 2, are observed the results of the synergy between corrosion and wear (tribocorrosion), both the substrate and the [HfN/VN]n multilayer coatings (n=1, 30, 50 and 80 bilayers). For the accomplishment of the tests it was used as a pin (counterbody) a steel sphere of 100Cr6 with 6mm of diameter. The applied load was 5N with a travel distance of 80m. Also, the Figure 2 shows that at the begging of the test the coefficient of friction has increased because at this point the force of static friction is overcome between the two contact surfaces, but once you start the relative movement is reduced this ratio reaching a point in which it is approximately constant, this reduction is attributed to the lubricating effect of the corrosive environment that softens the contact between the pin and the surface of the coating, furthermore, due to the movement generated by the solution makes tribometer wear particles ejected out of the contact zone. In addition, it can be see that as the number of bilayers increases, the coefficient of friction decreases, being smaller for n =80 bilayers, due to the function or contribution of each layer that constituting the coating, providing, either as hardness, adhesion, particle size and other surface properties, which help to improve the wear resistance.

The electrochemical impedance spectroscopy assays were performed in a frequency range of 100kHz to 0.001Hz, as a result of these tests, in Figure 3 the shows Nyquist plots for the substrate and the multilayer coatings [HfN/VN]n. As increases the number of bilayers the semicircles of Nyquist plots have increased amplitude as compared to uncoated steel diagram, this also indicates a significant increase in polarization resistance increase being higher for the system with n=80 bilayers. Therefore wear in the multilayer coatings [HfN/VN]n decreases; improving the protection of AISI 4140 steel, corrosion caused by the NaCl solution.

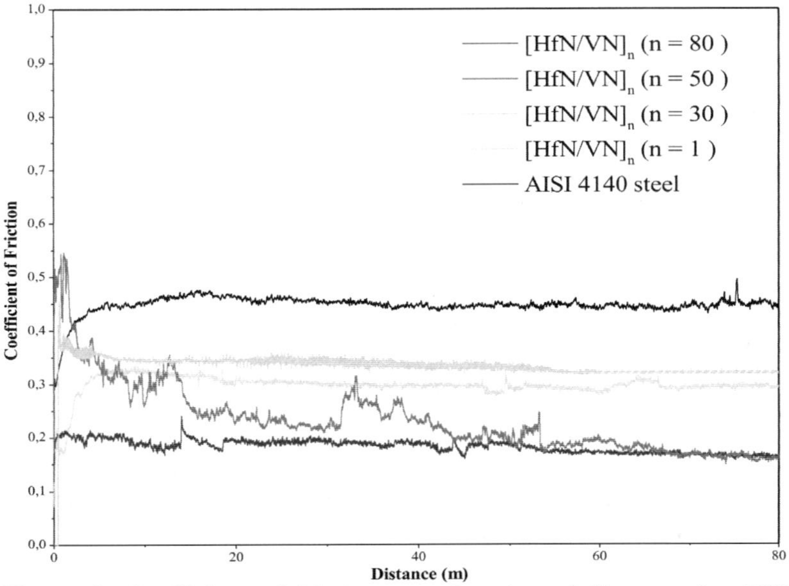

Figure 2. Coefficient of friction as a function of distance; for AISI 4140 steel and multilayer coatings [HfN/VN]n, in presence of NaCl 3.5% saline.

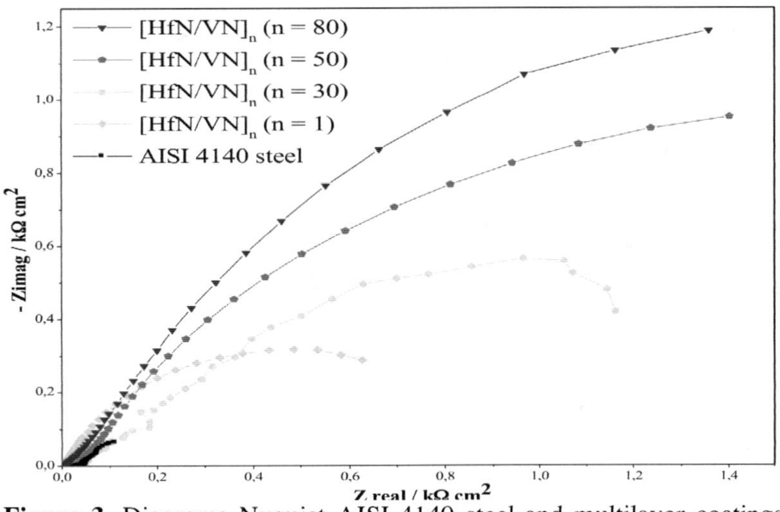

Figure 3. Diagrams Nyquist AISI 4140 steel and multilayer coatings [HfN/VN]n.

In Figure 4 are observed the polarization curves in a potential range of -0.25mV to 1V, with the cathodic and anodic slopes was determined corrosion rate of the coated and uncoated steel. After analyzing the results in Figure 4, it can be seen that as the number of bilayers increase, the coating have a better performance against degradation phenomena, because they exhibit noble corrosion potential (positive) and low density, indicating that the corrosion rate is less for multilayer coatings especially for n=80 bilayers; in comparison with the uncoated steel which has a higher value.

This value is related to the friction coefficient obtained in Figure 2, since it could generate a greater amount of wear particles; and when they are in contact with saline medium increases the corrosion rate significantly.

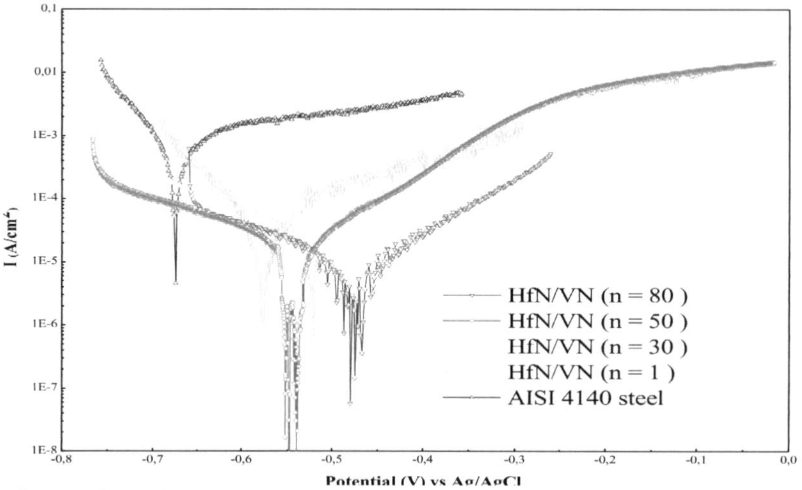

Figure 4. Tafel polarization curves of the AISI 4140 and multilayer coatings [HfN/VN]n.

4. Conclusion

The multilayer coatings [HfN/VN]n (n = 1, 30, 50 and 80 bilayers); give to the AISI 4140 industrial steel an enhancement in its properties. An improved resistance to polarization and a lower corrosion rate, ensuring their use in mechanical systems exposed to the phenomena of tribocorrosion.

The number of bilayers has great influence on the tribo-chemical properties of the [HfN/VN]n coatings and which give suitable tribological and electrochemical properties; creating good corrosion barrier between the substrate and the corrosive medium.

Acknowledgments

W Aperador, thanks the financial support from the Universidad Militar Nueva Granada, contract number ING-1775.

References

[1] Bienk E J, Reitz H and Mikkelsen N J 1995 *Surf Coat Technol* **76** 475
[2] Caicedo J C, et al 2011 *Appl Surf Sci* **258** 312
[3] Mischler S, et al 1997 *Matériaux et Techniques* **HS** 5
[4] Fischer A and Mischler S 2006 *J Appl Phys D* **39**
[5] Ponthiaux P, et al 2004 *Wear* **256** 459
[6] Mischler S 2008 *Tribol Int* **41** 573
[7] Mischler S, Debaud S and Landolt D 1998 *J Electrochem Soc* **145** 750
[8] Stemp M, Mischler S and Landolt D 2001 *Tribol Series* **39** 539
[9] Wood Robert 2007 *J Appl Phys D* **40** 5502
[10] Berg G, et al 1995 *Surf Coat Technol* **74** 135
[11] Atar E, Kayali E S and Cimenoglu H 2004 *Wear* **257** 633
[12] Nowak R and Li C L 1997 *Thin Solid Films* **305** 297
[13] Quintero V, et al 2011 *RCF* **43** 707
[14] Tomasz S, et al 2011 *Mater Lett* **65** 2146

Chromatographic analysis of phytochemicals components present in mangifera indica leaves for the synthesis of silver nanoparticles by $AgNO_3$ reduction

D Martínez-Bernett[1], A Silva-Granados[1], S N Correa-Torres[2] and A Herrera[1]
[1] Universidad de Cartagena, Cartagena, Colombia.
[2] Universidad Pontificia Bolivariana, Floridablanca, Colombia.

Email: aherrerab2@unicartagena.edu.co , sandra.correa@upb.edu.co

Abstract. It was studied the green synthesis of silver nanoparticles (AgNPs) from the reduction of a silver nitrate solution (1 and 10mM) in the presence of an extract of mangifera indica leaves. Phytochemicals components present in extracts of mango leaves were determined using a GC-MS chromatograph. The results showed the presence of the phenolic compound pyrogallol (26.9% wt/5mL of extract) and oleic acid (29.1% wt/5mL of extract), which are useful for the reduction of the metallic salt $AgNO_3$ and the stabilization of silver nanoparticles. The synthesized nanoparticles were characterized by UV visible spectroscopy (UV-vis), evidencing absorbances at wavelengths of 417nm (AgNPs-1) and 414nm (AgNPs-10), which are characteristic peaks of this metallic nanoparticles. Scanning Electron Microscopy (SEM) was used to determine the size of the synthesized nanoparticles. A particle size of about 28±7nm was observed for the AgNPs-1 sample and 26±5nm for the AgNPs-10. This suggests the advantages of green chemistry to obtain silver nanoparticles with a narrow size distribution.

1. Introduction

Synthesis of silver nanoparticles represents a matter of great interest within the scientific community nowadays thanks to their chemical, optical and morphological properties. Their large volumetric surface area allows the liberation of silver ions, which inhibit bacterial growth, adding microbial resistance to the material [1,2]. These nanoparticles can be synthesized by green chemistry methodologies using plant extracts with phytochemical components that promote the reduction of the silver nitrate metallic salt [3]. One of the advantages of this method is that it avoids the use of organic solvents such as hydrazine, which can produce volatile toxic residues like toluene and chloroform [4].

The purpose of this research was the synthesis of silver nanoparticles by using mango (mangifera indica) leaves extracts as reducing agent for the silver nitrate salt ($AgNO_3$). For that reason, a chromatographic analysis of the extract was performed, aiming to identify the phytochemical components that are present in the biomass. Such substances could promote both reduction of the metallic salt and stability of the suspended nanoparticles. A gas chromatograph coupled to a mass spectrometer was used for this analysis. Synthesis of the silver nanoparticles was carried out at room temperature, varying the concentration of silver nitrate from 1.0 to 10mM. The synthesized nanoparticles were characterized by means of both UV-visible spectroscopy and Scanning Electron Microscopy (SEM) to determine the size distribution of the nanoparticles as a function of the concentration of the precursor reagent.

2. Materials and methods

Mango (mangifera indica) fruit trees were selected from the rural population of Arroyo Grande (Bolívar, Colombia) for the gathering of leaves. This biomaterial was used for the preparation of the aqueous extract and the characterization of phytochemical components. Infusion technique was used when preparing the extract. In order to do that, the leaves were washed, dried and grinded. Then, 100g of biomass were added to 1L of distilled water that had been previously heated at 95°C. The infusion was left to cool down to 60°C and then it was filtered to remove fibrous material. Finally, this aqueous solution was concentrated until achieving a volume of 100mL.

The synthesis of silver nanoparticles involved the preparation of an initial concentrated solution of 100mM silver nitrate ($AgNO_3$ 99%, purchased from PanReac AppliChem) by diluting 1.7g of the metallic salt in 100mL of distilled water. Afterwards, 1.0mM and 10mM diluted solutions were obtained from the original. Green chemistry methodology was implemented for the synthesis of silver nanoparticles by mixing 50mL of $AgNO_3$ solution (1.0mM and 10mM) with 5mL of the mangifera indica extract. Reaction mixture was carried out at room temperature with magnetic agitation for 1 hour. Then, silver nanoparticles were centrifuged at 5500rpm, washed with distilled water and ethanol, and dried at room temperature.

3. Results

3.1. Phytochemical components in the mangifera indica extract

Phytochemical components of mangifera indica leaves extracts were determined by using a gas chromatograph coupled to a mass spectrometer (GC/MS). This analysis allowed the identification of chemical compounds that could promote the reduction of the metallic salt during the synthesis of the silver nanoparticles. This measurement required an organic extraction of the phytochemical components using a HPLC grade ethanol. This organic extract was concentrated by using a Stuart Vertical Condensor rotary evaporator at 60°C for 15 minutes until obtaining a volume of 5mL. Chromatographic analysis was carried out by direct injection of the extract into organic phase, using an AT6890 Series Plus Agilent Technologies gas chromatograph coupled to a AT MSD 5975 Inert XL selective mass detector operating in radiofrequency full scan mode. An additional DB-5MS (J&W Scientific) 60m×0.25mm×0.25μm column with 5% phenyl-poly(dimethylsiloxane) was used for this purpose. Injection was performed in split mode (30:1) with an injection volume of 2μL.

Figure 1 displays the chromatogram obtained from this measurement and the Table 1 shows a list of the main components that were detected in this analysis. Chromatographic analysis indicated the main presence of pyrogallol (26.9%) and hydroquinone (5.1%), which can promote the reduction of silver ions, facilitating the formation of nanoparticles [5]. In the case of pyrogallol, (1,2,3-benzenoitriol) accounts for a highly reducing agent for metals. The properties of this phenol derivative allow the transformation of metallic materials into their basic electronegativity state, which corresponds to a nanometric scale. This electrochemical change causes free ions to attract each other and form nanoparticles (clusters) [6]. On the other hand, the presence of oleic acid (29.1%) can contribute to the stabilization of AgNPs by acting as a primary surfactant during the nucleation stage of this nanomaterial [7]. It is an organic monosaturated acid that is primarily found in plants. This fatty acid can be coupled to the external structure of metallic nanoparticles through carboxylate bonds (COO^-) which can limitate size growth of the material [8].

In addition, it is important to note that chromatographic analysis showed the presence of 23.7% unidentified oxygenated compounds. This result makes it possible to hypothesize the presence of other phytochemicals such as flavonoids, aldehydes, or ketones that can be involved in the formation of the silver nanoparticles. Finally, the presence of diethyl phthalate (4.9%) was also detected. This phenomenon is associated to a plastic component. It is possible that a sample contamination had occurred as mango leaves were collected in plastic bags, and they were also stored in plastic Ziploc bags after the grinding process.

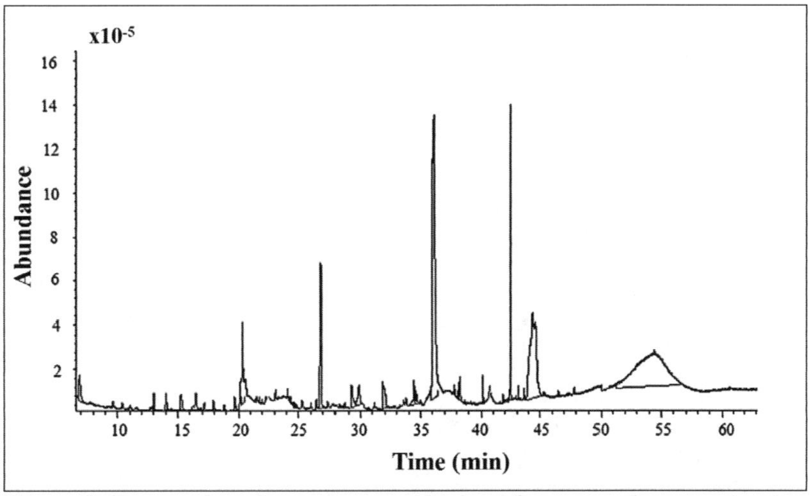

Figure 1. Sequencing chromatogram obtained for mango leaves extract.

Table 1. Chromatography analysis for the mangifera indica leaves extract.

t_R (min)	Component identification	Rel. Quantification %wt/5 mL
6.87	Acetaldehyde	1.5
13.0	Furfural	0.6
14.03	2-Furanometanol	0.8
20.32	Phenol	3.1
26.83	2,3-Dihidro-3,5-dihifroxi-6-metil-4H-Piran-4-ona	5.1
29.24	*o*-Catechol	1.3
31.89	Hydroquinone	1.7
36.12	Pyrogallol	26.9
38.27	2-Hydroxyacetophenone	1.0
42.46	Diethyl phthalate	4.9
54.43	Oleic acid	29.1

3.2. Characterization of silver nanoparticles

The synthesized nanoparticles were characterized by UV-visible spectroscopy using a Labomed, Inc. UV 2650 equipment. Figure 2 shows the results of this analysis. There, absorption peaks in the wavelengths of 417nm (AgNPs-1) and 414nm (AgNPs-10), were observed. Such peaks are characteristic of these metallic nanoparticles [9,10].

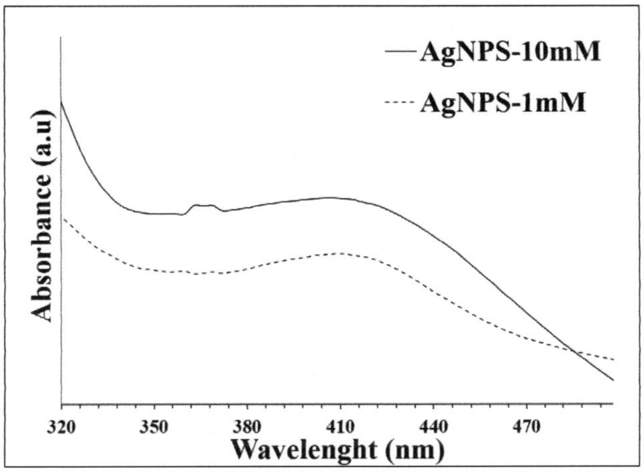

Figure 2. UV-vis spectroscopy of the silver nanoparticles that were synthesized from 1.0 and 10mM silver nitrate solution, using a mango leaves extract as reducing agent.

Figure 3 shows images of the nanoparticles that were obtained with a Quanta FEG 650 Scanning Electron Microscope (SEM). The ImageJ software was used for the determination of size distribution. Average sizes of 28±7nm and 26±5nm were obtained for the AgNPs-1 and AgNPs-10 samples, respectively. This indicates that the concentration of metallic salt did not influence the size distribution of the nanoparticles.

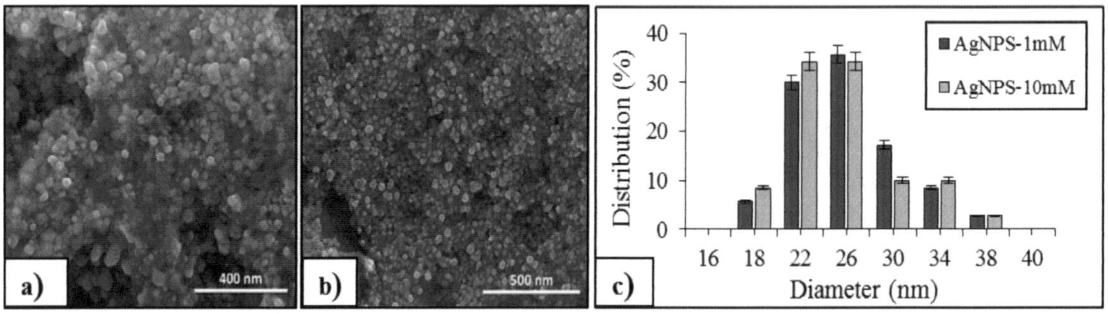

Figure 3. SEM images of silver nanoparticles synthesized by green chemistry from silver nitrate at concentrations of (a) 1.0mM, (b) 10mM, and (c) frequency histogram of the average particle size.

4. Conclusions

This research proved that the phytochemicals components present in the mangifera indica leaves extract (pyrogallol, hydroquinone, and oleic acid) can promote the formation of silver nanoparticles at room temperature with a uniform size distribution. This methodology is characterized by its easy performance and its low cost, which contributes to the environmental wealth that is achieved when no toxic compounds are used during the synthesis stage.

Acknowledgments

We would like to thank the University of Cartagena for the financial support of this project (grant No. 064-2013) and Pontifical Bolivarian University Bucaramanga sectional for the financial support of this research (DGI-BI050B project).

References

[1] Ahmed S, Ahmad M, Babu L and Saiqa I 2015 *Journal of Advanced Research* **1** 1
[2] Prakash P, Gnanaprakasam P, Emanuel R, Arokiyaraj S, Saravanan M 2013 *Colloids and Surf B Biointer* **108** 255
[3] Rajaram K, Aiswarya C and Sureshk Jumar P 2015 *Materials Letters* **138** 251
[4] Varma R 2012 *Current Opinion in Chemical Engineering* **1** 123
[5] Raghua P, Madhusudana T, Reddaiaha K, Jaidev L and Narasimha G 2013 *Enzyme and Microbial Tech* **52** 377
[6] Ghanshyam U, Prakash G, Prakash O, Mahendra P 2010 *Chemico Biological Interactions* **183** 333
[7] Wang Y, Wong J, Teng X, Zhang X, Yang H 2003 *Nano Letters* **3** 1555
[8] Wen J, Li J, Liu L and Chen Q 2011 *Colloids and Surfaces A Physicochem Eng Aspects* **373** 29
[9] Zahran M, Ahmed H and El-Rafie M 2014 *Carbohydrate Polymers* **111** 971
[10] Fonseca A, Martinez D, Meister S, Mazarin A, Brandelli A 2014 *Colloids and Surf B Biointerfaces* **113** 115

Evaluation of the photocatalytic activity of iron oxide nanoparticles functionalized with titanium dioxide

A Herrera[1], A Reyes[1] and J Colina-Márquez[1]
[1] Universidad de Cartagena, Cartagena, Colombia.

Email: aherrerab2@unicartagena.edu.co , jcolinam@unicartagena.edu.co

Abstract. Photocatalytic activity of iron oxide (IO) nanoparticles functionalized with TiO_2 was evaluated through photodegradation of phenol under UV irradiation. For this, magnetic nanoparticles were synthesized by co-precipitation method obtaining aggregates with a size of 46nm. The IO nanoparticles were encapsulated in a polysiloxane matrix and then functionalized with TiO_2 at 25°C (sample A: 0.1g TiO_2 and B: 0.3g TiO_2). Photodegradation experiments were carried out for six hours at pH 3.0 using concentrations of IO-TiO_2 nanoparticles of 0.2, 0.5, and 1.0g/L. A maximum amount of 89% of phenol photodegradation was achieved by using 0.2g/L of the IO-TiO_2-B sample. In addition, it was evaluated the possibility to re-using the nanomaterial after magnetic separation. For this, 0.2g/L of B sample were submitted for five cycles of photodegradation. A stable photocatalytic activity was observed as well as the nanoparticles were regenerated by calcination among cycles, which suggests the versatility of these nanoparticles for the photodegradation of organic pollutants.

1. Introduction

Elimination of phenolic compounds in wastewaters has been widely studied by different processes such as biological treatments, photocatalysis, adsorption, electrochemical oxidation, and Photo-Fenton oxidation [1-4]. Despite their effectiveness, many of them present disadvantages like transfer of pollutants, long degradation times and loss of valuable materials (catalysts) [1,2]. Development of new technologies to achieve the degradation of phenol in a suitable way becomes then necessary.

Titanium dioxide nanoparticles represent a promising alternative regarding contaminated wastewater treatment, thanks to their excellent photocatalytic activity and large surface area [3-5]. However, the separation of the nanoparticles from the aqueous solution once the process has finished represents a major drawback for their application. Supported titanium dioxide can face this inconvenient by allowing an easy separation of the catalysts from the reagents and products, as well as a lesser product contamination and high adaptability to continuous processes [1,2,6].

This study focuses on the evaluation of the photocatalytic activity of iron oxide (IO) nanoparticles modified with TiO_2. In order to achieve this objective, the photodegradation of phenol was studied in an aqueous solution (50ppm of initial concentration of phenol and a pH of 3.0) under constant UV irradiation. In addition, it was studied the possibility of reusing the nanomaterial by magnetic separation and calcination. From these experiments, a stable photocatalytic behaviour was observed after five photodegradation cycles.

2. Methodology

2.1. Synthesis of iron oxide nanoparticles and functionalization with titanium dioxide (IO-TiO_2)

Iron oxide (IO) nanoparticles were synthesized from an aqueous solution of ferric chloride (0.36M) and ferrous chloride (0.18M) [7]. This mixture was heated up to 70°C and then 10mL of NH_4OH were added. The synthesis was carried out at 80°C and a pH of 8.0. Afterwards, nanoparticles were washed with distilled water and dried at 80°C. On the other hand, TiO_2 was synthesized by reduction of titanium isopropoxide (5mM) with 15mL of an aqueous lemongrass leaves extract. This process was carried out at 25°C at 175rpm for 24 hours. Then, nanoparticles were centrifuged at 5000rpm for 15min, washed with distilled water/ethanol, and calcinated at 550°C for 3 hours.

For the production of the IO-TiO_2 nanocomposite, 0.7g of IO nanoparticles were suspended in 20mL of dimethyl sulfoxide (DMSO) using an ultrasound at 8000rpm. Then, it was added 3mL of tetraethyl orthosilicate (TEOS). Reaction was carried out at 100rpm for 48 hours [2]. IO nanoparticles encapsulated in the TEOS silica matrix were later separated by centrifugation and suspended again in DMSO (1%w/v). In order to obtain multifunctional nanomaterials, TiO_2 nanoparticles were added to the suspension of IO nanoparticles in DMSO. For this purpose, it was added 0.1g of TiO_2 (sample IO-TiO_2-A) and 0.3g of TiO_2 (sample IO-TiO_2-B). The functionalization was carried out with mechanical agitation during 12 hours. Afterwards, the multifunctional material was separated by centrifugation at 5000rpm during 15 minutes, washed with ethanol and dried at room temperature.

2.2. Evaluation of the photodegradation of phenol using the iron oxide-TiO_2 nanomaterials

Photocatalytic activity was evaluated as a function of nancomposite concentration by adding 0.2, 0.5 and 1.0g/L of the IO-TiO_2 nanomaterials to a synthetic phenol solution, prepared at 50ppm and a pH of 3.0. The photodegradation of phenol was performed in a UV irradiation chamber for 6 hours at 100rpm [1]. Phenol absorbance was read in an UV spectrophotometer at a wavelength of 271nm [1]. Additionally, it was tested the performance of IO-TiO_2 nanomaterials for continuous photodegradation applications. Thus, after finish a photodegradation cycle, IO-TiO_2 nanomaterials were decanted using a magnet, washed three times with distilled water/ethanol (2:1), and calcinated again at 550°C.

2.3. Characterization

Nanoparticles were characterized by dynamic light scattering (DLS) using a ZetaSizer NanoSeries/Nano-Zs90 equipment. Crystalline structure was determined through X-ray diffraction using an XPert PRO equipped with a PW3050/60 goniometer. A JEOL JSM-6490LV scanning electron microscope coupled to an EDS was used to take images of these nanomaterials. The photodegradation of phenol was determined using a Labomed Inc. UV 2650 spectrophotometer.

3. Results and discussion

3.1. Determination of size and composition

Figure 1 shows the hydrodynamic size obtained for IO nanoparticles suspended in DMSO, from which it was estimated a size of 46±2.0nm (Figure 1(a)). In the case of TiO_2 nanoparticles, DLS measurement was not performed because it was not possible obtain a stable colloidal suspension of this nanomaterial in DMSO media. X-ray diffraction was used to determine the crystal structure of the synthesized nanoparticles (Figure 1(b)). From this measurement it was observed the characteristic diffraction patterns of the cubic inverse spinel structure of IO nanoparticles (maghemite/magnetite) and the presence of the anatase crystal phase in the TiO_2 nanoparticles, both results are in accordance with the information reported in literature [8,9]. Scherrer's equation [2] was used to estimate an average crystal size of 30±20nm for IO nanoparticles and 19±4nm for TiO_2 nanoparticles. These results are in agreement with the crystal size reported by Alvarez and co-workers (2010), whose calculated a crystal size of 25.8nm for IO nanoparticles and 14.5nm for TiO_2 nanoparticles [2].

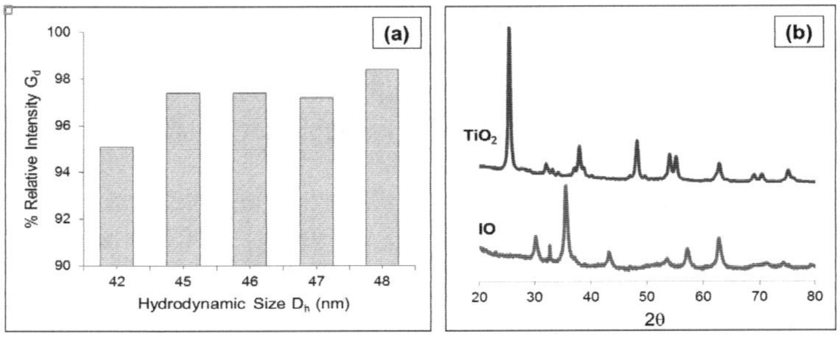

Figure 1. (a) Hydrodynamic size measurements of the IO nanoparticles suspended in DMSO. (b) X-ray diffraction patterns of IO and TiO_2 nanoparticles.

Figure 2 shows images of the nanoparticles using a SEM. There, it can be observed that the IO and TiO_2 nanoparticles have an amorphous shape with a wide size distribution. Once the modification was performed, the elemental composition of iron, oxygen, silica and titanium atoms in the samples was analysed by EDS. Table 1 presents a summary of these results, from it was estimated a molar ratio Fe:Ti equal to 10 for sample IO-TiO_2-A and a molar ratio Fe:Ti equal to 55 for sample IO-TiO_2-B.

Figure 2. SEM images of the synthesized nanoparticles. (a) IO, (b) TiO_2, (c) IO-TiO_2-A, and (d) IO-TiO_2-B.

Table 1. Elemental composition of the synthesized nanoparticles.

Element	IO-TiO_2-A, weight %	IO-TiO_2-B, weight %
O	40.54	46.88
Si	6.21	6.61
Ti	4.84	0.83
Fe	48.41	45.69

3.2. Evaluation of the photocatalytic activity of the IO-TiO_2 nanomaterials

Figure 3 shows the results obtained after the photodegradation of phenol by using the IO-TiO_2 nanocomposites. From this, it was observed a maximum photodegradation of 89% by using 0.2g/L of the IO-TiO_2-B sample. This can be attributed to the elemental composition of this nanomaterial, which displayed a Fe:Ti ratio of ~55. Despite the protective layer of siloxanes surrounding the magnetic nanoparticles, it is possible that Fe (III) ions can be released to the solution [9]. These ions can increase the photocatalytic activity of titanium dioxide at pH of 3.0 by acting as electron acceptors and preventing the occurrence of an electron-hole recombination process [2]. The other reason can be the probable occurrence of a Fenton-like process, which could lead to the free hydroxyl radicals formation [4,9]. These additional oxidant species can increase the photodegradation rate of the phenol.

3.3. Magnetic removal and photodegradation cycles

The lifespan of IO-TiO_2 nanomaterial, along with its photocatalytic sensitiveness, was evaluated as a function of the operating time. For achieving this, phenol photodegradation was studied by submitting sample IO-TiO_2-B (0.2g/L) to five continuous cycles of UV irradiation at a pH of 3.0 for 6 hours. The reuse of the IO-TiO_2 material was performed applying magnetic decantation and redispersion in a fresh phenol solution at 50ppm. Table 2 presents the obtained results. A ~20% decrease in the photocatalytic activity of the nanomaterial was observed. This can be attributed to the deactivation by catalyst poisoning, which causes a decrease in the active sites that are available for photodegradation. This is why the calcination of the material at 550°C was considered and evaluated after each photodegradation cycle, with the purpose of removing all of the impurities that could deactivate the catalyst. Table 2 shows these results, from which a stable photocatalytic activity was observed for the regenerated nanomaterial, suggesting its possible reutilization.

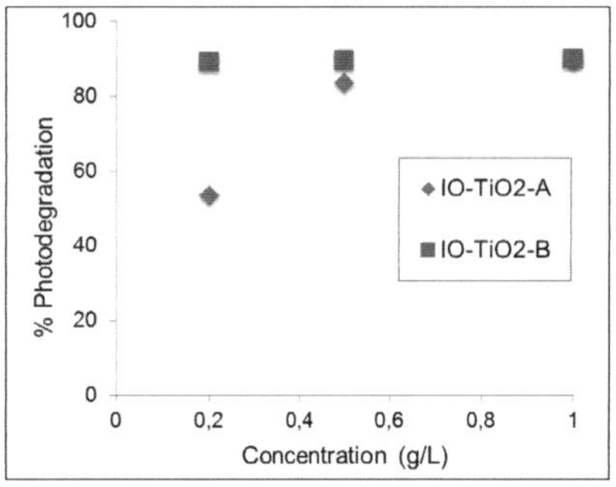

Figure 3. Phenol (50ppm) photodegradation using the IO-TiO_2 nanomaterials under ultraviolet irradiation by 6h.

Table 2. Photocatalyic sensitiveness of the IO-TiO_2-B sample.

Cycle	Photodegradation	Photodegradation/calcination
1	89.1	89.7
2	86.6	86.4
3	85.0	86.8
4	79.5	84.6
5	72.6	83.5

4. Conclusions

The synthesis of iron oxide nanoparticles and their modification with titanium dioxide was presented by executing a simple and easily implemented method. The synthesized nanomaterial (IO-TiO_2-B) achieved a high photodegradation percentage of phenol by using a low concentration (0.2g/L), which contributes to the cost effectiveness of the photodegradation process. It was necessary to implement an additional regeneration process which was found to be necessary for maintaining a stable photocatalytic behaviour. All of the above characteristics show how attractive the use of this nanomaterial can be when it is intended for wastewater treatments, due to their magnetic/photocatalytic duality.

Acknowledgments

We would like to thank the University of Cartagena and COLCIENCIAS for the financial support of Ana Reyes through the Young Researcher Program (Grant No.110752128546).

References

[1] Laoufi N, Tassalit D and Bentahar F 2008 *Global NEST Journal III* **10-3** 404
[2] Alvarez P, Jaramillo J, López F and Plucinski K 2010 *Applied catalysis B Environmental* **100** 338
[3] Akpan U and Hameed B 2009 *Journal of Hazardous Materials* **170-2-3** 520
[4] Asiri A, Al-Amoudi M, Al-Talhi T and Al-Talhi A 2011 *Journal of Saudi Chemical Society* **15-2** 121
[5] Pazokifarda S, Farrokpay S, Mirabedni M and Esfandec M 2015 *Progress in Organic Coatings* **87** 36
[6] Sood S, Ahmad U, Surinder K and Sushil K 2015 *Journal of Colloid and Interface Science* **450** 213
[7] Wei Y, Han B, Hu X, Lin Y, Wang X and Deng X 2012 *Procedia Engineering* **27** 632
[8] Wei Sh, Hu X, Liu H, Wang Q He Ch 2015 *Journal of Hazardous Materials* **294** 168
[9] Zhana J, Zhangb H, Zhu G 2014 *Ceramics International* **40** 8547

Ultrasonic cavitation for obtainment of nanometric sized particles

A Santos[1], R Guzmán[1], J Espinosa[1] and J Estrada[1]
[1] Universidad Pontificia Bolivariana Seccional Bucaramanga, Floridablanca, Colombia.

E-mail: alfonso.santos@upb.edu.co

Abstract: This project aims to determine the possibility of obtaining nanometric size particles of aluminium oxide (Al_2O_3) and titanium dioxide (TiO_2) from commercial micron-sized powders, through the physical principle of ultrasonic cavitation, in order to be used as supply material in coatings made through a process of thermal spray by flame. The tests are performed on a Hielscher UIP 1000hd Ultrasonics equipment, in a 20 micron wave amplitude and in times of 6, 8, 12, 18 and 24 hours. The determination of the particle size is done through image processing using ImageJ software, obtained by the technique of scanning electron microscopy (SEM); while the elemental composition of the processed samples is analyzed through the technique of energy dispersing spectroscopy (EDS). The results show that Al_2O_3 and TiO_2 have a reduction behaviour of the particles size after being subjected to ultrasonic cavitation, however is only reached the nanometric size in the TiO_2 samples.

1. Introduction
In recent years have been developed a series of methods for obtaining of nanoparticles, which are usually grouped into two categories, the approaches "from top to bottom" which consists of the division of solid mass into smaller portions [1]. Recently, it has been reported results of the use of a method based on the use of ultrasound for the reduction of size from micrometer-scale particles to Nanometric dimensions, mainly in elements such as carbon, gallium, zinc, gold, etc. [2-4]. The principle searches through the application of ultrasonic waves in a medium, usually water, generate a phenomenon known as cavitation, which can locally cause temperatures of up to 5000K, pressures up to 2000MPa and speeds up to 280m/s.

The size of the particles used on the accomplishment of coatings, significantly influences the mechanical properties related to its behaviour to the wear phenomenon; it has been shown that the hardness and the yield strength of a material increases with decreasing grain size of the crystal until a critical value of 20nm; from this value, these properties decrease with decreasing grain size [5].

This project wants to apply the technique of ultrasonic cavitation for the processing of Al_2O_3 particles and TiO_2 micron-sized, contained in water, to determine the behaviour of size depending on the time of exposure to the phenomenon and determine whether it is possible or not to reach nanometric sizes of these oxides from this principle, for its use in obtaining coatings through thermal spray by flame process.

2. Conceptual framework

2.1. Synthesis of nanoparticles
Various physical methods are currently used for synthesis and commercial production of nanostructured materials. The first and most widely used technique involves the synthesis of single-phase metals and ceramic oxides by the technique of inert gas evaporation [6]; the cathodic pulverization is another

technique used to produce groupings of nanostructured materials, as well as a variety of thin layers. The third physical method involves the generation of nanostructured materials through a severe mechanical deformation [7].

Chemistry has also played an important role in the development of new materials with innovative and technologically important properties [8]. The advantage of chemical synthesis is its versatility in designing and synthesizing new materials that can be refined in the final product. The main advantage offered by chemical processes on the other methods is its good chemical homogeneity, since it enables the mixture at the molecular level [9]. The general procedure involves reactions in aqueous or non-aqueous solutions containing soluble or suspended salts. Once the solution is super saturated with product, the precipitate is formed both by homogeneous or heterogeneous nucleation [10]. Nanostructured materials are also prepared by chemical vapour deposition (CVD) or chemical vapour condensation (CVC) [11]. CVC method is used to produce a variety of powders, metal fibres and compounds, whereas the CVD method has been employed in the synthesis of ceramic metals, intermetallic and composites [12].

2.2. Ultrasonic cavitation

Ultrasound has evolved in the last ten years and has become a processing technology completely commercial of high reliability, scalability, and energy efficiency. Ultrasonic processor generates cavitation, ultrasonic basic effect that allows new results on physical and chemical processes.

While low intensity or high frequency ultrasound is mainly used for the analysis of non-destructive testing and imaging; high intensity ultrasound is used for the treatment of liquids, in other words what is meant to achieve is shearing, in order to cause cracks in the surface of the particles.

Recent researches have used ultrasonic cavitation method for obtaining nanometric size particles of metals of low melting point and alloys [2]; ferrite of zinc [3] and gallium, among others, showing promising results that allow to expect positive results with other materials.

3. Methodology

3.1. Preparation of Al_2O_3 and TiO_2 samples

The weight of the material used for the tests was 5 grams, which are diluted in 500mL of sterile water; the homogenization of the mixture was made through a process of stirring with a glass rod. Finally the sample enters the flow cell of the system UIP 1000hd used.

3.2. Testing

The equipment used for testing was a processor of ultrasound UIP 100hd, which gives a power of 1000W at a frequency of 20 kHz; amplitude used was 20 microns, which is equivalent to 80% of the maximum allowed by the equipment. The mixing process was carried out continuously and processing time of them was 6, 8, 12, 18 and 24 hours. After the test time, the mixture is brought to an electric oven, which is set at a constant temperature of 250°C for a period of 30 minutes, causing evaporation of the water.

3.3. Morphological characterization of processed materials

In order to obtain a value for the size of the particles with a reliability greater than 95%, a statistician analysis is made with a pre-sample of 10 magnitudes, for both the Al_2O_3 as for the TiO_2. From images obtained of SEM and a public domain digital image processing software called ImageJ, it was proceeded to dimension 10 particles randomly, always taking the lower dimension of them. The procedure consisted of the following steps: load the image into ImageJ; scale the image in order to obtain a constant dimension during all performed measurements, as shown in Figure 1; perform the conversion of the value of the indicated scale on the label of the image, to obtain its corresponding value in nanometers; check the scaling and last perform the measurement of the size of the particles, through dimensioning of the lower section of the same.

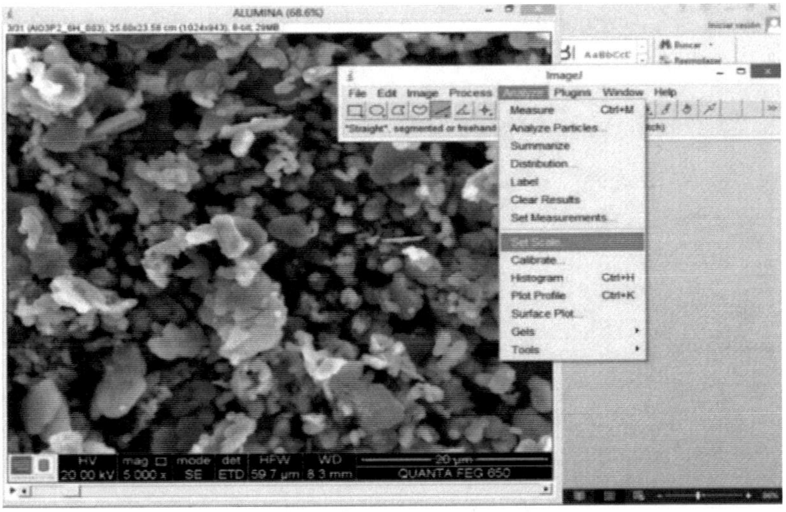

Figure 1. Image of scanning electron microscopy of Al_2O_3 used to obtain the size of the particles.

4. Results

The total number of required measurements to achieve a reliability of 95%, with a 5% error rate was 115 for the alumina and 100 for titanium dioxide; the results of the average values for each material in each of the defined procedures, are listed in Figure 2.

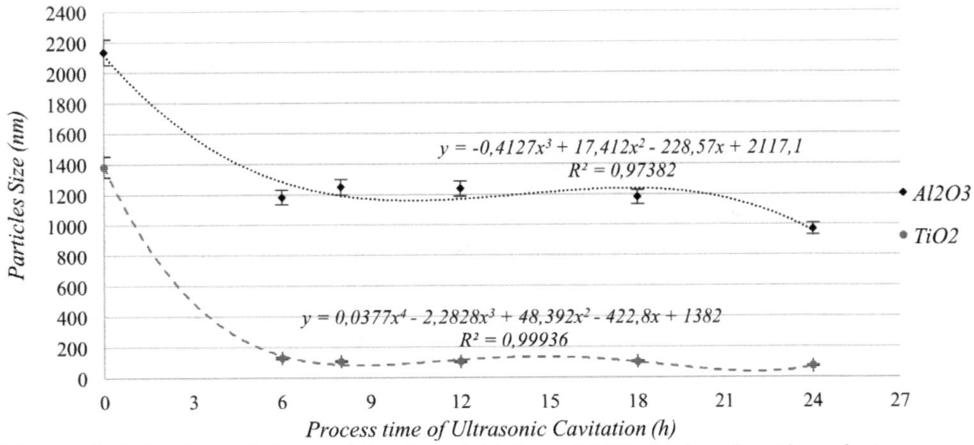

Figure 2. Behaviour of the particle size of Al_2O_3 and TiO_2 in time function of process.

The results obtained of the grinding process through ultrasonic cavitation embodied in Figure 2, show a concordance with those reported by Vargas-Hernandez et al [13], where is evidence of an effect of acoustic cavitation in the grinding of particles. The curve of the behaviour of the particle size of Al_2O_3 depending on the time of ultrasonic cavitation process shows a 4 degree polynomial relationship. There is evidence of an approximate decrease of 45% of the size in the first 6 hours of process; in processing times of 8, 12, 18 and 24 hours, no significant change is seen in the average particle size. The average size of the particles obtained by ultrasonic cavitation process is 969.64nm, which represents a decrease of the 55.4% of the initial value; however nanometric sizes (< 100 nm) are not reached.

The curve of the behaviour of TiO_2 particles size on the basis of the time of ultrasonic cavitation process also shows a 3 degree polynomial relationship. There is evidence of an approximate decrease of 92.76% of the size in the first 6 hours of process; processing time of 8, 12 and 18 hours does not show a significant difference with the results obtained in a time of 6 hours; there is again a tendency to reduce

Journal of Physics: Conference Series **687** (2016) 012035 doi:10.1088/1742-6596/687/1/012035

size for 24 hour processing time. The average size of the smallest particles of TiO_2 obtained through the process of ultrasonic cavitation is 70.8nm, which represents a decrease of the 94.9% of the initial value.

4.1. Chemical characterization of the materials processed through the process of ultrasonic cavitation
Processed powders chemical analysis was carried out through the obtained values of microanalysis by energy dispersion spectroscopy of x-rays (EDS), which are listed in Figure 3.

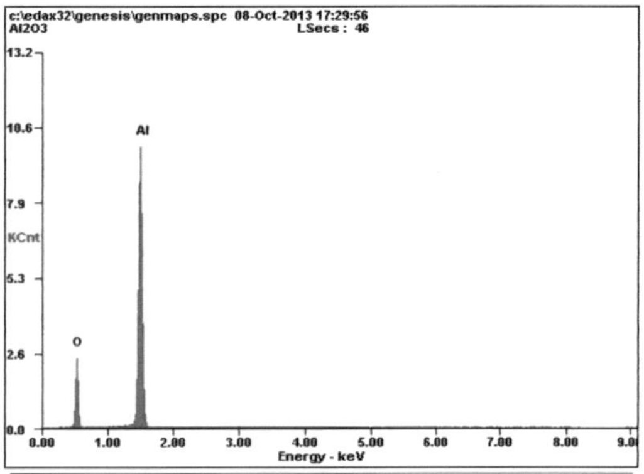

Element	Wt%	At%
OK	38.35	51.20
AlK	61.65	48.80

(a)

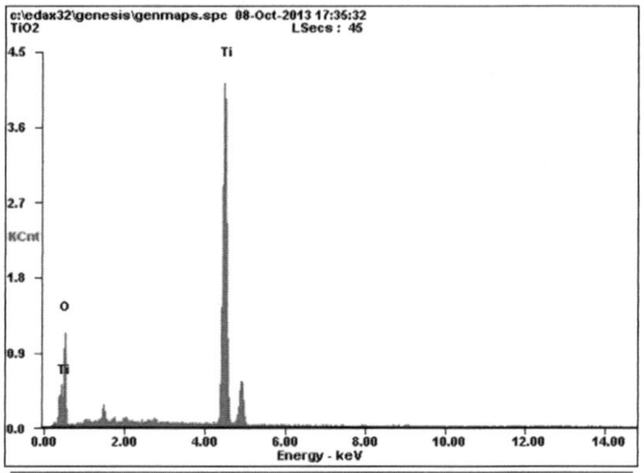

Element	Wt%	At%
OK	31.46	43.65
TiK	68.54	56.35

(b)

Figure 3. Spectrum obtained in EDS analysis. (a) Alumina powder; (b) titanium dioxide powder.

The percentage of error that was obtained between the calculated theoretical atomic weight and the experimental retrieved by EDS analysis, for Al_2O_3 and TiO_2 was 3.3% and 9% respectively; which are consistent with the values obtained from the material without processing through ultrasonic cavitation.

5. Conclusions

The ultrasonic cavitation process originates in the tested materials an average decrease of the size of the particles that comprise it; reaching nanometric values on the titanium dioxide.

The ultrasonic cavitation process does not affect the chemical composition of the materials processed when the medium used to dissolve the material and propagate sound waves is distilled water.

References

[1] R Zanella 2012 Mundo nano **5-1** 69
[2] H Friedman et al 2012 *Ultrasonics Sonochemistry* **20** 432
[3] M Sivakumar et al 2005 *Current Applied Physics* **6** 591
[4] V Bhooshan Kumar et al 2013 *Ultrasonics Sonochemistry* **21** 1166
[5] S Tjong and H Chen 2004 *Materials Science and Engineering R. Reports* **45-1** 1
[6] R W Siegel 1993 *Materials Science and Engineering A* **168-2** 189
[7] S K Ganapathi and D A Rigney 1990 *Scripta Metallurgica et Materialia* **24 9** 1675
[8] P Psaras and D Langford 1987 *Advancing materials research, U.S. national academy of engineering and national academy of sciences* (Washington: National Academy Press) p 203
[9] C N R Rao 1993 *Materials Science and Engineering B* **18-1** 1
[10] J Th G Overbeek 1982 *Advances in Colloid and Interface Science* **15** 521
[11] W Chang et al 1994 *Nanostructured Materials* **4-5** 507
[12] G Cao and Y Wang 2014 *Nanostructures and nanomaterials synthesis properties and applications 2nd ed* (Singapore: World Scientific Publishing)
[13] Y Vargas–Hernández et al 2000 *Efectos de la cavitación acústica sobre la molienda de partículas en suspension* (Madrid: XXXI Congreso Nacional de Acústica–TecniAcustica Ultrasonidos ULT 19)

Raman spectroscopy of ZnMnO thin films grown by pulsed laser deposition

S Orozco[1], H Riascos[1] and S Duque[1]
[1] Universidad Tecnológica de Pereira, Pereira, Colombia.

E-mail: sorozco@utp.edu.co

Abstract. ZnMnO thin films were grown by Pulsed Laser Deposition (PLD) technique onto Silicon (100) substrates at different growth conditions. Thin films were deposited varying Mn concentration, substrate temperature and oxygen pressure. ZnMnO samples were analysed by using Raman Spectroscopy that shows a red shift for all vibration modes. Raman spectra revealed that nanostructure of thin films was the same of ZnO bulk, wurzite hexagonal structure. The structural disorder was manifested in the line width and shape variations of E_2(high) and E_2(low) modes located in 99 and 434cm^{-1} respectively, which may be due to the incorporation of Mn ions inside the ZnO crystal lattice. Around 570cm^{-1} was found a peak associated to E_1(LO) vibration mode of ZnO. 272cm^{-1} suggest intrinsic host lattice defects. Additional mode centred at about 520cm-1 can be overlap of Si and Mn modes.

1. Introduction

ZnO is a semiconductor with wurtzite hexagonal structure and belong to space group $P6_3mc$. This material promises applications in optoelectronics, transparent electronics, gas sensors [1], and very interesting material for spintronics [2,3]. ZnO hexagonal lattice is characterized by two interconnecting sublattices of Zn^{2+} and O^{2-}, such that each Zn ion is surrounded by tetrahedral of O ions and vice-versa [4]. The irreducible representation of the optical phonons is $\Gamma opt = 2A_1 + 2B_1 + 2E_1 + 2E_2$. In this expression, the A_1 and E_1 modes are polar, and may split into transverse (TO) and longitudinal (LO) optical phonon, both are active in Raman and infrared spectroscopy. The non-polar E_2 modes are only Raman active and the B_1 modes are Raman inactive. The Raman spectra can be interpreted to estimate the degree of disorder in a crystal. The structural disorder is manifested in the line shape (line width and asymmetry variation) of the E_2 mode [2]. The properties of the grown thin films depend on the deposition process and deposition parameters. In ZnMnO films, formation of defects or introduction of extrinsic doping can cause stress, which can modify the physical properties [5]. In this work, ZnMnO thin films were grown by PLD technique. The optical properties of the grown samples were characterized by Raman spectroscopy; we report some features of Mn-doped ZnO thin films produced by PLD, and we discuss the effects of the oxygen pressure, substrate temperature and Mn concentration on structural and optical properties of ZnMnO films.

2. Experimental details

Mn-doped ZnO thin films were grown by conventional pulsed laser deposition (PLD) using a stainless steel HV deposition chamber and a pulsed Nd:YAG laser (1064nm wavelength, 9ns pulse duration). The laser was operated at 10Hz and the laser beam was incident at an angle of 45 grades with respect to the target surface. The targets were prepared from high-purity powders of ZnO (99.999%), and

MnO (99.999%) serving as the doping agent. The pressed targets were sintered at 1000°C for 12h in air. The targets were fabricated with a nominal composition of 1, 5, 10, 15 and 20wt.% Mn, and were continuously rotated during laser ablation to renew the irradiated surface and to prevent crater formation. The ablated material was collected onto a Si (100) substrate, over a temperature range of RT to 600°C, placed in front and at 6cm from the target. Prior to an experiment, the chamber was evacuated to 7×10^{-4} Pa. During deposition, oxygen (99.999%) was flown through the chamber and the background pressure was varied between 2.67 and 26.66Pa. The laser fluence was kept at 3.0 ± 0.4J/cm^2. Three sets of ZnMnO samples were prepared by PLD. In the first set, ZnO thin films were deposited at different Mn-doping ranging from 1 to 20wt.% Mn, preserving substrate temperature and oxygen pressure in room temperature and 6.13Pa respectively. In the second set, the substrate temperature was varied from room temperature to 600°C, with 5wt.% of Mn concentration and 6.13Pa of oxygen pressure. The third set was deposited at different oxygen gas pressure in the range 2.67Pa-26.66Pa at room temperature and 5wt.%. of Mn content. Optical analysis of the films was carried out in a Jobin–Yvon micro-Raman spectrometer.

3. Results and discussions

The different ZnMnO Raman patterns were analysed in the range 40-900cm^{-1}. Figures 1, 2 and 3 show the spectra of ZnMnO thin films varying Mn concentration, substrate temperature and oxygen pressure respectively. The spectra in Figure 1 show several peaks at 100cm^{-1}, 434cm^{-1} and 570cm^{-1} associated to E_2(low), E_2(high), and E_1(LO) vibration modes respectively. This peaks are typical of ZnO bulk. These spectra are very similar for Figure 2 and 3. The high frequency E_2 mode, E_2(high), involves predominantly the lighter oxygen atoms and ZnO wurzite structure [6]. The line width are assigned to defects, structural damages, residual stress, and impurities of the sample [2]. E_2(low) mode is associated with the non-polar vibration of heavier cations Zn sublattice [6,7]. E_1(LO) Raman line intensity varies with oxidation of excess zinc in the ZnO thin film [8,9]. This mode is caused by defects of O vacancy and Zn interstitial [8]. An additional peak appears at 273cm^{-1} and it indicates host lattice defects in the ZnO structure [10]. The peak at 520cm^{-1} is associated with overlapping of vibration modes from Si and Mn.

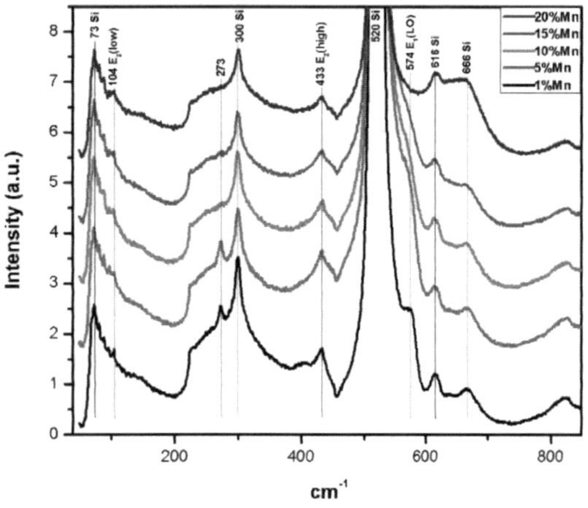

Figure 1. Raman patterns of ZnO thin films deposited by PLD at different Mn-concentration ranging from 1wt.% to 20wt.% of Mn, at room temperature and 6.13Pa.

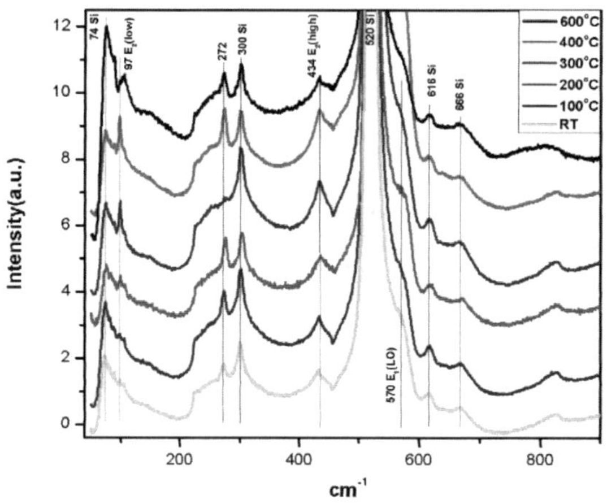

Figure 2. Raman patterns of ZnMnO thin film deposited by PLD with the substrate temperature varying from room temperature to 600°C. The Mn-doping and oxygen pressure were kept constant, 5wt.% Mn and 6.13Pa, respectively.

3.1. Influence of Mn concentration

The peaks intensity decreases with increasing of Mn content as is shown in Figure 1. For E_2(high) its line width increases with Mn content increase, while the intensity decreases. However E_2(low) peak intensity is very weak for all Mn concentration. For 1wt.% Mn, a broad peak appear associated with E_1(LO); as Mn concentration increases the broadening is higher till is overlap for the Si and Mn modes at 520cm^{-1}. This peak (at 520cm^{-1}) became more wide and its intensity low with increase Mn concentration. In addition, the peak at 273cm^{-1} appears with low intensities for 1wt.% and 5wt.% of Mn only.

3.2. Influence of substrate temperature

The variations of substrate temperature are shown in Figure 2. E_2(low) and E_2(high) appear with low line width and higher intensity at 300°C. As the substrate temperature decreases, E_2(high) is red shift so that the tensile stress increases [11]. The shoulder related with E_1(LO) has same behaviour that in Figure 1. The peak associated with defects on the lattice appear with very weak intensity at 300°C. According to these spectra characteristics, the thin film grown at 300°C has best crystalline quality and optical properties.

3.3. Influence of oxygen pressure

E_2(high) mode at 9.33Pa is very sharp peak as is shown in Figure 3. For this peak the maximum intensity was observed at 9.33Pa. This behaviour can be understood by two competing processes; the increase of oxygen pressure in the chamber improves the stoichiometry of the films and the crystal quality. On the other hand, the kinetic energy of the reactive particles in the plasma decreases due to high oxygen pressure, which limits surface diffusion of the growing atoms and thus degrades the film quality [12,13]. E_2(low) is very weak peak for all range of oxygen pressure. E_1(LO) mode has same behaviour that for Mn concentration and substrate temperature. The peak at 272cm^{-1} has very weak intensity at 9.33 and 13.33Pa.

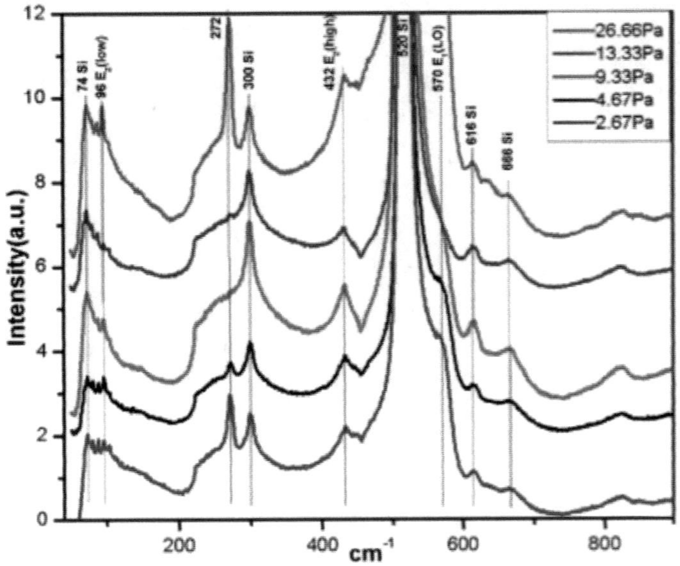

Figure 3. Raman patterns of ZnMnO thin film deposited by PLD, the oxygen gas pressure was varied from 2.67Pa to 26.66Pa. The Mn-doping and substrate temperature were kept constant, 5wt.% and room temperature, respectively.

4. Conclusion

The optical properties have a dependence on Mn-doping concentration, Oxygen gas pressure and substrate temperature. All prepared films present a typical hexagonal wurtzite ZnO structure and a peak own to Mn doped was present. Any extra peaks were found related with other impurities independent of the deposition parameters. It is seen that the crystalline quality of the films are varying with processing conditions (Mn concentration, substrate temperature and Oxygen gas pressure). Defects and residual stress of the samples were found for high Mn concentration, also for high and low substrate temperature and oxygen pressure. The films with better optical properties and less structural disorder have been obtained at relatively low Mn-concentration (1-5wt.% Mn), an intermediate substrate temperature (300–400°C) and relatively intermediate oxygen gas pressure (4.67-9.33Pa).

References

[1] Ferro R, Rodríguez J A and Bertrand P 2008 *Thin Solid Films* **516** 2225
[2] Kharroubi B, Baghdad R, Abdiche A, Bousmaha M, Bousquet M, Zeinert A, Marssi M El, Zellama K and Hamzaoui S 2012 *Physica Scripta* **86** 015805
[3] Duan L, Zhao X, Liu J, Geng W, Xie H and Chen S 2011 *Journal of Magnetism and Magnetic Materials* **323** 2374
[4] Coleman V A and Jagadish C 2006 *Basic properties and applications of ZnO* ed Jagadish C and Pearton S Australia (Australia: Elsevier Limited) chapter 1 p 2
[5] He H P, Zhuge F, Ye Z Z, Zhu L P, Wang F Z, Zhao B H and Huang J Y 2006 *Journal of Applied Physics* **99** 23503
[6] Chai G Y, Chow L, Lupan O, Rusu E, Stratan G I, Heinrich H, Ursaki V V and Tiginyanu I M 2011 *Solid State Sciences* **13** 1205
[7] Chen L C, Tien C H and Fu C S 2012 *Materials Science in Semiconductor Processing* **15** 80
[8] Atanas J P, Asmar R A, Khoury A and Foucaran A 2006 *Sensors and Actuators A Physical* **127** 49
[9] Exarhos G J and Sharma S K 1995 *Thin Solid Films* **270** 27
[10] Bundesmann C, Ashkenov N, Schubert M, Spemann D, Butz T, Kaidashev E M, Lorenz M and Grundmann M 2003 *Applied Physics Letters* **83** 1974
[11] Shan F K, Liu G X, Liu Z F, Lee W J, Lee G H, Kim I S, Shin B C and Yu Y S 2004 *Journal of the Korean Physical Society* **45** 771
[12] Zhu B L, Sun X H, Zhao X Z, Su F H, Li G H, Wu X G, Wu J, Wu R, Liu J 2008 *Vacuum* **82** 495
[13] Riascos H, Orozco S, Uzuriaga J 2015 *Procedia Materials Science* (Approved)

Growth of PEO ceramic coatings on AA 2024-T3 aluminium alloy

S Forero Sotomonte[1], C Blanco Pinzon[2] and S García Vergara[1]
[1] Universidad Industrial de Santander, Bucaramanga, Colombia.
[2] Instituto Colombiano del Petróleo (ICP), Piedecuesta, Colombia.

E-mail: sagarver@uis.edu.co

Abstract. The growth of PEO ceramic coatings on AA 2024-T3 aluminium alloy in an aqueous Na_2SiO_3 (10.5g/l), KOH (2.8g/l) solution at 310 and 400V for 500 and 710s, was investigated. The morphology, roughness and thickness of the coatings were determined by SEM, digital microscopy, XRD diffraction analysis and thickness measuring instrument. The results show that thicker coatings are produced with longer process times and high applied voltages. Due to the nature of the PEO process, the roughness of the surface coatings increases as the coating become thicker, due to the development of sparks. The coatings are porous, with a crater like morphology and they are mainly amorphous.

1. Introduction

Aluminium alloys are widely used in aerospace and automotive industry due to their excellent properties such as high strength, low density, non-magnetic properties and good formability [1,2]. However, the corrosion performance of these alloys is low due to presence of alloying elements like copper. Plasma Electrolytic Oxidation (PEO), a relatively new surface modification technique, can be used to provide the corrosion resistant of these aluminium alloys [3,4]. Additionally, the PEO coatings can improve wear, thermo-optical, dielectric and thermal barrier properties [5-7]. PEO coatings exhibit a superior performance, cost effectiveness and environmental friendliness, compare to conventional methods used on aluminium-copper alloys [8-9]. The PEO coatings formation mechanism starts when the natural, very thin and passive film of the substrate begins to dissolve. With increasing the applied voltage, gas bubbles are produced which result in a porous film with a columnar structure perpendicular to the substrate. Plasma micro-discharge begins when the applied voltage exceeds the breakdown voltage of the oxide film [10,11]. The properties of the PEO coating mainly depend on electrolyte concentration and composition [12], the alloying elements of substrate [13], the applied current density [14] and type of power source [15].

In the present study, PEO treatment was performed on AA 2024-T3 aluminium alloy in an aqueous Na_2SiO_3-KOH electrolyte under DC conditions, in order to determine the influence of applied voltage and treatment time on the morphology, thickness and crystallinity of the coatings.

2. Experimental

Rectangular samples with dimensions of 35x20x1,76mm of AA 2024-T3 (wt%: 4.476 Cu, 1.207 Mg, 0.582 Mn, 0.189 Fe, 0.163 Si, 0.037 Zn, 0.039 Ti, 0.013 Cr) aluminium alloy were used in the present study. The surface of samples was ground with 400#, 800#, 1000# and 1200# SiC abrasive papers consecutively, cleaned with distilled water and acetone, and then dried by cool air flow.

The electrolyte was an aqueous solution of Na_2SiO_3 (10.5g/l) and KOH (2.5g/l) in distilled water. DC power supply with approximate 500V of the maximum output voltage was used for PEO treatment. PEO treatment was performed under constant sparking voltages of 310 and 400V for 500 and 710s. A stainless steel electrode sheet was used as cathode. After coating, the samples were removed from the electrolyte, washed in distilled water and acetone and then dried by cool air flow.

Microstructure and surface roughness were studied by scanning electron microscopy (SEM) (Quanta FEG 650) and digital microscopy (Hirox KH-7700). The coating thickness was determined using a gauge (PosiTector 6000). XRD diffraction analysis was used to study the crystallinity of the coatings.

3. Results

The formation of PEO coatings observed on the current density-time curves, showed an initial increase of the current density followed by a current drop, which corresponds to uniform establishment of "soft sparking" over the entire sample surface. Such current drop is attributed here, to an increased thickness and density of the coating over the entire sample surface and consequently the higher resistance of the oxide material to mass transfer.

The thickness and roughness of the PEO coatings are presented in Table 1. The PEO treatment was applied at 310 and 400V for 500 and 710s. It is clear, that with the increase in the treatment time and applied voltage, the thickness of the PEO coating increases. The roughness of the sample surface increases for the ticker coatings, as reported for other researchers [10,16]. This increase in roughness is due the fact that the sparking on the surface produces many discharge channels and craters.

Table 1. Thickness and average roughness of the PEO coatings formed on AA 2024-T3 aluminium alloy in Na_2SiO_3 (10.5g/l) and KOH (2.5g/l) in distilled water solution.

	Coating thickness (μm)	Average roughness (μm)
A1 (310V, 500s)	1,62	6,949
A2 (310V, 710s)	2,72	5,838
B1 (400V, 500s)	9,48	9,387
B2 (400V, 710s)	13,90	27,741

Figure 1 presents the SEM micrographs of the surface of PEO treated AA 2024-T3 aluminium alloy at conditions used in the present work. Pores of corresponding discharge channels distributed all over the surface can be distinguished. The surface of the coatings formed at higher conditions (410V, 710s) presents a crater-like morphology. The discharge channels are the result of localized discharge, and the pores are possibly the result of escaping gases during PEO process [5,17]. Through these channels, the molten alumina flows out of the channel and rapidly solidifies, and at the same time anionic components such as $SiO3^{-2}$ enter the channels, leaving the sharp and distinctly visible boundaries. The craters are the result of cooling and deposition of molten material on the surface. As the voltage of the process increases and also the time, the surface of the coatings showed the presence micro-cracks.

The X-ray diffraction patterns of PEO coatings formed on AA 2024-T3 aluminium alloy are presented in Figure 2. The X-rays can easily penetrate through the coating because of its porosity, thus the diffraction peaks of substrate are strongly detected. It can be seen that coating formed at 400V consists mainly of γ-Al_2O_3, which is a typical phase in PEO coatings of aluminium alloys [12]. There is not presence of α-Al_2O_3. The oxide layer that ejected out from the channel is objected to high cooling rate in immediately contact with surrounding electrolyte and promotes the formation of γ-phase during solidification of alumina droplets. However, the coatings formed at 310V are mainly amorphous. Signals corresponding to mullite and cristobalite are also detected. The presence of these phases is due to the electrolyte used for the PEO treatment. It has been reported that the amorphous phase can be formed as a result of rapid localized quenching around each individual discharge during the formation of the layer. The presence of such amorphous material influences the thermal conductivity of the coating.

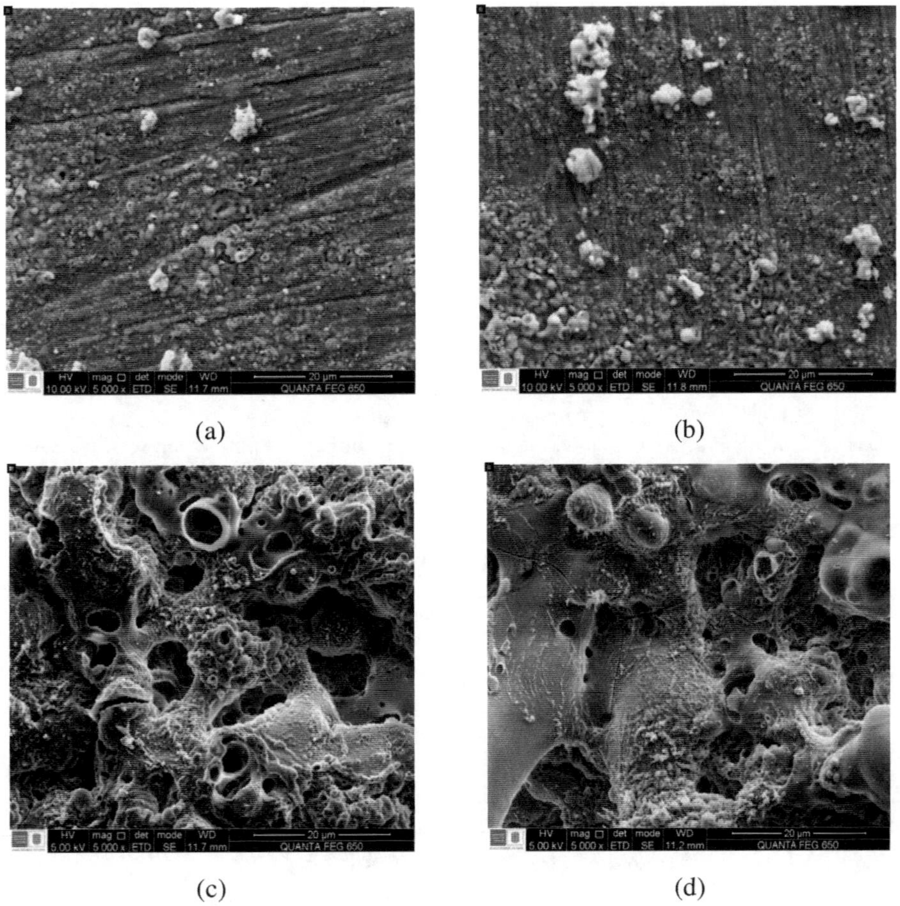

Figure 1. SEM micrographs of PEO coatings formed on AA 2024-T3 aluminium alloy in Na_2SiO_3 (10.5g/l) and KOH (2.5g/l) in distilled water solution. (a) 310V-500s, (b) 310V-710s, (c) 400V-500s, (d) 400V-710s.

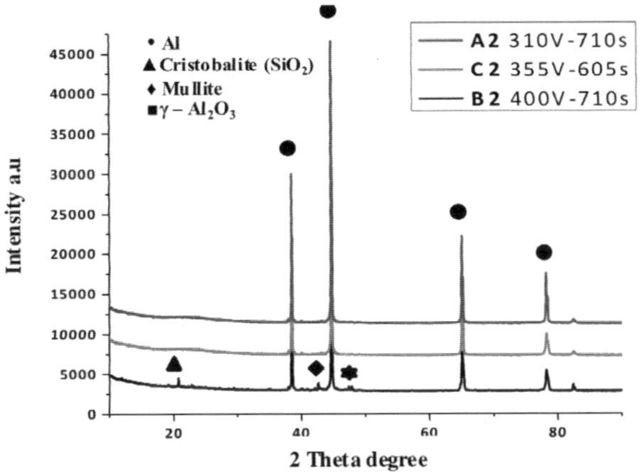

Figure 2. The XRD pattern of PEO treated 2024-T3 Al alloy in Na_2SiO_3 (10.5g/l) and KOH (2.5g/l) in distilled water solution.

4. Conclusions

Thick porous ceramic coatings are developed on AA 2024-T3 aluminium alloy by PEO process in a Na_2SiO_3 (10.5g/l) and KOH (2.5g/l) in distilled water solution. The thickness of the PEO coatings is about 13.9 μm when a 400 V is applied for 710 s. At these conditions, the coatings are composed by γ-Al_2O_3, mullite and cristobalite phases. Generally the thickness and the roughness of the coatings increase as the time and the voltage applied increase. The surface of the coatings showed a crater-like morphology.

Acknowledgments

The authors are grateful to COLCIENCIAS (Convocatoria 566-2012, Jóvenes Investigadores e Innovadores) for the financial support.

References

[1] Wen L, Wang Y, Zhou Y, Ouyang J H, Guo L and Jia D 2010 *Corros Sci* **52** 2687
[2] Ding H Y, Dai Z, Skuiry S C and Hui D 2010 *Tribol Int* **43** 868
[3] Bierwagen G, Brown R, Battocchi D and Hayes S 2010 *Prog Org Coat* **68** 48
[4] Wen L, Wang Y, Zhou Y, Ouyang J H, Guo L and Jia D 2010 *Corros Sci* **52** 2687
[5] Curran J A and Clyne T W 2005 *Surf Coat Technol* **199** 168
[6] Gnedenkov S V, Khrisanfova O A, Zavidnaya A G, Sinebrukhov S L, Kovryanov A N, Scorobogatova T M, et al 2000 *Surf Coat Technol* **123** 24
[7] Blawert C, Heitmann V, Dietzel W, Nykyforchyn H M and Klapkiv M D 2007 *Surf Coat Technol* **201** 8709
[8] Liu R, Wu, Xue W, Qu Y, Yang C, Wang B and Wu X 2014 *Materials Chemistry and Physics* **148** 284
[9] Mecuson F, Czerwiec T, Belmonte T, Dujardin L, Viola A and Henrion G 2005 *Surf Coat Technol* **200** 804
[10] Stojadinovic S, Vasilic R, Belca I, Petkovic M, Kasalica B, Nedic Z and Zekovic L 2010 *Corros Sci* **52** 3258
[11] Gu W C, Lv G H, Chen H, Chen G L, Feng W R and Yang S Z 2007 *Mater Sci Eng A* **447** 158
[12] Wang K, Koo B H, Lee C G, Kim Y J, Lee S H and Byon E 2009 *Trans Nonferrous Met Soc China* **19** 866
[13] Oh Y J, Mun J I and Kim J H 2009 *Surf Coat Technol* **204** 141
[14] Su P, Wu X, Guo Y and Jiang Z, 2009 *J Alloys Compd* **475** 773
[15] Arrabal R, Matykina E, Viejo F, Skeldon P, Thompson G.E and Merino M C 2008 *Mater Sci* **43** 1532
[16] Kalkanc H and Kurnaz S C 2008 *Surf Coat Technol* **203** 15
[17] R H U Khan, A Yerokhin, X Li, H Dong, A Matthews 2010 *Surf Coat Technol* **205** 1679

Electrochemical preparation and structural characterization of $CuGaSe_2$ thin films

L Manfredy[1], O P Marquez[1], S A Lopez-Rivera[1], J Marquez[1], Y Martínez[1] and D A Miranda[2]
[1] Universidad de los Andes, Mérida, Venezuela.
[3] Universidad Industrial de Santander, Bucaramanga, Colombia.

E-mail: manfredy@ula.ve

Abstract. $CuGaSe_2$ thin films have been synthesized by one-step electrodeposition on titanium and SnO_2/glass substrates from aqueous solutions that contain thiocyanate ion as a complexing agent. According to composition, and morphology analysis at deposition potential of -0.55V vs SCE (Standard Calomel Electrode) for 20 minutes from aqueous solutions of 4.6mM CuCl, 25.0mM $GaCl_3$ y 9.2mM SeO_2 in 2.0M de KSCN at pH 2.5 was found the optimum condition for obtain the $CuGaSe_2$ thin films with good morphology and good stoichiometric. The samples were annealed at temperature of 520^{0}C in inert atmosphere for 30 minutes to improve their crystallinity. The obtained $CuGaSe_2$ thin film has a direct band gap (E_g) of 1.65eV. In addition, the cyclic voltammetry (CV) studies were performed in order to elucidate the electrodic processes that occur for the formation of semiconductor $CuGaSe_2$. Therefore, the electrochemical behaviour for each ion (Cu^+, Ga^{3+} and Se^{4+}), for binary (Cu-Se, Ga-Se, and Cu-Ga) and for ternary (Cu-Ga-Se) systems are described. This material has potential application in solar cells.

1. Introduction

Thin solar cells based on $Cu(In,Ga)Se_2$ (CIGS) absorber layers have demonstrated 21.7% conversion efficiencies in laboratory devices [1]. Moreover, $CuGaSe_2$ (E_g=1.65eV and α=1,2x10^{4}cm^{-1}) is one of the most-promising wide-gap chalcopyrite absorber and a strong candidate material high open-circuit voltage in thin films solar cells [2]. The current record for thin films solar cell devices with this ternary compounds prepared by physical vapour deposition (PVD), have reached 9.7% efficiencies [3]. This technique produces high-quality materials, but involves a process that is difficult to increased to large-scale. Electrodeposition is an alternative method to obtain good quality and large-scale thin films. However, to obtain high quality layers, parameters such as pH, reactant concentrations, reduction potential, etc., are critical and need optimization, since these parameters influence the deposits stoichiometry and the semiconductor electrical properties [4]. Previously, several research groups have reported the preparation of $CuGaSe_2$ using electrochemical techniques [5-8]. In this paper, we present the detailed procedure to obtain $CuGaSe_2$ thin films grown by one-step electrodeposition on Ti and SnO_2/glass from aqueous solutions containing thiocyanate ions.

2. Experimental procedure

The CV studies and the electrodeposition of $CuGaSe_2$ thin films, were performed in a three-electrode cell, connected to a BAS Epsilon potentiostat. Pt foil was the reference electrode, SCE (Saturated Calomel Electrode) was the reference electrode, and Ti foil or SnO_2/glass was the working electrode.

The electrolytes solutions consists of 4.6-5.0mM CuCl, 5.0-9.2mM SeO_2, 25.0 $GaCl_3$, and 2.0M KSCN at pH 2.5. The films were grown at different potentials (Table 1) and concentration ratios in the electrolytic medium (Table 2). Annealing at 520^{0}C in inert atmosphere for 30 minutes improved the the $CuGaSe_2$ thin films crystallinity. The chemical composition, morphology and crystalline properties of the films were analysed by scanning electron microscopy (SEM, Hitachi Model S-2500) equipped with energy dispersive spectrometer (EDS, Noran System Six Company). The crystal structure was analysed by X-ray diffraction (XRD, Siemens model D5005). The E_g was determined by transmittance measurement using UV-VIS-IR spectrophotometer (Cary 17I).

3. Results and discussions

3.1. Electrochemical studies

The electrochemical behaviour of the individuals constituents, binary and ternary systems were systematic studied in order to elucidate the electrodic processes that occur for the formation of semiconductor $CuGaSe_2$. Figure 1 shows the CV reduction potentials for the three ions. The thiocyanate ion effect is to shift the Cu^+ reduction potential in the negative direction to about -0,78V, where metallic gallium is also deposited. Moreover, a reduction peak at -0.68V probably corresponds to Se^{4+} reduction to Se^0.

The Cu-Ga-Se system voltammogram (Figure 2) presents a multiplicity of cathodic peaks that by comparing with the electrochemical behaviour of the binary systems each process was analysed. At cathodic potentials between -0.35 to -0.75V a peak is observed which is attributed to the formation of binary phases $Cu_{2-X}Se$ [9,10]. At more negatives reduction potentials, beyond -0.75V, two reduction peaks were observed which are assigned respectively to Cu and Ga metallic, respectively. Furthermore, in the Ga-Se system, a cathodic peak was observed at -0.52V corresponding to the formation of Ga_2Se_3, which could to gallium incorporation in the film, increasing its concentration (Table 1). These studies suggest that the formation of $CuGaSe_2$ could occur through the reaction [8]:

$$Cu_{2-X}Se + Ga_2Se_3 \rightarrow 2CuGaSe_2 \quad (1)$$

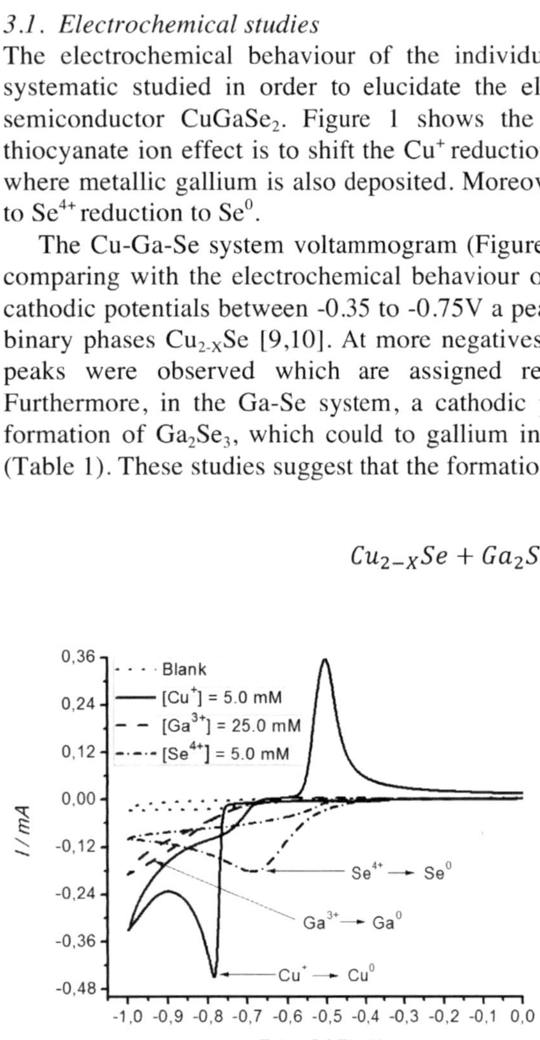

Figure 1. CV obtained on Ti from the precursor ions Cu^+, Ga^{3+} and Se^{4+}, respectively.

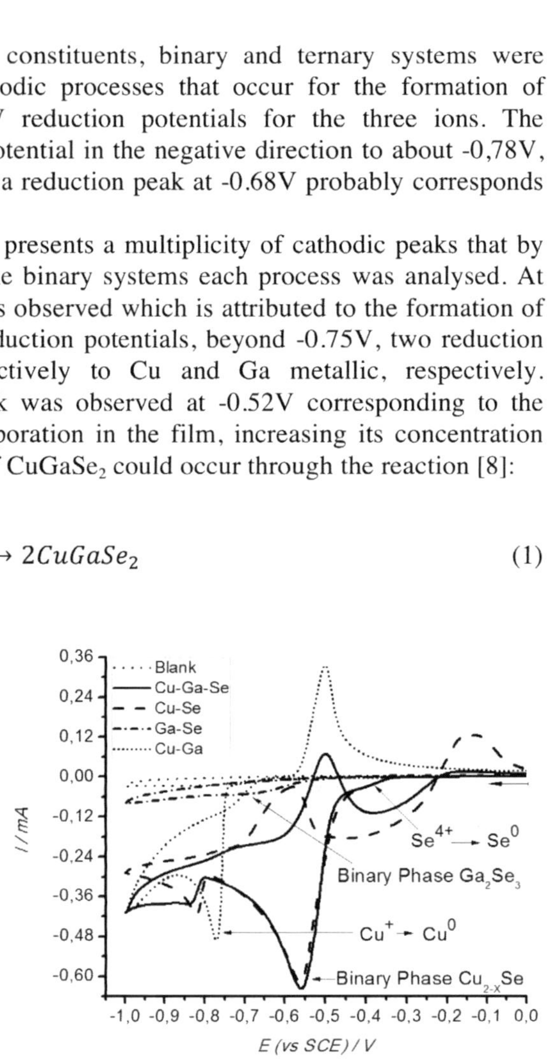

Figure 2. CV of binary and ternary systems obtained on Ti.

3.2. Compositional and morphology studies

The atomic composition and morphology of the Cu-Ga-Se films were analysed by EDS and SEM, respectively. Table 1 summarizes the influence of deposition potential on the film compositions and morphology. In all cases, a solution containing 5.0mM CuCl, 25.0mM $GaCl_3$ and 5.5mM SeO_2 was used. In the range between -0.35 and -0.45V, the amount gallium is too low, this is because the reduction reaction of Ga^{3+} to Ga^0 cannot take place, the gallium insertion into the film proceed by formation of gallium oxide or hydroxide [7]. In addition, in this range a $Cu_{2-x}Se$ phase is also deposited, giving an induced deposition mechanism [10]. The micrograph image (Figure 4) showed dispersed deposits attributed to copper selenide phases.

At -0.55V cathodic potentials, it is observed that gallium atomic concentration increases due to gallium, incorporation in the film by metal selenides formation (Figure 2). At this potential, $Cu_{2-X}Se$ formation is observed (Figure 2). In the range between -0.55 and -0.75V it can be seen that the Ga/Cu ratio remains almost constant. This demonstrates that the stoichiometry of the films is independent of the applied deposition potential in this range. Additionally, the synthesized films have a stoichiometric ratio very close to the ternary compound $CuGaSe_2$ (Figure 3, Table 1) but with $Cu_{2-X}Se$ excess. Moreover, the micrograph image (Figure 4) showed a homogeneous granular morphology.

Table 1. EDS analysis of Cu-Ga-Se layers obtained at different potentials in aqueous solution with 25.0mM Ga^{3+}, 5.0mM Cu^+, 5.5mM SeO_2 and 2.0M de KSCN at pH 2.5 (5.0:25.0:5.0).

Potentials (V vs SCE)	Atomic percent (%)			Ga/Cu ratio	Stoichiometry
	Cu	Ga	Se		
-0,35	34,86	7,45	57,69	0,21	$CuGa_{0,21}Se_{1,65}$
-0,45	31,36	14,11	54,53	0,45	$CuGa_{0,45}Se_{1,74}$
-0,55	27,05	17,78	55,17	0,66	$CuGa_{0,66}Se_{2,04}$
-0,65	27,74	18,06	54,20	0,65	$CuGa_{0,65}Se_{1,97}$
-0,75	26,40	17,77	56,33	0,67	$CuGa_{0,67}Se_{2,13}$
-0,85	28,84	14,96	55,39	0,52	$CuGa_{0,52}Se_{1,92}$

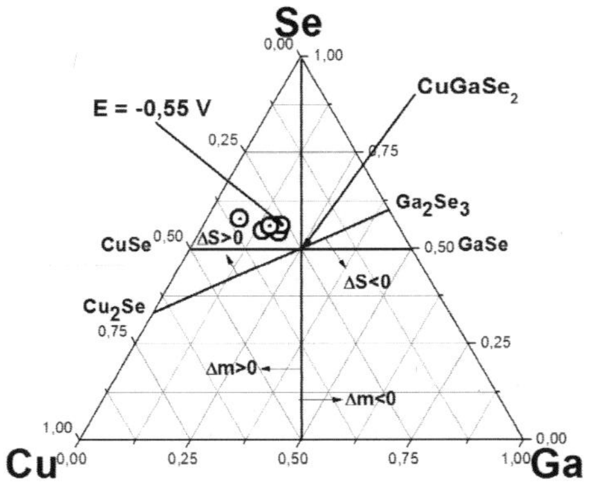

Figure 3. Ternary phase diagram of $CuGaSe_2$ electrodeposited at different potentials.

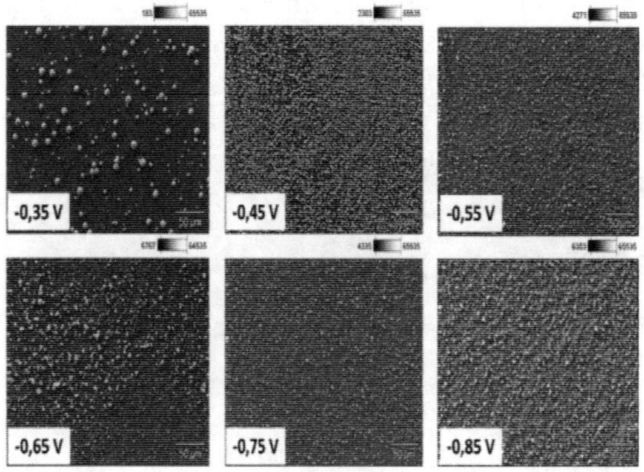

Figure 4. Morphology of $CuGaSe_2$ electrodeposited at different potentials.

The compositional data obtained in the Table 1 have been expressed into deviation from molecularity (Δm) and deviation from stoichiometry (ΔS) in Figure 3 and take into account the role of the deviation from the ideal composition. In Figure 3 such a schematic ternary phase diagram for $CuGaSe_2$ is shown. To describe the $Cu_xGa_ySe_z$ composition of a point within the ternary field relative to the composition of the chalcopyrite, Groenick and Janse [11] have introduced two independent parameters, i.e. the deviation in molecularity Δm and the deviation in valence stoichiometry ΔS:

$$\Delta m = \frac{X}{Y} - 1,\ \Delta S = \frac{2Z}{X+3Y} - 1 \tag{2}$$

where x, y, and z are the atomic concentrations of the individual atoms. As indicated in the Figure 3, ΔS describes compositional deviations from the pseudo binary Cu_2Se-Ga_2Se_3 line, where ΔS<0 corresponds to anion poor and ΔS>0 to anion rich material. Δm divides the triangle into Cu-rich and Cu-poor compositions. In our case, it can be seen that the films synthesized at potentials between -0.35 to -0.85V have ΔS>0 and Δm>0, indicating that these films contain an excess of selenium and copper, respectively. This means that these potentials would form the ternary semiconductor with an excess of the $Cu_{2-x}Se$ phase.

Based on the morphologic analysis a -0.55V potential was selected as the optimum for the $CuGaSe_2$ deposition, since it requires less energy to produce uniform and homogeneous films. However, it is necessary to do further studies changing the concentration of the ions, trying to reduce the binary $Cu_{2-x}Se$ phase formation. In Table 2, is shown the atomic percent of the deposits, changing the concentration of the precursor ions. It is shown that with the 4.6:25.0:9.2 ion concentrations give a stoichiometric relation close to the expected ternary.

Table 2. EDS analysis of electrodeposited Cu-Ga-Se thin films at -0.55V for 20 minutes from different concentration ratios.

Cu:Ga:Se	Atomic percent (%)			Stoichiometry
	Cu	Ga	Se	
5,0:25,0:5,5	27,05	17,78	55,17	$CuGa_{0,66}Se_{2,04}$
4,6:25,0: 5,5	26,55	25,36	44,09	$CuGa_{1,11}Se_{1,50}$
4,6: 25,0: 9,2	23,91	23,53	52,53	$CuGa_{0,98}Se_{2,20}$

3.3. X-ray diffraction analyses

The XRD analyses of $CuGaSe_2$ thin films (Cu:Ga:Se=4.6:25.0:9.2) electrodeposited on Ti and SnO_2/glass substrates are shown in Figure 5 and Figure 6, respectively. Figure 5(a) shows XRD of as deposited on Ti substrate, in which only shows peaks due to the substrate. After the thermal treatment at 520^{0}C, the peaks corresponding to the chalcopyrite $CuGaSe_2$ are observed (112), (200, 204) and (312, 116), (JCPDS 75-0104) (Figure 5(b)). Similar results are observed for the films on SnO_2/glass under similar conditions (Figure 6(a), (b), (c)). In both substrates the chalcopyrite is the only phase. This is very important results.

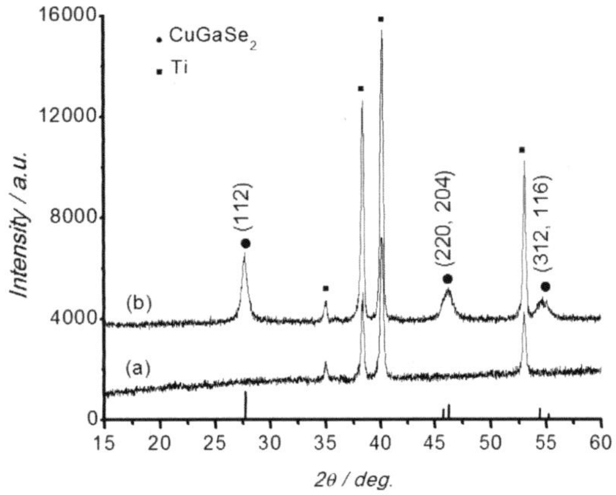

Figure 5. XRD patterns for (a) as-deposited on Ti substrate and (b) with annealing at 520^{0}C for 30 min in inert atmosphere.

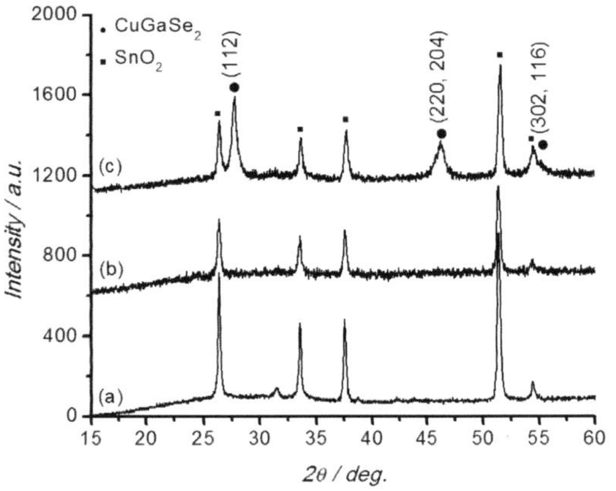

Figure 6. XRD patterns for (a) SnO_2/glass, (b) as-deposited and (c) with annealing at 520^{0}C for 30 min in inert atmosphere.

3.4. Optical absorption spectroscopy analyses

Using the transmittance (T) values and film thickness (5.6 microns) the absorption coefficient (α) it was obtained. The band gap energy was estimated by extrapolating the linear portion of $(\alpha h\nu)^2$ vs

$(h\nu)$ plot to the intercept on the x-axis. Figure 8 reveals that the obtained $CuGaSe_2$ thin film has a direct band gap (E_g) of 1.65eV.

Figure 7. SEM cross-section image of the $CuGaSe_2$ (5.0:25.0:9.2) with annealed.

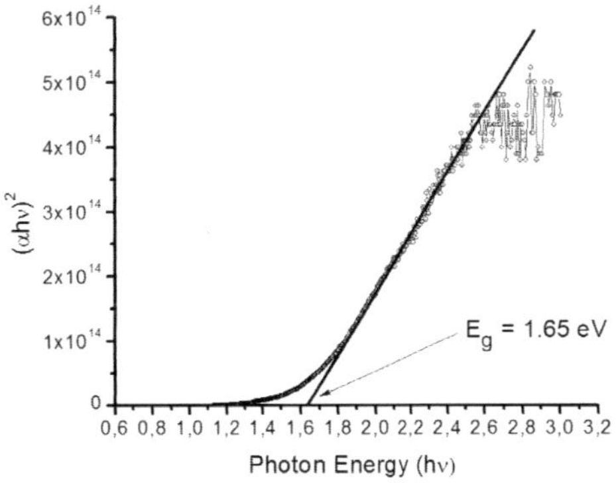

Figuras 8. Plot of $(\alpha h\upsilon)^2$ against photon energy for the $CuGaSe_2$ thin film on SnO_2/glass.

4. Conclusions

In this paper, it was found that we could obtain by electrochemical method good thin films of $CuGaSe_2$ over Ti and SnO_2/glass substrate with applying at potential of -0.55V during 20 minutes of deposition. The electrolytic medium was 4.6mM of CuCl, 25.0mM of $GaCl_3$ y 9.2mM of SeO_2 in 2.0M of KSCN at pH 2.5. Annealing at 520^{0}C in inert atmosphere for 30 minutes improved the $CuGaSe_2$ thin films crystallinity. The crystal structure of thin films was chalcopyrite type with space group $I\bar{4}2d$, with the direct energy gap of 1.65eV and high optical absorption coefficient. This material has potential application in solar cells.

Acknowledgements

The authors acknowledge Consejo de Desarrollo Científico Humanístico y de Arte (CDCHTA, projects C-1477-07-08-A and C-1477-07-06-E), and the Venezuelan National Science and Technology Foundation (FONACIT, project PEII-2011001099 and Mision Ciencias) for their financial support.

References

[1] Jackson P, Hariskos D, Wuerz R, Kiowski O, Bauer A, Frieldmeier T and Powals M 2014 *Phys Status Solidi RRL* **9** 28

[2] Rusu M, Wiesner S, Fuertes-Marron D, Meeder A, Doka S, Bohne W, Indner S, Schedel-Niedrig T, Giesen C, Heuken M and Lux-Steiner M 2004 *Thin Solid Films* **556** 451

[3] Saad M, Riazi H, Bucher E and Lux-Steiner M 1996 *Applied Physics A* **62** 181

[4] Kapur V, Basol B and Tseng E 1987 *Ternary and Multinary Compounds: Proceedings of the 7th International Conference, Snowmass, Colorado* (Pittsburgh: Materials Research Society)

[5] Oda Y, Minemoto T and Takakura H 2008 *J Electrochem Soc* **155** H292

[6] Muthuraj J, Rasmussen D and Suni I 2011 *J Electroanal Soc* **158** D54

[7] Liu F, Yang J, Zhou J, Lai Y, Jia M, Li J and Liu Y 2012 *Thin Solid Films* **520** 2781

[8] Lee H, Lee J, Hwang Y and Kim Y 2014 *Current Applied Physics* **14** 18

[9] Vedel J 1997 *Inst Phy Conf Serial* **152** 152

[10] Kemell M, Saloniemi H, Ritala M, Leskela M and Rauhala E 2000 *J Electrochem Soc* **147-3** 1080

[11] Groenick and Janse 1978 *Z for Pysikalische Chemie N F* **110** 17

[12] Manfredy L 2014 *Preparación electroquímica de semiconductores ternarios $CuInS_2$ y $CuGaSe_2$* (Mérida: Universidad de los Andes)

Analysis of the influence of the applied voltage and the scan speed in the atomic force microscopy local oxidation technique

C Mendoza[1,2], A Plata[1], Z Lizarazo[1] and C A Chacón[1,2]
[1] Universidad Industrial de Santander, Bucaramanga, Colombia.
[2] Laboratorio de Microscopia, Parque Tecnológico Guatiguará, Piedecuesta, Colombia.

E-mail: carolinamendozaluna@gmail.com

Abstract. The local nanolithography oxidation technique is implemented by using an atomic force microscope (AFM) for the fabrication of nanoscale patterning structures on a silicon substrate covered whit a thin film of silicon nitride. During the fabrication process, the microscope is operated on air and contact mode utilizing a silicon tip covered whit a hard Cobalt-Chromium coat. The dependence of the oxide growth with the applied voltage was investigated varying this parameter in a range of 1 to 10V to constant scanning speed; the influence of the writing speed in the dimensions of the oxide formed is also analysed varying the speed values between 0.1 to 1μm/s. Is found that the dimensions of lines depend of scanning speed and voltages applied.

1. Introduction

The fabrication of micro-scale and nano-scale devices is a fundamental part of the nanoscience, nanotechnology research and some of their new applications. However, this introduces a technological challenge due to the requirement of the high-resolution nanolithography technique that allows defining whit precision the geometry of the desired pattern. For this purpose, there is a range of lithography techniques to transfer the patterns onto different kind of substrates such as the Scanning Probe Lithography technique (SPL) and electron, ion and light beams [1,2]. Thus, compared whit other micro and nano-scale fabrication techniques, the Scanning Probe Lithography is a low cost option; the AFM techniques offer the possibility to write directly patterns under flexible environmental conditions and can be implemented through different mechanisms. Thus, the surface of a sample can be modified under the influence of an intense localized electric field, similarly used for the anodic oxidation, surface decomposition, modification by thermal effect or by the tip indentation and others [1-4].

Among the SPL techniques, the local oxidation has been widely used in conductive and semiconductive materials for the direct fabrication of different kind of devices such as: dielectric barriers, mask for selective etching, and the fabrication of a wide range of electronic and mechanical devices/structures at nanoscale that includes conductive nanowires, transistors, quantum rings and dots, microlenses, etc [1,2,5,6]. Due to the versatility of this technique and with the purpose of continuity to previous works realized by the laboratory inn fabrication of diffractive elements using AFM [7]. In this work the anodic local oxidation technique is implemented to create the pattern desired. The first step in the implementation of the local anodic oxidation lithography is the analysis of the experimental parameters that influence the oxide height and width [8]. Such, different scanning speeds and voltage values, to find the minimum lateral dimension possible to achieve for the experimental conditions described in this paper.

2. Local Anodic Oxidation

The local oxidation SPL technique consist in the application of a positive voltage between the AFM tip and the surface of the sample in an atmosphere with relative humidity greater than 50%, this will induce the local oxidation on the surface of the substratum. The humidity in the ambient generates a meniscus of water between the tip and the surface of the sample, the applied voltage ionizes the molecules producing OH^- ions, which are accelerated through the surface where the oxidation reaction occurs; the required local electric field must be higher than the critical electric field whose value is around 1V/m [9,10] then the OH^- ions are transported to the surface of the sample guided by the electric field forming oxide structures while the reaction with the atoms of the surface take place. The oxide growth will depend on experimental conditions like scanning speed, applied voltage, ambient humidity and others.

The oxidation process reveals an electrochemical nature due to the presence of currents of order of picoamps that induce the oxidation along with the presence of a meniscus of water creating basically an electrochemical nanocell. For the silicon nitride the oxidation reaction on the anode (surface of the sample) occurs based on $Si_3N_4 + 6(OH^-) \rightarrow 3SiO_2 + 6e^-$ [9, 11].

3. Experimental parameters

For the local oxidation process and the acquisition of images a di-CPII Veeco Atomic Force Microscope is used, to generate oxide lines the Nanolithography 1.9 software is used and for image treatment and its analysis the SPMLab Ver. 6.02 software is used. Then, the Voltage-Pulse operation mode allows applying a voltage difference between the AFM tip and the sample up to ±10V.

For the experimental study of this parameters in the local oxidation nanolithography process, lines with a separation of 1μm are created on a substrate of nitride on silicon wafer p type <100>, resistivity 0.001-0.005Ohm*cm with a RMS roughness of 0.4nm, using a silicon AFM tip with a hard Co-Cr coat, with a nominal radius of 35 nm; the microscope is operated under environmental conditions of humidity and temperature (65% y 20°C, respectively). Then, each of the parameters is varied as follows: scanning speed values between 0.1 and 1μm/s are taken applying a constant voltage of 10V. The variation of the voltage from 1 to 10V is perform to constant speed of 0.5μm/s and finally the oxide dimensions, height and width, are measured along the line to obtain an average value.

4. Results

The local oxidation was induced on a silicon wafer with a thin layer of silicon nitride, varying the scanning speed with a constant voltage of 10V. For the characterization of the oxide dimensions an AFM image was taken in contact mode and the height and width are measure form the profile Figures 1 and 2. The average height and lateral dimensions in function of the scanning speed are showed in Figures 3(a) and 3(b). For the voltage analysis, the same procedure is used varying the applied voltage between 1 to 10V forming lines with a step of 1μm for the scanning speed of 0.5μm/s, the results are showed in Figures 4(a) and 4(b).

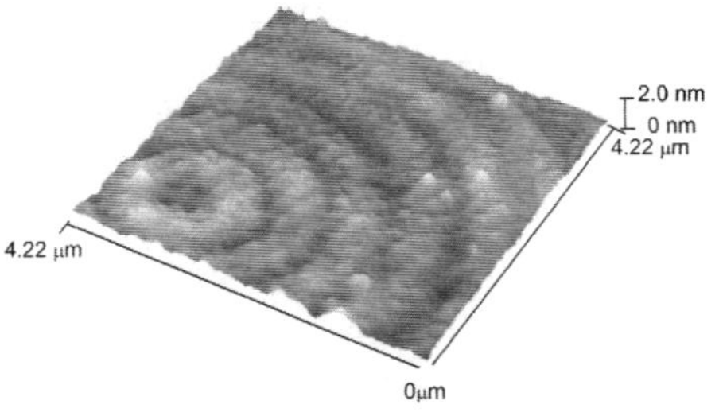

Figure 1. 3D AFM image showing the results for anodic oxidation for six of the ten values of scanning speed (0.1 to 1μm/s from centre to edge).

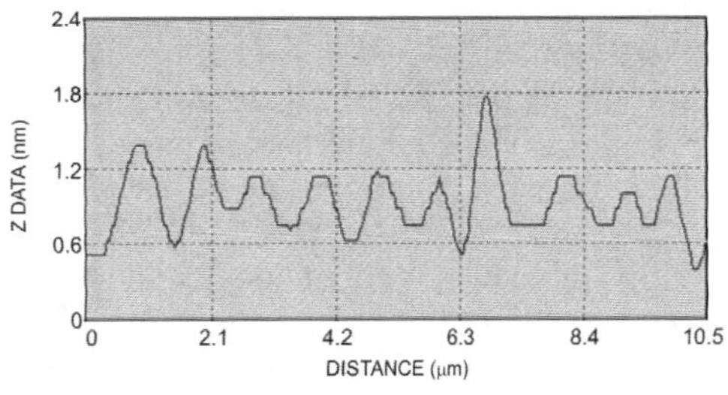

Figure 2. Profile of the oxide lines generated using different values of scanning speed between 0.1μm/s to 1μm/s (from left to right).

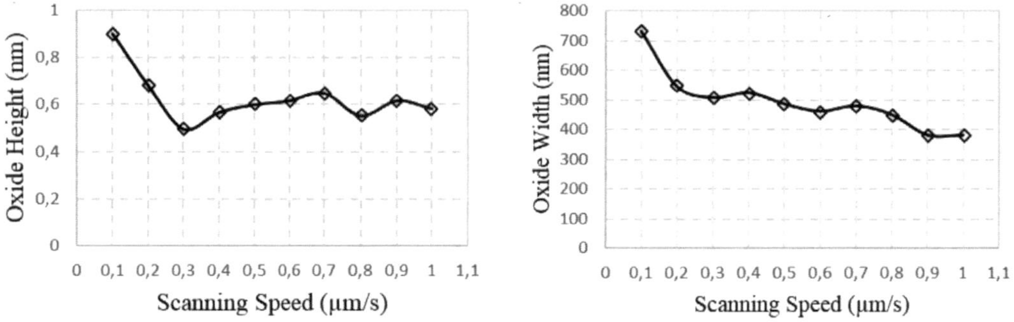

(a) (b)

Figure 3. (a) Average oxide height in function of the scanning speed, (b) Average oxide width in function of the scanning speed.

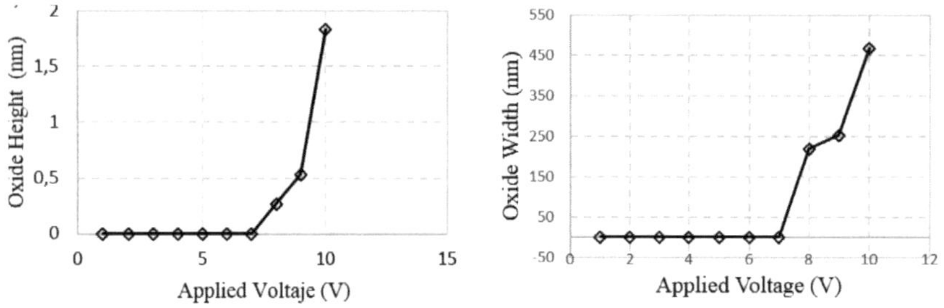

(a) (b)

Figure 4. (a) Average oxide height in function of the applied voltage, (b) Average oxide in function of the applied voltage.

5. Conclusions

The implementation of the local anodic oxidation lithography technique allow produce homogeneous lines with high fidelity. An inverse relation in the oxide height compared with the scanning speed is found with low variation in this dimension for speeds greater than 0.3μm/s. The same relation is found between the lateral dimension of the generated line and the scanning speed, obtaining a minimum width for a speed of 1μm/s. Also, a direct relationship between the lines dimensions and the applied voltage, finding that is required a voltage greater than 7V to induce the local oxidation of the surface. Such that, these are the optimal values for the implementation of the local anodic oxidation lithography technique with an Atomic Force Microscope.

References

[1] Barnett M R, Nave M D, Bettles C 2004 *J Mater Sci Eng A* **386** 205
[2] B Bhushan 2010 *Handbook of nanotechnology 3rd ed* (Berlin: Springer)
[3] Hans-Eckhardt Schaefer 2010 *Nanoscience the science of the small in physics engineering chemistry biology and medicine* (Berlin: Springer)
[4] Zheng Cui 2008 *Nanofabrication principles capabilities and limits* (US: Springer)
[5] A Majumdar, P I Oden, J P Carrejo, L A Nagahara, J J Graham, et al 1992 *Appl Phys Lett* **61** 2293
[6] Luis Guillermo Villanueva, Oscar Vazquez-Mena, Cristina Martin-Olmos, Veronica Savu, Katrin Sidler and Juergen Brugger 2013 *Micromachines* **4** 370
[7] J A Dagataa, F Perez-Murano and G Abadal K Morimoto, T Inoue, J Itoh and H Yokoyama 2000 *Applied Physics Letter* **76-19** 2710
[8] Bharat Bhushan 2010 *Scanning probe microscopy in nanoscience and nanotechnology* (Berlin: Springer)
[9] Marta Tello, Fernando García and Ricardo García 2002 *Journal of Applied Physics* **92-9** 5539
[10] P Avouris, T Hertel, R Martel 1997 *Appl Phys Lett* **71**, 285
[11] Avouris, R Martel, T Hertel, R Sandstrom 1998 *Appl Phys A Materials Science & Processing* **66** S659

Structural and impedance analysis of Co-doped $SrTiO_3$ perovskite

E Echeverri[1] and O Arnache[1]
[1] Universidad de Antioquia, Medellín, Colombia.

E-mail: oscar.arnache@udea.edu.co

Abstract. $SrTi_{1-x}Co_xO_3$ ($0.2 \geq x \geq 0$) polycrystalline samples were prepared by solid-state reaction. X-ray diffraction (XRD) analysis shown the perovskite type structure is conserved for all samples without impurities. A small increase in the lattice parameters were observed for $x \geq 0.05$. Morphology and composition were analysed by scanning electron microscopy (SEM-EDX). Impedance spectroscopy measurements form 50Hz to 1MHz were made at different temperatures (25–400°C). The spectra were analysed by Z' vs Z'' plots, which reveal 3 contributions associated to electrodes, grain boundary and grain. From the fits the frequency (*f*) and times relaxation (τ) of the grain were estimated, with values of f~15KHz and τ~67μs for 20% Co samples at RT. A dispersion in the permittivity ε^* at low frequency (<300Hz) were observed, which increase with the temperature. Each component of ε^* converge for frequencies up to ~300kHz. Finally, a behaviour and activation energy analysis of the electrical conductivity is presented from ln(σ) vs 1/T plots.

1. Introduction

$SrTiO_3$ are materials widely used in microelectronic devices (multilayer capacitors, DRAM, thermistors, piezo actuators, varistors, etc) by its dielectric properties and thermodynamically stable at high temperature. Beyond its technological relevance, $SrTiO_3$ (STO) is a useful model material [1]. It exhibits both ionic and electronic conduction, and its bulk behaviour is well understood. However, bulk and grain boundary resistivity in $SrTiO_3$ ceramics changes with acceptor doping (Ni, Fe, and Al) [2-4]. These resistive grain boundary barriers are reduced and field-enhanced leakage currents though the material, especially at grain boundary barrier interface a detailed dielectric studies in the perovskite-type titanates is required. The impedance spectroscopy has been a powerful tool to distinguish the grain and grain boundary electrical contribution of many oxide ceramic materials [4, 5]. Data from impedance can be analysed using different complex formalism: $Z^*=Z'-jZ''$, where Z' and Z'' are the real and imaginary components of impedance, respectively. Electrical modulus $M^*=j\omega C_0 Z^*$, where angular frequency $\omega=2\pi f$, f is the applied frequency (in Hz) and C_0 is the empty cell capacitance. On the other hand, the induction of ferromagnetism in non-magnetic materials by doping with magnetic impurities (Co, Mn) [6,7] has recently gained considerable attention because potential applications in spintronic. In this way, since evidenced ferromagnetism in cobalt-doped anatase thin films, opened an avenue for exploration of other titanium-based oxides [7]. Therefore, new mechanisms have to be proposed to explain the transport and magnetic behaviour. In this paper, $SrTi_{1-x}Co_xO_3$ (x=0, 0.3, 0.5, 0.1 and 0.2) ceramics have been prepared by solid state reaction. The temperature and frequency dependence on the impedance and conductivity for STO and Co-dope samples in the frequency range 50Hz to 1MHz and in the temperature range RT-400°C have been investigated and analysed to determine some Co effect in the conduction mechanisms.

2. Experiments

Strontium titanate ($SrTiO_3$) and $SrTiO_3$Co-doped (Co-3%, Co-5%, Co-10% and Co-20%) powder used in this study were obtained by solid state reaction, prepared in air by mixing $SrTiO_3$, TiO_2, and Co_3O_4 precursors in stoichiometric according to the formula $SrTi_{1-x}Co_xO_3$ (x=0, 0.03, 0.05, 0.1 and 0.2). For each concentration all powder were ball-milled in a conventional planetary ball milling (Fritsch Pulverisette P5) using agate vials at 100RPM by 2h. Powder as milled were heated to 1100°C at a rate of 2°C/min, held at this temperature for 12h, and furnace-cooled to room temperature. The calcined powders were pressed into discs specimens with diameter ϕ=6 mm and a thickness of t~1.2mm. The discs were sintered at 1150°C for 12h. X-ray diffraction experiments were performed after the powders were obtained in order to identify the main crystalline phase of $SrTiO_3$ powders and to determine the average crystallite size, recording the diffractogram from 10° to 100° 2θ, 0.013 step sizes in a Empyre (PANalytical), with a Cu Ka radiation. Rietveld's whole X-ray powder fitting method was used to analyse the diffraction patterns. The Rietveld quantitative analyses reported here used the FULPROF program. The Williamson–Hall method was used to evaluate the crystallite size; this method is based on the line broadening of the diffraction lines due to the strain and crystallite size. The microstructure of the powders was studied by a JEOL JSM 6490LV Scanning Electron Microscopy (SEM). Energy dispersive X-ray microanalysis (EDS) of samples was carried out to estimate the phase composition. All the discs were polished and thin layer of silver paste was applied as the electrode and baked at 650°C for 30min to dielectric measurements. The dielectric properties of $SrTiO_3$ were determined by using a FLUKE PM6306 RCL Meter in the frequency from 50Hz to 1MHz, connected with a PC. Capacitance was measured in the range from RT to 400°C at 100°C interval.

3. Results and discussion

Figure 1(a) shows the XRD pattern for the cobalt doped samples at 3, 5, 10 and 20%, which were similar respect to the pure $SrTiO_3$ sample. The results revealed that all samples are single phase Perovskites without any detectable impurity or secondary phase. The data were refined using the Rietveld method assuming a structural model based on a Pm3m cell. The lattice parameters were plotted as a function of Co doping (Figure 1(b)). The lattice parameter decreases as the cobalt content increase until 5%, above this concentration the lattice parameter increase. This behaviour probably is attributed by two factors: first, possible disproportionation of Co ions (Co^{+2} and Co^{+3}) in the Ti sites and second, possible oxygen vacancy in the perovskite structure. The Williamson-Hall analysis and size-strain plot method were used to study the individual contributions of crystallite sizes on the peak broadening of the samples. SEM images of the powders are shown in the inset in Figure 1(c). From images, the grain are similar for all samples, there were agglomerations of powder bonded with grains (dendritic type morphology). On the other hand, Co substitution for all samples was confirmed by EDS analyses, the average Co compositions of the samples by elementary analysis are similar to the nominal values. Figure 2(a) shows the Cole-Cole plots at RT. Clearly two semicircles were observed in Co-doped samples at different temperatures. These can be assigned to charge transport within the grain and grain boundary effect, respectively. In general, the grain boundary effect on electric properties may originate from a grain boundary potential barrier, which should be related to Ti-rich boundary [8]. In the case of STO sample we did not see any semicircles in the frequency and temperature ranges considered in this study. The large semicircle at low frequencies being due to space depletion layer at grain boundaries and the small one at high frequencies being due to the bulk. In order to quantitatively analyse resistances at grains (R_g) and grain boundaries (R_{gb}), experimental impedance spectra, were simulated using a corresponding equivalent circuit with two parallel RC elements in series. Therefore, R_g and R_{gb} at each test temperature were extracted from the intercept of the semicircle on the real axis (Z′) in complex plot (Figure 2(b)). The bulk signal is particularly pronounced in the Z′ and Z′′ plot in Co-10% and Co-20% (Figure 2(c)).

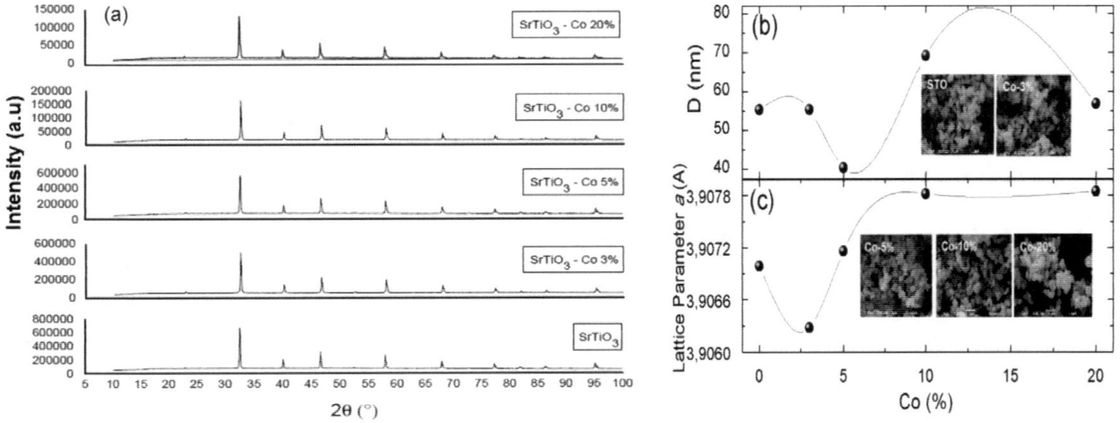

Figure 1. (a) XRD patterns of $SrTi_{1-x}Co_xO_3$ (b) Crystallite Size as a function Co doping in STO structure by Williamson-Hall analysis. (c) Lattice parameters as a function Co-doping. Continuous line is a visual guide. Inset in (b) and (c) show SEM surface morphologies of $SrTi_{1-x}Co_xO_3$ samples.

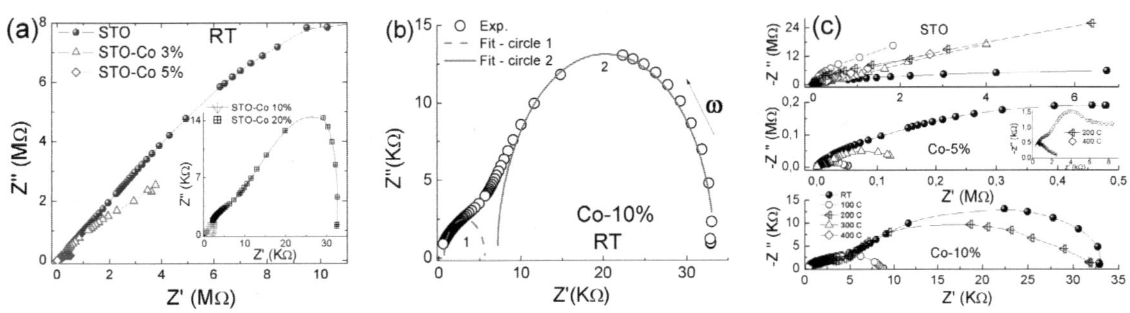

Figure 2. (a) Cole-Cole plots of STO, Co-3%, Co-5%, Co-10% and Co-20% (b) Complex impedance plane plot of Co-10% at RT. Continuous line are the fits (1) and (2) corresponding to grain and grain boundary contributions, respectively. The arrow address means the sense of increasing the frequency (ω). (c) Complex impedance plot (Z′ and Z′′) of the STO, Co-5% and Co-10% samples at different temperatures from RT to 400C. Inset show complex impedance for Co-5% in a less scale for 200 and 400°C.

Figure 3(a) shows the imaginary part of the dielectric modulus, M′′, as a function of frequency for STO, Co-5% and Co-10% samples at different temperatures. It is clear that the maximum in M′′ peak shifts to higher frequency with increasing temperature. This effect is most evident in STO Co-20% sample. The frequency region below peak maximum M′′ is associated to the range in which charge carriers are mobile over long distances. For a frequency above peak maximum M′′, the carriers seem to be confined to potential well, thus becoming mobile over a short distance [9]. It is noted that in Co-doped samples (3, 5 and 10%), M′′ curves do not shown a characteristic peak associated to relaxation time over the temperature range investigated in this work. For the purpose of a more clarity and quantitative treatment, we consider the Bode plots in Figure 3(b) which better resolve the contributions from the more conductive grain interior. As seen from Figure 3(b), one relaxation time is clearly identified for Co-20% sample. In order to estimated conductivity of total bulk, the R_g at each test temperature was converted into conductivity (σ_{dc}) using $\sigma = t/(A*R_g)$, where t is the sample thickness and A is the electrode area in the sample surface. Figure 3(c) shows the values of σ_{dc} for grains as a function of temperature in Co-3%, Co-5% and Co-20% samples. The results demonstrate that the conductivity of grains follows the Arrhenius law, and activation energy can be calculated from the slope of the linear portion of $\ln(\sigma_{dc})$ vs $10^3/T$ graph. This energy is important because plays a major role in leakage currents behaviour. The activation energy for grain (E_g) conduction in Co-3%, Co-5%

and Co-20% samples, according to the fit shown in Figure 3(c) were calculated. The activation energies of grain samples were 1.895, 1.891 and 1.08eV, respectively. This result shows that the activation energy of the grain in Co-20% doped sample is lowest. A possible explanation for this is the fact that cobalt content reduces the height of the moving barrier of charges in grains and therefore Co-20% can exhibits a larger leakage current. Finally, these experimental results imply that the natures of charge carriers responsible for dielectric relaxation peaks and dc conduction in Co-20% belong to the same category, which indicate that the polarization relaxation has a strong relation to the conductivity in the grain interior. The relaxation times estimated in this work for Co-20% were 66.7, 0.25 and 0.17μs for RT, 100 and 200°C, respectively.

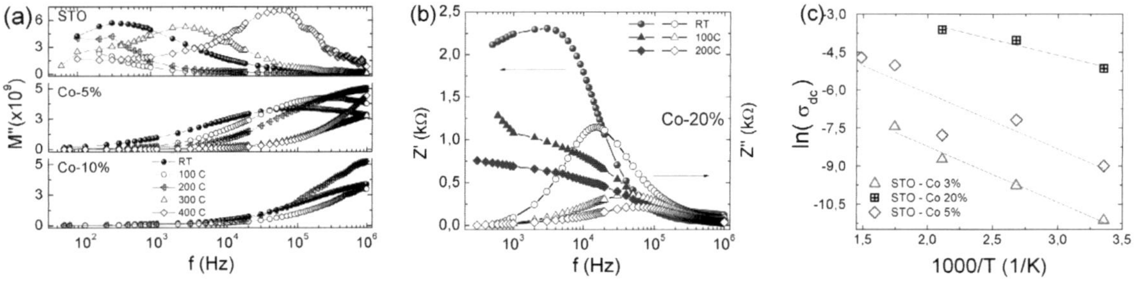

Figure 3. (a) Frequency dependence of the dielectric modulus, M´´, for pure and Co-doped STO (5 and 10%) at different temperatures (RT, 100, 200, 300 and 400°C). (b) Complex impedance (Z´ and Z´´) as a function frequency of STO Co-20% samples at RT, 100 and 200C. (c) Variation in total conductivity with inverse of temperature (1000/T) for Co-3%, Co-5% and Co-20% samples. Line is the fit using Arrhenius law.

4. Conclusions

In this work $SrTi_{1-x}Co_xO_3$ ceramics were synthetized by solid state reaction and characterized by XRD, SEM-EDS. It was found that samples show a perovskite single phase. Impedance spectroscopy studies have been performed to determine the correlation between Co-doped effect and electrical conduction behaviour. Co-doped ceramics exhibit grain and grain boundary evidenced by two semicircles in complex impedance plots. The activation energy of grain conductions is low in Co-20% samples, this indicate that Co doping can induce an easy conduction in the grains, where the charged particles participate in the conduction process at high temperatures. Into account the relaxation process, the results indicate that measures from lower frequencies are required to study the relaxation phenomena in STO and Co-doped samples.

Acknowledgments

This work was supported by Universidad de Antioquia under Solid State Group – GES sustainability strategy 2014-2015.

References

[1] Vollman M and Waser R 1994 *J Am Ceram Soc* **77** 235
[2] Waser R 1991 *J Am Ceram Soc* **74** 1934
[3] Denk I, Munch W and Maier J 1995 *J Am Ceram Soc* **77** 3265
[4] Rodewal S, Fleig J and Maier J 2001 *J Am Ceram Soc* **84** 521
[5] Jurado J R, Colomer M T and Frade J R 2000 *J Am Ceram Soc* **83** 2715
[6] Pascanut C, Dragoe N and Berthet P 2006 *J Magn Magn Mater* **305** 6
[7] Malo S and Maignan A 2004 *Inorg Chem* **43** 8168
[8] Mogus-Milankovic A, Santic A, Licina V and Day D E 2005 *J Non-Cryst Solids* **351** 3235
[9] Balaya P, Jamnik J, Fleig J and Maier J 2006 *Appl Phys Lett* **88** 062109

Study of optical and structural properties of CZTS thin films grown by co-evaporation and spray pyrolysis

R Moreno[1], E A Ramirez[1] and G Gordillo Guzmán[1]
[1] Universidad Nacional de Colombia, Bogotá, Colombia.

E-mail: rmorenom@unal.edu.co

Abstract. Results regarding optical and structural properties of Cu_2ZnSnS_4 (CZTS) thin films prepared by co-evaporation using a novel procedure are compared with those obtained with CZTS films grown using a solution based route. The lattice strain ε and crystallite size D of CZTS films prepared by co-evaporation and by spray pyrolysis were estimated through X-ray diffraction (XRD) measurements using Williamson–Hall-isotropic strain model. The results of estimated average crystallite size of CZTS films by Scherrer and Williamson–Hall plot methods were compared with AFM (atomic force microscopy) measurements. It was found that the average crystallite size measured by Williamson–Hall plot methods agree quite well with AFM results. Further, information regarding the influence of preparation method on both, crystalline phases and the formation of structural defects was achieved through Raman and Urbach energy measurements.

1. Introduction

Cu_2ZnSnS_4 is emerging as a potential material that has several promising attributes for efficient, inexpensive solar cells made from abundant and nontoxic elements [1]. The first CZTS solar cell was reported by Katagiri et al. in 1997 with an efficiency of 0.66% [2]. In 2008, Katagiri's group reported an energy conversion efficiency of 6.7% [3]. Until recently state of the art CZTSSe thin-film solar cells were limited to 11.1% power conversion efficiency, being achieved via a hydrazine slurry approach [4]. Other vacuum and non-vacuum-based deposition techniques have also been successful in fabricating CZTSSe solar cells with efficiencies above 8% [5]. In 2013 Solar Frontier announced a world record for CZTS cell efficiency, reaching 12.6% [6].

This paper reports results concerning optical and microstructural properties of Cu_2ZnSnS_4 thin films prepared by both co-evaporation of precursors and spray pyrolysis from a solution containing chlorides of metallic precursors and thiourea, dissolved in dimethyl sulfoxide (DMSO). The influence of the preparation method on the crystallite size and formation of micro structural defects was studied through X-ray diffraction measurements using the scherrer and Williamson-Hall equations. In addition, we report results related to the influence of the preparation method on the crystalline phase and formation of structural defects, obtained through measurements of Raman spectroscopy and Urbach Energy.

2. Experimental

The Cu_2ZnSnS_4 films were prepared by co-evaporation of Cu, Sn and ZnS in presence of sulphur (evaporated from a Knudsen cell) using an evaporation source consisted of a cylindrical graphite crucible, which includes two coaxial chambers. The purpose of using the coaxial source is to improve the homogeneity in chemical composition of the CZTS films both laterally and in depth, which is not possible using laterally separated crucibles to evaporate the precursors. Cu and Sn evaporate from the same camera considering that they have similar evaporation temperature, while the ZnS evaporates in a separate chamber due to the fact that this evaporates at a lower temperature. To achieve simultaneous evaporation of Cu, Sn and ZnS, the coaxial source was built with a special design, which allows generating a temperature profile such that the chamber containing the ZnS is at a temperature around 980°C and the chamber containing Cu+Sn at a temperature around 1300°C.

The preparation of Cu_2ZnSnS_4 films following a solution based route was done using a solution containing copper chloride (0.01M), zinc chloride (0.005M), tin chloride (0.005M) and thiourea (0.04M) as source of metals and sulphur, respectively and dimethyl sulfoxide (DMSO) was used as solvent. A spray pyrolysis deposition system was used to deposit the CZTS thin films; this consist in a pneumatic spray nozzle that employ nitrogen as carrier gas and a solution flow rate of 15L/min. CZTS films were deposited on soda lime glass substrates held at a temperature of 350ºC. Post deposition annealing was also carried out in a furnace installed inside a vacuum chamber, kept at a temperature of 500°C for 30 minutes. The camera is initially evacuated to a pressure of 10^{-4} mbar, then nitrogen is introduced until it reaches a pressure of 20mbar and finally the furnace temperature was ramped up at 10K/min until the annealing temperature is reached.

The transmittance and reflectance measurements were performed using a Varian–Cary 5000 spectrophotometer and the film thickness was determined using a Veeco Dektak 150 surface profiler. The XRD measurements were performed with a Shimadzu-6000 diffractometer and Raman spectroscopy with a Thermo Scientific DXR Raman spectrometer with a DPSS laser of 785nm, 15mW focused with a 50X objective.

3. Results

3.1. Optical properties

In Figure 1 are compared transmission spectra of CZTS deposited by co-evaporation with those of a CZTS film grown by spray pyrolysis. It is observed that both type of CZTS films exhibit high transmittances indicating that the photon absorption occurs predominantly via band to band transitions. Curves of $(\alpha h\nu)^2$ vs hν are also depicted in Figure 1. The absorption coefficient α was calculated through the relation [7]: $\alpha = -\frac{1}{d}\ln\frac{T(\lambda)}{1-R(\lambda)}$. Where T(λ) and R(λ) are values of transmittance and reflectance experimentally obtained and d the film thickness. The Eg values obtained from the intercept with the axis hν of the curve of $(\alpha h\nu)^2$ vs hν were 1.45eV for samples prepared by co-evaporation and 1.48 for samples prepared by SP.

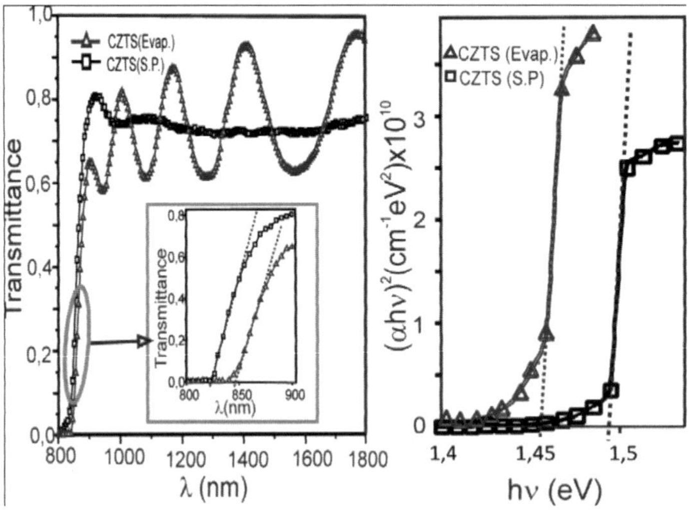

Figure 1. Curves of spectral transmittance and of $(\alpha h\nu)^2$ vs hν, corresponding to typical CZTS films deposited by co-evaporation and spray pyrolysis.

Structural defects give rise to distortion of the bands generating tails of states that extend the bands within the gap causing a decrease of the slope of the transmittance curves (see insert of Figure 1). Absorption via transitions between states of tails of bands was evaluated through the calculation of the absorption coefficient near the band edge α_E that has an exponential dependence on photon energy and Urbach energy [8] ($\alpha_E = \alpha_0 exp\left[\frac{hv-E_i}{Eu}\right]$), where E_u is the Urbach energy, E_1 and α_0 are constant. Thus, a plot of lnα vs. hv should be linear and Urbach energy can be obtained from the slope. The Eu values obtained from the slope of the transmittance curves, were 22.98meV for samples prepared by co-evaporation and 43.17meV for samples prepared by SP. Considering that Urbach energy is related to density of localized states in the band gap induced by structural disorder (stress and dislocation) [9], can be concluded that the co-evaporated CZTS films present an improved crystallinity, as compared with those grown by SP.

3.2. Structural characterization

In Figure 2 are compared XRD pattern of typical CZTS thin films deposited by both co-evaporation and spray pyrolysis. It is observed that the CZTS samples present reflexions corresponding only to the Cu_2ZnSnS_4 phase (AMC cart #99-100-5143). The formation of single phase CZTS compounds was also proven from Raman spectroscopy measurements.

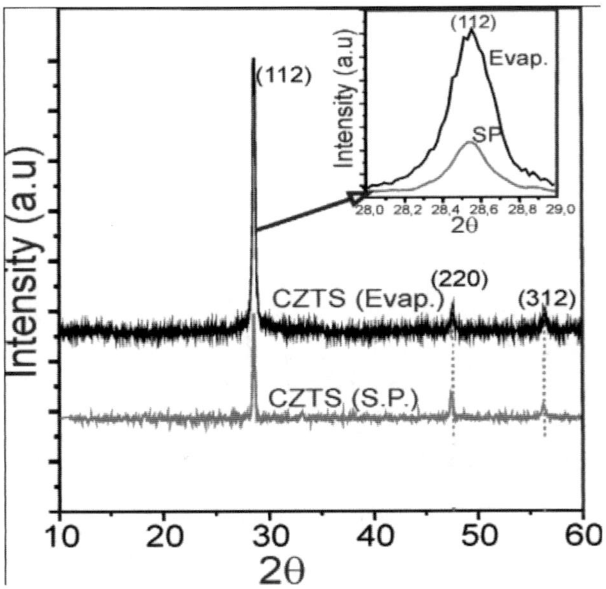

Figure 2. XRD spectra of CZTS film prepared by co-evaporation and by spry pyrolysis. Inset shows a scan around the (112) reflexion.

The scans around the (112) reflexion shown in inset of Figure 2 show that the full width at half maximum (FWHM) value corresponding to the studied CZTS films is different, indicating that crystallite size of CZTS films is affected by the deposition method.

Initially the crystallite size value is calculated using Scherrer's formula to compare with the values obtained from the Williamson-Hall (W–H) models and AFM analysis results. The crystallite size D of CZTS films is estimated from the Scherrer's equation [10] as: $D = \frac{K\lambda}{\beta_{hkl} cos\theta_{hkl}}$. Where D is the crystallite size (nm); K is the shape factor (K=0.9), λ is the wave length of the X-rays (for CuKa radiation); θ_{hkl} is Bragg diffraction angle and β_{hkl} is the instrumental corrected broadening of the hkl diffraction peak measured a half of its maximum intensity. In addition to the instrumental X-ray peak broadening, lattice strain and crystallite size are the other two independent factors that contribute to the total peak broadening. The strain induced broadening β_S is given by the relation $\beta_S = 4\varepsilon tan\theta_{hkl}$. Now the total peak broadening is represented by the sum of the contributions of crystallite size and strain present in the material. Assuming that the strain present in the material is uniform, the W–H equation [11] for the total peak broadening is given by: $cos\theta_{hkl} = \frac{K\lambda}{D} + 4\square sin\theta_{hkl}$.

A plot taking $4\sin\theta_{hkl}$ along X-axis and $\beta_{hkl}\cos\beta_{hkl}$ along Y-axis allow to extract the strain present in the material and the crystallite size from the slope and the intercept of the linear fit made to the plot.

In Table 1 are listed values of crystallite size estimated from Scherrer's formula and Williamson Hall Equation as well as from AFM measurements, for typical CZTS films prepared by the mentioned methods; values of lattice strain ε present in the CZTS films are also listed.

Table1. Values of crystallite size calculated from Scherrer's formula and W-H equation, compared with values obtained from AFM measurements.

Deposit. Method	Sherrer´s Equation	Williamson Hall Equation		AFM
	D (nm)	D (nm)	Lattice strain (ε x10^{-3})	D (nm)
Evap	127,3	130,8	0,0076	130
SP	78,0	80,2	0,0086	78

It is observed that the lattice strain ε of evaporated CZTS films is significantly less than those of CZTS films grown by SP, indicating that evaporated CZTS films grow with improved crystallinity. The results also revealed that the crystallite size of evaporated CZTS samples is much larger than those of Spray pyrolysis deposited CZTS films.

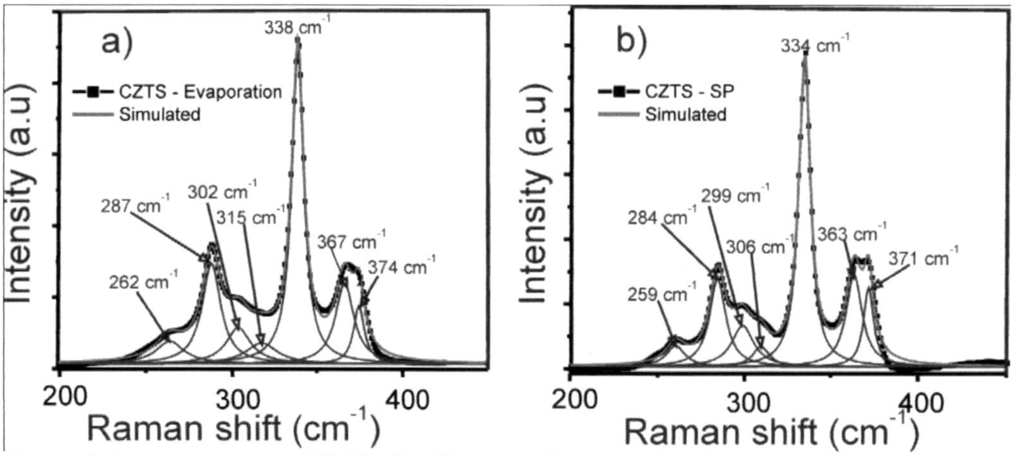

Figure 3. Raman spectra of CZTS thin films and fitting peaks with Lorentzian functions for (a) co-evaporated sample and (b) SP deposited sample.

Besides X-ray analysis, Raman spectroscopy measurements were done on thin films of CZTS prepared by both, spray pyrolysis and co-evaporation. In Figure 3 the Raman spectra of a typical CZTS film grown by co-evaporation is compared with that of a sample prepared by spray pyrolysis. The fitting with Lorentzian functions curves of the Raman spectrum of the evaporated sample gives evidence of peaks presence at 262, 287, 302, 338, 367 and 374cm^{-1} which have been attributed to the Cu_2ZnSnS_4 phase [12]. An identical Raman Spectrum was reported elsewhere for stoichiometric Cu_2ZnSnS_4 films grown with structure kesterite, measured using a 785nm laser [13]. The Raman spectrum of the sample prepared by SP exhibit the same peaks of the sample evaporated but slightly shifted towards lower values of wave number, indicating that this sample contain only peaks associated with the Cu_2ZnSnS_4phase. The Raman shift has been attributed to the confinement of optical phonons in a small crystalline particle [13]. In both spectra the dominant peaks are located at 338 (334) and 287 (284)cm^{-1}

which arises from A1 vibration mode, where only S-anions are involved. The bands at 367cm^{-1} and 375cm^{-1} correspond to B1 and B2vibrational modes with cations moving only along the z direction.

4. Conclusions

Single phase Cu_2ZnSnS_4 thin films with kesterite type structure were grown by both co-evaporation and dimethyl sulfoxide (DMSO) based spray pyrolysis, and their structural properties evaluated through X-ray diffraction measurements using Williamson–Hall-isotropic strain model and Urbach energy estimated from absorption coefficient calculated near the band edge. It was found that evaporated CZTS thin films have values of lattice strain and Urbach energy lower than those obtained for CZTS films grown by SP, indicating that the first type of CZTS samples exhibit a lower density of structural defects than the CZTS films prepared by spray pyrolysis. Considering that Urbach energy is related to density of localized states induced by structural defects, can be concluded that the coevaporated CZTS films present an improved crystallinity.

It was also found that the CZTS films prepared by co-evaporation and SP present Eg values of 1.45eV and 1.48eV respectively and that both type of CZTS films exhibit high transmittances indicating that the photon absorption occurs predominantly via band to band transitions.

Acknowledgements

This work was supported by Colciencias (Contract #038/2013) and Universidad Nacional de Colombia, Sede Bogotá, Facultad de Ciencias, Grupo GMS&ES, Bogotá D.C., Colombia (Proy. 20287 supported by DIB).

References

[1] H Katagiri, K Saitoh, T Washio, H Shinohara, et al 2001 *Sol Energy Mater & Sol Cells* **65** 141
[2] H Katagiri, N Sasaguchi, S Hando, S Hosino, et al 1997 *Sol Energy Mater & Sol Cells* **49** 407
[3] H Katagiri, K Jimbo, S Yamada, T Kamimura, W S Maw, et al 2008 *Appl Phys Express* 104 1201
[4] T K Todorov, J Tang, S Bag, O Gunawan, T Gokmen, Y Zhu, D B Mitzi 2013 *Adv Energy Mater* **3** 34
[5] I Repins, C Beall, N Vora, C DeHart, D Kuciauskas, P Dippo, B To, J Mann, W Chsu, A Goodrich, R Noufi 2012 *Sol Energy Mater Sol Cells* **101** 154
[6] Martin A Green, Keith Emery, Yoshirhiro Hisnikawa, et al 2015 *Progress in* Photovoltaics **23** 1
[7] J I Pankove 1971 *Optical processes in semiconductors* (New York: Dover Publications Inc)
[8] M V Kurik 1971 *Status Solidi A* **8** 9
[9] P Schrrer, Phis Z 2011 *Nature Nanotecnology* **6** 534
[10] G K Williamson,W H Hall 1953 *Acta Metall* **1** 22
[11] P A Fernandes, P M Saloméa, A F da Cunhaa 2011 *Journal of Alloys and Compounds* **509** 7600
[12] X Fontane 2013 *Caracterización por espectroscopía Raman de semiconductores Cu_2ZnSnS_4 para nuevas tecnología fotovoltaicas* (España: Universidad de Barcelona)
[13] M Yang, D Huang, P Hao, F Zhang, X Hou and X Wang 1994 *J Appl Phys* **75** 1

Synthesis of an air-working trilayer artificial muscle using a conductive cassava starch biofilm (manihot esculenta, cranz) and polypyrrole (PPy)

Y E Núñez D[1], Á A Arrieta A[2] J A Segura B[3] and S D Bertel H[1]
[1] Fundación Universitaria Los Libertadores, Colombia.
[2] Universidad de Sucre, Sincelejo, Colombia.
[3] Universität Hamburg, Hamburg, Germany.

E-mail: yenunezd@libertadores.edu.co

Abstract. In this study, a methodology for obtaining a conductive cassava starch biofilm doped with lithium perchlorate ($LiClO_4$) is shown, as well as the electrochemical technique for the synthesis of polypyrrole films, which are used for developing the trilayer artificial muscle PPy/Biopolymer/PPy designed to operate in air. Furthermore, results from the trilayer movement using chronoamperometric techniques are shown.

1. Introduction

The constant progress of material science has allowed to develop actuators capable of converting chemical reactions into motion by electrical stimulation (reason why are known as electrochemomechanic actuators). Since these devices perform functions of contraction and expansion similar to natural muscles, are also called artificial muscles [1-3]. Nowadays, artificial muscles have been developed in a vast variety of materials. in the laboratory of electrochemistry at the Basque University in San Sebastián, Spain, a bilayer device of polypyrrole (PPy) with another neutral flexible polymer substrate was devised and patented, where the metal electrode, covered with a PPy film electrogenerated, was peeled using a commercial nonconductive polymeric adherent film and the peeled bilayer was used then as new electrode in electrolyte solution. The macroscopic result was a remarkable inflection of the bilayer [4]. Actuators based on conducting polymers have become more interesting due to their characteristics such as high mechanical tension, large deflection values, high reversibility and the possibility of precise control using small voltages [5]. However, the reported artificial muscles are made from polymers and most of them work in aqueous media, only a few works reported movements in air [6-8]. Currently, implementation of biopolymers conductors on the generation of artificial muscles has not yet been reported. In this work is presented the development of a polypyrrole/biopolymer conductor/polypyrrole trilayer muscle able to move in air.

2. Experimental procedure

2.1. Preparation of the conductive biofilm

Initially cassava starch (3 g), which was obtained from core operations of washing, peeling, grating, filtering, decanting, drying and grinding was taken. Subsequently, in order to provide flexibility and conductive properties to the biofilm, it was necessary to use glycerin (3*g*), glutaraldehyde (2*g*), and $LiClO_4$ (3*g*) [9].

2.2. Synthesis of polypyrrole

To prepare PPy films, electrochemical polymerization techniques were used [10]. In which a solution containing 0.1*M* $LiClO_4$ and 0.2*M* pyrrole gauged in 25*ml* of acetonitrile was prepared. Using a system consisting of a working electrode (EW) in the center, two counter-electrodes at the ends (EA), and a reference electrode (ER) of calomel saturated with KCl, a potential of 0.8*V* was applied by a potentiostat for a time of 2500*s*.

2.3. Obtainment of the trilayer

After obtaining the PPy layers, a cut of 4,5cm^2 (3*cm* x1,5*cm*) conductive biofilm was made, and placed on one of the layers of PPy which was in the steel sheet (EW), due to the adhesion properties acquired by the biofilm, it was possible to detach the layer of PPy. Then, using the same method, the second layer of PPy is removed, thus obtaining the three layers constituent (PPy/conductive biopolymer/PPy).

3. Results and discussion

3.1. Movement of the trilayer actuator

In order to evaluate the movement of the trilayer, potential steps of 2*V* and -2*V* were set during 20*s* and 40*s*, for oxidizing and reducing respectively the PPy layers.

Figure 1 shows the chronoamperometric response of the trilayer device, subject to periodic oxidations and reductions. In the first 20*s* the movement is from the reference point (0°) to positive angles (see Figure 2), consuming an electrical charge of 95.44*mC*, which is higher than in the movement toward negative angles, between 20*s* and 40*s*, where the electrical charge consumption was 77.17*mC*.

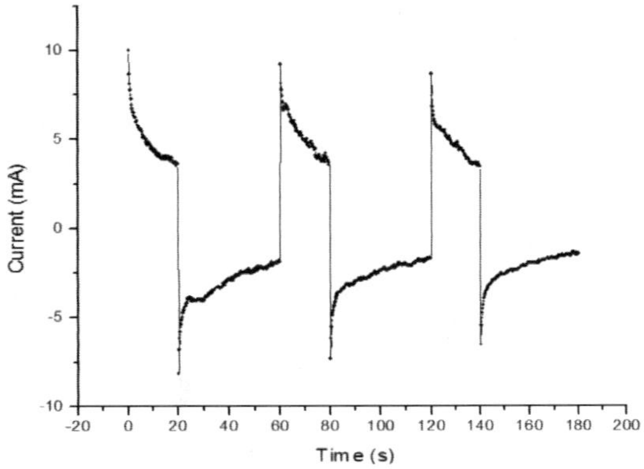

Figure 1. Chronoamperometric response of the trilayer actuator.

Figure 2 shows the connection scheme and movement of the trilayer artificial muscle, where is sketched the movement and its relationship to the redox activity of the PPy layers, originating the response of the device. In this case, the oxidation corresponds to the movement of perchlorate ions from the biopolymer layer which acts as conductive solid electrolyte toward the PPy layer, causing an increase in volume, while the other layer of PPy is reduced, implying a reduction in volume of the PPy layer and therefore causing the device to bend towards it.

The record of the movement, is presented in Figure 3. This figure shows the sequential photographs of the movement of the trilayer device operating in air from the position 0° to positive and negative angles, where it can be noticed the movement due to the increase and decrease in volume of the PPy layers.

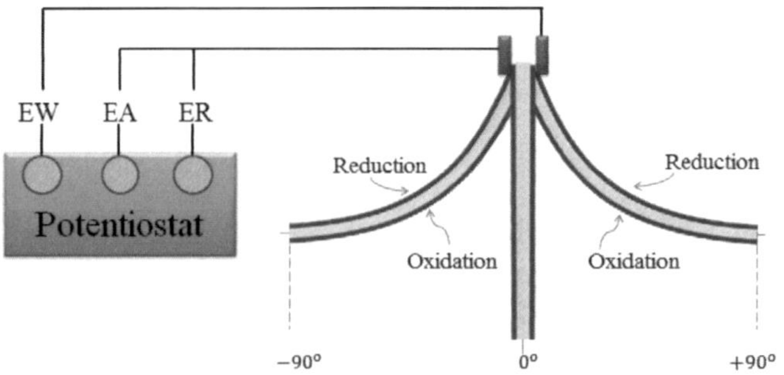

Figure 2. Movement scheme of the trilayer actuator.

(a)

(b)

Figure 3. Photo sequence of the trilayer artificial muscle movement: a) to positive angles, b) to negative angles.

4. Conclusions

The developed conductive biopolymer, proved to be very stable physically and in its favorable conductive properties. Which is, the plasticizers in combination with the $LiClO_4$ added, in the aforementioned amounts, granted favorable electrical and mechanical properties for its use as an artificial muscle.

Developing the trilayer actuator (PPy/biopolymer/PPy), showed that the biopolymer was effective when used as solid polymer electrolyte, allowing the successful exchange of perchlorate ions from and towards the PPy layers. Allowing oxidation and reduction of the PPy layers generating a significant change in volume, which allow the smooth flowing movement of the actuator in the air in both directions using currents with the order of milliamps.

References

[1] T Otero 2003 *Revista Iberoamericana de Polímeros* **4** 26
[2] M Cortés 2002 *Músculos artificiales de tricapa, caracterización y diseño de dispositivos* (Bilbao: Universidad del País Vasco)
[3] T Otero and M T Cortés 2003 *Sensors and Actuators B* **96** 152
[4] T Otero, E Angulo, J Rodríguez y C Santamaría 1992 *Dispositivos laminares que emplean polímeros conductores capaces de provocar movimientos mecánicos* (Patente: EP–9200095, EP–9202628)
[5] U L Zainudeen, M A Careem and S Skaarup 2008 *Sensors and Actuators B Chemical* **134** 467
[6] D F Parra, C C Tadini, P Ponce and A B Lugão 2004 *Carbohydrate Polymers* **58** 475
[7] Nicho, M and H Hu 2000 *Solar Energy Materials & Solar Cells* **63** 423
[8] J M Sansiñena, V Olazàbal, T F Otero, C N Polo da Fonseca and Marco-A De Paoli 1997 *Chem Commun.* **22** 2217
[9] A Arrieta y A Jaramillo 2014 *Revista Colombiana de Materiales* **5** 42
[10] M Diaz, A Rojas, E Vera y M Cortes 2005 *Revista Colombiana de Física* **37-2** 357

First principles study of structural, electronic and magnetic properties of magnesium

G P Abdel Rahim[1], J A Rodríguez M[1] and M G Moreno-Armenta[3]
[1] Universidad Nacional de Colombia, Bogotá, Colombia.
[2] Universidad Nacional Autónoma de México, México D.F., México.

E-mail: pabdelrahim@gmail.com , moreno@cnyn.unam.mx

Abstract. We investigated the structural, electronic, and magnetic properties of Mg, in the CS (simple cubic), NiAs (Nickel arsenide), FCC (rock-salt), R (Rhombohedral), Diamond and WZ (wurtzite) phases. Calculations were performed using the first-principles pseudo-potential method within the framework of spin-density functional theory (DFT).

1. Introduction

Previous studies with Mg in the hcp phase show that it is a fragile metal at room temperature with a strong surface irregularity which to date has not been surpassed [1-5]. However, in the NaCl and CsCl phases it has been mixed with Li at room temperature, obtaining an alloy that can change from ductile to brittle, depending upon the amount of Li in the mixture [5]. Today, the industry is interested in this alloy, Mg-Li, due to its low density and dent resistance for the construction of automobile parts [6]. Several theoretical and experimental papers have shown that Mg is stable in the CsCl and NaCl phases, both in bulk and in the form of thin films. Mg in the CsCl phase is stable at pressures P ~50GPa, but it is unstable at room temperature. The stability of the alloy Mg-Nb in the CsCl phase at room temperature allows forming multilayer films of Mg-Nb. Mg is combined with other materials to form ions of Mg and achieve the construction of rechargeable and non-rechargeable batteries, in order to replace Li batteries with Mg ones, since Mg is 10^4 times more abundant in the earth's crust than Li [7].

2. Method

Calculations were performed using the first-principles pseudo-potential method in the framework of spin density-functional theory (DFT) by means of Quantum Espresso Code [8]. Exchange and correlation effects were dealt with using the generalized gradient approximation (GGA) implemented in the Perdew–Burke–Ernzerhof functional (PBE) [9]. Ultrasoft pseudo-potentials were employed [10]. The wave function was expanded in plane waves up to a cutoff of 40Ry. A gamma-centered Monkhorst–Pack [10] k-point grid of 10×10×10 was used in the First Brillouin zone. The convergence for the energy was 0.0001Ry. For determining the properties of Mg in this configuration, a supercell model was implemented. For example, in WZ a 1×1×2 supercell was used, and a $\sqrt{2} \times\sqrt{2} \times 2$ supercells was used in the simple cubic phase.

3. Structural properties

Figure 1 show the energy of the compound versus volume (E vs. V) by unit formula of Mg. It can be seen that each curve has minimum energy, which indicates that the phase is metastable. The energies

Journal of Physics: Conference Series **687** (2016) 012043 doi:10.1088/1742-6596/687/1/012043

and volumes are given by unit formula.

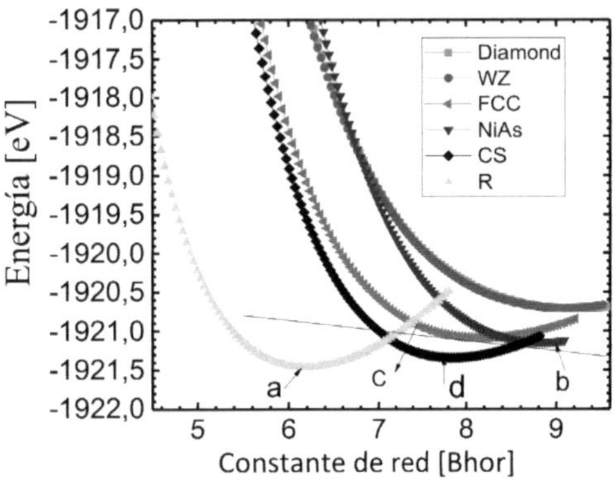

Figure 1. Energy vs. volume for the Mg compound in six structures.

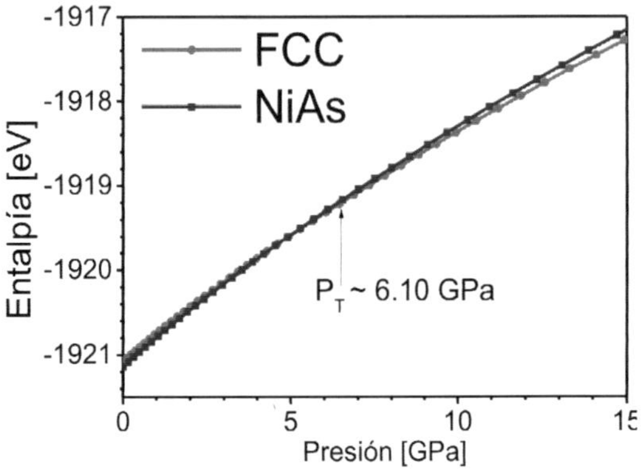

Figure 2. Enthalpy vs. pressure for the Mg compound in two structures.

These graphics (Figure 1 and 2) are fit to the Murnaghan equation of state [11] for each phase. In this way, the equilibrium lattice constant, the bulk modulus, and other structural parameters can be obtained (see Table 1).

Figure 1 shows the cohesion energy as a function of the Mg for the R, CS, NiAs, FCC, Diamond, and WZ structures. In this superimposition it can be seen that the ground state crystallization is R (Rhombohedral) in a hexagonal cell, since it is the lowest energy with respect to the other phases (Table 1), but there is also a local minimum in the CS structure with energy of ~0.12eV/unit formula. In this Figure 1, four points, labeled a, b, c, and d, correspond to: a) R phase in conditions of minimum energy P=0. b) NiAs phase before a phase transition FCC, P<PT. c) FCC phase after the phase transition from NiAs, P>PT. d) CS phase in conditions of minimum energy P=0.

Figure 2 depicts enthalpy versus pressure and shows an intersection between the phases NiAs and FCC at P_T ~6.10GPa. We found that there is no transition of the R phase nor of the other five phases, but possible transitions from the NiAs phase to the WZ and diamond phases were found.

Table 1 The structural parameters of Mg in the R, CS, NiAs, FCC, Diamond and WZ-like structure. Cohesive energies and volumes are given per unit cell (containing eight atoms). The parameter a_o (Bhor) in the cubic structure is taken as a (Bohr)=a_0(Bohr)/√2 where a (Bohr) is the lattice constant.

Structural Parameter	R Our results	CS Our results	NiAs Our results	FCC Our results	Diamond Our results	WZ Our results
a_{min} [Bohr]	6.18	7.85	8.83	8.15	9.23	9.09
Volume [Bhor^3]	258.87	342.51	374.91	191.43	277.78	266.53
c/a	5.05	2.8284	2.51	2.8282	2.8284	3.28
B [GPa]	19.00	15.91	16.73	22.08	13.06	15.96
B'	2.95	2.98	2.68	2.99	2.67	2.53
E cohesion [eV]	-1921.46	-1921.34	-1921.14	-1921.08	-1920.71	-1920.71

Table 1 reports the structural parameters obtained. These parameters are determined by fitting the curves of E vs. V to Murnaghan's state equation [11]. In the tables, c/a is the relationship between lattice parameters, V is the volume of the unitary cell, B is the bulk modulus, and E_0 is the cohesion energy.

4. Electronic properties

The electronic properties of Mg in this configuration were studied for the R and CS phases. Figures 3 and 4 show the total and partial density of states for each of these phases at zero pressure, respectively. The first thing we observe is that the Mg is composed of orbitals 3*s* and 2*p*. The presence of states near the Fermi level indicates that this material behaves as a metal. In the calculations, the polarization of the spin was taken into account, but it was not necessary to include how the density of up-spin states was equal to that of the down-spin states. Therefore, we can also say that the material is not magnetic in any of these phases.

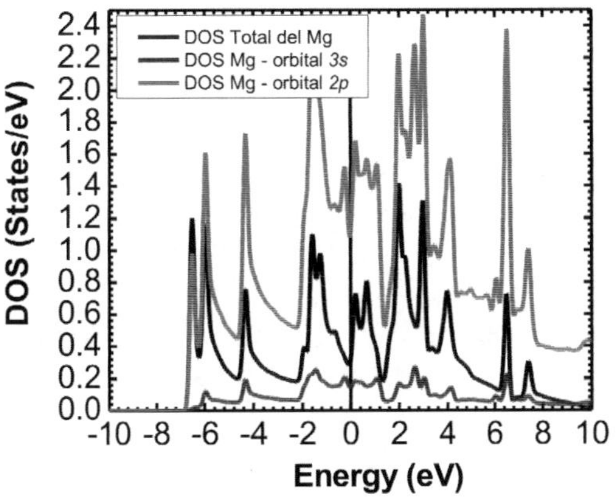

Figure 3. Partial and total DOS of Mg in the R phase in conditions of minimum energy P=0.

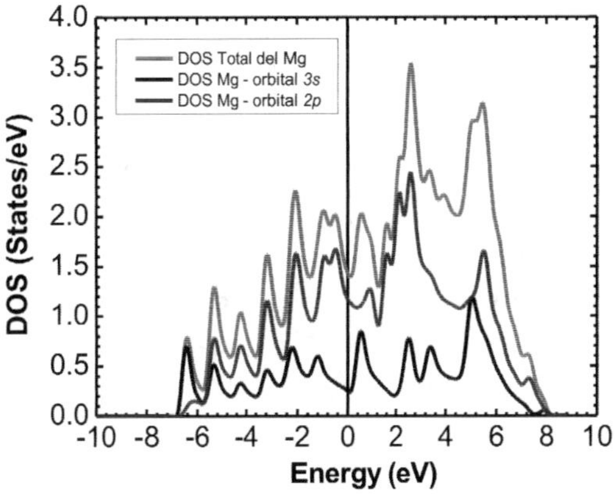

Figure 4. Partial and total DOS of Mg in the CS phase in conditions of minimum energy P=0.

5. Phase transition

We studied the electronic properties Mg in dots immediately before and after the transition point. Figure 5 shows the partial and total DOS in the NiAs phase before a phase transition FCC, $P<P_T$, and as in the two previous figures a metallic behavior is observed. Figure 6 shows the partial and total DOS of Mg in the FCC phase after the phase transition NiAs, $P>P_T$, showing a metallic behaviour. For both, electronic contributions are due to hybridization of orbitals s and p of the Mg.

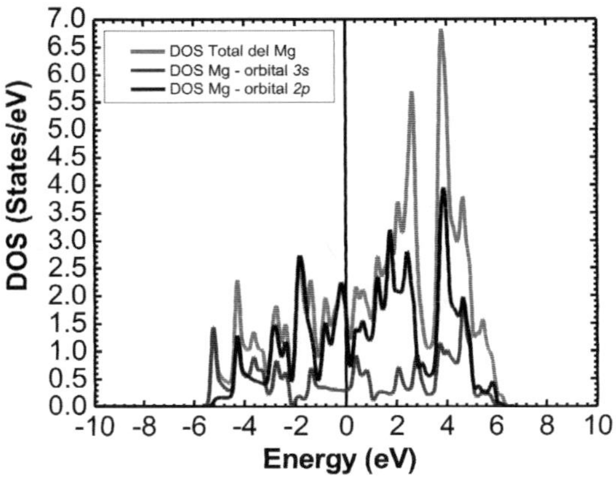

Figure 5. Partial and total DOS of Mg in the NiAs phase before a phase transition FCC, $P<P_T$.

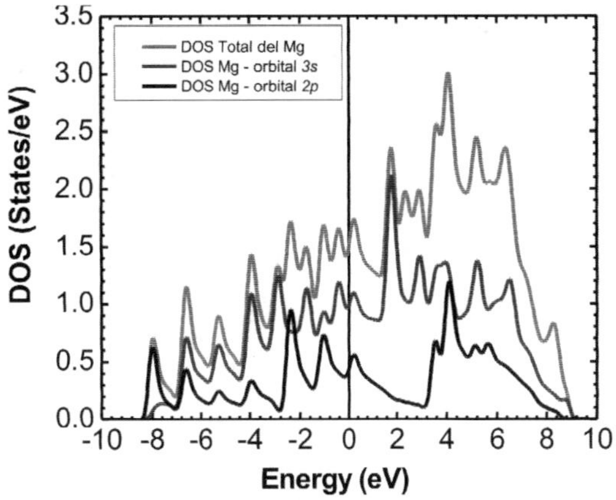

Figure 6. Partial and total DOS of Mg in the FCC phase after the phase transition NiAs, $P>P_T$.

6. Conclusions

We calculated the structural, electronic and magnetic properties of Mg, carried out by means of DFT using the Wu-Cohen GGA approximation. Our calculations predict that in a hexagonal 1×1×2 supercell and √2×√2×2 supercell in the simple cubic phase, Mg crystallizes in the R (Rhombohedral) phase in a hexagonal supercell. The volume module in the six phases is not greater than ~22.08GPa, indicating that the metal is not very rigid compared with Si in the diamond structure. Furthermore, for a transition pressure of ~6.10GPa, Mg changes from NiAs to FCC. Finally, Mg in this configuration behaves as a metallic, nonmagnetic material.

Acknowledgments

This research had partial support from the Universidad Nacional de Colombia, Faculty of Engineering, Universidad Antonio Nariño by means of project title "Propiedades físicas de la adsorción de bismuto sobre la superficie/(001) cubica-MgO en un slab 2x2, number 20141086 de la UAN"; project number 20141086 in Colombia and Universidad Nacional Autónoma de México, UNAM by means of DGAPA project, IN102714-3. Calculations were performed at the Dirección General de Cómputo y de Tecnologías de la Información y Comunicación de la Universidad Nacional Autónoma de México (DGCTIC-UNAM) supercomputing center project SC14-1-I-2. Other calculations were carried out at the cluster of Scientific Calculation Support Center of Universidad Antonio Nariño and the cluster of Grupo de Estudio de Materiales GEMA, of Dept. of Physics of Universidad Nacional de Colombia, Bogotá, Colombia.

References

[1] Barnett M R, Nave M D, Bettles C 2004 *J Mater Sci Eng A* **386** 205
[2] Nave M D, Barnett M R 2004 *Scripta Mater* **51** 881
[3] Styczynski A, Hartig C, Bohlen J, Letzig D 2004 *Scripta Mater* **50** 943
[4] Gehrmann R, Frommert M M, Gottstein G 2005 *Mater Sci Eng A* **395** 338
[5] W A Counts, M Fria´k, D Raabe, J Neugebauer 2009 *Acta Materialia* **57** 69
[6] Ilgyou Shin and Emily A Carter 2014 *Acta Materialia* **64** 198
[7] Matthew M Huie, David C Bock, Esther S Takeuchi, Amy C Marschilok, Kenneth J Takeuchi 2015 *Coordination Chemistry Reviews* **287** 15
[8] P Giannozzi, S Baroni, N Bonini, M Calandra, R Car, C Cavazzoni, D Ceresoli, G L Chiarotti, M Cococcioni, I Dabo, A Dal Corso, S de Gironcoli, S Fabris 2009 *J Phys Condens Matter* **21** 395502
[9] J P Perdew, K Burke, M Ernzerhof 1996 *Phys Rev Lett* **77** 3865
[10] K Laasonen, A Pasquarello, R Car, C Lee, D Vanderbilt 1993 *Phys Rev B* **47** 10142
[11] F D Murnaghan 1944 *Proc Natl Acad Sci USA* **30** 5390

Journal of Physics: Conference Series **687** (2016) 012043 doi:10.1088/1742-6596/687/1/012043

First principles study of structural, electronic and magnetic properties of magnesium

G P Abdel Rahim[1], J A Rodríguez M[1] and M G Moreno-Armenta[2]

[1] Universidad Nacional de Colombia, Bogotá, Colombia.
[2] Universidad Nacional Autónoma de México, México D.F., México

E-mail: pabdelrahim@gmail.com , moreno@cnyn.unam.mx

CORRIGENDUM TO: G P Abdel Rahim, J A Rodríguez M and M G Moreno-Armenta 2016 *J. Phys.: Conf. Ser.* 687 012043

The authors would like to correct an institution name in the affiliations of this work. The correct list of affiliations for the paper 'First principles study of structural, electronic and magnetic properties of magnesium' is:

First principles study of structural, electronic and magnetic properties of magnesium

G P Abdel Rahim[1], J A Rodríguez M[1] and M G Moreno-Armenta[2]

[1] Universidad Antonio Nariño, Bogotá, Colombia..
[2] Universidad Nacional Autónoma de México, México D.F., México

Quicklime as an alternative in the photodegradation of contaminants

M R Joya[1], J H Bautista Ruiz[2] and A M Raba[2]

[1] Universidad Nacional de Colombia, Bogotá, Colombia.
[2] Universidad Francisco de Paula Santander, San José de Cúcuta, Colombia.

E-mail: mrinconj@unal.edu.co

Abstract. Chemical oxidation is one of the many different methods of removing contaminants that has emerged recently, is an alternative method to traditional techniques. According to this research calcium peroxide is suitable choice for contaminant biodegradation in soil and ground water, but it happens at a very low speed. We hope that the method of synthesis of calcium peroxide in nano size by increased ratio of surface to volume can increase the speed of reaction and solve the problem. In the first synthesis of the material we obtained CaO instead of CaO_2. Calcium oxide (CaO), commonly known as quicklime or burnt lime, is a widely used chemical compound in obtaining hypochlorite and to neutralize acidic soils. In this study we characterize the material by X-ray scanning electron microscopy, and ultraviolet-visible spectroscopy.

1. Introduction

CaO-based sorption material is most abundantly available from natural resource such as limestone and dolomite. The sources of CaO are not limited to these two sources only. There are other researchers that proposed to use biomaterial sources such as from crab shell and oyster shell as the source of CaO [1]. Calcium oxide (CaO) is an important inorganic compound, which is used across various industries as catalyst, toxic-waste remediation agent, adsorbent, etc [2].

Among the heterogeneous catalysts that are being used in transesterification, calcium oxide (CaO) has a promising place, and many reports have been published on CaO-catalyzed transesterification using laboratory grade. It is cheap, abundantly available in nature (as limestone), and some of the sources of this compound are renewable (waste material consisting of calcium carbonate, $CaCO_3$). However, the utilization of waste materials as heterogeneous catalysts has been of recent interest in the search for a sustainable process [3, 4].

Calcium Oxide is a off white coloured bulky powder. It is also called lime, quicklime or caustic lime. Calcium oxide is highly reactive with moisture or water. It releases considerable heat when it gets hydrated. As it can react with the moisture in the air as well, it is often used as a dessicant. The reaction takes place with carbon dioxide generating water insolube carbonates. Due to this property of Calcium Oxide, it is used in plastic industry as well for the removal of moisture. Besides this, other uses of calcium oxide include manufacturing porcelain, glass, bleaching powder, calcium carbide, calcium cynamide and purifying sugar. It is used in water softners and also in mortar and cement. It is also used to treat acidic soil [5], in Figure 1, some applications are observed in the limestone. Calcium Oxide is used in metallurgy industry with different purposes. In Steel is incorporated during iron and steel process as clinker agent,

Journal of Physics: Conference Series **687** (2016) 012044 doi:10.1088/1742-6596/687/1/012044

for then be able to eliminate impurities existing in metal parts. Calcium Oxide (lime) must contain of a high purity to improve its reactivity and performance [6].

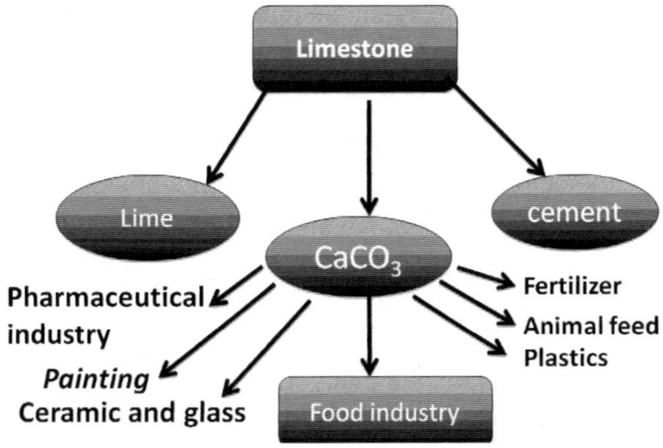

Figure 1. The processing and major uses of limestone [7].

2. Theoretical Formalism

Calcium oxide powders were prepared from polymeric precursors by the Pechini method [8]. The metal calcium Sulphate 2-hydrate solution was mixed with a stoichiometric amount of citric acid. The resulting solution was stirred for about 1 hour on a hot plate and the temperature was stabilized at 70^oC. The mixture was heated to 90^oC, at which point ethylene glycol was added at a mass ratio of with respect to citric acid. The temperature was maintained constant up to resin formation, which polymerized at 300^oC. The precursor powders were then calcined for 2 hours at various temperatures, ranging from 350^oC, 450^oC, 550^oC to 650^oC. The two processes are the same, unlike a process that ammonia was added and the other not.

Calcium Oxide (CaO) has a potential to be developed as adsorbent material to capture CO_2 at high temperature. CO_2 could be capture using some adsorbent such as activated carbon and molecular sieve. However, these types of adsorbent are only active at low temperature and require high pressure. In example of gasification process, the effluent gases produced were usually in the form of hot gas stream (over 600^oC). CaO-based sorbent has been identified as the suitable candidate for this job [9]. In recent years, many researchers worked on enhancing the performance especially for multi-cycles of the CaO through various modification methods such as hydration, acid treatment and chemical doping. The ethanol-water hydration method was observed the simplest and economically to apply [10].

3. Results and discussion

In Figure 2, the SEM sample of CaO is observed, samples with ammonia have size less particle Figure 2(a) and (b). In Figure 2(c) and (d), have samples without ammonia, grain shape are small bars, the bar size increases with increasing temperature.We hope to make a comparison with these samples photodegradation; this is a job for the future. The photocatalytic removal of organics and its degradation efficiency will be evaluated by determination of reduction in the chemical oxygen demand values. The organic pollutant degradation efficiency is expected in a wide range of pH values, irradiation time and optimum catalyst load. In additional effects like aeration and stirring increase the photodegradation efficiency of organic.

The XRD pattern of CaO nanoparticles has been shown in Figure 3. All reflection peaks in Figure 3 can be readily indexed to a pure cubic phase of CaO with a Fm-3m space group. XRD patterns showed broadening of the peaks, indicative of the ultra-fine nature of the crystalline material. Figure (a), (b) are ammonia samples due to particle size, we have more interest in these.

Figure 2. SEM images of CaO NPs. Figure (a) and (b) with addition of ammonia at 450 and 650^0C, (c) and (d) without added ammonia in the process; 450 and 650^0C respectively.

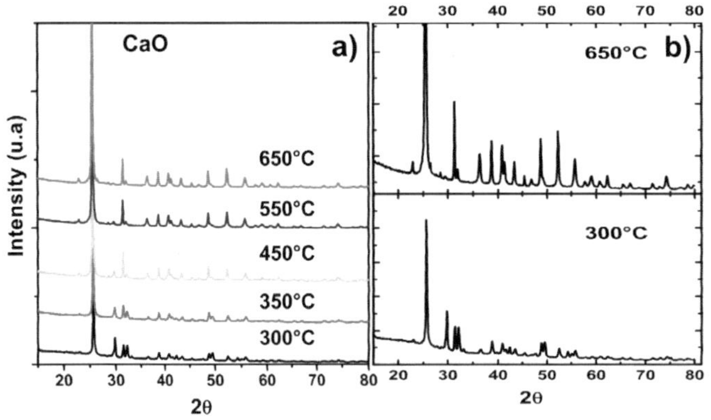

Figure 3. The XRD patern of CaO nanoparticles.

We intend to do tests with these nanoparticles in degradation of pesticides and as a bactericide. According to some works of literature the pesticides and herbicides, intentionally released into the environment, are ubiquitous in aquatic systems; they are often detected at low levels and commonly occur in the form of complex mixtures [11]. Leaching of chemical fertilizers and pesticides, applied to agricultural and forest land, is one of the main reasons for organic pollution in several water streams. Pesticides and herbicides are harmful to life because of their toxicity, carcinogenicity and mutagenicity [11]. Therefore toxicity of pesticides and their degradation products is making these chemical substances a potential hazard by contaminating the environment. Others author say that samples of CaO respond well as antibacterial. With increasing calcination time, antibacterial activity of CaO nanoparticles was also increased [12].

4. Acknowledgements

Authors acknowledge financial support from the agencies Colciencias Colombia and CNPq Brazil.

References

[1] M T Arpin and S Yusup 2011 *Canad J Appl Sci* **5** 1391
[2] S C Ameta, P B Punjabi, R Ameta, C Ameta 2014 *Microwave-assisted organic synthesis a green chemical approach 1st ed* (New York: CRC Press)
[3] A Buasri, N Chaiyut, V Loyuenyong, P Worawanitchaphong and S tronyong 2013 *Sci Worl J Hind* **2013** 7
[4] R M Nageeb 2013 *Adsorption Technique for the Removal of Organic Pollutants from Water and Wastewater* (Croatia: InTech) chapter 7 p 167
[5] P Patnaik 2002 *Handbook of inorganic* (New York: McGraw-Hill)
[6] European Commision Joint Research Centre 2001 *Integrated pollution prevention and control* (Seville: European Commission)
[7] D J Harrison 1993 *Indistrial minerals laboratory limestone* (Nottingham: British Geological Servey)
[8] M YGaceram, M C Pujol, M Aguilo and F Diaz 2007 *J Sol-Gel Sci Techn* **42** 79
[9] C Yu, C Huang and C Tanr 2012 *Aeros Air Qual Reser* **12** 745
[10] V Manovic, E J Anthonyr 2010 *Int J Environ Res Public* **8** 3129
[11] H Liu, X Cai, Y Wang, J Chen 2011 *Water Research* **45** 3499
[12] Z X Tangi, Z Yu, Z L Zhang, X Y Zhang, Q Pan and L E Shi 2013 *Quim Nova* **36** 933

Coal tar, material used in soil improvement for use in road engineering

R Ochoa Díaz[1], A Montañez[1] and J Cuentas[1]
[1] Universidad Pedagógica y Tecnológica de Colombia, Tunja, Colombia.

E-mail: ricardo.ochoa@uptc.edu.co

Abstract. Coal tar is a by-product of coal distillation in the absence of oxygen to obtain metallurgical coke; its colour varies from dark coffee to black, slightly viscous and its density is greater than that of water. Taking into account the previous characteristics, this document presents a study of the feasibility of using coal tar for the improvement of physical properties, mechanics and dynamics of materials used in road engineering. In this way, the origin, characteristics, and properties of tar are first described. Next, its combination with which granular-based material is evaluated through the CBR test procedure to determine its resistance and to compare it with the non-stabilized material. Finally, the behaviour of the material when subjected to dead loads by means of resistant modules found with the NAT (Nottingham Asphalt Tester) is explored. As a result, the option of using coal tar as a stabilizer was identified due to its use under specific conditions.

1. Introduction

Upon seeing the current state of roads in the country, in which high investments are made and the pavement structures do not support the period of time for which they are constructed, it becomes necessary to search for new construction and stabilization methods for pavement which comply with construction specifications.

The fulfilment of granular layer specifications is not always obtained through available sources; this has led to the continuous search for and development of methodologies of improvement based on the amendment of an external agent such as stabilizing treatments with lime, cement and asphalt, among others [1].

Is for that reason that this document shows the study of an alternative in granular bases stabilization, as the coal tar, it is a bituminous product semisolid or liquid that is obtained as a distillation remainder of coal [2], in lack of air. The coal tar is a binder in cold, water resistant, insoluble in lubricant oils and fuels derivatives from petroleum [3]. The coal tar used in this study is produced in the steel Acerías Paz del Río S.A., at the coker, that process is made to 1000°C [4]. The material to be stabilized should be product from the mechanical grinding and must meet with specifications for a granular-based material type; to make that test were performed based in the testing standards for material cart of the INVIAS [5].

2. Materials and methodology

They made some tests to coat tar, there were tending to stablish physical-chemical characteristics, the results were: moisture: 7%, specific gravity: 1.18g/cm^3, calorific value: 8.9Kcal/Kg, melting point: 80°C. The granular- based material corresponds to BG- 40 from a quarry that belongs to the company CSS Tunja.

The method used to develop the research was proposed in three stages: the first one was the making of the tests needed to know the characteristics of the materials. The second one was about making the mix of the granular-base type material with the coal tar, to establish the resistance characteristics (CBR) [6] and to determine the optimal percentage of coal tar. And the third one is about to make the comparison with the parameters of material resistance without alter with the mixed material with the optimal percentage and the analysis of the convenience of using the tar as a stabilizer.

3. Experimental design

For this study material with the granular base type BG-40, article 330 of the general specifications for INVIAS road construction [7] was used, and mixed with a material of four different percentages of tar (1, 3, 5 y 7%), and through the CBR laboratory test (rule INV E-148-13) the resistance of each mix was determined in order to obtain the optimal percentage of tar to obtain the best characteristics of the material. A mix was made with this optimal percentage and the mechanical and dynamic processes of the material were determined and verified with the parameters of the non-modified material.

4. Results

4.1. Optimal percentages of coal tar

With the CBR from each percentage of tar we made a graphic of CBR Vs. tar percentage with which we determine the optimal percentage to stabilize the granular-base material, the optimal percentage selected 3%, with which we obtained the most material resistance.

4.2. Parameters of mechanical behaviour

In the lineal elastic analysis, one sole value of the elastic modular value and a Poisson relation representative of the material were considered. To conduct the statistical analysis of the mechanical behaviour of the coal tar as a stabilizer in granular bases the CBR test, the index used to grade the quality of materials in accordance with INVIAS, was used.

As Figure 1 show, the stabilized material with optimal percentage coal tar improved by 19 units of CBR corresponding with an increment of 32% in respect to the unstabilized material. It demonstrates that coal tar is effective in the stabilization of granular bases.

Figure 2 shows the averages of the three cylinder expansions for each case, unchanged and stabilized material with the optimal percentage, in which it is evident that one of the main properties of coal tar is in to reduce the spread, reducing by approximately 80 %.

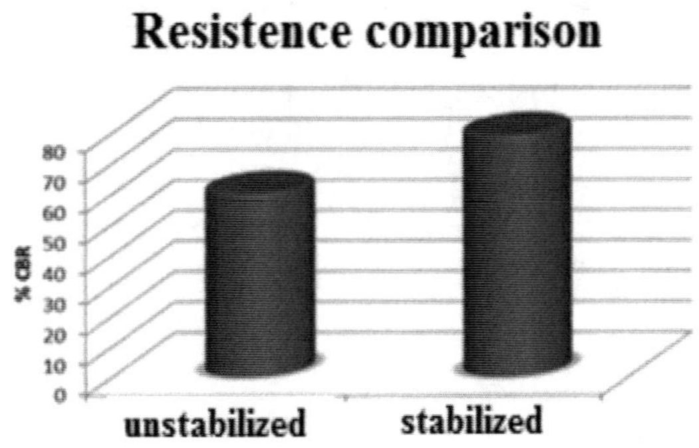

Figure 1. Comparison of resistance.

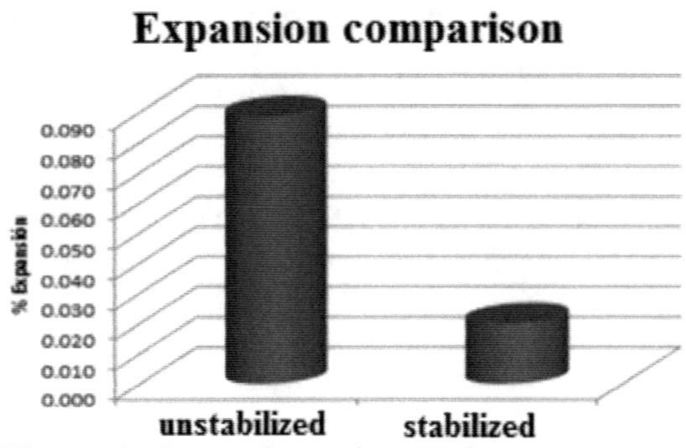

Figure 2. Comparison of expansion.

4.3. Parameters of dynamic behaviour

In Figure 3, an increasing tendency is observed: as the state of force grows the resilient modulus of the material grows. For the case of the specimens with an unstabilized base, the modulus of the material diminishes when the percentage of humidity in the specimen grows.

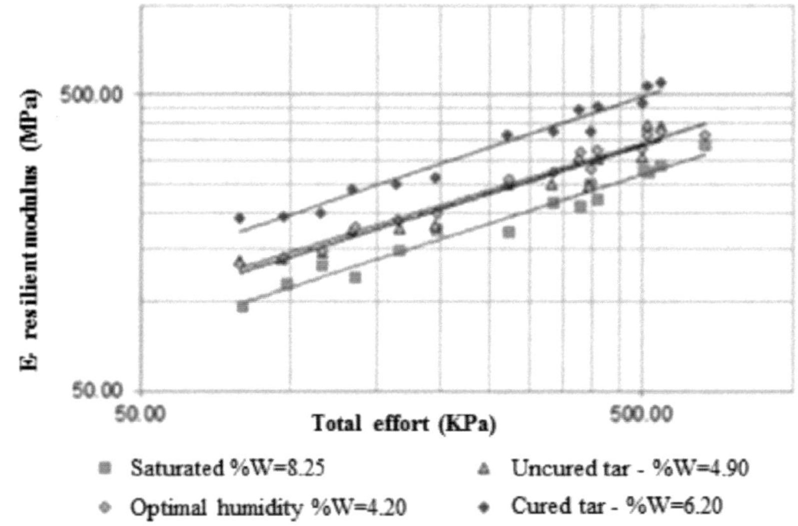

Figure 3. Modular resistance tendency.

5. Conclusions

The optimal percentage to stabilize the granular-base material is 3% in weight. Although at the moment of stabilize the granular-base we do not reached to meet the minimum specifications stipulated by the INVÍAS 2013 standards, for a granular- base type A, for a transit level NT3, we can to increase the resistance by approximately 35% getting a CBR of 79%, which we conclude that coal tar from the company Acerías Paz del Río has effectiveness as a stabilizer of granular - bases, improving their resistance.

It was determined that one of the main improvements of stabilizing base material with coal tar is expansion, achieving an 80% reduction for which the expansion percentage is 0.02. The expansion percentage found in the material is considered low or non-critical according to the specifications of INVIAS, in which the minimum percentage of a material is 2%.

The resilient modulus test is considered the most accurate way to characterize the resistance of a material. In the case of the tests made in the chamber triaxial, it is observed that the moisture content in the specimen affects directly the resistance of the material; in that way, to higher humidity less resilient modulus or less resistance.

Coal tar is classified as a physical-chemical stabilizing agent, where its main support is due to the coating of fine particles, considered noxious when they are sandy silt, and with the presence of water it varies volumetrically.

References

[1] Márquez S 2005 *Estabilización de suelos* (Argentina: Universidad Nacional de la Patagonia San Juan Bosco)
[2] Ochoa R 2013 *IOP Publishing Journal of Physics Conference Series* **466**
[3] Acerías Paz del Río SA 2014 *Alquitrán (Tar Crude)* (http://www.pazdelrio.com.co/es-es/Productos /coproductos/Paginas/ alquitran. aspx)
[4] Gómez A 2002. *Proceso siderúrgico planta belencito* (Colombia: Acerías Paz del Río S.A.)
[5] Instituto Nacional de Vías 2013 *Normas de ensayos de materiales para carreteras* (Bogotá: INVIAS)
[6] Higuera C 2011 *Nociones sobre métodos de diseño de estructuras de pavimento para carreteras* (Colombia: Universidad Pedagógica y Tecnológica de Colombia)
[7] Instituto Nacional de Vías 2013 *Especificaciones generales de construcción de carreteras* (Bogotá: INVIAS)

Synthesis of praseodymium doped cerium oxides by the polymerization-combustion method for application as anodic component in SOFC devices

A F Cruz Pacheco[1], J A Gómez Cuaspud[1] and E Vera López[1]
[1] Universidad Pedagógica y Tecnológica de Colombia, Tunja, Colombia.

E-mail: andresfelipe.cruz@uptc.edu.co

Abstract. This work reports the synthesis and the characterization of six oxides; it is based on $Ce_{1-x}Pr_xO_2$ (x=0.0, 0.2, 0.4, 0.6, 0.8 and 1.0) system, which is obtained by the polymerization-combustion technique for potential applications on design of advanced electrodic components, for solid oxide fuel cells (SOFC). Initially the solid precursors are characterized by infrared spectroscopy (FTIR) and thermal analysis (TGA-DTA), allowing to determine the formation of prevalent citrate species and the optimal temperature for the consolidation of the desired crystalline phases. The X-ray diffraction (XRD) and the transmission electron microscopy analysis (TEM) are performed over calcined samples which provided information about the formation of a fluorite phase with grain distribution, surface, textural and morphological properties consistent with the nanometric obtaining crystallites (30nm), it is oriented along the (1 1 1) facet, with d spacings of 0.31nm for the main diffraction signal. These results indicate the effectiveness of the proposed synthesis method for potential applications in the design of advanced anodic materials for solid oxide fuel cells.

1. Introduction

The operating principle of a fuel cell is based on its three basic components, an anode (fuel electrode), a cathode (air electrode) and an electrolyte (polymer or ceramic) [1,2]. Depending on the type of electrolyte, the fuel cells can be classified in systems of low, medium and high operation temperature. The first group is integrated by the proton exchange (PEMFC), direct methanol (DMFC) and phosphoric acid fuel cells (PAFC), which operates directly with hydrogen and oxygen, through the catalytic effect that can be achieved by an expensive platinum catalyst in the fuel electrode. In the second group, the molten carbonate fuel cells (MCFC), employ a metallic electrocatalyst of low cost. However, these systems are characterized by a low CO tolerance, which causes a poisoning phenomenon on the anodic catalyst, causing a blockage of active sites of the catalyst and hence, the anode degradation. On the other hand, the solid oxide fuel cells (SOFC), have an incomparable advantage with respect to the systems that works at low and medium temperatures, since these devices have the possibility to use ceramic electrocatalysts (Ni-YSZ), economically more affordable, with possibilities of use for a wide range of fuels without high levels of processing or use of noble metals. However, the carbon deposition, represents the main drawback of this cell, recently, the cerium oxides modified with praseodymium oxide have had a growing interest as anodic components, since fluorite structure allows the movement of cations and charge carriers along structure, producing good levels conductivity and structural stability [3,4], reason of which, the manufacture of anodic components using rare earth oxides for fuel cells has increased in order to produce more efficient devices that help to reduce the emissions of greenhouse gases and the consumption of fossil fuels [5,6].

However, it is clear that chemical routes used to obtain the ceramic electrodes of these devices do not allow the generation of acceptable properties, at textural, morphological and surface levels for developing catalytic processes. So, it is clear that in the materials science's context, it is necessary the search of new synthesis methods that allow the obtain of new and improved materials for the applications in solid oxide fuel cells technology [7,8]. In this context, the present work is focused on the synthesis and characterization of 6 cerium oxides based on the $Ce_{1-x}Pr_xO_2$ (x=0.0, 0.2, 0.4, 0.6, 0.8 and 1.0) system, by a chemical route of polymerization-combustion, with the aim of studying the effect on structural, morphological, electrical and catalytic properties for use as potential anodic components in SOFC technology.

2. Experimental

For the synthesis of cerium oxides, it is started from corresponding nitrates of $Ce(NO_3)_3$ 1.0molL^{-1}, $Pr(NO_3)_4$ 1.0molL^{-1} and citric acid monohydrate 99.99%. The dissolutions of metal ions were measured in glass reactors equipped with magnetic stirring (150rpm), temperature control and reflux. In this way, the proposed systems remained on simultaneous heating for 2 hours at 80°C, after which it is added a solution of citric acid at a ratio of 1:0.5, with respect to the total content of metal cations, in order to enable the formation of the corresponding coordination compounds [9]. The solutions were treated between 100 to 250°C to evaporate the solvent and obtain gels, which were treated at 350°C until the beginning of a combustion process, which generates a cluster of heterogeneous solids. The obtained precursors are treated at 500°C under oxygen flow (10mLmin^{-1}) for three hours in order to remove carbonaceous residues from previous combustion stage and to form the crystalline phase [10].

The metal-organic precursors (before calcination), were analysed by FT-IR technique in a Shimadzu prestige-21 spectrophotometer to confirm the presence of citrate species [11]. The relationship between weight change and heat exchange in function of temperature (TGA-DTA) were carried out in a simultaneously Mettler Toledo equipment TG-APMS TGA/SDTA851e/LF/1600. under air flow conditions (50mLmin^{-1}), in platinum crucibles with a heating rate of 5°Cmin^{-1} between 25 and 1000°C to evaluate the optimal temperature for the consolidation of the crystalline phases [12]. The chemical composition and crystallographic structure of calcined solids were determined by X-ray diffraction, on a PANalytical X'Pert PRO MPD diffractometer equipped with an Ultra fast X´celerator detector and a Bragg-Brentano configuration, using the Cu Kα radiation (λ=1.54186Å), between 10 and 90° with steps of 0.02°. The measurements are developed with a voltage of 40kV and a current of 20mA. The analysis by transmission electron microscopy was performed on a JEOL 2100 microscope equipped with a LaB_6 thermo-ionic gun, operated with accelerating voltage of 200kV equipped with a CCD image acquisition system crystalline phase.

3. Results and discussion

The Figure 1 presents the FT-IR spectra of one of the precursors with their corresponding absorption bands, which point out the formation of a polymeric citrate complex for all synthesized precursors. Showing the same absorption bands, it is possible to identify the main signals as follow: The band located at 3431cm^{-1} is associated with vibrational tension modes of O-H group of dimers intermolecular bonded to hydroxyl groups, which can be moved on slightly by the NH_4OH addition effect to the reaction medium. In the same region at 3241cm^{-1}, it is clear, a tension band associated with the vibrational mode of O-H bond in carboxylic acid dimers linked to hydrogen bridges, which contain information from the N-H bond in a tension mode [4]. The band located at 2360cm^{-1}, corresponds to the stretching modes associated with the presence of occluded CO_2 in the material pores, derived of thermal treatment. The bands associated with asymmetric stretching and bond deformation (COO-M) are in the region at 1300-1500cm^{-1} and a band that arises from the symmetric deformation of the groups (COO-M) at 1653cm^{-1} [13]. The signal at 1040cm^{-1} corresponds to the C-O bond tension associated with acetals and primary alcohols, as well as bends out of plane of C-H bond, which are present in the precursors and that contain information about the NH_4OH addition during the process of adjusting the pH [14]. The bands located at low wavenumber correspond to the bonds of the different metal cations of Ce^{3+} y Pr^{3+} with OH groups and oxygen atoms [15,16].

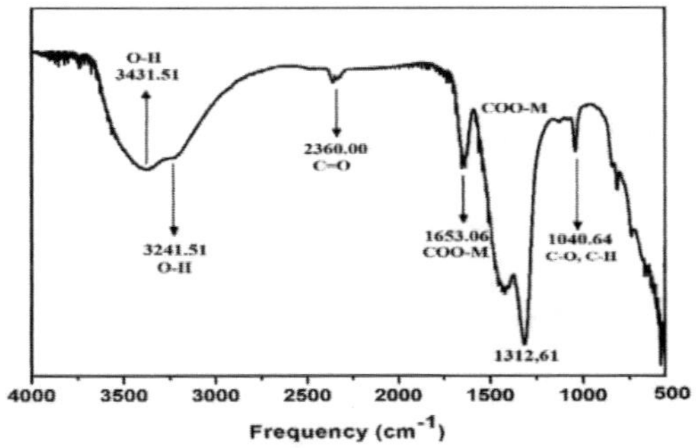

Figure 1. Representative infrared spectrum of the $Ce_{1-x}Pr_xO_2$ (x=0.0, 0.2, 0.4, 0.6, 0.8 and 1.0) precursor in KBr pellet obtained at 25°C.

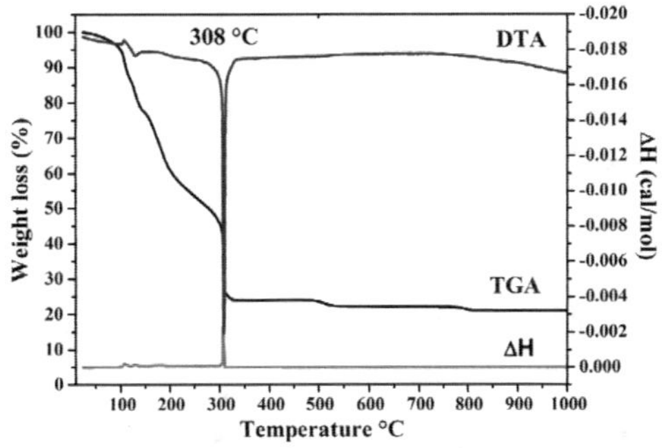

Figure 2. Representative thermal analysis (TGA-DTA) of $Ce_{1-x}Pr_xO_2$ (x=0.0, 0.2, 0.4, 0.6, 0.8 and 1.0) precursor.

The TGA-DTA analysis is shown in Figure 2. In the first stage, occurs the de-protonation of OH groups between 50 and 60°C, while in the second stage occurs the elimination of water molecules present until achieving a temperature of 150°C. Dehydration continues until 210ºC and additionally in the case of the citrate precursor, it starts the formation of C=C double bonds, with partial conversion of citrate to aconitate, while ammoniacal species formation begins to appear between 200 and 230°C, together with the elimination of R-O radical species, it is accompanied by an endothermic event until 285ºC [4]. At that moment, the system reach a maximum level of exothermicity, which is related to the combustion of organic matter between 285 and 340°C and the weight loss upper to 70%. At this point, it is initiated the removal of formed carbonates and consolidation of the crystalline phase, in this way, most of organic species are removed when the temperature reaches 505°C [17].

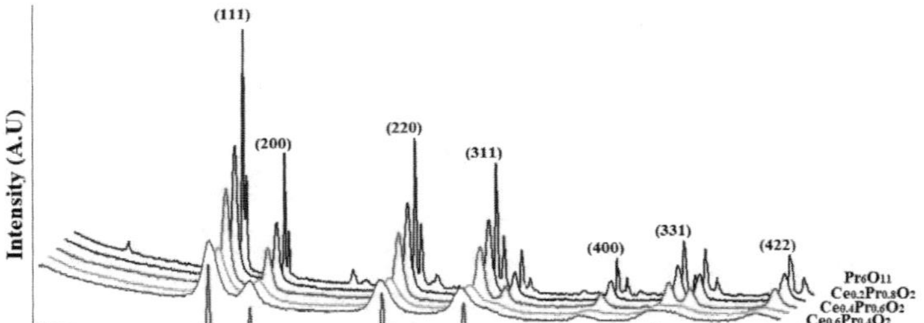

Figure 3. Comparison between X-ray patterns of $Ce_{1-x}Pr_xO_2$ (x=0.0, 0.2, 0.4, 0.6, 0.8 and 1.0) samples.

The X-ray diffraction analysis (XRD), confirms that synthesized oxides of the $Ce_{1-x}Pr_xO_2$ (x=0.0, 0.2, 0.4, 0.6, 0.8 and 1.0), which has a homogeneous composition and a size of crystallite specific of the synthesis method, with a preferential crystal orientation along (1 1 1) facet, as it is illustrated in Figure 3 [10]. The estimation of crystallite size in all samples are made using the most intense signals by use of the Debye-Scherrer equation which gives as result an average of 30nm in all compositions, results that corroborate to those obtained from transmission electron microscopy, in order to validate the obtention of nanometric oxides. [18]. The search performed by X'Pert High Score software in the ICCD databases, leads to a phase classification according to the reference compound of CeO_2, ICDD code 00-034-0394, with space group Fm-3m (225), cubic crystal system with a cell parameters a=5.4113Å and a cell volume of 158.46Å^3 [19].

The analysis by means of transmission electron microscopy (TEM), confirms the presence of nanometric agglomerates of regular morphology, with an average size of 30nm, outer areas of 1853nm^2, and interplanar distances of 0.31nm for the (1 1 1) facet [4,18] (see Figure 4).

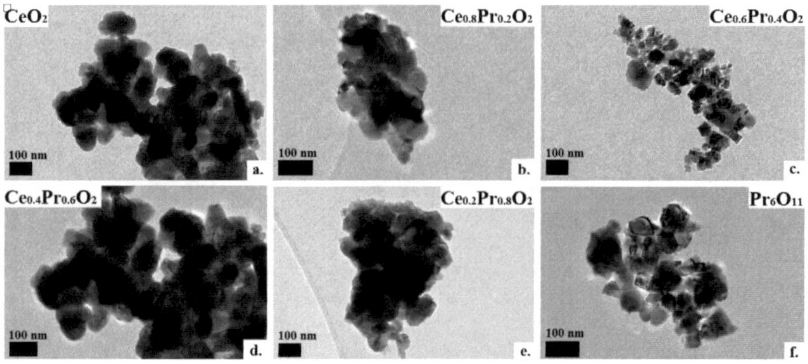

Figure 4. Transmission electron microscopy images of the $Ce_{1-x}Pr_xO_2$ (x=0.0, 0.2, 0.4, 0.6, 0.8 and 1.0) systems respectively.

4. Conclusions

The FT-IR and TGA-DTA analysis performed over citrate precursors allows evaluating the formation of intermediate coordination complexes and the optimal temperature for the consolidation of the crystalline phase, in accordance with the synthesis method used. The characterization by X-ray diffraction and transmission electron microscopy on calcined samples provide information about the formation of a prevalent fluorite phase with morphological and surface properties which consistent with the obtaining of nanometric crystallites (30nm) oriented along (1 1 1) facet and interplanar distances of 0.31nm for the main diffraction signal. The effect of citric acid is a critical parameter to establish the compounds coordination, which favours the development of highly homogeneous materials with improved properties for potential applications, since it involves the preservation of surface properties of relevance in the development of catalytic properties for eventual applications as anodic electrocatalysts in solid oxide fuel cells.

References

[1] Minh N Q 1993 *J Am Ceram Soc* **76** 563
[2] Zhu W Z and Deevi S C 2003 *Materials Science and Engineering* **362** 228
[3] Tuller H L, Bishop S R, Chen D, Kuru Y, Kim J J and Stefanik T S 2012 *Solid State Ionics* **225** 194
[4] Gomez J A, Valencia J S and Carda J B 2010 *J Chil Chem Soc* **55-4** 445
[5] Shaikh S P, Muchtar A and Somalu M R 2015 *Renewable and Sustainable Energy Reviews* **51**1
[6] Patil B B and Basu S 2014 *Energy Procedia* **54** 669
[7] Raj E S and Irvine J T S 2010 *Solid State Ionics* **180** 1683
[8] Deganello F, Marca G and Deganello G 2009 *Journal of the European Ceramic Society* **29-3** 439
[9] Gomez J A and Schmal M 2014 *Applied Catalysis B: Environmental* **148-149** 1
[10] Gomez J A and Valencia J S 2009 *Rev Colomb Quim* **38** 303
[11] Yu H F, Wang J, Wang S S and Kuo Y M 2009 *J Phys Chem Solid* **70** 218
[12] Varma H K, Kumar K P, G WK and Damodaran A D 1998 *J Mater Sci Lett* **11** 1313
[13] Weil K S, Hardy J S and Kim J Y 2007 *J Adv Mater* **2** 84
[14] Gomez J A and Valencia J S 2009 *Rev Colomb Quim* **38** 289
[15] Vanhoyland J P, D'Haen J, Mullens S and Mullens S J 2005 *Solid State Chem* **178** 166
[16] Pinheiro F, Machado de Souza F, Martins T S, Abreu Fantini M C and Isolani P C 2010 *J Therm Anal Calorim* **99** 385
[17] Titov A A, Klimenko M A, Goryacheva E G, Opolchenova N L, Stepareva N N and Sokolova N P 2008 *Inorg Mater* **44** 1101
[18] Esther Jeyanthi C, Siddheswaran R, Kumar P, Siva Shankar V and Rajarajan K 2014 *Ceramics International* **40** 8599
[19] Gomez J A and Schmal M 2013 *International Journal of Hydrogen Energy* **38** 7458

Structural properties, electric response and magnetic behaviour of $La_2SrFe_2CoO_9$ triple complex perovskite

F Casallas[1], E Vera[1], D Landínez[2], C Parra[1] and J Roa[2]
[1] Universidad Pedagógica y Tecnológica de Colombia, Tunja, Colombia.
[2] Universidad Nacional de Colombia, Bogotá, Colombia.

E-mail: francy.casallas@uptc.edu.co

Abstract. The triple perovskite $La_2SrFe_2CoO_9$ was prepared by the solid state reaction method from the high purity precursor powders La_2O_3, $SrCO_3$, Fe_2O_3, Co_2O_3 (99.9%). The crystalline structure was studied by X-ray diffraction experiments and Rietveld refinement analysis. Results reveal that this material crystallizes in an orthorhombic triple perovskite belonging to the space group Pnma (#62) with lattice constants a=5.491978(2)Å, b=7.719842(2)Å and c=5.436260(3)Å. The granular surface morphology was studied from images of Scanning Electron Microscopy. The electric response was studied by the Impedance Spectroscopy technique from 10.0mHz up to 0.1MHz, at different temperatures (77–300K). Measurements of magnetization as a function of temperature permitted to determine the occurrence of a paramagnetic – ferromagnetic transition for a Curie temperature of 280K, which suggests it application in nanoelectronic devices. From the fit of the magnetic response with the Curie-Weiss equation it was concluded that the effective magnetic moment is particularly large due to the contribution of La, Fe and Co cations.

1. Introduction

Ceramic materials represent a great percentage of systems which are actually investigated by the physics and chemistry of solids because their interesting properties. Particularly, the perovskite family has concentrated important attention in last decades. From the point of view of chemical composition, perovskites have been characterized by the ideal formula ABX_3, where A generally is an alkaline earth element, B represents a transition metal or rare earth element and X, more of times, is the oxygen [1]. Modifications of atomic radii of A and B, introduces structural distortions and new crystalline phases. Inclusions of rare earth elements instead the alkaline earth in the A site, and magnetic elements in the B site, give the possibility to produce materials with exotic electric and magnetic properties [2,3]. Partial substitutions of the A and B cations give rise to complex materials as the triple perovskites $A_2A'B_2B'O_9$ [4]. It chemical configuration supplies multiple chances to combine different elements, generating the possibility to synthesize new materials, which involve a more large gamma of physical properties. Depending on magnetic and electric characteristics of A, A', B and B' it is relatively simple to create new perovskite systems with half-metallic properties [5], magnetoelectric response [6] or magnetic ordering [7], which offer promissory perspectives in the new spintronics technology [8]. In order to analyse the possibility to create new magnetic perovskite materials, in this work we report the synthesis, the structural characterization and the electronic properties of the new triple perovskite $La_2SrFe_2CoO_9$, which was idealized as the introduction of the lanthanide La in the A site, the alkaline earth Sr in A', and the magnetic transition metals Fe and Co in the B and B' locations of the $A_2A'B_2B'O_9$ formula, in order to construct a triple complex perovskite.

2. Experimental

The samples were synthesized by the solid state reaction recipe. The precursor powders La_2O_3, $SrCO_3$, Fe_3O_3 and Co_3O_4 (Aldrich 99.9%) were mixed in stoichiometric proportions. Mixture was ground to form a pellet and annealed at 1100°C for 48h. The sample was then regrinded, repelletized and sintered at 1200°C for 48h. X-Ray diffraction (XRD) experiment was performed by means a PW1710 diffractometer with $\lambda_{CuK\alpha}$=1.5406Å. Rietveld refinement of diffraction pattern was carried out by the GSAS code [9]. Scanning Electron Microscopy (SEM) images and Electron Dispersive X-ray spectra were obtained by the utilization of a VEGA 3 electronic microscope. Electrical impedance spectroscopy measurements were performed by the application of frequencies from 10.0mHz up to 0.10MHz and AC voltage of 10mV rms in the range of temperatures between 77K up to 300K. Measurements of magnetic susceptibility in the Zero Field Cooling (ZFC) and Field Cooling (FC) recipes and curves of magnetization as a function of applied fields were carried out by means a VersaLab Quantum Design equipment.

3. Results and discussion

The analysis of XRD pattern showed in Figure 1 reveals the presence of characteristic peaks of complex perovskite systems.

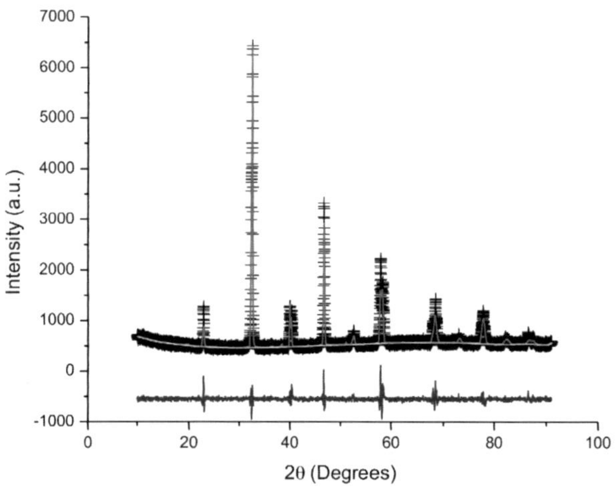

Figure 1. XRD pattern for the $La_2SrFe_2CoO_9$ triple perovskite.

In Figure 1, crosses represent the experimental data and line corresponds to simulated pattern by means of GSAS code. Base line is the difference between theoretical and experimental results. Rietveld refinement permitted to establish that this material crystallizes in an orthorhombic triple perovskite, space group Pnma(#62), lattice constants a=5.491978(2)Å, b=7.719842(2)Å and c=5.436260(3)Å. These results show deviation of no more than 3% when compared with the theoretical values obtained from the Structure Prediction Diagnostic Software SPuDS [10], which predicts that lattice constants a=5,5286 Å, b=7,7805Å and c=5,4756Å for the $La_2SrFe_2CoO_9$ material. Parameters of refinement are: RF^2=9.82% and x^2=2.196. The numeric results of the Rietveld analysis are shown in the Table 1.

A qualitative approximation to the surface structure of the $La_2SrFe_2CoO_9$ triple perovskite is obtained from the SEM technique as exemplified in Figure 2. The surface of material is constituted by sub-micrometric grains which form groups with appearance of clusters. These have a mean size of 5μm and appear strongly sintered with a texturized morphology in the external surface and granular, as gravel, with weak inter-granular coupling in the deepest regions.

Table 1. Atomic positions of cations and anions on the unit cell for $La_2SrFe_2CoO_9$.

Atom	Wyckoff Site	x	y	z
La	4c	0.5000	0.2500	0.4978
Sr	4c	0.5000	0.2500	0.5035
Fe	4b	0.0000	0.0000	0.5000
Co	4b	0.0000	0.0000	0.5000
O	4c	1.0120	0.2500	0.4745
O	8d	0.2739	0.0382	0.7401

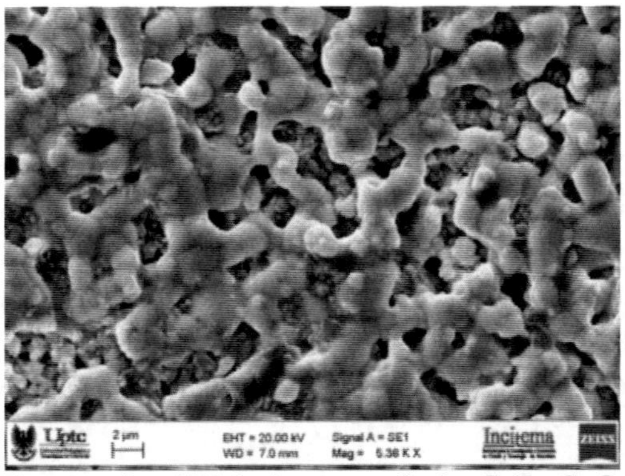

Figure 2. SEM image of $La_2SrFe_2CoO_9$ material.

Measurements of Impedance Spectroscopy permit to establish the electric feature of the $La_2SrFe_2CoO_9$ material. The impedance plots Z(ω) or Nyquist plot are shown in Figure 3(a). This picture not only derives the information about the electrical properties of our compound but also gives an idea of the relaxation mechanism and the conductance in the material. The impedance spectrum is characterized by the appearance of semicircle arc whose patterns changes, but not its shape, when the temperature is decreased [11].

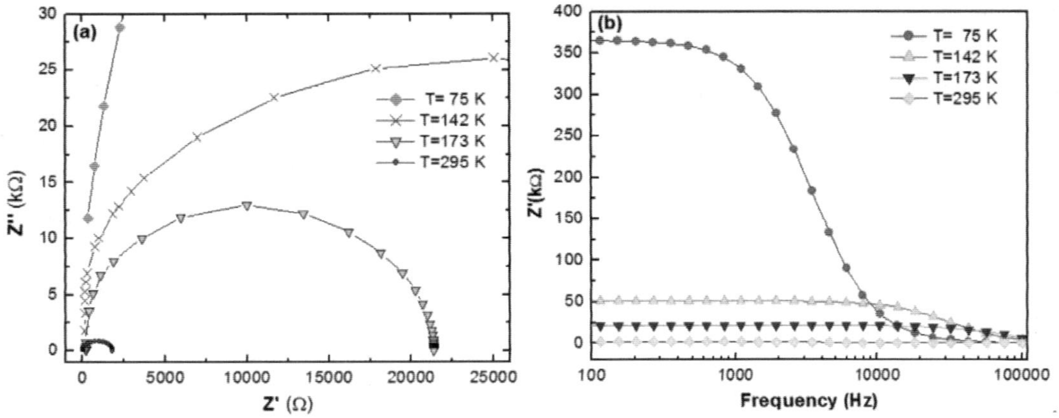

Figure 3. (a) Curve of Nyquist and (b) Bode diagram for the $La_2SrFe_2CoO_9$ material.

The Nyquist plot is related to the electrical processes occurring within the sample and their correlation with the microstructure of material. Figure 3(b) shows the variation of Z' as a function of the frequency for several temperatures. The impedance value is typically higher at lower temperatures in the low-frequency region and decreases gradually with increasing frequency. Also, Z' decreases with increasing temperature, indicating an increase of the ac conductivity. The value of Z' appears to merge in the high-frequency region irrespective of temperature. This result may possibly be related to the release of space charge as a result of the reduction in the barrier properties of material with rise in temperature, and may be a responsible factor for the enhancement of conductance of the material with temperature at high frequencies. The merge of the value of Z' for all temperatures at higher frequencies can be interpreted by the presence of space charge polarization [12]. The magnetic response was studied in curves of susceptibility as a function of temperature on the application of constant fields H=50, 600, 1200, and 5000 Oe, as shown in figure 4(a). As observed in the picture, the temperature of ferromagnetic ordering is T_C=280K. Close to this temperature value an irreversibility occurs between the ZFC and FC susceptibility curves. This behaviour, which is characterized of disordered magnetic systems as spin glasses [13], is attributed to the cationic disorder of the La and Sr ions in the A and A' sites of the triple perovskite, and Fe and Co in the B and B' sites, because their values of ionic radii are very similar and its individual magnetic spins can have essentially different orientations, in groups or clusters of spins that respond in different ways under the application of an external magnetic field. In the ZFC procedure, when the temperature is decreased in absence of external field, these clusters of spins are frozen, so that the magnetic field and low temperature are not sufficient to break the correlation of exchange between them to orientate them in the direction of the external field. In the FC procedure, the temperature is decreased in the presence of magnetic field and all spin clusters are effectively oriented in the direction of the applied field. Therefore, in the ZFC-FC curve a bifurcation of the magnetic response has place, which results in a lower susceptibility to ZFC and higher for the other. Furthermore, magnetic irreversibility temperature Ti is sensitive to the intensity of the applied magnetic field. As it is increased, the value of T_i increases from T_i = 320K for H=50Oe to T_i=370K for H=1200Oe. Subsequently, the value of T_i begins to decrease until T_i=275K for H=5000Oe. This behavior may be due to the occurrence of a value limit of applied field, below which the sizes and the grain boundaries can contribute to increasing the value of T_i.

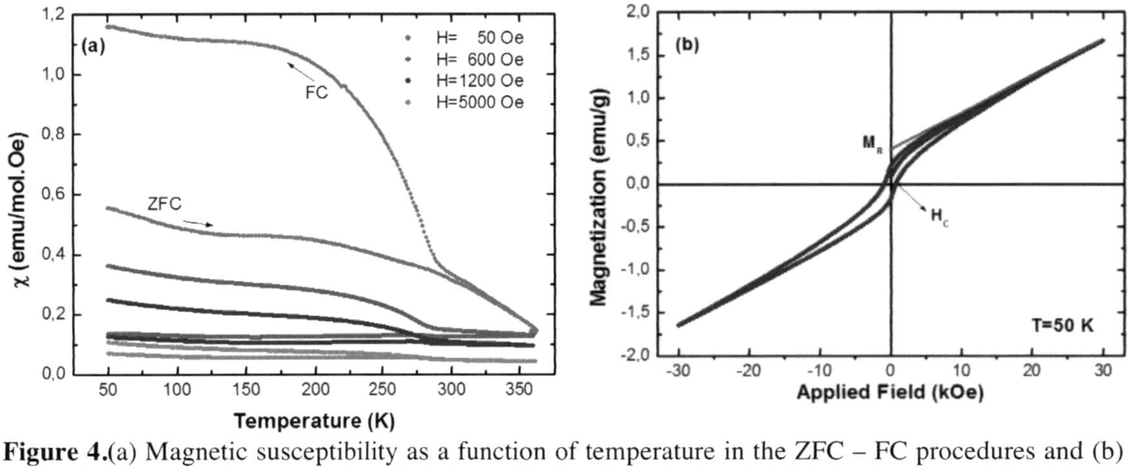

Figure 4.(a) Magnetic susceptibility as a function of temperature in the ZFC – FC procedures and (b) magnetization as a function of applied fields H=50, 600, 1200 and 5000 Oe for the La2SrFe2CoO_9 perovskite.

On the other hand, Figure 4(b) shows the hysteretic ferromagnetic behaviour of this material at low temperature values, below T_C. Magnetic parameters as coercive field HC=1.1kOe and remnant magnetization M_R=0,37emu/g were obtained from the picture. We notice that no magnetization saturation is observed in this experimental result. However, a quasi-linear behaviour is observed in the moderately high field regime, which is typical of granular systems with sub-micrometric crystallite sizes, in accordance with the SEM images showed in Figure 2.

4. Conclusions

We have synthesized the La2SrFe2CoO$_9$ complex perovskite by the standard solid reaction method. Structural characterization from XRD experiments and Rietveld refinement reveals that this material crystallizes in an orthorhombic triple perovskite belonging to the Pnma (#62) space group. SEM images showed the strongly sub-micrometric granular morphology of the material with crystallite clusters in the surface and weak coupled grains in the deepest regions. Impedance spectroscopy measurements reveal a systematic increase of the conductivity in this material as a function of the temperature. Results of the magnetic response in curves of susceptibility as a function of temperature permitted to determine the ferromagnetic character of this material with ordering temperature about 280K. Irreversibility between the ZFC and FC susceptibilities were observed and attributed to the cationic disorder of the Fe and Co magnetic cations. Eventually, the cationic disorder of La and Sr cations can be relevant due the strongly paramagnetic character of the La^{3+} ion. At last, from the hysteretic behaviour of magnetization as a function of applied fields the coercive field and remnant magnetization was characterized for this interesting material.

Acknowledgments

This work was partially supported by Division of Investigations of the National University of Colombia

References

[1] Hazen R 1988 Perovskites *Scientific American* **258-6** 74

[2] Sharma P, Ahn J, Hur N, Park S, Kim S, Lee S, Park J, Guha S, Cheong S 2004 *Physical Review Letters* Lett **93** 177202

[3] Moreno L, Valencia J, Landınez D, Arbey J, Rodrıguez M, Martınez M, Roa-Rojas J, FajardoF 2008 *Journal of Magnetism and Magnetic Materials* **320** 19

[4] Albornoz J, Téllez D, Roa-Rojas J, Munévar J, Baggio-Saitovich E 2013 *Journal of Superconductivity and Novel Magnetism* 26 pp 2313-2317

[5] Bonilla M, Téllez D, Rodríguez J, Aguiar J, Roa-Rojas J 2008 *Journal of Magnetism and Magnetic Materials* **320** 397

[6] Roa-Rojas J, León-Vanegas A, Téllez D, Pureur P, Dias F, Vieira V 2008 *Journal of Magnetism and Magnetic Materials* **320** 104

[7] Viola M, Martinez-Lope M, Alonso J, Velasco P, Martinez J, Pedregosa J, Fernandez-Diaz M 2002 6 *Chemistry of Materials* **14** 812

[8] Majewski P, Geprägs S, Sanganas O, Opel M, Gross R, Wilhelm F, Alff, L. 2006 *arXiv* preprint cond-mat/0602071

[9] Larson A, Von Dreele R 1994 *Los Alamos National Laboratory Report* (USA: LANSCE MS-H805)

[10] Lufaso M, Woodward P 2001 *Structural Science* 57 725

[11] Prasad K, Chandra K, Kulkarni A 2011 *Journal of materials science* **46** 2077

[12] Rahmouni H, Nouiri M, Jemai R, Kallel N, Rzigua F, Selmi A, Alaya S 2007 *Journal of Magnetism and Magnetic Materials* **316** 23

[13] Mantilla J, Ter Haar E, Coaquira J, Bindilatti V 2007 *Journal of Physics Condensed Matter* **19** 386225

Theoretical investigation of GaN carbon doped

M J Espitia Rico[1], M G Moreno Armenta[2], J A Rodríguez[3] and N Takeuchi[2]
[1] Universidad Distrital Francisco José de Caldas, Bogotá, Colombia.
[2] Universidad Nacional Autónoma de México, Ensenada B.C., México.
[3] Universidad Nacional de Colombia, Bogotá, Colombia.

E-mail: mespitiar@udistrital.edu.co

Abstract. In this work we used first principles calculations in the frame of density functional theory (DFT) in order to study the structural and electronic properties of GaN doped with carbon. The computational calculations were carried out by a method based on plane waves pseudopotentials, as implemented in the Quantum Espresso code. In the wurtzite type GaN supercell the nitrogen atoms were replaced by carbon atoms (C by N) and then also the gallium atoms by carbon atoms (C by Ga). The carbon concentrations in the GaN volume was set as x=25, 50 y 75%. For each concentration x of carbon the formation energy was calculated for the substitutions C by N and CxGa. We found that it is more energetically favourable that the carbon atoms occupy the positions of the nitrogen atoms (C by N), because in all the x concentrations of carbon the formation energies were lower than that in the substitutions (C by Ga). It was found that the new compounds C_xGaN_{1-x} have higher bulk moduli. So they are very rigid. This property makes them good candidates for applications in hard coatings or devices for high power and temperatures. Analysis of the density of states show that the new C_xGaN_{1-x} ternary compound have metallic behaviour that comes essentially from the hybridization states N-p and C-p cross the Fermi level.

1. Introduction

Gallium nitride (GaN) is one of the wide direct band gap semiconductors that normally crystallizes in the wurtzite structure [1] it can be used in optoelectronic components, like blue light emitting diodes and lasers [2]. Due to its notable thermal and chemical stability, it is also ideally suitable for the applications in harsh environments, such as high pressure and temperature [2,3]. Large piezoelectric constants of GaN point out to possible applications of GaN-based materials in sensors. Due to wide band gap, these sensors are expected to operate in a broad temperature range and/or in a harsh environment. On the other hand, the carbon in the GaN had not been sufficiently studied. For many years carbon was considered an atmospheric impurities that could contaminate the semiconductor devices based on GaN during the manufacturing process [4,5]. In particular, the carbon polluting behaviour was reported for GaAs, AlAs and InAs grown by Molecular Beam Epitaxy [5]. However, in recent years some theoretical and experimental studies have concluded that the introduction of carbon in the GaN volume strongly alters their electronic and magnetic properties. H. Tang et al [6] at 2001 and R. P. Vaudo et al [7] at 2003 found that the incorporation of carbon in the GaN films generate in an insulating material with higher resistance $>10^8\Omega$ [8,9], for this reason GaN films are becoming increasingly popular in their highly resistive form to be used as insulating buffer layers or substrates in GaN-based device technology.

In this work we execute first principles calculations to investigate the effects of the incorporation de C in the GaN volume on structural and electronica properties of C:GaN system. First, in the GaN structure wurtzite we replaced gallium atoms by carbon atoms (C by Ga) and then nitrogen atoms by carbon atoms (C by N), whit carbon concentration of 25, 50 y 75%. We calculate the formation energy of each substitution to find the energetically most favourable configurations.

2. Computational method

To determine the equilibrium geometry of the $C_xGa_{1-x}N$ y C_xGaN_{1-x} (x=0.25, 0.50, 0.75), compounds, in the wurtzite structure the volume and the c/a ratio was optimized. We used the pseudopotential plane-wave method, as implemented in the Quantum espresso code [10]. The correlations and exchange effects are treated using the generalized gradient approximation (GGA). The formation energy of the ternary phase is defined as [11].

$$\Delta E = E_{C:GaN} - N_C E_C - N_{Ga} E_{Ga} - N_N E_N \tag{1}$$

Where $E_{C:GaN}$ is the total energy of the ground state for the corresponding concentration of C by Ga or C by N in the ternary compound. E_C is the total energy of carbon in the ground state (graphite), N_C is the number of carbon atoms in the compound. E_{Ga} is the total energy of ground state of Ga, N_{Ga} is the number of carbon atoms, E_N is the total energy of ground state of N and N_C is the number of nitrogen atoms in the ternary compound.

3. Results and discussions

3.1. Structural properties

The Table 1 show the formation energy calculated with the Equation (1) for ternary $C_xGa_{1-x}N$ and C_xGaN_{1-x} (x=0.25, 0.50, 0.75) compounds.

Table 1. Formation energy.

	%	Formation energy (eV)
C by N	0.25	-3.714
C by N	0.50	-0.286
C by N	0.75	2.345
C by Ga	0.25	-0.612
C by Ga	0.50	2.586
C by Ga	0.75	10.612

For all concentrations, the formation energy is lower when a carbon atom occupies the position of a nitrogen atom, therefore, it is energetically more favourable that the carbon atoms occupy the positions of the nitrogen atoms in the GaN wurtzite type. Additionally, the ternary $C_{0.25}GaN_{0.75}$ y $C_{0.05}GaN_{0.50}$ has negative formation energy, therefore this compound are thermodynamically stable, whereas that formation energy of $C_{0.75}GaN_{0.25}$ compound is positive, which implies that this compound is thermodynamically unstable, this means that this compound cannot grow under equilibrium conditions, so in order to grow them, it is necessary to supply power to the system, in agreement with the result obtained by Zhang y S. Veprek [12] in their first principles study of metastable phases and thermodynamic modelling of $Ti_{0.5}Al_{0.5}N$ compound.

The Table 2 shows the main structural parameters of GaN and the allowed ternary compounds. The structural parameters of the binary GaN compound calculated in this work accords well with values reported theoretically [1] and experimentally [13], since it differs by less than one percent. The Table

1 show that by increasing the concentration of carbon atoms, the lattice constants of ternary compounds only change slightly with respect to GaN, this may be because the radius of the C atom (0.914Å) is very close to the atomic radius of N (0.92Å). The Figure 1(a) show the variation of the bulk modulus as a function of concentrations of C, whereas the Figure 1(b) show the variation of total energy as a function concentrations of C. The bulk modulus decreases with the increasing amount (x) of the concentration of C atoms with respect to bulk modulus of GaN, whereas that energy total increases linearly (is less negative) with an increasing of the concentration of C atoms in the GaN compound.

Table 2. Structural parameters lattice constant (a_0), ratio c/a, bulk modulus (B_0) and total energy (E_0) of the compounds.

Compound	a_0 (Å)	c/a	V_0 (Å^3)	B_0 (GPa)	E_0 (eV)
	3.2048	1.633	23.286	175.33	
GaN	3.1986[a]	1.6339[a]	23.153[a]	176.54[a]	-2693.128
	3.2210[b]	1.6226[b]	22.838[b]	188.00[b]	
$C_{0.25}GaN_{0.75}$	3.2019	1.6312	47.41	164.50	-2693.451
$C_{0.50}GaN_{0.50}$	3.2120	1.6311	47.41	153.35	-2633.743
$C_{0.75}GaN_{0.25}$	3.2141	1.6360	56.10	146.04	-2604.074

[a] Theoretical reference [1], [b] Experimental reference [13].

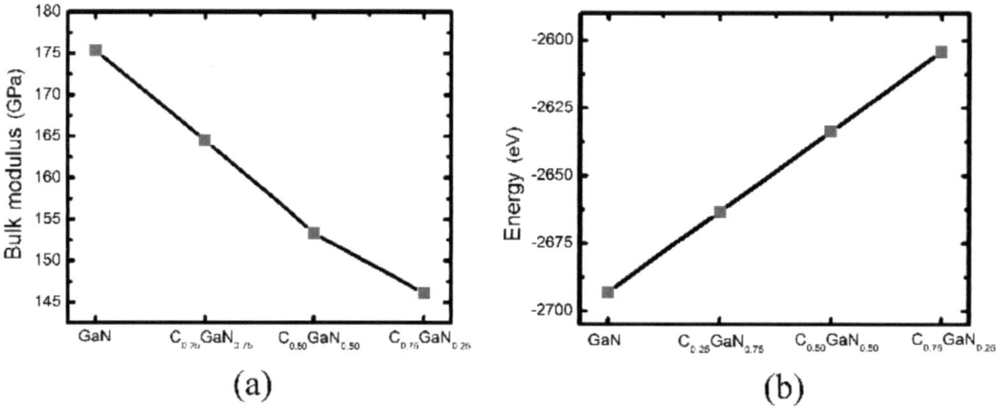

Figure 1. (a) Bulk module GPa) as a function of concentration of carbon atoms. (b) Total energy as a function of concentration of carbon atoms for C_xGaN_{1-x} (x=0.25, 0.50 y 0.75) compounds. In the figures the lines are only a guide to the eye.

3.2. Electronic properties

The Figure 2(a) show the total and partial density of states of the of GaN clean, this compound has a semiconductor behaviour with a forbidden energy gap of 2.0eV, The magnitude of this gap is smaller than the experimentally reported for GaN in wurtzite structure, this occurs because the GGA approximation underestimates the forbidden energy gap in semiconductors. The Figures 2(b), 2(c) and 2(d) shows the total and partial spin-polarized density of states of the ternary $C_{0.25}GaN_{0.75}$, $C_{0.50}GaN_{0.50}$ y $C_{0.5}GaN_{0.25}$ compounds. Due to the introduction of the C atoms in the structure of GaN, it loses its semiconductor nature, there are a penetration towards prohibited energy zone of the states C-p in greater proportion and the states N-p in lesser proportion, therefore, the allowed ternary compounds has metallic behaviour determined by such states.

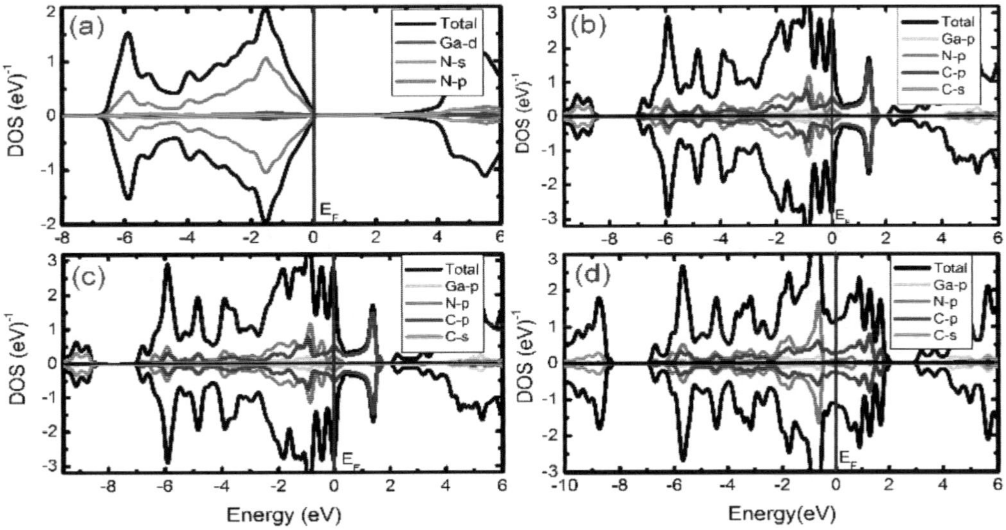

Figure 2. Total and partial density of state of (a) GaN, (b) $C_{0.25}GaN_{0.75}$, (c) $C_{0.50}GaN_{0.50}$ and (d) $C_{0.75}GaN_{0.25}$ in the wurtzite phase.

4. Conclusions

In this work we executed a study of GaN carbon doped, with concentrations of carbon atoms of x = 25%, 50% y 75%, using the Density Functional Theory DFT in the frame pseudopotential plane wave. The formation energy calculations reveal that is more favourable that the carbon atoms occupy the positions of the nitrogen atoms in the GaN wurtzite type. We found that the allowed ternary $C_{0.25}GaN_{0.75}$, $C_{0.50}GaN_{0.50}$ y $C_{0.5}GaN_{0.25}$ have higher bulk moduli, so they are very rigid. This property makes them good candidates for applications in hard coatings or devices for high power and temperatures. The density states studies reveal that there are a penetration towards prohibited energy zone of the states C-p in greater proportion and the states N-p in lesser proportion, therefore, the allowed ternary compounds has metallic behaviour.

Acknowledgments

The Centro de Nanociencias y Nanotecnología of the UNAM for their support through the project DGAPA IN102 713, Supercomputo DGCTIC-UNAM, project SC 15-1-IR-18.

References

[1] Arbouche O, Belgoumène B, Soudini B and Driz M 2009 *Computational Materials Science* **47** 432
[2] Nakamura S 1997 *Solid State Commun* **102** 237
[3] Nakamura S, Senoh M, Iwasa N and Nagahama S I 1995 *Applied Physics Letters* **67** 1868
[4] Kimura T and Hashizume T 2009 *Journal of Applied Physics* **105** 014503
[5] Ito H and Ishibashi T 1991 *J Appl Phys* **30** 994
[6] Tang H, Webb J B, Bardwell J A, Raymond S, Salzman J and Uzan S S 2001 *Appl Phys Lett* **78** 757
[7] Vaudo R P, Xu X P, Salant A, Malcarne J and Brandes G R 2003 *Phys Status Solidi* **200** 18
[8] Chen Z, Cao C B and Li W S 2009 *Cryst Growth Des* **9** 792
[9] Lopatiuk O, Chernyak L, Feldman Y and Gartsman K 2007 *Thin Solid Films* **515** 4365
[10] Giannozzi P, Baroni S and Bonin N 2009 *J Phys Condens Matter* **21** 395502
[11] Hu X L, Zhang Y F, Zhuang N F and Qian J Q 2010 *Journal of Solid State Chemistry* **183** 2741
[12] Zhang R F and Veprek S 2007 *Acta Materialia* **55** 4615
[13] Zheng Gu, Edgar J H, Pomeroy J, Kuball M and Coffey D W 2004 *Journal of Materials Science: Materials in Electronics* **15** 555

Synthesis of (azelaic-co-dodecanedioic) polyanhydride by microwave technique

M Gutiérrez[1], C Sierra[1], M Acevedo Morantes[1] and A P Herrera[1]
[1] Universidad de Cartagena, Cartagena, Colombia.

Email: aherrerab2@unicartagena.edu.co , macevedom@unicartagena.edu.co

Abstract. A polyanhydride was synthesized through microwave radiation using azelaic acid and dodecanedioic dicarboxylic acid at concentrations of 75:25, 50:50, and 25:75%w/w with acetic anhydride as crosslinking agent. Polymerization was carried out during 3 and 5 minutes. The copolymer with the highest molecular weight was selected using the intrinsic viscometry technique and by Huggin/Kraemer and Solomon/Ciuta methods. Based on these measurements, the 50:50 copolymer was selected with a polymerization time of 3 minutes in the microwave. This sample displayed the highest intrinsic viscosity (41.82cm^3/g), demonstrating the relevance of the microwave technique for the synthesis of biopolymers.

1. Introduction

Polyanhydrides are biodegradable polymers with suitable properties for biomedical applications. The products of decomposition are non-toxic diacid monomers that can be metabolized and eliminated by the body [1]. These biopolymers have been studied for the delivery of chemotherapeutic drugs including antibiotics and anesthetics [2,3]. Policondensation is the method of synthesis of this type of biopolymer. This process is carried out under rigorous conditions to obtain high molecular weight polymer with very long reaction times [4]. Microwave radiation offers significant benefits when compared to conventional heating since this technology allows a solvent-free and a fast polymerization by reaching high reaction temperatures [5,6].

The present work evaluates the synthesis of polyanhydrides from two dicarboxylic acids: azelaic acid and dodecanedioic acid, using the microwave technique and acetic anhydride as a crosslinking agent. The intrinsic viscosity method was used to determine the molecular mass of the synthesized copolymers as a function of the reaction time (3 and 5 minutes) and at different monomer concentrations (75:25, 50:50 and 25:75% w/w). Samples of the copolymers (0.1g/mL) were prepared and suspended in a chloroform/ethanol solution (60/40% v/v). The Hugging and Kraemer equations were used to calculate intrinsic viscosity by lineal extrapolation, as well as the Solomon-Ciuta equation by single-point.

2. Materials and methods

2.1 Synthesis of the (azelaic-co-dodecanedioic) polyanhydride

All chemical reagents were used as received. Azelaic acid 90% and dodecanedioic acid 99% were used as reacting monomers. Microwave irradiation was performed using 1120W microwave. Pre-polymerization of both monomers was performed in covered boron-silicate vials of 10mL capacity by microwave irradiation (5 minutes) using a 1:3 relation of solid dicarboxilic acid to acetic anhydride. The copolymerization step was subsequently performed using concentrations of 75:25, 50:50 and

25:75% w/w of azelaic acid:dodecanedioic acid, respectively. This stage was performed by heating the pre-polymers mixture in the microwave and varying reaction time between 3 and 5 minutes.

2.2 Determination of the intrinsic viscosity of the synthesized copolymers

A chloroform/ethanol (60/40% v/v) solution was used as solvent for these measurements. At a first stage, 1g of the copolymer was dissolved into 10mL of solvent. Dilutions were then prepared from this solution to obtain samples with concentrations of 0.005, 0.0075 and 0.01g/mL. Intrinsic viscosity measurements were carried out using a capillary viscometer, which was vertically submerged in a thermal bath at 25°C [7]. A sample volume of 7 mL was taken, and a standby time of 10 minutes was applied as factor for achieving thermic equilibrium. In a typical procedure, the sample was suctioned through the capillary until that the polymer solution reached the upper mark. Then, a measurement of the time was registered when the fluid reached the bottom mark of the capillary. Afterwards, the viscometer was washed with ketone and dried. This procedure was repeated for each copolymer solution, as well as for the chloroform/ethanol solution. All measurements were conducted in triplicate.

3. Results

Intrinsic viscosity of the copolymers was determined as a function of their relative viscosity, according to Huggins and Kraemer equations (Equation 1) [8],

$$\frac{\eta_{sp}}{c} = [\eta] + k_H[\eta]^2 c \quad ; \quad \frac{\ln \eta_r}{c} = [\eta] - k_K[\eta]^2 c \qquad (1)$$

where η_{sP} is the specific viscosity, η_r is the relative viscosity and c is the concentration in g/mL, k_H and k_K are dimensionless constants, called as Huggins constant and Kramer constant, respectively. Table 1 shows the results that were obtained with the capillary viscometer. It can be observed that the sample of the 50:50 copolymer with 3 minutes of microwave irradiation presented the highest values of relative viscosity, which were between 1.2 and 1.25 for the studied range of concentrations [9].

Table 1. Relative and intrinsic viscosities of the copolymer samples (azelaic-co-dodecanedioic) prepared in the microwave.

	Polymerization time of 3 min in the microwave					
	Relative viscosity			Intrinsic viscosity		
	0.005g/mL	0.0075g/mL	0.010g/mL	Solomon-Ciuta	Huggins-Kraemer	Standard deviation
75:25	0.99	1.12	1.16	10.65	-	-
50:50	1.20	1.21	1.25	29.44	41.82	7.30
25:75	1.05	1.08	1.11	11.39	12.48	0.41
	Polymerization time of 5 min in the microwave					
	Relative viscosity			Intrinsic viscosity		
75:25	1.01	1.049	1.08	6.04	4.07	2.80
50:50	1.02	1.034	1.04	4.69	4.98	0.54
25:75	1.02	1.028	1.05	4.48	2.83	0.90

Determination of intrinsic viscosity was achieved by plotting the values of η_{SP}/c and $\ln \eta_r/c$ versus concentration, according to Huggins and Kraemer equations. Figure 1 shows these plots. This analysis did not display an adequate lineal adjustment for the 75:25% w/w relation of azelaic acid:dodecanedioic acid that was prepared under 3 minutes of polymerization time (Figure 1(a)). The same happened for the sample with a relation of 25:75%w/w with a polymerization time of 5 minutes (Figure 1(f)). This can be attributed to the low solvent-homopolymer affinity, since it is known that adequate solvents lead high intrinsic viscosities in polymers solutions. Solubility parameters (SP) are

Journal of Physics: Conference Series **687** (2016) 012049 doi:10.1088/1742-6596/687/1/012049

used when describing this behaviour; the closer the values of SP for both solvent and solute, the solution is more suitable [10]. This implies that for relative viscosities lower than 1.0, there is no lineal correlation; it is only for values higher than 1.2 that a suitable accuracy and linearity are achieved with use the Huggins and Kramer equations through extrapolation of data for deducting intrinsic viscosity [11]. The copolymer synthesized with a ratio of 50:50% w/w azelaic acid:dodecanedioic acid at 3 minutes in the microwave (Figure 1(b)) presented higher intrinsic viscosity, since the intercepts of the Huggins and Kraemer straight lines exhibited a higher value when it is compared to the ones achieved in the remaining samples. This leads to the deduction that such values of monomer concentrations and polymerization time allowed the formation of a polyanhydride with a higher molecular mass.

The Solomon-Ciuta (SC) equation (Equation 2) was used to validate the data of intrinsic viscosities that were graphically obtained from the Huggins and Kramer equations [12]. Table 1 shows a summary of the results in this comparison.

$$[\eta] = \frac{[2(\eta_{SP} - \ln \eta_r)]^{\frac{1}{2}}}{c} \quad (2)$$

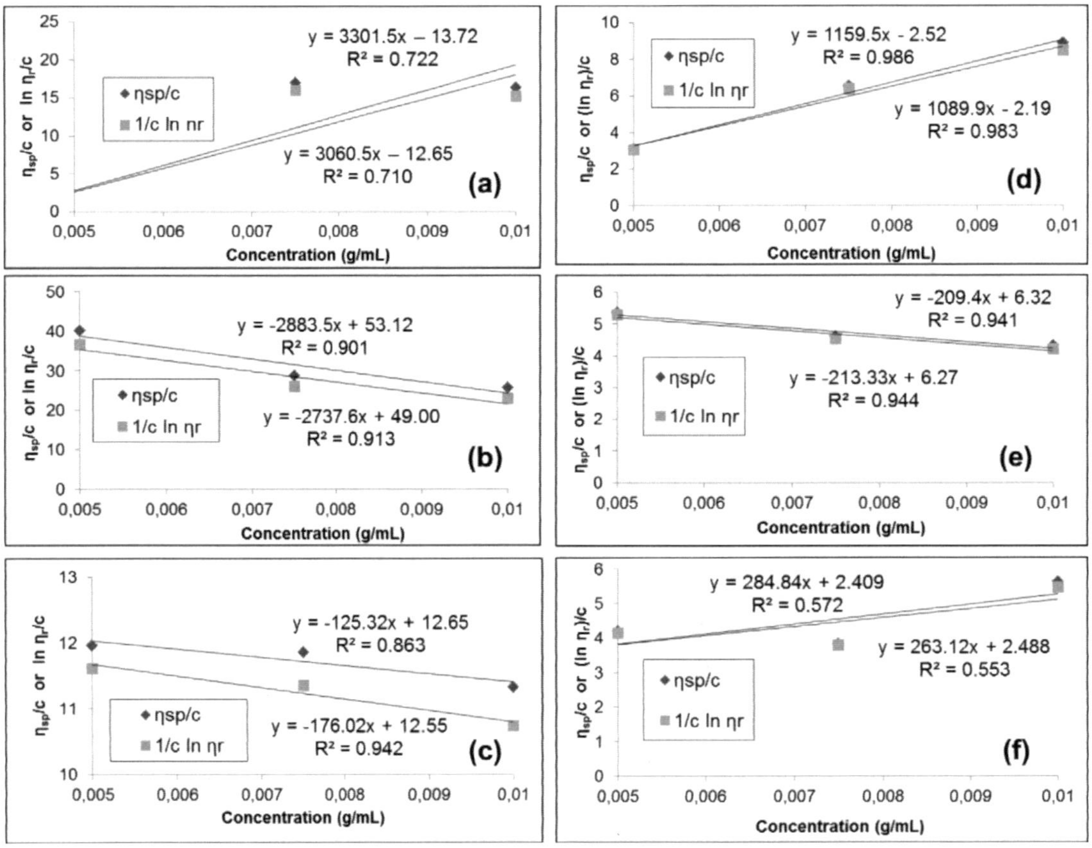

Figure 1. Determination of intrinsic viscosity by Huggins (♦) and Kraemer (■) methodology for the (azelaic-co-dodecanedioic) copolymers synthesized by microwave irradiation. Polymerization time of 3 minutes for the (a) 75:25, (b) 50:50, and (c) 25:75 proportions. Polymerization time of 5 minutes for the (d) 75:25, (e) 50:50, and (f) 25:75 proportions.

A JEOL JSM-6490LV scanning electron microscope (SEM) coupled to an energy dispersive spectroscopy (EDS) was used to get images of the synthesized (azelaic-co-dodecanedioic) polyanhidride that was prepared with a concentration of 50:50%w/w after 3 minutes of microwave irradiation. Figure 3 shows this photograph. There, the formation of a non-regular, porous morphology with microchannels can be observed. Additionally, EDS allowed the determination of an elemental composition of about 72% weight of carbon and 28% weight of oxygen for this copolymer.

Figure 3. SEM image of the (azelaic-co-dodecanedioic) polyanhidride that was prepared with a concentration of 50:50%w/w after 3 minutes of microwave radiation.

4. Conclusions

Microwave radiation proved to be a suitable method for the synthesis of biomaterials. It allows a simple, fast and economic procedure for the production of polyanhydrides from dicarboxylic acids. Capillary viscometry represents a low cost as useful technique for the determination of intrinsic viscosities and the deduction of molecular masses for polymeric materials. The highest intrinsic viscosity (41.82cm^3/g) was achieved with a copolymer that presented a 50:50 proportion of monomers and a polymerization time of 3 minutes. This reflects the polymer-solvent affinity and evidences that the resulting concentration of the copolymer depends on the relative reactivity of each monomer, as well as on their proportions in the reaction mixture. Thus, materials with combined properties can be produced by copolymerization, instead of using separate homopolymers.

Acknowledgments

We would like to thank the University of Cartagena for the financial support of this project (grant No. 052-2013). The authors are grateful with Chem. Eng. Leandro Marrugo for his collaboration during the development of this work, through the financial support program Young Researcher from the Department of Science and Technology of Colombia COLCIENCIAS (grant No. 0216-2014).

References

[1] Kumar N, Langer R and Domb A 2002 *Advanced Drug Delivery Rev* **54** 889
[2] Kost J and Langer R 2012 *Advanced Drug Delivery Rev* **64** 327
[3] Eameema M, Duvvuri L, Khan W and Domb A 2014 Polyanhydrides *natural and synthetic biomedical polymers* (Burlington, MA: Elsevier) chapter 10 pp 181–92
[4] Sosnik A, Gotelli G and Abraham G 2011 *Progress in Polymer Sci* **36** 1050
[5] Brandon M, Surya K and Balaji N 2004 *Macromolecular Rapid Communications* **25** 330
[6] Kappe O, Dallinger D and Murphree S 2009 Microwave theory *practical microwave synthesis for organic chemists* (Germany: Wiley-VCH) chapter 2 pp 11–20
[7] Seuvre A and Mathlouthi M 2010 *Food Chemistry* **122** 455
[8] Masuelli1 M and Sansone M 2012 *Products and applications of biopolymers* (Croatia: InTech) pp 88–95
[9] Khan M and Qazi R 2010 *The Arabian Journal for Science and Engineering* **35-2A** 67
[10] Han K, Jeon G, Hong I and Lee S 2013 *J of Industrial and Engineering Chemistry* **19** 1130
[11] Mehrdad A, Shekaari H and Niknam Z 2013 *Fluid Phase Equilibria* **353** 69
[12] Delpech M and Oliveira C 2005 *Polymer Testing* **24** 381

Mercury removal in wastewater by iron oxide nanoparticles

E Vélez[1], G E Campillo[1], G Morales[1], C Hincapié[1], J Osorio[2], O Arnache[2], J I Uribe[2] and F Jaramillo[2]
[1] Universidad de Medellín, Medellín, Colombia.
[2] Universidad de Antioquia, Medellín, Colombia.

E-mail: evelez@udem.edu.co

Abstract. Mercury is one of the persistent pollutants in wastewater; it is becoming a severe environmental and public health problem, this is why nowadays its removal is an obligation. Iron oxide nanoparticles are receiving much attention due to their properties, such as: great biocompatibility, ease of separation, high relation of surface-area to volume, surface modifiability, reusability, excellent magnetic properties and relative low cost. In this experiment, Fe_3O_4 and γ-Fe_2O_3 nanoparticles were synthesized using iron salts and NaOH as precipitation agents, and Aloe Vera as stabilizing agent; then these nanoparticles were characterized by three different measurements: first, using a Zetasizer Nano ZS for their size estimation, secondly UV–visible spectroscopy which showed the existence of resonance of plasmon at λ_{max}~360 nm, and lastly by Scanning Electron Microscopy (SEM) to determine nanoparticles form. The results of this characterization showed that the obtained Iron oxides nanoparticles have a narrow size distribution (~100nm). Mercury removal of 70% approximately was confirmed by atomic absorption spectroscopy measurements.

1. Introduction

The toxicity and persistence in environment of heavy metals released by industrial effluents have generated a significant threat to public health; these effluents have been treated through various conventional techniques (precipitation, electrochemical processes and/or membrane processes) [1]. Mercury is one of these persistent pollutants in wastewater and its removal has become a necessity. It shows its most toxic form as the highly reactive mercury (II) salts [2], which are often converted by bacteria into a neurotoxic substance, methylmercury, which increases the health risk for humans and wildlife through the aquatic food chain [3,4]. Nanomaterials such as iron oxide nanoparticles (IONPs) are efficient adsorbents because they combine magnetic separation with ionic exchange capacity for heavy metals removal [5,6]. Also, magnetic properties of iron oxide nanoparticles allow, for simple dispersion, the removal from an aqueous system [5-7]. Specifically the magnetite (Fe_3O_4) and maghemite (γ-Fe_2O_3) nanoparticles have been used in the adsorption of arsenic and chromium due to their surface properties [8], in this research we used it for mercury removal. In the present work, we report an efficient removal of Hg (II) ions from polluted water by using IONPs. The nanoparticles were synthesized by co-precipitation method, and characterized using different techniques to verify nanoparticles size and form.

2. Experiment

γ-Fe_2O_3 nanoparticles (IONPs) were prepared by co-precipitating Fe^{2+} and Fe^{3+} ions (in a molar ratio 1:2) in aqueous medium. Chemical precipitation was achieved at room temperature under vigorous stirring while adding NaOH solution (1.5M), Aloe Vera solution (6.25mL) was used as stabilizing agent [9], and in this step a black precipitate is formed. During the reaction process, the pH was maintained at about 11-12. The precipitate was isolated after decanting the solution, then washed with deoxygenated water and finally centrifuged at 4000rpm (this process was done three times). Lastly, 0.01M solution of HCl is added to neutralize the ionic charges of nanoparticles. Cationic colloidal nanoparticles are then separated by centrifugation at 4000rpm, resulting in a translucent colloidal solution. Synthesized samples were characterized by using dynamic light scattering measurements (for this purpose we use both: a Nanozeta S - Malvern device), UV–vis absorption spectroscopy (UV-Visible Cary-100 VARIAN) in the wavelength range from 200 to 900nm, was also used to identify the presence of metal nanoparticles, these small particles exhibit an absorption of visible electromagnetic waves by the collective oscillation of the electrons in the conduction band at the surface (this is known as the surface plasmon resonance effect) [10]. The absorption bands for iron oxide nanoparticles have been reported to be in the range between 330–450nm [11,12]. The size and morphology of iron oxide nanoparticles were observed by Scanning Electron Microscopy (SEM) from JEOL JSM-6490LV equipment. Samples for SEM measurements were suspended in ethanol and ultrasonically dispersed. The Atomic Absorption Spectroscopy (AAS) measurements were carried out by a Thermo CientificTM iCE 3000.

3. Results and discussion

3.1. Synthesis and characterization

To identify the presence of iron oxide nanoparticles (IONPs) we used UV-visible spectrometer, and the Figure 1 represents the UV-vis spectra, which shows a broad absorption peak at ~360nm, due to the absorption and scattering of light by IONPs, indicating nanoparticles presence. The Figure 2 shows the size distribution versus intensity of the synthesized IONPs, which were mono-disperse with a diameter of ~99.56nm.

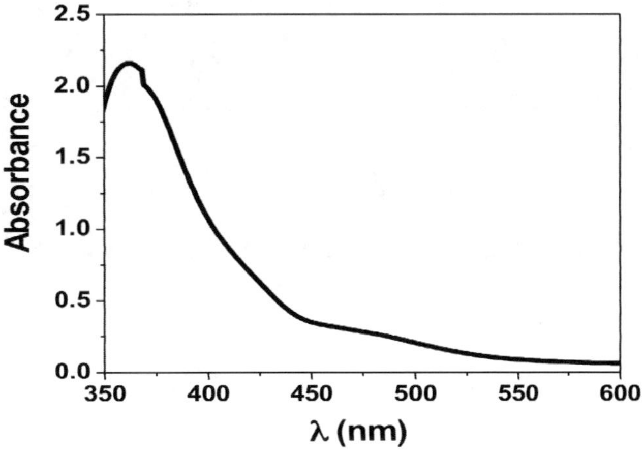

Figure 1. UV-vis spectra of IONPs.

In this work, we used Aloe Vera as stabilizing agent because of its low cost, and non-toxicity. 6.25mL of solution of Aloe Vera was used to obtain the size and shape of synthesized nanoparticles [9]. The usage of surfactant is necessary to avoid nuclei aggregation, in order to decrease the total surface energy. This aggregation, which can be a consequence of electrical charges in the surface (produced by dipolar moment) between atoms of the crystals, should be inhibited or limited to restrict the final particle size at the nanometric scale. One way to prevent nanoparticles from aggregation is by using substances that lead to steric repulsion between individuals [13]. The shape and size distribution of

IONPs were characterized by SEM and is shown in Figure 3, here the presence of nanoparticles with a size approximately 100nm is confirmed, as well as the presence of particles of greater size, but in minimum quantities.

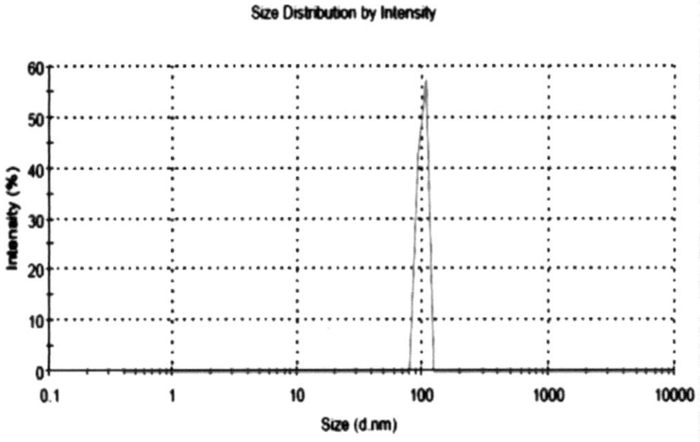

Figure 2. Particle size distribution analysis of IONPs.

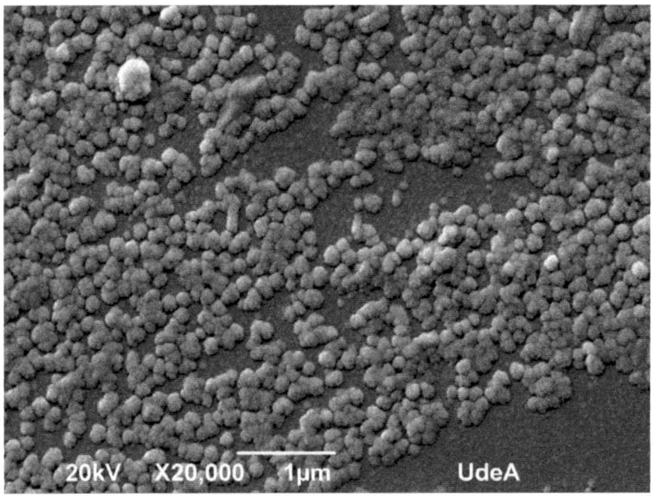

Figure 3. Micrograph SEM of IONPs with Aloe Vera.

3.2. Mercury removal

In order to evaluate the mercury removal capacity, the IOPNs synthesized in this study were used. A series of dilutions of nanoparticles in an Agar Plate Count matrix were prepared, on which a fixed volume of IOPNs solution was incorporated. When each system has been solidified, 5mL of standard solution Hg (II) is added to each different concentrations (4, 6, 8, 10, 25, 50mg/L). The experiment was performed in triplicate, and was left in contact by 24 h. The quantification of Hg (II) in the solution was carried out by atomic absorption spectroscopy measurements, complexed with NH_4Br. To evaluate the efficiency of IOPNs for Hg (II) removal, a removal percentage (%) is calculated. Figure 4 shows the effect of initial Hg (II) concentration on the removal efficiency of Hg (II) using IONPs.

A maximum removal percentage of about 87% was observed from an initial Hg (II) concentration of 8mg/L; for concentrations of 6mg/L and 4mg/L the removal percentages (%) have values close to 73 % and 68% respectively. At 10mg/L and 25mg/L, removal percentages (%) decreased with the increase of initial Hg (II) concentration, this is because the IOPNs sites would eventually become saturated with adsorbed Hg (II), at this point further addition of Hg (II) to the solution hardly increase the amount of adsorbed Hg (II). At 50mg/L, removal percentage (%) was below 50% and its value is not reported in Figure 4(a). Keeping the highest Hg (II) concentration of 8mg/L with the same contact time, different concentrations of IONPs was applied to explore the removal percentage of Hg (II), Figure 4(b). It is observed that at any concentration value of IONPs the removal percentage of Hg (II) is above 60%, therefore it is not necessary to use high concentrations of nanoparticles for effective removal of mercury.

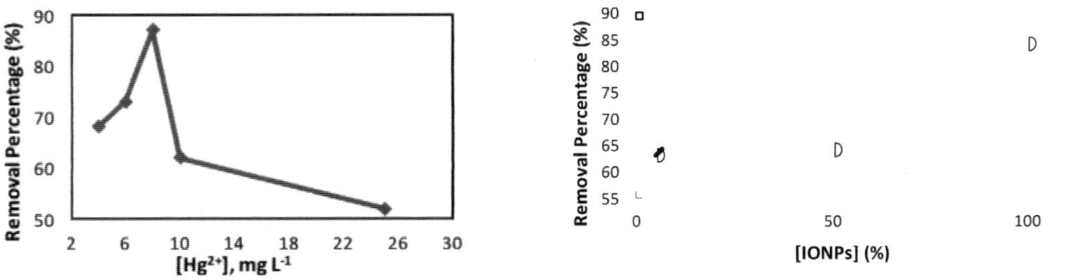

(a) (b)

Figure 4. Hg (II) removal percentage (a) from Hg (II) solutions of different concentrations, [IOPNs]=100%. (b) With different concentration of IONPs, [Hg (II)]=8mg/L. Adsorption time of 24h.

4. Conclusions

In this paper the synthesis and characterizations of Iron Oxides Nanoparticles, stabilized with Aloe Vera was carried out. The nanoparticles showed a monodisperse behaviour, with an average size ~100nm, which was corroborated by UV-Vis, DLS and SEM measurements. Mercury removal more than 85% was achieved (70% average), using low concentrations of IOPNs solution; therefore these can be used as an alternative for treating wastewater.

Acknowledgments

This work was supported by Project 738 of Vicerrectoría de Investigaciones, Universidad de Medellín and the Solid State Group (GES), Sustainability Strategy 2014-2015 CODI, Universidad de Antioquia. The authors also gratefully acknowledge to Centro de Laboratorios, Universidad de Medellín.

References

[1] Gavrilescu M 2004 *Eng Life Sci* **4** 2190
[2] Penka I Girginova; et al 2010 *J Colloid Interf Sci* **345** 234
[3] P Miretzky, A Fernandez Cirelli 2009 *J Hazard Mater* **167** 10
[4] X Wang, S Pehkonen, A Ray 2004 *Electrochim Acta* **49** 1435
[5] Wang X Z, C Dou P, Ding Y and Xu P 2009 *Bioresource Technol* **100** 2301
[6] Mayo J T, Yavuz C, Yean S Cong L, Shipley H, Yu W, Falkner J, Kan A, Tomson M, Colvin V L 2007 *Sci Technol Adv Mater* **8** 71
[7] Dixit S, Hering J G 2003 *Environ Sci Technol* **37** 4182
[8] Chowdhury S R, Yanful E K 2010 *J Environ Manage* **91** 2238
[9] S. P. Chandran, M Chaudhary, R Pasricha, A Ahmad and M Sastry 2006 *Biotechnol Prog* **22** 577
[10] S Link, M A El-Sayed 2000 *Int Rev Phys Chem* **19** 409
[11] O U Rahman, S C Mohapatra, S A 2012 *Materials Chemistry and Physics* **132** 196
[12] P Periyathambi, W S Vedakumari, S Bojja, S B Kumar, T P Sastry 2014 *Materials Chemistry and Physics* **148** 1212
[13] Le T T T, Fribourg-Blanc E, Dang M C 2011 *Adv Nat Sci Nanosci Nanotechnol* **2** 1

Structural and magnetic properties $CuAl_{1-x}Cr_xS_2$ alloys

C Ortega López, G Casiano Jiménez[1] and M J Espitia[2]
[1] Universidad de Córdoba, Montería, Colombia.
[2] Universidad Distrital Francisco José de Caldas, Bogotá, Colombia.

E-mail: cortegal@unicor.edu.co

Abstract. In this paper computational calculations were performed based on the density functional theory DFT, to investigate the structural, electronic and magnetic properties of $CuAl_{1-x}Cr_xS_2$ (x=0.0, 0.50 and 1.00) alloys. Pseudopotential method was used, as is implemented in the Quantum-Espresso code. We found that the alloys crystallize in a tetragonal structure belonging to space group 122 (I-42d) with lattice constants a=5.290Å, c=10.378Å for x=0.5 and a=5.283Å, c=10.366Å for x=1.00. These values are in good agreement with experimental results. Additionally, we found that the alloys possess an antiferromagnetic behaviour with magnetic moments 4,20μ_β/cell and 4,05μ_β/cell, respectively. From the analysis of the density of states, it is clear that the alloys have a half-metallic behaviour due to the Cr-d and Cu-d states crossing the Fermi level. This compounds can be used in spintronic.

1. Introduction

The I-III-VI2 chalcopyrite compounds were I=Cu; III=Al, Ga, In; and VI=S, Se, the have been attracting considerable attention due to their potential applications in solar-cells, detectors, light-emitting diodes, optoelectronic and nonlinear laser devices applications. Among the I-III-VI2 chalcopyrite semiconductors, $CuAlS_2$ compounds have a wide direct band gap around 3.49eV that useful for ultraviolet absorption [1-8]. This is a typical example of the chalcopyrite structure, space group I-42d with four formula units per one unit cell, which is a ternary analog of the diamond structure and essentially a superlattice o superstructure of zinc blende. Like the atoms in diamond and zinc blende structure, each constituent in these ternary compounds, is tetrahedrally coordinated to four neighbors atoms: every metal ion is coordinated by four sulphur ions, every sulphur ion has two Al y two Cu nearest neighbors (see Figure 2(a)). In this paper we study the structural and electronics properties of $CuAl_{1-x}Cr_xS_2$ (x=0.0, 0.50 and 1.00) alloys, due of possible applications in diluted magnetic semiconductors, spin injectors and other spintronics applications.

2. Computational method

Calculations were performed using periodic density functional theory (DFT), such as is implemented in the plane-wave self-consistent field (PWscf) code in the Quantum ESPRESSO package [9]. The exchange and correlation energies were modelled according to the generalized gradient approximation (GGA) with the Perdew Burke Ernzerhof (PBE) gradient-corrected functional [10]. Electron–ion interactions were treated with the pseudopotential method [11,12]. The electron wave functions were expanded into plane waves with a kinetic-energy cutoff of 40Ry. For the charge density, a kinetic energy cutoff of 400Ry was used. A 6×6×3 Monkhorst-Pack mesh [13] was used to generate the k-points in the supercell. To calculate the lattice constant, the bulk modulus and the total energy of each studied compound, the calculated data are fitted with the Murnaghan equation of state [14], equation (1).

$$E(V) = E_0 + \frac{B_0 V}{B_0'}\left[\frac{\left(V_0/V\right)^{B'}}{B_0'-1}+1\right] - \frac{B_0 V_0}{B_0'-1} \qquad (1)$$

Where B_0 is the bulk modulus and its first derivative is B'_0, V_0 is the equilibrium volume of the cell, and E_0 is the binding energy.

3. Results and discussions

3.1. Structural properties

In order to investigate the structural properties of $CuAl_{1-x}Cr_xS_2$ alloys (x=0.0, 0.50 and 1.00), the total energies of different compounds as functions of volume have been calculated in both ferromagnetic (FM) and antiferromagnetic (AFM) states and fitted by Murnaghan's equation of state equation (1). For x=0.0, namely, the $CuAlS_2$ not have magnetics properties. In the $CuAl_{0.50}Cr_{0.50}S_2$ and $CuCrS_2$ alloys, the relative stability of the FM phase with respect to the AFM one has been investigated using various AFM structures to obtain the most stable AFM structure. For these purpose, the supercells of 1x1x2 were used to get even numbers of Cr atom for switching spin state up and down.

The calculated total energy differences between the FM and AFM states ($E=E_{AFM}-E_{FM}$) are about -0.075eV and -0.008eV for the $CuAl_{0.50}Cr_{0.50}S_2$ and $CuCrS_2$ alloys, respectively. For the two cases, the AFM states are more favourable in energy than FM states. The Figure 1 shows the total energy-volume curves for three alloys.

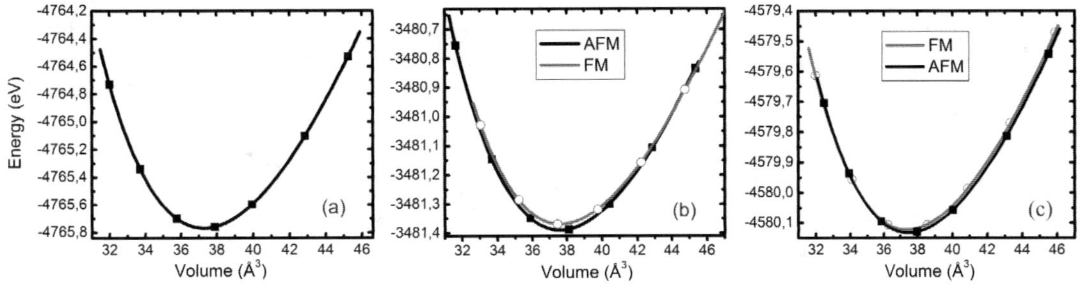

Figure 1. Total energy as a function of the volume: (a) $CuAlS_2$, (b) $CuAl_{0.50}Cr_{0.50}S_2$ and (c) $CuCrS_2$.

Table 1. Structural parameters: lattice constants (a_0) and (c), bulk modulus (B_0), total energy (E_0) and magnetic moment.

compound	a_0 (Å)		c (Å)		B_0 (GPa)	E_0 (eV)	μ (μ_β)	
$CuAlS_2$	5.338	5.334[a]	10.391	10.444[a]	165.31	-4765.769	0.0	0.0[a]
$CuAl_{0.50}Cr_{0.50}S_2$	5.290	5.312[a]	10.378	10.389[a]	155.23	-3481.331	~ 4.20	4.10[a]
$CuCrS_2$	5.283	-	10.366	-	155.60	-4580.122	~ 4.05	3.90[a]

[a] Experimental reference [15].

The calculated volume parameters including equilibrium lattice constant, bulk modulus B_0, total energy E_0 in AFM phase and magnetic moments, for all alloys are listed in Table 1, our results are compared with experimental values reported in reference [15]. The calculated lattices constant is good agreement with values reported experimentally, since it differs by less than one percent.

The Figure 1 shows that the alloys are stable or metastable due there are a minimum energy in the corresponding curve. The Figures 2 show the crystal structure of the alloys $CuAl_{1-x}Cr_xS_2$ (x=0.0, 0.50 and 1.00) obtained after structural relaxation. In all cases, the space group obtained is the same, the tetragonal structure 122 (I-42d).

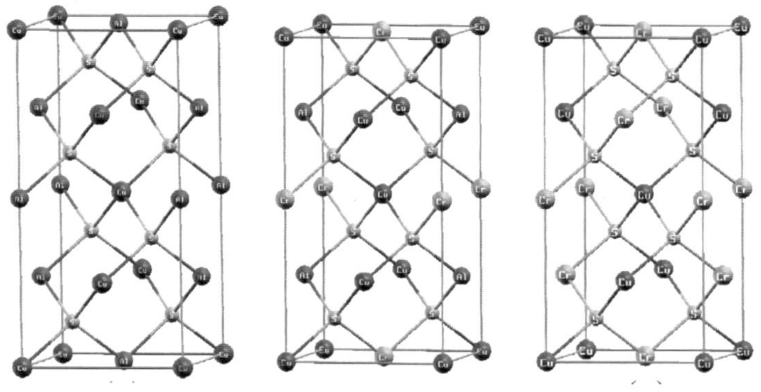

(a) (b) (c)

Figure 2. Unit cell of the alloys: (a) $CuAlS_2$, (b) $CuAl_{0.50}Cr_{0.50}S_2$ and (c) $CuCrS_2$ after structural relaxation.

3.2. Electronic properties

The Figures 3(a), 3(b) and 3(c) shows the total density of states (TDOS) and partial density of states (PDOS) of the orbitals that more contribute near the Fermi level of the alloys $CuAl_{1-x}Cr_xS_2$ (x=0.0, 0.50 and 1.00), respectively. The calculations were performed with spin polarization up and down. The Figure 3(a) confirm the semiconductor nature of $CuAlS_2$ with a forbidden energy gap of ~2.0 eV, The magnitude of this gap is smaller than the experimentally reported for $CuAlS_2$ in chalcopyrite structure (~3.5eV), this occurs because the GGA approximation underestimates the forbidden energy gap in semiconductors. The Figures 3(b) and 3(c) shows the total and partial spin-polarized density of states of the $CuAl_{0.50}Cr_{0.50}S_2$ and $CuCrS_2$ alloys. Due to the introduction of the Cr atoms in the structure of $CuAlS_2$, it loses its semiconductor nature, there are a penetration towards prohibited energy zone of the states Cr-d in greater proportion and the states Cu-d and S-p minor proportion, therefore, the allowed ternary compounds has half-metallic behavior determined by such states. These last two alloys have magnetic behavior with magnetic moments of 4.20 and 4.05μ_β respectively.

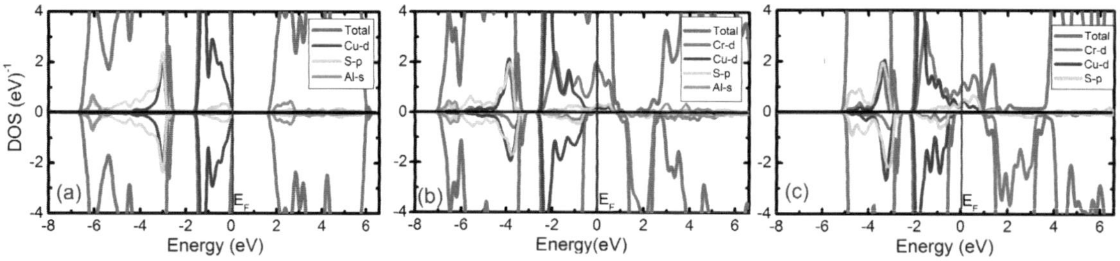

Figure 3. Total and partial density of state of (a) $CuAlS_2$, (b) $CuAl_{0.50}Cr_{0.50}S_2$ and (c) $CuCrS_2$ in the chalcopyrite structure.

4. Conclusion

We executed a study of the alloys $CuAl_{1-x}Cr_xS_2$ (x=0.0, 0.50 and 1.00) using the Density Functional Theory (DFT) in the frame pseudopotential. The calculated lattices constant calculated are good agreement with values reported experimentally, since it differs by less than one percent. The density states study reveal that due to the introduction of the Cr atoms in the structure of $CuAlS_2$, it loses its semiconductor nature, there are a penetration towards prohibited energy zone of the states Cr-d in greater proportion and the states Cu-d and S-p minor proportion, the alloys $CuAl_{0.50}Cr_{0.50}S_2$ and $CuCrS_2$ have a magnetic moment of 4.20 and 4.05μ_β respectively, these properties show that the alloys are good candidates for possible applications in diluted magnetic semiconductors, spin injectors, and other spintronics applications.

References

[1] Kodigala S 2010 *Cu(In1-xGax)Se2 Based thin film solar cells* (San Diego: Academic Press Elsevier)
[2] Harichandran G and Lalla N P 2008 *Materials Letters* **62** 1267
[3] Caglar M, LLincan S and Caglar Y 2008 *Optics Communications* **281** 1615
[4] Yue G H, Wang X, Wang L S, Wang W and Peng D L 2008 *Physics Letters A* **372** 5995
[5] Hsiang Y, Wang A, Bao N and Gupta A 2010 *Solid State Sciences* **12-3** 387
[6] Chen S, Gong Z G, Walsh A and Wei S H 2009 *Appl Phys Lett* **94** 041903
[7] Bid S and Pradhan S K 2002 *J Appl Crystallogr* **35** 517
[8] Bose P, Pradhan S K and Sen S 2003 *Mater Chem Phys* **80** 73
[9] Giannozzi P, Baroni S and Bonin N 2009 *J Phys Condens Matter* **21** 395502
[10] Perdew J P, Burke K and Ernzerhof M 1996 *Phys Rev Lett* **77** 3865
[11] Vanderbilt D 1990 *Phys Rev B* **41** 7892
[12] Laasonen K, Pasquarello A, Car R, Lee C and Vanderbilt D 1993 *Phys Rev B* **47** 10142
[13] Monkhorst H and Pack J 1976 *Phys Rev B* **13** 5188
[14] Murnaghan F D 1944 *Proc Natl Acad Sci USA* **30** 244
[15] Villarreal M A, Grima P, Quintero M, Moreno E, Fernández J, Silva P and Villegas J 2013 *Revista Mexicana de Física* **59** 521

Electronic and magnetic properties GaN/MnN/GaN and MnN/GaN/MnN interlayers

C Ortega López[1], G Casiano Jímenez[1] and M J Espitia[2]
[1] Universidad de Córdoba, Montería, Colombia.
[2] Universidad Distrital Francisco José de Caldas, Bogotá, Colombia.

E-mail: cesarorlo@gmail.com

Abstract. In this work we execute computational calculations to investigate the structural, electronic and magnetic properties of the GaN/MnN/GaN and MnN/GaN/MnN interlayers. The calculations were carried out by a method based on pseudopotentials, as implemented in the Quantum ESPRESSO code. For the description of the electron-electron interaction, generalized gradient approximation (GGA) was used. The total energy calculation reveal that the GaN/MnN/GaN interlayer is energetically most favourable that the MnN/GaN/MnN. Analysis of the density of states show that the interlayers have metallic behaviour that comes essentially from the hybridization and polarization states Mn-d and N-p cross the Fermi level. The interlayers have magnetic properties with a magnetic moment of $8\mu_\beta$/cell. Due these properties the superlattices can be potentially used in the field of spintronic.

1. Introduction

MnN has been extensively studied, both theoretically and experimentally, due to its excellent mechanical, physical and chemical properties such as high hardness, high melting point and high resistance to wearing conditions, corrosion and oxidation [1]. On the other hand, GaN is a direct-gap semiconductor that normally crystallizes in wurtzite structure [2]. Due to its excellent properties, GaN has been used in a range of applications like radiation emitting and sensing devices as well as in high-power/high-temperature electronics; moreover, its efficiency in blue, green and yellow LEDs, injection lasers and ultraviolet detectors is truly extraordinary [3-6]. Furthermore, the combination of the semiconductor properties of GaN and the magnetic properties of the transition metal ions is of great current interest because it could be used at applications in diluted magnetic semiconductors (DMS) [7,8]. Such an interest is also placed on various applications in spintronics (e.g. electro-optical switches and spin injectors) [9]. In this work we study the structural and electronics properties of GaN/MnN/GaN and MnN/GaN/MnN interlayers in the wurtzite structure.

2. Computational method

The Calculations were performed within the framework of Density Functional Theory (DFT) such as is implemented in the Quantum ESPRESSO package [10]. The exchange and correlation energies were modelled according to the generalized gradient approximation (GGA) with the Perdew Burke Ernzerhof (PBE) gradient-corrected functional [11]. Electron–ion interactions were treated with the pseudopotential method [12,13]. The electronic wave functions were expanded into plane waves with a kinetic-energy cutoff of 40Ry. For the charge density, a kinetic energy cutoff of 400Ry was used. A 6x6x3 Monkhorst-Pack mesh [14] was used to generate the k-points in the supercell. To calculate the lattice constant, the bulk modulus and the total energy of each studied compound, the calculated data are fitted with the Murnaghan equation of state [15]. In other to study the relative stability of GaN/MnN/GaN and MnN/GaN/MnN interlayers the energy of formation was calculated. For the ternary compound, the formation energy is defined as the difference between the total energy of the ternary K=GaN/MnN/GaN (or K=MnN/GaN/MnN) and the total energy of the binary compounds in their ground state MnN-NaCl and GaN-wurtzite, respectively. Therefore, the formation energy is given by Equation (1) [16,17].

$$\Delta E_f = E_K^{wurtzita} - (1-x)E_{MnN}^{NaCl} - xE_{GaN}^{wurtzita} \quad (1)$$

The interlayers were modelled according to special quasirandom structures approach [18] and the disorder aspects was ignored.

3. Results and discussions

3.1. Structural properties

To determine the structural properties in the ground state, such as the lattice constant (a_0), bulk modulus (B_0) and total energy (E_0) of the GaN/MnN/GaN and MnN/GaN/MnN interlayers, in the wurtzite structure, the total energy was calculated as a function of the volume, the results were fit to the Murnaghan equation of state. Figure 1 shows the energy-volume curves for the GaN/MnN/GaN and MnN/GaN/MnN interlayers.

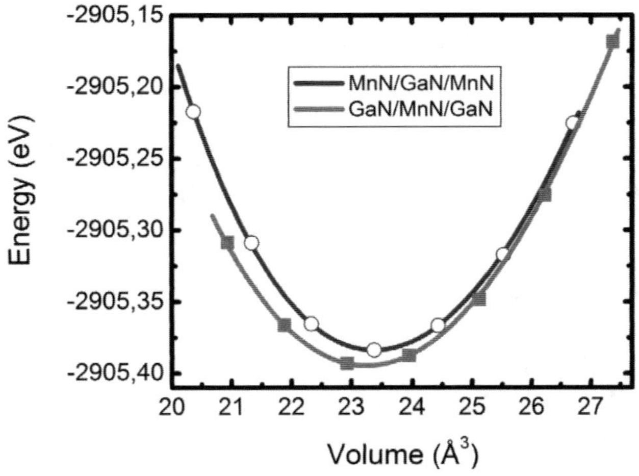

Figure 1. Total energy as a function of the volume of the interlayers.

Table 1. Structural parameters of the interlayers.

Interlayer	a_0 (Å)	c/a	V_0 (Å^3)	B_0 (GPa)	E_0 (eV)	μ (μ_β/cell)	l_{Ga-N} (Å)	l_{Mn-N} (Å)
MnN/GaN/MnN	3.194	3.291	46.488	119.67	-2905.394	~8.0	2.887	2.102
MnN/GaN/MnN	3.191	3.321	46.752	120.75	-2905.383	~8.0	2.802	2.098

We observe that the GaN/MnN/GaN interlayer is more energetically favourable that the MnN/GaN/MnN. The calculated values for the lattice constants of the interlayers, the c/a ratio, the equilibrium volume (V_0), the bulk modulus (B_0), the total energy (E_0), the magnetic moment and bond length of the Ga-N and Mn-N (see Figure 2(a)) of the GaN/MnN/GaN and MnN/GaN/MnN interlayers in the wurtzite structure are listed in Table 1. The values of the bulk modules of interlayers are higher, which confirms that they are quite rigid, making them good candidates for possible applications in devices operated at high temperature and high power, as well as hard coatings.

In order to verify the relative stability of the interlayers, we calculated the energy of formation of each interlayer. For this purpose, we calculated the total energy E_0 of the binary compounds GaN and MnN, in their ground states. The values are -3505.20eV and -2307.7eV, respectively. Table 2 shows the values of formation energy ΔE_f calculated using Equation 1.

Table 2. Formation energy.

Superlattice	ΔE_f (eV)
MnN/GaN/MnN	0.906
MnN/GaN/MnN	0.917

In the Table 1 we observed that the total energy of the GaN/MnN/GaN and MnN/GaN/MnN interlayers is negative, in the same way the total energy of the binary compounds MnN and GaN in their ground state is negative, however according to the results of Table 2, the value of the energy of formation of each interlayer is positive. Therefore, GaN/MnN/GaN and MnN/GaN/MnN interlayers are metastable. This means that the interlayer cannot grow under equilibrium conditions, so in order to grow them, it is necessary to supply energy to the system. The Figure 1 confirm that the interlayers are metastable due there are a minimum energy in the corresponding curve. The Figure 2 show the crystal structure of the GaN/MnN/GaN and MnN/GaN/MnN interlayers obtained after structural relaxation. In all cases, the space group obtained is the same, the hexagonal structure whit space group N° 156. It show the bond length Ga-N and Mn-N.

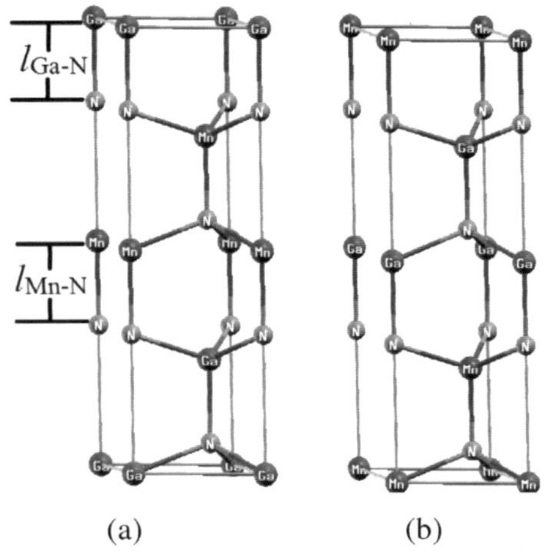

Figure 2. Unit cell of the: (a) GaN/MnN/GaN (b) MnN/GaN/MnN interlayers after structural relaxation. It show the bond length.

3.2. Electronic properties

The Figure 3 shows the total density of states (TDOS) and partial density of states (PDOS) of the orbitals that more contribute near the Fermi level of the interlayer most energetically favourable GaN/MnN/GaN. The interlayer exhibit a metallic behaviour due that of valence orbitals cross the Fermi level. This metallic behaviour it is determined by the states Mn-d in greater proportion and the states N-p in minor proportion. The interlayer has magnetic properties with a total magnetic moment of $8\mu_\beta$/cell. This value in the magnetic moment value can understood as follows: the magnetic moments of $8\mu_\beta$/cell is due Mn^{3+} configuration, whose electronic configuration Mn^{3+}=:[Ar]$3d^4$; when the Mn atom is in the interlayer, the Mn atom gives three electrons. Then, the Mn atom remain four valence electrons (configuration d^4). These valence electrons couple ferromagnetically, as result the four electrons produce a total magnetic moment of $8\mu_\beta$/cell, as there are two Mn atoms in the supercell (see Figure 2(a)), each Mn atom contribute with $4\mu_\beta$/atom-Mn.

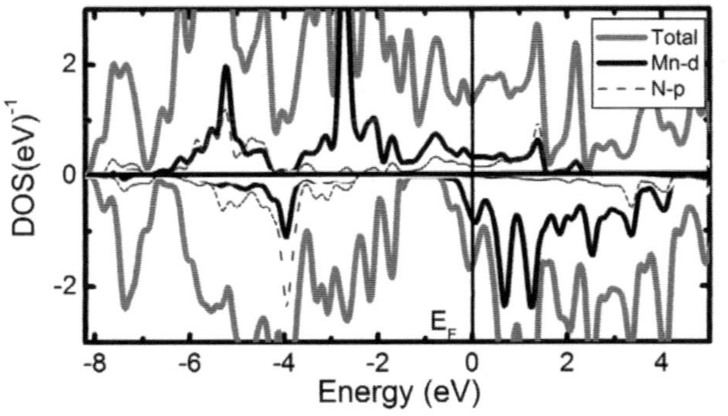

Figure 3. Total and partial density of state of GaN/MnN/GaN.

4. Conclusions

We executed a study of the GaN/MnN/GaN and MnN/GaN/MnN interlayers using the Density Functional Theory DFT in the frame pseudopotential. The total energy calculation reveal that the GaN/MnN/GaN interlayer is energetically most favourable that the MnN/GaN/MnN. The values of the bulk modules of interlayers are higher, which confirms that they are quite rigid, making them good candidates for possible applications in devices operated at high temperature and high power, as well as hard coatings. Analysis of the density of states show that the interlayers have metallic behaviour that comes essentially from the hybridization states Mn-d and N-p cross the Fermi level. The interlayers have magnetic properties with a magnetic moment of $8\mu_\beta$/cell. Due these properties the superlattices can be potentially used in the field of spintronics.

References

[1] Janotti A, Wei S H, Bellaiche L 2009 *Appl Phys Lett* **82** 766
[2] Arbouche O, Belgoumène B, Soudini B, Driz M 2009 *Computational Materials Science* **47** 432
[3] Mei-Xin F, Shu-Ming Z, De-Sheng J, Jian-Ping L, Hui W, Chang Z, Zeng L, Huai W, Feng W, Hiu Y 2012 *Chin Phys B* **21** 084209
[4] Wagner J M, Bechstedt F 2002 *Physical Review B* **66** 115202
[5] Nakamura S, Senoh M, Nagahama S, Iwasa N, Yamada T, Matsushita T, Sugimoto Y, Kiyoku H 1997 *Appl Phys Lett* **70** 1417
[6] Nakamura S, Senoh M, Iwasa N, Nagahama S 1995 *Applied Physics Letters* **67** 1868
[7] Hu B, Man B Y, Yang C, Lui M, Cheng S C, Gao X, Xu S, Wang S C, Sung Z C 2011 *Applied Surface Science* **258** 525
[8] Gao C X, Yu F C, Kin D J, Ihm Y 2009 *Journal of the Korean Physical Society* **54** 633
[9] Gonzalez R, López W, Rodriguez J 2009 *Materials Science and Engineering B* **163** 633190
[10] Giannozzi P, Baroni S, Bonin N 2009 *J Phys Condens Matter* **21** 395502
[11] Perdew J P, Burke K, Ernzerhof M 1996 *Phys Rev Lett* **77** 3865
[12] Vanderbilt D 1990 *Phys Rev B* **41** 7892
[13] Laasonen K, Pasquarello A, Car R, Lee C, Vanderbilt D1993 *Phys Rev B* **47** 10142
[14] Monkhorst H and Pack J 1976 *Phys Rev B* **13** 5188
[15] Murnaghan F D 1944 *Proc Natl Acad Sci USA* **30** 244
[16] Sheng S H, Zhang R F, Veprek S 2008 *Acta Materialia* **56** 968
[17] Vargas-Hernández C, Espitia M, Báez R E 2015 *Computational Condensed Matter* **4** 1
[18] Zunger A, Wei S H, Ferreira L, Bernard J E 1990 *Physical Review Letters* **65** 353

Structural and electronic properties of $Er_{0.5}Lu_{0.5}B_2$ and LuB_2 compounds by using the PBE0 hybrid functional

C Ortega[1], D Rasero[2] and L C Sánchez[1]
[1] Universidad de Córdoba, Montería, Colombia.
[2] Universidad de Sucre, Sincelejo, Colombia.

E-mail: cortegal@gmail.com

Abstract. We report first-principles calculations to determine the structural and electronic properties of $Er_{0.5}Lu_{0.5}B_2$ and LuB_2 compounds, by using Density Functional Theory (DFT) and Full Potential Linearized Augmented Plane-Waves (FP-LAPW) method. For the description of the electron-electron interaction was used the Local Density Approximation (LDA), Generalized Gradient Approximation of Perdew-Burke-Ernzerhof (GGA-PBE) and PBE0 hybrid functional. From the density of states (DOS), it is found that the addition of a fraction of the exact Hartree-Fock exchange energy in PBE approximation, evidence the localization of the 4f-Er and 4f-Lu orbitals, which favours the polarization of the electronic spins of these orbitals in LuB_2 and $Er_{0.5}Lu_{0.5}B_2$ compounds. The PBE0 scheme is justified because it describes of more appropriate manner the electronic and magnetic properties of strongly correlated systems than the LDA and PBE approaches.

1. Introduction

The rare earth diborides, REB_2, ie. compounds with the presence of rare-earth metals in the boride lattice leads to a wide diversity of thermodynamic, structural, electronic and magnetic properties. These rare-earth borides such as diborides (XB_2), tetraborides (XB_4), hexaborides (XB_6), and dodecaborides (XB_{12}) have attracted great attention over the last years as a rich class of materials that exhibit various phenomena related with electronic spins of f-type orbitals. Within this family of materials, XB_2 diborides constitute a group of compounds with AlB_2-type crystal structure, integrated by hexagonal layers composed of rare earth element atoms alternating with layers of hexagonal boron atoms [1-3]. From the theoretical point of view, as pointed Fan et al. [4] in their work, it should be noted that in this type of compounds, the presence of f orbitals of rare earth element implies strong electronic correlation that theoretically, is not adequately described with the local density approximation (LDA) or generalized gradient (GGA). In this case, it becomes necessary to employ exchange-correlation functionals, such as LDA + U or GGA-PBE0 [5], taking into account this feature of the f orbitals.

In this work, we present a study of the electronic properties of the LuB_2 and $Er_{0.5}Lu_{0.5}B_2$ compounds, using the FP-LAPW method employing exchange-correlation functionals with the GGA-PBE and GGA-PBE0 parameterizations, within the formalism of DFT.

2. Computational details

Calculations are performed within the framework of DFT, using the FP-LAPW method. Exchange-correlation effects are treated using the LDA and GGA-PBE approximations [6-7], and PBE0 hybrid functional [8-10].

The charge density and potential are expanded in spherical harmonics up to l_{max}=10 and the wave function is expanded in plane waves with cutoff parameter K_{max}=8.00/R_{mt-min}; where R_{mt-min} is the smallest radius of the atomic sphere in the unit cell and K_{max} limits the kinetic energy. To ensure convergence in the integration of the first Brillouin zone, are used 95 k-points in the irreducible part of the first brillouin zone. The integrals over the brillouin zone are solved using the k-point special approximation of Monkhorst-Pack [11]. The self-consistency is achieved by requiring that the convergence of the total energy is less than 10^{-4}Ry. For expanding the potential in the interstitial region is considered G_{max}=12.00. Muffin-tin radius used for Er, Lu, and B were R_{mt-Er}=2.20, R_{mt-Lu}=2.10 and R_{mt-B}=1.60, respectively. The hybrid functional PBE0 is applied to the 4f electrons of the rare earth elements Er and Lu, similar to what was done in a previous work of Casiano et al [10].

3. Results and discussion

The crystalline structures of the compounds were optimized in AlB_2-type hexagonal structure with space group P6/mnm(#191). The total energies are computed for different values of cell volume. The lattice parameters are obtained by minimizing the energy as a function of volume. To perform this, we fit the obtained energy by the equation of state of Murnaghan [12]. The calculated values in this work, the theoretical values reported by other authors and some available experimental values are shown in Table 1. Our results are in good agreement with the experimental values reported.

Table 1. Lattice parameters, minimum energy E_0, minimum volume V_0, bulk modulus B_0, and the magnetic moment per cell μ. Other results are used for comparison.

	Reference	Method	a_0 (Å)	c/a	V_0 (Å^3)	B_0 (GPa)	E_0 (eV)	μ (μ_0/cell)
ErB_2	[10]	FP-LAPW:LDA	3.168	1.191	32.78	168.81	-11.085	2.659
	[10]	FP-LAPW:GGA-PBE	3.247	1.182	35.02	148.71	-14.896	2.697
	[10]	FP-LAPW:GGA-PBE0	3.260	1.166	34.98	168.33	-13.713	3.072
	[4]	FP-LAPW:LDA	3.280	1.155	-	-	-	-
	[13]	LDA	3.171	1.192	-	169.47	-	-
	[14]	PP-PW:GGA	3.275	1.161	-	172.9	-	-
	[15]	Experimental	3.268	1.157	-	-	-	-
$Er_{0.5}Lu_{0.5}B_2$	Our work	FP-LAPW:LDA	3.175	2.325	32.29	193.09	-37.519	2.642
	Our work	FP-LAPW:GGA-PBE	3.241	2.323	34.26	163.03	-32.991	2.644
	Our work	FP-LAPW:GGA-PBE0	3.251	2.308	34.33	169.62	-22.617	3.067
LuB_2	Our work	FP-LAPW:LDA	3.181	1.137	31.69	277.52	-18.721	0.000
	Our work	FP-LAPW:GGA-PBE	3.236	1.143	33.52	217.88	-16.141	0.000
	Our work	FP-LAPW:GGA-PBE0	3.237	1.141	33.56	189.27	-16.064	0.000
	[4]	PP-PW:LDA	3.241	1.141	-	-	-	-
	[13]	LDA	3.181	1.144	-	204.35	-	-
	[16]	PP-PW:GGA	3.242	1.146	-	180.7	-	-
	[3]	Experimental	3.244	1.142	-	-	-	-

From the B_0 values shown in Table 1, is concluded that the compounds are rigid. This property, due to the strong bonds that exist between boron atoms present in the AlB_2-type hexagonal structure, makes it attractive for potential applications as coatings on cutting tools.

For the analysis of the electronic properties only is our interest to compare the results of ErB_2 [10] and $Er_{0.5}Lu_{0.5}B_2$ compounds.

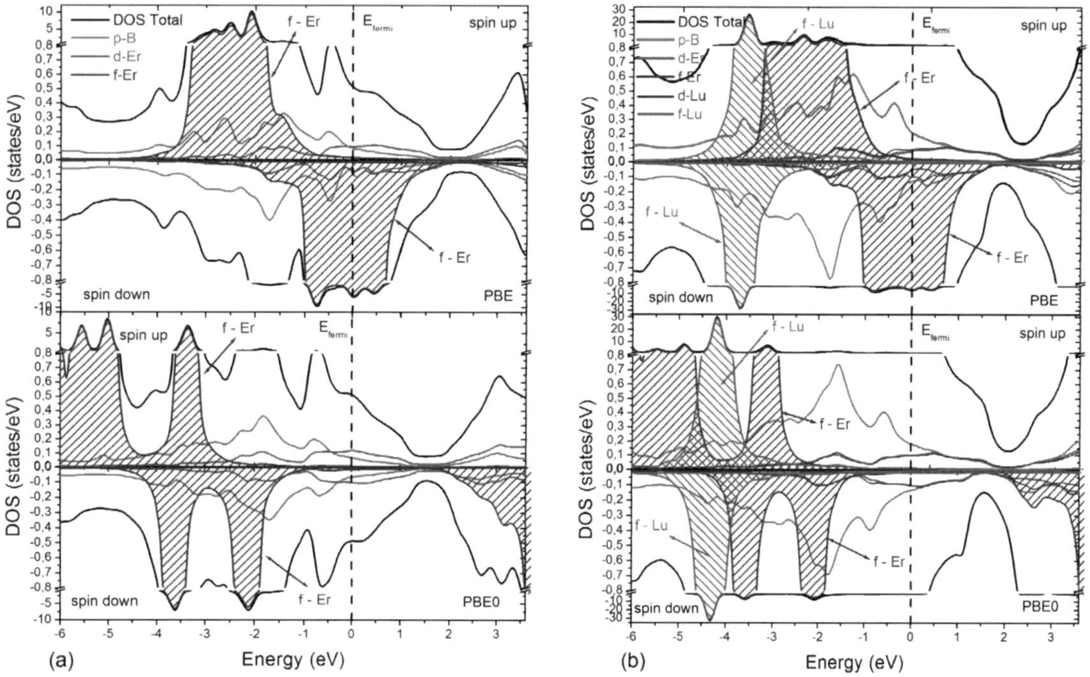

Figure 1. Total and partial density of states for ErB_2 (taken of reference [10], only to compare the electronic properties) (a) and $Er_{0.5}Lu_{0.5}B_2$ (b) with the GGA-PBE (top) and GGA-PBE0 (bottom) parameterizations. Are shown the contributions of the p-B, d-Er, f-Er, d-Lu and f-Lu states with red, green, blue shaded area, violet and olive shared area, respectively. The black color line represents the total density of states. The zero of the energy is taken as the Fermi level.

The total (DOS) and partial (PDOS) density of states calculated for ErB_2 and $Er_{0.5}Lu_{0.5}B_2$ are shown in Figure 1. The calculations were carried out considering spin polarizations. In Figure 1 we consider only the orbitals that have the greatest contribution to the Fermi level. For our purpose we only analyze the DOS results for the ErB_2 and $Er_{0.5}Lu_{0.5}B_2$ compounds (LuB_2 not presents interesting properties: metallic, non-magnetic) in the PBE and PBE0 approximations (PBE is a refinement of the PBE).

In the case of ErB_2 [10], the GGA-PBE result indicates that for spin up polarization, the largest contribution at the Fermi level is mainly due to p-B and d-Er states, while f-Er states are located between energy ~-3.5eV and ~-1.0eV. For spin down polarization, the largest contribution at the Fermi level is due to f-Er states in the energy region ~-1.5eV and ~0eV. For $Er_{0.5}Lu_{0.5}B_2$ is observed that, in the GGA-PBE case with polarization of spin up, the largest contributions due to p-B, d-Er and d-Lu and f-Er states are located in the region of energies ~-3.5eV and ~-1.0eV. For polarization with spin down, the mayor contribution is due to f-Er states. It also notes that GGA-PBE result, f-Lu states are anchored in the same region of energy (~-5.0eV and ~-2.5eV) which can be explained considering that in Lu the f orbitals are completely occupied by electrons (Lu:[Xe] $4f^{14}5d^{1}6s^{2}$). For the GGA-PBE0 calculations we found notable differences with respect to GGA-PBE. The density of electronic states GGA-PBE0, both ErB_2 and $Er_{0.5}Lu_{0.5}B_2$, shows the f-Er localized states before the Fermi level in the region from ~-6.0eV to ~-2.5eV for spin up polarization, and between ~-4.0eV to ~-1.5eV for spin down polarization.

It is further noted that as GGA-PBE case, for $Er_{0.5}Lu_{0.5}B_2$ f-Lu states are located within the range ~-5.0eV and ~-2.5eV with spin up and down polarizations almost symmetric. This is because, as was indicated earlier, f-Lu states do not possess unpaired electrons. This characteristic behaviour of f-Er and f-Lu states is the reason because the difference between the values of the magnetic moment of ErB_2 and $Er_{0.5}Lu_{0.5}B_2$ are relatively small ($0.005\mu_0$/cell). Is not observed a signicant mixing with other states of the valence band outside this region. Thus, f-Er and f-Lu states, particularly the 4f are highly localized and can only interact via a RKKY (Ruderman-Kittel-Kasuya-Yosida) interaction as noted by Mori et al. [3] on a theoretical and experimental study of ferromagnetism and electronic structure of TmB_2. In the region -4eV<E <E_{Fermi}, for ErB_2, there are contributions from 2p states of B and 5d of Er. The latter, crossing the Fermi level and are responsible for the metallic character of the material. For $Er_{0.5}Lu_{0.5}B_2$ in addition to 2p-B and 5d-Er states are observed the presence of 5d-Lu states, which also contribute to metallic behaviour of this compound. This feature distinguishes the calculated DOS with GGA-PBE0 those obtained from GGA-PBE in which f-Er states also cross the Fermi level. For the energy values of ~-1.0eV and ~-2.0eV two peaks are observed that are characteristic of the diboride compounds based on rare-earth elements. This behaviour is consistent with the results of Mori at al. [3] for TmB_2 and YbB_2. It can be noted, moreover, that the differences between the derived DOS from the GGA-PBE and GGA-PBE0 parameterizations, must be, as explained Nourmohammadi et al. [17], in which the exchange splitting using PBE0 increases, causing that the localization of magnetic electrons (mainly the 4f of Er) is greater, and that the values of the lattice constant at equilibrium, and magnetic moment predicted by GGA-PBE0 be slightly greater than the result of GGA-PBE as shown in Table 1.

4. Conclusions

It was found that the effect of adding a fraction of the exact exchange energy of Hartree-Fock in PBE approximation, evidence the localization of 4f-Er and 4f-Lu orbitals, which favors electronic spin polarization of these orbitals in the ErB_2 and $Er_{0.5}Lu_{0.5}B_2$compounds. This explains the increase in the total magnetic moment PBE0 ($3.072\mu_0$/cell in ErB_2 and $3.067\mu_0$/cell in $Er_{0.5}Lu_{0.5}B_2$) compared to the total magnetic moment in PBE ($2.697\mu_0$/cell in ErB_2 and $2.664\mu_0$/cell in $Er_{0.5}Lu_{0.5}B_2$).

References

[1] Kosolapova T Y 1990 *Handbook of high-melting compounds properties production and applications* (New York: Metallurgiya Hemisphere)
[2] Cannon J, Cannon M and Hall H 1997 *J Less Common Met* **56** 83
[3] Mori T, Takimoto T, Leithe-Jasper A, Cardoso-Gil R, Schnelle W, Auermann G, Rosne H and Grin Y 2009 *Phys Rev* B **79** 104418
[4] Fan Y, Ru-Shan H, Ning-Hua T and Wei G 2002 *Chin Phys Lett* **19** 1336
[5] Vallars P 1997 *Crystallographic data for intermetallic phases* (Ohio: Pearson's Handbook, ASM International)
[6] Hohenberg P and Kohn W 1964 *Phys Rev* **136** B864
[7] Perdew J P, Burke K and Ernzerhof M 1996 *Phys Rev Lett* **77** 3865
[8] Ernzerhof M and Scuseria G E 1999 *J Chem Phys* **110** 5029
[9] Adamo C and Barone V 1999 *J Chem Phys* **110** 6158
[10] Casiano G, Sánchez L, Rodriguez J A, Rasero D and Ortega C 2014 *Rev Mex Fis* **60** 276
[11] Monkhorst H J and Pack A T 1976 *Phys Rev B* **13** 5188
[12] Murnaghan F 1944 *P Natl Acad Sci* **30** 244
[13] Zazoua F, Kacimi S, Djermouni M and Zaoui A 2011 *J Appl Phys* **110** 014908
[14] Ozisik H, Deligoz E, Colakoglu K and Surucu G 2013 *Chin Phys* B **22** 046202
[15] Novikov V, Chukina T and Verevkin A 2010 *Physics of the Solid State* **52** 364
[16] Deligoz E, Ozisik H, Colakoglu K, Surucu G and Ciftci Y O 2011 *J Alloys Compd* **509** 1711
[17] Nourmohammadi A and Abolhasani M R 2010 *Solid State Communications* **150** 1501

A study ab-initio of relative stability and electronic properties of substitutions of C by B and N atoms in the w-BN compound

G Casiano Jiménez[1], C Ortega López[1] and M J Espitia[2]
[1] Universidad de Córdoba, Montería, Colombia.
[2] Universidad Distrital Francisco José de Caldas, Bogotá, Colombia.

E-mail: glacaji@hotmail.com

Abstract. In this work we study relative stability, structural and electronic properties of pure BN and with substitutions of carbon by boron and nitrogen atoms, e.i. for $B_{0.9375}C_{0.0625}N$ and $BC_{0.0625}N_{0.9375}$, respectively. Using first-principles calculations in the Density Functional Theory (DFT), we found that it is more energetically favourable that the carbon atoms occupy the positions of the boron atoms (C by B). It was found that the new compounds $BC_{0.0625}N_{0.9375}$ (C by N) and $B_{0.9375}C_{0.0625}N$ (C by B) have high bulk modulus, so they are very rigid. This property makes them good candidates for applications in hard coatings or devices for high power and temperatures. The density of state calculations shows that compound has metallic behaviour due to the hybridization of the orbitals C-2p and N-2p that cross the Fermi level.

1. Introduction

The boron nitride (BN) is a semiconductor material. Due to different modifications in structure, zincblende (cubic) boron nitride (c-BN), wurtzite boron nitride (w-BN), hexagonal boron nitride (h-BN) and rhombohedral boron nitride (r-BN), it exhibits some very fascinating physical properties such as high temperature stability, chemical inertia, high hardness, low coefficient of thermal expansion and high thermal conductivity [1-4]. BN has a broad range of potential applications, in high temperature devices, radio frequency and high-frequency, high-power laser diodes, light emitting diodes operating in the ultraviolet region, solar detectors, field effect transistors and high electron mobility transistors [4-8]. In this paper, we present first-principles total energy calculations of the structural and electronic properties of pure BN and BN:C system. First, we calculated the structural parameters of pure BN and the two ternary compounds considered in this study (C by B and C by N). Finally, we made a detailed analysis of the density states of BN:C most favourable system. We found that the ternary compounds exhibit metallic behaviour and have a high bulk modulus, therefore can be used in hard coatings.

2. Computational method

The calculations are performed in the DFT framework using the Quantum Espresso package [9]. The correlation and exchange effects of the electrons are treated using the generalized gradient approximation (GGA) of Perdew, Burke, and Ernzerhof (PBE) [10]. Electron–ion interactions were treated with the pseudopotential method [11,12]. The electron wave functions were expanded into plane waves with a kinetic-energy cutoff of 40 Ry. For the charge density, a kinetic energy cutoff of 400 Ry was used. A 6x6x6 Monkhorst-Pack mesh [13] was used to generate the k-points in the unit cell. The calculations are performed taking into account the spin polarization. To calculate the structural, electronic, and magnetic properties of pure BN, a 32-atom *2a* x *2b* x *2c* wurtzite supercell was considered. First, we study a carbon located in the N-substitutional site (C by N), i.e. the concentration $BC_{0.0625}N_{0.9375}$ and finally, we analyse a carbon located in the B-substitutional site (C by B), the $B_{0.9375}C_{0.0625}N$. In order to calculate the lattice constant, the bulk modulus and the total energy of boron nitride BN and the substitutions C by N and C by B, the calculated data are fitted with the Murnaghan equation of state [14], Equation (1).

$$E(V) = E_0 + \frac{B_0 V}{B_0'}\left[\frac{\left(V_0/V\right)^{B'}}{B_0'-1}+1\right] - \frac{B_0 V_0}{B_0'-1} \tag{1}$$

3. Results and discussions

3.1. Structural properties

The Figure 1 shows the wurtzite supercell of pure BN, the concentrations $BC_{0.0625}N_{0.9375}$ (C by N) and $B_{0.9375}C_{0.0625}N$ (C by B) respectively, after of structural relaxation. In three cases, the space group obtained is the same, the tetragonal structure p3m1 (#156). The structural properties in the ground state, such as lattice constant (a_0), a/c ratio, bulk modulus (B_0) and total energy (E_0) of BN and BN:C (C by B and C by N). The structural properties were obtained adjusting the results of isotropic variation of the volume to the Murnaghan equation of state, Equation (1). Figure 2 shows the energy-volume curves for BN and with substitutions C by B and C by N.

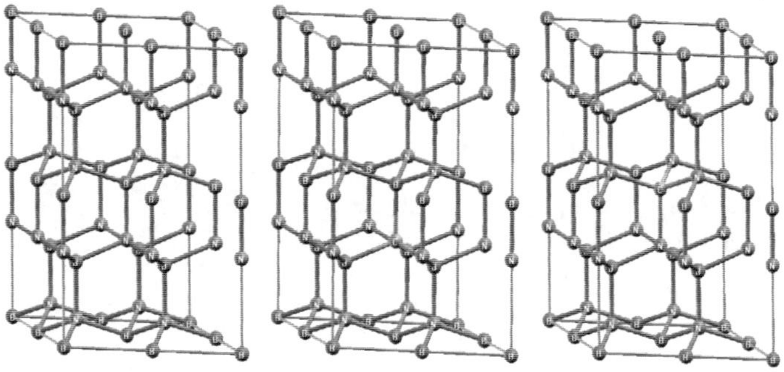

Figure 1. Unit cell of the ternary compound: (a) pure BN, (b) $BC_{0.0625}N_{0.9375}$ and (c) $B_{0.9375}C_{0.0625}N$ after structural relaxation.

In the Figure 2 we can observe that, with respect to pure-BN, the $BC_{0.0625}N_{0.9375}$ (C by N) is energetically unfavourable because, in the ground state the total energy difference between pure-BN and $BC_{0.0625}N_{0.9375}$ ($\Delta E=E_{BN:C}-E_{BN}$) is 232.07eV, difference is positive because the minimum of energy of pure-BN is lower than the $BC_{0.0625}N_{0.9375}$ compound, while that $B_{0.9375}C_{0.0625}N$ (C by B) is energetically most favourable, because the difference between pure-BN and $B_{0.9375}C_{0.0625}N$ ($\Delta E=E_{BN:C}-E_{BN}$) is -112.94eV. These results indicate that the $B_{0.9375}C_{0.0625}N$ compound is energetically most stable, because it has the lowest energy. Additionally, table 1 shows the main structural parameters of pure-BN and $BC_{0.0625}N_{0.9375}$ and $B_{0.9375}C_{0.0625}N$ concentrations.

As we can see in Table 1, the calculated lattice constant of the binary BN compound accords well with both experimentally [2] and theoretical [3] values, since it differs by less than one percent. Bulk modulus of w-BN agrees with theory work [3], with an error of about 1.0%. Calculated bulk modulus for w-BN is close with the diamond (442GPa). This may be due to the strong sp3 B-N bonding (similar structure to diamond). We note that values of the bulk modules of the pure BN, BC0.0625N0.9375 (C by N) and B0.9375C0.0625N (C by B) concentrations are higher, which confirms that they are quite rigid, making them good candidates for possible application in devices operated at high temperature and high power, as well as in hard coatings.

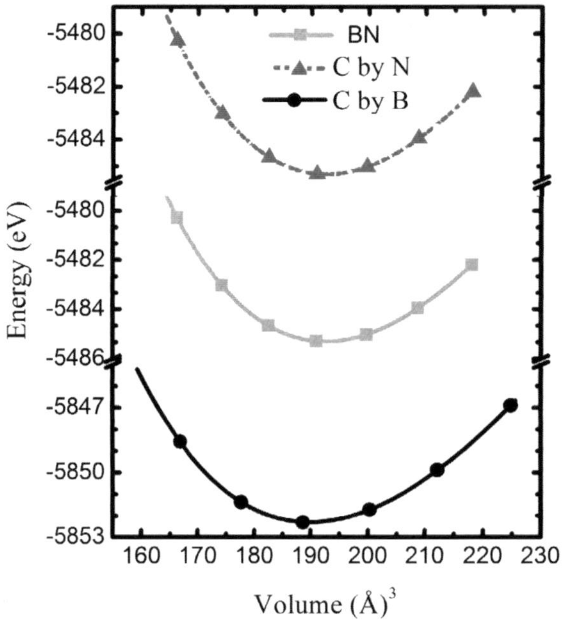

Figure 2. Energy versus volume BN, C by B and C by N substitutions.

Table 1. Structural parameters lattice constant (a_0), bulk modulus (B_0) and total energy (E_0) of the BN, C by B y C by N.

	a_0 (Å)	c/a	V_0 (Å^3)	B_0 (GPa)	E_0 (eV)
BN	2.550	1.640	189.80	398.95	-5717.36
	2.530[a]	-	-	-	-
	2.556[b]	1.650[b]	-	403[b]	-
C by B	2.540	1.649	189.80	389.78	-5852.30
C by N	2.550	1.648	192.14	395.80	-5485.29

a: Experimental reference [2], b: Theoretical reference [3].

3.2. Electronics properties

The Figure 3 shows the total density of states (TDOS) and partial density of states (PDOS) of the orbitals that more contribute near the Fermi level for most energetically favourable substitution, namely, $B_{0.9375}C_{0.0625}N$ (C by B). We note that the compound exhibit a metallic behaviour due to that valence orbitals (the two spin states, spin up and down) cross the Fermi level. This metallic behaviour it is determined by the states C-2p in greater proportion and the states N-2p and B-2p in minor proportion. Moreover, the partial density of states belonging to the C-2p orbital exhibits a high peak near of Fermi level, about - 0.3eV, thus indicating localized states in that region. Additionally, we can see (Figure 3) that in the valence band around the Fermi level, the C-2p states overlap with another two peaks of smaller amplitude belonging N-2p and B-2p. Therefore, there is strong hybridization between the C-2p, N-2p, and B-2p orbitals that cross the Fermi level resulting in a strong covalent bond that is responsible for the high rigidity of the ternary compounds. A similar result was obtained by Murillo et al [15] in their first-principles study of the C-doped AlN.

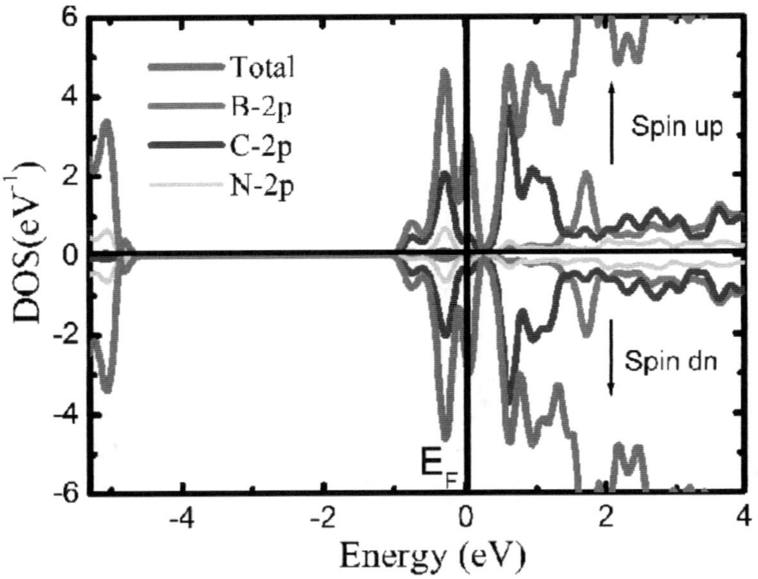

Figure 3. Density of states for substitution C by B ($B_{.9375}C_{0.0625}N$) w-BN compound

4. Conclusions

We found energetically more favourable that C atoms occupy the atomic positions of B in the w-BN supercell. Additionally, we found that the compounds are rigid quite due to that the bulk modulus are high, therefore are ultra-hard and can be potentially used in hard coatings. The calculations of the density of states, show that compound has metallic behaviour due to the hybridization C-2p, B-2p and N-2p orbitals that cross the Fermi level.

References

[1] Ahmeda R, Aleema F, Hashemifarb S J, Akbarzadeh H 2007 *Physica B* **400** 297
[2] Soma T, Sawaoka S, Saito S 1974 *Mater Res Bull* **9** 755
[3] Ohba N, Miwa K, Nagasako N, Fukumoto A 2001 *Physical Review B* **63** 115207
[4] Wang H, Taychatanapat T, Hsu A, Watanabe K 2011 *Elec Device Letters IEEE* 32 **9** 1209
[5] Pittroff W, Erbert G, Beister G, Bugge F 2001 *Adv Packaging IEEE Transactions* **24** 434
[6] Subramani S, Devarajan M 2014 *Electron Devices IEEE Transactions on* **61** 3213
[7] Takahashi S, Yusuke I, Akinori K, Yuji H 2014 *Journal of Alloys and Compounds* **615** 141
[8] Ishida H, Rimdusit S 1998 *Thermochimica Acta* **320** 177
[9] Giannozzi P, Baroni S, Bonin N 2009 *J Phys Condens Matter* **21** 395502
[10] Perdew J, Burke K and Ernzerhof M 1996 *Physical Review Letter* **77** 3865
[11] Vanderbilt D 1990 *Phys Rev B* **41** 7892
[12] Laasonen K, Pasquarello A, Car R, Lee C, Vanderbilt D 1990 *Phys Rev B* **47** 10142
[13] Monkhorst H and Pack J 1976 *Phys Rev B* **13** 5188
[14] Murnaghan F D 1944 *Proc Natl Acad Sci USA* **30** 244
[15] Murillo J F, Ortega C, Espitia M J 2015 *Results in Physics* **5** 250

Feasibility for the use of coal tar as a new material for road surfaces (pavement) construction

M A Romero Farfán[1], H E Murillo Vega[1] and F A Trujillo Pinto[1]
[1] Universidad Pedagógica y Tecnológica de Colombia, Tunja, Colombia.

E-mail: miguel.romero@uptc.edu.co

Abstract. The stabilization products often used to improve the support of granular layers in the construction of road surfaces may be expensive and difficult to get. Therefore, it is necessary to test different materials, which are cheap and easy to obtain, and which will enhance the physical and mechanical properties of pavement layers. This document evaluates the use of coal tar, as a stabilizer for granular subbase. Initially, with a description of tar properties, determining the optimal conditions for the granular subbase material compaction, by means of modified proctor tests and the calculation of the resistance of the unaltered material by using CBR lab tests (California Bearing Ratio). Afterwards, with the design and development of granular material mixes with different percentages of coal tar and determining its CBR as comparative parameter with that of the unaltered material. Finally, by calculating the optimal coal tar percentage in order to stabilize the subbase granular, the results showed an improvement in the resistance of the granular material and a decrease in its expansion due to the use of coal tar.

1. Introduction

When suitable granular materials are not available for pavement construction [1], it becomes necessary to decide the most convenient stabilization treatment for the soil, in order to make them useful for the construction of the layers forming such structure [2]. Tar is often used in asphalt layers to embed them with higher resistance, imperviousness to water and to prolong their lifetime [3]. Nevertheless, granular soil stabilization using tar has not been applied so far.

This research aims to determine the feasibility of coal tar as a granular subbase stabilizer, by means of characterizing such materials, making granular soil mixes with tar, and the analysis of such mix with the determination of its resistance method (CBR) [4].

2. Materials and methodology

The applied methodology involves obtaining the materials, the characterization of such materials by lab tests according to INVIAS-2007 norms [5], the resistance determination (CBR) for the unaltered material, design and making of mixes with different coal tar percentages and its resistance determination (CBR) and choosing the optimal coal tar percentage. The subbase material corresponds to SBG-2, coming from a quarry belonging to the company CSS constructors in Tunja. The coal tar was obtained from enterprise Acerías Paz de Río, as a sub-product of coke [6]. The research was performed at laboratory floors and pavements of the Universidad Pedagógica y Tecnológica de Colombia, known as Uptc, in Tunja (Boyacá).

3. Experimental design and results

It was decided to use four test percentages in weight for coal tar: 3%, 5%, 7% and 9%, in order to check any variation of soil resistance with the increasing content of coal tar and to find an optimal percentage of it [7]. The characterization of coal tar coming from Siderúrgica Paz de Río was performed according to the parameters given by INVIAS for bituminal materials, with the aim of simulating the use of this stabilizer as asphalt, according to results shown in Table 1. In the same way, the physical characterization of the granular material modified with clay fine material, based on the test norms for subbase materials given by Instituto Nacional de Vías–2007, article 300-07.

Table 1. Coal tar characterization.

Trial	Test Standard	Obtained Value
Ignition point and flame point using Cleveland's open coup (°C)	E-709-07	90 – 138
Specific gravity (g/cm3)	E-707-07	1.20 – a 25°
Viscocity Saybolt Furol (SSF)	E-714-07	245
		392 a 60°C
		170 a 70°C
Viscosity using rotational viscosimeter (mPa)	E-717-07	92 a 80°C
		60 a 90°C
		43 a 100°C

Later on, different tar percentage mixes were designed (3%, 5%, 7% and 9% tar) in order to calculate its resistance using submerged and not submerged CBR tests, according to proceedings of norm INV E-148-07. The mixes were elaborated outside during sunny days and were compacted in open air, this due to the type of solvents coming out of tar, in moulds for CBR tests, with compression energies of 10, 25 and 56hits/layer, subject to a tanning period of 72 hours, with the results given in Table 2.

Table 2. CBR test results.

Unaltered subbase CBR			Unaltered modified subbase CBR		
CBR (%)		45.0	CBR (%)		17.0
Submerged tar mix CBR					
% TAR	3%	5%	6.7%	7%	9%
CBR (%)	14.0	17.2	20.3	20.2	10.2
Non-submerged tar mix CBR					
% TAR	3%	5%	6.5%	7%	9%
CBR (%)	16.5	19.7	21.0	20.8	15.5

The expansion of the material was also evaluated. Samples were immersed with overload and its respective perforated disc stem in a water tank and then expansion initial readings were taken with the dial gauge and tripod. The immersion process lasted about four days, which the dissipation of pressure and saturation was achieved compacted soil. At the end of the period of immersion readings they were taken final expansion, to calculate the percentage of soil expansion subbase and determine the susceptibility of the material to volumetric change in saturated condition.

4. Analysis of results

The results of experiments show that the resistance of the mixes increase with the percentage variation up to a certain point, after which, resistance starts decreasing with higher amounts of binder. According to this behaviour, the inflection point at which the subbase hits the maximum resistance with the addition of tar was calculated to be 6.5% in weight, as the optimal amount of coal tar to stabilize the granular subbase material.

Mixes with 6.5% tar presented a remarkable increase in resistance, compared to the unaltered material, showing CBR values higher than 20%, which represents a profit of more than 3 CBR units. It was also evident that low tar percentages in the subbase do not improve its resistance; on the contrary, the CBR value decreased in mixes with 3% tar; in the same fashion mixes with 9% got CBR values of 10.2% and 15.5%. This loss in resistance could be the case because the binding substance in the mix does not cover all the components; therefore the mix is not compact in a uniform fashion. On the other hand, a high amount of tar corresponds to more tar in the interspaces between particles, thus, losing consistency and becoming highly deformable.

Another factor to be taken into account is material expansion. The tearless material gave an expansion percentage close to 0.2%, whereas the mixes with tar showed negative expansions, meaning, they compressed their volume. This could be explained due to the bonding character of the tar, that will not allow water to separate the material particles, thus inducing a settlement. Taking this into account, we can conclude that tar reduces totally the volume expansion.

5. Conclusions and recommendations

It was proven the possibility of stabilize a subbase granular material for road construction using coal tar coming from steel industry Acerías Paz del Río, since mixing the right amount of it with the subbase granular increased the resistance, showing CBR values beyond 3% of those for the unaltered material.

With the analysis of the CBR values for subbase-tar mixes, the optimal value for coal tar was found, with which it was obtained the higher resistance increase, being 6.5% in weight for the stabilization of granular subbases.

Since the CBR increase for the granular subbases produced by the coal tar stabilization is not too big, it can be considered as a stabilization method for materials with a CBR that does not behave as specified by INVIAS, but nevertheless, close to the acceptance limit.

With the stabilization of subbases with coal tar there is a profit in terms of volume expansion, since the mixes did not show any expansion, and on the contrary, they contracted slightly, which means a decrease in this factor as a whole.

Coal tar was characterized and manipulated as liquid asphalt; nonetheless, it was observed that the material is viscous and its volume changes drastically as temperature increases, behaving in a different way than asphalt and thus, there is a necessity to get a technical norm for lab tests with coal tar as well as for its use as a stabilization agent, since there is nothing of that sort as of now. It is recommended to use personal protection gear in order to manipulate the coal tar (gloves, anti-gas mask, eyeglasses, among others), and to do it in places well ventilated or outside if possible, since it is a bituminous substance that emits toxic gases, potentially harmful for people.

It is recommended also for the granular subbase stabilization with coal tar to be performed in a sunny day, mixing the compound with the bituminous binder and letting it rest for about a week period for its proper sealing before open it for traffic use or building the superior layer. Also, during the sealing period, the layer should be protected from rain, and preferably during nice weather conditions, so that all the tar solvents can be evaporated.

Journal of Physics: Conference Series **687** (2016) 012055 doi:10.1088/1742-6596/687/1/012055

References

[1] H S Carlos Hernando 2011 *Nociones sobre métodos de diseño de estructuras de pavimentos para carreteras* (Tunja: Universidad Pedagógica y Tecnológica de Colombia)

[2] M Santiago 2005 *Estabilización de suelos* (Argentina: Universidad Nacional de la Patagonia San Juan Bosco)

[3] D Laura Marcela and H Diego Andrés 2013 *Análisis del comportamiento del alquitrán de hulla como estabilizante de subbases granulares* (Tunja: Universidad Pedagógica y Tecnológica de Colombia)

[4] Ochoa R 2013 *Journal of Physics Conference Series* **466**

[5] Instituto Nacional de Vías 2007 *Normas de ensayo para materiales y especificaciones generales de construcción de carreteras* (Bogotá: INVIAS)

[6] Acerías Paz del Río SA 2014 *Alquitrán (Tar Crude)* (http://www.pazdelrio.com.co/es-es/Productos /coproductos/Paginas/alquitran.aspx)

[7] A B Siervo Andrés and G V Rubén Darío 2012 *Evaluación del comportamiento de mezclas bituminosas MDC-2 con agregados pétreos, escoria granulada, alquitrán y cemento portland* (Tunja: Universidad Pedagógica y Tecnológica de Colombia)

Effect of sulfonation and diethanolamine addition on the mechanical and physicochemical properties of SEPS copolymer

D Patiño[1], E Correa[1] and M Acevedo-Morantes[1]
[1] Universidad de Cartagena, Cartagena, Colombia.

E-mail: macevedom@unicartagena.edu.co , dpatinor@unicartagena.edu.co

Abstract. Modification techniques have been developed to achieve changes in the processing of polymers, and modification of their mechanical, thermal and morphological properties, as well as their hydrophobicity and conductivity. Sulfonation improves ion conductivity, antistatic behaviour, hydrophilicity and solubility of the polymers. These characteristics are related to the presence of sulfonic groups in the polymer matrix. This research project focuses on the evaluation of mechanical, physical and chemical properties of membranes that are based on a sulfonated Styrene-Ethylene-Propylene-Styrene (SEPS) copolymer. The membranes were functionalized with diethanolamine at 5, 15 and 30% w/w, to separate carbon dioxide. FTIR and XRD analyses were used to characterize the membranes. The sulfonated-loaded membrane with 15% of diethanolamine showed the best results in each characterization.

1. Introduction

Sulfonation processes are used to introduce sulfonic acids into polymers and modify properties such as ion conductivity, to increase their hidrophilicity and to improve polymer solubility. Sulfonated polymers have a wide range of applications. They are used as membranes for fuel cells, for electrodialysis, to exchange resins and in gas separation. [1,2]. Incorporation of polar groups by sulfonation favors water uptake, thus increasing conductivity and inducing stronger interactions with polar feed gases such as CO_2. These properties are typical of polymers that are commonly used in fuel cells and gas separation, respectively. Polymers with aromatic rings are more susceptible to experience sulfonation. This has been demonstrated with sulfonated polyethertherketones (SPEEKs), polystyrene copolymers [3], polyimides [4], polyphenyllenesulfides, polycarbonate [5], Poly (phenylene oxide) [6], and polysulfones. The properties of the modified polymer are strongly affected by the degree of sulfonation. For instance, selectivity for CO_2 separation of Poly (2,6-dimethyl-1,4-phenylene oxide) (PPO) can be increased from 16.7 to 33.2in sulfonated brominated PPO membrane with a 19.7% degree of sulfonation (DS) [7].

The copolymer that is used for this application is composed by a soft, rigid block known as block copolymer, which confers stability and rigidity. The Styrene-Ethylene-Propylene-Styrene (SEPS) is composed by two styrene blocks and an ethylene and propylene block that confers impact resistance and elasticity, respectively. SEPS is resistant to UV radiation at high temperature and it is stable during its production process. The purpose of this work is to study the effect of sulfonation and diethanolamine on the mechanical and physicochemical properties of SEPS membranes to evaluation of material stability in carbon dioxide separation [8].

2. Methodology

2.1. Materials
Styrene-Ethylene-Propylene-styrene (SEPS) copolymer was kindly donated by Kraton®. Diethanolamine was purchased from Sigma Aldrich to be used as a carrier agent to functionalize the membranes and improve carbon dioxide separation.

2.2. Preparation of the membrane
A solution of 5 grams of SEPS copolymer was prepared in 100mL of dichloromethane and then mixed with a sulfonating agent. The sulfonating agent was prepared in 100mL of dichloromethane at 0°C in an ice bath; 4.73mL of acetic anhydride were then added, along with 2.68mL of sulphuric acid after 10 minutes. The solution was left to rest for an additional 10 minutes. The solution was reacted for a certain time and then stopped with methanol. The obtained product was removed and washed with de-ionized water until neutralization. The sample was dried in an oven at 75°C [9,10].

The sample was dissolved in toluene. The solution was loaded with 5, 15 and 30% w/w (DEA) and stirred until obtaining a stable, clear solution. They were then transferred to a petri-dish until it was completely dry [9,10]. Polymer membranes were obtained with DEA (Loaded membranes LM) at different concentrations and with sulfonation (Sulfonated membranes SM). Moreover, the effect of DEA was determinate about sulfonated membranes (Sulfonated-loaded membranes SLM). All samples obtained were evaluated with respect to unmodified membranes (UMM).

2.3. Characterization of the membranes
Contact angle was determined from images that were previously taken when measuring the angle of a water drop on the surface of each membrane. This analysis was carried out at room temperature [11]. Mechanical properties like tensile strength, deformation resistance and Young's modulus were determined for each membrane using a EZ-S Shimadzu [11]. FTIR analysis was used to identify the sulfonic groups that are present in the membranes at different times. It was also used to establish the different functional groups, such as sulfonic and amines. A Scanon FTIR-8400 equipment was used to perform Fourier Transformed Infrared Spectroscopy (FTIR) in a frequency range between 4000 and 400cm^{-1} to determine the composition of each type of modified membrane [11]. An X-Ray Diffraction (XRD) analysis was carried out using a PW3373/00 Cu LFF DK332001 equipment with a copper anode, 45kW voltage and 40mA current. It was used for determining the structural characteristics of the unmodified and 15% sulfonated-loaded DEA membranes [11].

3. Analysis and discussion of results

3.1. Effect of sulfonation at different times using FTIR analysis
Figure 1(a) shows the infrared spectrum of three sulfonated samples. There, it can be observed that the sample exhibits the highest concentration of sulfonic groups after 60 minutes at 700, 1030 and 1150cm^{-1}. The sulfonic acid group $(SO_3H)^-$ is regularly observed at 709, 1028, 1080, 1252 and 3440cm^{-1}. The peak at 1252cm^{-1} is overlapping in the intensity of 1150cm^{-1} and the peak at 3440cm^{-1} overlaps at frequency 3420cm^{-1} because of the presence of moisture. The peak at 1080cm^{-1} is shifted to 1150cm^{-1} due to the presence of the aromatic ring that reduces the symmetry and directly influences vibration frequencies. Asymmetric effort of the C=CH group caused by the benzene ring is observed between 3000cm^{-1} and 3100cm^{-1}. Straining of the HCH group which is part of the polymer chain becomes evident at 1460 and 1490cm^{-1} [3].

In addition, spectrum of the sulfonated-loaded membrane with 15% of DEA in Figure 1(b) shows the presence of sulfonic groups in the peaks at 888, 1067 and 1239cm^{-1}. Amino group is evidenced in slight peaks at 3500cm^{-1}. The same behaviour is observed in the membrane that was loaded with 30% of DEA. Figure 1(b) showed the same behaviour of each functional group in Figure 1(a). However,

FTIR analysis of the modified membranes can prove the presence of the different functional groups in their structures [3].

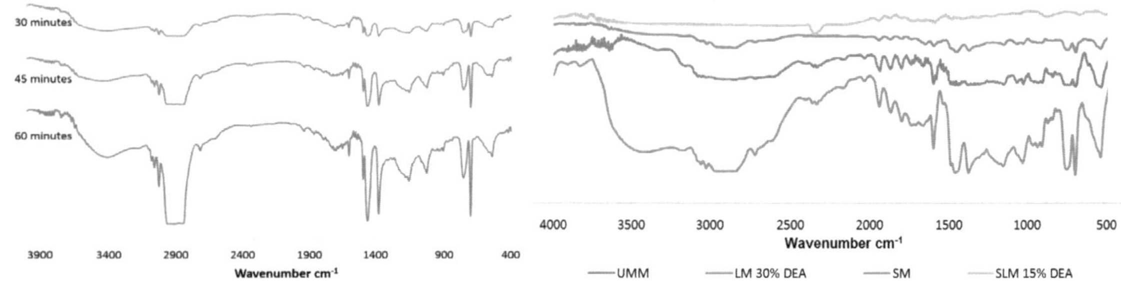

(a) (b)

Figure 1. (a) FTIR of the sulfonated samples at 30, 45 and 60min (b) FTIR of the modified membranes.

3.2. Contact angle

Images of each type of membrane were taken for recognizing the contact angles. A water drop was placed on the membrane surface in order to identify the hydrophobic or hydrophilic properties of the material [12].

In Table 1, it is observed that all the membranes presented hydrophilic properties, since all angles are lower than 90 degrees. The angles decrease as concentration DEA increases. The sulfonated-loaded membranes have the smallest angle. Water molecules create more interaction with the amine groups on the membrane surface. This interaction is amplified by the porosity of the membranes, which allows a greater immersion of the drops into the material [12].

Table 1. Contact angles for each membrane.

Membrane	Angle (°)
UMM	70
LM 5% DEA	62
LM 15% DEA	61
LM 30% DEA	60
SM	58
SLM 5% DEA	54
SLM 15% DEA	53
SLM 30% DEA	53

3.3. Mechanical properties

The membranes were subjected to mechanical properties tests. Table 2 shows that the unmodified membrane presented the highest results because of its high density. Moreover, loaded membranes decreased their mechanical properties values.

Table 2. Average values obtained from the mechanical properties tests for each sample.

Sample	Max. Strength (N)	Max. Tension (N/mm2)	Máx. Deform (%)
UMM	138.5	11	2466.9
LM 5% DEA	81.5	6.5	1381.4
LM 15% DEA	120.9	9.6	1551.8
LM 30% DEA	76.6	6.1	1458.4
SM	112.2	8.9	2433
SLM 5% DEA	93.2	7.4	2071.7
SLM 15% DEA	91.9	7.3	2125.1
SLM 30% DEA	77.8	6.2	1308.3

Nevertheless, the loaded membrane with 15% of DEA showed the best results, followed by the sulfonated one. In the case of sulfonated-loaded membranes, the behaviour corresponds to a more notorious decrease in mechanical properties. By increasing the DEA load DEA, porosity of the membrane increases, leading thus to a decrease in the mechanical characteristics because of the decrease in density [11]. When DEA loads are added, it significantly benefits elongation of the material, which in turn causes a greater effect on elasticity of the membranes [13].

3.4. Analysis by X-ray diffraction (XRD)

The unmodified and 15% DEA sulfonated-loaded membranes were subjected to XRD test to evaluate the structural behaviour between unmodified and modified membranes by DEA impregnation.

As it is shown in Figure 2, the copolymer is amorphous and does not change with sulfonation and DEA load. At 18.0° the intensity of the peak changes in both membranes; the 15% DEA sulfonated-loaded membrane showed a decrease in intensity because of the distance between the polymeric chains and the free fractional volume (FFV) according to Bragg's law [11]. The sulfonated-loaded membrane would present stronger interactions between polymeric chains that generate contraction in the polymer and increase porosity.

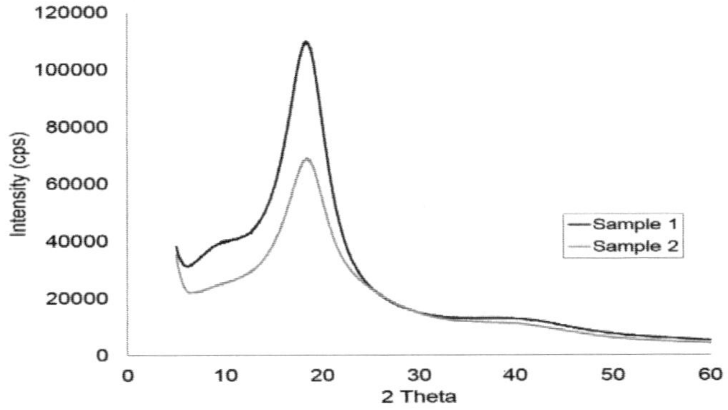

Figure 2. X-Ray of an unmodified (sample 1) and 15% sulfonated-loaded membranes (sample 2).

4. Conclusion

Copolymer of SEPS displayed excellent physical properties with positive values that are characteristics of membrane elongation. However, DEA sulfonated-loaded membranes exhibited the lowest results due to formation pores that decrease their resistance. DEA addition and sulfonation of SEPS copolymer increases the hydrophilic properties and the elongation characteristics of the material due to the porosity. All the same, it decreases the resistance to mechanical testing. Although the physic-chemical and mechanical properties were modified, the amorphous structure of the copolymer remained unmodified.

References

[1] Seulgi Kim, Hyunjung Lee, Dahee Ahn, Hae Woung Park, et al 2013 *J of Membrane Science* **427** 85
[2] Ji-Won Rhim, Geeta Chowdhury and Takeshi Matsuura 2000 *Journal of Applied Polymer Science* **76** 735
[3] S M Javaid Zaidi 2003 *The Arabian Journal for Science and Engineering* **28-2B** 183
[4] C Genies, R Mercier, B Sillion, N Cornet, G Gebelb, M Pineri 2001 *Polymer* **42** 359
[5] S Sridhar, Tejraj M Aminabhavi and M Ramakrishn 2007 *Journal of Applied Polymer Science* **105** 1749
[6] S Sridhar, B Smitha, M Ramakrishna, Tejraj M Aminabhavi 2006 *Journal of Membrane Science* **280** 202
[7] F Hamad, T Matsuura 2005 *Journal of Membrane Science* **253** 183
[8] Hudson John Michael, Graulus Hendrik, Hammoond Philip James 1994 *Composicion de gel que comprende polimeros de bloques de estireno etileno propileno* (Chile: Patent C08L053/02)
[9] D Suleiman, Y Elabd, E Napadensky, J Sloan, D Crawford 2005 *Thermochimica Acta* **430** 149
[10] Y Elabd, E Napadensky 2004 *Polymer* **45** 3037
[11] Shahab Saedi, Sayed S Madaeni, keyumars Hassanzadeh, et al 2014 *J of Indus and Eng Chem* **20** 1916
[12] R Naim, A F Ismail 2013 *Journal of Hazardous Materials* **250-251** 354
[13] Lei Zhang, Bin Zhao, Xianying Wang, Youxuan Liang, HanxunQiu, et al 2014 *Carbon* **66** 11

A structural study of the hydrogen absorption properties by replacing vanadium with zirconium in metal alloys

D Bellon[1], A Martinez[1], D Barreneche[1] and D S dos Santos[2]
[1] Universidad de Santander, Bucaramanga, Colombia.
[2] Universidade Federal do Rio de Janeiro, Rio de Janeiro, Brasil.

E-mail: alejandrom@udes.edu.co

Abstract. Hydrogen storage and microstructure the $TiCrV_{0.9}$, $TiCrV_{0.9}Zr_{0.2}$ and $TiCrV_{0.7}Zr_{0.4}$ alloys were investigated. The alloys were melted in arc furnace, the structure was analysed by X- ray diffraction and hydrogen absorption times were analysed by Sievert's type apparatus. The results showed that the addition of Zr replacing the V decreases in the absorption capacity passing of 3.8%*wt* to 2.0%*wt* in contrast the speed of hydrogen absorption increases considerably.

1. Introduction
Eighty-seven percent (87%) of global energy demand is covered by fossil fuels such as carbon, petroleum and gas. From an environmental point of view, the use of fossil fuels is the main culprit in global warming, as well as causing serious respiratory problems in major cities due to pollution [1].

This state of affairs is unsustainable over the medium-term and as such it is necessary to find different forms of energy production and consumption that are clean, safe and dependable. One possible response to this crisis is to implement the use of hydrogen as a fuel source. This falls within a framework known as the "hydrogen economy", a term currently used to refer to an alternative economic model not based on using fossil fuels for powering transportation.

Using hydrogen as an energy vector could revolutionize our immediate future [2]. The main advantage of hydrogen is that it is the most abundant element on Earth: it is present in water as well as fossil fuels; it is found in the Universe; and its combustion does not result in harmful emissions such as carbon dioxide which results from burning petroleum. The main challenge, however, is that hydrogen is rarely found in pure form making it difficult to exploit it effectively.

In this sense, the storage process for hydrogen is key. Traditional storage methods are conceptually simple using pressurized tanks or cylinders, or liquid hydrogen. Currently, researchers are investigating the use of metallic hydrides to store it in solid form.

One safe storage method for hydrogen in solid state is to use what are called H_2 storage alloys that result from forming metallic hydrides. The main advantages that make metallic hydrides very attractive, and are worth mentioning, include the high volumetric storage density, stability, safety and ease of handling. [3]

Ti-based hydrides have long been studied due to their relatively high hydrogen absorption capacities at room temperature, such as TiCrV which can store between 3,5 and 3,7% in weight, and presents optimum stability and cyclability properties meaning that it can be reused many times over [4].

In this work I seek to structurally characterize and study the hydrogen absorption TiCrV based alloys by adding Zr to replace V.

2. Experiment methodology

The alloys were formed using arc melting in atmosphere of argon using high-purity materials (>99.95). For this purpose, 20g ingots respectively made of $TiCr_{1.1}V_{0.9}$, $TiCr_{1.1}V_{0.9}Zr_{0.2}$, $TiCr_{1.1}V_{0.7}Zr_{0.4}$, were used and the samples were smelted 3 times in order to achieve a homogenized microstructure.

For structural study was necessary to use the sample metal powder which passed through a mesh of 270 microns, following which it was subjected to XRD (X-Ray Diffraction) using a Shimadzu 5000 with Cu-Kα radiation and the lattice parameters were calculated.

The hydrogen absorption properties of the as-cast sample were measured in a Sievert's type apparatus (PCT pro 2000). The activation of the samples was carried out in an autoclave where a vacuum was maintained for 2 hours at 350°C, the main objective being to remove any oxide layer and surface water ensuring the initial activation of the sample.

3. Analysis and discussion of results

During the first stage of the process, an alloy was cast based on it is high hydrogen absorption capacities, with the stoichiometry $TiCr_{1.1}V_{0.9}$. Two other alloys were cast: $TiCr_{1.1}V_{0.9}Zr_{0.2}$ y $TiCr_{1.1}V_{0.7}Zr_{0.4}$. One of the objectives is to diminish the amount of V and increase the Zr content since the use of V in technological applications is very expensive, and Zr also has a high affinity for producing hydrides.

3.1. XDR analysis

In Figure 1(a) it can be observed that the diffractogram corresponds with a body-centred cubic structure (bcc) characteristic of this kind of alloy, giving results that correspond with those reported by [4].

This body-centred cubic structure can be expected given that V and Cr have a bcc, and Ti crystalized into this same structure at temperatures above 800°C. As such only a single-phase was expected.

These kinds of compounds based on solid solutions with bcc structures demonstrate high hydrogen absorption rates at room temperature.

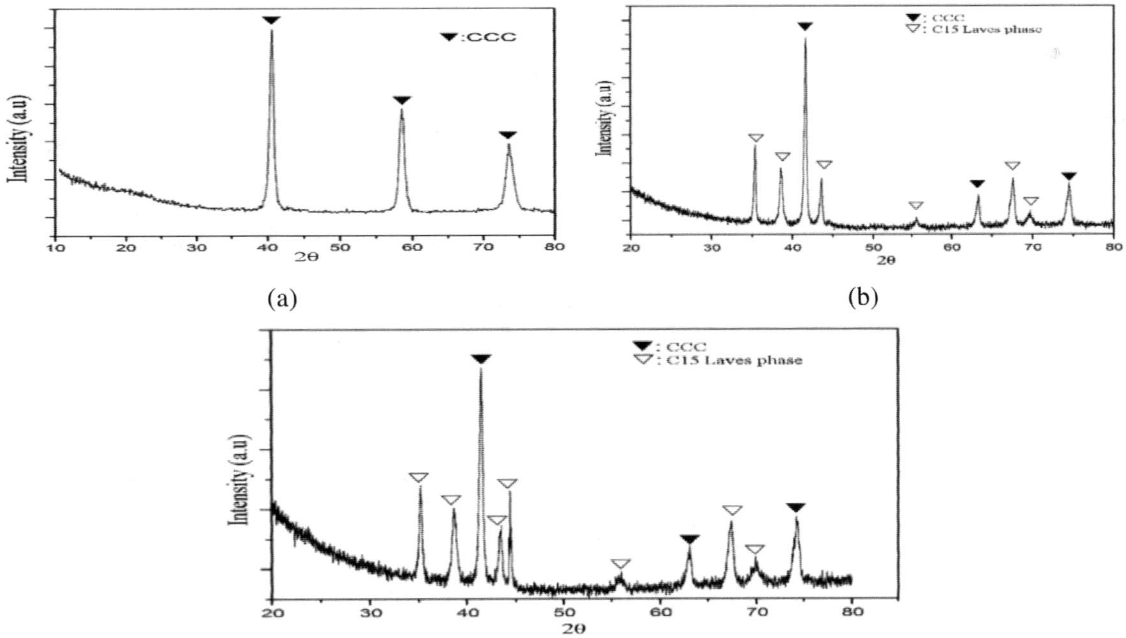

Figure 1. Diffractogram for the different alloys (a) $TiCr_{1.1}V_{0.9}$ alloy, (b) $TiCr_{1.1}V_{0.9}Zr_{0.2}$ alloy (c) $TiCr_{1.1}V_{0.7}Zr_{0.4}$ alloy.

The diffractogram for the $TiCr_{1.1}V_{0.9}Zr_{0.2}$ alloy can be seen in Figure 1(b). Here a double-phase can be observed; one which is a bcc phase and the other a Laves phase with a cubic-type C15 compound $ZrCr_2$, which arises from the addition of Zr. This result corresponds with those of K. Shashikala et al [5] who added Zr as a substitute for Ti in the alloy TiCrV.

The formation and stability of this Laves phase depends upon the ideal ratio of atomic radii between neighbouring atoms A and B being $r_A / r_B = \sqrt{3} / \sqrt{2} = 1,225$. As such, it is possible to experimentally obtain a ratio between 1,06 and 1,67. This property corresponds with the atoms geometrical compression factor since they are able to contract and expand as they approach the ideal ratio [6]. For this reason, the geometrical rule is important for predicting the appearance of the Laves phases.

Since the pairs of elements approach the ideal ratio ≈1,225 $(R_{Zr} / R_{Cr} = 1.26)$, it can be expected that the alloys in the present study form Laves phases. The atomic ratio for Zr is 1.60Å and for Cr it is 1.27Å.

The diffractogram for the $TiCr_{1.1}V_{0.7}Zr_{0.4}$ alloy can be seen in Figure 1(c) where the appearance and growth of a new peak corresponding with the Laves phase C15 can be seen. This is due to increasing the amount of Zr in the alloy as explained above.

The lattice parameters for each of the alloys are shown in Table 1 along with the phases that occurred for each one.

Table1. Phases and lattice parameters for each of the alloys studied.

Alloy	Phases	Lattice parameter (Å)	Capacity (*%wt*)
$TiCr_{1.1}V_{0.9}$	BCC	3.06	3.6
$TiCr_{1.1}V_{0.9}Zr_{0.2}$	BCC	3.14	3.0
	Laves phase C15	7.19	
$TiCr_{1.1}V_{0.7}Zr_{0.4}$	BCC	3.18	2.0
	Laves phase C15	8.03	

3.2. Absorption properties

Figures 2 and 3 show the time taken for hydrogen to be taken up and the storage capacity of each alloy. Figure 2(a) for the $TiCr_{1.1}V_{0.9}$ alloy shows a high sorption capacity of 3.6%wt (% percentage by weight), however, the process takes approximately 58 minutes before hydrogenation begins.

This is characteristic of these types of alloys that produce a heavy oxide layer that cannot be easily removed without using high temperatures. Such a process of removing any kind of surface water or oxidation is called activation and is carried out in a vacuum at approximately 350°C during an hour. Even so, the oxide layer remains intact and could be removed at higher temperatures but with the result that the sample carrier would be destroyed. The absorption process begins at the point at which the hydrogen penetrates the oxide layer.

In Figures 2(b) the hydrogen absorption capacity decrease to 3.0%wt for the $TiCr_{1.1}V_{0.9}Zr_{0.2}$ alloy and Figure 3 the hydrogenation time decreases significantly to 2.0%wt for the $TiCr_{1.1}V_{0.7}Zr_{0.4}$ for to a matter of minutes due to the formation of laves phases thanks to increasing the amount of Zr.

The activation process is the same for all of the samples with the difference that the formation of the laves phases acts as a catalyst for the hydrogen reaction with the surface of the metal. Given that these phases make up a small percentage of the main matrix and their mechanical properties are weaker than those of the preponderant alloy TiCrV, the hydrogen is first diffused in this phase creating preferred pathways or short-circuits until the main bcc structure is broken. This process, which concurs with the results reported by [7], occurs in a question of minutes until the alloy is saturated with hydrogen.

However, the hydrogen uptake rate drops from 3.8%wt to 3.0% and 2.0%wt respectively. This is due to the fact that the Laves phases increase their content and compete with the bcc phase, and in so doing impede the number of available places that can serve as hydrogen storage. This matches the studies by [7,8].

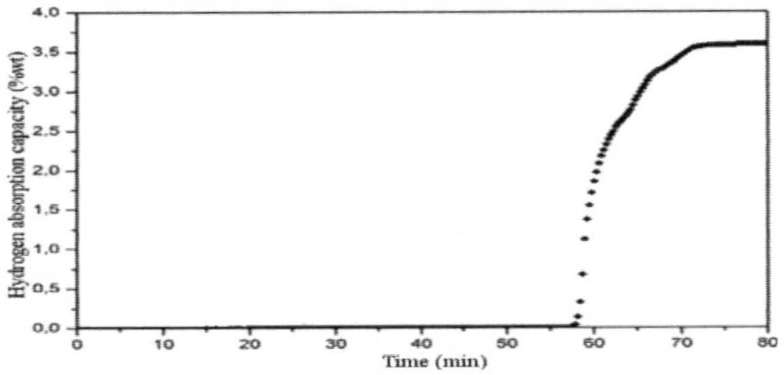

(a)

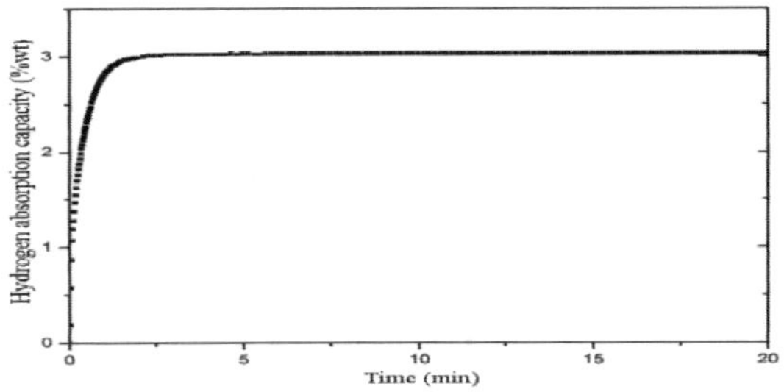

(b)

Figure 2. Hydrogenation curve for the alloys (a) $TiCr_{1.1}V_{0.9}$ and (b) $TiCr_{1.1}V_{0.9}Zr_{0.2}$.

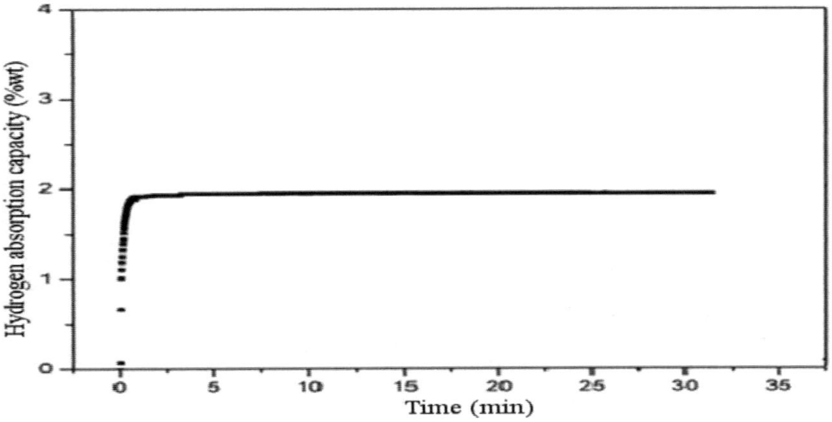

Figure 3. Hydrogenation curve for the alloy $TiCr_{1.1}V_{0.7}Zr_{0.4}$.

4. Conclusions

Substituting Zr for V in the alloy TiCrV was studied. It was verified that the compounds $TiCr_{1.1}$ $V_{0.9}Zr_{0.2}$ and $TiCr_{1.1}V_{0.7}Zr_{0.4}$ give rise to a double-phase structure made up of primarily a bcc structure and a smaller Laves C15 cubic-type structure.

The appearance of a Laves C15 phase in the TiCrV alloys can be controlled by varying the number of Zr and V atoms.

Substituting Zr augments the lattice parameters and decreases the hydrogen absorption capacity due to the increase in laves phase structures that compete with the bcc structures. The laves phase acts as a catalyst for the absorption reaction.

References

[1] A K Akella, R P Saini, M P Sharma 2009 *Renewable Energy* **34** 390
[2] Andrews J and Shabani B 2012 *International Journal of Hydrogen Energy* **37** 1184
[3] Andreas Zuttel 2007 *Mitig Adapt Strat Glob Change* **12** 343
[4] E Akiba, H Iba 1998 *Intermetallic* **6** 461
[5] K Shashikala, Seemita Banerjee, Asheesh Kumar, M R Pai, C G S Pillai 2009 *International Journal of Hydrogen Energy* **34** 6684
[6] Katherine C Chen, Eric J Peterson, Dan J Thoma 2001 *Intermetallics* **9** 771
[7] A Martinez, D S dos Santos 2012 *Journal of alloys and compounds* **536** S231
[8] Jeong-Hyun Yoo, Gunchoo Shim, Sung-Wook Cho, Choong-Nyeon Park 2007 *International Journal of Hydrogen Energy* **32** 2977

Development of a web application for estimate the total charge in an electric discharge

F H Vera-Rivera[1], V D Dugar-Zhabon[2], H J Dulce-Moreno[1] and E D V-Niño[2,3]
[1] Universidad Francisco de Paula Santander, San José de Cúcuta, Colombia.
[2] Universidad Industrial de Santander, Bucaramanga, Colombia.
[3] Universidad Politécnica de Madrid, Madrid, España.

E-mail: fredyhumbertovera@ufps.edu.co , deydannv@gmail.com

Abstract. This research seeks to implement a web application to estimate the total charge in an electric discharge from the experimental parameters established in reactors such as the JUPITER-MOSMET [1-5], used to perform surface modifications of metallic and nonmetallic species in solids substrates by hybrid discharges pulsed of high voltage and electric arc at low pressures [6-13]. Estimate the total charge generated in an electric discharge will allow to know the approximate value of the dose of ions implanted in the material surface [14-15], which are of great importance in each one of the experiments performed in the plasma laboratory of the Universidad Industrial de Santander (UIS), where the reactor is installed. Currently there are no applications developed to ease the process optimization and estimate the total charge generated in a pulsed electric discharge.

1. Introduction

The Laboratory of Plasma of the Research Group in Physics, Plasma and Technology and Corrosion (FITEK) adjoint to the school of Physics at the Universidad Industrial de Santander (UIS) in conjunction with the Development Research Groups of Software Engineering (GIDIS) and Ceramic Technology (GITEC) of the Universidad Francisco de Paula Santander, are performing experimental research in the area of surface modification of metallic materials by the three-Dimensional Ion Implantation technique (3DII) [2,3,9,10] generated by hybrid discharges of high voltage and electric arc at low pressures [6-13]. The Ion implantation produces defects in an extension of material volume, which are very difficult to observe directly [16-18]. Most of the superficial characterization experimental techniques employed to measure or observe the defects and the implanted ions concentrations only provide information on the surface, instead of examining the structure below it [19], where the ion amount and types of defects produced are of importance in the analysis of changes in the surface properties of the treated material [16-19].

For researchers in the area of surperficial modification, especially the one performed by electric discharges of high voltage at low pressures [6-13], it is very important to characterize the estimation of the total charge and the implanted particles in metallic solids as a function of depthness, moreover, there is a need to understand the complex phenomena that occur during the process. Therefore, the development of a web application and a computer simulation are of great importance, especially because we can obtain information that today could not be obtained by other means. That is why it is necessary to develop computational tools that estimate more quickly, accurately and precisely the total charge generated in an electric discharge and the ion concentration implanted in metallic materials.

2. Experiment methodology
In Figure 1 is present the three phases executed according to the methodological model proposed in the present investigation.

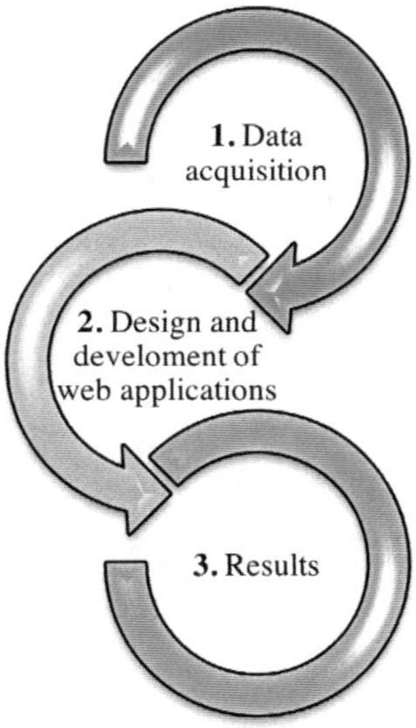

Figure 1. Methodological model.

2.1. Phase 1: Data acquisition
The data acquisition (current pulses) was performed from the experimental parameters that were established in a surface modification process (ion implantation) of metal substrates by means of an electric discharge [4,12,13]. The current pulses registered in the electric discharge, represent the total electric charge generated in funsion of the time.

2.2. Phase 2: Design and development of web application
The design and development of web applicative is a computational tool made in order to calculate quickly and accurately the total charge estimated in an electric discharge from the data acquired of a set of current pulses. The obtained data were digitized and graphed in order to calculate the area under the curve representing the total electrical discharge estimated.

2.3. Phase 3: Results
The results of the area under curve estimated (total electric charge estimated) are reported by the web applicative.

3. Results
A web application was implemented in order to estimate quickly and accurately the total electrical charge from a pulse of electric current in funsion of the time. Figure 2 shows the use case diagram describes the cua features that the user is available on the web application is presented.

The web application allows the user to upload image files and data in CVS format (see Figure 3) corresponding to the pulse of current that can be captured on an oscilloscope during ignition of the electric discharge (in our case we use the oscilloscope Tektronix TDS 2002 B); where the computational tool performs a process of digitization of data and graphic. After presented the graphic along with the image of the pulse, the user sets the maximum and minimum values that determine the pulse width necessary to calculate the area under the curve and estimate the total electrical charge (see Figure 4).

Journal of Physics: Conference Series **687** (2016) 012058 doi:10.1088/1742-6596/687/1/012058

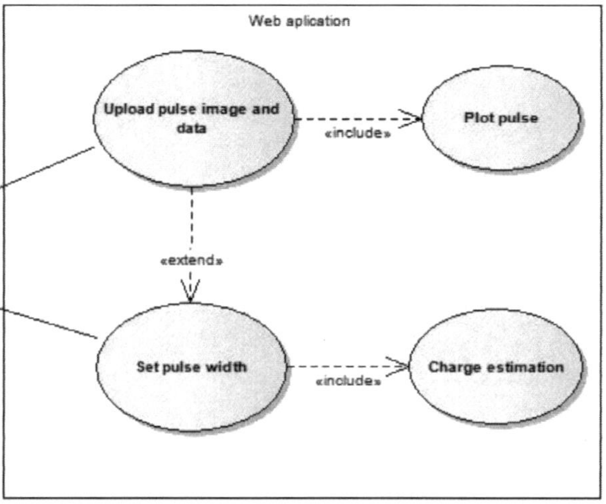

Figure 2. Functionalities of the web application.

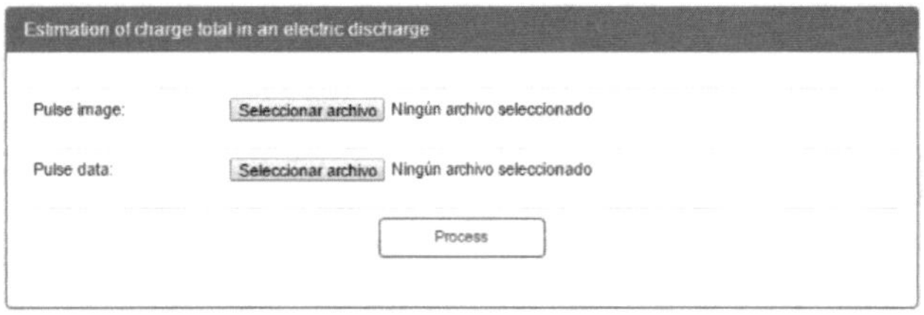

Figure 3. Section to load the image and the pulse data.

In Figure 3 after the data is loaded, it is clicked on process and the system automatically processes the data, the graphics and shows the estimated total charge of the electric current pulse (see Figure 4). At the top of Figure 4 is reported the pulse image obtained of the oscilloscope Tektronix TDS 2002 B and at the bottom the pulse graphed from the web application (as shown in the Figure 4, the graph obtained is very similar the acquired with the oscilloscope).

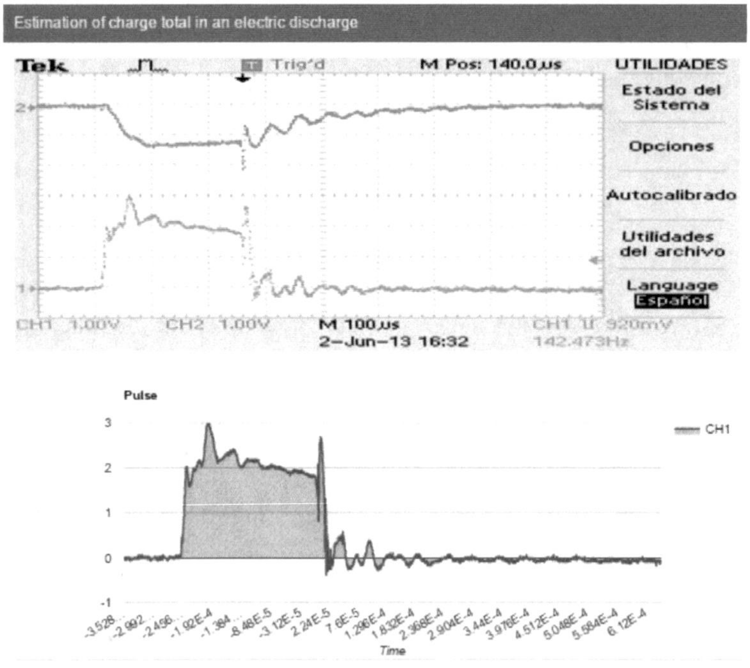

Figure 4. Image and graphic of pulse in the web application.

Finally, at the bottom of Figure 5 the estimated calculation of the total charge in the electric discharge is shown.

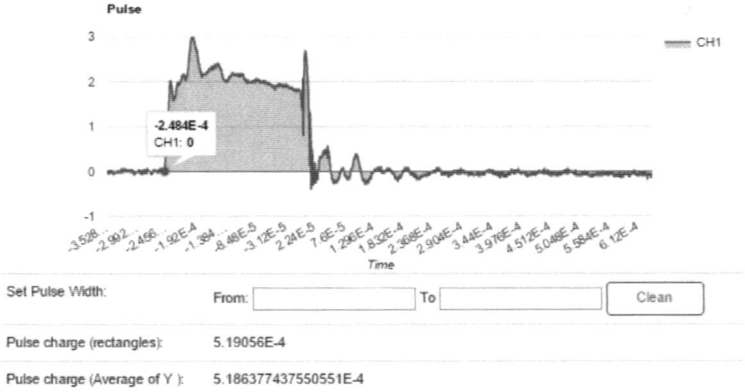

Figure 5. Results of the estimate total charge.

4. Conclusions

A computational tool to streamline and estimate the calculation of the total charge generated in a process of surface modification by electric discharge, bringing benefits to experimental researchs in the plasma laboratory of the UIS was built.

The web application developed can be used anywhere, at any time and without any special requirements, just need an Internet connection and the current pulse data to be processed. The format of the data received by the system can be downloaded from the web application.

The processing of data obtained from the web application are very similar to those current pulses reported experimentally during the ignition of the electric discharge, ensuring that the implemented web application works correctly to estimate the total charge.

References

[1] Conrad J R 1988 *Method and apparatus for plasma source ion implantation* (USA: Patent No 4764394)
[2] Dougar-Jabon V D, Castro B J, Dulce Moreno H J, Tsygankov P A 1999 *Rev Colom de Física* **31-2** 181
[3] Dougar-Jabon V D, Dulce Moreno H J and Tsygankov P A 2003 *Revista Respuestas* **8** 19
[4] Dugar-Zhabon V D, Dulcé-Moreno H, Garnica-Villamizar H A and V-Niño E D 2012 *Materials Research* **15-6** 969
[5] Valbuena Niño E D, Tsygankov P, Dugar-Zhabon V, Jaimes Ochoa C J, Parada Becerra F F and Plata A 2011 *Revista Colombiana de Física* **43-2** 458
[6] Conrad J R, Radtke J L, et al 1987 *J Appl Phys* **62** 4591
[7] J R Conrad 1987 *Mater Sci Eng* **116A** 197
[8] Raizer Y P 1991 *Gas discharge physics* (Belin: Springer)
[9] V I Khvesyuk, P A Tsygankov 1997 *Surface & Coatings Technology* **96** 68
[10] V Dougar, J Dulcé and P A Tsygankov 2002 *Rev Sci Instr* **73** 828
[11] H J Dulcé-Moreno, P A Tsygankov, V Dugar-Zhabon, E D V-Niño, F F Parada-Becerra 2011 *Puente Revista Científica* **5-2** 21
[12] F F Parada Becerra, R Cabanzo Hernandez, V Dugar-Zhabon, P Tsygankov, E Mejia-Ospino and E D V-Nino 2012 *Journal of Physics Conference Series* **370** 012054
[13] E D V-Niño, V Dugar-Zhabon, H J Dulce-Moreno, G Peña-Rodríguez, H A Garnica, P A Tsygankov 2012 *Revista Iteckne* **9-1** 14
[14] Valbuena Niño E D and Dugar-Zhabon V 2006 *Revista Colombiana de Física* **38-1** 61
[15] E D V-Nino, H J Dulce Moreno and V Dugar-Zhabon 2010 *Revista Colombiana de Física* **42-3** 387
[16] Yu V Gott 1978 *Interaction of Particles with substances in plasma research* (Moscow: Atomizdat)
[17] J F Ziegler, J P Biersack, U Littmark 1985 *The stopping and range of ions in matter* (New York: Pergamon)
[18] Edmanuel Torres, Valeri Dougar-Jabon 2006 *Revista Colombiana de Física* **38-1** 89
[19] K Nordlund, J Keinonen, M Ghaly and R S Averback 1999 *Nucl Instr Meth Phys Res B* **148** 74

PIC numerical study of ECR plasmas confinement in a minimum-B and zero-B magnetic traps with GPU

M T Murillo Acevedo[1,2] **and V D Dugar-Zhabon**[2]

[1] Universidad Santo Tomás, Bucaramanga, Colombia.
[2] Universidad Industrial de Santander, Bucaramanga, Colombia.

E-mail: plasmao@gmail.com

Abstract. This work analyzes through computational methods the phenomenon of confinement and heating of plasmas, in open magnetic traps, Minimum-B, and zero-B under conditions of resonance electron cyclotron (ECR). This simulation is made using electrostatic particle in cell method. First, it simulates the minimum-B trap, which has been studied both numerically and experimentally, by which is accomplished the confrontation of 6 different types of results that help us to validate our code. In the same way the zero-B trap is analysed. Proposed by Dr. Dugar-Zhabon, the main characteristic of the trap is the nullity of the magnetic field in the centre of the trap. The results show the detailed behaviour of the electronic component in the initial stage of the formation of plasma. Given the computational cost of the used model that allowed us to simulates fine details of the dynamics of plasma. Results were only reached in the time of half-life of the electrons. During this period the minimum-B trap proved to be better for the production of ions than the zero-B trap. Due to the huge amount of equations needed to solve the motion equations and the charge density, they are calculated in a Parallel way by GPU clustering.

1. Introduction

ECR plasmas are using in several scientific, industrial and technologist research areas [1,2]. In almost all practical applications are required of high density plasmas that need that the ions and electrons have to be confined to a limited space where plasma can stay and increase their energy.

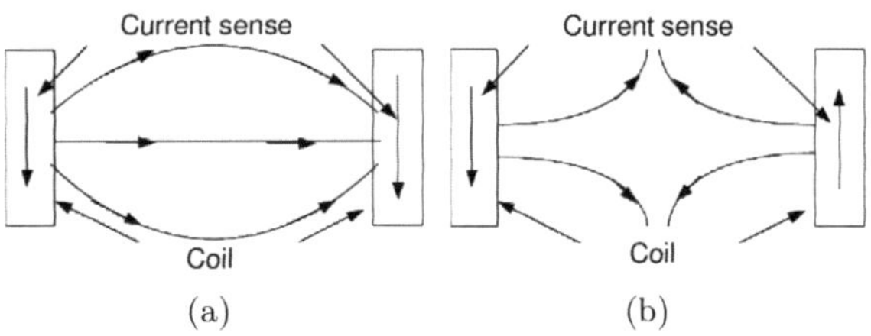

Figure 1. View of fields from coils for (a) minimum-B and (b) zero-B.

The magnetic trap system called minimum-B is used in ECR ion sources. This system is used to produce intense, high charge state ion beams. In order to produce higher multi-charged ions it is necessary to increase the density and time confinement of the plasma. In the minimum-B trap the magnetic field is minimum in the geometrical centre, that makes the plasma be pressured toward this region. So, a way to increase this effect could be achieved making the whole magnetic field in the centre as zero. This can be possible if the sense of the currents that feed the coils placed at ends of chamber (Figure 1) have an opposite sense. This idea was proposed by Dr. Valeriy Dougar-Zhabon [3]. In this work, we present a comparative study between zero-B and minimum-B configurations through computational simulation using PIC method. Which runs during 30000 microwave cycles. Due to the huge amount of equations to solve by cycle, the code has been performed in parallel by using the Cuda C library.

2. Physical and computational modeling

The plasma is modelled by electrostatic PIC method on a uniform mesh, which step length is 0.2cm. The charge weighting is performed in three-linear way. The plasma is confined inside cylinder discharge chamber, it has 4cm in radius and 26cm in length, the z-axis coincide with the symmetry axis of the chamber and the XY plane coincide with the transverse centre plane. The plasma is considered as a hydrogen one, it is heated by 14 GHz microwaves for a TE_{114} mode. The temporal mesh step is taken as the microwave period divided by 160. We have taken 5×10^6 superparticle for electrons and the same amount for ions, each superparticle represents to 1×10^6 real particles. The static magnetic field of the trap is formed by two coils to produce the longitudinal magnetic field, a hexapole system to generate a transversal magnetic field. The ECR zone has the shape like an ellipsoid . The evolution of the plasma is made as individual interacting particles in electromagnetic fields by solving the Newton-Lorentz equation as is present in our before work [4]. The discretized Newton-Lorentz equation for electrons 1 was calculated in one GPU using CUDA C library. While the corresponding Newton-Lorentz equation for ions was make using the CPU cores with OpenMP library. The discretized Poisson equation was solved in the CPU through iterative method. In case of a particle go out from chamber it is returned inside of chamber in aleatory way.

$$\frac{\vec{u}^{n+1/2} - \vec{u}^{n-1/2}}{\Delta\tau} = \vec{g}^n + \frac{\vec{u}^{n+1/2} + \vec{u}^{n-1/2}}{2\gamma_B^n} \times \vec{b}^{n-1/2} \quad (1)$$

where $\vec{u} = \frac{\vec{v}}{c\gamma}$ relativistic moment in units of m_0c, $\vec{g} = \frac{\vec{E}_T}{B_0}$ total electric field, $\vec{b} = \frac{\vec{B}_T}{B_0}$ total magnetic field; both of them are divided by the magnitude of the magnetic resonance field $B_0 = \frac{m_q\omega c}{q}$; $\gamma_B = \sqrt{1+u^2}$ relativistic factor and $\tau = \omega t$.

3. Minimum-B system

The ECR surface has a shape like an ellipsoid which volume is 131.63cm^{-3}.

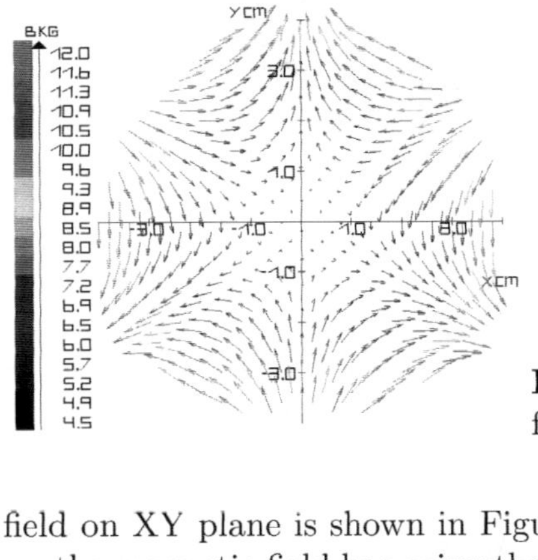

Figure 2. The magnetic field profile on z-axis.

The magnetic field on XY plane is shown in Figure 2, this configuration can form the local plasma tramps since the magnetic field has azimuthal and longitudinal gradients, the azimuthal ones is due to two factors. The first is that the radial component from the coils reinforcement the field from south poles and reduce field from north poles in the half first part of the chamber ($z < 0$). The opposite phenomenon is then on in the second part of the chamber ($z > 0$). This magnetic configuration make that electronic component of the plasma form a six-bar shape in the central region, its transversal section view can be shown in the Figure 3. These bars can unite by couples near the planes $z = 3.5$ cm and $z = -3.5$ as is shown in the Figure 3.

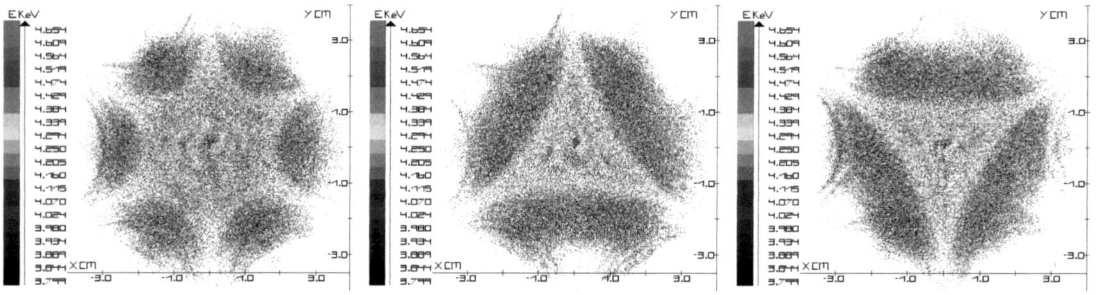

(a) (b) (c)

Figure 3. Spatial distribution of electrons confinement in the local trap for minimum-B system. (a) XY view of the electrons located in the local magnetic trap (around plane z=0 with 2cm in width, (b) XY view of the electrons located in the local magnetic trap (around plane z=3.5cm with 2cm in width and (c) XY view of the electrons located in the local magnetic trap (around plane z=-3.5cm with 2cm in width.

The asymmetry in the connections can be explain as a consequence of asymmetry in the magnetic field before mentioned, since in the regions where the magnetic field was weakened, the electrons can leakage from plasma and to impact the cylinder chamber walls. This phenomenon can be shown in the plasma marks on the wall, which ones are shown in the Figure 4(a). This result is agree with the experimental data [5].

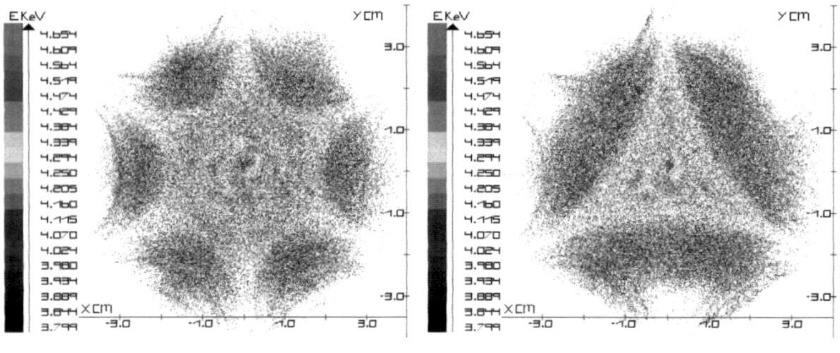

(a) (b)

Figure 4. Results for minimum-B system. (a) Plasma marks leaved for the electronic component on the cylinder wall of the chamber and (b) Electronic distribution function.

The particles in this region has a energy range of 3.8-4.61keV. This local trap represents an accumulation zone, it can be shown in its energy spectrum (Figure 4(b)), which has a peak around 4.5keV. The average electronic density inside of ECR zone is $8.88 \times 10^9 \text{cm}^{-3}$.

3.1. Software validation

For validate our software we using some experimental results and we also compare with other computational results. First, we make a plasma photo in the light visible spectrum,

Journal of Physics: Conference Series **687** (2016) 012059 doi:10.1088/1742-6596/687/1/012059

corresponding to cold electron emissions [6, 7]. To do it, we calculate the energy density in the same instant, and draw it in a plane without consider the deep. The image of this is shown in Figure 5.

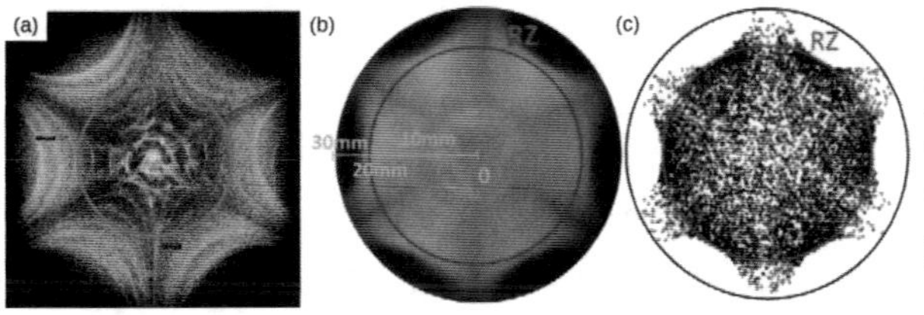

Figure 5. XY distribution of the cold electrons: (a) our results, (b) light visible photo of a real plasma [6] and (c) TrapCAD code [6].

Although this result is not exactly (Figure 5), however can be consider as good since the results do not corresponding at the same magnetic configuration; but its general behaviour can be showed, Like this one, we make another 3 results, for warm electrons 3.5-10keV (Figure 6) which are in the X-ray spectrum [6, 8–10], our results look like the real ones, for hot electrons we take the range of 5-10keV see [11, 12].

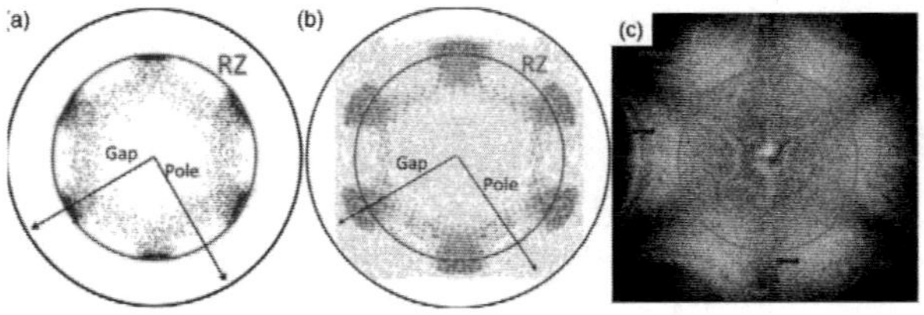

Figure 6. XY distribution of the cold electrons: (a) TrapCAD code [6], (b) X-ray photo of a real plasma [6] and (c) our results.

We also present the spatial distribution of warm electrons which energy range were taken as 10-50keV in Figure 7 where it was shown with other similar results. We can appreciate that our result is agree with the real ones, in the gradients of intensity.

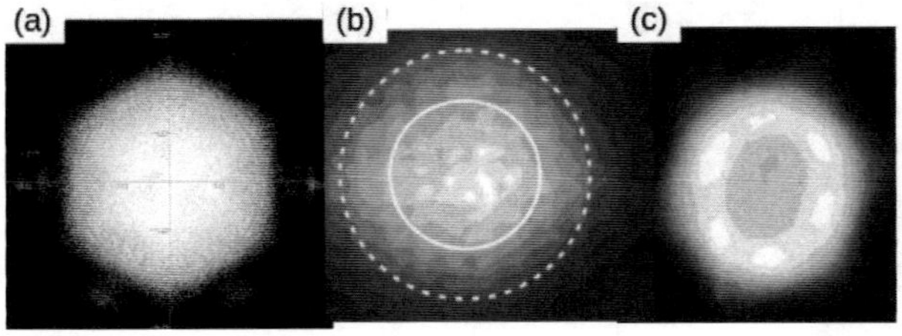

Figure 7. XY distribution of hot electrons: (a) our result and (b) experimental result [11] and (c) another simulation [11].

To measure the cold electron lifetime, we allow the particles leakage from chamber. We suppose an exponential decay then making an semi-logarithmic graph, we get two lifetimes the first 1.184855×10^{-5} s. This value is agree with [13]. The second was 1.184855×10^{-5} s, this can be possible according to [7]. The period of free decay was on during 1000 microwave cycles. In this time the electron component of the plasma was reduced by 8.48%, the 38.4% leave the plasma by cylinder walls, while the 61.16% do it for the end walls.

4. Zero-B system

For zero-B trap we only change the sense of the coil currents, the ECR zone has a volume of 400.88cm^{-3}. In this trap we also can find local traps the spatial distribution is shown in Figure 8. These particles have an energy range of 3.8-4.61keV. This zone also is a accumulation region as is shown in Figure 9(b). However during the period of free decay the electron component of the plasma was reduced by 20.56%, the 91.6% leave the plasma by cylinder walls, while the 8.4% do it for the end walls. The fact that the most particles go out from cylinder walls can be partly explain by the longitudinal magnetic cusp guide the electrons to cylinder walls and their radial component always reduce north poles which are converted in leakage zone greater than zones next to south poles (Figure 9(a)). The average electronic density inside of ECR zone is $7.52 \times 10^9 \text{cm}^{-3}$.

We also find two lifetimes for the cold electrons the first was 1.02×10^{-6}s and the second was 3.78×10^{-6}s.

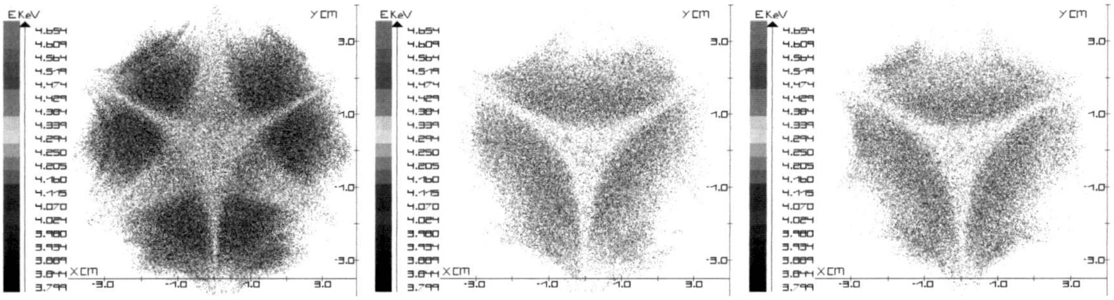

(a) (b) (c)

Figure 8. Spatial distribution of electron trapped in the local traps for zero-B system. (a) XY view of the electrons belong to local magnetic trap (around plane z=0 with 2cm in width), (b) XY view of the electrons belong to local magnetic trap (around plane z=5cm with 2cm in width) and (c) XY view of the electrons belong to local magnetic trap (around plane z=-5cm with 2cm in width).

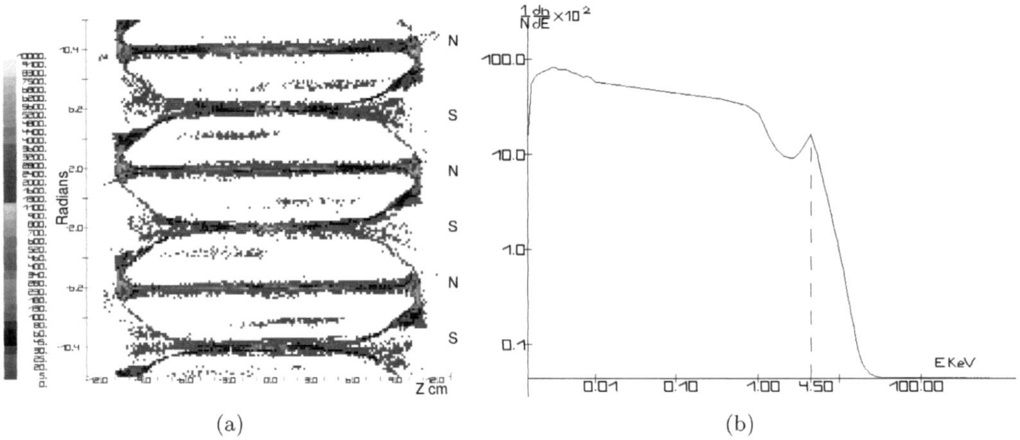

(a) (b)

Figure 9. Results for zero-B system. (a) Plasma marks leaved for the electronic component on the cylinder wall of the chamber and (b) Electronic distribution function.

5. Conclusions

At this stage, the minimum-B has better parameters, the first lifetime for cold electrons is 1.84 times higher than zero-B; the second lifetime for cold electrons is 3.12 times higher than zero-B. The amount of leakage particles in the zero-B trap was 2.42 times higher that minimum-B trap.

The density in the minimum-B was a little greater, it was 1.18 times higher that zero-B trap. Therefore the minimum-B trap showed advantages front zero-B trap, it can be possible for the absence of longitudinal magnetic field in the zero-B trap, specially in centre region where the plasma is pushed longitudinally but radially can be addressed on the cylinder walls. Although is necessary to reached the lifetime of ions to have a definitive results, since only at that moment the Lawson parameter of plasma can be measurement. Then our work simply show one small deficiency in the design of the zero-B trap. However, the biger volume of the ECR zone represent a significative advantage since can be increase the probability of ionization process.

Acknowledgment

Experiments presented in this paper were carried out using the GridUIS-2 experimental testbed, being developed under the Universidad Industrial de Santander (UIS) High Performance and Scientific Computing Unit development action with support from UIS Vicerrectoria de Investigacin y Extension (VIE-UIS) and several UIS research groups as well as other funding bodies (see https://grid.uis.edu.co).

References

[1] R Geller 1996 *Electron cyclotron resonance ion sources and ECR plasmas* (Bristol: Institute of Physics)
[2] Yoshinobu Kawai, Hideo Ikegami, Noriyoshi Sato, Akihisa Matsuda, Kiichiro Uchino, Masayuki Kuzuya and Akira Mizuno 2010 *Industrial plasma technology applications from environmental to energy technologies* (Germany: WILEY-VCH)
[3] V D Dougar-Jabon, F A Vivas Mejia and A M Umnov 2000 *Physica Scripta* **62** 183
[4] V D Dugar-Zhabon and M T Murillo Acevedo 2009 *IEEE Transaction on Plasma Science* **38** 3449
[5] S Biri, A Derzsi, Fekete, I Ivn 2007 *High Energy Physics and Nuclear Physics* **31** 165
[6] R Rcz, S Biri and J Plinks 2011 *Plasma Sources Sci Technol* **2** 025002
[7] Z Q Xie 1998 *Review of Scientific Instruments* **69-2** 625
[8] S Biri, A Valek and T Suta, E Takcs, Cs Szab and L T Hudson 2004 *Review of Scientific Instruments* **75-5** 1420
[9] Sndor Biri, Endre Takcs, Richrd Rcz, Larry T Hudson and Jzsef Plinks 2011 *IEEE Transactions on Plasma Science* **39-11** 2494
[10] E Takcs, B Radics, C I Szab, S Biri, L T Hudson, J Imrek, B Juhsz, T Suta, A Valek, J Plinks 2005 *Nuclear Instruments and Methods in Physics Research B* **235** 120
[11] P Grbling, J Hollandt and G Ulm 2002 *Review of Scientific Instruments* **73-2** 614
[12] U Lehnert and G Zschornack 1996 *Review of Scientific Instruments* **67-3** 1264
[13] Z Q Xie, C M Lyneis 1995 *Rev Sci Instrum* **66** 4218

Influence of thermal residual stress on behaviour of metal matrix composites reinforced with particles

R E Guzmán[1] and E Hernández Arroyo[1]
[1] Universidad Pontificia Bolivariana, Floridablanca, Colombia.

E-mail: rolando.guzman@upb.edu.co

Abstract. The properties of a metallic matrix composites materials (MMC´s) reinforced with particles can be affected by different events occurring within the material in a manufacturing process. The existence of residual stresses resulting from the manufacturing process of these materials (MMC´s) can markedly differentiate the curves obtained in tensile tests obtained from compression tests. One of the themes developed in this work is the influence of residual stresses on the mechanical behaviour of these materials. The objective of this research work presented is numerically estimate the thermal residual stresses using a unit cell model for the Mg ZC71 alloy reinforced with SiC particles with volume fraction of 12% (hot-forging technology). The MMC´s microstructure is represented as a three dimensional prismatic cube-shaped with a cylindrical reinforcing particle located in the centre of the prism. These cell models are widely used in predicting stress/strain behaviour of MMC´s materials, in this analysis the uniaxial stress/strain response of the composite can be obtained through the calculation using the commercial finite-element code.

1. Introduction

Metal matrix materials MMC´s show, besides light weight, excellent mechanical behaviour (raised Modulus of Elasticity), also has good wear characteristics and a lower coefficient of thermal expansion than magnesium alloys [1-4]. These materials are generally composed by a metal matrix (Mg or Al alloy based) and reinforcements of fibres or ceramic particles. The Matrix provides the shape to the structural element and transmit stress to the reinforcement, which in turn bears the stress transfer by the matrix [5-7]. This paper reports the results of a numerical investigation of the influence of thermal residual stress on behaviour of metal matrix composite reinforced with ceramic particles (SiC, 12% vol). The global response of a unit cell model [8-10] is then studied under four load conditions; (1) uniaxial tensile loading, with residual stress/strain via a temperature drop; (2) uniaxial tensile loading, without residual stress (3) uniaxial compressive loading, with residual stress, (4) uniaxial compressive loading, without residual stress.

2. Material

Table 1 shows the coefficients of dilatation and specific heats of SiC for both materials.

Due to particles addition in the ZC71-matrix, it is observed a change of the coefficients of thermal expansion and conductivity K. These coefficients will change the temperature increase depending of the material used. The temperature drop that occurs when cooling from the processing temperature down to the working temperature results in large residual thermal stresses/strains due to the large difference in the coefficients of thermal expansion.

The following parameters have been derived from experimental curves σ-ε: Young's modulus *(E)* and elastic limit *(σ_y)* (Figure 1).

Table 1. Thermal properties (source: MELRAM composite, U.K.).

Material	$\alpha\,(K^{-1})$	$c\,(Jkg^{-1}K^{-1})$
SiC	4.3×10^{-6}	840
Mg	26×10^{-6}	1025

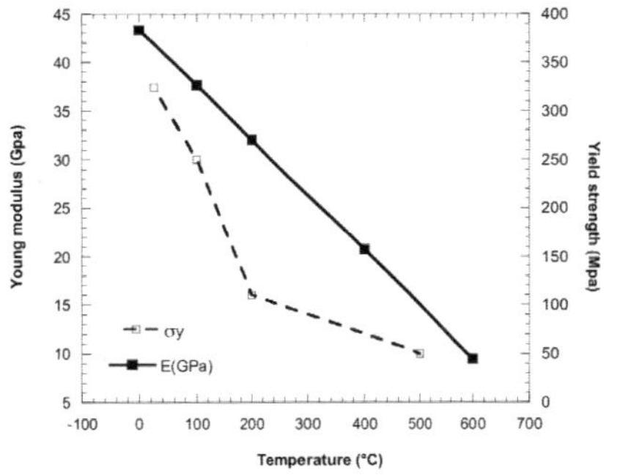

Figure 1. Young's modulus and elastic limit of base alloy.

3. Numerical simulations

A three-dimensional numerical model was used to simulate the deformation process by cooling from the processing temperature 728K (455°C) (the annealing temperature) down to 296K (23°C) (room temperature). The numerical model includes the specimen geometry "cubic unit cell" with the two parts corresponding to the matrix Mg and the particle in contact "Tie". With this numerical model were simulated four load conditions (1) uniaxial tensile loading, without residual stress; (2) uniaxial tensile loading, with residual stress/strain via a temperature drop; (3) uniaxial compressive loading, without residual stress and (4) uniaxial compressive loading, with residual stress. The numerical simulations were performed using the commercial finite-element code ANSYS. Different mesh sizes were considered in order to analyse the mesh sensitivity and to define the optimum one (2240 element and 2745 nodes for the matrix; 7328 element and 3971 nodes for the SiC particle, reduced integration hexahedral elements with eight nodes) (Figure 2(a)). The calculation is performed using a commercial code of finite elements, by implementing the Matrix constitutive law, following a thermoelasto-viscoplastic behaviour.

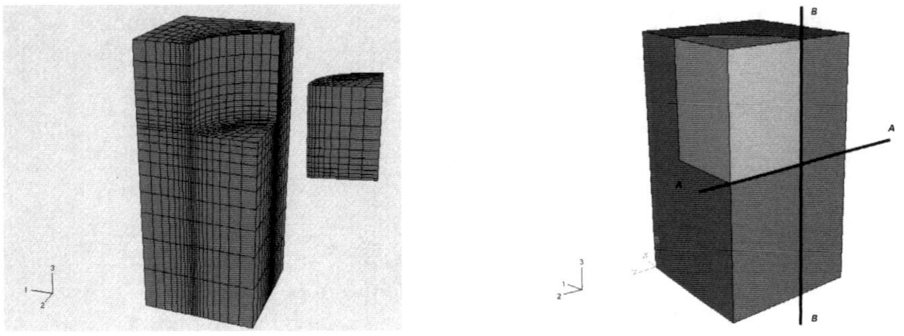

(a) (b)

Figure 2. (a) Meshing numerical model. (b) Part Assembly the unit cell model.

4. Results and discussions

4.1. Residual stress

Figure 2(a), 2(b), displays the zones that were considered to have a more relevant stress distribution. Sections A-A and B-B are in the most singular zone of the unit cell, which is the reinforcement-matrix interface. This zone produces a high concentration of stresses. Figure 3 shows the variation of the equivalent stress in section AA, as a function of the distance to the cell axis. On the left half side of the Figure 3, the corresponding curves of the equivalent stress in the reinforcement and matrix are overlapped. It can also be observed a rapidly increase in the stress of the reinforcement up to the sharp edge, while the variation of stress in the matrix remains almost constant with changes to the distance from the corner, except far from the reinforcement, where there is a steep drop.

Figure 4 shows the variation of the equivalent stress in section BB, as a function of the distance to the cell axis. On the left half side of the figure, the corresponding curves of the equivalent stress in the reinforcement and matrix are overlapped. It can also be observed a rapidly increase in the stress of the reinforcement up to the sharp edge, while the variation of stress in the matrix remains almost constant with changes to the distance from the corner, except far from the reinforcement, where there is a steep drop.

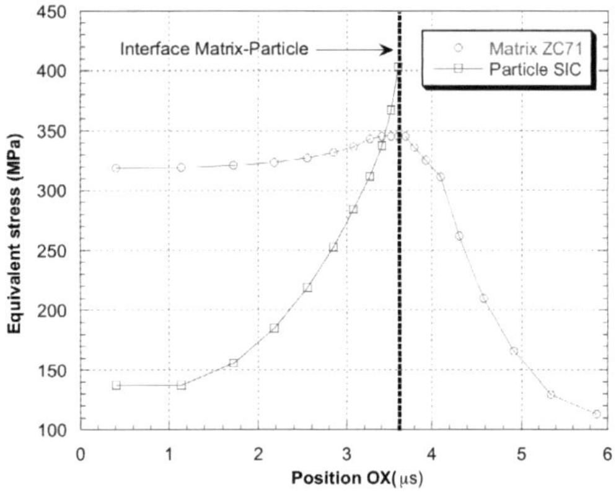

Figure 3. Equivalent stress distribution across AA.

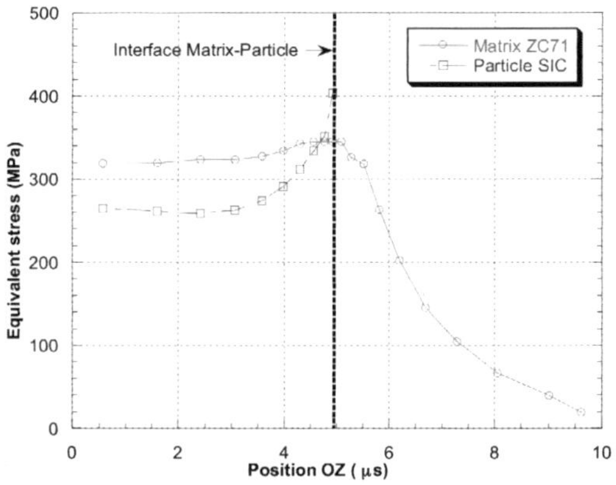

Figure 4. Equivalent stress distribution across BB.

Figure 5 shows the stress σ_{zz} in section AA, as a function of the distance to the cell axis. On the left half side of the graph, the corresponding curves of the stress σ_{zz} in the reinforcement and matrix are overlapped. It can be observed that the stress in the reinforcement and matrix are very similar, and correspond to the applied mutual compression. The stress in the matrix experiences a steep drop from the negative numbers, associated to the matrix-particle compression, to the positive numbers.

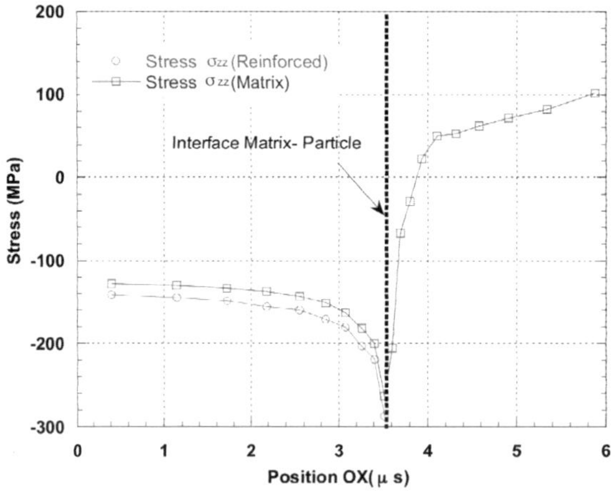

Figure 5. Stress σ_{zz} across AA.

4.2. The influence of the loads conditions

First, a field of residual stresses with thermal origin has been obtained as a result numerical of the cooling phase. Second, it has been applied to the particle model with and without residual stresses of thermal origin, a load condition (load condition 1 and 2) with a value lower than the elastic limit of the material after obtaining the enlargement ΔL of the particle-matrix (cell) set. Table 2 presents values for the elastic modulus obtained in the two mentioned situations (load condition 1 and 2): With and without previous residual stresses (equivalent to have carried out a thermal treatment that could eliminate them). It can be observed that the difference between both values of the elastic modulus is 13%, as a consequence, this difference to the mesoscopic residual stresses of thermal origin.

Table 2. Young´s modulus with without previous residual stresses.

	E (Gpa) tensile	*E (Gpa) compression*
With residual stresses	47	43
Without residual stresses	54	54

In order to provide all the information about the influence of residual stresses on the elastic modulus E, this was also calculated based on the deformation reached after applying to the cell a compression stress with a value lower than the elastic limit. In order to isolate the effect of the field of residual stresses from other mechanical effects (possible difference of the behaviour in tension and compression), the same stress-strain law has been applied to the matrix material. Table 2 shows the values for the elastic modulus obtained in the two mentioned situations (load condition 3 and 4): With and without previous residual stresses (equivalent to have carried out a thermal treatment that could eliminate them). It can be observed that in the absence of residual stresses, the obtained Young's modulus are the same. However, if a previous state of residual stress exists, the values obtained for the elastic modulus are different, thereby obtaining a lower value during a compression test.

The following conclusions of the cell model are drawn from the results presented here:

- Allow to obtain, with notable accuracy, the field of residual stresses in the particle-matrix interface.
- Allow to reflect specific conditions like particle geometry, the proportion in volume, and interface characteristics.
- The numerical values for the mechanical parameters obtained using cell models, clearly show trends and influences of other parameters and conditions. However, there is some disagreement with experimental values. These differences could be due to an incorrect selection of the boundary conditions applied to the model.

References

[1] Y El-Saeid Essa 2000 *Caracterización mecánica de materiales compuestos de base aleaciones de magnesio* (España: Universidad Carlos III de Madrid)

[2] AA VV 1999 *Composite materials hand book. Metal matrix composites vol 4* (USA: Department of Defence)

[3] Clyne T and Withers P 1993 *An introduction to metal matrix composites* (Cambridge: Cambridge University Press)

[4] J L Pérez-Castellanos, R Guzmán-López, A Rusinek and J Meléndez 2010 *Materials and Design* **31-7** 3259

[5] G Meijer, F Ellyin, Z Xia 2000 *Composites Part B* **31** 29

[6] Corbin S and Wilkinson D 1994 *Acta Metall Mater* **42-4** 1311

[7] Wang Z, Chen T, Lloyd D 1993 *Metall Trans A* **24** 197

[8] Llorca J and González C 1998 *Mech Phys Solids* **46-1** 1

[9] Levy A and Papazian J 1990 *Metall Trans A* **21** 411

[10] Levy A 1991 *J Am Ceram Soc* **9-74** 2141

Numerical modeling of high temperature fracture of metallic composites

E D Cendales[1] **and A García**[1]
[1] Fundación Universitaria Los Libertadores, Bogotá, Colombia.
E-mail: edcendalesl@libertadores.edu.co , agarciam04@libertadores.edu.co

Abstract. Mechanical properties of materials are strongly affected by increasing temperature, showing behaviors that could cause failure as creep. This article provides a brief theoretical description about fracture of materials, deepening on creep and intergranular creep. Some parameters as creep strain, strain rate, time to failure and displacement of the crack tip of a metallic glass selected at high temperature were studied. This paper shows a computer numerical model that permits establish mechanical behavior of a metal composite material $Zr_{52.5}Cu_{18}Ni_{14.5}Al_{10}Ti_5$, bulk metallic glass. In the presence of cracking when the material is subjected to temperatures exceeding 30% of the melt temperature of material. The results obtained by computer simulation show correlation with the results about the behavior of the material viewed through the creep test. From the results we conclude that the mechanical properties of the material generally do not undergo major changes at high temperatures. However, at temperatures greater than 650C, the effect of the application of stress during creep entails failures in this kind of material.

1. Introduction

At low temperatures metallic material is strained only if the stress state at a point reaches the yield stress of the material. However, at high temperatures plastic strain is developed with the application during some period of lower levels of stress than yield stress [1]. This process called Creep is developed in crystalline and non-crystalline materials and involves some strain mechanisms mainly involving movement of dislocations in the material [2]. The main aim of this research is to evaluate stress, creep strain, strain rate, and displacement of the crack tip a metal composite material $Zr_{52.5}Cu_{18}Ni_{14.5}Al_{10}Ti_5$, bulk metallic glass.

2. Computational implementation

In this article for the construction of the mesh tetrahedral elements were used, because this type of elements coupled better to the geometry near to the tip of the crack, compared with hexahedral elements of similar sizes. On the tip of the crack mesh refinement was performed, this refinement has the shape of circular rings, which permits to find more realistic values around the tip of the crack. All algorithms used in the mesh are available in the software Comsol Multiphysics, that was used to resolve the state of the stress in each node. Table 1 presents the material propierties used in the simulation. Some of these propierties has been extracted from Shakur et al. [3], and others has been taken from the materials propierties database of the software. Interactions and the type of material used as second phase in the composite material was not considered because it generates higher computational complexity. Similarly, this model differs from CTOD (crack tip opening displacement) due the simulation developed in this study represents a condition of plastic strain at high temperature, aversely to CTOD method that requires the plasticity remains in a small zone in the front of the crack tip.

Table 1. Propierties of the material model used in the simulation.

Mechanical propierties	Value	Units
Elasticity module in X axis	200000	N/mm^2
Poisson coeficient in XY	0.29	$\sim$
Stiffness module in XY	77000	N/mm^2
Mass density	7900	kg/m^3
Yield stress in X	420.51	N/mm^2
Elastic limit	351.57	N/mm^2
Thermal expansion coeficient in X	1.5e-005	K^{-1}
Thermal conductivity in X	47	$W/(m{\cdot}K)$
Specific Heat	420	$J/(kg{\cdot}K)$

2.1. Thermal-elastic model
Computational model is focused in the analysis of stress and strain behavior of the material under study as shows the Equation 1, which relates the external force F aplied on the volume of material V with the variation in the time of the rate of deformation and the internal stresses σ into the the bulk material.

$$\rho \frac{\partial^2 [\mathbf{u}]}{\partial t^2} - \nabla \cdot \sigma = F \cdot V \tag{1}$$

Equation 2 is the general heat conduction equation in the cracked plate coupled with the thermal-elastic deformation gradient. In this equation ρ_o and C_p are the density of the material and the specific heat at constant pressure of the material, respectively. Similarly, the term $\mathbf{K}$ is the conductivity of the material and the heat transfer during deformation Q is coupled with the term for the variation of the thermal deformation in the material under constant external stress applied.

$$\rho_o C_p \frac{dT}{dt} = \nabla \cdot (\mathbf{K} \nabla T) + Q - T_o \left(\frac{\partial \epsilon}{\partial T}\right)\Big|_\sigma \tag{2}$$

Nabarro-Herring high temperature fracture theory is given in the Equation 3, which shows the relation between the nodal displacements and the temperature in the node when the domain is discretized in the single edge crack plate.

$$\epsilon_{\mathrm{NH}} = \frac{1}{2}\left[(\nabla \cdot \mathbf{u})^T + \nabla \cdot \mathbf{u}\right] \tag{3}$$

3. Results and analysis
The displacement of the crack tip was established based on the computational simulation and is shown in Figure 1(a) in which is noted as this parameter varies as a time function measured in hours at different temperatures for the same condition of remote stress applied over cracked flat plate. From the results shown in Figure 1(a) can be established that under stress $\sigma = 0.5\sigma_y$ at temperatures of 600°C and 650°C presents the highest values of displacement for the crack

tip in function of the time due to at this temperature the material flows more readily than to lower temperatures. Figure 1(b) illustrates the behavior of the level of applied stress and the time required for breaking the cracked metallic material. From the Figure 1(b) can be inferred that for a established value of stress applied on the cracked plate at a higher temperature, the material reaches the most unfavorable condition; because for a given level of stress, the material require less time to reach additional cracking of the plate.

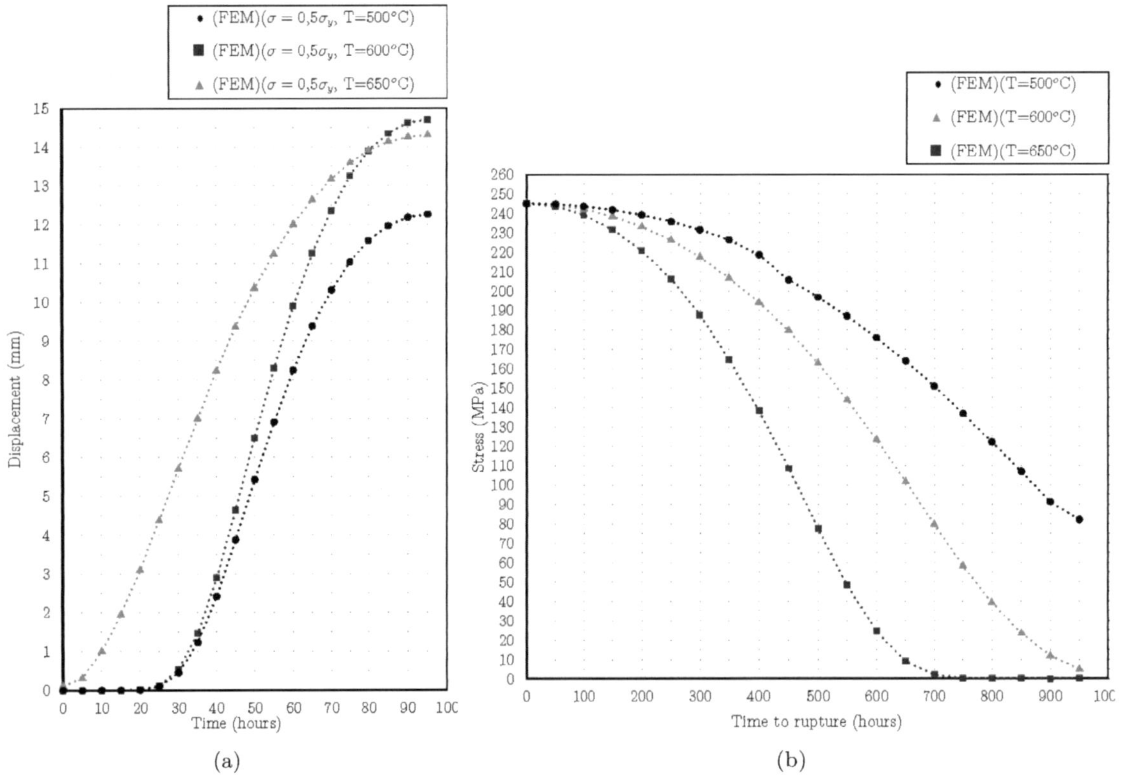

Figure 1. (a) Crack tip displacement and (b) Stress level σ_{xx} in front of the crack tip ($\theta = 0$) as a time function.

Temperatures near 600°C present load conditions more unfavorable, the analysis was developed to establish the strain behavior due action of stress and temperature on the material. Thus, Figure 2(a) illustrates the behavior of the creep strain of the crack tip at a specific temperature (600°C) and stress values of $\sigma = 0.5\sigma_y$, $\sigma = 0.75\sigma_y$ y $\sigma = 0.95\sigma_y$. Based on these results can be established that at this temperature and higher applied stress the highest values for the creep strain of the crack tip are reached. Moreover, Figure 2(b) shows the strain rate of the crack tip as a function of the temperature with respect to the applied stress. In Figure 2(b) is observed than higher temperature a higher strain rate is obtained for several remote stresses applied on the cracked material.

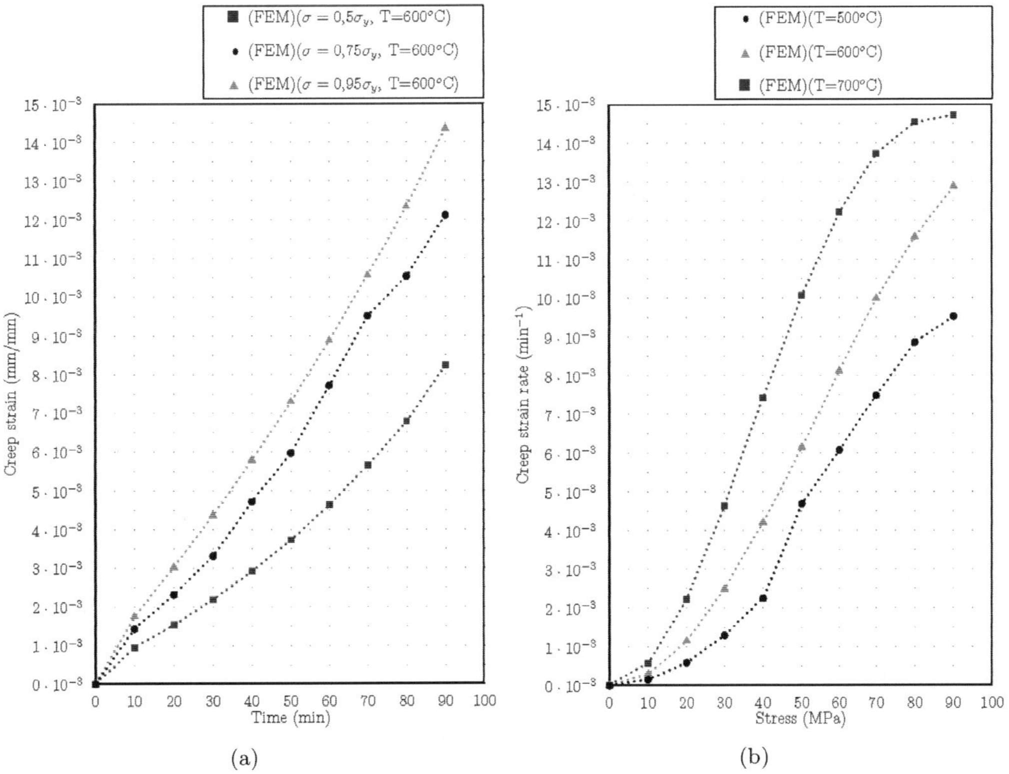

Figure 2. (a) Strain as a time function for several stress levels and (b) Strain rate as a function of material temperature.

4. Conclusion

This study presented a finite element method that has been implemented based closely on the Nabarro-Herring high temperature fracture theory [2]. The implementation is able to evaluate CTOD (crack tip opening displacement), stress, creep strain and creep strain rate directly without any requirement for aditional calculations for 3D simulations. Results shows convergency between experimental results established in the literature, [4] and the results obtained through FEM method. However is evident that more refinement in the FEM model must be established as those present in a XBEM implementation which presents more accurancy than FEM method on the evaluation of mechanical parameters. Nevertheless XBEM present higher complexity in the computational implementation for that reason FEM could be used as a first estimation for these parameters used in dimensioning and desing of mechanical systems involving composite materials.

References

[1] Gilman J 1969 *Micromechanics of flow in solids* (New York: McGraw-Hill)
[2] Nabarro F 1987 *Theory of crystal dislocations* (New York: Dover Publications)
[3] Shakur H, Scudino S, Khn U and Eckert J 2014 *Materials and Design* **59** 241
[4] Wang J, Cao W, Wang L, Guan S, Huang L and Li R 2015 *Journal of Alloys and Compounds* **637** 5

Phase diagram of the Kondo lattice model with a superlattice potential

J Silva-Valencia[1], R Franco[1] and M S Figueira[2]

[1] Universidad Nacional de Colombia, Bogotá, Colombia.

[2] Universidade Federal Fluminense, Niterói, Rio de Janeiro, Brasil.

E-mail: jsilvav@unal.edu.co

Abstract. We study the ground state of a Kondo lattice model where the free carries undergo a superlattice potential. Using the density matrix renormalization group method, we establish that the model exhibits a ferromagnetic phase and spiral phase whose boundaries in the phase diagram depend on the depth of the potential. Also, we observed that the spiral to ferromagnetic quantum phase transition can be tuned by changing the local coupling or the superlattice strength.

1. Introduction

Over the last few decades, cold atom setups have allowed the observation of Bose-Einstein condensates [1], BEC-BCS crossover of an attractive Fermi gas [2], and quantum phase transitions [3]. Now it is possible to emulate several condensed matter Hamiltonians with these setups, with the advantage of having full control over the parameters (kinetic energy, interactions, and density) and access to various dimensions [4].

The interplay between orbital and spin degrees of freedom is a crucial issue that determines the physical properties of manganese oxide perovskites and heavy-fermion materials [5]. Recently there has been interest in studying this competition in cold atom setups. The first proposal was made by Gorshkov et al. [6], who suggested the use of alkaline-earth atoms confined in optical lattices. Bauer et al. [7] proposed the use of a mixture of ^{40}K and ^{23}Na to generate a Kondo-correlated state, showing that the Kondo energy scale is accessible through current experimental techniques. Also, a new scheme for obtaining a SU(3) orbital Kondo effect with ultracold atoms was suggested by Nishida [8].

Recently, using ^{173}Yb atoms loaded into a 3D optical lattice, it was possible to observe a SU(6) Mott insulator state [9]. Also, Zhang et al. [10] achieved the first spectroscopic observation of SU(10) symmetry with ^{87}Sr atoms confined in a 2D optical trap. A one-dimensional realization of strongly-correlated liquids of ultracold fermions interacting repulsively within SU(N) symmetry, with a tunable number N of spin components, was reported by Pagano et al. [11].

Recently, Scazza et al. [12] confined ^{173}Yb atoms in two meta-stable electronic states and measured the spin-exchange contact interactions. This development will allow the emulation of two-orbital problems with cold atoms, such as for example the Kondo lattice model. An extension of this model with a harmonic potential for the free carries was studied recently [13,14], and it was found that in the ground state different insulating and metallic regions can coexist.

Journal of Physics: Conference Series **687** (2016) 012062 doi:10.1088/1742-6596/687/1/012062

2. Model

In the present paper, we consider ^{171}Yb atoms, which have a total spin of 1/2. Our setup consists of confining these atoms in the $^1S_0$ (g) and $^3P_0$ (e) states independently in two different optical lattice potentials with the same periodicity. Furthermore, it is reasonable to consider that there is a high Coulomb repulsion between the e atoms and that the collisions do not cause spin changes. Hence the delocalized g atoms experience an external potential due to confinement, while they interact locally with the localized e atoms through a Heisenberg-type term (see the setup in Figure 1). The Hamiltonian of this new Kondo lattice system is given by,

$$H = -t \sum_{i=1,\sigma}^{L-1} \left(\hat{c}^\dagger_{i,\sigma} \hat{c}_{i+1,\sigma} + \text{h.c.} \right) + \sum_{i=1}^{L} \left(J \mathbf{s}_i \cdot \mathbf{S}_i + V_i \hat{n}_i \right) \tag{1}$$

Where, $\hat{c}^\dagger_{i\sigma}$ creates an atom at site i in the electronic state $^1S_0$ and (nuclear) spin state σ. $\mathbf{S}_i$ is a localized spin-1/2 operator, $\hat{n}_i = \sum_\sigma \hat{c}^\dagger_{i\sigma} \hat{c}_{i\sigma}$, and $\mathbf{s}_i = \frac{1}{2} \sum_{\alpha\beta} \hat{c}^\dagger_{i,\alpha} \boldsymbol{\sigma}_{\alpha\beta} \hat{c}_{i,\beta}$ is the delocalized atom spin-density operator, $\boldsymbol{\sigma}$ being the vector of Pauli matrices. L is the lattice size, and J is the antiferromagnetic local exchange coupling. We fix our energy scale taking $t = 1$.

To observe new ingredients in the magnetic phase diagram of the Kondo lattice model, we consider that the free carries undergo a superlattice potential, which has the value $V_i = V$ for odd sites and $V_i = 0$ for even sites. Here V denotes the shift in the energy levels for each site.

For half-filling, this confinement potential leads to a quantum phase transition from a Kondo insulator spin liquid state without potential to a charge-gapped antiferromagnetic state with nonzero potential [15]. Now we will concentrate on fillings far away from half-filling and explore the ground state for low local couplings as a function of the superlattice potential.

To study the Hamiltonian (1), we considered $m = 600$ states per block within the finite-size algorithm of the density matrix renormalization group method (DMRG) [16] taking into account open boundary conditions. The discarded weight was around 10^{-6} or less.

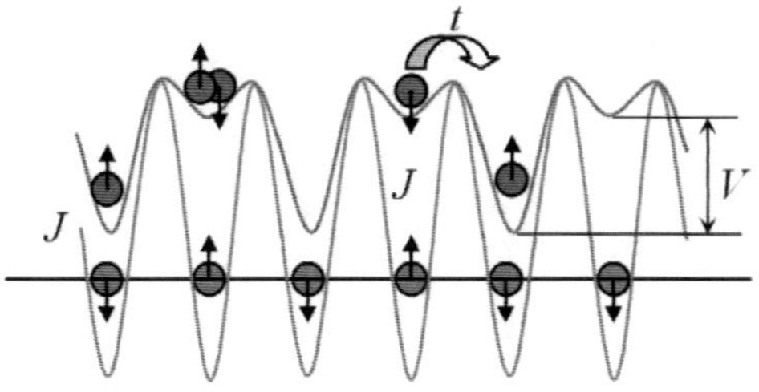

Figure 1. (color online) Schematic of the set-up. g atoms are red, and undergo a superlattice potential; e atoms are blue, and are fixed at each site.

3. Results

Today, it is well known that the one-dimensional Kondo lattice model ($V_i = 0\ \forall\ i$ in Hamiltonian (1)) exhibits a very rich phase diagram away from half filling. For a large local coupling, the ground state is ferromagnetic, whereas for small couplings the local spins exhibit a special magnetic order called the "island" phase by Garcia et al. [17]. Between these phases there is a spiral phase, where the magnetic order is unclear. The main difference between these phases is the magnetic order, which can be determined by calculating the spin structure factor $S(q) = \frac{1}{L} \sum_{j,k} e^{iq(j-k)} \langle \mathbf{S}^T_j \cdot \mathbf{S}^T_k \rangle$ of the ground state, where $\mathbf{S}^T$ is the total spin (free plus local) operator. This quantity exhibits a maximum at $q = \rho\pi$ ($q = 0$) in the island (ferromagnetic) phase, ρ being the density of the free carries. If we start with a certain phase and begin to change the parameters of the system, a critical point will be the parameter value for which $S(q)$ reaches its maximum value at the characteristic position of a different phase.

In Figure 2 (a), we show the spin structure factor for the original Kondo lattice model for $J = 1$, and it exhibits a maximum at $q/\pi < \rho$; therefore the ground state of the system is spiral. If we keep the local coupling and turn on the superlattice potential, for instance $V = 1$, we note that the maximum of $S(q)$ moves to zero, indicating that the ground state is ferromagnetic (Figure 2(b)). The superlattice potential causes a phase transition from the spiral to the ferromagnetic phase. Fixing the superlattice potential and decreasing the local coupling, we expected that the size of the singlet formed by delocalized and localized atoms would increase and the ferromagnetic order would not be favored, which we can observe in Figure 2(c) for $J = 0.8$, where the spin structure factor indicates a spiral ground state. In this case, another small peak appears, indicating an additional underlying magnetic order in the system. Now, the phase transition is due to the local coupling. The superlattice potential favors the localization of the atoms; therefore, the formation of singlets, i. e. a further increase in the superlattice potential, brings the system back to a ferromagnetic phase (Figure 2(d)). The above discussion allows us to conclude that the system undergoes a quantum phase transition from the spiral to a ferromagnetic state, which can be tuned by varying the local coupling or the potential strength. We believe that this phase transition is continuous.

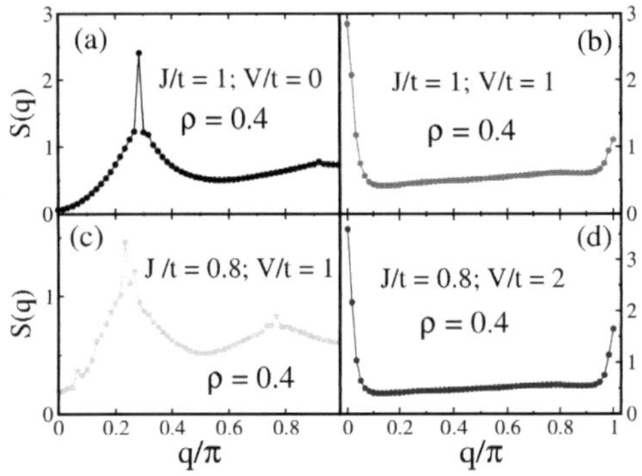

Figure 2. (color online) Spin structure factors for $\rho = 0.4$ with different parameters J/t and V/t.

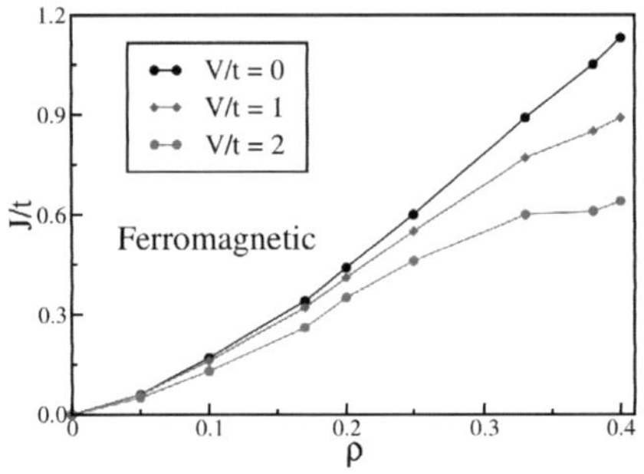

Figure 3. Phase diagram as a function of the density for $V = 0, 1$ and 2. The lines are visual guides.

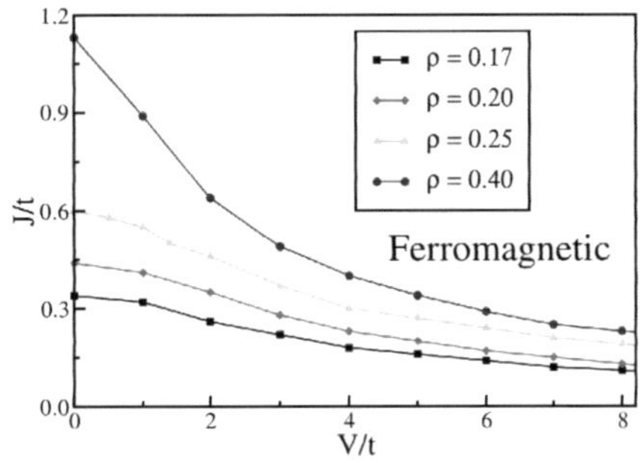

Figure 4. Phase diagram as a function of the strength. The lines are visual guides.

The boundary between the spiral and the ferromagnetic phase of the original Kondo lattice model is shown in Figure 3 (black points). As is well known, the critical coupling increases with the density [18]. Also, we show the boundary between these phases for $V = 1$ (green points) and $V = 2$ (red points). For both superlattice potentials, we observed that the critical points take low values for any density. This tendency increases as the overall density rises. In Figure 4, we plot the local critical coupling as a function of the superlattice strength for some densities.

We observe that the critical points take lower values as the potential increases; i.e., for a fixed density the critical points decrease monotonously, separating the ferromagnetic and the spiral phases. This behavior is expected, due to the localization generated by the potential, which reduces the spatial extension of the singlets and leads to a lower value of the local coupling in order to obtain ferromagnetism. For any density, we note that the decrease of the critical points is greater for low- and intermediate -strength potentials; however, for higher values, the critical points vary very slowly, indicating the difficulty of redistributing the carries in the system.

4. Conclusions

Using the density matrix renormalization group method, we studied the ground state of a Kondo lattice model with a superlattice potential, which can be emulated in an optical lattice setup. For densities lower than $\rho = 0.4$, we found that the spiral to ferromagnetic phase transition can be tuned by varying the local coupling or the superlattice potential. We observed that all phases are robust enough to survive at the thermodynamic limit.

We conclude that the external superlattice potential will determine the ground state once the density and the local coupling are defined. The critical points that separate the different phases of this model decrease monotonously as a function of the potential strength, reflecting the localization effect. Our calculations lead us to think that these transitions are of second order, although a bosonization calculation can help to confirm this.

Acknowledgments

The authors are thankful for the support of DIB-Universidad Nacional de Colombia, and the Brazilian agency CNPq.

References

[1] Anderson M H, Ensher J R, Matthews M R, Wieman C E and Cornell E A 1995 *Science* **269** 198
[2] Navon N, Nascimbène S, Chevy F and Salomon C 2010 *Science* **328** 729
[3] Greiner M, Mandel O, Esslinger T, Hänsch T W and Bloch I 2002 *Nature* **415** 39
[4] Bloch I, Dalibard and Nascimbéne J S 2012 *Nat Phys* **8** 267
[5] Tsunetsugu H, Sigrist M and Ueda K 1997 *Rev Mod Phys* **69** 809
[6] Gorshkov A V, Hermele M, Gurarie V, Xu C, Julienne P S, Ye J, Zoller P, Demler E, Lukin M D and Rey A M 2010 *Nat Phys* **6** 289
[7] Bauer J, Salomon C and Demler E 2013 *Phys Rev Lett* **111** 215304
[8] Nishida Y 2013 *Phys Rev Lett* **111** 135301
[9] Taie S, Yamazaki R, Sugawa S and Takahashi Y 2012 *Nat Phys* **8** 825
[10] Zhang X, Bishof M, Bromley S L, Kraus C V, Safronova M S, Zoller P, Rey A M and Ye J 2014 *Science* **345** 1467
[11] Pagano G, Mancini M, Cappellini G, Lombardi P, Schäfer F, Hu H, Liu X-J, Catani J, Sias C, Inguscio M and Fallani L 2014 *Nat Phys* **10** 198
[12] Scazza F, Hofrichter C, Höfer M, De Groot P C, Bloch I and Fölling S 2014 *Nat Phys* **10** 779
[13] Foss-Feig M, Hermele M and Rey A M 2010 *Phys Rev* A **81** 051603
[14] Silva-Valencia J and Souza A M C 2012 *Phys Rev A* **85** 033612
[15] Silva-Valencia J, Franco R and Figueira M S 2013 *Phys Lett A* **377** 643
[16] White S R 1992 *Phys Rev Lett* **69** 2863
[17] Garcia D J, Hallberg K, Alascio B and Avignon M 2004 *Phys Rev Lett* **93** 177204
[18] Gulacsi M 2004 *Advances in physics* **53** 769

IMRMPT2015 IOP Publishing
Journal of Physics: Conference Series **687** (2016) 012063 doi:10.1088/1742-6596/687/1/012063

Spiral to ferromagnetic transition in a Kondo lattice model with a double-well potential

R C Caro[1], R Franco[1] and J Silva-Valencia[1]
[1] Universidad Nacional de Colombia, Bogotá, Colombia.
E-mail: jsilvav@unal.edu.co

Abstract. Using the density matrix renormalization group method, we study a system of ^{171}Yb atoms confined in a one-dimensional optical lattice. The atoms in the $^1S_0$ state undergo a double-well potential, whereas the atoms in the $^3P_0$ state are localized. This system is modelled by the Kondo lattice model plus a double-well potential for the free carries. We obtain phase diagrams composed of ferromagnetic and spiral phases, where the critical points always increase with the interwell tunneling parameter. We conclude that this quantum phase transition can be tuned by the double-well potential parameters as well as by the common parameters: local coupling and density.

1. Introduction

Heavy-fermion materials exhibit a rich variety of ground states, some of these being magnetic and others not. To explain some properties of these materials, the scientific community has used the periodic Anderson model and the Kondo lattice model [1]. The main problem for the experimental study of these materials is the absence of control of the parameters, which has stimulated a search for new ways to test the ideas involved in the above models, where the charge and spin degrees of freedom compete. Taking into account the success of cold-atom setups for studying quantum phenomena such as the superfluid-insulator quantum phase transition, Gorshkov et al. proposed using alkaline-earth atoms confined in cold-atom setups to emulate the Hamiltonians that describe the physical properties of heavy-fermion materials [2].

Recently, Nishida [3] suggested a new scheme for obtaining a SU(3) orbital Kondo effect with ultracold atoms, and Bauer et al. [4] proposed the use of a mixture of ^{40}K and ^{23}Na to generate a Kondo-correlated state. Also, the realization of a one-dimensional system of fermions interacting repulsively with SU(N) symmetry, where N can be tuned [5] and the spin-exchange contact interactions between confined ^{173}Yb atoms can be measured [6], has stimulated study of this subject.

Foss-Feig et al. [7] and Silva-Valencia et al. [8] follow Gorshkov's proposal and consider that ^{171}Yb atoms in $^1S_0$ (g) and $^3P_0$ (e) states can be confined independently in two different optical lattice potentials with the same periodicity. If the collisions do not cause spin changes and the strong repulsion between e atoms is taken into account through a unit-filling constraint, then it is possible to simulate the Kondo lattice model [9] plus an external potential to the free carries. Considering a quadratic potential, they show that metallic regions and Kondo insulator domains with one g atom at each site coexist.

2. Model and results

^{171}Yb atoms in a double-well potential [10] can be described by considering a kinetic term for the free atoms, which undergo the external potential, and a local interaction term between local and free atoms. Therefore the Hamiltonian is given by

$$H = -\sum_{i,\sigma}\left(\hat{c}^{\dagger}_{2i-1,\sigma}\hat{c}_{2i,\sigma} + t\hat{c}^{\dagger}_{2i,\sigma}\hat{c}_{2i+1,\sigma} + \text{h.c.}\right) + \sum_{i}\left(J\mathbf{s}_i\cdot\mathbf{S}_i + V_i\hat{n}_i\right) \quad (1)$$

Where, $\hat{c}^{\dagger}_{i\sigma}$ is the creation operator for an atom at site i in the electronic state $^1S_0$ and (nuclear) spin state σ. $\mathbf{s}_i$ and $\mathbf{S}_i$ are spin-1/2 operators for free and localized atoms, respectively, and $\hat{n}_i = \sum_\sigma \hat{c}^{\dagger}_{i\sigma}\hat{c}_{i\sigma}$ is the density operator. The shift in the energy levels between neighbor sites V_i is zero for odd sites and V for even sites. The intrawell tunneling is our energy scale and equal to one, whereas the interwell tunneling is t. J is the antiferromagnetic local exchange coupling and the lattice size is L.

Figure 1 shows a picture of our experimental setup, where the double-well potential for the free atoms can be created through the superposition of two optical lattices of different frequencies, with the frequency of one optical lattice being double that of the other. Note that V can be tuned by shifting one laser beam with respect to the other, and the inter and intrawell tunnelings can be adjusted independently through careful control of the intensities [11, 12]. Here we consider $t \leq 1$.

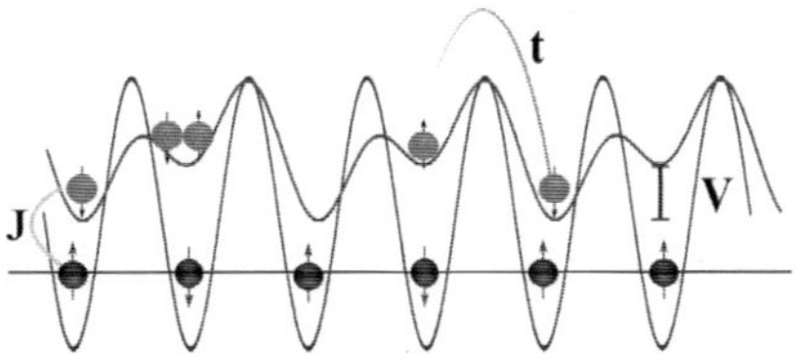

Figure 1. Schematic of the set-up. g atoms are red, and undergo a double-well potential; e atoms are blue, and are fixed at each site. t is the interwell tunneling parameter.

The ground state of the Hamiltonian (1) was studied using the density matrix renormalization group method (DMRG) method [13], taking into account open boundary conditions. We considered $m = 400$ states per block, and the discarded weight was around 10^{-6} or less.

To characterize the different phases of our model, we calculate the spin structure factor $S(q) = \frac{1}{L}\sum_{j,k} e^{iq(j-k)}\langle \mathbf{S}^T_j \cdot \mathbf{S}^T_k\rangle$, where $\mathbf{S}^T_j$ is the total spin at site j. This quantity gives us information about the order of the system. For instance, when the total spin at the sites without singlets points in the same direction, we have a ferromagnetic order, which is reflected by a maximum at the wavenumber zero. An antiferromagnetic order leads to a maximum at π. Our main goal in this paper is to determine the evolution of the quantum phases of the model with the intrawell tunneling t when the other parameters of the model are fixed. The Kondo lattice model Hamiltonian is obtained by equation (1) considering $t = 1$ and $V = 0$. It is well known that for a global density $\rho = 0.25$, the critical point is located around $J_c/t \approx 0.60$, i. e. for values larger than J_c the ground state is ferromagnetic and the spin structure factor has a maximum at $q = 0$. For smaller values, the ground state is "spiral", which means that the magnetic order is unclear, and the localized spins can show diverse alignments along the lattice, characterized by a non-definite position of the maximum of $S(q)$ [14].

Now we consider the double-well potential and the spin structure factor as a function of the wave number, shown in Figure 2, where the global density is constant and equal to $\rho = 0.25$. Considering a local coupling $J = 0.35$ and with $V = 0$, we know that the ground state will be of the spiral type when $t = 1$, and this type remains for $t = 0.6$, as shown in Figure 2(a). When the potential is more deeper ($V = 2$), we observe that the maximum of the spin structure factor is at $q = 0$, which indicates that the ground state is ferromagnetic (see Figure 2(b)).

Fixing the parameter V and decreasing the local coupling, we observe in Figure 2(c) that the maximum at $S(q)$ is offset from zero, indicating a spiral state. Note that the spiral states in Figure 2(a) and (c) are different, i.e. the magnetic order is not the same, and the maximum positions are different. If the depth of potential increases, for instance $V = 4$ in Figure 2(d), the maximum of the spin structure factor is at $q = 0$, indicating that the ferromagnetic state is recovered. We observe that the local coupling and the depth of potential generate localization of the free atoms, which allows the formation of singlets, leading to the possibility of tuning the ferromagnetic to spiral quantum phase transition by changing the local coupling or V when the interwell tunneling and the density are fixed.

Maintaining the depth of potential at $V = 0$, we expect that as t increases the free atoms will tend to move to the neighbor sites, which hinders the formation of singlets, leading to the need to consider larger values of the local coupling for obtaining a ferromagnetic state. This fact is corroborated in Figure 3, where we show the local coupling versus the interwell tunneling for a global density $\rho = 0.25$. For $V = 0$, we note that the ferromagnetic phase decreases as the interwell tunneling grows. The critical points that separate the ferromagnetic and spiral phases for each value of the interwell tunneling grow monotonously. This behavior remains the same for other values of the depth of potential, $V = 2$ and 4. However, the localization effect of the depth of the potential causes the area of the ferromagnetic phase in the phase diagram to increase with the depth of the potential.

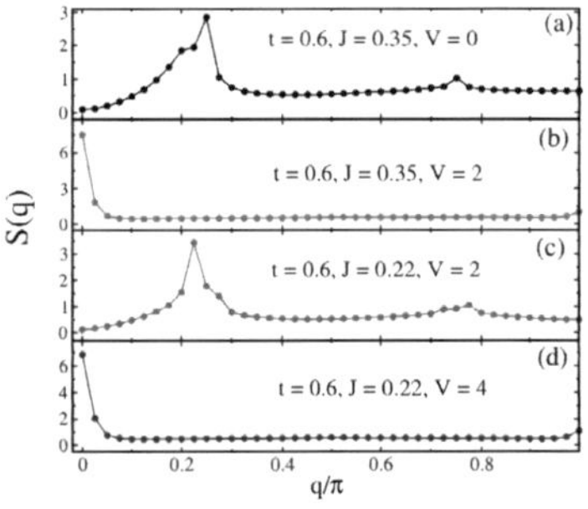

Figure 2. The spin structure factor for different parameters. The global density is $\rho = 0.25$, and $L = 80$.

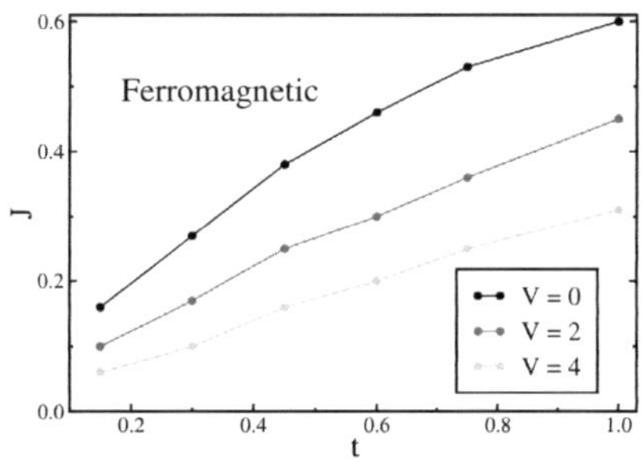

Figure 3. Phase diagram as a function of the interwell tunneling for three values of $V = 0, 2, 4$.

The spin structure factor for atoms in a double-well potential with parameters $t = 0.3$ and $V = 2$ is shown in Figure 4 for different values of the local coupling and the global density. For a constant global density $\rho = 0.25$, we observe that for local couplings $J = 0.1$ and $J = 0.2$ the ground state of the system is spiral and ferromagnetic, respectively, as shown in Figure 4(a) and (b). Keeping the same local coupling as in Figure 4(b) but increasing the number of free atoms, we obtain a larger density, for instance $\rho = 0.40$, and we note that the ground state changes (see Figure 4(c)). Now we have a spiral with a very complex structure: note that none of wavevectors dominate, as happened before for other parameter values. To obtain a ferromagnetic state, we need a larger value of the local coupling when the density is $\rho = 0.40$; for instance, we show in Figure 4(c) that for $J = 0.4$, the maximum of $S(q)$ is at $q = 0$, and the ground state is ferromagnetic.

In Figure 5, we show the phase diagram of the model for the densities $\rho = 0.25$ and $\rho = 0.40$ when the depth of potential is fixed at $V = 2$. We observe that the critical points increase monotonously as a function of the interwell tunneling regardless the global density. For each value of the interwell tunneling, we see that for larger densities a larger local coupling is necessary in order to obtain a ferromagnetic state, because more atoms lead to a larger double occupation probability, which makes the formation of singlets difficult, and we have to localize the atoms by increasing the local coupling. Above the lines in Figure 5, we have a ferromagnetic phase, and below them, we have a spiral phase.

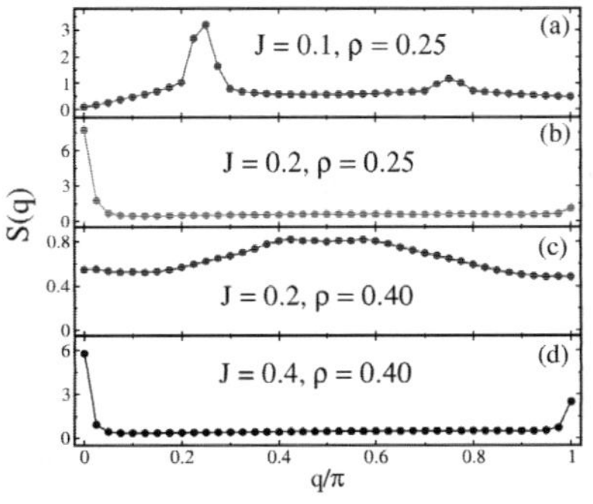

Figure 4. The spin structure factor for different parameters. The double-well parameters are $t = 0.3$ and $V = 2$.

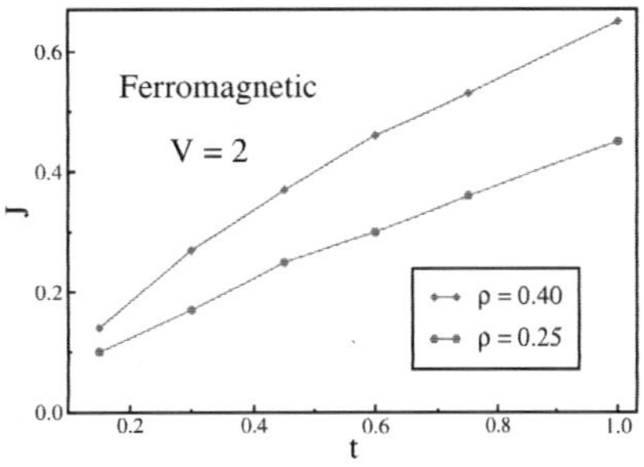

Figure 5. Phase diagram as a function of the interwell tunneling. The double-well parameters are $t = 0.3$ and $V = 2$.

3. Conclusions

Yterbium atoms in the $^1S_0$ (g) state that undergo a double-well potential interacting with localized Yterbium atoms in the $^3P_0$ (e) state were studied using the DMRG method. This one-dimensional system was modeled by the Kondo lattice model plus a double-well potential for the free atoms. For the parameters considered here, we found a ferromagnetic and a spiral phase, which compete in the phase diagram. The ferromagnetic area increases as the density decreases, and the depth of potential grows. The quantum phase transition between these phases can be tuned by varying the local coupling, the depth of the potential, the interwell tunneling, or the density when three of these parameters are fixed.

Acknowledgments

The authors are thankful for the support of DIB-Universidad Nacional de Colombia and COLCIENCIAS (grant No. FP44842-057-2015).

References

[1] Gulacsi M 2004 *Advances in physics* **53** 769
[2] Gorshkov A V et al. 2010 *Nat Phys* **6** 289
[3] Nishida Y 2013 *Phys Rev Lett* **111** 135301
[4] Bauer J, Salomon C and Demler E 2013 *Phys Rev Lett* **111** 215304
[5] Pagano G, et al 2014 *Nat Phys* **10** 198
[6] Scazza F, et al 2014 *Nat Phys* **10** 779
[7] Foss-Feig M, Hermele M and Rey A M 2010 *Phys Rev A* **81** 051603
[8] Silva-Valencia J and Souza A M C 2012 *Eur Phys J B* **85**
[9] Shibata N and Ueda K 1999 *J Phys Condens Matter* **11** R1
[10] Qian Y, Gong M and Zhang C 2013 *Phys Rev A* **87** 013636
[11] Sebby-Strabley J, et al 2007 *Phys Rev Lett* **98** 200405
[12] Trotzky S, et al 2008 *Science* **319** 295
[13] White S R 1992 *Phys Rev Lett* **69** 2863
[14] Garcia D J, Hallberg K, Alascio B and Avignon M 2004 *Phys Rev Lett* **93** 177204

IMRMPT2015 IOP Publishing
Journal of Physics: Conference Series **687** (2016) 012064 doi:10.1088/1742-6596/687/1/012064

Density matrix renormalization group study of the Anyon-Hubbard model

J Arcila-Forero[1], R Franco[1] and J Silva-Valencia[1]
[1] Universidad Nacional de Colombia, Bogotá, Colombia.
E-mail: jsilvav@unal.edu.co

Abstract. Recently optical lattices allow us to observe phase transition without the uncertainty posed by complex materials, and the simulations of these systems are an excellent bridge between materials-based condensed matter physics and cold atoms. In this way, the computational physics related to many-body problems have increased in importance. Using the density matrix renormalization group method, we studied a Hubbard model for anyons, which is an equivalent to a variant of the Bose-Hubbard model in which the bosonic hopping depends on the local density. This is an exact mapping between anyons and bosons in one dimension. The anyons interlope between bosons and fermions. For two anyons under particle exchange, the wave function acquires a fractional phase $e^{i\theta}$. We conclude that this system exhibits two phases: Mott-insulator and superfluid. We present the phase diagram for some angles. The Mott lobe increases with an increase of the statistical. We observed a reentrance phase transition for all lobes. We showed that the model studied is in the same universality class as the Bose-Hubbard model with two-body interactions.

1. Introduction

Experimental setups of cold atoms using optical lattices have the unique ability to tune interactions and density, and because of that they have emerged as unusual laboratories for the realization of Hubbard boson and fermi models, as well as for observing phase transitions without the intrinsic uncertainty posed by materials. The simulations are an incredible bridge between materials and cold atoms: the scale of modern numerical simulations is able to match the scale of experimental cold atom systems, thereby allowing a direct comparison [1].

We used the density matrix renormalization group (DMRG) method. This method was developed by S. White in 1992 [2] and has become a powerful numerical method that can be applied to low-dimensional strongly-correlated fermionic and bosonic systems. Its field of applicability has now been extended beyond condensed matter, and it is successfully used in statistical mechanics and high-energy physics. The DMRG allows for a systematic truncation of the Hilbert space by keeping the most probable states that describe a wave function [3].

Anyons are the third fundamental category of particles. For two anyons under particle exchange, the wave function acquires a fractional phase $e^{i\theta}$, giving rise to fractional statistics with $0 < \theta < \pi$. Greater interest for the study of anyons emerged when the fractional quantum Hall effect observed experimentally had a natural explanation in term of anyons [4]. Another discovery that reinforced this interest was the evidence of superconductor anyon gas [5]. Various experimental setups have been proposed for the creation and detection of anyons, for example creating bosons with conditional-hopping amplitudes using assisted Raman tunneling in an optical lattice [6, 7]. Through other means, anyons in rotating Bose-Einstein condensates consisting of a small number of atoms have been achieved [8].

2. Model and results

Anyons in 1D are defined by the generalized commutation relations $a_j a_k^\dagger - e^{-i\theta sgn(j-k)} a_k^\dagger a_j = \delta_{jk}$ and $a_j a_k = e^{i\theta sgn(j-k)} a_k a_j$, where the operators $a_j^\dagger$, a_j create or annihilate an anyon on site j. We study the Anyon-Hubbard model, which can be described by the Hamiltonian

$$H = -t \sum_j^L \left(a_j^\dagger a_{j+1} + h.c \right) + \frac{U}{2} \sum_j^L n_j (n_j - 1) \quad (1)$$

Where t is the tunneling amplitude connecting two neighboring sites and U is the on-site interaction energy, h.c. denoting the hermitic conjugate. This Hamiltonian describes an interacting gas of anyons in a one-dimensional optical lattice. Let us define the fractional version of a Jordan-Wigner transformation $a_j = b_j exp \left(i\theta \sum_{i=1}^{j-1} n_i \right)$, with $n_i = a_i^\dagger a_i = b_i^\dagger b_i$ the number operator for both particle types, provided that the particles of type b are bosons (b are the boson operator) with $[b_j, b_j^\dagger] = \delta_{ji}$ and $[b_j, b_i] = 0$.

By inserting the mapping, the Hamiltonian can be rewritten in terms of bosonic operators.

$$H = -t \sum_j^L \left(b_j^\dagger b_{j+1} e^{i\theta n_j} + h.c \right) + \frac{U}{2} \sum_j^L n_j (n_j - 1) \quad (2)$$

The mapped, bosonic Hamiltonian thus describes bosons with an occupation-dependent amplitude $e^{i\theta n_j}$ for hopping processes. If the target site j is unoccupied, the hopping amplitude is simply t. If it is occupied by one boson, the amplitude reads $te^{i\theta}$, and so on [6].

We present the results for the Anyon-Hubbard model using the DMRG method. We used the finite-size algorithm for sizes up to $L = 256$; we considered a truncated Hilbert space with five states by site and fixed the density $\rho = N/L = 1$. We kept up to $m = 200$ states per block and obtained a discarded weight around 10^{-8} or less. The first aim was to determine the quantum phase. We considered the chemical potential, the energy for adding (removing) a particle to the system, which is given by $\mu^+ = E_0(L, N+1) - E_0(L, N)$ and $\mu^- = E_0(L, N) - E_0(L, N-1)$ respectively, where $E_0(L, N)$ is the ground-state energy for L sites and N particles. In this we defined the gap by $\Delta\mu = \mu^+ - \mu^-$.

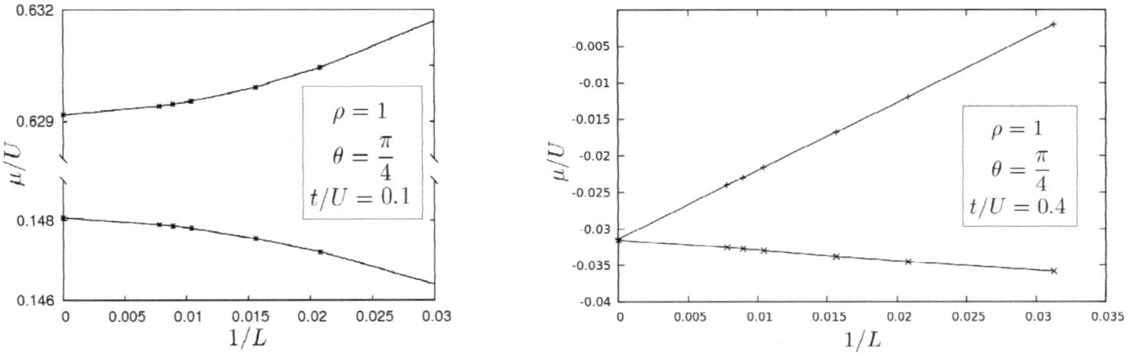

Figure 1. System size dependence on the chemical potential. The upper set of data points corresponds to the particle excitation energy, and the lower one to the hole excitation energy. Points are DMRG results.

The Mott insulating phase has a finite gap for single-particle excitations. At the thermodynamic limit, $N, L \to \infty$ and $\rho = N/L$ integer. A finite gap is expected, i. e., $\Delta\mu = lim_{L,N\to\infty}\Delta\mu(L) > 0$. On the other hand, the superfluid phase is gapless. This is a criterion for determining the type of phase, given the parameters t, U and θ.

We show in Figure 1 (left panel) the energy for adding and removing particles with $\rho = 1$, $\theta = \pi/4$ and $t/U = 0.1$. We can see that the particle (hole) excitation energy decreases (increases) as the lattice size increases; we also observe that the size functional dependence is quadratic. At the thermodynamic limit ($L \to \infty$), we determine their values $\Delta\mu(L) = \mu^+(L \to \infty) - \mu^-(L \to \infty) = 0.63 - 0.15 = 0.48$ considering that the gap is nonzero. We can conclude that the state of the system is Mott insulator. In Figure 1 (right panel), we show the particle and hole excitation energies for the same angle and density but now with $t/U = 0.4$; in this case the functional dependence is lineal and we observe that both quantities coincide at the thermodynamic limit, i.e. the gap disappears, and the ground state is superfluid. In this context, we observe the closing of the gap with an increase of t/U.

We show the phase diagram for different angles in Figure 2. The values of the particle and hole excitation energies were extrapolated at the thermodynamic limit. We observe a Mott insulator phase surrounded by a superfluid phase and we reproduce the well-known phase diagram of the Bose-Hubbard model ($\theta = 0$) [9]. The Mott lobe grows in both directions with an increase of the angle (leading to anyonization of the gas), implying that the critical point position is shifted to the right. Thus the state-dependent hopping helps to localize the particles. This presents the possibility of inducing a quantum phase transition from the superfluid into the Mott insulator phase by changing the statistical angle and not just by the competition between t and U.

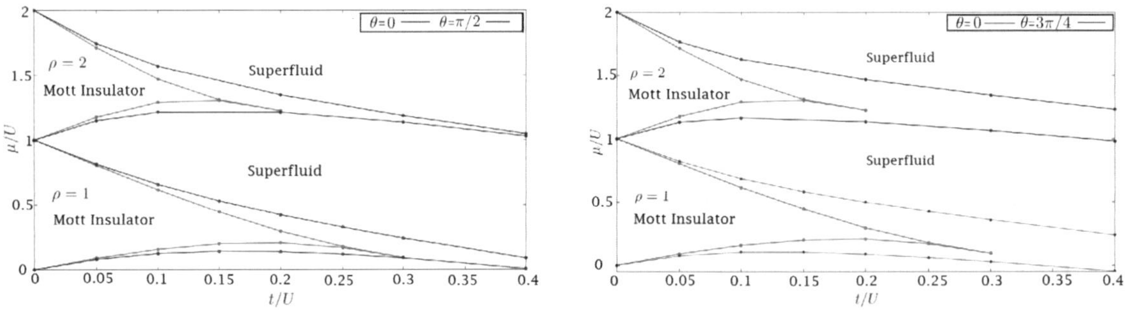

Figure 2. Phase diagram of anyons with statistical angle $\theta = \pi/2$ (Left) and $\theta = 3\pi/4$ (right) for the densities $\rho = 1$ and $\rho = 2$. The lines are visual guides and the points are DMRG results.

For first time using DMRG, we calculate the lobes with $\rho = 2$ for angles $\theta = \pi/2$ and $\theta = 3\pi/4$ (Figure 2). We show the comparison with the bosonic case in the same phase diagram. In order to ensure the high accuracy needed for the considerations below, we truncate the local Hilbert space at $\rho + 5$ bosons in local operators [10]. When we consider a density greater than one, the Mott region increases and the critical point position is shifted to the left for both angles considered. The increase of the density favors the emergence of the superfluid phase. This result contrasts with the previous mean field calculations [6].

We observe a reentrance phase transition in Figure 3. Reentrance is an unusual feature where the phase boundaries of a system exhibit a succession of transitions between two phases A and B, such as $A - B - A - B$, when just one parameter is varied monotonically [11]. In our model, at some suitable constant chemical potential (for example $\mu = 0.1$), the model displays a sequence of quantum phase transitions between the Mott insulator and superfluid phases. In the sequence, we have (A: Mott Insulator and B: Superfluid) a change from an insulator region with $\rho = 1$ to a superfluid region with $\rho < 1$ and then again to an insulator region, and finally the system remains in a superfluid state.

We discussed how the quantum phase transition happens in a similar study of Silva-Valencia and Souza [12]. We can use the critical point and try to fit the gap to a special function. One possibility is the function that describes the Kosterlitz-Thouless transition, for which the gap follows $\Delta\mu = Aexp\left[-k/\sqrt{t_c - t}\right]$, where A and k are constants.

In Figure 4, we show the gap as a function of $t_c - t$ with density $\rho = 1$ and statistical angle $\theta = \pi/4$. In the inset, we observe a linear tendency, which indicates that the Kosterlitz-Thouless behavior is suitable for describing the closing of the gap. We conclude that the model studied is in the same universality class as the usual Bose-Hubbard model.

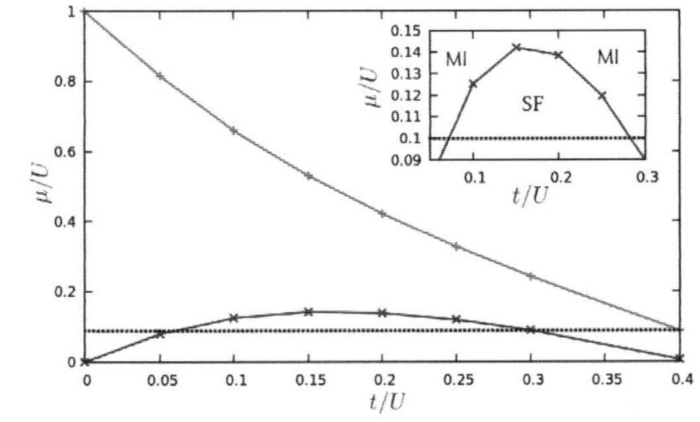

Figure 3. Reentrance behavior at constant $\mu/U = 0.1$ with $\rho = 1$ and $\theta = \pi/2$. Inset: sequence MI-SF-MI.

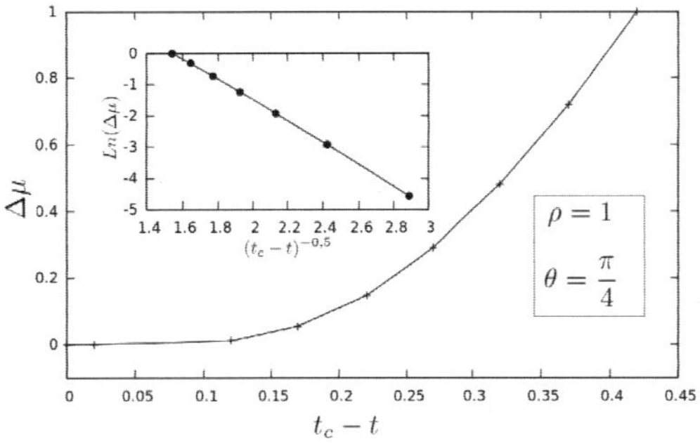

Figure 4. Gap as a function of $t_c - t$. Inset: Adjustments to the Kosterlitz-Thouless transition.

3. Conclusions

We used the density matrix renormalization group method to study the Anyon-Hubbard model. This model exhibits a Mott insulator and a superfluid phase, and we determined the phase diagram for different angles, which increases the Mott insulator region in both directions when increasing the statistical. We observed a reentrance phase transition. The position of the critical point decreases with an increase in the density, in contradiction to previous mean-field results. We show that the gap of the model closes, following the Kosterlitz-Thouless function.

Acknowledgments

The authors are thankful for the support of DIB-Universidad Nacional de Colombia.

References

[1] Duchon E, Loh Y and Triveldi N 2014 *Optical lattice emulators Bose and Fermi Hubbard models* (Novel Superfluids vol 2) ed K Bennemann and J Ketterson (London: Oxford University Press) chapter 17
[2] White S R 1992 *Phys Rev Lett* **69** 2863
[3] Hallberg K 2010 *Adv Phys* **55** 5
[4] Tsui D C, Stormer H L and Gossard A C 1982 *Phys Rev Lett* **48** 1559
[5] Fetter A, Hanna C and Laughlin R 1989 *Phys Rev Lett* **39** 9679
[6] Keilmann T, Lanzmich S, McCulloch I and Roncaglia M 2011 *Nat Commun* **2** 361
[7] Greschner S and Santos L 2015 *Phys Rev Lett* **115** 053002
[8] Paredes B, Fedichev P, Cirac J and Zoller P 2001 *Phys Rev Lett* **87** 010402
[9] Ejima S, Fehske H and Gebhard F 2011 *Europhys Lett* **93** 30002
[10] Carrasquilla J, Manmana S and Rigol M 2013 *Phys Rev A* **87** 043606
[11] Pino M, Prior J, Somoza M, Jaksch D and Clark S R 2012 *Phys Rev A* **86** 023631
[12] Silva-Valencia J and Souza A M C 2011 *Phys Rev A* **84** 065601

Journal of Physics: Conference Series **687** (2016) 012065 doi:10.1088/1742-6596/687/1/012065

Mott insulator and superfluid phases in bosonic superlattices

G J Cruz[1], **R Franco**[2] **and J Silva-Valencia**[2]
[1] Universidad Santo Tomás, Bogotá, Colombia.
[2] Universidad Nacional de Colombia, Bogotá, Colombia.
E-mail: greiscruz@usantotomas.edu.co

Abstract. We study the ground-state phase diagram of boson chains on a 2-period superlattice using the density matrix renormalization group method. New insulators for commensurate densities were found, differentiated by the arrangement of the particles in the unit cell, which was corroborated by analysis of the density versus the potential strength. Also, phase transitions between insulators for $\rho \geq 1$ were seen, and a maximum in the behavior of the von Neumann entropy in the critical region was revealed, which suggests a superfluid phase between the insulators.

1. Introduction

A lot of research in recent years has focused on the ground state of bosonic chains due to the advantage of being able to emulate them in optical lattices. These allow controlling the parameters, such as the kinetic energy, the density, and the particle interactions [1, 2]. Some experiments using optical lattices have established that the phase diagram exhibits a transition from superfluid to Mott insulator [3], which was observed by Greiner et al. in 2002 [4]. An interesting problem is to study the chains of particles in l-period superlattices, whose arrangement is characterized by a unit cell with l sites, where for instance there is an energy difference (λ) between one site and the other $l-1$ sites [5, 6].

Through mean-field approximation, Buonsante and Vezzani [7] described an ultracold system and found that at zero temperature, insulator domains appeared for fractional densities. More recently, Dhar et al. [8] showed new insulators in the 2-period superlattice. They demonstrated insulators at $\rho = 1/2$ and $\rho = 1$, and a phase transition between the Mott insulator and the new insulator phase when λ is near the strength of the local interaction U at $\rho = 1$. However, although the above results describe the behavior for particular densities of one-dimensional boson systems on superlattices, the phases for densities larger than one, the localization of the particles, and the state of the system around the critical point remain unclear.

In the present paper, we show the phases of the ground state of boson chains on 2-period superlattices. Applying the density matrix renormalization group method at the thermodynamic limit, we determined the phase diagram, and new insulators for commensurate densities were obtained. Also, we calculated the local density and identified periodic arrangements of the particles in the superlattice for some values of λ, when the system is in an insulator phase. Finally, we explored the critical region through calculations of the von Neumann entropy, obtaining a maximum at the transition point, which suggests a superfluid phase in that region.

2. Model and results

One-dimensional systems of bosons in superlattices are described by the modified Bose-Hubbard model, defined as:

$$H = -t\sum_i (a_{i+1}^\dagger a_j + H.c.) + \frac{U}{2}\sum_i \hat{n}_i(\hat{n}_i - 1) + \sum_i \lambda_i \hat{n}i \quad (1)$$

Where t is the hopping parameter, $a_i^\dagger$ (a_i) creates (annihilates) a boson at site i, U represents the local interaction, $\hat{n}_i = a_i^\dagger a_i$ is the number operator, and λ_i denotes the shift in the energy levels of the sites in each unit cell. We set our energy scale taking $t = 1$ and the interaction parameter $U/t = 10$. The 2-period superlattice has two sites per unit cell, which is represented in Figure 1.

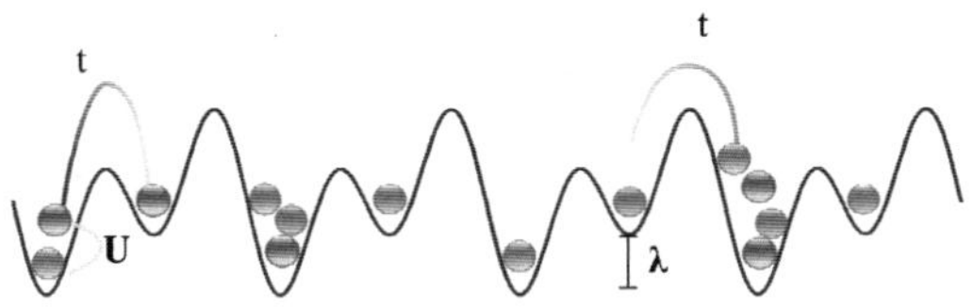

Figure 1. (color online) Schematic representation of the 2-perid superlattice. This is formed by the superposition of two waves, where the first has double frequency than the other. We consider that the difference of the hopping parameter between neighboring states is very small.

In the present paper, $\lambda_i = \lambda$ for even sites, and $\lambda_i = 0$ for odd sites. The current investigation involved calculating the energies $E(N, L)$ for lattices with different lengths L and N, $N + 1$ and $N - 1$ particles such that we obtain the chemical potential for increase (μ_p) and decrease (μ_h) the number of particles by one, where the general expressions for chemical potentials are: $\mu_p = E(N+1, L) - E(N, L)$ and $\mu_h = E(N, L) - E(N-1, L)$. To obtain the energies, we used the density matrix renormalization group method for lattices from 24 to 84 sites, obtaining an error of around 10^{-13}. Then we extrapolated our results at the thermodynamic limit and repeated the process for a wide range of values of the parameter λ, as illustrated by Figure 2. In Figure 2(a), we show the chemical potential at the limit $L \longrightarrow \infty$ for $\rho = 1$ and $\lambda/t = 20$, and we observe an insulator phase due to the presence of the finite gap. In Figure 2(b), we obtain a superfluid phase because the gap is zero when $L \longrightarrow \infty$.

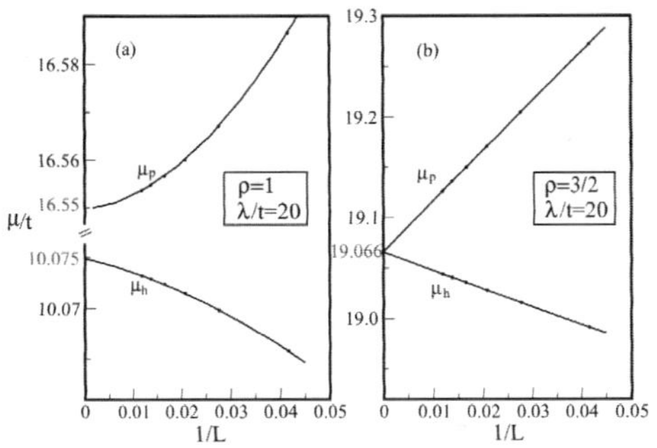

Figure 2. The chemical potential μ/t versus $1/L$, with L the size of the lattice. a) exhibits an insulator state and b) exhibits a superfluid phase at the thermodynamic limit.

Data obtained with the previous methodology allow us to study this system and obtain the following results: Figure 3 illustrates the findings of the density ρ against the chemical potential μ for the 2-period superlattice at values of λ/t between 0 and 24. $U/t = 10$ is taken for comparison with the previous results. The homogeneous system of bosons is represented in Figure 3(a), where plateaus at densities $\rho = 1$ and $\rho = 2$ are exhibited, indicating the size of the energy gap in the Mott insulator phase. For different values of μ/t that do not imply an insulator phase, a compressible phase is achieved and the system is superfluid. Figure 3(b) for $\lambda/t = 10$ shows plateaus at semi-integer densities, while the plateaus at integer densities are fewer than in the previous case. In fact, when λ is near U, the findings suggest a phase transition from a Mott insulator to a new insulator. In Figure 3(c), we show the density for $\lambda/t = 20$. The plateau at $\rho = 3/2$ is not present, and this happens when $\lambda/t \approx 2U$. Finally, the presence of insulators at all commensurate densities (integer and semi-integer) for $\lambda/t = 24$ are given in Figure 3(d). From the above results, it could be inferred that for $\rho > 1$ there may be phase transitions between two types of insulators, passing through a superfluid phase.

Now, we explore the charge distribution for some insulator states by observing the local density along the superlattice as detailed in Figure 4. It can be observed that the boson configuration $\{2, 1, 2, 1, ...\}$ for $\rho = 3/2$ and $\lambda/t = 10$ is shown in Figure 4(a). Data in Figure 4(b) suggest a configuration $\{3, 0, 3, 0\}$ when λ increases. From Figure 4(c), it can be seen that the configuration of particles is around two when $\rho = 2$ and $\lambda/t = 2$, which is itself in a Mott insulator phase. It can be seen in Figure 4(d), with $\rho = 2$ and $\lambda/t = 24$, that the system is organized as $\{3, 1, 3, 1, ...\}$. Accordingly, the boson organization depends on the competition between the parameter λ and the particle interaction U.

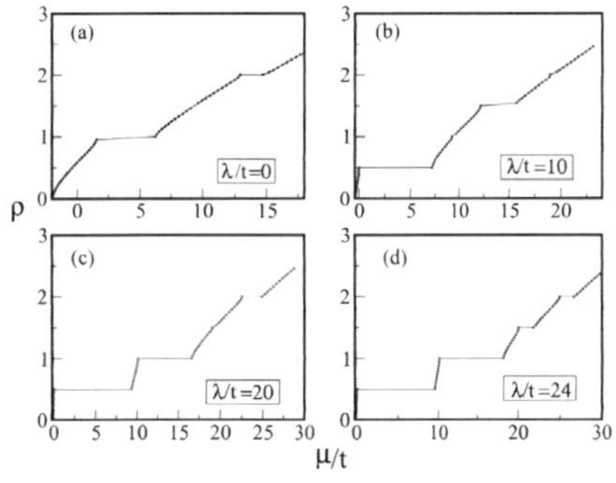

Figure 3. (color online). Density profile ρ versus chemical potential μ for bosons in a 2-period superlattice, with $U = 10$.

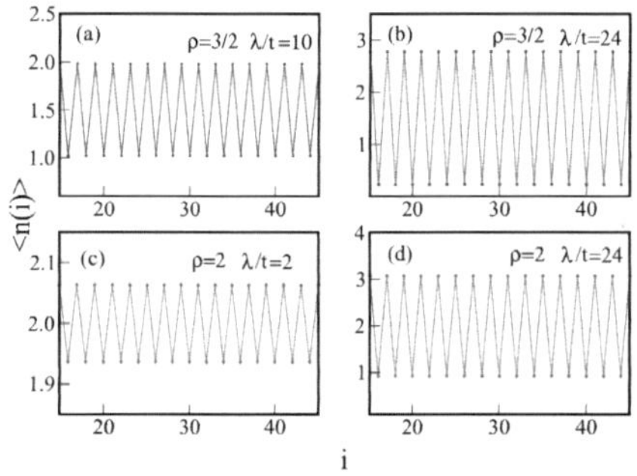

Figure 4. (color online). On-site number density plotted against lattice site index with $L = 50$ sites.

As mentioned earlier, we calculated the phase diagram, and this is shown in Figure 5. In this case, we take the values of λ and μ in U scale. We can observe that the gap size for $\rho = 1/2$ remains constant for large values of λ/U, and the remaining integer and semi-integer densities have insulator regions and transitions between them for values of densities $\rho \geq 1$ with λ values near U (for $\rho = 1$ and $\rho = 2$) and λ near $2U$ (for $\rho = 3/2$).

França and Capelle [9] showed that the average of local von Neumann entropy is able to indicate the critical points in an inhomogeneous system, which is given by: $\epsilon^* = \frac{1}{L}\sum_i \epsilon_{\nu N}(i)$, where $\epsilon_{\nu N}(i) = -Tr\sigma_i \log_2 \sigma_i$ is the local von Neumann entropy, and $\sigma_i = Tr_B\sigma$ is the density matrix of a single site located at i, where B represents the environment with $L - 1$ sites, and σ is the density matrix of the whole system [10]. We can infer a superfluid region between the insulators, as is corroborated by Figure 6. In this plot, the average entropy achieves a maximum for values around the transition region, which implies a significant growth in the degrees of freedom and a superfluid phase. This is shown in Figure 6(a), where the average entropy is calculated for the density $\rho = 1$ with a superlattice of $L = 80$ sites, and it can be observed that the maximum is near $\lambda \approx U$. In Figure 6(b), the average entropy is shown for $\rho = 3/2$ and the maximum is given for λ near $2U$.

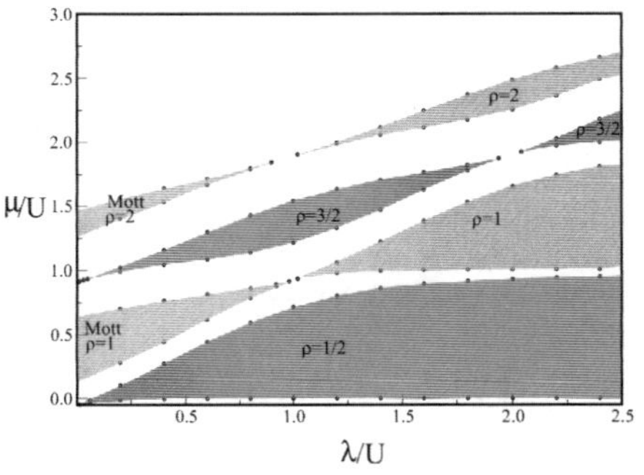

Figure 5. (color online). Phase diagram of the AB chain. The points are DMRG data.

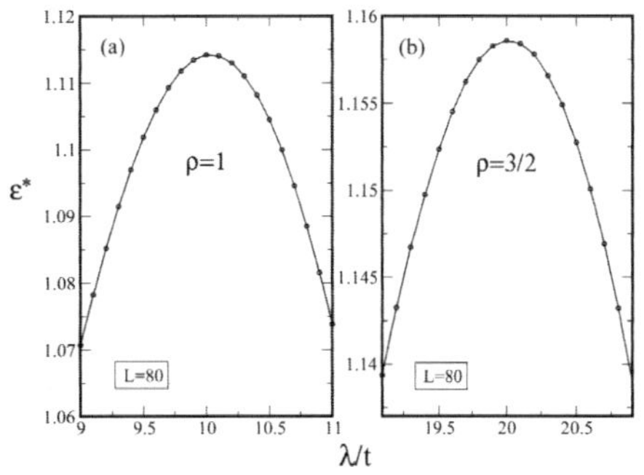

Figure 6. Average von Neumann entropy for a lattice with size $L = 80$ and density (a) $\rho = 1$ and (b) $\rho = 3/2$.

3. Conclusions

Using the density matrix renormalization group method, we determined the chemical potential of the 2-period superlattice at the thermodynamic limit and found the phase diagram. For a small energy difference (λ), we observe that insulator regions with a peculiar charge distribution appear for semi-integer densities, which doesn't occur for integer densities, for which the Mott insulator phase still appears. The size of the new insulator phases for densities less than one increases with λ but then stabilizes, and these phases remain in the phase diagram. On the other hand, for densities $\rho \geq 1$, we always found a superfluid region that separates two insulator phases. For integer densities, the system starts in a Mott insulator phase, but for $\lambda > U$ the system exhibits a new insulator phase. The most surprising result is that the new insulator for $\rho \geq 1$ is unstable with increasing λ, and we obtain an insulator phase for values of $\lambda > U$, which depends on the density. Between the insulator phases, we always have a superfluid phase, a result that was confirmed using von Neumann entropy.

Acknowledgments

Franco and Silva-Valencia are thankful for the support of DIB-Universidad Nacional de Colombia. Cruz acknowledges the support of the Departamento de Ciencias Básicas of the Universidad Santo Tomás.

References

[1] Bloch I, Dalibard J and Zwerger W 2008 *Rev Mod Phys* **80** 885
[2] Bloch I, Dalibard J and Nascimbne S 2012 *Nat Phys* **8** 267
[3] Jaksch, Bruder C, Cirac J, Gardiner C and Zoller P 1998 *Phys Rev Lett* **81** 3108
[4] Greiner M, Mandel O, Esslinger T, Hänsch T W and Bloch I 2002 *Nature* **415** 39
[5] Roati G, D'Errico C, Fallani L, Fattori M, Fort C, Zaccanti M, Modugno G, Modugno M and Inguscio M 2008 *Nature* **453** 895
[6] Atala M, Aidelsburger M, Barreiro J T, Abanin D, Kitagawa T, Demler E and Bloch I 2013 *Nat Phys* **9** 795
[7] Buonsante P and Vezzani A 2004 *Phys Rev A* **70** 033608
[8] Dhar A, Mishra T, Pai R and Das B P 2011 *Phys Rev A* **83** 053621
[9] França V V and Capelle K 2008 *Phys Rev Lett* **100** 070403
[10] Silva-Valencia J and Souza A M C 2012 *Phys Rev A* **85** 033612
[11] Torio M E, Aligia A A, Japaridze G I and Normand B 2006 *Phys Rev B* **73** 115109

Quantum phases of AB_2 fermionic chains

L S Murcia-Correa[1], **R Franco**[1] **and J Silva-Valencia**[1]
[1] Universidad Nacional de Colombia, Bogotá, Colombia.
E-mail: `jsilvav@unal.edu.co`

Abstract. A fermionic chain is a one-dimensional system with fermions that interact locally and can jump between sites in the lattice, in particular an AB_n chain type, where A and B are sites that exhibit a difference in energy level of Δ and site B is repeated n-times, such that the unit cell has $n+1$ sites. A limit case of this model, called the ionic Hubbard model ($n = 1$), has been widely studied due to its interesting physics and applications. In this paper, we study the ground state of an AB_2 chain, which describes the material $R_4[Pt_2(P_2O_5H_2)_4X] \cdot nH_2O$. Specifically, we consider a filling with two electrons per unit cell, and using the density matrix renormalization group method we found that the system exhibits the band insulator and Mott correlated insulator phases, as well as an intermediate phase between them. For couplings of $\Delta = 2, 10$ and 20, we estimate the critical points that separate these phases through the structure factor and the energy gap in the sector of charge and spin, finding that the position of the critical point rises as a function of Δ.

1. Introduction

The effect of the interactions of many particles in quantum systems and superconductivity applications has been shown to be one of the most fascinating problems in physics, because it has allowed understanding the behavior of the electrons in solids, using some models such as the Hubbard model (HM). This model, in general, includes a Coulomb interaction term at short range and another term to quantify the kinetic energy. An example of a material that could be described by this model is vanadium oxide [1]. The Hubbard model has an exact solution in one dimension [2], through the *Bethe Anzatz* method, but it has also been studied using several analytic and numerical methods. Recently, this model has been emulated in optical lattices, allowing precise control of the hopping and interaction parameters, and filling in the Hamiltonian [3].

Other materials, such as tetrathiafulvalene-p-chloranil, can be modelled with a modification of the Hubbard model that considers that there exists a change in energy level of Δ between sites in the lattice. This model is called the ionic Hubbard model, which exhibits the phases: band insulator and Mott insulator, with a ferroelectric phase between them [4].

Trying to generalize the ideas of the ionic Hubbard model, in 2006 Torio et al. proposed a AB_n fermionic chain model, where A is a type of atom with a certain energy level and B is another type of atom at a different site that repeats n times [5]. This type of chain describes systems like halogen-bridged binuclear metal $R_4[Pt_2(P_2O_5H_2)_4X]{\cdot}nH_2O$ [6]. They used different techniques, such as bosonization, perturbation theory, and the exact diagonalization method for small clusters, and determined the phase diagram for AB_2 and AB_3 chains (which exhibits a band insulator phase, a correlated insulator phase, and a ferroelectric insulator phase) for a system of L sites by following the crossing of appropriate excited energy levels, which turns out to be equivalent to the method of jumps in Berry phases. However, their phase diagram does not clearly indicate the critical values where transitions occur, and this is an open question in the literature yet.

Given that experimentalists in the area of cold atoms can construct real one-dimensional lattices and that the ionic Hubbard model in an optical honeycomb lattice has been studied measuring the double occupation [7], this motivates us to study AB_n fermionic chains, in particular the $n = 2$ case, using the density matrix renormalization group (DMRG) method for large lattice sizes.

In this paper, we focus on the transition points that separate the band and Mott insulator phases, using the spin and charge gaps and the structure factors.

2. Model

The Hamiltonian for the AB_2 chain is defined by:

$$\mathcal{H} = -t \sum_{i\sigma} (\hat{c}^{\dagger}_{i+1,\sigma} \hat{c}_{i,\sigma} + h.c.) + U \sum_{i} \hat{n}_{i\uparrow} \hat{n}_{i\downarrow} + \sum_{i} \Delta_i \hat{n}_i \tag{1}$$

Where $\hat{c}^{\dagger}_{i,\sigma}$ creates an electron at site i with spin σ, $n_{i\sigma} = \hat{c}^{\dagger}_{i,\sigma}\hat{c}_{i,\sigma}$, and $n_i = n_{i\uparrow} + n_{i\downarrow}$. The parameter t is the nearest-neighbour hopping amplitude (we take $t = 1$), U is the on-site Hubbard interaction, and Δ_i is the difference in on-site energies. When we talk about the AB_2 chain, we have a unit cell with three sites, so two sites have an energy difference of $\Delta_i = 0$ and the other of $\Delta_i = -\Delta$, and this structure is repeated periodically.

On the other hand, for a fill equal to two electrons per unit cell, we have some limit cases for the AB_2 chain: for $U < \Delta$, the site A is doubly occupied and the site B is empty. This means that physically the lower band is filled and the others are empty, so the system behaves like a band insulator (BI). For $U > \Delta$ and for any value of t, in the Hubbard model we take a Mott insulator (MI) with $n = 1$. In the case of $n = 2$ and $U - \Delta \gg t$, the system adopts a form of modified MI state, called "correlated insulator" (CI), with almost exactly one particle per site and a charge gap of approximately $U - \Delta$ [5].

To study the ground state energy of our system, we calculate the spin (Δ_s) and the charge (Δ_1) gaps defined by

$$\Delta_s = E_0(N,1) - E_0(N,0), \qquad \Delta_1 = E_0(N+1,1/2) + E_0(N-1,1/2) - 2E_0(N,0) \tag{2}$$

Where $E_0(N, S_z)$ is the ground state energy for a lattice of size L with N particles in a spin sector S_z, S_z being the z component of the total spin [4]. With density matrix renormalization group, Δ_s can be calculated by the ground state energy in different subspaces S_z in two different sweets, whereas for Δ_1 we calculate the ground state energy in the subspaces with the corresponding particle numbers. In total, we work with four states.

Also, we decided to calculate the spin $S(k)$ and charge $N(k)$ structure factors, which indicates how the spin and charge are ordered in real space. For the ground state, the spin and charge structure factors can be defined by $S(k) = \frac{1}{L}\sum_{j,l} e^{ik(j-l)} \langle \mathbf{S}_j \cdot \mathbf{S}_l \rangle$ and $N(k) = \frac{1}{L}\sum_{j,l} e^{ik(j-l)} \langle \hat{n}_j \cdot \hat{n}_l \rangle$ respectively, where $\mathbf{S}$ is the spin operator, and $\hat{n}$ is the charge density operator. As for the spin ordering, if the spin structure factor exhibits a maximum at 0, the ordering is ferromagnetic; if the maximum is at π, it is antiferromagnetic. Otherwise, there will be a different ordering, maybe paramagnetic. Finally, for the charge ordering, the first peak in the charge structure factor indicates how much charge there is in the system, and the second peak shows how many electrons there are per unit cell. Therefore, we can determine the charge and spin distribution for our whole system.

3. Results

In order to numerically study the ground state of the AB_2 chain for two electrons per unit cell, we used the density matrix renormalization group method with 7 sweets and 300 states per block, and we obtained an error on the order of 10^{-9} or less, for system sizes $L = 18, 36, 54, 72, 90$, and 108.

In Figure 1, we show the gap difference, $\Delta_1 - \Delta_s$, as a function of U for three different values of Δ. For all Δ values, we observe that there first exists a left-hand region in which the gap difference is equal to zero. This region grows as Δ increases. So the system is in a region type BI (as was noted by Torio [5]), because the spin and charge gaps are equal. Thus we obtain one site with double-occupation and the others with zero occupation. Also, we see that the gap difference increases rapidly within a short range of values of U, and after that there is a small increase in the gap difference, tending to a constant value, which indicates that we are in a zone in which gaps are different and there is a particular ordering in the system. According to this Figure, we will have three phases in the system: a band insulator region, an Mott insulator region, and an intermediate phase between them.

For $\Delta = 10$ and $U = 1$, in Figure 2(a) we are in the band insulator region, which indicates that there are two electrons at one site (site A) and zero electrons at the other (site B), which characterizes this region. In Figure 2(b), we are in the Mott insulator region (with $U = 18$), and we have an electron at site A and another electron between two sites B. Thus we know that the system has two phases with different charge and spin ordering.

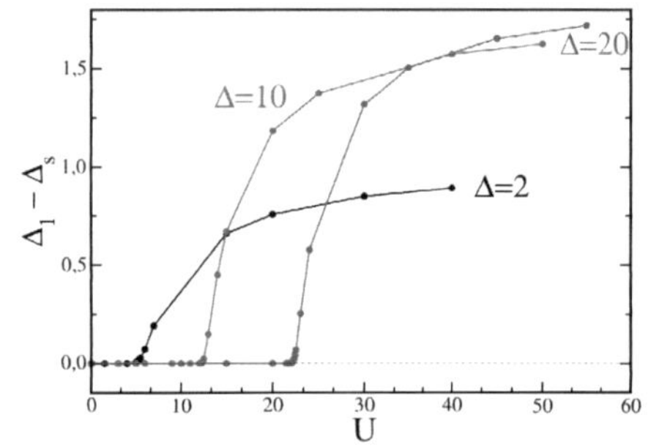

Figure 1. Gap difference $\Delta_1 - \Delta_s$ as a function of U for $L = 72$.

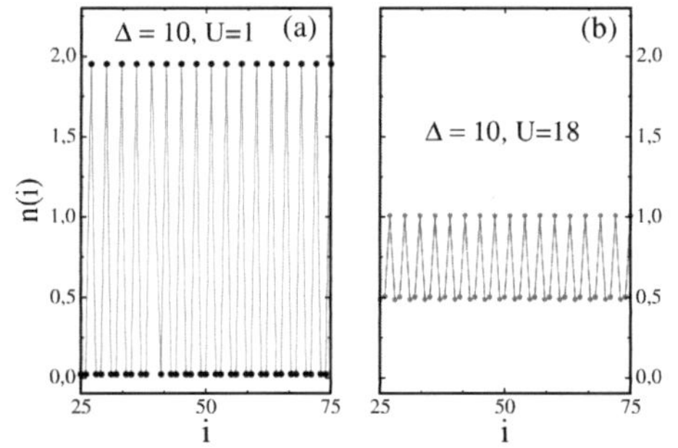

Figure 2. Charge Density for: (a) BI and (b) MI.

In order to separate the phases, we need to find the critical point where the phase transitions occur. So we calculate the spin and charge structure factors. However, the charge structure factor is not shown in our results, because this does not give clear information about where the phase transition is. In Figure 3, we show the spin structure factor as a function of the momentum k/π for $\Delta = 2$ and $\Delta = 20$ with two values of U. We can see in Figure 3(a) that the structure factor shows an increasing and continuous behavior, indicating that there does not exist a particular ordering in the system. If we increase the value of U (for instance $U = 5.3$ in Figure 3(b), we see that the spin structure factor exhibits a maximum at $k = 2\pi/3$, which indicates a particular charge ordering for equal or larger values of U than this. If we consider $\Delta = 20$ for $U = 18$, we are in the intermediate phase, where the spin structure factor $S(k)$ does not exhibit a maximum value (Figure 3(c)), so we don't have a correlated phase. However, if we increase the value of U to 22.4 or larger, the behavior changes, indicating that we have a maximum value of $S(k)$, as we see in Figure 3(d).

We observe that the point where we have a maximum in the spin structure factor $S(k)$ depends on the size of the lattice. In order to find the transition point, we have to extrapolate to the

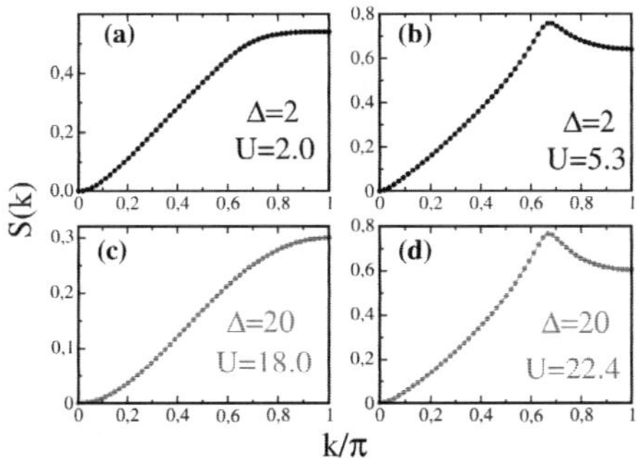

Figure 3. The spin structure factor $S(k)$ of the spin-spin correlation function for system size $L = 108$.

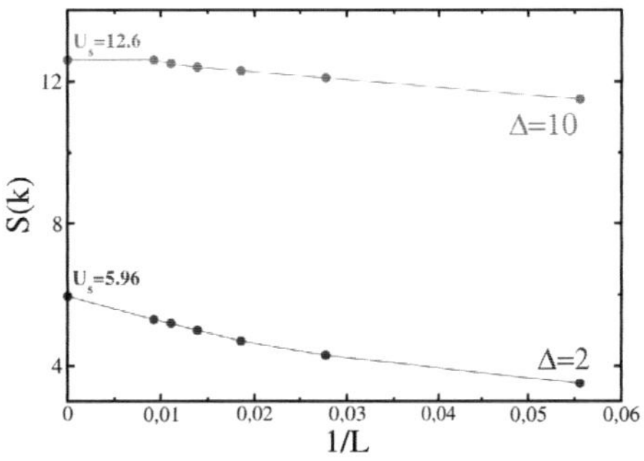

Figure 4. The spin structure factor $S(k)$ as a function of $1/L$ for $L = 18, 36, 54, 72, 90$.

thermodynamic limit, as in Figure 4. We can clearly see that as the size of the lattice grows, the transition point rises, and finally, using the extrapolation, we found, for $\Delta = 2$, that the critical point is $U_s = 5.96$. However, the value determined by Torio et al. was $U_s \approx 8.2$ for $\Delta = 2$. As can be seen, our values and theirs diverge, which reflects the finite size effects, because they used small lattices, while our results indicate the transition points at the thermodynamic limit. Also, the values for $\Delta = 10$ behave in a manner similar to $\Delta = 2$. We found the transition point at $U_s = 12.6$, so we can say that at the thermodynamic limit, the critical points will exhibit a crescent behavior as function of Δ.

4. Conclusions

We studied the AB_2 chain model using the density matrix renormalization group method and we were able to clearly identify two phases: a band insulator and a Mott correlated insulator. Therefore, we speculated that there is an intermediate phase between them.

We corroborated that the gap difference is zero for the band insulator phase. Also we showed that, the spin structure factor allows calculating the transition point where the Mott correlated phase begins.

Finally, we determined the transition points at the thermodynamic limit, and we saw that the position of the points rises as a function of Δ.

Acknowledgments

The authors are thankful for the support of DIB-Universidad Nacional de Colombia.

References

[1] Demishev S V, Chernobrovkin A L, Glushkov V V, Grigorieva A V, Goodilin E A, Ohta H, Okubo S, Fujisawa M, Sakurai T, Sluchanko N E, Samarin N A and Semeno A V 2011 *Phys Rev B* **84** 094426
[2] Bethe H 1931 *Zeitschrift für Physik* **71** 205
[3] Esslinger T 2010 *Annu Rev Condens Matter Phys* **1** 129
[4] Manmana S R, Meden V, Noack R M and Schönhammer K 2004 *Phys Rev B* **70** 155115
[5] Torio M E, Aligia A A, Japaridze G I and Normand B 2006 *Phys Rev B* **73** 115109
[6] Yamamoto S 2001 *Phys Rev B* **64** 140102(R)
[7] Messer M, Desbuquois R, Uehlinger T, Jotzu G, Huber S, Greif D, Esslinger T 2015 (ArXiv:1503.05549)

Computational modeling of the mechanism of hydrogen embrittlement (HE) and stress corrosion cracking (SCC) in metals

E D Cendales[1], F A Orjuela[1] and O Chamarraví[1]
[1] Fundación Universitaria Los Libertadores, Bogotá, Colombia.
E-mail: edcendalesl@libertadores.edu.co , ochamarravig@libertadores.edu.co

Abstract. In this article theoretical models and some existing data sets were examined in order to model the two main causes (hydrogen embrittlement and corrosion-cracking under stress) of the called environmentally assisted cracking phenomenon (EAC). Additionally, a computer simulation of flat metal plate subject to mechanical stress and cracking due both to hydrogen embrittlement and corrosion was developed. The computational simulation was oriented to evaluate the effect on the stress-strain behavior, elongation percent and the crack growth rate of AISI SAE 1040 steel due to three corrosive enviroments (H_2 @ 0.06MPa; HCl, pH=1.0; HCl, pH=2.5). From the computer simulation we conclude that cracking due to internal corrosion of the material near to the crack tip limits affects more the residual strength of the flat plate than hydrogen embrittlement and generates a failure condition almost imminent of the mechanical structural element.

1. Introducción

Many processes in the oil and gas industry involve the use of rawmaterials and environments rich in hydrogen, main cause of failure by embrittlement and stress corrosion cracking in pipes and pressure vessels manufactured usually with low carbon steel. The result is normally a catastrophic fragile fracture and occurs unexpectedly, in some cases after a long time in service [1]. In this industry the evaluation of the susceptibility to stress corrosion cracking is a basic aspect that is required for safe and economical operation of many types of equipments [2]. On the other hand, machines and structural components in permanent contact in hydrogen environment can fail prematurely. Due the efect of external loading or even internal stresses some damage can occur in materials through cracking, blistering, reduction of ductility, and/or by formation of brittle hydrides [3].

2. Model for hydrogen and acid diffusion due to mechanical stress

Duda et al. [4], developed a one-dimensional model in the context of the mechanics of the continuous media to simulate mechanical strain, degradation and hydrogen diffusion through free surfaces in elastic solids. Hyodo et al. [5], evaluated the growth of a crack in high strength steel subjected to environmental and mechanical effects by coupling numerical models of linear elastic fracture mechanics and continuous damage theory (see Equation 1).

$$\rho\frac{\partial^2[\mathbf{u}]}{\partial t^2} - \nabla \cdot \sigma = F \cdot V \tag{1}$$

Hydrogen is concentrated in a zone called fracture process zone which is near to the crack tip. Due to high stress triaxiality near the crack tip the crystal lattice expands, which increases the hydrogen solubility locally. The high local concentration of hydrogen causes embrittlement in both, the crack tip and the fracture process zone. Embrittlement along with the high local stresses, results in microcracking in the process zone. Within the metal, hydrogen atoms are dissolved at interstitial sites remain trapped at lattice vacancies and dislocations. Both, Hydrogen diffusion and thermal desorption measurements suggest the importance of trapping by vacancies and dislocations. The equilibrium distribution of hydrogen among traps is governed by the hydrogen chemical potential (μ_H) or fugacity in the metal and the binding energy of H to a dislocation E_{bD}. Additionally, the concentration of H trapped at dislocations is taken as $C_{HD}=K_{DH}C_H$, where the dislocation trapping equilibrium constant is $K_{DH}=\exp(-E_{bD}/RT)$ [6].

Hydrogen diffusion can be modeled through the term of effective diffusivity D_{eff} into the Equation 2, in the which V_H is the partial molar volume of hydrogen in iron-based alloys, R is the gas constant, T the actual temperature, T_Z the absolute zero temperature both in Kelvin and σ_h is the hydrostatic stress [7].

$$\frac{\partial C}{\partial t} = D_{\text{eff}}\nabla^2 C + D\frac{V_H}{R\cdot(T-T^z)}C\cdot\nabla^2\sigma_h \tag{2}$$

3. Computational implementation

Tetrahedral elements were used for the construction of the mesh, because this type of elements coupling better to the geometry near to the tip of the crack, compared with hexahedral elements of similar sizes. On the crack tip, mesh refinement was performed. This refinement has the shape of circular rings, which permits to find more realistic values around the tip of the crack. All algorithms used in the mesh are available in the software Comsol Multiphysics, that was used to resolve the state of the stress in each node. Material propierties of low carbon steel AISI SAE 1040 used in the simulation were taken from the software material database. However, some of these propierties has been extracted from the literature available about the behavior of metals under the stress corrosion cracking and hydrogen fragilization conditions [8]. The computational simulation is oriented to evaluate the effect on the stress-strain behavior, elongation percent and the crack growth rate of the material under study due to three corrosive enviroments: (H_2 @ 0.06MPa; HCl, pH=1.0; HCl, pH=2.5).

4. Results and analysis

Figure 1 presents the results of simulations to evaluate the corrosion under stress and embrittlement due to the presence of hydrogen. These graphs illustrate schematically the behavior of the stress concentration factor about the remote applied stress on the plate as well as the first and second order for the stress σ_{xx} at the crack tip $\theta = 0$.

Figure 2(a) shows the behavior of the applied stress and strain based on the results obtained from the computational simulation for metallic material cracked in the presence of three corrosive environments at a rate strain of $3\cdot 10^{-6}s^{-1}$. From the graph it can be inferred that under the condition of remote stress on the cracked flat plate with a crack that starts from one end, the presence of corrosive environment HCl (pH=1.0) presents the most unfavorable loading condition because according to the results of the simulation stress levels reached before the sudden failure of the material is lower than those obtained for other corrosive environments.

Figure 2(b) also presents the stress-strain relation for the same material under the same conditions of corrosion and embrittlement, however, the strain process is developed to a higher strain rate which is in the order of $6 \cdot 10^{-6}s^{-1}$. Based on the results shown in this graph, the load condition in presence of HCl (pH=1.0) again presents the lower stress values, before that material reaches the level of K_{IEAC}.

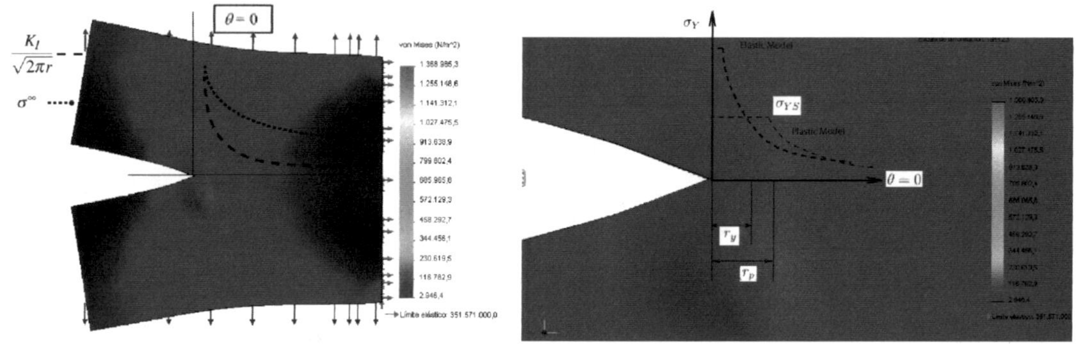

(a) (b)

Figure 1. (a) Representation of normal stress in the plane of the crack and the curve of remote applied stress on the plate, and (b) Representation of the estimates of first and second order (elastic and plastic model) to the size of the plastic area at the tip of the crack in a flat plate.

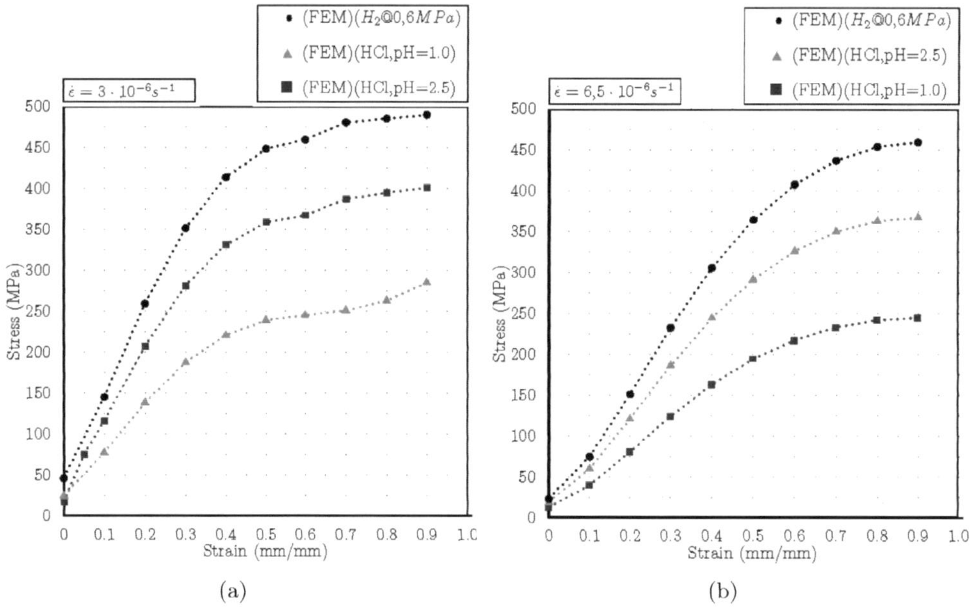

(a) (b)

Figure 2. (a) Stress-strain relation in the presence of corrosive medium at a rate of $3 \cdot 10^{-6}s^{-1}$ and (b) at a rate of $6 \cdot 10^{-6}s^{-1}$.

The numerical model implemented allows the evaluation of the results for elongation at the tip of the crack in the presence of corrosive environments so the results illustrated in Figure 3(a) presents the elongation percentage evaluated to a load condition of $\sigma = 0.75\sigma_y$. Based on the results for this load condition is established that HCl (pH=1.0) environment has the highest elongation percentage of the crack tip as a function of the time compared with others corrosive environments. This condition reaches values near to 4% in a few hours after started the process of degradation of the surface the crack.

Figure 3(b) shows the rate of crack growth as a function of the level of remote stress applied to the cracked flat plate. In this figure is evidenced that the highest crack growth rate is achieved under the condition of stress and corrosion for the environment HCl to pH=1.0, which reaches values of $9 \cdot 10^{-7}$cm/s.

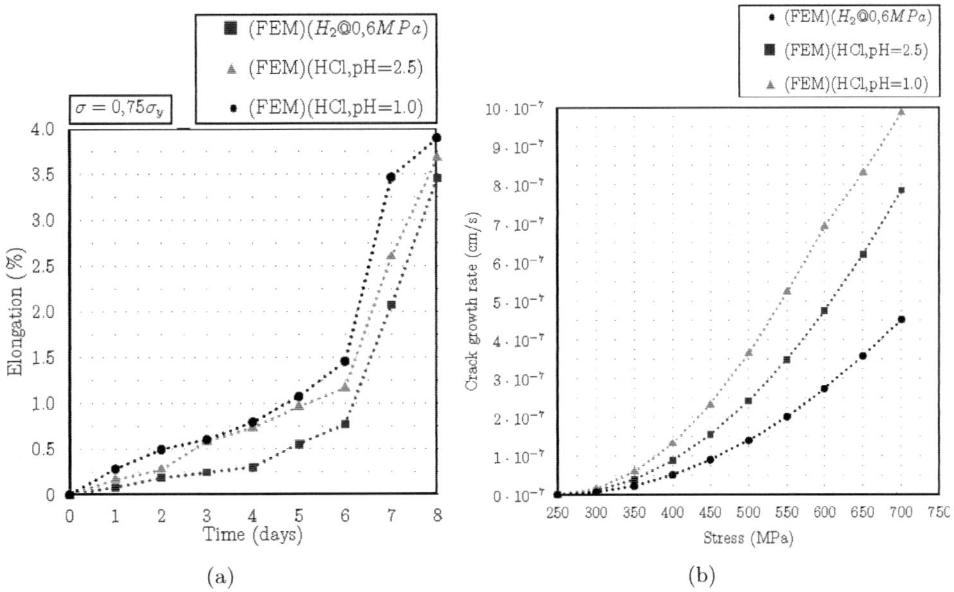

Figure 3. (a) Crack tip opening elongation in presence of corrosive environment and (b) Crack growth rate as a function of the applied stress.

5. Conclusion

In this study has been established the behavior of the strength of the material, crack tip plastic strain and crack growth rate considering the influence of hydrogen and corrosion damage. Finite element method has been implemented based closely on the stress intensity factors in Linear Elastic Fracture Mechanics theory. Two dimensional analytical solution of diffusion combined with FE analysis of crack growth permits evaluate experimental predictions as those determined by Scheider et al. [8], in the sense that values obtained in the present research through simulation for stress and strain of the material under stress corrosion cracking are similar to the values determined through the simulation developed in that research.

References

[1] Gangloff R 2003 *Hydrogen assisted cracking in high-strength alloys* (Oxford: Elsevier)
[2] Costa Mattos H, Bastos I and Gomes C 2014 *Corrosion Science* **80** 143
[3] Raykar N, Maiti S and Singh R 2011 *Engineering Fracture Mechanics* **78** 3153
[4] Duda F, Souza L, Guimaraes L and Barbosa J 2007 *Mechanics of Solids in Brazil* **1** 199
[5] Hyodo T and Kimura M 1989 *Corrosion Science* **27** 340
[6] Hebert K 2015 *Electrochimica Acta* **168** 199
[7] Alvaro A, Olden V and Akselsen O 2014 *International Journal of Hydrogen Energy* **39** 3528
[8] Scheider I, Pfuff M, Dietzel W and Besson J 2008 *Engineering Fracture Mechanics* **75** 4283

Mathematical modeling and validation of the carburizing of low carbon steels

A García Mariaca[1], E D Cendales[1] and O Chamarraví[1]
[1] Fundación Universitaria Los Libertadores, Bogotá, Colombia.
E-mail: agarciam04@libertadores.edu.co , ochamarravig@libertadores.edu.co

Abstract. This paper shows the mathematical modeling of heat and mass transfer in transient state of cylindrical bars of low carbon steel subjected to carburizing process. The model solution for the two phenomena was performed using a one-dimensional analysis in the radius direction, using the numerical method of finite differences; also a sensitivity analysis by varying the coefficient of convective heat transfer (h) is performed. The modeling results show that this carburization steel is strongly dependent on h. These results suggest that if it can increase the value of h in this kind of process could reduce the time of process for this heat treatment. Additionally, an experimental procedure was established by carburization of a steel AISI SAE 1010, which develops cementing solid phase and the specimen steel and micrographic hardness profiles obtained from samples of the specimen analysis was performed, to determine the penetration depth of the carbon and validate this result over the values obtained by the computer model.

1. Introduction
This paper presents the mathematical model of the carburizing at $1173K$ over a cylindrical bar made of AISI SAE 1010. This model is made based on the coupling phenomena of heat and mass transfer, on the other hand an experimental validation of the results obtained in the model is made. The results of the model and the experiment show great similarities in concentrations for different carburizing times, also a sensitivity analysis is performed by varying the coefficient of convective heat transfer (h) getting as a result that this process is strongly dependent on this parameter.

2. Mathematical modeling
The mathematical model is taking into account the phenomena of heat and mass transfer. In order to solve this, the finite difference numerical method, specifically implicit method in the radial direction of the bar is used [1–4].

2.1. Thermal modeling
To solve the thermal model, the physical properties like thermal conductivity (K) and specific heat (Cp) were taken as variabales and temperature dependent [5,6]. It is proceeded to propose the general equation for heat transfer with variables properties:

$$\frac{1}{r}\left(\frac{\partial}{\partial r}kr\frac{\partial T}{\partial r}\right) = \rho Cp\frac{\partial T}{\partial t} \tag{1}$$

2.1.1. Boundary conditions As a one-dimensional analysis is performed, the model needs two boundary conditions, these are in the center and the outer surface of the bar. For the center, isolated border mirror image is used.

$$\frac{\partial}{\partial r}k\frac{\partial T}{\partial r}+k\frac{\partial^2 T}{\partial r^2}=\rho Cp\frac{\partial T}{\partial t} \tag{2}$$

Furthermore, the surface has a convection boundary condition equal to the conductive heat transfer in the bar [7].

$$-k\frac{\partial T\,(r,t)}{\partial r}=h(T_\infty-T\,(r,t)) \tag{3}$$

2.1.2. Initial condition The heat transfer begins with an initial uniform temperature across the bar setted as $293K$ and for this case h is fixed in $20W/m^2K$.

$$T_0=293K \quad \forall \quad (r,0) \tag{4}$$

2.2. Mass transfer
Mass transfer is described by Fick's second law, with variable diffusion coefficient in function of the temperature [8]. The equation that describes this process is:

$$\frac{1}{r}\left(\frac{\partial}{\partial r}Dr\frac{\partial \rho_c}{\partial r}\right)=\frac{\partial \rho_c}{\partial t} \tag{5}$$

In terms of the mass fraction of carbon.

$$\frac{1}{r}\left(\frac{\partial}{\partial r}Dr\frac{\partial w_c}{\partial r}\right)=\frac{\partial w_c}{\partial t} \tag{6}$$

2.2.1. Boundary conditions There are two boundary conditions: these are in the center and the outer surface of the bar. For the center, isolated border mirror image is used.

$$\frac{\partial}{\partial r}D\frac{\partial w_c}{\partial r}+D\frac{\partial^2 w_c}{\partial r^2}=\frac{\partial w_c}{\partial t} \tag{7}$$

On the surface, the initial condition is similar to the heat transfer, mass diffusion in the bar is equals to the mass transfer convection boundary.

$$-D\frac{\partial w_c\,(r,t)}{\partial r}=h_m(w_\infty-w_c\,(r,t)) \tag{8}$$

Where h_m is convection mass transfer coefficient, which is obtained maching the Nusselt and Sherwood numbers.

2.2.2. Initial condition The initial condition for the mass transfer is an uniform carbon concentration in the bar at time zero:

$$w_c=0.1\% \quad \forall \quad (r,0) \tag{9}$$

3. Experimental validation

The experimental validation of the model was performed with samples of steel AISI SAE 1010, of $12.7mm$ diameter, which were subjected to a carburizing process carried out with mineral carbon at a temperature of $1173K$ for three different times (18000, 21600 and $25200s$). With the carburizing made, it is proceeded to polish the samples, which are attacked with a reagent that highlights the case-hardened layer. The reagent used was Nital 0.2%. Subsequently it was measured in the microscope the depth of this layer using a graduated ocular. Metallographic study was conducted in order to supplement the data obtained through the microhardness test that was done on the cross-sectional area of the specimen. The procedure for making microhardness was according to the standard ASTM E92, ASTM E384 and ISO R81 for hardness vickers (HV) [9].

4. Results

The results of microhardness and depth of case-hardened layer for different test times are presented in Table 1. The HV decreases with depth of the case-hardened layer.

Table 1. HV and hardness rockwell C (HRC) of the specimens at different depth.

Time test $[s]$	0.2 $[mm]$	0.4 $[mm]$	0.6 $[mm]$	0.8 $[mm]$	depth $[mm]$	w_c [%]
18000	827-62	802-61	778-60	766-60	1.1	0.92
21600	850-63	812-62	795-61	778-60	1.35	1.05
25200	884-64	820-62	800-62	780-60	1.5	1.1

In Figure 1, the variation of the carbon concentration is observed at different carburizing times with respect to the radius of the specimen obtained with the mathematical model. In this Figure a match with the experimental measurements shown in Table 1 was observed [10, 11].

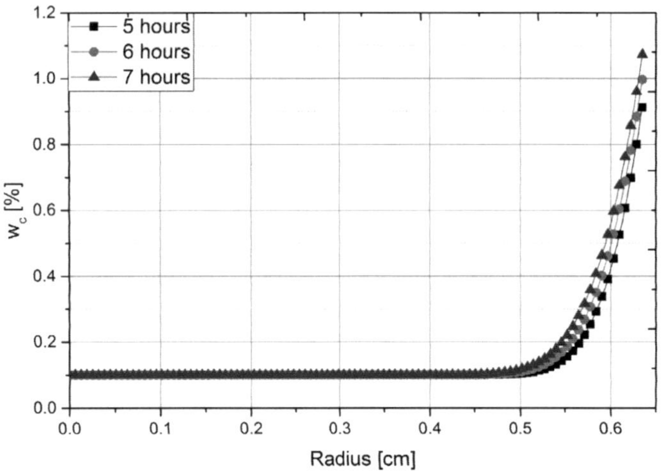

Figure 1. variation of the carbon concentration at different times carburizing.

In Figure 2 are shown results of the mathematical model varying h. These results show that carburizing process improves with increasing this parameter.

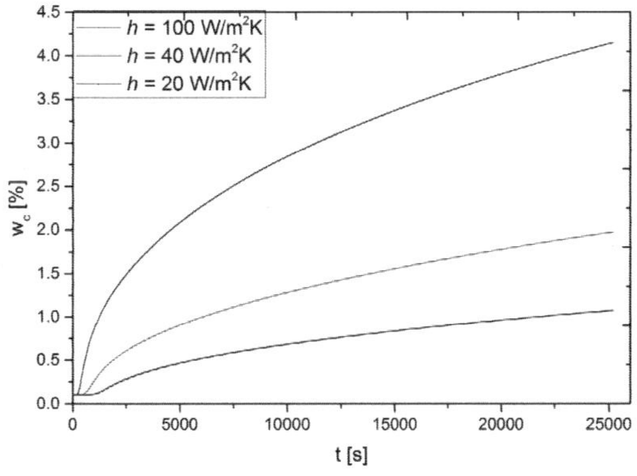

Figure 2. Behavior of the carbon concentration on the surface for different heat transfer coefficients.

In Figure 3, it is observed the difference in the carbon concentration at the surface between the mathematical model and experimental validation.

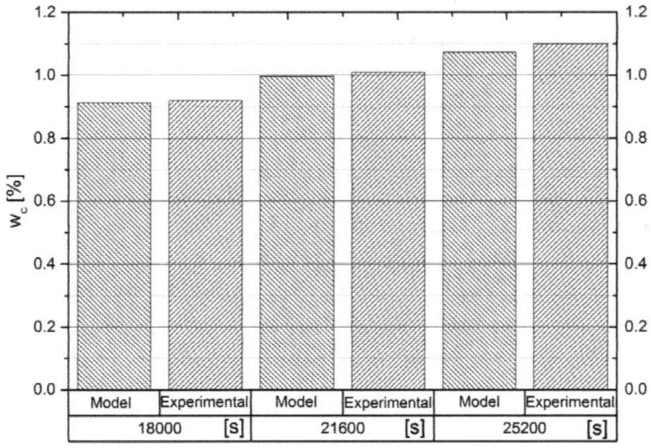

Figure 3. Variation of the carbon concentration in the surface between the mathematical model and experimental validation.

5. Conclusions

The results from the mathematical model show that the carburizing process is strongly dependent on the coefficient of convective heat transfer h. In Figure 2, it can be observed that an increase of 100% of this parameter could reduce the process time by 6 and it could be seen in cost reduction. The mathematical model applied for each carburizing time has a percentage relative error (E_t) less than 2% for concentrations on the surface, as shown in Figure 3. For this reason, the proposed mathematical model describes accurately the behavior of mass transfer and heat transfer occurring in the carburizing process.

6. References

[1] Morral J, Dupen B and Law C *Metallurgical Transactions A* **23** 2069
[2] Keddam M, Djeghlal M, Barrallier L and Salhi E *Computational Materials Science* **29** 43
[3] Ozisic M N 1994 *Computational methods in heat transfer* (London: CRC)
[4] Goldstein J I and Moren A E *Metallurgical Transactions A* **9** 1515
[5] Incropera F P and DeWitt D P 1999 *Fundamentos de tansferencia de calor* (México: Pearson Educación)
[6] Cengel Y A and Pérez H 2004 *Heat transfer a practical approach* (México: McGraw-Hill)
[7] Mukai R, Matsumoto T, Dong-ying J, Suzuki T, Saito H and Ito Y *Transactions of Nonferrous Metals Society of China* **16** 566
[8] Sun Y and Bell T *Materials Science and Engineering A* **224** 33
[9] Lyman and Taylor 1967 *Metals handbook* (USA: American Society for Metals)
[10] Wei Y, Zhang L and Jr R S *Journal of Materials Engineering and Performance* **22** 1886
[11] Rodionov A V, Ryzhov N M, Fakhurtdinov R S and Zhidkov E N *Metaliovedenie i Termicheskaya Obrabotka Metallov* **7** 28

Journal of Physics: Conference Series **687** (2016) 012069 doi:10.1088/1742-6596/687/1/012069

Ab initio calculations of half-metallic ferromagnetism in $(VN)_1/(AlN)_1$, $(VN)_1/(GaN)_1$ and $(VN)_1/(InN)_1$ superlattices

M J Espitia Rico[1], J H Díaz F[1] and C Ortega López[2]
[1] Universidad Distrital Francisco José de Caldas, Bogotá, Colombia.
[2] Universidad de Córdoba, Montería, Colombia.

E-mail: mespitiar@udistrital.edu.co

Abstract. We employed density functional theory (DFT) in order to study the electronic and magnetic properties of $(VN)_1/(AlN)_1$, $(VN)_1/(GaN)_1$ and $(VN)_1/(InN)_1$ superlattices, in the wurtzite structure. The calculations were carried out using the pseudopotential method, employed exactly as implemented in Quantum Espresso code. For the description of the electron-electron interaction, generalized gradient approximation (GGA) was used. We found that the superlattices exhibit a half-metallic ferromagnetic behaviour and all the superlattices have magnetic properties with a magnetic moment of $2\mu_\beta$/cell. Analysis of the density of states show that ferromagnetic behaviour of the superlattices can be explained by the strong hybridization between states V-d and N-p crossing of the Fermi level. Due these properties the superlattices can be potentially used in the field of spintronics or spin injectors.

1. Introduction

At the present time the understanding of magnetic behaviour of system with reduced dimensionality is one of the most important challenges in condensed matter physics, due the prospect of using the spin y developing devices in the field of spintronics. Spintronic devices use both the electrical charge and the spin of electrons to improve considerably the functionality of devices designed with conventional semiconductors [1–5]. For this reason, current research in spintronics is focused in the study of magnetic behaviour of new materials, for example, magnetic ordering of dopants in semiconductors, the magnetic properties of alloys formed by semiconductors with other compounds. Due potential applications in diluted magnetic semiconductors with high-temperature ferromagnetism, spin injectors, in magnetic memories or ultrahigh-density hard disks. Of particular interest are alloys with semiconductor having a half-metallic behaviour and have magnetic properties. This new materials one of the two spin channels is metallic and the other is semiconducting, leading to 100% spin polarization in the ground state at the Fermi level and can thus be used as spin injectors for magnetic random access memories or in other spin-dependent devices. This fact shows that it is worthwhile to carry out theoretical studies of $(VN)_1/(AlN)_1$, $(VN)_1/(GaN)_1$ and $(VN)_1/(InN)_1$ superlattices that will provide information on the structural, electronic, and magnetic properties of these superlattices and enable the design of new devices that will contribute to the development of current semiconductor technology.

2. Computational method

The calculations are performed within the DFT framework using the Quantum Espresso package [6]. The correlation and exchange effects of the electrons are treated using the generalized gradient approximation (GGA) of Perdew, Burke, and Ernzerhof (PBE) [7]. Electron–ion interactions were treated with the pseudopotential method [8,9]. The electron wave functions were expanded into plane waves with a kinetic-energy cutoff of 40Ry. For the charge density, a kinetic energy cutoff of 400Ry was used. A 6x6x4 Monkhorst-Pack mesh [10] was used to generate the k-points in the unit cell.

The calculations are performed taking into account the spin polarization due to the presence of the V atom. To calculate the lattice constant, the bulk modulus and the total energy of each studied superlattice, the calculated data are fitted with the Murnaghan equation of state, equation (1).

$$E(V) = E_0 + \frac{B_0 V}{B_0'}\left[\frac{\left(V_0/V\right)^{B'}}{B_0'-1}+1\right] - \frac{B_0 V_0}{B_0'-1} \quad (1)$$

Where B_0 is the bulk modulus and its first derivative is B'_0, V_0 is the equilibrium volume of the cell, and E_0 is the total energy. The $(VN)_1/(AlN)_1$, $(VN)_1/(GaN)_1$ and $(VN)_1/(InN)_1$ superlattices were moudeled according to special quasirandom structures approach [11] and the disorder aspects were ignored. The $(VN)_1/(AlN)_1$, $(VN)_1/(GaN)_1$ and $(VN)_1/(InN)_1$ superlattice a hexagonal unit cell with alternating [0001] layers of (AlN, GaN and InN) and VN in conventional wurtzite structure was employed, as show the Figure 1. Where a and c are the lattice constants, u denotes the dimensionless parameter of the internal structure and the positions of the atoms are: for V or M (0,0,0), (1/3,2/3,1/2) and N (0,0,u), (1/3,2/3,u+1/2).

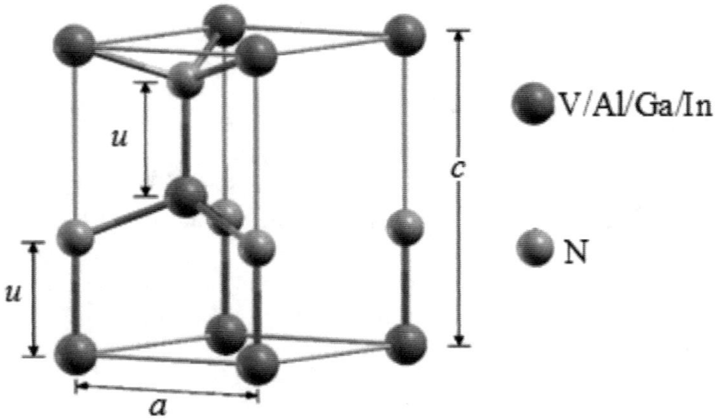

Figure 1. Unit cell $(VN)_1/(AlN)_1$, $(VN)_1/(GaN)_1$ and $(VN)_1/(InN)_1$ superlattices.

3. Results and discussions

3.1. Structural properties

To determine the structural properties in the ground state, such as the lattice constant (a_0), bulk modulus (B_0) and total energy (E_0) of the $(VN)_1/(AlN)_1$, $(VN)_1/(GaN)_1$ and $(VN)_1/(InN)_1$ superlattices, in the wurtzite structure, the total energy was calculated as a function of the volume, the results were fit to the Murnaghan equation of state, Equation (1). Additionally, the total energy variation was calculated as a function of the volume of the ferromagnetic (FM) and antiferromagnetic (AFM) phases to find the most favourable magnetic phase of the superlattices. Figure 2 shows the energy-volume curves for $(VN)_1/(AlN)_1$, $(VN)_1/(GaN)_1$ and $(VN)_1/(InN)_1$ superlattices in the FM and AFM states.

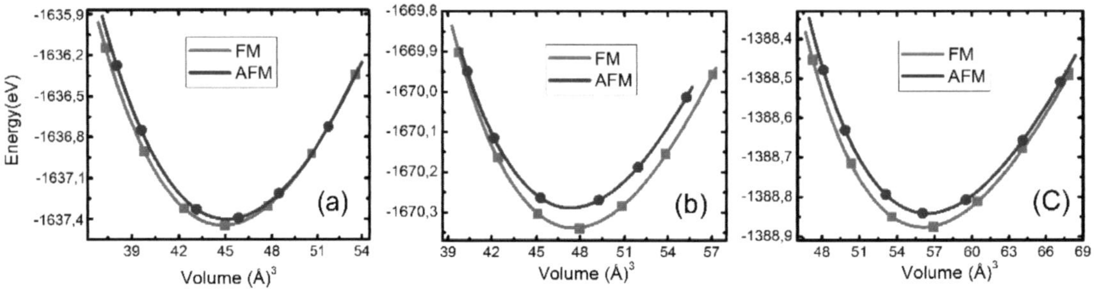

Figure 2. Total energy as a function of the volume of the superlattices (a) $(VN)_1/(AlN)_1$, (b) $(VN)_1/(GaN)_1$ (c) $(VN)_1/(InN)_1$ in the FM and AFM states.

In the ground state, the total energy differences between the FM and AFM states ($E = E_{AFM} - E_{FM}$) were - 0.029, - 0.051 and - 0.043eV for $(VN)_1/(AlN)_1$, $(VN)_1/(GaN)_1$ and $(VN)_1/(InN)_1$, respectively. In the three cases, the FM state is more energetically favourable that the AFM state. Our results and other selected theoretical results are provided in Table 1.

Table 1. Structural parameters lattice constant (a_0), bulk modulus (B_0) and total energy (E_0) of the superlattices.

Superlattice	a_0 (Å)	c/a	V_0 (Å^3)	B_0 (GPa)	E_0 (eV)	μ (μ_β)
$(VN)_1/(AlN)_1$	3.171	1.628	44.82	174.87	-1637.44	~ 2.0
$(VN)_1/(GaN)_1$	3.227 3.267[a]	1.631	47.41	176.07	-1670.33	~ 2.0
$(VN)_1/(InN)_1$	3.406	1.636	56.10	165.20	-1388.87	~ 2.0

[a] Theoretical reference [12].

The calculated lattice constant for $(VN)_1/(GaN)_1$ accord well with theoretical value reported, since it differs by less than one percent. The values of the bulk modules of the superlattices are higher, which confirms that they are quite rigid, making them good candidates for possible applications in devices operated at high temperature and high power, as well as hard coatings. On the other hand, despite the difference in the crystalline structure between $(VN)_1$ NaCl and (AlN), $(GaN)_1$, $(InN)_1$ wurtzite, joining of the layers of the VN with AlN, GaN and InN to form a superlattice had not change the AlN, GaN and InN wurtzite structure, as seen in Table 1 in the value of the c/a ratio of the superlattices, which are very close to the value ideal of the c/a (1.63) of AlN, GaN and InN in the wurtzite structure.

3.2. Electronic properties

Figure 3 shows the total and partial spin-polarized density of states of the $(VN)_1/(AlN)_1$, $(VN)_1/(GaN)_1$ and $(VN)_1/(InN)_1$ superlattices. The Figure 3 shows that they are half-metallic and ferromagnetic; this result occurs because in the valence band close to the Fermi level, the majority spins (spin-up) are metallic, and the minority spins (spin-down) are semiconductors. These compounds have a spin polarization of 100% of the conduction carriers in the ground state, which is a requirement of the spin injectors [5].

This finding suggests that these ternary compounds can be efficiently used as spin injectors. Figure 3 shows that in the valence band near the Fermi level, the spin-up density (the majority spins) is mainly dominated by the V-3d states and, to a lesser extent, by the N-2p states, which cross the Fermi level. The magnetic properties comes essentially of the vanadium atom and is attributed at V^{3+} configuration, because the V atom is the superlattice gives three electrons, the V atom remain two valence electrons (The electronic configurations V^{3+}=:[Ar]$3d^2$). This valence electrons couple ferromagnetically, as result the two electrons produce a total magnetic moment of $2\mu_\beta$/atom-V.

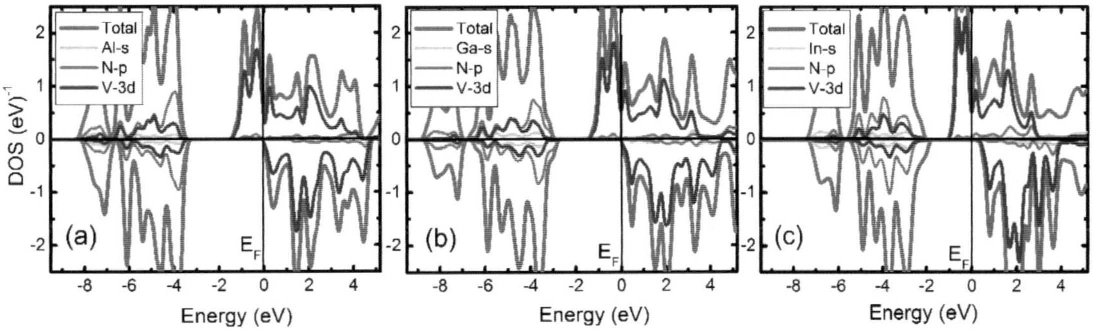

Figure 3. Total and partial density of state of (a) $(VN)_1/(AlN)_1$, (b) $(VN)_1/(GaN)_1$, (c) $(VN)_1/(InN)_1$ superlattices.

4. Conclusions

We reported first principles calculations to determine the structural, electronic, and magnetic $(VN)_1/(AlN)_1$, $(VN)_1/(GaN)_1$ and $(VN)_1/(InN)_1$ superlattices. The calculated values of the bulk modules are quite high; therefore, the superlattices are quite rigid, which makes them attractive for potential applications at high temperatures and for hard coatings. On the basis of the density of states, we found that the superlattices exhibit a half metallic behaviour, due to the orbital V- and N-*p* that cross the Fermi level. Finally, we found that $(VN)_1/(AlN)_1$, $(VN)_1/(GaN)_1$ and $(VN)_1/(InN)_1$ superlattices exhibit magnetic properties with magnetic moment $2\mu_\beta$. These properties show that superlattices are good candidates for possible applications in diluted magnetic semiconductors; spin injectors, and other spintronics applications.

References

[1] González A, López W and González R 2011 *Solid State Communications* **151** 1794
[2] Ohno H 1998 *Science* **281** 951
[3] Ohno Y, Young D, Beschoten B, Matsukura F, Ohno H and Awschalom D 1999 *Nature* **402** 790
[4] Wolf S A, Awschalom D D, Buhrmann, Daughton J M, Von-Molnar S, Roukes M L, Chtchelkanova A Y and Treger D M 2001 *Science* **294** 1488
[5] Heddar H, Zaoui A and Ferhat M 2013 *Superlattices and Microstructures* **53** 16
[6] Giannozzi P, Baroni S and Bonin N 2009 *J Phys Condens Matter* **21** 395502
[7] Perdew J, Burke K and Ernzerhof M 1996 *Physical Review Letter* **77** 3865
[8] Vanderbilt D 1990 *Phys Rev B* **41** 7892
[9] Laasonen K, Pasquarello A, Car R, Lee C and Vanderbilt D 1993 *Phys Rev B* **47** 10142
[10] Monkhorst H and Pack J 1976 *Phys Rev B* **13** 5188
[11] Zunger A, Wei S H, Ferreira L and Bernard J E 1990 *Physical Review Letters* **65** 353
[12] González R, López W and Rodríguez J 2009 *Materials Science and Engineering B* **163** 190

Journal of Physics: Conference Series **687** (2016) 012070 doi:10.1088/1742-6596/687/1/012070

Computational two-dimensional modeling of the stress intensity factor in a cracked metallic material

J E Rolón[1], E D Cendales[1] and I M Cruz[2]

[1] Fundación Universitaria Los Libertadores, Bogotá, Colombia.
[2] Universidad ECCI, Bogotá, Colombia.

E-mail: jerolonr@libertadores.edu.co , icruzb@ecci.edu.co

Abstract. Cracking of metallic engineering materials is of great importance due cost of replacing mechanical elements cracked and the danger of sudden structural failure of these elements. One of the most important parameters during consideration of the mechanical behavior of machine elements having cracking and that are subject to various stress conditions is the stress intensity factor near the crack tip called factor Kic. In this paper a computational model is developed for the direct assessment of stress concentration factor near to the crack tip and compared with the results obtained in the literature in which other models have been established, which consider continuity of the displacement of the crack tip (XBEM). Based on this numerical approximation can be establish that computational XBEM method has greater accuracy in Kic values obtained than the model implemented by the method of finite elements for the virtual nodal displacement through plateau function.

1. Introduction

Geometrical discontinuities such as cracks or notches are found usually in different structures and machine elements [1]. These imperfections in the material acts as stress concentrators so that the material is subject to lower values of stress than the yield stress can increase the size of the crack to reach the critical size at which the failure of the material occurs suddenly [2, 3]. Because this condition, fracture study in materials represents a fundamental area the enginnering analysis and design of structures [4, 5].

2. Computational implementation

In this article for the construction of the mesh tetrahedral elements were used, because this type of elements coupled better to the geometry near to the tip of the crack, compared with hexahedral elements of similar sizes. On the tip of the crack mesh refinement was performed, this refinement has the shape of circular rings, which permits to find more realistic values around the tip of the crack. All algorithms used in the mesh are available in the software used to resolve the state of the stress in each node. According to this computational implementation developed in this research, it was not required the second mesh having a size of crack increased. It was enough to recalculate the stiffness matrix considering new nodal coordinates.

3. Results

During the development of finite element analysis for the case illustrated in this paper, it was established the comparison with results from other investigations in order to validate the results obtained by implementing performed. Figure 1 shows the results obtained for the simulations in which number of degrees of freedom of the nodes of the mesh in the tip of the crack change and the simulation was established for different values of this parameter. The same figure illustrates the expected behavior of parameter K_I according to a study by Rooke et al. [6]. Also the results obtained by Alatawi et al. [7], are presented in order to make comparison due they developed computer simulations to evaluate the relative stress intensity factor K_I/K_o, for a similar geometry that was used in the present study.

Based on the results shown, it is observed that the results obtained by Alatawi et al. [7], present convergence to increase the number of degrees of freedom in the nearby area the crack tip when a number of degrees of freedom greater than 150 was used, so that this implementation allows direct evaluation of parameter K_I. Moreover, Figure 1 shows that the results obtained in present study for the direct evaluation the stress intensity factor through computational implementation of FEM have relative convergence regarding the results established by implementing of XBEM. However considerable discrepancy is observed when the number of degrees of freedom is less than 200, approximately.

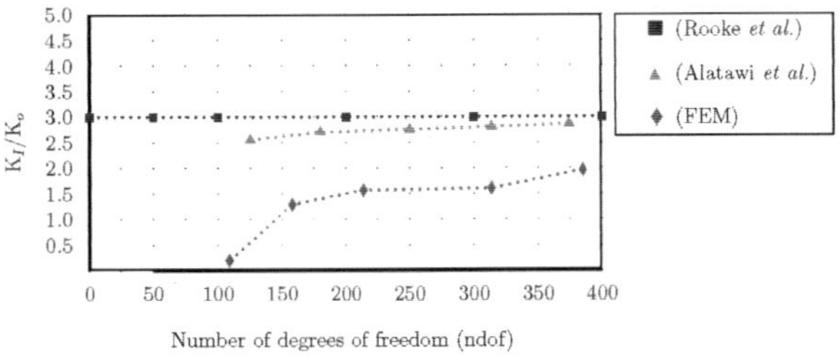

Figure 1. Variation of standardized stress intensity factor as a function of dnof.

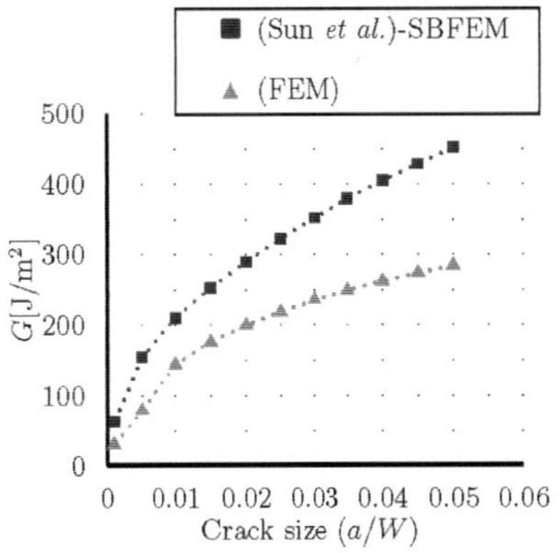

Figure 2. Variation of the driving force for crack growth as a function of the relative crack size.

Figure 2 shows the variation of the driving force (G) for the increase of the crack size obtained by Sun et al. [8], through the implementation of the method SBFEM for relative stress intensity factor K_I/K_o, for the same geometry. Sun et al., report that there is convergence between the values for G parameter obtained by computational simulation which was calculated by the Equation 1 in order to validate the simulation. From Figure 2 can be established that the FEM computational implementation developed in the present work shows subestrimacion for the driving force during advancing of the crack respect to the results obtained by computational implementation based SBFEM.

$$G = -\frac{1}{l}\int_0^l \Delta\mathbf{u}(\varepsilon_{crack}(r,\theta),\eta(\theta))^T\sigma(\varepsilon_{initial}(r,\theta),\eta(\theta)) \quad (1)$$

The behavior of stress was evaluated by FEM simulation and were subsequently compared with the results reported by Sun et al. [8]. As Figure 3 illustrates, the values for the same parameter obtained in this study through FEM simulation present overestimation of the normal stresses on the front of crack ($\theta = 0$) for crack sizes smaller than 0.01mm. However, for higher values there is a considerable difference between the two computational estimates, presenting underestimation of stresses at points near the crack tip.

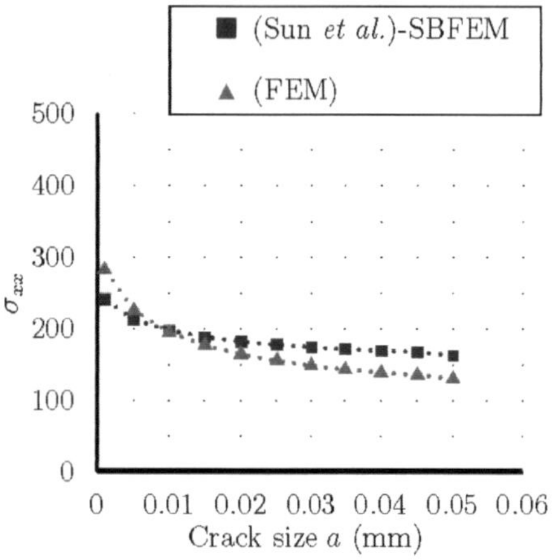

Figure 3. σ_{xx} in front of the crack ($\theta = 0$) depending on the crack size.

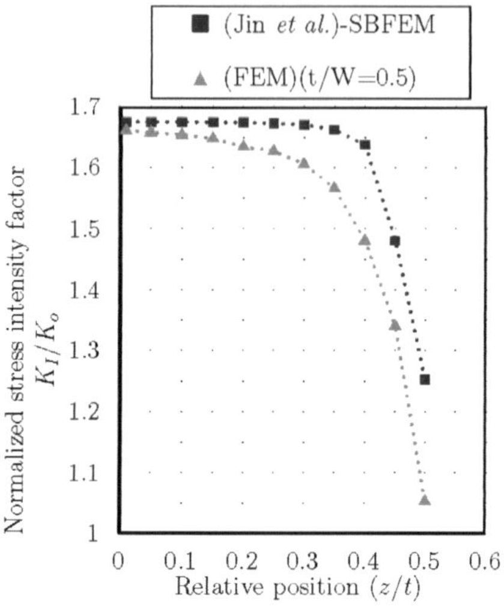

Figure 4. Variation of standardized stress intensity factor as a function of the relative position at the crack tip.

In order to evaluate the stress triaxiality in the crack tip the results of computational simulations obtained by Jin et al. [9] and the results obtained by simulation implemented in this study are presented in Figure 4. The results in both simulations illustrate a purely triaxial behavior at the center of the crack ($z/t = 0$) in which the parameter value K_I has the highest value. Moreover, in distant symmetrical positions at the crack tip ($z/t > 0$) this parameter decreases sharply to a minimum value as expected at the free surface of the plate in which the stress state has a distinctly a plane stress condition.

4. Conclusions

Finite element method has been implemented based closely on the stress intensity factors in linear elastic fracture mechanics theory. The implementation is able to evaluate SIFs directly without any requirement for aditional calculations for 2D simulations. Results shows convergency between results established using XBEM method and FEM method. However is evident that XBEM implementation present more accurancy than FEM method on the evaluation of SIFs and G (driving force). Nevertheles XBEM present higher complexity in the computational implementation for that reason FEM could be used as a first estimation for these parameters used in dimensioning and desing of mechanical systems.

References

[1] Williams M and Ellinger G 1953 *Welding Journal* **32** 498
[2] Sih G 1966 *International Journal of Fracture Mechanics* **2** 628
[3] Hull D and Bacon D 1953 *Introduction to dislocactions* (Woburn: Betterworth-Heinemann)
[4] Eshelby J 1975 *The Physics of Metals* **2** 10
[5] Westergaard H 1939 *Journal of Applied Mechanics* **6** 49
[6] Rooke D and Cartwright D 1976 *Compendium of Stress Intensity Factors* (London: Her Majesty's Stationery Office)
[7] Alatawi I and Trevelyan J 2015 *Engineering Analysis with Boundary Elements* **52** 56
[8] Sun Z, Ooi E and Song C 2015 *Engineering Fracture Mechanics* **134** 330
[9] Jin Z and Wang X 2013 *Theoretical and Applied Fracture Mechanics* **132** 110

IMRMPT2015 IOP Publishing
Journal of Physics: Conference Series **687** (2016) 012071 doi:10.1088/1742-6596/687/1/012071

Mott lobes evolution of the spin-1 Bose-Hubbard model

A F Hincapie-F[1], R Franco[1] and J Silva-Valencia[1]

[1] Universidad Nacional de Colombia, Bogotá, Colombia.

E-mail: `jsilvav@unal.edu.co`

Abstract. We study spin-1 bosons confined in a one-dimensional optical lattice, taking into consideration both ferromagnetic and antiferromagnetic interaction. Using the density matrix renormalization group, we determine the phase diagram for the two firsts lobes and report the evolution of the first and second Mott lobes with respect to the spin-exchange interaction parameter (U_2). We determine that for the antiferromagnetic case, the first lobe is suppressed while the second grows as $|U_2|$ increases. For the ferromagnetic case, the first and second Mott lobes are suppressed by the spin-exchange interaction parameter. We propose an expresion to describe the evolution of the critical point with the increase in $|U_2|$ for both cases.

1. Introduction

The confining of ultracold atoms in optical lattices has been a very important tool for the study of condensed matter. The precision and the easy control of the various couplings makes of this a natural quantum simulator. Fermionic and bosonic systems have been explored in these experimental setups. The most well-known experiment was carried out in 2002 by Greiner *et* al. [1]. They show the phase transition from a superfluid state to the Mott insulator state, which has been much studied theoretically [2, 3].

In conventional magnetic traps, the spins of atoms are frozen so effectively that they behave like spinless particles. Now, recent progress in cooling techniques has opened the door to new studies of these systems. For instance, using purely optical traps, alkali atoms such as ^{87}Rb or ^{23}Na can be condensed and loaded into an optical lattice, preserving the spin component [4]. The size of the effects of the spin interaction can be irrelevant when the number of atoms is relatively large, so, it's natural to think about the idea of taking these interactions into consideration in optical lattices where the number of atoms is about 3 or 4 per lattice site [1].

The spin-1 Bose-Hubbard model shows the superfluid to Mott insulator phase transition [5,6], and the phase diagram was presented and discussed by Rizzi et al. [7] using the density matrix renormalization group (DMRG) method for the antiferromagnetic case, and by Batrouni *et* al. [8] using quantum Monte Carlo for the ferromagnetic case for small lattices. They show the location of the phase boundary, and Rizzi shows the asymmetry present with the odd-even filling discussed in [6]. But evolution of the critical point with the change in the spin-exchange interaction parameter has never been shown. This is the aim of this paper.

2. Model

The Hamiltonian for describing spin-1 bosons in an optical lattices is given by

$$H = -t \sum_{i=1,\sigma}^{L-1} \left(\hat{b}_{i,\sigma}^{\dagger} \hat{b}_{i+1,\sigma} + \text{h.c.} \right) + \frac{U_0}{2} \sum_{i=1}^{L} \hat{n}_i \left(\hat{n}_i - 1 \right) + \frac{U_2}{2} \sum_{i=1}^{L} \left(\mathbf{S}_i^2 - 2\hat{n}_i \right) \tag{1}$$

Where $\hat{b}_{i\sigma}^{\dagger}$ creates a boson with spin component σ at site i in a lattice of L sites, h.c denotes the hermitian conjugate, $\hat{n}_i = \Sigma_\sigma \hat{b}_{i,\sigma}^{\dagger} \hat{b}_{i,\sigma}$ and $\mathbf{S}_i = \Sigma_{\sigma,\sigma'} \hat{b}_{i,\sigma}^{\dagger} \mathbf{T}_{\sigma,\sigma'} \hat{b}_{i,\sigma'}$ are the number and spin operators for site i where $\mathbf{T}$ are the spin 1 Pauli matrices [5]. The first term in the Hamiltonian represents the kinetic energy and t is the hopping parameter, and the second and third term represent the on-site interaction (U_0) and spin exchange interaction (U_2), respectively. The local strengths are given by

$$U_0 = \frac{4\pi\hbar^2}{M} \frac{a_0 + 2a_2}{3} \text{ and } U_2 = \frac{4\pi\hbar^2}{M} \frac{a_0 - a_2}{3} \tag{2}$$

Where a_n are the scattering lengths and M is the mass of the atoms. These coupling parameters obey the constrain $-1 < U_2/U_0 < 1/2$, U_2 can be positive (antiferromagnetic case) or negative (ferromagnetic case) [4]. We fix our energy scale taking $U_0 = 1$.

3. Results

To study the Hamiltonian (1), we used the finite-size density matrix renormalization group method. The local Hilbert space for the on-site part of the Hamiltonian is fixed, imposing a maximum occupation number $\hat{n}_{max} = 4$. We carried out five iterations, considering $m = 250$ states per block, and the discarded weight was around 10^{-8} or less.

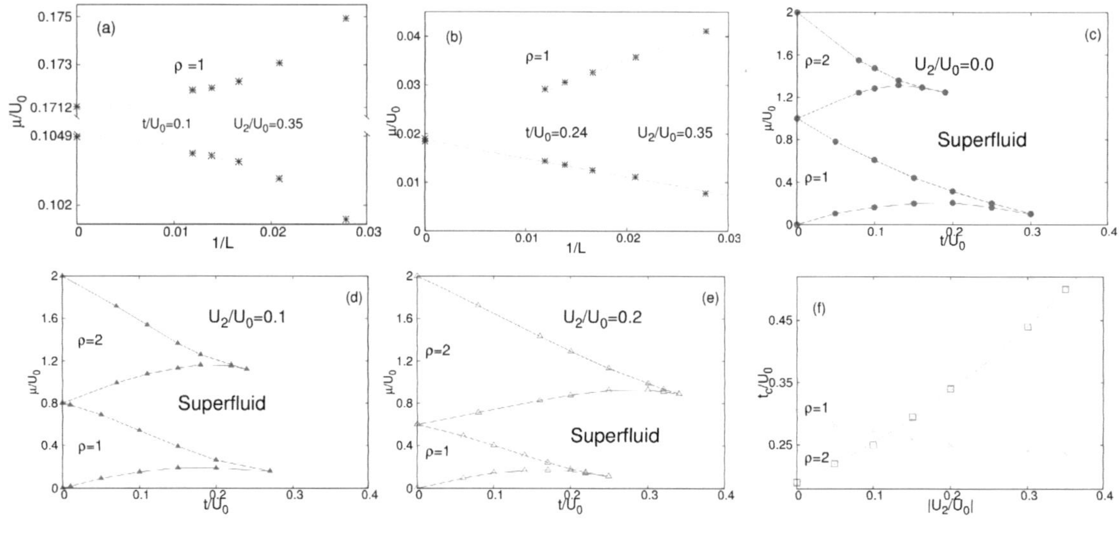

Figure 1. (Color online) Antiferromagnetic case. Chemical potential versus $1/L$ for $U_2/U_0 = 0.1$, density $\rho = 1$ and (a) $t/U_0 = 0.1$ (b) $t/U_0 = 0.24$. (c-e) Phase diagrams for the first two Mott lobes of the model case for $U_2/U_0 = 0{,}0.1$ and 0.2.

For a small kinetic energy, we expected that the bosons would be localized in the lattice and the system would be in a Mott-insulator phase. In order to show this, we calculated the energy for adding or removing a boson, given by $\mu^a(L) = E(L, N+1, S_z) - E(L, N, S_z)$ and

$\mu^r(L) = E(L,N,S_z) - E(L,N-1,S_z)$ respectively, where $E(L,N,S_z)$ is the ground state energy for L sites, N particles and S_z spin component. Following Katsura, our results are obtained considering $S_z = 0$ for the ground state [10]. The dependece of the size of the system on adding or removing energy is shown in Figure 1(a) for $\rho = 1$, $t/U_0 = 0.1$ and $U_2/U_0 = 0.35$; we can see that the added energy decreases and the removed energy increases as the lattice size increases. It can be seen that the dependence is quadratic, and we can determine its values at the thermodynamic limit. We calculated the gap given by $\Delta = \mu^a(L) - \mu^r(L)$, and obtained a nonzero finite value, showing the existence of a Mott insulator state. In Figure 1(b), we show the addition and removal energies for $\rho = 1$, $t/U_0 = 0.24$ and $U_2/U_0 = 0.35$. It can be seen that the dependence now becomes linear, and at the termodynamic limit the gap is zero, showing the existence of a superfluid state.

In Figure 1(c), we show the phase diagram for the two first Mott lobes ($\rho = 1$ $\rho = 2$) of the model, taking $U_2/U_0 = 0$. We observe that our results are in accordance with the reported results for the spinless case [11]. On the other hand, if we turn on the spin interaction when considering the antiferromagnetic case, we obtain that the Mott insulator area for $\rho = 1$ decreases and for $\rho = 2$ increases with the spin-exchange interaction parameter, confirming the odd and even asymmetry discussed by Imambekov [5]. This behavior is shown in Figure 1(a) for $U_2/U_0 = 0.1$ and Figure 1(b) $U_2/U_0 = 0.2$. Also, at the no-hopping limit ($t/U_0 = 0$) the espin-1 Bose-Hubbard model exhibits features that are in contrast with the spinless model. In Figure 1(d-e), we show that the antiferromagnetic interaction favors the singlet state, where the total local spin is $S_i = 0$ and the energy is set by $E_n = n_i U_0 - 2U_2$. In Figure 1(f), we report the evolution of the critical point. It can be seen that the critical value increases for $\rho = 2$ and decreases for $\rho = 1$ monotonically, and the best possible fit would be quadratic for both cases. For $\rho = 2$, we see that the spin interaction favors the Mott insulator state, and in order to evolve into superfluid statea large kinetic energy is necessary as the spin-exchange interaction parameter increases.

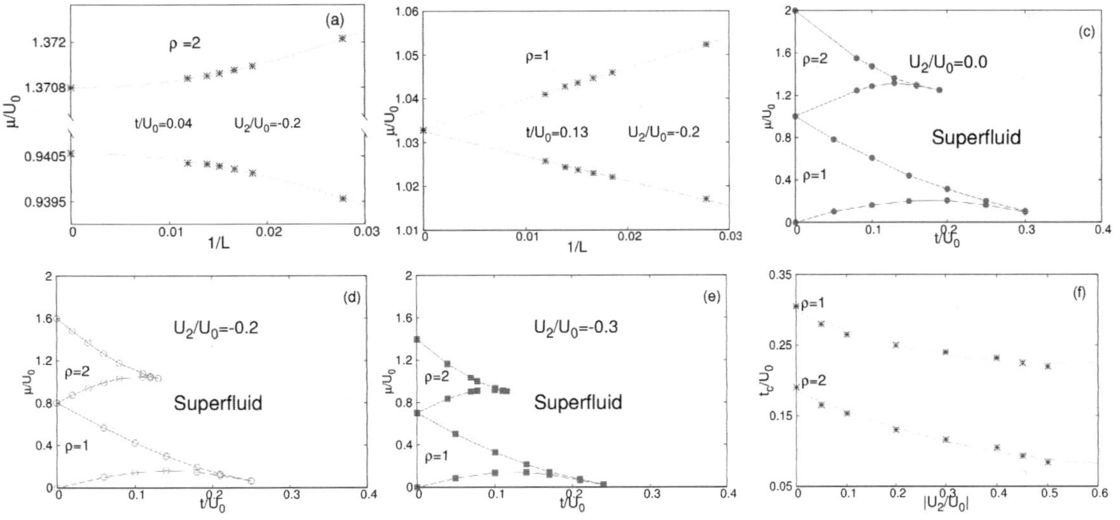

Figure 2. (Color online) Ferromagnetic case. The chemical potential versus $1/L$ for $U_2/U_0 = -0.2$, density $\rho = 2$ and (a) $t/U_0 = 0.04$ (b) $t/U_0 = 0.13$. (c-e) Phase diagrams for the first two Mott lobes of the model for $U_2/U_0 = 0, -0.2$ and -0.3.

For the ferromagnetic case, which has not been much studied, we performed a numerical study at the thermodynamic limit, in contrast with Batrouni *et al.* [8] who show the phase diagram considering finite small lattices and using quantum Monte Carlo (QMC). Based on the power of our DMRG code, we extended their calculations to the thermodynamic limit. Then, considering ferromagnetic interaction, we show that for small kinetic energy the system has a nonzero finite value for the gap, as is shown in Figure 2(a), and the system will be in a Mott insulator state, as in the antiferromagnetic case. If the hopping parameter increases, the gap will vanish at a critical value and the system evolves into the superfluid state, as is shown in Figure 2(b). Again it is possible to see that in the insulator region the best fit is quadratic, whereas in the superfluid region it is linear. In Figure 2(c-e) we show the phase diagrams for the ferromagnetic case for the two first Mott lobes ($\rho = 1$ $\rho = 2$) of the model. We show that the area of the Mott lobes decreases with an increase in the spin-exchange interaction, in contrast with the antiferromagnetic case, which accord with Batrouni. Also, at the no-hopping limit, the energy is set as $E_n = n_i(U_0 + U_2)$, implying that the interaction favors on-site ferromagnetic states where the total local spin is $S_i = 2$. In Figure 2(f), we report the evolution of the critical point. It can be seen that the critical value decreases monotonically for $\rho = 1$ and $\rho = 2$, and the best possible fit would be quadratic for both cases. We see in both cases that the spin interaction favors the superfluid state, and to evolve into a Mott insulator state, a small kinetic energy is necessary as the spin-exchange interaction parameter increases. Furthermore, if we compare the behavior of the critical point for the first Mott lobe in each case, ($U_2 < 0$ and $U_2 > 0$) in Figure 2(f) and Figure 1(f), we can see that the evolution is almost the same, and the corresponding fits would be similar.

4. Conclusions

We performed a numerical study of the ground state of the spin-1 bosons confined in a one-dimensional optical lattice. Using the density matrix renormalization group method, we determined the phase diagram for the first lobes ($\rho = N/L = 1$ and 2), reporting the evolution of the critical point of the transition, calculated by means of the chemical potential with respect to the spin-exchange interaction parameter (U_2), for ferromagnetic and antiferromagnetic interactions, and showed that the best fit would be quadratic in both cases. For the antiferromagnetic case, we found that the area of the first lobe decreases while that of the second increases as the $|U_2|$ increases, and for the ferromagnetic case, we calculated the phase diagram at the thermodynamic limit, reporting that the area of the first and second Mott lobes decreases as $|U_2|$ increases. We conclude that the behavior of the critical point for the first Mott lobe is almost the same for both cases considered.

Acknowledgments

The authors are thankful for the support of DIB-Universidad Nacional de Colombia.

References

[1] Greiner M, Mandel O, Esslinger T, Hänsch T W and Bloch I 2002 *Nature* **415** 39
[2] Jaksch D, Bruder C, Cirac J I, Gardiner C W and Zoller P 1998 *Phys Rev Lett* **81** 3108
[3] Fisher M P A, Weichman P B, Grinstein G and Fisher D S 1989 *Phys Rev B* **40** 546
[4] Stamper-Kurn D M, Andrews M R, Chikkatur A P, Inouye S, Miesner H-J, Stenger J and Ketterle W 1998 *Phys Rev Lett* **80** 2027
[5] Imambekov A, Lukin M and Demler E 2003 *Phys Rev A* **68** 063602
[6] Demler E and Zhou F 2002 *Phys Rev Lett* **88** 163001
[7] Rizzi M, Rossini D, De Chiara G, Montangero S and Fazio R 2005 *Phys Rev Lett* **95** 240404
[8] Batrouni G G, Rousseau V G and Scalettar R T 2009 *Phys Rev Lett* **102** 140402
[9] White S R 1992 *Phys Rev Lett* **69** 2863
[10] Katsura H and Tasaki H 2013 *Phys Rev Lett* **110** 130405
[11] Kühner T D and Monien H 1998 *Phys Rev B* **58** 14741

2D electrostatic PIC algorithm for studying laser induced plasma in vacuum

C A Álvarez[1], H Riascos[1] and C Gonzalez[1,2]

[1] Universidad Tecnológica de Pereira, Pereira, Colombia.
[2] Universidad de Buenos Aires, Buenos Aires, Argentina.

E-mail: caalvarez@utp.edu.co

Abstract. Particle-In-Cell(PIC) method is widely used for simulating plasma kinetic models. A 2D-PIC electrostatic algorithm is implemented for simulating the expansion of a laser-induced plasma plume. For potential and Electric Field calculation, Dirichlet and periodic boundary conditions are used in the X (perpendicular to the ablated material) and Y directions, respectively. Poisson-solver employs FFTW3 library and the five-point Laplacian to compute the electric potential. Electric field calculation is made by central finite differences method. Leap-frog scheme updates particle positions and velocities at each iteration. Plume expansion analysis is done for the Emission and Post-Emission stages. In the Emission phase (while the laser is turned on), fast electron expansion is observed and ion particles remain near the surface of the ablated material. In the post-emission stage (with the laser turned off) the charge separation produces an electric field that accelerates the ions leading to the formation of a KeV per particle Ion-Front. At the end of the expansion, fastest electrons escape from the simulation space; an almost homogeneous ion-electron distribution is observed, decreasing the electric field value and the Coulomb interactions.

1. Introduction

In the procces known as PLD, a pulsed laser is focused onto a target of the material to be deposited. For sufficiently high laser energy density, each laser pulse vaporizes or ablates a small amount of the material creating a plasma plume [1]. The ablated material is ejected and a plasma cloud forms in front of the surface which expands into the vacuum above it [2]. Laser ablation cloud consist of excited or ground-state neutrals, electrons and ions [3].

Collective interaction between the species in the plume represents a complicated theoretical problem which can be studied through computational models. Particle-In-Cell method [4–6] is used for simulating plasma from a kinetic approach. It consists on solving the differential equations involved in plasma evolution (Electric Potential, Electric Field and kinetic equations are solved on every step). Computational space is taken as a grid point collection where finite size particles phase-space change in discrete time-steps. Preliminary results for a 2D PIC model for Silver (Ag) laser-induced plasma expanding in vacuum are presented in this paper with emphasis on the ion front behaviour.

2. Model

We will consider an isothermal initial expansion for the species at temperature T_0, with a constant evaporation flux $J_{e,0} = n_{e,0}v_{e,0} = J_{i,0} = n_{i,0}v_{i,0}$. $n_{i,0}$ and $n_{e,0}$ are the densities outside of the surface for ions and electrons respectively, and the flow velocities are given by:

$$v_{i,0} = \left(\frac{2k_bT_0}{\pi m_i}\right)^{1/2}, \quad v_{e,0} = \left(\frac{2k_bT_0}{\pi m_e}\right)^{1/2} \tag{1}$$

When particles are ejected from the surface ($x = 0$), we assume a half Maxwellian distribution ($v_x > 0$ in front of the surface, obviating the negative velocities) and a full Maxwellian distribution for velocities in X (perpendicular to the surface) and Y (parallel to the surface) directions, respectively [2, 7, 8]:

$$f_0^\alpha(v_x) = n_0\sqrt{\frac{2m_\alpha}{\pi k_bT_0}}exp\left(\frac{-m_\alpha v_x^2}{2k_bT_0}\right), \quad f_0^\alpha(v_y) = n_0\sqrt{\frac{m_\alpha}{2\pi k_bT_0}}exp\left(\frac{-m_\alpha v_y^2}{2k_bT_0}\right) \tag{2}$$

α represents plasma species and k_b is the Boltzmann constant.

In PIC method, the plasma is represented by a finite number of "computational particles", whose dynamics are followed by self-consistent fields. N_k real particles are represented by a computational particle. The expansion space is divided in a spatial grid where each grid point has size Δx; computational steps advance time on Δt, solving field equations on each iteration. To compute the charge densities on the grid and for projecting forces that are found on the grid back to the particles, we use the weighting scheme known as Cloud in Cell [9]. Charge distribution on grid cells is given by:

$$\rho_{jk} = eN_k(n_{i,jk} - n_{e,jk})W(x - x_{jk}) \tag{3}$$

The electric potential and field equations are solved by five-point discrete version of the Poisson equation using Fast Fourier Transform and central finite difference scheme respectively (see equations 4 and 5).

$$\frac{\phi_{j+1,k} + \phi_{j-1,k} + \phi_{j,k+1} + \phi_{j,k-1} - 4\phi_{j,k}}{\Delta x^2} = -\frac{\rho_{j,k}}{\epsilon_0} \tag{4}$$

$$E_{x,i} = -\frac{(\phi_{x,i+1} - \phi_{x,i-1})}{\Delta x} \tag{5}$$

The particle trajectories are advanced with the "leap-frog" method, where the position and velocity values are staggered in time. The leap-frog method for k-th particle yields:

$$\vec{r}^{k+1} = \vec{r}^k + v^{k+0.5}\Delta t, \quad \vec{v}^{k+0.5} = \vec{v}^{k-0.5} + \frac{q_\alpha}{m_\alpha}\vec{E}_p\Delta t \tag{6}$$

where $\vec{r}^k$, $\vec{v}^k$, m_α are the position, velocity and mass of k-th particle respectively, $\vec{E}_p$ is the electric field acting on the particle, and Δt is the computational time step.

3. Parameters

Laser and solid surface parameters used in this work are the same employed in [10]. Laser pulse $t_0 = 100\,fs$, fluence $1.0\,J/cm^2$ and Ag solid surface (ion-electron mass relationship $m_i/m_e = 1.98 \times 10^5$). Under these conditions we have a particle flux $J_0 = 4.5 \times 10^{33}\,particles/m^2s$ with $v_{i,0} = 10^3\,m/s$. Initial temperature for both species is $T_0 = 1.77\,eV$.

We choose $\Delta t = 10^{-5}t_0 = 10^{-18}\,s$ and $\Delta x = \lambda_D \approx 10^{-11}\,m$ in order to satisfy Courant stability criterion ($\Delta t \leq \Delta x/v_{max}$) which avoids that the particles go through more than a cell per iteration [2, 11].

In our model, 10000 superparticles are used with a weight $N_k = 10$. Simulation space will be $L_x = 4097\Delta x$, $L_y = 8\Delta x$. Since the same number of ions and electrons are ejected on emission time, the boundary conditions are: at $x = 0$ electric potential is zero $\phi(x = 0) = 0$, at $x = L_x$ $\phi = 0$ and in Y-direction we assume periodical conditions due to L_y small size.

In order to achieve a higher computational efficiency, dimensionless parameters are used, taking for normalization the initial values $v_0 = v_{i,0}$, t_0, $x_0 = v_0 t_0$, the mass m_i, $n_0 = 10000/x_0^3$ and $E_0 = k_b T_0/\pi e x_0$ [12]. The algorithm is implemented in a C++ code and the electric potential is computed using the modern FFTW3 Fast Fourier Transform library.

4. Results

At the beginning of the expansion, the electrons velocity is higher than the ions velocity; this leads to a charge separation on the simulation space, wich generates an electric field that accelerates ions. Figure 1 shows ion velocity distribution for different times of the expansion. At $t/t_0 = 1$ most ion superparticles have $v = 0$; the number of superparticles decrease quickly up to $v = 5v_0$. At $t/t_0 = 2$ superparticle number with $v = 0$ is smaller than at $t = t_0$. However, the most superparticles have high velocities with the highest values around $15v_0$; this suggest that these superparticles are the ion front [4]. At $t/t_0 = 4.5$ the superparticles have an almost homogeneous velocity distribution. Fastest ions present $T > 1\,keV$. Figure 2 shows ion phase-space at $t = 2t_0$; the superparticles with velocities near $v = 16v_0$ and $x = 12x_0$ are the ion front of plasma [13].

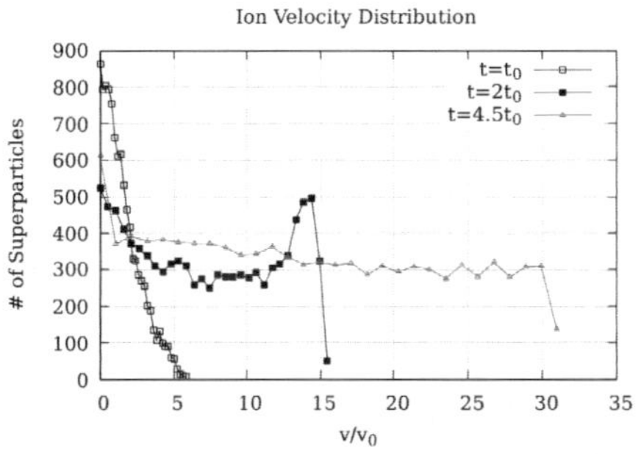

Figure 1. Ion velocity distribution for different times.

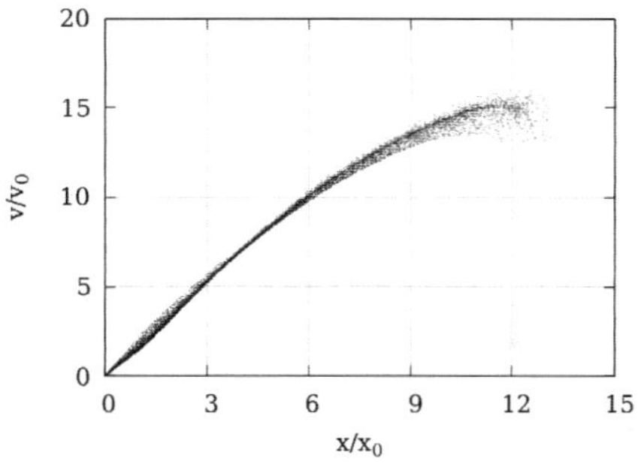

Figure 2. Ion Phase-space, $t = 2t_0$.

Figures 3 and 4 show ion and electron number density at $t = 5t_0$. An almost uniform distribution is observed for both species; therefore, electric field becomes zero, but in $x \approx 90x_0$ it is maximum as it is shown in (Figure 5).

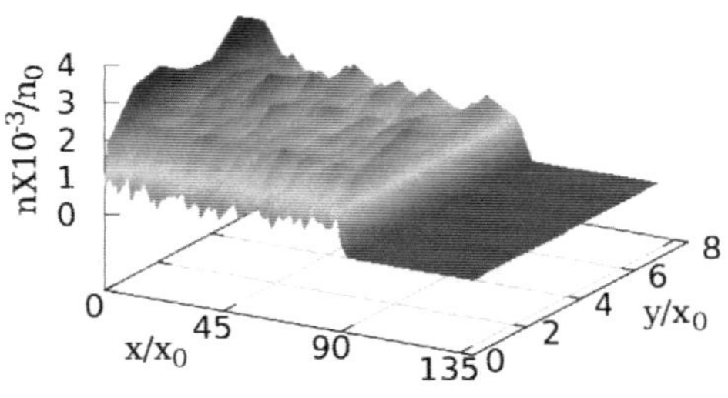

Figure 3. Ion Number Density at t=5t0.

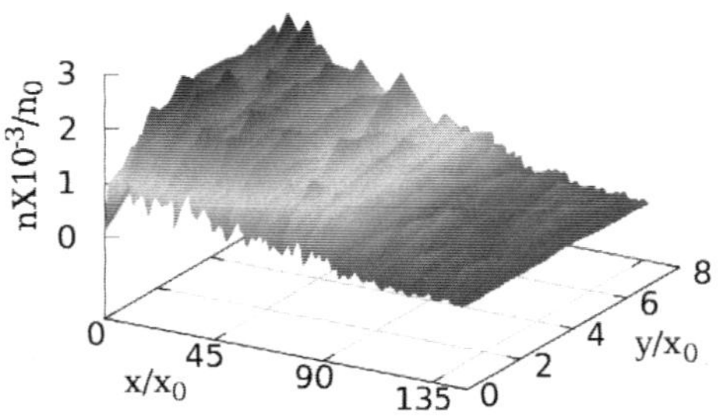

Figure 4. Electron Number Density at t=5t0.

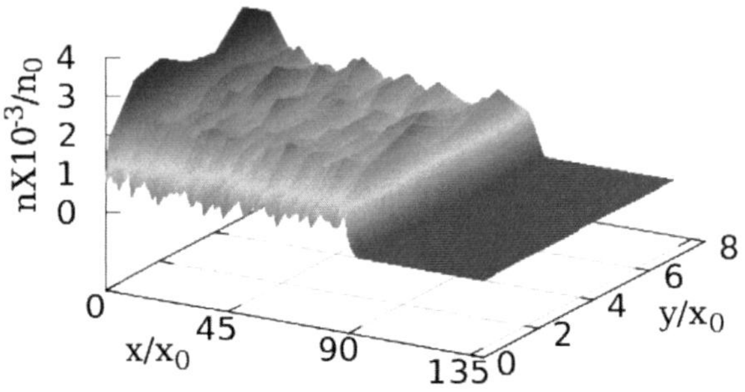

Figure 5. Electric field on X direction at t=5t0.

5. Conclusions
Electrostatic expansion of the laser-induced plasma plume is generally studied as a one-dimensional problem, due to the symmetry of the Coulomb forces in the directions that are parallel to the ablated surface. Results from 1D electrostatic studies were corroborated, allowing us to continue with the full 2D electromagnetic case.

References
[1] Eason R 2007 *Pulsed laser deposition of thin films applications-led growth of functional materials* (New Jersey: John Wiley & Sons Inc)
[2] Nedelea T and Urbassek H M 2002 *Physics of Plasmas* **9** 3209
[3] Phipps C 2007 *Laser ablation and its applications* (New Mexico: Springer Science Bussines Media LLC)
[4] Gurevich A and Meshcherkin A 1981 *Zh Éksp Teor Fiz* **53** 1810
[5] Gurevich A and Meshcherkin A 1981 *Zh Éksp Teor Fiz* **81** 1295
[6] Dawson J 1983 *Reviews of Modern Physics* **55** 403
[7] Gonzalez C A, Arteaga J A, Gomez Y H, Osorio J, Jaramillo J and Riascos H 2012 *Journal of Physics: Conference Series* **370** 012033
[8] Kato T N and Takabe H 2010 *Physics of Plasmas* **17** 032114
[9] Birdsall C K and Fuss D 1997 *Journal of Computational Physics* **135** 141
[10] Ellegaard O, Nedelea T, Schou J and Urbassek H 2002 *Applied Surface Science* **197-198** 229
[11] Birdsall C K and Langdon A B 1991 *Plasma physics via computer simulation* (Bristol: IOP Publishing Ltd)
[12] Perego M, Howell P D, Gunzburger M D, Ockendon J R and Allen J E 2013 *Physics of Plasmas* **20** 052101
[13] Allen J E and Perego M 2014 *Physics of Plasmas* **21** 034504

Journal of Physics: Conference Series **687** (2016) 012072 doi:10.1088/1742-6596/687/1/012073

2D electrostatic PIC algorithm for laser induced studying plasma in vacuum

C A. Álvarez[1], H Riascos[1] and C Gonzalez[1,2]

[1] Universidad Tecnológica de Pereira, Pereira, Colombia
[2] Universidad de Buenos Aires, Buenos Aires, Argentina.

CORRIGENDUM TO: C A. Álvarez, H Riascos and C Gonzalez 2016 *J. Phys.: Conf. Ser.* 687 012072

The authors would like to correct the title of this work. The correct title for the paper '2D electrostatic PIC algorithm for laser induced studying plasma in vacuum' is:

2D electrostatic PIC algorithm for studying laser induced plasma in vacuum

C A. Álvarez[1], H Riascos[1] and C Gonzalez[1,2]

[1] Universidad Tecnológica de Pereira, Pereira, Colombia
[2] Universidad de Buenos Aires, Buenos Aires, Argentina.

Control of atom-atom entanglement by cavity detuning

O Calderón[1], M R Joya[2] and K M Fonseca Romero[2]

[1] Universidad de los Andes, Bogotá, Colombia.
[2] Universidad Nacional de Colombia, Bogotá, Colombia.

E-mail: kmfonsecar@unal.edu.co

Abstract. Using the atomic levels previously employed to demonstrate a two-photon maser, we show that the atom-atom entanglement produced by the successive passage of two three-level Rydberg atoms across a single-mode lossless cavity can be enhanced using the Stark shift. The atoms are assumed to be prepared in their excited states and to interact with the field during the same amount of time. Employing a physically motivated perturbation-theory approach, we obtain an effective two-level Hamiltonian. We show that, within the limits of validity of the approximation, atomic entanglement can be controlled by changing the frequency of the cavity field, and can be enhanced up to a maximum where the squared concurrence attains the value 16/27.

1. Introduction

Entanglement was initially discussed as a strange phenomenon which would demonstrate the lack of completeness of quantum mechanics [1]; today it is, without question, one of the most important concepts in the fields of quantum communication and quantum information [2]. Entanglement has been produced in several different systems, most of them involving matter-radiation interaction; in particular, in cQED (Cavity Quantum Electrodynamics) experimental setups [3]. In this context, one-photon processes have been extensively discussed [3]. Multiphoton processes, especially two-photon processes [4–7] are well known; however, the investigation of the production and control of entanglement employing these processes have received little attention. A remarkable exception is a recent report [8] in which, using ideas previously discussed in other systems [9, 10], two-photon precesses are proven useful to control the entanglement between two atoms which do not interact directly.

In this contribution, we reconsider the atomic states used in the reference [11] to demonstrate a two-photon maser, and show that the atomic entanglement of two atoms sent through a high-Q cavity can be controlled by an appropriate choice of detuning of the electromagnetic field with respect to the allowed atomic transitions. In section 2, we present the physical system and show why the analysis of [8] fails in this case. Using a physically motivated perturbation theory, we find an effective Hamiltonian; we show that the atomic concurrence can be controlled between zero and $4/3\sqrt{3}$ under a particular choice of interaction time (section 3).

Journal of Physics: Conference Series **687** (2016) 012073 doi:10.1088/1742-6596/687/1/012073

2. Physical system

We consider a situation in which two Rubidium atoms sequentially traverse a high-Q microwave superconducting cavity. The relevant atomic states of the ^{85}Rb atoms are the ground state $39S_{1/2}$ ($|g\rangle$), the intermediate state $39P_{3/2}$ ($|i\rangle$) and the excited state $40S_{1/2}$ ($|e\rangle$) (see Figure 1). The transition frequency between the excited (intermediate) and ground state is $\omega_{eg}/2\pi=$ 136.83174GHz ($\omega_{ig}/2\pi=$ 68.37687GHz), and the frequency of the single-mode cavity field is ω. The atom-field coupling between the intermediate level and the other states is described by the one-photon Rabi angular frequencies $\Omega_{ei}\approx\Omega_{ig}\approx 0.7\text{MHz}=\Omega$.

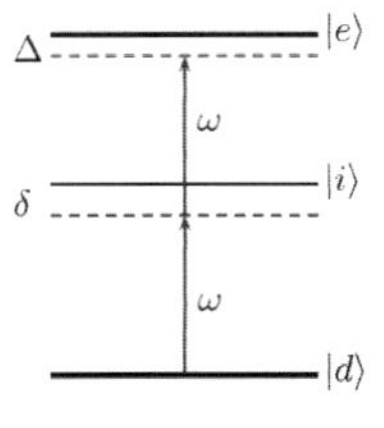

Figure 1. Relevant energy levels of the Rubidium atom.

The Hamiltonian of the system in the dipole- and rotating-wave approximations is

$$\widehat{H} = \hbar\omega\hat{a}^\dagger\hat{a} + \sum_{k=e,i,g} \hbar\omega_k\,|k\rangle\,\langle k| + \hbar\Omega\left(\hat{a}\hat{S}^+_{ig} + \hat{a}^\dagger\hat{S}^-_{ig} + \hat{a}\hat{S}^+_{ei} + \hat{a}^\dagger\hat{S}^-_{ei}\right), \tag{1}$$

Where the operators S^+_{kl} are the raising operators between levels $|k\rangle$ and $|l\rangle$, and the S^-_{kl} are the corresponding lowering operators. The transitions $|g\rangle \leftrightarrow |i\rangle$ and $|i\rangle \leftrightarrow |e\rangle$ are assumed to be almost resonant, i.e., the one-photon detunings are small in the sense that $\delta_g\equiv\omega_i-\omega_g-\omega\ll\omega_i-\omega_g$ and $\delta_e\equiv\omega_e-\omega_i-\omega\ll\omega_e-\omega_i$. We also define the two-photon detuning $\Delta=\delta_e+\delta_g=\omega_{ei}+\omega_{ig}-2\omega$.

Employing a canonical transformation method the authors of [8] find the effective Hamiltonian

$$H_{\text{GMN}} = [\Delta + (\beta_e + \beta_g)a^\dagger a]S_z + \frac{1}{2}(\beta_e - \beta_g)a^\dagger a + g(S^+_{eg}a^2 + S^-_{eg}a^{\dagger 2}), \tag{2}$$

Where $S_z=(|e\rangle\langle e|-|g\rangle\langle g|)/2$. Here, $g=\Omega^2(\delta_g-\delta_e)/(\delta_e\delta_g)$, and the Stark shifts associated with the levels e and g are, respectively, $\beta_e=\Omega^2/\delta_e$ and $\beta_g=\Omega^2/\delta_g$. The authors restrict themselves to the case $\beta_e=\beta=\beta_g$, and show interesting possibilities of control of atom-atom concurrence for the parameters $\beta=\Delta=2g$ and $\beta=-\Delta=g$. We show that the particular choice used in Reference [8] breaks down in the case of a single intermediate level.

The equality of the Stark shifts of the levels $|e\rangle$, β_e, and $|g\rangle$, β_g, imply that the one-photon detunings are smaller than the two-photon detuning Δ, $\delta_g=\frac{1}{2}\Delta=\delta_e$. In this case it is inconsistent to ignore one-photon transitions, an implicit assumption in the derivation of Hamiltonian (2).

Taking as reference values those of zero Stark shift and exact two-photon resonance, striking differences on the values of atom-atom entanglement were obtained in [8] for $\beta/g=1$ and $\Delta/g=-1$. We consider the slightly more general form $\beta/g=x^2$ and $\Delta/g=-y^2$, for x and y real parameters. If we multiply the dimensionless Stark parameter and the dimensionless two-photon detuning, we get $\beta\Delta=-\,(xyg)^2<0$, which is incompatible with the equality $\beta\Delta=\frac{\Omega^2}{\delta_g}\times 2\delta_g=2\Omega^2>0$. Our results, to be presented in the following sections, complement those of Reference [8].

We find an effective two-level approximation by defining a physically motivated perturbation theory. Efficient two-photon processes between the ground and the excited state, and almost no population transfer to the intermediate state, are expected for large one-photon detunings ($|\delta_e|, |\delta_g|\gg\sqrt{n+1}$) and a small two-photon detuning ($|\Delta|\ll\sqrt{n+1}$). We define the one-photon detuning to be of zeroth order, the terms proportional to the dipole moments to be of first order, and the two-photon detuning to be of second order. The effective Hamiltonian, up to second order and ignoring the contribution of the intermediate level, is given by

$$H_{\text{eff}} = \hbar\Big\{\Delta - \frac{\Omega^2}{\delta_g}\Big(\hat{a}^\dagger\hat{a}+1\Big)\Big\}\,|e\rangle\langle e| - \frac{\hbar\Omega^2}{\delta_g}\Big(\hat{a}^\dagger\hat{a}+2\Big)\,|g\rangle\langle g| - \frac{\hbar\Omega^2}{\delta_g}\Big\{\hat{a}^2\widehat{S}^+_{eg} + \hat{a}^{\dagger 2}\widehat{S}^-_{eg}\Big\}. \tag{3}$$

3. Atomic entanglement

We consider a situation in which two atoms, initially prepared in their excited states, traverse the initially empty cavity. The atoms, which are never simultaneously present in the cavity, interact with the cavity field for the same amount of time. Though the atoms never interact directly, they become entangled. If the interaction time is set to $\tau{=}\frac{\pi}{\omega_2}$, the atomic concurrence is $C(s){=}2s\sqrt{1-s}$ where $s{=}\sin^2(\varphi)\sin^2\!\Big(\pi\frac{\omega_0}{\omega_2}\Big)$, $\tan\varphi{=}\frac{\Omega^2}{2\delta}\sqrt{2}/\Big(\frac{\Delta}{2}+\frac{\Omega^2}{2\delta}\Big)$, and

$$\omega_n \;=\; \sqrt{\left(\frac{\Delta}{2}+\frac{\Omega^2}{2\delta}\right)^2+\left(\frac{\Omega^2}{2\delta}\right)^2(n+1)(n+2)}, \quad n=0,2. \tag{4}$$

The interaction time, of the order of hundreds of microseconds, is reasonable in cQED experiments. When $s = 2/3$, concurrence attains its maximum value (under the condition of equal interaction times) $C_{\max}{=}\frac{4}{3\sqrt{3}}{\approx}\ 0.77{<}1$. When the two-photon detuning is varied between -20kHz and 20kHz, atomic concurrence varies from almost zero to $C_{\max}$, then to a local minimum and again to $C_{\max}$, before decreasing to almost zero (as shown in Figure 2).

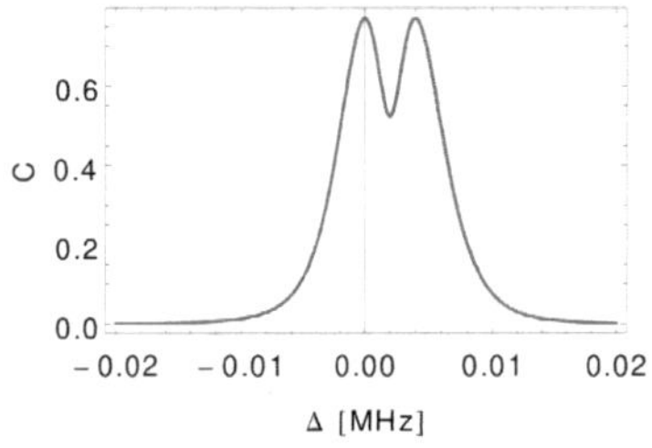

Figure 2. Atomic concurrence as a function of two-photon detuning.

4. Conclusions

We considered the sequential passage of two Rubidium atoms, prepared in specific Rydberg states, through a high-Q cavity, initially prepared in its vacuum state. We have shown that, choosing a particular interaction time, the two-photon detuning can be used to control the atomic entanglement. To assess the feasibility of the present scheme, it is necessary to include dissipation and small changes on the interaction time in the mathematical model.

Acknowledgments

We acknowledge partial support from Universidad Nacional de Colombia under project 9366.

References

[1] Einstein A, Podolsky B and Rosen N 1935 *Phys Rev* **47** 777
[2] Horodecki R, Horodecki P, Horodecki P M and Horodecki K 2009 *Rev Mod Phys* **81** 865
[3] Raimond J M, Brune M and Haroche S 2001 *Rev Mod Phys* **73** 565
[4] Brune M, Raimond J M and Haroche S 1987 *Phys Rev A* **35** 154
[5] Yoo H I and Eberly J H 1985 *Phys Rep* **118** 239
[6] Alsingh P and Zubairy M S 1987 *J Opt Soc Am B* **4** 177
[7] Puri R R and Bullough R K 1988 *J Opt Soc Am B* **5** 2021
[8] Ghosh B, Majumdar A S and Nayak N 2008 *J Phys B At Mol Opt Phys* **41** 065503
[9] Halder M, Beveratos A, Gisin N, Scarani V, Simon C and Zbinden H 2007 *Nature Phys* **3** 692
[10] León J and Sabín C 2009 *Phys Rev A* **79** 012301
[11] Brune M, Raimond J M, Goy P, Davidovich L and Haroche S 1987 *Phys Rev Lett* **59** 1899

From sequential to correlated tunneling of two bosons

R Avella[1], R R Rey González[1] and K M Fonseca Romero[1]
[1] Universidad Nacional de Colombia, Bogotá, Colombia.
E-mail: kmfonsecar@unal.edu.co

Abstract. The sequential to correlated-tunneling transition of two identical bosons confined in a symmetrical double well potential is studied using a time-dependent variational formalism. Increasing the strength of the short-range boson-boson interaction, the probability of finding both bosons in the same well allow to identify three regimes. For small values of the boson-boson coupling constante, this probability display beatings. Then, the probability of finding both bosons in the same well becomes quasi-periodical in time, for almost all values of the coupling constant; however, there is a zero-measure set, which contains an infinity of points, for which this probability is time-periodic. Finally, for larger values of the coupling constant, this probability changes periodically. Based on the dynamics of the probability of finding one boson on each well and increasing the boson-boson interaction, we find that tunnelling changes from a regime where one boson tunnels after the other to a regime in which they tunnel together.

1. Introduction

Tunneling, one of the most characteristic quantum effects, has been studied in many systems. In the context of cold atoms, Josephson tunneling of Bose-Einsteins Condensates (BEC) in multi-well systems have received considerable attention [1–3]. Recent experimental techniques have allowed the exploration of the tunneling of a few bosons [4]. In fact, optical lattices permit a high control of both the geometry and the depth of confining potentials [5]. In this contribution, we study the tunneling of two bosons in a double-well potential using a time-dependent variational approach.

2. Physical system

We consider two ^{87}Rb atoms, with mass of 144.42×10^{-27}kg, to be confined in a double well potential (Figure 1, dashed line). The width of each well and of the $h\times3$kHz potential barrier are 10μm and 2μm [6], respectively. Experimental reports of tunneling in these systems describe a sequential to correlated-tunneling transition [4, 7]. Previous theoretical treatments [8–11] show sequential tunneling for weak boson-boson interactions and correlated tunneling near the fermionization limit.

We use dimensionless quantities, distances in units of $\xi=1\mu$m, energies in units of $\epsilon=10^{-31}$J, and times in units of $\tau=\frac{\hbar}{\epsilon}$. For example, the height of the potential barrier is $v_0\approx19.88$ in these units. The single-particle Hamiltonian then becomes

$$h_0(x) = -\frac{\hbar^2}{2m\xi^2\epsilon}\frac{d^2}{dx^2} + v(x). \tag{1}$$

The (single-particle) eigenvectors can be given in analytic form, but in terms of the eigenvalues of $h_0(x)$, which are the solutions of a trascendental equation. The two smallest eigenenergies, $E_0 \approx 0.03548$ and $E_1 \approx 0.03552$ are well separated from the other eigenvalues, $E_n > 0.14$ for $n \geq 2$.

We choose the first eigenfunction $\psi_0(x) = \langle x|0\rangle$ to be non-negative everywhere, and the second, $\psi_1(x) = \langle x|1\rangle$ to be positive for $x \geq 0$. Thus, the wavefunction $\langle x|L\rangle = \psi_L(x) = \frac{1}{\sqrt{2}}(\psi_0(x) - \psi_0(x))$ ($\psi_R(x) = \frac{1}{\sqrt{2}}(\psi_0(x) + \psi_0(x))$) describes a particle in the left (right) well. The tunneling time from one well to be other is 7.5×10^4.

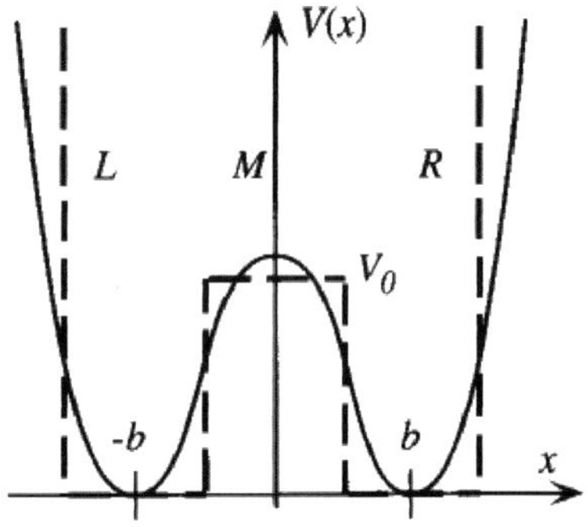

Figure 1. Realistic (solid line) and square double well potentials (dashed line).

The two-particle Hamiltonian is $H(x_1, x_2) = h_0(x_1) + h_0(x_2) + \lambda\delta(x_1 - x_2)$, where x_1 and x_2 are the dimensionless positions of the two bosons, and λ is the interaction coupling constant.

3. Time-dependent variational principle

The time-dependent variational principle, which provides and approximation of the exact dynamics, corresponds to the minimization of the action functional $S = \int_0^T dt\,(i\,\langle\Psi|\Psi\rangle - \langle\Psi|H|\Psi\rangle) = \int_0^T dtL(t)$. In order to investigate the tunneling dynamics, we use the trial wave function $\Psi(x_1, x_2, t) = c_1(t)\psi_0(x_1)\psi_0(x_2) + \frac{c_0(t)}{\sqrt{2}}\,(\psi_1(x_1)\psi_0(x_2) + \psi_0(x_1)\psi_1(x_2)) + c_{-1}(t)\psi_1(x_1)\psi_1(x_2)$.

The "Lagrangian" $L(t)$ depends on the value of the integrals $I_{ij} = \int dx\phi_i^2(x)\phi_j^2(x)$, $i,j = 1,2$. By setting $E_0 = \frac{E_M - \delta}{2}$ and $E_1 = \frac{E_M + \delta}{2}$, and employing the approximation $I_{ij} \approx I = 0.07235$, with an error of about 3 parts per ten thousand, we obtain the Lagrangian,

$$\begin{aligned} L(t) &= ic^*_{-1}(t)\dot{c}_{-1}(t) + ic^*_0(t)\dot{c}_0(t) + ic^*_1(t)\dot{c}_1(t) - \Big\{E_M + \delta(-|c_1(t)|^2 + |c_{-1}(t)|^2) \\ &\quad + \lambda I\Big[1 + |c_0(t)|^2 + c_{-1}(t)c^*_1(t) + c^*_{-1}(t)c_1(t)\Big]\Big\} \end{aligned} \tag{2}$$

The tunneling process depends on the coefficients $c^*_m(t)$, $m = 0, \pm 1$, whose equations of motion are given by $\frac{d}{dt}\left(\frac{\partial L}{\partial \dot{c}^*_m(t)}\right) = \frac{\partial L}{\partial c^*_m(t)}$.

4. Return probability

In order to investigate the tunneling process, we assume that both bosons are initially on the left well, i.e., the initial state of the system is $|LL\rangle$. The probability that they remain on the left well is ($g = \lambda I$),

$$P_{LL} = \frac{1}{2}\left(1 + \cos(gt)\cos\left(t\sqrt{g^2+\delta^2}\right)\right) + \frac{g\sin(gt)\sin\left(t\sqrt{g^2+\delta^2}\right)}{2\sqrt{g^2+\delta^2}} - \frac{\delta^2\sin^2\left(t\sqrt{g^2+\delta^2}\right)}{4\,(g^2+\delta^2)} \tag{3}$$

The probability of finding one boson on each well, $P_{LR}=\frac{\delta^2}{2(g^2+\delta^2)}\sin^2\left(\sqrt{g^2+\delta^2}t\right)$, reaches a maximum of $\frac{1}{2}$ for vanishing boson-boson interaction and decreases to zero as the interaction is increased. Thus, the tunneling is sequential for small values of the interaction constant and correlated for large values.

The behaviour of the quantum return probability P_{LL} present three distinct regions. The first region, corresponds to very small values of the effective coupling constant g, when the return probability can be approximated as $P_{LL}\approx\frac{1}{2}\left(3+4\cos(gt)\cos(\sqrt{g^2+\delta^2}t)+\cos(2\sqrt{g^2+\delta^2}t)\right)$. In this region, the most important characteristic is the presence of beatings: a low-frequency envelope $\cos(gt)$, which modulates the high-frequency periodic oscillations $\cos(\sqrt{g^2+\delta^2}t)$. The term $-1+\cos(2\sqrt{g^2+\delta^2}t)$ becomes important when the amplitude of the beating, $\cos(gt)$, is small. These details are shown in Figure 2.

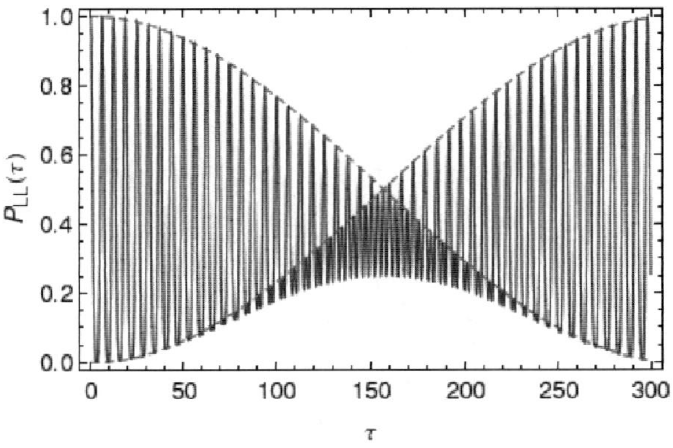

Figure 2. Population of the state $|LL\rangle$ with time measured in units of $1/\delta$, for $g=\frac{\delta}{100}$.

The effective interaction coupling constant g and the energy δ are of the same order in the second region, where the behaviour of the return probability is quasi-periodic (See Figure 3). In this region there are infinitely many values of the coupling constant where the return probability becomes periodic: when $g=\frac{\alpha\delta}{\sqrt{1-\alpha^2}}$, where α is a rational in $[0,1)$ (see Figure 4(a)). In fact, in this case the return probability has contributions corresponding to the angular frequencies $\omega,\omega(1\pm\alpha)$, with $\omega=\sqrt{g^2+\delta^2}$.

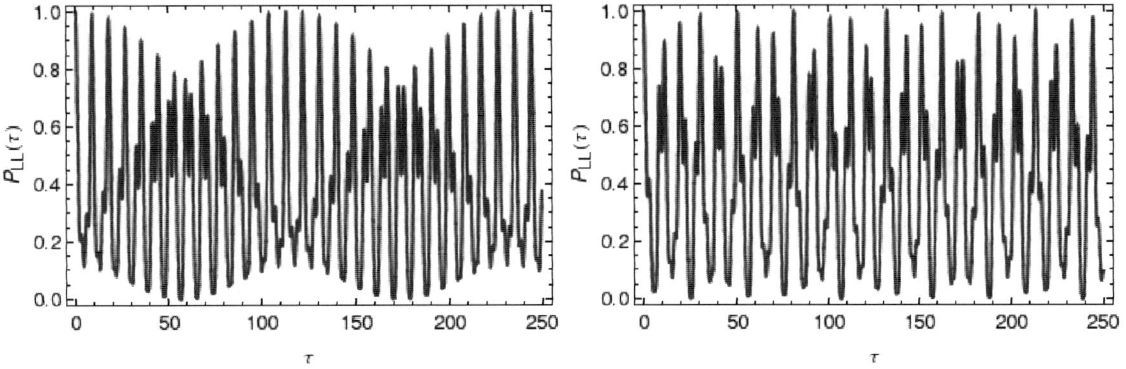

(a) (b)

Figure 3. Quase-periodic evolution of the population of the state $|LL\rangle$ with time measured in units of $1/\delta$, for (a) $g=\frac{\delta}{3}$ and (b) $g=\frac{\delta}{2}$.

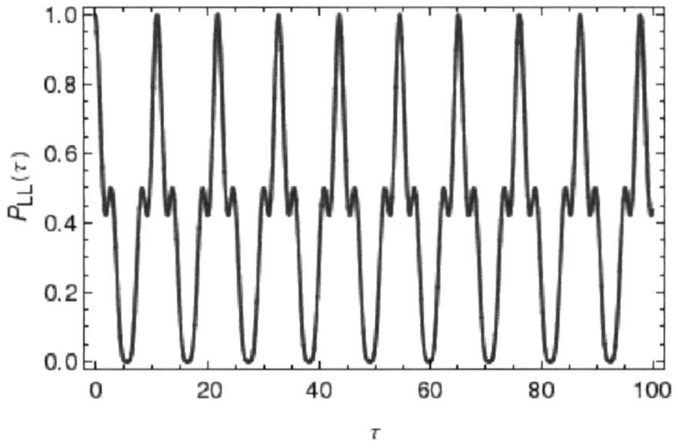

(a)

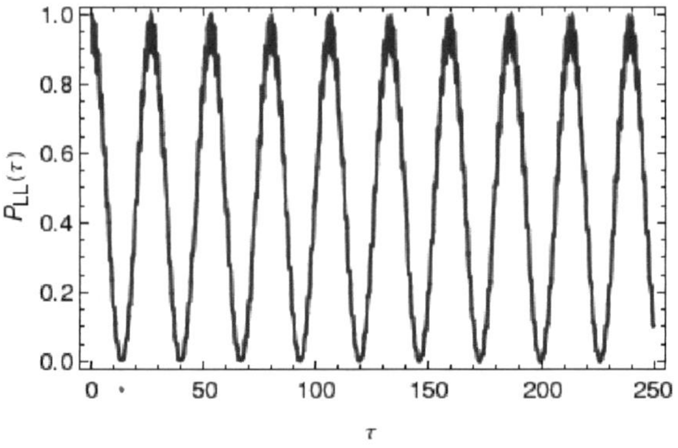

(b)

Figure 4. Periodic evolution of the population of the state $|LL\rangle$ with time measured in units of $1/\delta$, for (a) $g=\frac{\delta}{\sqrt{3}}$ and (b) $g=2\delta$.

The last region corresponds to an effective coupling constant much larger than the energy difference δ, when the return probability becomes periodic $P_{LL}(t)\approx\frac{1}{2}\left(1+\cos\left(\frac{\delta^2 t}{2g}\right)\right)$, (see Figure 4(b)).

5. Conclusions

We have considered the problem of tunneling of two interacting cold bosons in a double well potential. Using a simple trial function, we have used the time-dependent variational principle to show that there is a transition from sequential to correlated tunneling as the coupling constant is increased. For small values of the coupling constant, the return probability displays frequency beating; for intemediate values, the return probability is quasi-periodic except for an infinity of values when it becomes periodic. For large values of the coupling constant, the return probability is periodic; the larger the coupling constant the larger the tunneling period.

Acknowledgments

We acknowledge partial support from Universidad Nacional de Colombia under project 9366.

References

[1] Smerzi A, Fantoni S, Giovanazzi S and Shenoy S R 1997 *Phys Rev Lett* **79** 4950
[2] Javanainen J 1986 *Phys Rev Lett* **57** 3164
[3] Cataliotti F S, Burger S, Fort C, Maddaloni P, Minardi F, Trombettoni, Smerzi A and Inguscio M 2001 *Science* **293** 843
[4] Fölling S, Trotzky S, Cheinet P, Feld M, Saers R, Widera A, Müller T and Bloch I 2007 *Nature* **448** 1029
[5] Sebby J, Anderlini M, Jessen P and Porto J 2006 *Phys Rev A* **73** 033605
[6] Yong-Il S 2006 *Experiments with Bose-Einstein condensates in a double-Well potential* (USA: MIT)
[7] K Winkler K, G Thalhammer G and Lang F 2006 *Nature Lett* **441** 853
[8] Roy A and Reichl L 2008 *Phys Rev A* **77** 033418
[9] Analabha R and Reichl L 2009 *Time-of-flight signatures of two boson in a double well* arXiv:0903.5029v4
[10] Zöllner S, Dieter H and Schmelcher P 2008 *Phys Rev A* **78** 013621
[11] Zöllner S, Dieter H and Schmelcher P 2007 *Phys Rev Lett* **100** 040401

Morphological analysis of carbon steels using fractal geometry

D A Prada[1], C P González[1], P E Vera[1] and M A Álvarez[2]
[1] Universidad Pontificia Bolivariana, Floridablanca, Colombia.
[2] Universidad Industrial de Santander, Bucaramanga, Colombia.

E-mail: duwamg.prada@upb.edu.co

Abstract. In this paper we present the preliminary results of morphological analysis of $Fe_{59}Mn_{36.5}Al_{3.10}C_{3.56}Cu_{0.237}$ %Wt alloy. This alloy was prepared by mechanical alloying with various milling times, then hot compacted at various pressures and finally underwent presented a sintering process. The samples were characterized structurally by experimental techniques such as: X-ray diffraction and scanning electron microscopy. Each micrograph was analysed by fractal dimension using the Box Counting method in 2D; this type of tool allows you to assign a numerical value to the region which has been applied the technic. This numerical value is associated with the properties obtained by experimental techniques such as microhardness.

1. Introduction

The system Fe Mn Al disordered alloys, in the austenite phase [1], arouses great interest both scientific and technological due to its mechanical and corrosion resistance properties. In order to stabilize the austenitic phase, as is known, gives the alloy optimum mechanical properties, at both room temperature and high temperature, were varied preparation methods concerning adding some alloying powder metallurgy and [2-4], which are miscible with the powder metallurgy.

We prepared add samples 3.56% Wt of C and 0.237% Wt Cu Fe Mn Al to system with the powder metallurgy process: mechanical alloying (6 and 12h), uniaxial compaction in hot (T=250°C and pressures of 400, 600 and 800MPa) and sintering (T≈1150°C). The parts obtained are made of mirror finish them by metallographic process. The pieces were analysed with scanning electron microscopy (SEM) also obtaining energy spectra at various points in each sample. In order to analyse the morphology of each sample was developed Box- Counting 2D software.

2. Mathematical method

The Fractal Geometry is a branch of mathematics, very attractive fractal objects that are possible to construct the notion of self-similarity, dynamic systems, chaos, orbits and dimension, both topological and Hausdorff, which have become useful tools in numbers disciplinary fields such as mathematics, science, art, etc. [5-7].

Besides the self-similarity, the fractal objects have an idea out of the common; the fractal dimension. The dimension that is assigned by convention to certain geometrical and physical objects, is associated with a number infinite variables, for example, a cube is assigned a triple defined directly by the thickness, the width and height thereof, then the size of this object is three. This type of dimension is known as the topological dimension.

The fractal dimension, as properly its name suggests, is a fractional dimension and is determined by a rational number. This type of dimension has been widely used for example to measure the length of the coast of an island. Dimensionality value assigned to the objects of study under some technical spreadsheets fractal dimension is a value that ensures as Mandelbrot [5] but you do not find transitional areas without a clearly defined structure, I will identify areas with fractal whose dimension is either a fraction either an integer "abnormal", which also indicates an irregular or interrupted state. Furthermore, since many objects are systems such as comments, which Mandelbrot [5], in that they are composed of many different parts, hinged together, the fractal dimension describes one aspect of this rule joint.

The Box-counting method or by counting boxes, has been used to calculate the fractal dimension of certain objects that are represented in a plane [5]. The fractal dimension calculated using the Box-Counting method meets the slope of the linear fit between the points and the size of the grid with the relation $D=\ln(N)/Ln(k)$ where N is the finite number of subfigures or subsets all congruent k satisfies $r=1/k$ numerical value where r is a shrinkage factor. The accuracy of the method is improved with a greater number of iterations, the depth of these iterations fractal structures is not limited, that is, the size of the side of the case can be reduced many times as needed, but the processed images resolution it is the limiting factor to count points. Our goal is to apply this method of calculating fractal dimension for each micrograph obtained by SEM at the same points where energy spectra were obtained.

3. Results and discussion

Figure 1 shows a micrograph (a) for 6h and with a compaction pressure of 800MPa hot shown. To a point, was taken it samples the energy spectrum (b) which shows the alloying components and the existence of traces of oxygen attributed to air trapped in the compacted surface. Were taken spectra at several points and similar spectra were found not shown here, which shows a good homogeneity achieved during the process.

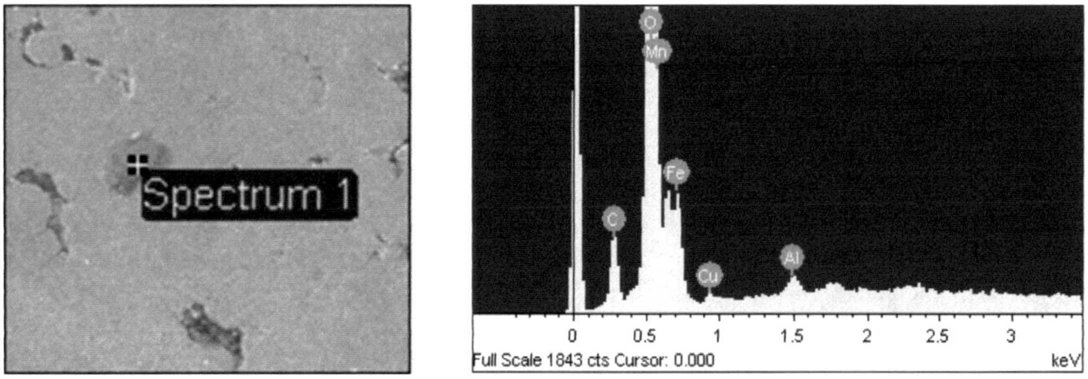

(a) (b)

Figure 1. Micrograph (a) of the alloy with 6 hours 800MPa and its corresponding energy spectrum spectrum (b) taken him in 1.

In Figure 2, occurs the dimensional analysis method achieved with Box Counting. The figure shows the region where the zoom is performed and calculated the fractal dimension with Box-Counting.

Figure 3, shows the micrograph taken for the sample with 12 hours 400MPa at the first point where the energy spectrum was obtained, doing on the image zoom and calculated the Box-Counting dimension shown in Figure 4.

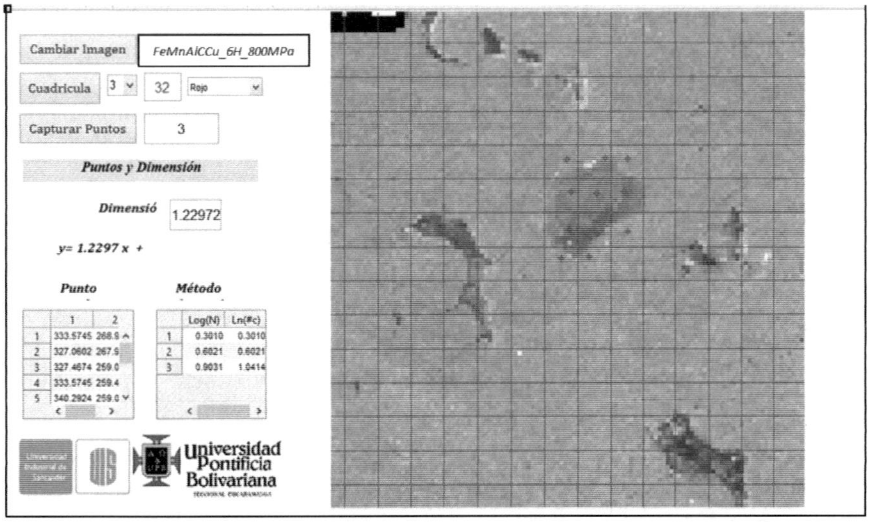

Figure 2. Fractal dimension micrograph 6 hours and 800MPa pressure spectrum one.

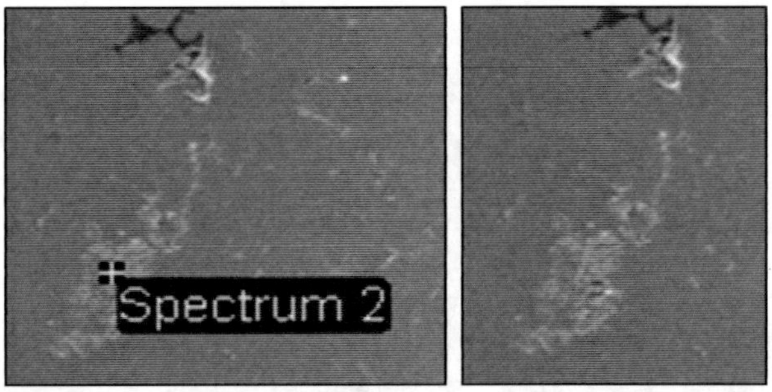

Figure 3. Fractal dimension micrograph 12 hours and 400MPa pressure spectrum 2.

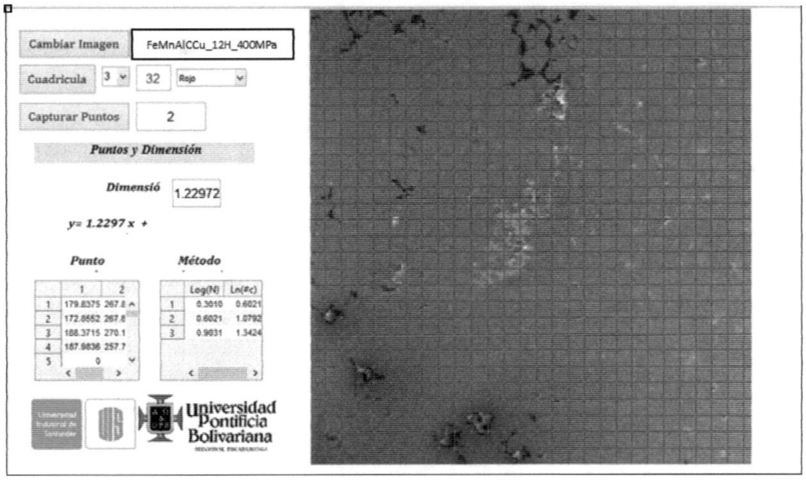

Figure 4. Fractal dimension shows 12 hours and 400MPa pressure spectrum 2.

Similarly was made to various points, the fractal dimension obtained in each case, are summarized in Table 1. It may be noted that the dimensionality is not affected by pressure changes or by compaction hours of mechanical alloying. With what might be said that there is a process of homogenization.

Table 1. Fractal dimension obtained for each spectrum.

Sign-spectrum	Fractal dimension
FeMnAlCCu_6H_800MPa – spectrum 1	1,22972
FeMnAlCCu_6H_800MPa – spectrum 3	1,16096
FeMnAlCCu_12_400MPa – spectrum 1	1,16096
FeMnAlCCu_12_400MPa – spectrum 2	1,22972

At the points where the spectra were taken and elsewhere in the sample data Vickers (HV) microhardness were taken with a 500g load and indentation time of 5s, finding an average for the samples with 6h 800MPa a microhardness of 244.52±0.5HV and for 12h and 400MPa shows a hardness of HV 224.8±0.5. It can be concluded that there is in each sample homogeniezación process by the average value and additionally microhardness results are better for 6h and 800MPa. However, it is important to note that the values obtained for 6 h and 800MPa to 400MPa 12h shows are better than those reported by Cruz B et al [8] almost certainly due to the conditions of sample preparation.

4. Conclusions

Each micrograph was analysed from one side taking energy spectra at various points of the sample, finding similar spectra in each. In all spectra, the presence of traces of alloying constituents, other trace oxygen. In addition, the fractal dimension analysis using the counting method 2D cases, it was possible to assign a numerical value to each of the regions where energy spectra were obtained; corresponding to the dimensionality of the samples, where that value was calculated by the relation $D=\ln(N)/\mathrm{Ln}\ (k)$ where N is the finite number of subfigures or subsets all congruent k satisfies $r=1/k$ numerical value where r is a shrinkage factor, and has been designated in relation to roughness, irregularity or interruption state condition, with the outline of each of the analysed micrographs. Dimensionality assigned at each point of each sample was very close, which could be associated with a homogeneity achieved in the development process of each sample. This homogeneity due to alloy diffusion process, gave better results than those reported by microhardness Cruz B et al [8] for such alloys.

References

[1] Rico Castro M M, Medina M H, and Pérez Alcázar G A 2000 *Hyp Int* **128** 503
[2] Rodríguez R and Pérez Alcázar G A 2000 *Phys Stat Sol b*. **220** 441
[3] Valderruten J F, Pérez Alcázar G A and Civale L 2004 *Journal of Materials Sci* **39** 5445
[4] Pérez Alcázar G A, González C and Cruz B 2003 *Hyp Int* **148-149** 295
[5] Mandelbrot B 2006 *Los objetos fractales forma, azar y dimension 6th ed* (España: Metatemas)
[6] Falconer K 2006 *Fractal geometry mathematical foundations and applications 2nd ed* (India: Wiley)
[7] Barnsley M 2012 *Fractals everywhere* (USA: Dover Publications)
[8] Cruz B and Peréz Alcázar G A 2005 *Scientia et Technica* **XI-28** 157

Self-consistent simulation of an electron beam for a new autoresonant x-ray generator based on TE_{102} rectangular mode

V D Dugar-Zhabon[1], E A Orozco[1] and A M Herrera[1,2]

[1] Universidad Industrial de Santander, Bucaramanga, Colombia.
[2] Heildelberg Institute for Theoretical Studies, Heildelberg, Germany.

E-mail: eaorozco@uis.edu.co

Abstract. The space cyclotron autoresonance interaction of an electron beam with microwaves of TE_{102} rectangular mode is simulated. It is shown that in these conditions the beam electrons can achieve energies which are sufficient to generate hard x-rays. The physical model consists of a rectangular cavity fed by a magnetron oscillator through a waveguide with a ferrite isolator, an iris window and a system of dc current coils which generates an axially symmetric magnetic field. The 3D magnetic field profile is that which maintains the electron beam in the space autoresonance regime. To simulate the beam dynamics, a full self-consistent electromagnetic particle-in-cell code is developed. It is shown that the injected 12keV electron beam of 0.5A current is accelerated to energy of 225keV at a distance of an order of 17cm by 2.45GHz standing microwave field with amplitude of 14kV/cm.

1. Introduction

X-ray radiation of traditional sources, where the electrons are accelerated by an electrostatic field, runs into the range of 50-150keV. The upper limit of the x-ray energy is determined by electric isolation problems. However, among the existing methods for producing x-rays, of great importance for our research are those that are based on the electron cyclotron phenomenon. A cyclic electron accelerator as an x-ray source is proposed in [1]. A compact x-ray source based on a mirror trap where electron heating is realized on the fundamental ECR frequency and its harmonics is described in [2]. Electron acceleration by microwaves of TE_{111} mode in a magnetostatic field under the electron cyclotron resonance conditions is studied theoretically [3]. An ECR acceleration of electrons in TE_{101} rectangular resonant cavity in the presence of a uniform magnetic field is simulated [4]. The results of this study make it possible to fabricate and test an x-ray source where the electrons are accelerated in spiral orbits in the cavity midplane [5]. It is worth noting that in the above mentioned accelerators the electron cyclotron frequency can be maintained equal to the microwave frequency within a very limited number of cyclotron periods, which severely constrains the maximum operating energy. Theoretically speaking, it is possible to circumvent the problem by increasing a homogeneous magnetic field at the time with a rate that compensates the relativistic electron mass rise. In such a manner the resonance can be maintained and large electron energies can be achieved [5, 6]. Recently, we have analyzed the cyclotron resonance speeding of electrons in the combined inhomogeneous magnetostatic and standing microwave fields both through numerical experiment and analytical analysis [7–9].

The 3D inhomogeneous magnetic field profile is determined by the necessity for sustaining the ECR conditions along all electron trajectories. This type of electron cyclotron resonance is named spatial autoresonance acceleration (SARA). An X ray source based on this concept has been certificated recently [10]. In the present paper, a physical scheme of an autoresonant accelerator system based on TE_{102} rectangular cavity is presented. This device can be classified as compact due to its characteristic dimensions of about 30cm. The autoresonant interaction between 0.5A electron beam and 2.45GHz standing microwave field of 14kV/cm of amplitude is studied. Taking into consideration that the problem is relativistic, a full electromagnetic particle-in-cell (PIC) code [12] is developed.

2. Theoretical formalism

The physical scheme of the autoresonant accelerator which can produce x-rays is shown in Figure 1. A magnetron emits 2.45GHz microwaves into TE_{102} rectangular cavity through a waveguide with a ferrite isolator and an iris window. The field amplitude in the cavity is $E_{0l} = 14kV/cm$. An axially symmetric magnetostatic field can be formed by a system of DC current coils or permanent magnetic rings. An electron gun injects the electron beam into the cavity along the magnetic field axis. The magnetic field parameters are those which maintain the electron beam in the space cyclotron autoresonance regime [7, 8]. To produce an x-ray flux, a molybdenum target is placed at the acceleration zone end. A linear polarized TE_{102} electromagnetic field is excited in the cavity.

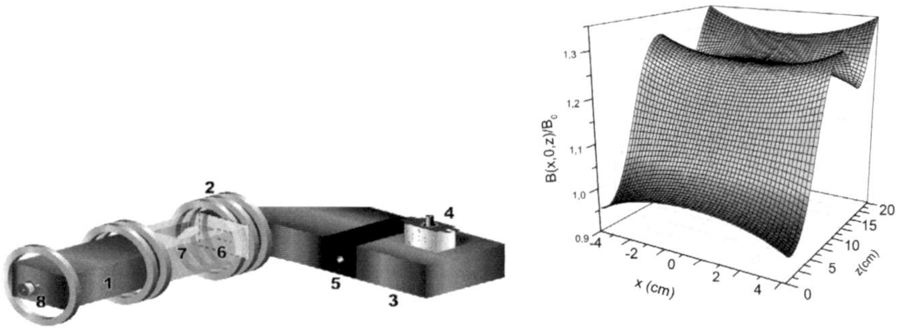

(a) (b)

Figure 1. (a) Physical scheme of the autoresonant system: 1-cavity, 2-current coils, 3-waveguide, 4-microwave generator, 5-ferrite isolator, 6-iris microwave window, 7-non-magnetic metal target, 8-electron gun. (b) magnetic field profile in the $y = 0$ plane.

For the electron motion analysis, the Cartesian coordinates, whose origin is set in the center of the cavity lateral side (Figure 1(a)), are applied. The z-axis coincides with the cavity axis along which the electron beam is injected; the other axes are shown in Figure 4. In these coordinates, the width, height and length of the cavity dimensions are Lx=7.74cm, Ly=3.82cm and Lz=20cm, respectively.

The magnetostatic field is generated by three axisymmetric coils using the same parameters given in [7]. This profile is shown in Figure 1(b). The self-consistent simulation of this system is studied by using the particle in cell method whose algorithm is presented in Figure 2 through a computational time cycle which involves the following steps:

(i) calculation of the current densities in the mesh points on the basis of the superparticle (SP) position and velocity data,

(ii) Computation of the fields on the mesh points,

(iii) calculation of the forces exerted on the *SPs*,

(iv) calculation of new positions and velocities of the *SPs* through integration of their equations of motion.

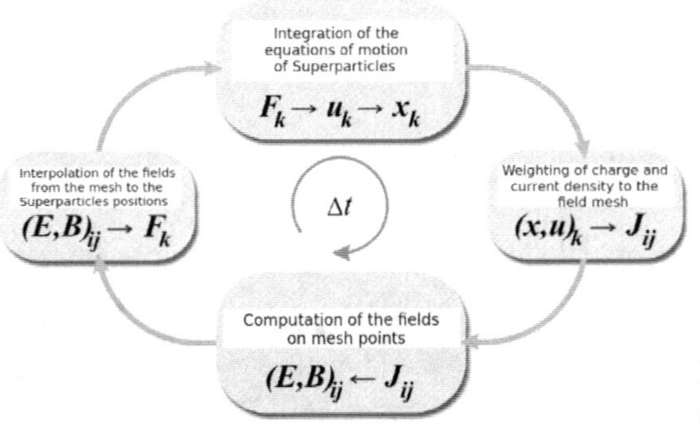

Figure 2. Particle in cell (PIC) algorithm.

In our simulations, the current density is found using the charge conservative method proposed by Umeda [11] and the beam electric and magnetic fields on the mesh points are found using the Yee method [13]. It is worth noting that the TE_{10} microwaves that propagate along the waveguide and are incident on the iris window excite in the cavity a linearly polarized TE_{102} mode which is calculated using the method described in [14]. The electric and magnetic field on the *SP* positions, $\vec{E}_P$ and $\vec{B}_P$, are calculated by a trilinear interpolation of the $\vec{E}$ and $\vec{B}$ mesh node data. To obtain the total magnetic field, fields produced by the current coils and by the beam are summed up. The *SP* dynamics is described by the relativistic Newton-Lorentz equation which is solved numerically for all the *SPs* using the Boris leapfrog procedure [12]. In order to accelerate the calculation process and reach faster the microwave steady state regime in the cavity, the Hanning method is applied [15]. To avoid non physic actions of the microwaves, we use the Uniaxial Perfectly Matched layer (UPML) technique [16]. The phase shift between the particle-velocity vector and the right-hand electric component of the microwave field, represented by φ, is calculated indirectly beginning from the phase of TE_{102} electric field component. The simulation time step Δt is taken according to the Courant stability condition [16].

3. Results and discussion

The numerical simulations on a spatial autoresonance interaction of an electron beam and microwaves are realized under the above-mentioned physical parameters. The electron beam of 0.5A is injected at the initial energy of 12keV through an orifice of 0.2cm into the cavity along its axis which taken as z-axis. The total length of 32.34cm, for which the simulations are fulfilled, is represented by the waveguide and the cavity. To excite a 14kV/cm field in the cavity, an input power is to be equal to 2.4MW in a continuous mode. Such a high level of the microwave power makes the system operate in a pulse regime. For example, the 2.4MW pulses of $1\mu s$ duration at 25Hz repetition frequency results in an acceptable consumed power of 60W. The electrons beam also needs to be injected in a pulsed regime at the same frequency. In these conditions the beam length will be of the order of 0.5m.

To execute the simulations during an appropriate time lapse the beam electrons which are injected during a microwave period are divided into 8.49×10^3 superparticles. The calculations are fulfilled on a rectangular mesh of $\Delta x = \Delta y = \Delta z = 0.07cm$ and using a time step

$\Delta t = 1.315\ ps$. The simulations are considered finished when the electrons impinge on a non-magnetic metal target whose longitudinal position is z=17.5cm. The dimensions of the target are $LT_x = 0.28cm$, $LT_y = 1.2cm$ and $LT_z = 1.0cm$.

The steady electric field distribution in the $y = 0$ cavity plane without the electron beam is shown in Figure 3(a). One can see the TE_{102} electric field image with distortions in the vicinity of the iris microwave aperture (the black line) and above the metal target (yellow spot) whose z-coordinate is 17.5cm. The first one is irrelevant because the beam does not appear there. Figure 3(b) evidences that the electric field strength is increased significantly at the target surface. But this distorted electric field does not prevent the beam from bombarding the target because it pointed perpendicularly to the target surface.

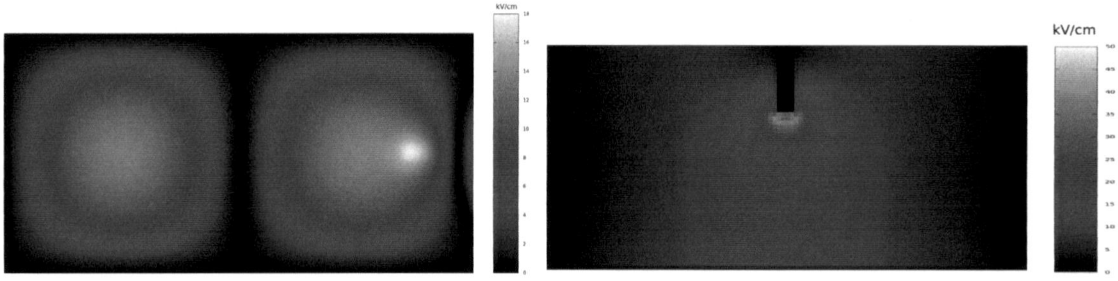

(a) (b)

Figure 3. Microwave electric field before the electron beam injection: (a) in the longitudinal $y = 0$ plane and (b) in the transverse $z = 17.5cm$ plane.

Figure 4 shows some snapshot of electrons beam evolution (red dots) from the injection point up to an impact on the target. Close to the injection point, the electrons move along an almost straight line due to low magnitudes of the microwave field in this region (see Figure 4(a)). In the longitudinal magnetostatic field the electrons move in a counterclockwise direction around the magnetic field axis. In Figure 4(b), this motion is indicated by the black arrows. In the position z=10cm the microwave field changes its phase to π. However, after some microwave periods the electrons regain their acceleration phases because of high frequency of the microwaves. Figure 4(d) shows the moment of the impact of electrons on the target.

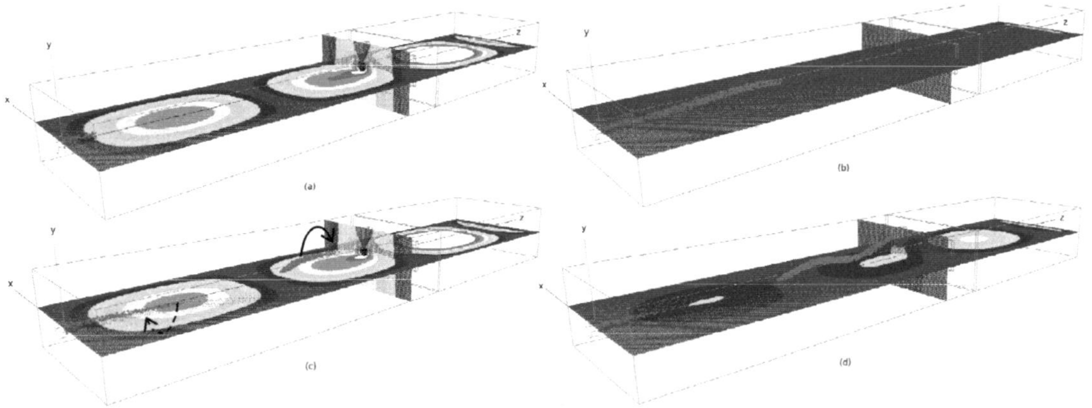

Figure 4. Some snapshot of electrons beam evolution simulation from its injection up to impact with the target.

Figure 5 shows that the self-consistent electric field on the beam surface field in the plane z=10cm is as high as 1.6kV/cm and thus cannot be ignored although this value remains significantly smaller than the microwave amplitude of 14kV/cm.

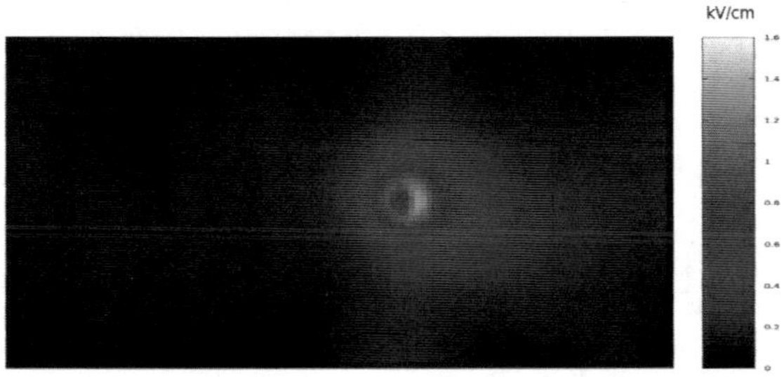

Figure 5. Transversal self consistent electric field strength in $z = 10cm$ plane.

The evolution of the phase shift between the particle velocity and the circular right-hand electric component of microwaves along the beam trajectory is shown in Figure 6(a). One can see that the electrons are injected with arbitrary phases with respect to the microwave field phase. Then, in the process of the interaction with the microwave field some electrons are accelerated, with some of them losing their energy, which results in a phase-focalization effect. The phase-focalization process is completed after 3cm run when the phase shift is observed practically equal to the exact cyclotron resonance value of π.

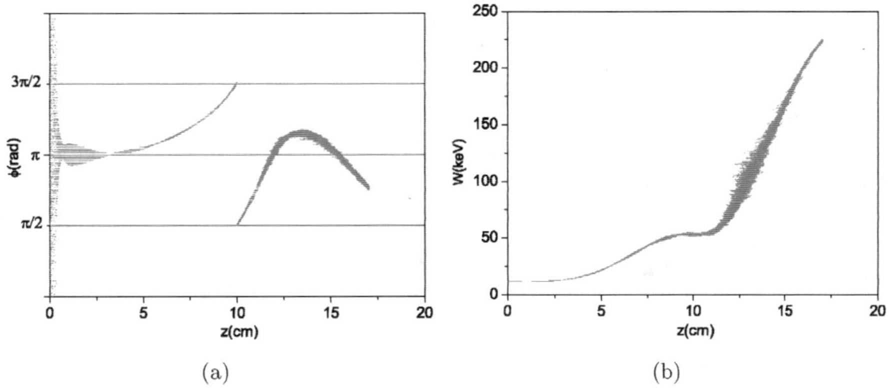

Figure 6. Phase shift between the right-hand circular polarized component of the electric microwave field and the transversal velocity of electrons (a) and the evolution of the beam energy (b) along the beam trajectory.

The rate of beam energy change as the beam moves into the increasing magnetostatic field is presented in Figure 6(b). The energy stop rising in the zone of z=10cm is due to the change of the microwave polarization vector to the opposite sense. In spite of the polarization direction jump all the electrons remain in the acceleration band because the electric field in this zone is very low at a local decrease of the magnetostatic field (see Figure 1). The latter provokes reduction of the cyclotron frequency magnitude. The evolutions of the transversal and longitudinal velocity components are shown in Figure 7.

One can see that in the zone 12cm<z<14cm the longitudinal electron velocities have their minimum values that are accompanied by an increase of the local density that causes rise of the self-consistent electric field and the dispersion of the energy in this zone (see Figure 6). Finally, the electrons impact on the target at a position of z=17cm with an energy of 225keV. In this position, the longitudinal velocity is so small that during the period of the cyclotron rotation the beam displaces in the longitudinal direction at a distance of 4-6mm which is smaller than the dimension of the target along z axis whose area is $1.2 \times 1.0 cm^2$. The transversal velocity and energy attain their maxima magnitudes in the impact position that guarantee the production of hard 170keV.

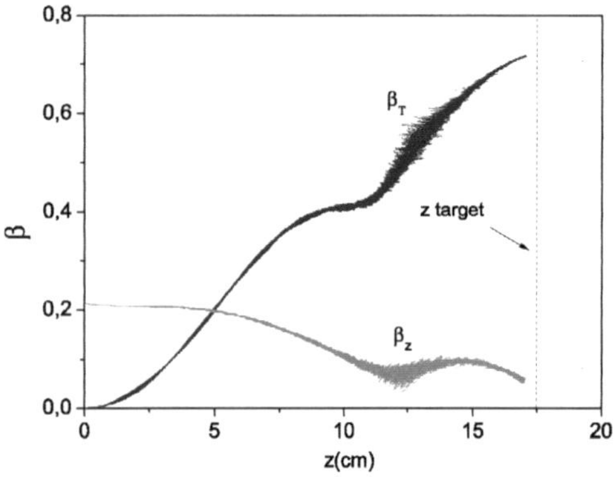

Figure 7. Evolution of the longitudinal (red dots) and transversal (blue dots) components of electron velocities along the beam trajectory.

4. Conclusions

The realized numerical experiment shows that the studied system based on a TE_{102} rectangular cavity can create the conditions for space cyclotron autoresonance acceleration of electron beams of order of 0.5A to energies of hundred electron-volts which can be used for an x-ray production. For the simulations, the full 3D electromagnetic software is developed. To enhance the effectiveness of the studied system in reference to a microwave power harnessing, an optimization of waveguide-cavity coupling system is scheduled.

Acknowledgments

This work is supported in part by the Fundación para la Promoción de la Investigación y la Tecnología (Bank of Republic) under contract No 201123. The authors would like to thank Orlando Duarte for his technical help.

References

[1] H R Gardner et al 1990 *Rev Sci Instrum* **61-2** 724
[2] K S Golovanivsky and V D Dugar-Zhabon 1992 *A microwave x-ray source and methods of sterilization* (US: Patent No WO 9317446)
[3] R Baskaran and T Selvakumaran 1998 *Rev Sci Instrum* **69-10** 3510
[4] R Baskaran and T Selvakumaran 2000 *Rev Sci Instrum* **71-2** 1203
[5] K S Golovanivsky 1983 *IEEE Trans Plasma Sci* **PS-11-1** 28
[6] O Gal 1989 *IEEE Trans Plasma Sci* **17-4** 622
[7] V D Dugar-Zhabon, E A Orozco and A Umnov 2008 *Phys Rev ST-AB* **11** 041302
[8] V D Dugar-Zhabon and E A Orozco 2009 *Phys Rev ST-AB* **12** 041301
[9] V D Dugar-Zhabon and E A Orozco 2010 *IEEE Trans Plasma Sci* **38-10** 2980
[10] V D Dugar-Zhabon and E A Orozco 2011 *Fuente compacta autoresonante de rayos x* (Colombia: Patent 11-112696)
[11] T Umeda et al 2003 *Comp Phys Com* **156** 73
[12] R W Hockney and W Eastwood 1988 *Computer simulation using particles* (Bristol: A Higler)
[13] K Yee 1966 *IEEE Trans Ant Propagat* **14** 302
[14] G Bellanca 1997 *IEEE Micr and Guid Wave Lett* **7-12** 411
[15] D Prescott and N Shuley 1994 *IEEE Trans Micr Theor Ech* **42-8** 1582
[16] A Taflove and S Hagness 2000 *Computational electrodynamics the finite-difference time-domain 2nd ed* (Norwood: Artech House)

3D electromagnetic simulation of spatial autoresonance acceleration of electron beams

V D Dugar-Zhabon[1], J D González[2] and E A Orozco[1]

[1] Universidad Industrial de Santander, Bucaramanga, Colombia.
[2] Universidad del Magdalena, Santa Marta, Colombia.

E-mail: jedgonzalezac@unal.edu.co

Abstract. The results of full electromagnetic simulations of the electron beam acceleration by a TE_{112} linear polarized electromagnetic field through Space Autoresonance Acceleration mechanism are presented. In the simulations, both the self-sustaned electric field and self-sustained magnetic field produced by the beam electrons are included into the elaborated 3D Particle in Cell code. In this system, the space profile of the magnetostatic field maintains the electron beams in the acceleration regime along their trajectories. The beam current density evolution is calculated applying the charge conservation method. The full magnetic field in the superparticle positions is found by employing the trilinear interpolation of the mesh node data. The relativistic Newton-Lorentz equation presented in the centered finite difference form is solved using the Boris algorithm that provides visualization of the beam electrons pathway and energy evolution. A comparison between the data obtained from the full electromagnetic simulations and the results derived from the motion equation depicted in an electrostatic approximation is carried out. It is found that the self-sustained magnetic field is a factor which improves the resonance phase conditions and reduces the beam energy spread.

1. Introduction

The self-sustenance of the cyclotron resonance interaction between the electron and the transverse electromagnetic wave propagated along a homogeneous magnetic field, has been known since 1962 as cyclotron autoresonance [1, 2] which can be realized in various ways [3]. An autoresonance accomplished in a homogeneous magnetic field which increases in time to compensate precisely a relativistic deflection of the electron cyclotron movement phase on the wave phase is known as GYRAC [4]. A cyclotron autoresonance acceleration of electrons in a guide inhomogeneous magnetostatic field by a travelling electromagnetic wave is analyzed in [5,6]. Electron acceleration to high energies by microwaves in inhomogeneous magnetostatic field found in the numerical experiments on plasma dynamics in an Electron Cyclotron Resonance (ECR) minimum-B trap [7] is attributed to a self-sustenance of the ECR conditions on some parts of the electron trajectory. An acceleration of protons by a circularly polarized electromagnetic wave propagates along a homogeneous guided field is studied in [8]. The results of self-sustenance of the ECR conditions in the static inhomogeneous magnetic field at a fixed microwave field frequency both through a numerical simulation and an analytical analysis were reported in [9–11]. This acceleration mode was named Spatial Autoresonance Acceleration (SARA). In the SARA concept, the determined magnetostatic field profile along the electron beam trajectory makes possible a compensation of the electron mass relativistic change by the corresponding variation of the magnetic field magnitude [10]. In such a manner the equality between the microwave frequency and the electron cyclotron frequency is maintained during the beam motion. An electrostatic (PIC) code [11] is used to analyze the influence of the self-consistent electric field on the space autoresonance interaction. For an X ray source certificated recently [12], the attainable electron energy has been calculated in the simulations based on this concept. To appreciate the significance of the magnetic field generated by electron beam, the SARA numerical simulations are fulfilled using a full electromagnetic (EM) PIC code, and the obtained results are compared with those calculated in the electrostatic approximation. In this code, the finite difference time domain (FDTD) method is put to use [13]. The beam current density is calculated by using the charge conservation method [14]. The relativistic Newton-Lorentz equation presented in the centered difference form is solved applying the Boris algorithm that provides visualization of the beam path and its energy, velocity and phase evolutions [15].

2. Theoretical formalism

The physical scheme of the autoresonant accelerator is shown in Figure 1. A magnetron (4) emits 2.45GHz microwaves into TE_{112} cylindrical cavity (1) through a waveguide (3), a ferrite isolator (5) and an iris window (6). To avoid non-physical reflections of the microwaves in the waveguide, we apply the Uniaxial Perfectly Matched Layer (UPML) method [16]. The perfectly matched layer is located before the TE_{10} excitation plane in the rectangular waveguide. The microwaves fill the cavity through a rectangular iris window of $7.65 \times 1.0cm^2$. The length and radius of the cylindrical cavity are $4.54cm$ and $20cm$, respectively. To attain the microwave steady-state regime in the cavity as soon as possible, the Hanning function is applied [17] on the assumption that the cavity and waveguide walls are made of the perfect electric conductor material. Once the microwave field achieves the steady state regime of amplitude of $E_0^l = 14kV/cm$, the cold electron beam of $1.0A$ or $3.0A$ is injected into the cavity through an orifice of $0.2cm$ in radius along the magnetic field axis which is taken as z-axis. The initial longitudinal energy of the beam electrons is $14keV$.

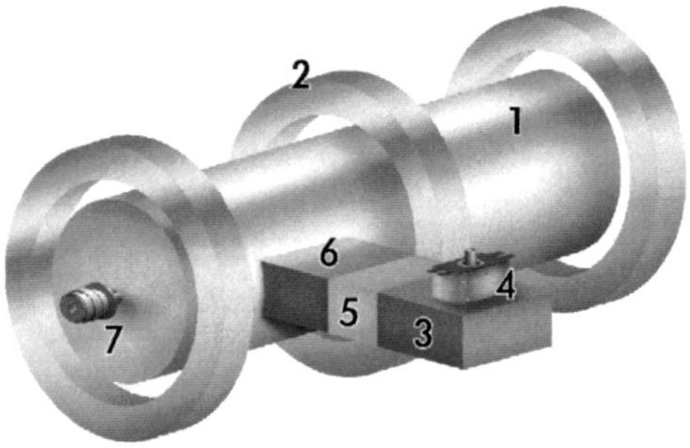

Figure 1. Physical scheme of the autoresonant system: 1-cavity, 2-magnetic coils system, 3-waveguide, 4-magnetrón, 5-ferrite isolator, 6-window, 7-electron gun.

The SARA space configuration of the axially symmetric magnetostatic field is formed by a system of three DC current coils (2). All the parameters referring to the coil currents are identical to those described in [11].

The beam motion is depicted in the framework of the Vlasov-Maxwell equation system:

$$\frac{\partial f_e}{\partial t} + \vec{v}\cdot\frac{\partial f_e}{\partial \vec{r}} - e(\vec{E} + \vec{v}\times\vec{B})\cdot\frac{\partial f_e}{\partial \vec{p}} = 0 \qquad (1)$$

$$\vec{\nabla}\times\vec{E} = -\frac{\partial \vec{B}}{\partial t}, \qquad \vec{\nabla}\times\vec{B} = \mu_0\vec{J} - \varepsilon_0\frac{\partial \vec{E}}{\partial t} \qquad (2)$$

where f_e is the electron distribution function and $\vec{E}$ and $\vec{B}$ are the electric and magnetic fields inside the cavity, respectively. The electric field $\vec{E}$ presents the superposition of TE_{112} electric component with the electric field generated by the beam electrons. The magnetic field $\vec{B}$ is composed of the TE_{112} magnetic component, the magnetic field generated by the beam electrons and the magnetostatic field produced by the coil system. The superparticle charge density and current density are determined by the expressions:

$$\rho(\vec{r},t) = -e\int f_e d\vec{p}, \quad \text{and} \quad \vec{J}(\vec{r},t) = -e\int \vec{v} f_e d\vec{p} \tag{3}$$

The equations system (1)-(3) is solved through using a (PIC) method [15]. The single superparticle (SP) dynamics is described with the following motion equations:

$$\frac{d\vec{r}_k}{dt} = \vec{v}_k, \qquad \frac{d(\gamma m_k \vec{v}_k)}{dt} = q_k(\vec{E}_P + \vec{v}_k \times \vec{B}_P) \tag{4}$$

where $\gamma = [1-(v_k/c)^2]^{-1/2}$ is the relativistic factor, $q_k(=-N_e e)$ and $m_k(=N_e m_e)$ are the charge and mass of a SP, $\vec{E}_p$ and $\vec{B}_p$ are the total electric and magnetic fields at the position of the k-th SP, respectively. The SP positions and velocities are found using the Boris leapfrog procedure [15]. In the electrostatic approximation (ES) the electric self-consistent field is determined in the mesh nodes by solving the Poisson equation at each time step, and a trilinear interpolation helps determine this field in the SP positions [11]. In order to obbtain the beam trajectory and its energy under the electromagnetic approach (EM), the Newton-Lorentz equation and the Maxwell equations are calculated jointly. The Maxwell equations in the finite difference form are:

$$\vec{B}^{n+1/2} = \vec{B}^{n-1/2} - \vec{\nabla} \times \vec{E}^n \tag{5}$$

$$\vec{E}^{n+1} = \vec{E}^n + \frac{\Delta t}{\varepsilon_0}[\mu_0 \vec{J}^{n+1/2} - \vec{\nabla} \times \vec{B}^{n+1/2}] \tag{6}$$

Where n is the index of the time step ($t_n = n \triangle t$). The self-consistent magnetic and electric fields are calculated on the mesh points using the Yee technique [13]. The fields $\vec{E}_p$ and $\vec{B}_p$ in the SP localization points are found with the help also a trilinear interpolation method. Starting from the initial data on the positions and velocities of the SPs, the current density is computed applying the charge conservative method [14]. In the simulations all the quantities are taken in dimensionless form: $\vec{E}^* = \vec{E}/(-B_0 c)$, $\vec{B}^* = \vec{B}/(-B_0)$, $\vec{J}^* = \mu_0 \vec{J}/(r_{l0} B_0)$, $\vec{r}^* = \vec{r}/r_{l0}$ and $t^* = \omega t$; where: $B_0 = m_e \omega / e$ is the magnetic induction corresponding to the classic cyclotron resonance for the electrons and $r_{l0} = c/\omega$ is the relativistic Larmor radius. The phase difference between the particle velocity vector and the right-hand electric field component of the microwave field is calculated indirectly. The simultaneous numerical solution of the motion equations (4) and Maxwell equations (5), (6) permits to calculate the electron positions, particle energy and longitudinal velocity at any time step.

3. Results and discussion

The simulations are fulfilled on a rectangular mesh with the following spatial steps: $\Delta x = \Delta y = 0.07cm$, $\Delta z = 0.2cm$ at a time step $\Delta t = 2.07ps$ chosen in accordance with the Courant stability condition [16]. Each superparticle consists of 3×10^5 electrons.

For the superparticles, the simulation is considered completed when it hits the lateral cavity wall or reaches a plane where the longitudinal velocity goes to zero which is due to the diamagnetic force. Figure 2 shows the image of the distribution of the microwave electric field component in the cross section $z = L_c/4$.

Figure 3 shows a snapshot of the electron beam trajectory in the SARA conditions. One can see that the Larmor radius of the electron whose guiding center which moves along the magnetic field axis grows smoothly because the energy of the electrons associated with the transversal motion is increased.

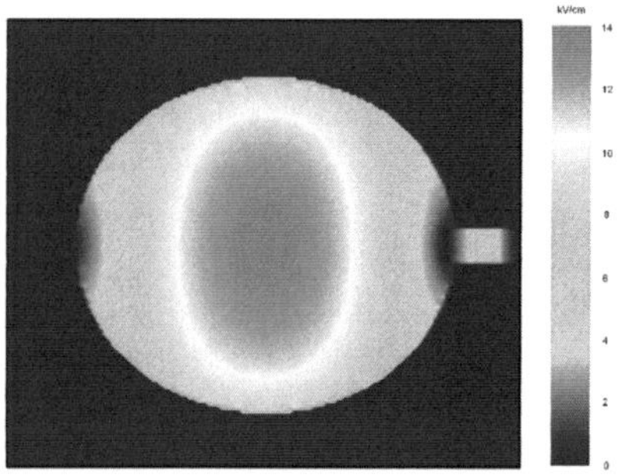

Figure 2. Microwave electric field component in the cross section $z = L_c/4$ before injection of electron beam.

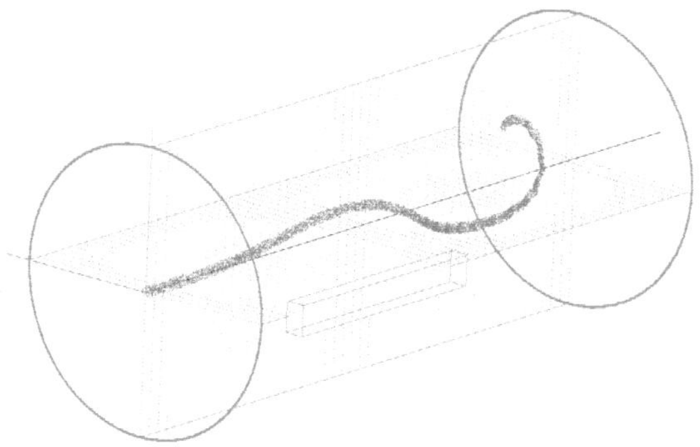

Figure 3. Snapshot of the electron beam in SARA conditions.

In Figure 4 (a) for $1A$ beam, the phase difference φ between the transverse electron velocity and the right-hand polarized microwave electric component after 9.72 microwave periods calculated for the ES and EM approaches is presented.

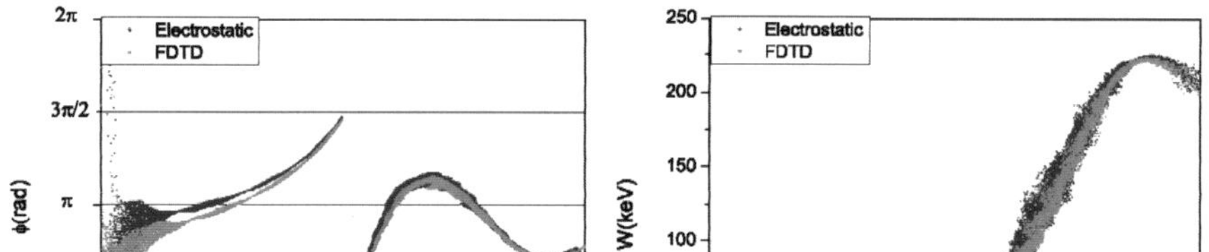

Figure 4. (a) Phase shift between the electric field of the right hand component of the microwaves field and the transversal velocities of electrons for $1A$ beam after 9.72 microwave cycles. Red dots correspond to FDTD simulations and blue dots to ES simulations. (b) Energy of a $1A$ beam after 9.72 microwave cycles.

The simulations show that in the EM case the beam occupies a smaller phase space in comparison with the case of the ES approximation. A direct consequence of this effect is a decrease of the energy spread along the beam trajectory, as compared with the ES (see Figure

4(b)). Since the only difference between these two approaches is the self-consistent magnetic field, we can conclude that the proper beam magnetic field opposes to the radial beam electric field and the diffusion of electrons in the phase space. It is important to note that the self-consistent magnetic field considerably diminishes the initial beam phase range as compared with the ES results and promotes the electron phase focalization. In the case of $1A$ beam, the focalization occurs at a distance of $1cm$ from the injection orifice where the electrons come into the exact autoresonance acceleration regime. The beam motion beyond the focalization point is found auto-maintained in the acceleration phase range $(\pi/2, 3\pi/2)$. The phase jump at $z = 10cm$ shown in Figure 4 (a) is due to the fact that this point is a TE_{112} microwave field node. Figure 4 indicates that the self-consistent magnetic field does not affect the maximum energy value, which means that its magnitude is insufficient to change the resonance conditions. This circumstance manifests itself through the practical equality of the longitudinal β_z and the transversal β_T electron velocity data obtained for the ES and EM simulations (see Figure 5).

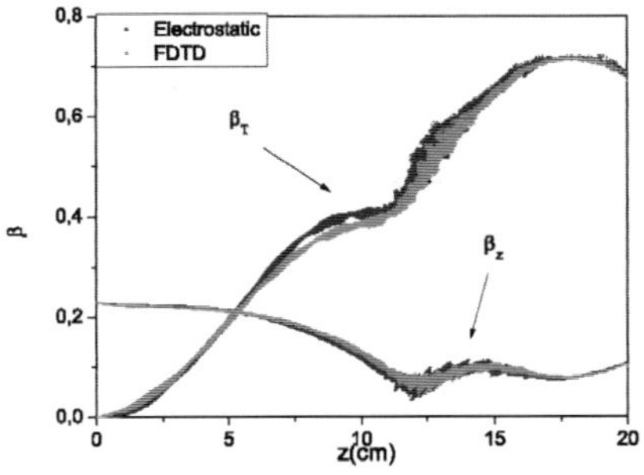

Figure 5. Transversal β_T and longitudinal β_Z components for $1A$ beam after 9.72 microwave cycles. Red dots correspond to FDTD simulations and blue dots to ES simulations.

A more significant influence of the self-consistent electric field on the beam characteristics is observed for the $3A$ beam. Figure 6(a) shows that the phase focalization length of $3.5cm$ is larger than in $1A$ case. This length increase can be attributed to a higher self-consistent electric field which opposes the focalization effect. For this reason, the energy spread of the $3A$ beam is found considerably larger in comparison with $1A$ case (see Figures 5 and 6(b)). It should be noted that the radial self-consistent electric field and longitudinal magnetostatic field provoke a motion of electrons around the beam center. This motion also results in longitudinal velocity decrease. The electron density in the neighborhood of $z = 10cm$ point increases due to decelerating of the diamagnetic force at diminishing of the microwave field (see Figure 7).

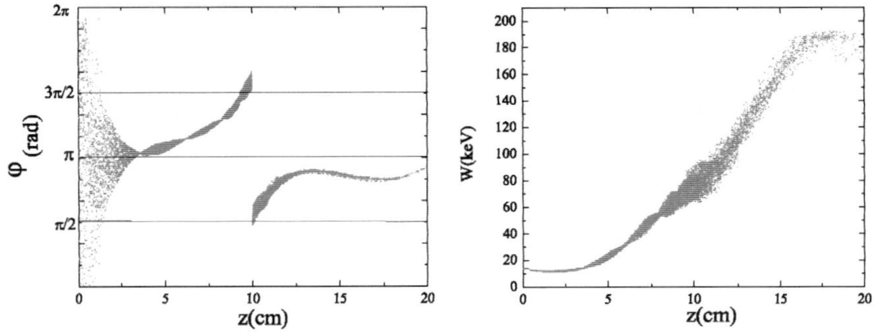

(a) (b)

Figure 6. (a) Phase shift between the microwaves and the transversal velocities of electrons for $3A$ beam after 8 microwave cycles. (b) Beam electron energy for $3A$ beam.

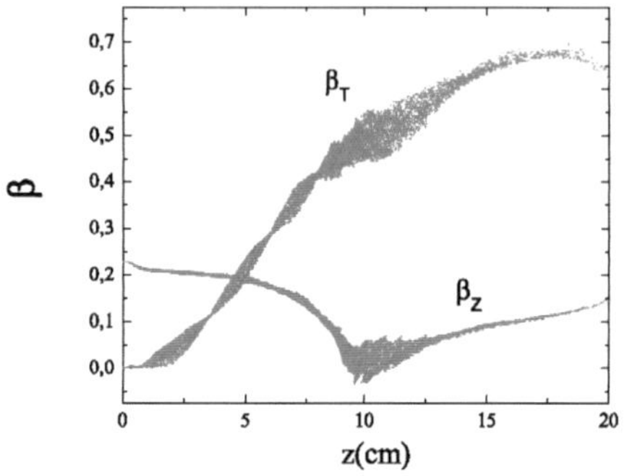

Figure 7. Transversal β_T and longitudinal β_Z components for $3A$ beam after 8 microwave cycles.

4. Conclusions

The executed space autoresonance 3D numerical simulations using an electromagnetic particle in cell code show that the studied system based on a TE_{112} cylindrical cavity can accelerate electron beams of $1A$ and $3A$ to energies of hundreds kilo-electron-volts. The different behaviour of $1A$ beam and $3A$ beam injected into the chamber through the same orifice is explained by the discrepancy in their densities which results in a difference in the self-consistent electric fields. It is shown that the self-consistent magnetic field reduces the beam phase area and energy spread. A further step to go into the system efficiency under study, namely the microwave power harnessing and variation of the microwave frequency is scheduled.

Acknowledgments

This work is supported by Foundation for Promotion for Research and Technology (Bank of Republic) under contract No 201123, Magdalena University (Fonciencias) and the Colombian Agency COLCIENCIAS through doctoral scholarships 567.

References

[1] A Kolomenskii and A Levedev 1962 *Dokl Akad Nauk USSR* **145** 1251
[2] V Davydovskii 1962 *Zh Eksp Teor Fiz* **43** 886
[3] V Milantiev 2013 *Physics Uspekhi* **56** 823
[4] K S Golovanivsky 1983 *IEEE Trans PLasma Sci* **11** 28
[5] R Shpitalnik, C Cohen, F Dothan and L Friendland 1991 *J Appl Phys* **70** 1101
[6] L Friendland 1994 *Phys Plasmas* **1** 421
[7] V Dougar-Jabon, A Umnov and D Suescún 2002 *Rev Sci Instrum* **73** 629
[8] A Holkundkar, G Brodin and M Marklund 2012 *Phys Rev ST-AB* **15** 091301
[9] V Dugar-Zhabon, E A Orozco and A Umnov 2008 *Phys Rev ST-AB* **11** 041302
[10] V Dugar-Zhabon and E A Orozco 2009 *Phys Rev ST-AB* **12** 041301
[11] V Dugar-Zhabon and E A Orozco 2010 *IEEE Transaction on Plasma Science* **38** 2980
[12] V Dugar-Zhabon and E A Orozco 2011 *Fuente compacta autoresonante de rayos X* (Colombia: Patent 11-112696)
[13] K Yee 1966 *IEEE Trans Ant Propagat* **14** 302
[14] T Umeda 2003 *Comp Phys Com* **156** 73
[15] R W Hockney and W Eastwood 1988 *Computer simulation using particles* (Bristol: A Higler)
[16] A Taflove and S Hagness 2000 *Computational electrodinamics the finite-difference time-domain 2d* (Norwood: M A Artech House)
[17] D Prescott and N Shuley 1994 *IEEE Trans on Micr Theor and Tech* **42** 1582

Virtual experimentation in electromagnetism, mechanics and optics: web-based learning

J H Escobar[1], H Sánchez[1], J R Beltrán[1], J De la Hoz[1] and J D González[1]

[1] Universidad del Magdalena, Santa Marta, Colombia.

E-mail: jedgonzalezac@unal.edu.co

Abstract. This paper present novel virtual laboratories of electromagnetism, mechanics and optics to be used in the teaching of physic for the students as well as to the teacher in the classroom. The model used is friendly with the users and also is responsible for displaying data and its manipulation. The Vista/model, implements the behavior of the view to respond to user actions and expose model data in a way that is easy to use bindings, it is the mechanisms by which we can link the elements of the user interface objects containing the information to be displayed in the view. During the process of application of virtual laboratories the students shown a increasing ability to learn this specifics physics topics using a Web-based learning environment and a guide of work with the virtual laboratories framed in the meaningful learning pedagogical model.

1. Introduction

The recent learning theories focusing on reform in science education is emphasizing the importance of rethinking the role and practice of experimentation in science teaching and learning. Important Authors in the educational research found that there is a significant difference between what a learner can do with and without an external help. This is a reason that is an imperative need the use of Virtual Experimentation (VE), it is according with the contemporary advantage of the communications especially through the web sites [1–4]. We expect to students abilities in physics will be improved and their learning and problem solving in Web-based learning environments. The computerized learning environments should balance giving and with holding information and instructional supports in order to maximize the learning outcome of students. Despite of that the Web-based learning environments give students a tool in their learning processes, if the students not exert enough cognitive efforts, if they receive information too early, resulting in failure to acquire a schema from the given learning tasks. Moreover, it has been found that learners often abuse instructional supports available in the computerized learning environment without actually engaging in learning. Additional researches shown the impact of apps online on students performance and attitudes in physics classes at a high school, and found that this kind of repetitive simulation without adequate pedagogies to deliver course content not offer the expected results in the students knowledge [5–10]. In this paper, we will present virtual simulations of physics laboratories which are in total 20 simulations in mechanics, electromagnetism and optics totally developed in the University of Magdalena. Only a few of them are present in this paper to show the way in which they are used for the students in the high-school and University of Magdalena, in the Engineer Faculty. These simulations of virtual laboratories are within the framework of meaningful learning through the use of a guide to work with the software. We know that meaningful learning is characterized by interaction between new knowledge and prior knowledge and the development of critical thought that lead to problematize the topic in study. For this process, the teacher acts as a facilitator and mediator between the student common knowledge and scientific knowledge, guiding reflection about the phenomenon studied, becoming a researcher of her own practice. We believe that the use of virtual laboratories should not be limited to observation of the physical phenomenon by using a virtual tool, but it must take the student to obtain a meaningful understanding of the physical phenomenon. Visit our web site to interact with the simulations (*http://simulacionfisica.unimagdalena.edu.co/*).

2. General architecture of experiences

MVVM (Model View View Model) without access to data (MVC) with Silverlight. It is the model responsible for all application data and business logic related view or views. Also responsible for displaying data to the user and allowing manipulation of application data. Vista model, implements the behavior of the view to respond to user actions and expose model data in a way that is easy to use bindings (mechanisms by which we can link the elements of the user interface objects containing the information to be displayed) in view.

2.1. Languages used on experiences

C#: Language used for Vista/Model. XAML: Language used to view (User Interfaces).

2.2. Background information for analysis of experiences

WPF application development with silverlight website development with ASP.Net and *C#*. Management of side events (delegates). Development of N-tier applications. XAP Deployment compiled on web sites.

2.3. Software requirements to manage the development of experiences

Silverlight SDK 4 and 5 Microsoft Visual Studio 2010

3. Virtual experimentation

We implement the new GUI design simulations for teaching physics that includes a pedagogical part Figure 1 The main objective is to develop generic and specific skills in the area of basic sciences (physics) by interacting with new technologies, achieving greater interest both teachers and students in this area and helping to improve learning environments used in the ownership, understanding and deepening knowledge deductively of the physical sciences. The research plan improvements in the level of interpretive, argumentative and purposeful students in the area of physical skills, through the development of laboratories using software, allowing students interested in a special way through this area, because Virtual labs are structured with interactivity, data processing area, friendly interface (display areas of the physical phenomenon, input and output data and aids). These learning tools become elements of great motivation for understanding natural phenomena and the development of skills. This research is relevant in institutions of secondary and higher education, as encourages the development of generic and specific skills in the areas of knowledge such as the theoretical and physical development to evaluate and propose equations to program in each experience, generation laboratory guidelines, developing simulations, application methodology practices and evaluation process before and after the implementation of simulations in the development of laboratories. Additionally in this research, we could count with the participation of the faculty of basic sciences of the schools of secondary and higher education in our city for the implementation of the virtual laboratories and the first experience of the student with the software.

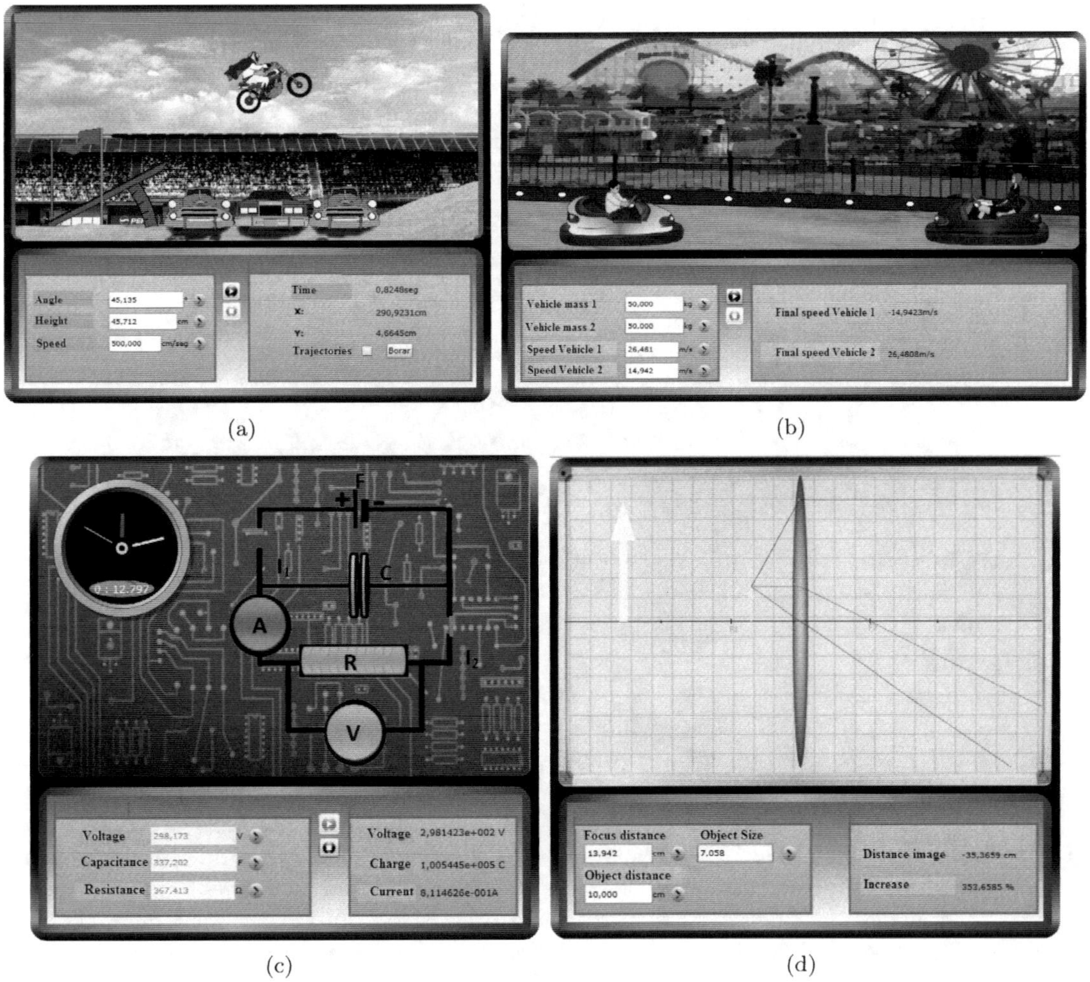

Figure 1. Snapshots of some virtual laboratories: (a) parabolic movement, (b) collision in one dimension, (c) discharge of a capacitor and (d) lenses.

4. Future works in virtual laboratories

The evolution of smartphones and tablets and use massively, offers a great opportunity to expand the use of virtual laboratories by students. Therefore, virtual laboratories will be implemented in a multi-platform that allows access from any Android, iOS or Windows device. We will update the graphical interface, creating a new 2D or 3D format depending on the type of experience Figure 2. However, neither the above logic equations simulator is not changed, the graphics engine also used the simulator will allow us to export any device by which one can be programmed once. Additionally, we propose that this new solution experiences are modifiable, ie, establish two or three scenarios for simulation giving the user a choice that is to your liking.

4.1. Technologies

For this new proposal we will use technology named Blender which is a free program that is used to create 3D modeling and animation, and also Unity is used which is an application used for game development and this can encourage logic programs modeled using the programming language C#.

Figure 2. Simulation experiment of parabolic motion using Unity 3D. Initial test runs on mobile devices.

5. Conclusions

We present virtual laboratories in specifics physics topics to improve the abilities of students in the basic sciences. The model vista model is used to respond to user actions and link the elements of the user interface objects containing the information to be displayed. We expect that the virtual laboratories improve the level of interpretive, argumentative and purposeful students in the area of physical skills. This research is relevant in institutions of secondary and higher education, as encourages the development of the physics. Finally, this research initiate the implementation of the virtual laboratories in IOS, Android and Windows to expand the use of virtual laboratories by students with news the graphical interface, creating a new 2D or 3D format, modifiable to establish two or three scenarios by simulation.

Acknowledgments

This work was financed by the University of Magdalena (Fonciencias).

References

[1] H Burton and R J Sinclair 2002 *Journal of Clinical Neurophysiology* **17** 575
[2] L B Flick 1993 *Journal of Science Teacher Education* **4** 1
[3] G E Glasson 1989 *Journal of Research in Science Teaching* **26** 121
[4] R L Goldstone and J Y Son 2005 *The Journal of the Learning Sciences* **14** 69
[5] M Reiner 1999 *Interactive Learning Environments* **7** 31e55
[6] M Resnick 1998 *Educational Technology Research and Development* **46** 43
[7] K Scheiter, P Gerjets,T Huk, B Imhof and Y Kammerer 2009 *Learning and Instruction* **19** 481
[8] J Swaak and T de Jong 1996 *Studies in Educational Evaluation* **22** 341
[9] J Swaak, W R van Joolingen and T de Jong 1998 *Learning and Instruction* **8** 235
[10] J Sweller 1994 *Learning and Instruction* **4** 295

Donor-assisted resonant electron tunneling in double-barrier heterostructures under tilted magnetic fields: A theoretical study

H Paredes[1], C L Beltrán Ríos[1] and J C Arce[2]

[1] Universidad Industrial de Santander, Bucaramanga, Colombia.
[2] Universidad del Valle, Cali, Colombia.

E-mail: hparedes@uis.edu.co

Abstract. A theoretical study of the current-voltage characteristics of a double-barrier heterostructure doped with a layer of shallow donors in the middle of the well and subjected to a magnetic field tilted with respect to the growth direction is reported. The parent materials are assumed to possess simple cubic lattices, the electronic structure of the system is modeled by means of a one-band tight-binding Hamiltonian, and the current is evaluated employing the Keldysh nonequilibrium formalism. The results reveal the magnetocompression of the donor wavefunction, in good qualitative agreement with the experimental report of Patanè et al [Phys Rev Lett **105** 236804 (2010)].

1. Introduction

High magnetic fields can be employed to probe sensitively and noninvasively features of the quantum states of low-dimensional structures embedded in a solid [1]. In particular, in magnetotunneling spectroscopy (MTS) an applied magnetic field perpendicular to the current permits mapping the momentum probability distributions of a nanostructure in the plane perpendicular to the current [1, 2], thereby yielding information about the shape and size of its wavefunctions. Recently, MTS has been applied to the study of the compression effected on the wavefunction of a donor by an applied magnetic field parallel to the current [3] and the stretching caused on the wavefunction of a quantum dot by a focused femtowatt laser beam [4]. Hence, this technique is a valuable aid in the design of nanostructures tailored for specific tasks [5].

In MTS experiments the target nanostructure is either located within the quantum well defined by two tunnel barriers [2, 3] or within a single tunnel barrier [2, 4]. In either configuration, the current-voltage ($I-V$) characteristic curves exhibit weak features associated with resonant electron tunneling (RET) through quasi-bound states of the nanostructure. Moreover, in the first configuration the $I-V$ curves exhibit strong peaks at higher voltage associated with RET through quasi-bound states of the metastable quantum well. Momentum probability density maps can be extracted from the behaviour of the current intensities associated with these features as functions of the perpendicular magnetic field strength.

In previous reports, it was shown that a simple theoretical methodology, based on a one-band tight-binding Hamiltonian and the Keldysh nonequilibrium formalism [6], can reproduce the features of the experimental $I-V$ characteristics of double-barrier heterostructures (DBHs) under perpendicular and parallel magnetic fields [7] and δ-doped DBHs under perpendicular magnetic fields [8]. In the latter study, it was demonstrated that the momentum probability density associated with the donor ground state can be quantitatively extracted from the theoretical data. In the present report, this methodology is applied to the case of a δ-doped DBH under a tilted magnetic field, with a view to reproducing the experimental observations of [3] on the magnetocompression of the donor wave function.

2. Methodology

The bare DBH is described by the one-band tight-binding Hamiltonian.

$$\hat{H} = \sum_i \epsilon_i^0 c_i^\dagger c_i + \sum_{i,j} v_{ij}^0 c_i^\dagger c_i \tag{1}$$

where ϵ_i^0 is the site energy, v_{ij}^0 is the hopping energy between sites i and j, and the summation is understood to be performed over neighboring sites only. Taking the z axis as the direction of the current, i.e. the DBH growth direction, the perpendicular component of the magnetic field $B_x\hat{\mathbf{i}}$ is introduced into the Hamiltonian (1) by means of the Peierls gauge transformation of the hopping energies [9].

$$v_{ij}^0 \rightarrow v_{ij}^0 \exp\left[i\frac{e}{\hbar}\mathbf{A}\left(\frac{\mathbf{l}+\mathbf{l}'}{2}\right)\cdot(\mathbf{l}-\mathbf{l}')a\right] \tag{2}$$

where $\mathbf{A} = B_x y\hat{\mathbf{k}}$ is the vector potential, $\mathbf{l}$ and $\mathbf{l}'$ are the position vectors of neighboring sites i and j, a is the lattice parameter, and e and $\hbar$ have the usual meanings. On the other hand, the presence of the parallel component $B_z\hat{\mathbf{k}}$ is taken into account by means of the substitution.

$$\epsilon_i^0 \rightarrow \epsilon_i^0 + (n+1/2)\hbar\omega_i \tag{3}$$

where $n = 0, 1, 2, \ldots$ and $\omega_i = eB_z/m_i^*$ are the Landau index and cyclotron frequency, respectively, with m_i^* being the in-plane (xy) electron effective mass.

Assuming that the parent materials possess simple cubic lattices and performing a Fourier transform employing the appropriate Cunningham points [10], the field-dressed three-dimensional (3D) Hamiltonian (1) is reduced to an effective 1D Hamiltonian with dressed energies that depend on the in-plane wave vector $\mathbf{k} = k_x\hat{\mathbf{i}} + k_y\hat{\mathbf{j}}$, given by,

$$E_i(\mathbf{k}) = \epsilon_i + 2v_i[\cos(k_x a) + \cos(k_y a)] \tag{4}$$

where ϵ_i and v_i are the site and hopping energies in the effective 1D lattice, with i =W,B,I for well, barrier or interface sites, respectively. The height of the potential barrier is given by [11] $\Delta E = (\epsilon_W - \epsilon_B) + 2(v_W - v_B)$.

Neglecting the band-bending effect due to the charge accumulation layers that arise at each interface in the real structure, the voltage bias is included in the site energies ϵ_i by means of a linear interpolation of the voltage V between the two barriers. The presence of the hydrogenic donor in the quantum well is taken into account by including in the site energies the Coulomb term $\epsilon_i^c = -e^2/4\pi\epsilon_{0,i}r_{i0}$, where $r_{i0} = |\mathbf{r_i} - \mathbf{r_0}|$ is the distance between the impurity located at the middle of the quantum well and the electron on site i, and $\epsilon_{0,i}$ is the value of the dielectric constant at site i.

Employing the Keldysh formalism [6] it can be shown that the average current induced in the system at 0K is given by [12].

$$I = \frac{4\pi^2 e v_W^2}{\hbar}\int_{\mu_R}^{\mu_L} \frac{\rho_L(\hbar\omega)\rho_R(\hbar\omega)}{|\Gamma(\hbar\omega)|^2} d(\hbar\omega) \tag{5}$$

where the subscripts L and R refer to the injector and collector located to the left and right of the structure, respectively, $\rho_L(\hbar\omega)$ and $\rho_R(\hbar\omega)$ are the densities of states, $\Gamma = (1 - g^a_{LL} g^a_{RR} v^2_W)(1 - g^r_{LL} g^r_{RR} v^2_W)$, with g^a_{LL} and g^r_{RR} being the advanced and retarded Green functions, and μ_L and μ_R are the chemical potentials, with $\mu_L > \mu_R$, μ_L will be referred to as the Fermi energy.

3. Results and discussion

The lattice constant, dielectric constant and electron effective mass were assumed to be the same at all sites and equal to the ones of GaAs, $a = 0.282$nm, $\epsilon_0 = 12.5$ and $m^* = 0.067m_0$, with m_0 being the free electron mass. The hopping energy between any two sites was evaluated by means of the formula $v^0 = \hbar^2/2m^*a^2$, yielding $v^0 = 7.179$eV. The Fermi energy was taken to be 100meV, which corresponds approximately to the injector doping chosen in ref. [3]. The structural parameters of the DBH were also taken from ref. [3] (barrier height of 300meV and barrier and well widths of 5.7nm and 9.0nm, respectively).

Each frame of Figure 1 displays the I-V characteristic curves of the structure for several values of B_x at a fixed value of B_z. The shoulder that appears at $B_x = 0$ T and develops into a broad peak as B_x increases, henceforth to be called D, is associated with RET through the ground state of the donor [3, 8]. At higher voltage there appears a strong peak (not shown) associated with RET through the lowest subband of the quantum well [3, 8]. It is observed that feature D shifts to lower bias and its current intensity I_D decreases monotonically with B_x. Each curve of Figure 2 shows the normalized dependence of I_D on B_x, extracted from the data in the corresponding frame of Figure 1 . It is clearly appreciated that I_D is the less sensitive to B_x the higher the strength of B_z, in qualitative accordance with the experimental results of [3].

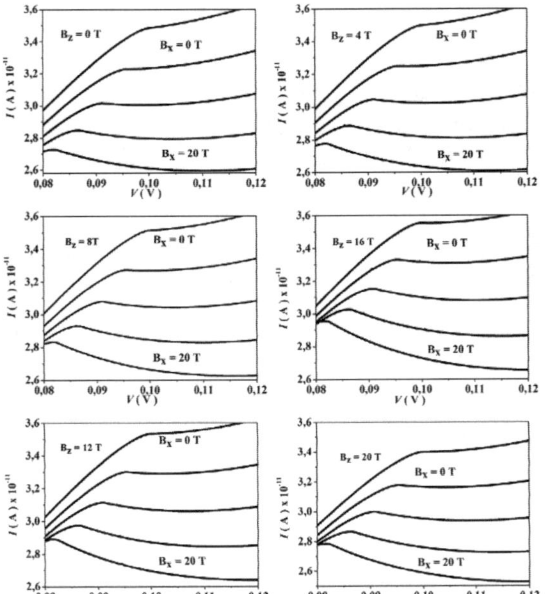

Figure 1. Current-voltage characteristic curves of the δ-doped double-barrier heterostructure under tilted magnetic fields. Each frame displays the curves for strengths of 0T, 5T, 10T, and 20T of the perpendicular component of the field at a fixed strength of the parallel component.

However, the decrease of I_D with B_x in the theoretical plots is observed to be much slower than in the experimental ones. As discussed in [7], the parallel field increases the hopping energy whereas the perpendicular field decreases it. Hence, the overestimation of the peak current produced by the model must be due to an underestimation of the effect on the hopping of the perpendicular field in relation with the one of the parallel field. This can be attributed, at least in part, to the nearest-neighbor approximation, which enters directly in Equation (2), but not in Equation (3).

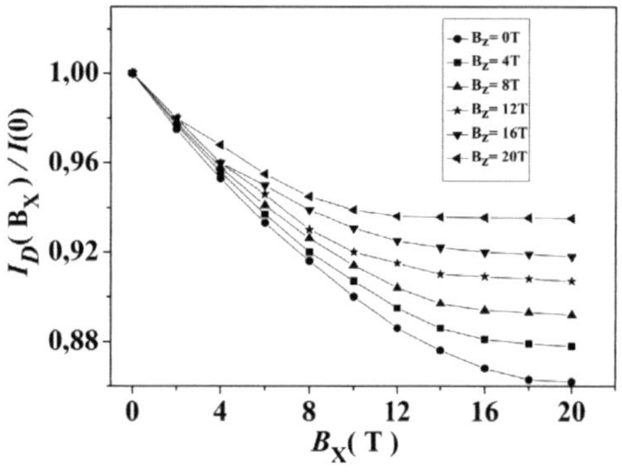

Figure 2. Dependence of the normalized donor peak current on the strength of the perpendicular component of the magnetic field. Each plot is extracted from the data in the corresponding frame of Figure 1.

At $B_z = 0$ T, the plot of I_D versus B_x should furnish a cut of the donor probability density in momentum space $|\Psi(k_y)|^2$ in the direction of the field [2, 3]. Indeed, in [8] it was shown that a plot of $\sqrt{I_D}$ versus B_x can be fitted accurately to a Lorentzian, revealing the 1s-like nature of the donor wavefunction. On the other hand, the model calculations presented in [3] indicate that the behaviour of $I_D(B_x)/I_D(0)$ at increasing strengths of B_z exhibited in Figure 2 can be attributed to the compression of the donor wavefunction effected by the field.

4. Concluding remarks

The present simplistic model is appropriate for the qualitative prediction of the effects of applied perpendicular and parallel magnetic fields, and combination thereof, on the I-V characteristics of undoped and δ-doped DBHs. In particular, this model is able to account for the magnetocompression of a donor wavefunction. This provides confidence in its more general utility for predicting the spatial characteristics of the quantum states of nanostructures under other applied perturbations.

Acknowledgments

Authors acknowledge support from the High Performance Computing Center at Universidad Industrial de Santander.

References

[1] A Patanè and L Eaves 2012 *Semiconductor Research* (Heidelberg: Springer Series in Materials Science) p 309
[2] A Patanè, R J A Hill, L Eaves, P C Main, M Henini, M L Zambrano, A Levin, N Mori, C Hamaguchi, Yu V Dubrovskii, E E Vdovin, D G Austing, S Tarucha and G Hill 2002 *Phys Rev B* **65** 165308
[3] A Patanè, N Mori, O Makarovsky, L Eaves, M L Zambrano, J C Arce, L Dickinson and D K Maude 2010 *Phys Rev Lett* **105** 236804
[4] O Makarovsky, E E Vdovin, A Patanè, L Eaves, M N Makhonin, A I Tartakovskii and M Hopkinson 2012 *Phys Rev Lett* **108** 117402
[5] G P Lansbergen, R Rahman, C J Wellard, I Woo, J Caro, N Collaert, S Biesemans, G Klimeck, L C L Hollenberg and S Rogge 2008 *Nature Phys* **4** 656
[6] L V Keldysh 1964 *Zh Eksp Teor Fiz* **47** 1515, L V Keldysh 1965 *Sov Phys JETP* **20** 1018
[7] H Paredes-Gutiérrez, N Porras-Montenegro and A Latgé 2003 *Phys Rev B* **68** 45311
[8] H Paredes-Gutiérrez, N Porras-Montenegro, J C Arce and A Latgé 2004 *J Appl Phys* **95** 4890
[9] R Peierls 1933 *Z Phys* **80** 763, F Elsholz, A Wacker, E Schöll, M Cast, G Strasser and E Gornik 2001 *Phys Rev B* **63** 033312
[10] S L Cunningham 1974 *Phys Rev B* **10** 4988
[11] A Nogueira and A Latgé 1998 *Phys Rev B* **57** 1649
[12] E V Anda and A Latgé 1994 *Phys Rev B* **50** 8559

Effect of a high intensity laser beam on impurity binding energy in a nanocone

H Paredes[1], C L Beltrán Ríos[1] and W Gutíerrez[1]
[1] Universidad Industrial de Santander, Bucaramanga, Colombia.
E-mail: hparedes@uis.edu.co

Abstract. This paper presents theoretical results of a study that analyzed the effect of a high-frequency laser in the ground state binding energy of a hydrogenic donnor impurity. For these results, the trigonometric sweep method and framework of the effective mass approximation is applied. The results showed that the binding energy changes depending on the laser intensity and the impurity position across of the nanocone axis. The results agree with previous results obtained in similar systems.

1. Introduction

There has been considerable interest in the study of the effect of an intense laser high frequency semiconductor devices, because it induces a nonlinear effect in the optical properties such as intersubband transitions, the peak refractive index change show a red shift [1], anisotropy in the absorption coefficients [2, 3] and capable of monitoring high-switching frequency optical signals [4].

A number of theoretical results on the study of the binding energy of a hydrogen-donor impurity in quantum well wires (QWWs) of GaAs-(Ga, Al)As under the laser-dressed effects both Coulomb potential and impurity confinement potential, respectively. Qu Fanyao et al. [5] , their results show a rapid decrease of the ground-state binding energy for different values of the wire radius with increasing field intensity laser. E. C. Niculescu and A. M. Radu [6], they calculated the binding energy of an off-center donor impurity in a cylindrical QWW, in the presence of a static magnetic field, the position of the donor impurity and two directions of polarization of the laser field. The binding energy behavior is only slightly affected by the magnetic field, however the strong confinement and the presence of linearly polarized laser radiation, a significant change is observed in the binding energy for various directions of polarization and partially solve the degeneration of impurity states. The binding energy was studied for shallow-donor impurity located off-axis in GaAs coaxial cylindrical quantum wire under the action of an intense high-frequency laser field. The dependence of the binding energy as a function of the donor position along and perpendicular to the direction of polarization of the radiation and for different values thicknesses of barrier; in this results was observed that as parameter of the laser is increased binding energy also increases and in the absence of laser radiation the binding energy is degenerate for impurities located symmetrically in the QWW and the laser field effect partially breaks the degeneracy of ground state in the binding energy [7]. Recently, C. A. Duque et al. [8], a theoretical study the effect of an intense laser field on the states of a donor impurity in a GaAs- (Ga, Al)As QWW of rectangular cross section, under the effect of a laser radiation field are applied independently oriented throughout horizontal and vertical directions in the xy plane. The application of the combined use of two fields of intense laser radiation oriented in different directions in the plane there are two effects induced by laser: i) two centers of Coulomb and ii) modifies the electron probability density due to the change of the confinement potential profile. Thus, the possibility of transition of one-dimensional electron coupled single, double and quadruple within the structure.

In particular, the research on solar cell in [9, 10] and photodetectors devices [4] as GaAs nanocone are important because they have an improved solar absorption, short-circuit current density and light-intense trapping. In this work was studied the effect of the intense high frequency laser on the ground-state binding energy of hydrogenic impurity located on the axis of a nanowire shape a nanocone of GaAs. The binding energy is determined according to the position of the impurities throughout the axis of the wire and the intensity of the laser field. This paper is organized as follows: in Section 2 we describe the theoretical framework. Section 3 contains the results and discussion, and finally, our conclusions appear in Section 4.

2. Theoretical model

A nanowire with cylindrical geometry and variable section was considered, the cylinder radii decrease with the nanowire height in agree with the next equation:

$$R(z) = R_1 - \left(\frac{R_2 - R_1}{H}\right) z \tag{1}$$

with R_1 and R_2 are the greater and lower radii, respectively, of the nanowire and H its height. z run between zero and H. In Figure 1 is showing a schematic picture of the nanowire with an impurity localized on axis.

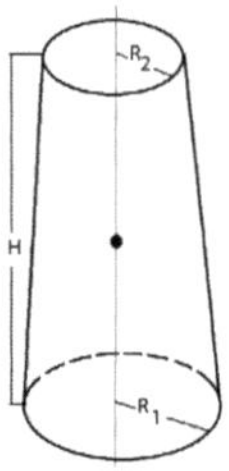

Figure 1. Nanocone schematic figure, with the geometrical parameters.

The impurity stationary schrödinger equation in arbitrary units (au), and considering the effective mass approximation framework, in cylindrical coordinates, is given by

$$-\nabla^2\Psi + \left(V(\rho, z) - \frac{2}{\sqrt{\rho^2 + (z - z_i)^2}}\right)\Psi = E\Psi \tag{2}$$

In the above equation the $V(\rho, z)$ term correspond to the geometrical confinement due to the structure, it is describe like infinite potential confinement give by,

$$V(\rho, z) = \begin{cases} 0 & \text{in of the nanocone} \\ \infty & \text{out of the nanocone} \end{cases} \tag{3}$$

And z_i is the impurity position, this potential is not modified by the laser. The Hamiltonian give by Equation 2 is modified when a laser beam is used. In order to include the effect of the laser we have consider that dimensions of the nanocone are in agree with the condition give by the Floquet method [11]. Under this method the Coulomb interaction is modified and new coulomb term, in au, is give by,

$$V_c = -\frac{1}{|\vec{r} - \vec{r_i} + \vec{\alpha}|} - \frac{1}{|\vec{r} - \vec{r_i} - \vec{\alpha}|} \tag{4}$$

Where $\vec{r_i}$ is the impurity position, and $\vec{\alpha}$ is a vector associated to laser dressing parameter ($\alpha = |\vec{\alpha}|$), this term is given by,

$$\alpha = \left(\frac{8\pi\, e^2 I}{m_e^2 c^2 \omega^4}\right)^{1/2} \tag{5}$$

Where e is the electron charge, I is the laser intensity, m_e is the electron mass, c light velocity and ω is the frequency. Again it was considered that the impurity is localized on the nanocone axis and the laser has a polarization in z, under above condition the coulomb term is writing, in cylindrical coordinates, as:

$$V_\alpha = -\frac{1}{\sqrt{\rho^2 + (z - z_i + \alpha)^2}} - \frac{1}{\sqrt{\rho^2 + (z - z_i - \alpha)^2}} \tag{6}$$

In agree with the above conditions the Equation 2 can be written as:

$$-\nabla^2\Psi + (V(\rho, z) + V_\alpha)\Psi = E\Psi \tag{7}$$

The shape of nanocone permit us considerate the adiabatic approximation [12] for to separate the particle axial and radial motion. By mean of this approximation, and considering the axial symmetry, the Schrödinger equation can be reduce from 3 to 1 dimension problem, and it can be written as:

$$-\frac{\partial^2 F(z)}{\partial z^2} + V_{eff}(z)F(z) = EF(z) \tag{8}$$

Where the effective potential V_{eff} is given by,

$$V_{eff}(z) = E_\rho(z) + \int_0^{R(z)} \rho\, V_\alpha\, \Phi^2(\rho)\, \mathrm{d}x \tag{9}$$

$\Phi(\rho)$ is the solution to the nanocone radial Schrödinger equation and $E_\rho(z)$ the ground state energy for the radial equation, $F(z)$ is the eigenfunction for z direction. The Equation 8 is solve numerically. The binding energy(E_b) for the impurity is obtained by the expression:

$$E_b = E_0 - E \tag{10}$$

The E_0 term correspond to the energy of a particle in the nanocone, without impurity.

3. Results and discussion

The theoretical calculation was realized considering the values for $R_1 = 15$nm, $R_2 = 10$nm, $H = 300$nm, for a GaAs structure.

In Figure 2 is showing the effective potential (V_{eff}), for a impurity localized in a $\frac{H}{2}$ and $\frac{H}{3}$ from the bottom of the structure, as a z function and different values of the α parameter. In this figure is observe that for $\alpha = 0$ there is a single well in V_{eff} caused by the coulomb interaction, and when the laser beam was considered ($\alpha \neq 0$) the single well is spliting in a double quantum well.

The V_{eff} in the double well structure has lower deep that the well in the case $\alpha = 0$. In Figure 2(a) it is observed that the barrier between wells are increasing when α is increasing. Also, in Figure 2(a) and (b), it is possible to observe that for a some value of $\alpha \neq 0$ the bottom of the left well is more deeper that the bottom right well, in Figure 2(a) it is observed that the left well for $\alpha = 50$nm is slightly deeper that in the case for $\alpha = 30$nm.

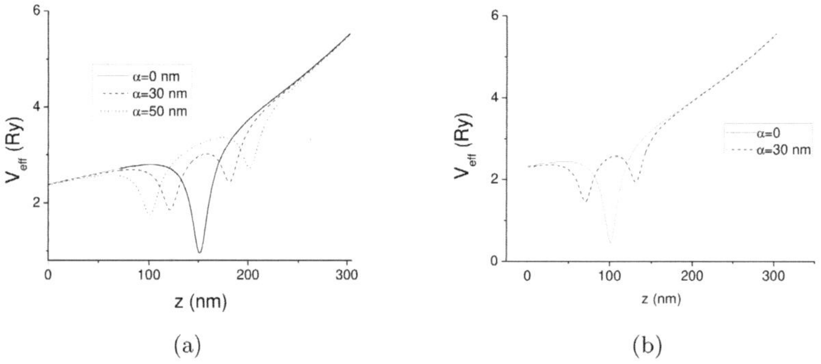

Figure 2. Effective potential for an impurity localized at (a) $z_i = H/2$ and (b) $z_i = H/3$ from the nanocone base, as a function of z, for different values of the α parameter.

In Figure 3(a) the binding energy is shown as a function of the impurity position and some values of α parameter. In all cases showed it can be observe a diminishing in the binding energy when the impurity from away of the wider region (R_1) to the narrow region (R_2). Additionally for the impurity localized in narrow region the binding energy is no modified by the laser intensity, in this case the geometrical confinement is greater than the laser effect.

In Figure 3(b), is showing the binding energy as a function of the α parameter and some values of the impurity position. In this figure is observed a diminishing in the energy for some values of α parameter and later the energy is increasing with α value. This behaviour can be understand with help of the V_{eff} Figure. For $z_i = H/2$ the electron is confined in the left well and if the α value is continued increasing the energy began to diminish, because the left well become deeper, and if we continued increasing the α value, the left well it get closer to the bottom of the structure and the electron is repel from the barrier, approaching it to impurity and the energy is increasing. For $z_i = H/3$ the electron has a higher energy that $z_i = H/2$ because the electron is localized in the wider region and more closer to the impurity, in this case the lateral confinement is lower that the case with $z_i = H/2$, but when α is increasing the energy has the same behaviour that in the case with $z_i = H/2$. When $z_i = 2\,H/3$ the impurity is localized in narrow region and the electron is far away from the impurity causing a lower energy that in $z_i = H/2$ case. In this last case the effect of the structural confinement is lower that the other cases and the variation in the energy is more slight that the other case when the electron is confined in the left well of the effective potential.

The increasing the binding energy with the laser field have been report by others author in similar system .

The increasing the binding energy with the laser field have been report by others author in similar system [6]. Radu [13] shown the effect upon impurity binding energy as a function of parameter and different impurity positions, for a cylindrical wire. Ungan et al. [14] studied the effect the laser in a quantum well for different position of the impurity, in this work the binding energy show a similar behavior that we report. Duque et al. [8] considered a wire with rectangular section and different laser beam polarizations to calculate the binding energy for a impurity in the wire. All these work show that the energy values have a high dependence with the impurity position and the parameter, and the values ranges for the binding energy are similar for all them.

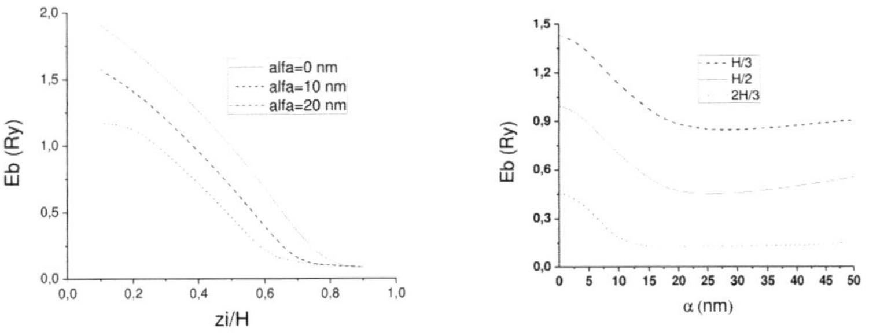

(a) (b)

Figure 3. (a) Binding energy of a impurity as a function of the impurity position and different values of the α parameter, (b) Binding energy for a impurity as a function of the α parameter and different values of the impurity position z_i.

4. Concluding remarks

Theoretical results obtained for an impurity in a GaAs nanocone with infinity confinement was presented in the effective mass frame work. It was found that the binding energy is diminishing when the impurity is localized from the wider region to narrow region, for the same impurity position the effect of a laser beam is increase the energy for some values of the parameter. The competition between geometrical confinement and laser beam effect cause a diminishing and later increasing in the binding energy, for different values of the parameter. The high dependence of the binding energy with the impurity position and the parameter associated with the laser beam has been observed in similar systems by others authors.

Acknowledgements

The authors express acknowledgements to PhD Ilia Mikahailov for the programming codes support. Similarly, to VIE of Universidad Industrial de Santander for financial support to develop a research project.

References

[1] L M Burileanu and A Radu 2011 *Opti Commun* **284** 2050
[2] E C Niculescu, L M Burileanu, A Radu and A Lapaşcu 2011 *J Lumin* **131** 1113
[3] H Sari, E Kasapoglu, I Sökmen and M Güneş 2003 *Phys Lett A* **319** 211
[4] LinBao Luo, JingJing Chen, MingZheng Wang, Han Hu, ChunYan Wu, Qiang Li, Li Wang, JianAn Huang, FengXia Liang 2014 *Adv Funct Mater* **24** 2794
[5] Qu Fanyao, A L A Fonseca and O A C Nunes 1996 *PRB* **54** 16405
[6] E C Niculescu, A M Radu 2007 *J Optoelectron Adv Mater* **9** 3598
[7] A Radu, E C Niculescu 2010 *Phys Lett A* **374** 1755
[8] C A Duque, M E Mora-Ramos, U Yesilgul, F Ungan, S Sakiroglu, E Kasapoglu, H Sari and I Skmen 2014 *Acta Phys Pol A* **125** 198
[9] Lei Hong, Rusli, Hongyu Yu, Xincai Wang, Hao Wang and Hongyu Zheng 2012 *Jpn J Appl Phys* **51** 10ND13
[10] Baomin Wang, Erica Stevens and Paul W Leu 2014 *Opt Express* **22** A386
[11] M Gavrila and J Z.Kaminski *Phys Rev Lett* **52** 613
[12] L F Garcia, W Gutiérrez and I D Mikhailov 2014 *Physica B Condensed Matter* **45** 14
[13] A Radu 2010 *Superlattices and Microstructures* **48** 114
[14] F Ungan, U Yesilgul, S Sakiroglu, E Kasapoglu, H Sari, I Sökmen 2010 *Phys Lett A* **374** 2980

Current density and state density in diluted magnetic semiconductor nanostructures

S T Pérez Merchancano[1], H Paredes Gutiérrez[2] and J A Zuñiga[1]

[1] Universidad del Cauca, Popayan, Colombia.
[2] Universidad Industrial de Santander, Bucaramanga, Colombia.

E-mail: sperez@unicauca.edu.co

Abstract. We study in this paper the spin-polarized current density components in diluted magnetic semiconductor tunnelling diodes with different sample geometries. We calculate the resonant JxV and the density of states. The differential conductance curves are analyzed as functions of the applied voltage and the magnetic potential strength induced by the magnetic ions.

1. Introduction

The spintronics employs both the electron charge and spin as information carriers. The design of a basic device uses the proximity of the first ferromagnet gate from the left to align the electron spins, which are unpolarized when entering the semiconductor from the source. Magnetic field control of the magnetization of the second ferromagnetic gate parallel (antiparallel) to the first gate makes the transit of the current easier (harder). The current can be measured when it emerges from the drain. Recent advances [1–3] in experimental studies of the spin transport effect in semiconductor structures have moved the state of the art closer to the realization of novel spintronic devices. The employment of spin-related phenomena is expected [4] to extend the functionality of conventional devices at the classical level and address fundamental problems of electronics in the quantum limit [1–3]. Different types of semiconductor and hybrid spintronic devices [5, 6] have been proposed. Spintronic devices allow us to control the functionality by spin-orbit and magnetic interactions and can be used as magnetic sensors or programmable logic elements. In general, transport in semiconductor spintronic devices can be characterized by the creation of a non-equilibrium spin polarization in the device (spin injection), the measurement of the final spin state (spin detection), the external control of spin dynamics by the electric (gate modulation) or magnetic fields, and uncontrolled spin dynamics leading to loss of information in the device (spin relaxation or spin dissipation).

In particular, the research on diluted magnetic materials is a highly interesting topic because these materials are important for the fabrication of high performance electronic devices [7–11]. We study in this paper the spin-polarized current density components in diluted magnetic semiconductor tunneling diodes with different sample geometries. The charge buildup and its fluctuation on the resonant levels of a given geometry can help us to project efficient low voltage spin filter devices working with a small barrier offset at a small magnetic field. The idea of exploring and manipulating the spin degrees of freedom for carriers in semiconductor devices, a field named spintronics, has shown promising trends as well as new challenges.

2. Theoretical model

In this case we consider a simple one band tight-binding model (TB) for the spin dependent resonant tunneling (neglecting the spin-orbit interaction) in a double barrier heterostructure (DBH) of $Cd_{1-x}Mn_xTe - CdTe - Cd_{1-x}Mn_xTe$ (see, Figure 1).

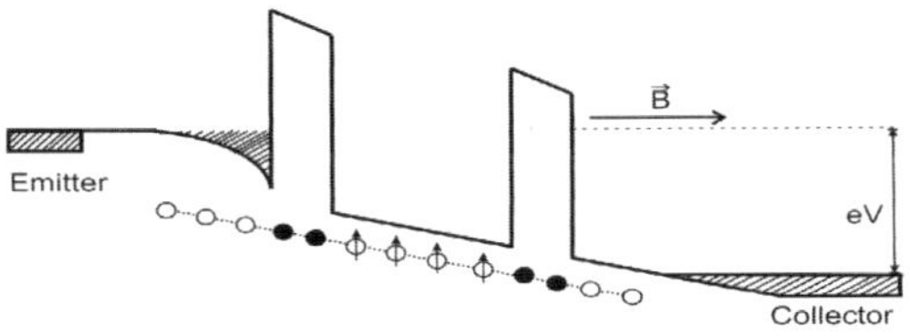

Figure 1. This scheme shows the configuration of the quantum dispositive or semiconductor nanostructure where the controlled impurities in the quantum well are affected by electric and magnetic fields. The different interactions between the nearest neighbors are studied by means of the tight-binding model.

The Hamiltonian of the system has the form:

$$H = \sum_{i,\sigma} \varepsilon_i c^{\dagger}_{i\sigma} c_{i\sigma} + \sum_{i,j,\sigma} V_{ij} c^{\dagger}_{i\sigma} c_{j\sigma} + \sum_{i,\sigma,\sigma'} A_{\sigma_i,\sigma'_i} c^{\dagger}_{\sigma_i} c_{\sigma'_i} \tag{1}$$

where ε_i is the diagonal energy, $c^{+}_{i\sigma}(c_{i\sigma})$ is the creation (annihilation) operator of an electron with spin up (S^+) and spin down (S^-), in either the barrier or the well, and V_{ij} is the hopping between neighboring sites i and j. In addition, $A_{\sigma,\sigma'} = g_0 \mu_B \overrightarrow{\sigma} \cdot \overrightarrow{B}$ in the Equation (1) is the Zeeman term, where g_0 is the gyromagnetic ratio of the electron, μ_B is the Bohr magneton, $\overrightarrow{\sigma}$ represents the Pauli spin-vector and $\overrightarrow{B}$ is the applied magnetic field parallel to the current direction.

In order to treat the Green functions (GF) in this nonequilibrium case, we use the Keldysh [12] scheme, the full system is decoupled into two equilibrium ones [right (R) and left (L)] and the associated GF are obtained. Renormalized dressed GF for the non equilibrium case can be derived through the Dyson equation, making the two subsystems interact by means of a perturbed Hamiltonian. The average current density induced to T=0K in the double barrier system is calculated through the following expression [13].

$$\langle J \rangle = \frac{2eT_\gamma}{\hbar} \int_{-\infty}^{+\infty} d\omega [G^{+-}_{01}(\omega) - G^{+-}_{10}(\omega)] \tag{2}$$

where G^{+-}_{01} G^{-+}_{01} are advanced and retarded Green functions and ω is the frequency. The above expression can be written in terms of the state density of the two subsystems in equilibrium, $\rho_{RL}(\hbar\omega)$, as:

$$J = 16\pi^3 J_0(B) T^2 \sum_{n}^{N} \int_{(n+1/2)\hbar\omega_c}^{E_F} \frac{\rho_L(\hbar\omega)\rho_R(\hbar\omega) d(\hbar\omega)}{|\Lambda(\hbar\omega|^2} \tag{3}$$

where $T = V_{01} = V_{10}$, $J_0(B) = \frac{e^2 B}{4\pi^2 \hbar^2 c}$, n runs over all Landau levels, E_F is the Fermi energy, $\omega_c = \frac{eB}{m^* c}$ is the cyclotronic frequency, m^* is the effective mass and

$$|\Lambda(\hbar\omega)|^2 = (1 - g^a_L g^a_R T^2)(1 - g^r_L g^r_R T^2) \tag{4}$$

where $g^{a(r)}_{L(R)}$ corresponds to the advanced (retarded) GF of the right and left subsystems.

3. Results

We make use of the Green Functions formalism shortly described in the preceding paragraph to study the polarized current density behaviour as a function of the magnetic field through a systematic analysis of magnetism in $CdTe - Cd_{1-x}Mn_xTe - CdTe$ magnetic diodes. We consider also the exchange interaction Hamiltonian between spins in localized diluted magnetic Mn^{++}-ions [3, 4, 6] and in free carriers as additional positive and negative uniform magnetic potentials, $V_m(x)=\pm N_0\alpha s_0(x)B_{5/2}[(5/2g_{_{Mn}}{_B}B)/(T - T_0(x))]$ [3, 5], for spin-up (S^+) and spin-down (S^-) carriers in the layer to the offset between layers, where $B_j(z)$ is the Brillouin function. For further details on this exchange-induced contribution to the energy separation of spin-split levels, see [6] and references therein. The parameters for $CdTe - Cd_{1-x}Mn_xTe$ layers are given in [3–5]. We have introduced a magnetic TB and considered the effects due to V_m(x) on spin-up, spin-down and total tunneling current densities in the magnetic DBH.

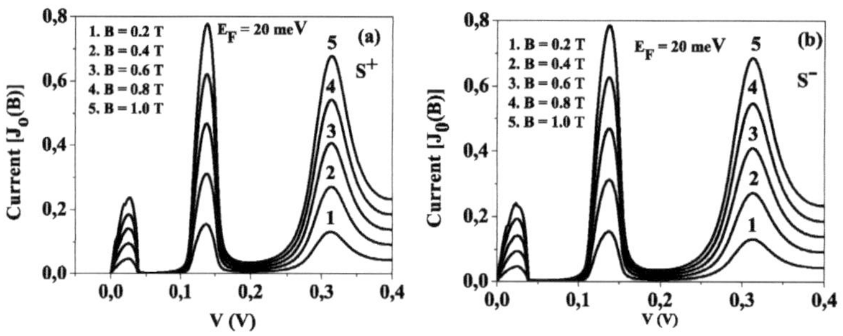

(a) (b)

Figure 2. Polarized current density with spin up (a) [spin down (b)] vs voltage for a $Cd_{1-x}Mn_xTe{-}CdTe{-}Cd_{1-x}Mn_xTe$ diode with different magnetic fields (0.2T, 0.4T, 0.6T, 0.8T and 1.0T) applied in the direction of the current. The barriers and the well are 30Å and 100Å respectively. The carrier density in the layer of the injector has a Fermi energy thick E_F=20meV.

In Figure 2 we show the behavior of the current density as a function of the bias voltage, for low and high magnetic fields strengths (0.2, 0.4, 0.6, 0.8 and 1.0T) in a quantum mechanical device. In this method we consider samples with fixed manganese concentration (Mn, x=0.00898) which allows us to increase the height of the potential barrier showing instantly a higher concentration of resonant states and instant magnetization. This substantially changes the current density in the system as the magnetic field strength increases and the voltage difference changes. Similarly, in this work, we also take into account that the barrier and well thicks are 30Å and 100Å respectively. It is important to note that the current density peaks (resonant states) are located near 0.03V, 0.15V and 0.32V for spin-up (Figure 2(a)). However, for the case of spin-down, the intensity peaks are located for values slightly lower, which implies a higher concentration of states that generates a carrier tunneling and show that the spin-down state in our system favors quantum transport information in this model. These resonances can be tuned by magnetic or electric fields since they determine both the energy and the spin-splitting between quantum well spin-polarized states in these devices. The strength peak is proportional to the tunnelling probability determined by the magnetic barrier height. In these figures too, we observe that the current intensity decreases in the intervals between 0.03V to 0.15V and between 0.32V and 0.40V showing a reversal in slope which implies a differential conductivity in the system. Also, in the second interval, the current density peak is lower than in the immediately preceding interval indicating saturation states because the supply voltage is large. Here, the last peak increases as the magnetic field strength increases, because the carriers are strongly localized in the nanostructure and favor the quantum tunneling of electrons.

In Figure 3 we show the results of the density of states with spin-up and spin-down for our DBH with conditions similar to those presented in Figure 2, as a function of the voltage with different magnetic fields applied in the direction of the current. We note that the number of quasi-bound states depends on the width of the quantum well, which is clearly correlated with the number of peaks in the Figure. The first peak, localized between 0.0 and 0.1eV, has a greater intensity which increases as the applied magnetic field increases. This implies a large concentration of localized states which increases the flow of a bigger number of carriers in the DBH. This indicates in turn growth in quantum information due to the spin. However, to the extent that the energy increases between 0.1 and 0.3eV, the height of the peaks decreases abruptly indicating that the quantum transport decreases due to the limited presence of the resonant states. In both figures we observe a similar behavior, except for the fact that in the case of the spin-down results, the quasi-bound states are localized in lower values of the energy that occur in lower voltages (see Figure 2(b)) favoring the passage of information.

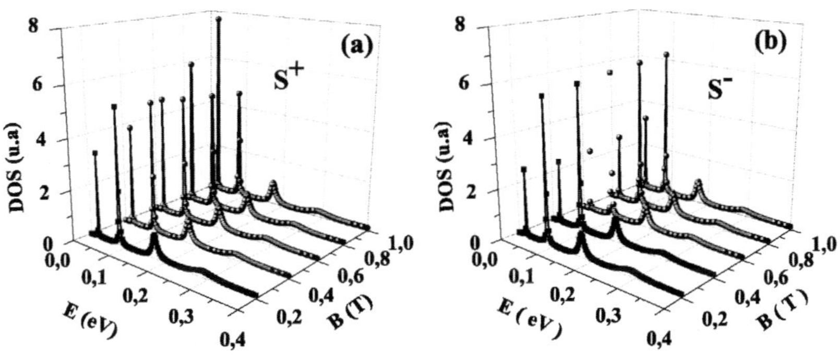

(a) (b)

Figure 3. State density (arbitrary units) with spin up (a) [spin down (b)] as an energy function for a diode system $Cd_{1-x}Mn_xTe - CdTe - Cd_{1-x}Mn_xTe$ with different applied magnetic fields (0.2T, 0.4T, 0.6T, 0.8T and 1.0T) in the direction of the current. The height of the potential barrier is Δ E=187meV, corresponding to a concentration of Mn (x=0.00898) and the barriers and the well are 30Å and 100Å thick, respectively.

In Figure 4(a) we show the differential conductance with spin up (spin down in Figure4(b)) vs voltage for the same structure shown in Figure 2. We observe that when the magnetic field amplitude increases, the differential conductance increases as a function of the voltage; this is because the values of the voltage applied to the device allow the electrons to tunnel through the conduction band of the structure $Cd_{1-x}Mn_xTe - CdTe - Cd_{1-x}Mn_xTe$ with spin up and spin down, since there are resonant states in the emission layer. The higher the magnetic field strength is, the more pronounced thepeaks in the figure are. However, for small values of the voltage between 0 to 0.05V, three peaks rise in the characteristic curve J_0(V), where there is a reversal of the conductance in the last of them for both spin up to spin down. Moreover, between 0.3 to 0.4V, we show a reversal in the differential conductance when the magnetic field increases which allows the tunnelling of carriers because the wave function is strongly localized and the cyclotron frequency is higher in the system. We can also see in these graphs that values of the voltage lower than 0.1V, the change in the differential conductance is appreciable for different strengths of the magnetic field for both spin up and spin down. This shows that resonant states are also produced for these voltage values allowing the passage of information via the spin system. We conclude that the information flow in diluted magnetic materials is produced via spin step at small potential differences.

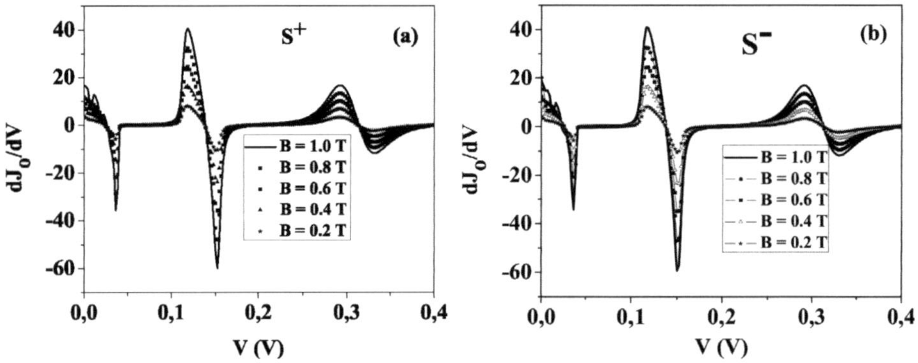

(a) (b)

Figure 4. Differential conductance with spin up (a) [spin down (b)] vs voltage for the same structure as in Figure 2.

In conclusion, in this paper we have shown the behavior of the current density in magnetic DBH tunnelling systems considering diluted magnetic impurities having ferromagnetic and antiferromagnetic orientations as a function of the magnetic field intensity, the magnetic ion concentration, the sample designed geometry and the potential bias. We have shown that the magnetic impurities have enhanced the spin-filter effect of quantum tunnelling in such systems with one of the spin-polarized tunnelling population being notably favored when different magnetic orientations are present.

Acknowledgments

The authors of this work acknowledge the partial financial support of the Cauca University, through the projects ID 3363 and 3088, and the Santander Industrial University (*UIS*) through the Programa de Movilidad de la Vicerrectoria de Investigaciones.

References

[1] G A Prinz 1998 *Science* **282** 1660
[2] A Slobodskyy, C Gould, T Slobodskyy, C R Beker, G Schmidt and L W Molenkamp 2003 *Phys Rev Lett* **90** 24601
[3] Mukesh Jain 1991 *Diluted Magnetic Semiconductors* (Singapore: World Scientific) chapter 11 pp 465-498
[4] J K Furdyna 1982 *J Appl Physics* **53** 7637
[5] Jan A Gaj and Jacek Kossut 2010 *Introduction to the Physics of Diluted Magnetic Semiconductors* (Heidelberg: Springer Series in Materials Science) chapter 3-4 pp 65-156
[6] A C R Bittencourt, J F Estanislau and G E Marques 2004 *Solid State Commun* **130** 253
[7] H Ohno 1999 *J Magn Magn Mater* **200** 110
[8] H Saito, A Yamamoto, S Yuasa and K Ando 2008 *Appl Phys Lett* **92** 192512
[9] M Elsen, O Boulle, J M George, H Jaffrs, R Mattana, V Cros, A Fert, A Lemaitre, R Giraud, G Faini and O Cedex 2006 *Phys Rev B* **73** 035303
[10] S H Chun, S J Potashnik, K C Ku, P Schiffer and N Samarth 2002 *Phys Rev B* **66** 100408
[11] W Wanga and C A Richterb 2006 *Appl Phys Lett* **89** 153105
[12] L V Keldysh 1965 *Sov Phys JETP* **20** 1018
[13] H Paredes-Gutiérrez, N Porras-Montenegro and A Latge 2003 *Phys Rev B* **68** 045311

Journal of Physics: Conference Series **687** (2016) 012082 doi:10.1088/1742-6596/687/1/012082

Biosynthesis and characterization of gold nanoparticles using extracts of tamarindus indica L leaves

S N Correa[1], A M Naranjo[1] and A P Herrera[2]
[1] Universidad Pontificia Bolivariana, Floridablanca, Colombia.
[2] Universidad de Cartagena, Cartagena, Colombia.

E-mail: sandra.correa@upb.edu.co , aherrerab2@unicartagena.edu.co

Abstract. This study reports the biosynthesis of gold nanoparticles using an extract of Tamarindus indica L. leaves. Phenols, ketones and carboxyls were present in the leaves of T. indica. These organic compounds that allowed the synthesis of nanoparticles were identified by gas chromatography coupled to mass spectrometry (GC/MS) and High Pressure Liquid Chromatographic (HPLC). Synthesis of gold nanoparticles was performed with the extract of T. indica leaves and an Au^{+3} aqueous solutions ($HAuCl_4$) at room temperature with one hour of reaction time. Characterization of gold nanoparticles was performed by UV visible spectroscopy, scanning electron microscopy (SEM) and EDX. The results indicated the formation of gold nanoparticles with a wavelength of 576nm and an average size of 52±5nm. The EDX technique confirmed the presence of gold nanoparticles with 12.88% in solution.

1. Introduction

A wide variety of physical and chemical procedures have been developed for the synthesis of nanoparticles with different compositions, sizes and shapes. However, physicochemical techniques for nanoparticles production like photochemical reduction, ablation laser, electrochemistry, aerosol technologies and ultrasound field, are highly expensive and require the use of toxic substances such as sodium borohydride, hydroxylamine, tetrakis (hydroxymetil) phosphonium chloride (THPC), and N, N-dimethylformaldehyde [1].

Synthesis through biological methods results attractive due to the environmentally friendly nature of the process. These biological procedures have been developed from several routes involving the use of metal salts precursors along with microorganisms, plants, fruits tissues and marine algae [2,3]. Such methodologies avoid the generation of negative impacts in the form of dangerous wastes, unlike mechanical, chemical and physical procedures [4-6].

Large amounts of agricultural residues including bagasse, leaves of trees and husks that are generated in monoculture regions of Colombia, could be used for the generation of nanoparticles through biological methods. This research was focused on the synthesis and characterization of gold nanoparticles from extracts of Tamarindus indica L. leaves as a biological method [7]. The main purpose of this study was to identify the compounds that allow the formation of gold nanoparticles. In order to do that, Ultraviolet-visible spectroscopy (UV-vis), Fourier Transform Infrared Spectroscopy (FTIR), Gas Chromatographic/Mass spectrometry (GC/MS) and High-Performance Liquid Chromagraphy (HPLC) techniques were implemented. Nanomaterials characterization was performed by Scanning Electron Microscopy (SEM) and Energy Dispersive X-ray Spectroscopy (EDX).

2. Methodology

Leaves of Tamarindus indica L. were obtained from the region of Arroyo Grande (Bolivar – Colombia). Biosynthesis of gold nanoparticles required the preparation of aqueous extracts from this raw material. To this end, infusion technique was performed by adding 1L of water to 100g of the plant, then heating the mixture to 90°C, subsequently filtering it and finally concentrating the obtained liquid until reaching a volume of 100mL. The extracts were then diluted in order to perform the UV-vis analysis in such a way that there was no interference with the wavelength of the gold nanoparticles. Consecution of functional groups belonging to reducing organic compounds (flavonoids, terpenes, tannins) was achieved by heating the extract samples to 80°C under constant agitation at 200rpm and concentrating them until 10mL of the extract were obtained. These samples were analysed by FTIR in an 8201 Shimadzu spectrophotometer.

A liquid-liquid extraction with HPLC grade ethanol was carried out on the organics extracts with the purpose of establishing the presence of organic compounds that act as bioreducing agents. The organic phase-extracts were concentrated in a Stuart Vertical Condensor rotary evaporator at 60°C for 15 minutes. The extracts were analysed by GC/MS and HPLC for the identification of flavonoids and terpenes. This was achieved by using an AT6890 Series Plus Agilent Technologies gas chromatograph coupled to a AT MSD 5975 Inert XL selective mass detector operating in radiofrequency full scan mode with a DB-5MS (J&W Scientific) 60m×0.25mm×0.25μm column with 5% phenyl-poly (dimethylsiloxane). The HPLC analysis was realized by Agilent Technologies 1200 Series (Palo Alto, California, USA), with a diode array detector (DAD) at λ=254nm using a DB-5MS (J&W Scientific) 60m×0.25mm×0.25μm column with 5% phenyl-poly (dimethylsiloxane).

Biosynthesis of the gold nanoparticles was carried out at room temperature on the basis of G.M. Nazeruddin methodology with some modifications [11]. It consisted of a simple, relatively fast procedure where 10mL of the chloroauric acid ($HAuCl_4$) precursor salt with a concentration of 0.27mM were mixed to sodium hydroxide (NaOH) at a concentration of 2M. After that, 1mL of the vegetal extract was added to the mixture, which proceeded to experience agitation for one hour and centrifugation at 6000rpm for 10 minutes. The settled material (gold nanoparticles) was washed with distilled water and submitted to an additional centrifugation. The solid material was extracted with 2mL of distilled water in order to generate a homogeneous solution of gold nanoparticles. A drop of NaOH 2M was added to the final solution, so that a pH of 10 was achieved, along with higher stability for the nanoparticles.

Formation of gold nanoparticles was verified by diluting the synthesized extracts of each biomass and using UV-Vis at a wavelength close to 550nm. Size and distribution of the nanoparticles were measured by SEM with a Quanta FEG 650 microscope (FEI, Netherlands). EDX analysis was performed using Apolo X equipment in high vacuum mode with a backscattered electron detector for secondary electrons with a resolution of 126.1eV (EDAX Inc, N.Y., USA). This allowed the identification of the elemental composition of T. indica extracts.

3. Results

3.1. Reducing agents identified by FTIR, GC/MS and HPLC

Functional groups in the leaves extracts of T. indica were determined by FTIR. Figure 1 shows the results of this analysis. There, it can be observed that a wide variety of functional groups were present in the extracts, including carbonyl compounds (1716.54cm^{-1}), aromatic rings (1559.37cm^{-1}), nitro compounds (1540.32cm^{-1}), alkanes (1394.81cm^{-1}), alkenes (1650.19cm^{-1}), amines (1254.59cm^{-1}), alcohols (3307.91, 1126.74, 1072.36cm^{-1}), phosphates (1072.36cm^{-1}) and alkyl halides (557.22cm^{-1}) X. These groups are commonly found as part of promoter agents for the bioreduction of gold nanoparticles, such as –OH and –COOH.

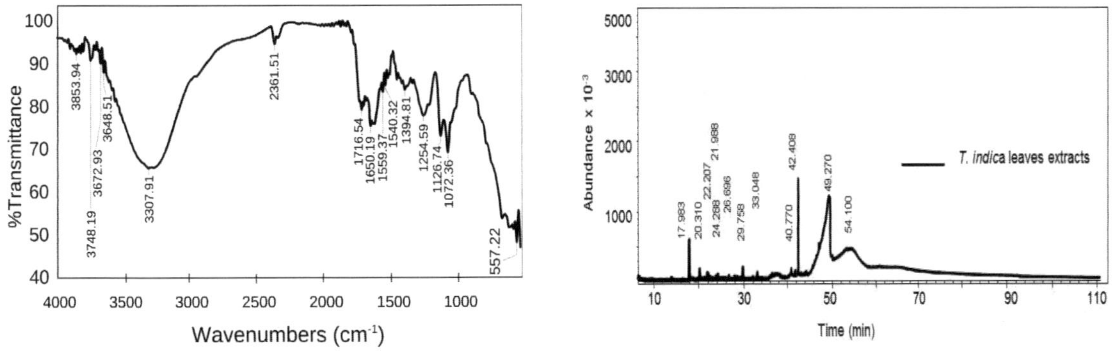

(a) (b)

Figure 1. (a) Infrared spectrum of the obtained Tamarindus indica L leaves extract. (b) Chromatogram of the samples of Tamarindus indica L leaves obtained by HPLC.

Figure 1(b) presents the identification of bioreducing compounds for the leaves extracts of Tamarindus indica L by GC/MS and HPLC/DAD. Organic compounds with aromatic rings such as phenols were identified in the T. indica sample. These included benzyl alcohol, o-guaiacol, 2,3-dihydrobenzofuran, p-vinilguaiacol and diethyl phthalate. Phenolic compounds act as bioreducing agents for nanoparticles by wrapping them and providing an excellent robustness in order to avoid agglomeration [7]. For the T. indica leaves extracts more compounds were detected using HPLC than the GC/MS.

3.2. Biosynthesis of gold nanoparticles

The synthesis of gold nanoparticles (AuNPs) was suggested by a pink coloration, and UV-Vis analysis confirmed the presence of this material at a wavelength of 576nm. Figure 2 shows the UV-Vis spectra that were obtained in the sample of original Tamarindus indica L. leaves extracts and the one with nanoparticles [7-9].

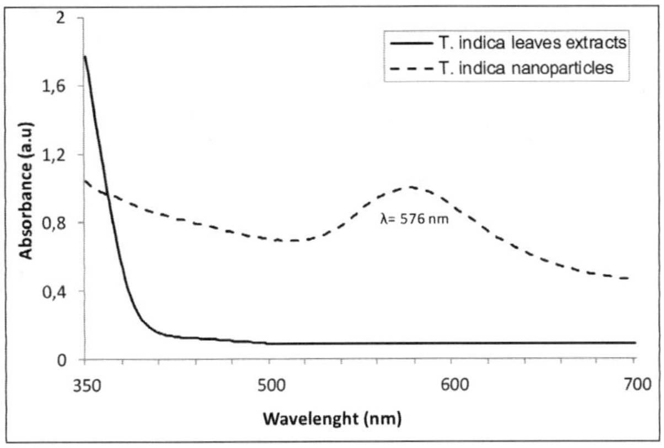

Figure 2. UV-Vis absorption spectra for 1mL of the original T. indica L. leaves extract and for nanoparticles that were synthesized with this extract.

3.3. Electron scanning microscopy (SEM)

The obtained nanoparticles exhibited an average size of 52±5nm according to SEM. An optimum size distribution was observed in the $HAuCl_4$ sample with an initial concentration as low as 0.27mM. Figure 3 shows the micrograph that was obtained by SEM, along with the analysis by EDX.

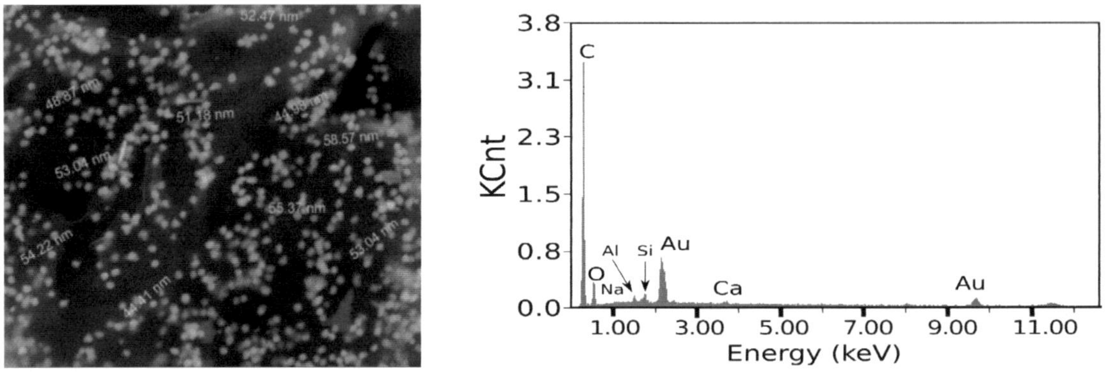

(a) (b)

Figure 3. (a) SEM micrograph of the gold nanoparticles that were synthesized from Tamarindus indica L. extracts (HAuCl4: 0.27mM). (b) EDX spectrum of gold nanoparticles synthesized with Tamarindus indica leaves.

These results show the achievement of uniform, spherical sizes, with some agglomeration sectors. EDX confirmed the presence of gold nanoparticles, which accounted for a 12.88% of the weight in the analysed sample. This recognition was made because of the registered energy, which is characteristic of gold nanoparticles.

4. Conclusion

Tamarindus indica L leaves extract is an efficient raw material for the synthesis of spherical gold nanoparticles with an average size of 52nm. FTIR analysis allowed the identification of carbonyl groups that can intervene in the reduction process and help stabilize the generation of nanoparticles. GC/MS and HPLC results confirmed the presence of these nanoparticles-producing compounds. UV-Vis, SEM and EDX techniques made it possible to identify the gold nanoparticles that were synthesized by biological methods. The resulting biomass from agricultural cultivation can be used to synthesize this material. Applications of gold nanoparticles continue to increase, and many of them play a vital in human and environmental health regarding mercury control and quality of water and air.

Acknowledgments

We would like to thank the Pontifical Bolivarian University of Bucaramanga sectional for the financial support of this project through funding for the improvement of science (DGI-BI050B project), and University of Cartagena for the financial support of this project (grant No. 064-2013).

References

[1] Kumar M A, Yusuf C and Chand B U 2013 *Biotechnology Advances* **31** 346
[2] Rajesh S, Raja D P, Rathi J M and Sahayaraj K 2012 *Sahayaraj J Biopest* **5** 119
[3] Iravani S 2011 *Green Chem* **13** 2638
[4] Dhillon G S, Brar S K, Kaur S and Verma M 2012 *Crit Rev Biotechnol* **32** 49
[5] Duran N and Seabra A B 2012 *Appl Microbiol Biotechnol* **95** 275
[6] Luangpipat T, Beattie I R, Chisti Y and Haverkamp R G 2011 *J Nanopart Res* **13** 6439
[7] Nazeruddin G M, Prasada N R, Prasadd S R, Shaikha Y I, Waghmare S R and Parag Adhyapak 2014 *Industrial Crops and Products* **60** 212
[8] Huang J L, Li Q B, Sun D H, Lu Y H, Su Y B, Yang X, et al 2007 *Nanotechnology* **18-10** 105104
[9] Sujitha M V and Kanna S 2013 *Spectrochim Acta Part A Mol Biomol Spectrosc* **102** 15

Borax as flux on sintering of iron Ancor Steel 1000® under glow discharge

H G Ariza Suarez[1], A Sarmiento Santos[2] and C A Ortiz Otálora[3]
[1] Universidad Pedagógica y Tecnológica de Colombia, Tunja, Colombia.

E-mail: geovannysy@hotmail.com

Abstract. This work studies the flux effect of borax (di sodium tetraborate decahydrate) on sintering of iron Ancor Steel 1000® in abnormal glow discharge. The incidence of the percentage by weight of borax and the sintering temperature in the process were observed. Samples of powder metallurgical iron were prepared with proportions of 0.50%, 2.0%, 4.0% and 6.0% by weight of borax using the procedures of powder metallurgy. The samples were sintered at 800 and 1100°C for 30min, by glow discharge at low pressure in a reducing atmosphere composed of 20% H_2+80% Ar. The samples in compact green-state were analyzed by TGA-DSC to determine the fusion process and mass loss during sintering. The analysis of microhardness and density, shows that at a sintering temperature of 800°C the sample density decreases and the sample microhardness increases with respect to sintered samples without borax. Sintered samples were analysed by DRX showing the absence of precipitates.

1. Introduction

Powder metallurgy is one of the processes used in industry for obtain chemical compositions in precise ratios in the manufacture of metallic materials [1] through the powder mixture which is subsequently compacted and sintered. The Sodium tetraborate pentahydrate (borax), with chemical formula: B_4O_7 $Na_2 \cdot 10H_2O$, is used in this study as metallurgical powder to analyse its effect as a lubricant and as a flux during sintering iron Ancor Steel 1000®. The fluxes have the property of lower the melting point of a mixture, and are widely used in welding. The Borax dissolves iron, tin, silver and nickel oxides [2], which can enhance the formation of metallurgical bonds in the sintering process. The oxide can be formed in the metal powder during the steps of: mixing, compaction and sintering.

2. Methodology

The powders Ancor Steel 1000® and Borax were weighed to have the desired amount of weight percentage; this process was performed using a microbalance. The Table 1 shows chemical composition of Ancor Steel 1000®. Metallurgical powders were mixed for 30 minutes and compacted in a steel matrix of diameter 10mm, 700MPa [3,6]. Sintering was performed in a plasma reactor by glow discharge with reductive gaseous atmosphere consisting of the mixture: 20% H_2+80% Ar for 30min with a pressure of 3Torr.

Sintering temperatures were 800°C and 1100°C that were achieved at a heating rate of 77°C/min. The porosity was analysed by micrographs at 100x magnification. DRX analysis was performed to detect the presence of precipitates. Density analysis was performed using a caliper and a microbalance.

Journal of Physics: Conference Series **687** (2016) 012083 doi:10.1088/1742-6596/687/1/012083

Microhardness measurement was obtained using a Vickers microhardness tester type, which was calibrated with a load of 100g for a time of 15s according to ASTM E384 [7].

Table 1. Chemical composition of Ancor Steel 1000®.

Element	Wt (%)	Element	Wt (%)
Carbon	0,004	Manganese	0,190
Sulfur	0,016	Copper	0,180
Oxygen	0,120	Nickel	0,080
Phosphorus	0,008	Molybdenum	0,050
Silicon	0,003	Chromium	0,008

3. Results

3.1. Compaction

The compacting step of Ancor Steel 1000 samples with different borax compositions was performed successfully; the samples were removed from the steel matrix without difficulty. This allows us to recommend the use of borax as a lubricant during the compaction step of powder metallurgical materials.

3.2. Thermogravimetric analysis

DSC-TGA analysis of Fe-6% mixture of borax are shown in Figure 1. Thermal processes show a dehydration process at 143.82° C due to the presence of borax and borax melting occurs at 745.16°C. Thermal dehydration process is presented between 140°C and 153°C, and causes a mass loss of 0.378mg.

The allotropic transformations of iron are also notorious, the allotrope β (Beta) with lattice parameter of 2.90Å and BCC crystal lattice is presented to 763.22°C and allotrope γ (Gamma), with FCC structure and 3.6Å lattice parameter, is presented around of 914.06°C.

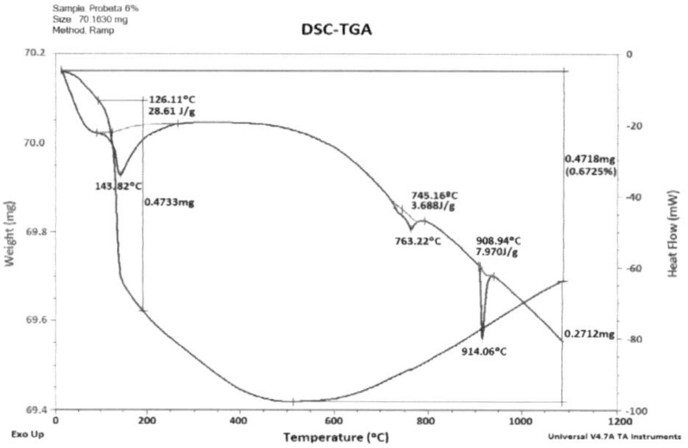

Figure 1. Thermogram of Fe-6% Borax.

3.3. Porosity

Samples of porosity increases with its borax content at 800°C sintering as shows in Figure 2. This increases because the regions occupied by borax during the forming step are then evacuated during the sintering step. Image analysis was performed at the micrographs in Figure 2 to quantify the porosity of the samples as a borax composition function, those results are summarized in Table 2.

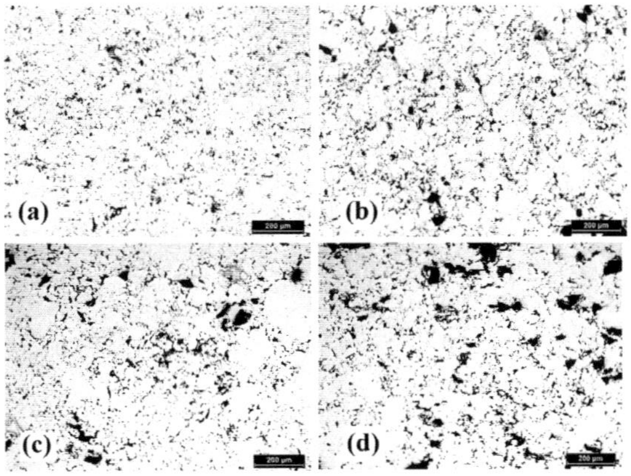

Figure 2. Micrographs at 100X: (a) 0.5% of borax, (b) 2% of borax, (c) 4% of borax and (d) 6% of borax.

Table 2. Porosity of Ancor Steel 1000® iron-borax samples sintered at 800°C.

Borax (% peso)	Porosity (%)
0,5	7,88
2,0	8,65
4,0	9,33
6,0	13,10

3.4. Density

The green and sintered samples show that density decreases with increasing addition of borax. Table 3 shows that the lowest density values were obtained with a percentage of 6% by weight of borax in both groups of samples.

Table 3. Iron-Borax samples density.

% of Borax	Density (g/cm^3)			
	Green state	Sintered to 800°C	Green state	Sintered to 1100°C
0.5	6,87	6,88	6,85	6,87
2,0	6,51	6,51	6,65	6,63
4,0	6,21	6,19	6,35	6,16
6,0	5,88	5,87	6,10	6,01

The sintered density should increases with respect to samples in green, but due to loss of mass during the heating step caused by borax output, sample density value is affected. The bombardment of the samples at the cathode of the glow discharge, also contributes to this mass loss.

3.5. Microhardness

Microhardness values are recorded in Table 4. Sintered samples at 800°C shown an increase in microhardness; it shows that the concentration of 2%, by weight of borax, has a microhardness value increase twice when sintered at 800°C. On the other hand, samples of 2.0%, 4.0% and 6.0% by weight of borax, sintered at 1100°C for 30 minutes, have a lower microhardness value than those sintered at 800°C. This is due to the rapid exit of borax during the heating step to reach the 1100°C temperature sintering. Because of flux properties of the Borax, the sintering is enhanced in its presence and then the mechanical properties of the material.

Table 4. Microhardness sintered 30min.

% of Borax	HV 0.1 Microhardness	
	Sintering to 800°C	Sintering to 1100°C
0,5	77,17	83,80
2,0	115,26	72,68
4,0	110,87	85,10
6,0	113,47	91,70

3.6. Analysis of DRX

DRX analysis carried out on the samples of 2.0%, 4.0%, 6.0% by weight of borax at 1100°C and the samples with 2.0% of borax sintering at 800°C, do not exhibit the formation of precipitates. The analysis only shows the characteristic peaks of ferritic iron. Figure 3 shows that DRX analysis is common to all samples.

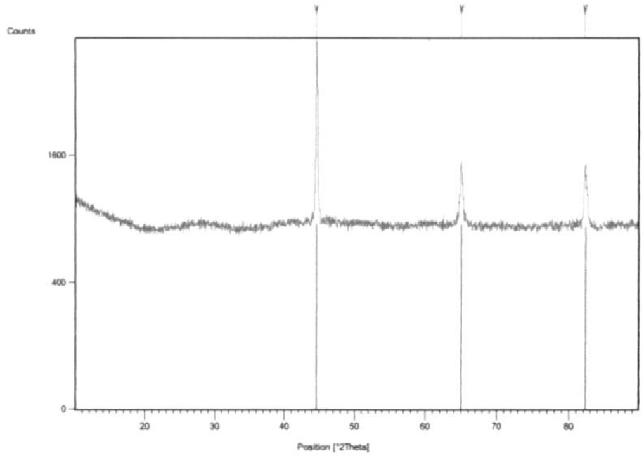

Figure 3. Analysis of DRX.

4. Conclusions

Ancor steel 1000® iron samples with 0.50%, 2.0%, 4.0% and 6.0% by weight of borax were manufactured using the standard procedure of the powder metallurgy. The sintering step was carried out in the cathode of abnormal glow discharge in a reducing atmosphere composed of 80% Ar and 20% H_2. The application of borax as a lubricant material during the forming step and application of the fluxing properties during sintering at 800°C was verified successfully.

The density obtained at 800°C sintering temperature is similar to that achieved at a sintering temperature of 1100°C. Samples sintered at 800°C, have higher microhardness than samples sintered at 1100°C. This is due to the oxide removing and flux properties of the borax at this temperature.

References

[1] Narasimhan K S 2001 *Mater Chem Phys* **67** 56
[2] Morral F R, Jimeno E y Molera P 1985 *Metalurgia general* (España: Editorial Reverté)
[3] Gural A, Tekeli S 2007 *Materials and Design* **28** 1224
[4] Gulsoy HO, Bilici MK, Salman S, Bozkurt Y 2007 *Materials and Design* **28** 2255
[5] Gulsoy HO, Salman S, Ozbek S 2004 *J Mater Sci* **39** 4835
[6] Selecká M, Šalak A, Danninger H 2003 *Journal of Materials Processing Technology* **141** 379
[7] ASTM E384-08 2008 *Standard test method for microindentation hardness of materials* (USA: ASTM International)

Synthesis and characterization of the oxide nanoparticles obtained by the polymeric precursor method

A M Raba[1], J Bautista[1] and E Murillo[1]
[1] Universidad Francisco de Paula Santander, San José de Cúcuta, Colombia.

E-mail: angelamercedesrp@ufps.edu.co

Abstract. Niobium pentoxide (Nb_2O_5) nanoparticles were prepared by the polymeric precursor method using citric acid as a chelating agent and ethylene glycol as a polymerizing agent. The powders obtained were characterized by X-ray diffraction and scanning electron microscopy. The results showed that the presence of HNO_3, HCl or NH_4OH in the employed aqueous solution favour the solubility of the used precursor salt, as well as also inside the oxide phase formation. The initial Nb_2O_5 powders were amorphous. The amorphous powders heated at $500\,°C$ contained Nb_2O_5 TT-phase, whereas at $650\,°C$ the Nb_2O_5 T-phase was obtained. In this way an increase in the synthesis temperature is related to the increase of the crystallinity, according to the values of the crystallite sizes that were estimated using the Scherrer method.

1. Introduction

The polymeric precursor method –Pechini method– is a chemistry synthesis method in which the general idea is to distribute the atomic cations through all polymeric structures [1]. Chelate formation takes place in an aqueous solution between cations and a carboxylic acid. Once this solution is mixed with a polyhydroxylated alcohol, the chelate is polyesterificated forming a resin. The decomposition of the amorphous resin is done by heating to temperatures below $350\,°C$ and then the material obtained is macerated. Powders are heated at temperatures between $500\,°C$ and $1100\,°C$ where the oxides with fine particles and suitable chemical composition are obtained, depending on the interest phase [2]. This method was chosen because it allows controlling factors such as porosity, surface area, particle size and shape which affect the microstructure, the reactivity and the final form of the oxide.

The oxide chosen is the Niobium pentoxide (Nb_2O_5). This is a n-type semiconductor with a band gap of about $3.4eV$ and is the most well-known and studied oxide material amongst all those reported oxides [3,4]. Many Nb_2O_5 polymorphs can be obtained depending on the temperature at which amorphous Nb_2O_5 is crystallized. The amorphous Nb_2O_5 begins to crystallize in the TT form (pseudohexagonal) around $500\,°C$. At higher temperatures T-, M- (orthorhombic) and H- (monoclinic) Nb_2O_5 forms are obtained [5]. The Nb_2O_5 crystalline behaviour also is influenced by the starting materials used, impurities that may be present and any interactions with other components.

Due to the excellent properties of the Nb_2O_5 (micro-textural, high refractive index, stability and corrosion resistance), this material has remarkable applications in optics, construction of ionic lithium batteries, electrochromic mechanisms and photovoltaic cells. The Nb_2O_5, particularly the T phase, is a promissory material since it provides good stability in aqueous medium for acid-catalysed reactions [6]. The Nb_2O_5 have significant advantages in catalytic performance because the nanochemical synthesis helps to control its structure and morphology [7].

In this work there will be an analysis the Pechini synthesis method of preparation for Nb_2O_5 nanostructures and its crystalline phase evolution.

2. Materials and methods

Nb_2O_5 nanoparticles were prepared by the polymeric precursor method -Pechini method. For this hydrated citric acid (Panreac 99.5%) was used as a chelating agent of niobium cations from $NbCl_5$ (Sigma-Aldrich 99%) and ethylene glycol (Panreac 99.8%) as a polymerizing agent. The employed solvent was distilled water (for the A samples) and solutions that contained HNO_3 (Panreac 65%), HCl (Panreac 37%) or NH_4OH (Panreac 30%) (B, C and D samples, respectively) where the objective was to dissolve better the niobium salt (see Table 1). The chelate was polyesterificated forming a resin. The resin decomposition was realized in a pre-calcination at 300 *°C* for 4 hours. The material obtained was macerated and the obtained powders were calcined at temperatures between 500 *°C* and 750 *°C* obtaining oxides with fine particles and of adequate chemical composition, which was controlled precisely during the process.

The characterization of the samples obtained was realized by X ray diffraction (XRD) with the objective being to determine the crystalline phases. For this an XPERT-PRO Panalytical was employed with radiation Co $K_{\alpha 1}$ (λ=1.78901$\mathring{A}$) operating at 40*mA*, at a velocity of 0.0133 °/*min* and varying 2θ between 10° and 90°. A scanning electron microscope (SEM) FEI QUANTA 200 operating at 5*kV* was used for analysing the morphology of the synthesized materials.

3. Results and discussion

Figure 1 presents the XRD patterns of the B samples as a function of the temperature. For all samples the Nb_2O_5 was identified using data contained in the Power Diffraction File, cards No. 28-0317 for the pseudohexagonal TT-phase (a=b=3,607$\mathring{A}$, c=3.925$\mathring{A}$) and cards No. 30-0873 for the orthorhombic T-phase, (a=6,175$\mathring{A}$, b=29,175$\mathring{A}$, c=3,930$\mathring{A}$). The crystallite sizes were estimated using the Scherrer method (see Table 1).

Table 1. Results of XRD analysis at room temperature[a].

Sample	Heating temperature (°C)	Solvent	Phase composition	Crystallite size (nm)
B-Nb500	500	H_2O - HNO_3	Nb_2O_5 (TT)	34 (9)
C-Nb500	500	H_2O – HCl	Nb_2O_5 (TT)	18 (6)
D-Nb500	500	H_2O - NH_4OH	Nb_2O_5 (TT)	45 (17)
B-Nb650	650	H_2O - HNO_3	Nb_2O_5 (T)	37 (5)
C-Nb650	650	H_2O – HCl	Nb_2O_5 (T)	39 (7)
D-Nb650	650	H_2O - NH_4OH	Nb_2O_5 (T)	35 (1)
B-Nb750	750	H_2O - HNO_3	Nb_2O_5 (T)	42 (2)
C-Nb750	750	H_2O – HCl	Nb_2O_5 (T)	41 (3)
D-Nb750	750	H_2O - NH_4OH	Nb_2O_5 (T)	44 (4)

[a] The heating time was 2 hours.
TT = low-temperature pseudohexagonal phase. T = orthorhombic phase.

The samples heated at 500 *°C* corresponded to Nb_2O_5 (TT) and showed broad diffraction lines. Nb_2O_5 (T) formed upon the heating of samples A, B, C and D at 650 *°C* and exhibited rather sharp diffraction lines. The single diffraction lines of the TT-phase were well resolved in the T-phase as seen in Figure 1. The analysis of these patterns indicates that an increase in the synthesis temperature leads to materials with higher crystallinity. In the Figure 2 another trend was observed at the oxide orthorhombic formation when the synthesis routes B and D are utilized. This is an indication that the HNO_3 and the NH_4OH produced the same crystalline structure (homogeneous Nb_2O_5 orthorhombic).

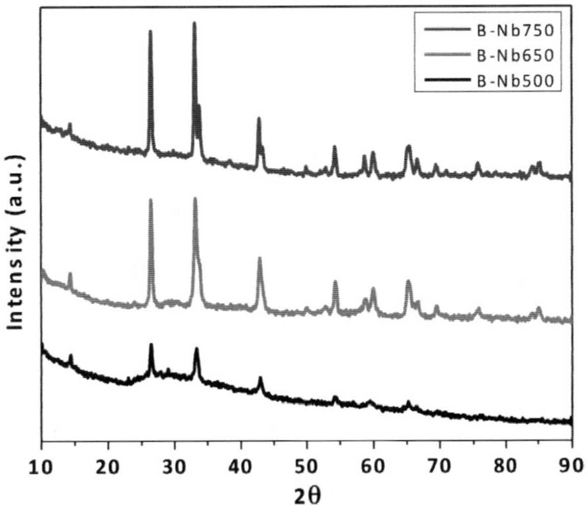

Figure 1. XRD patterns of the B samples as a function of the temperature.

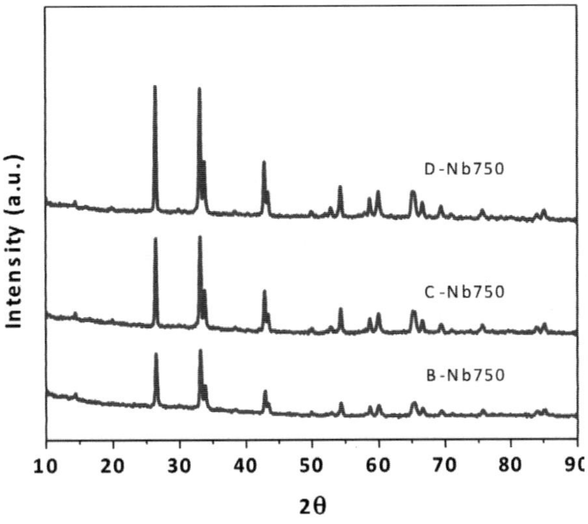

Figure 2. XRD patterns of the samples heated at 750 *°C*.

The morphological characterization samples are realized through SEM imagery analysis. In Figure 3 the morphological differences are observed into the samples A-Nb500 (a), A-Nb650 (b) and A-Nb750 (c). The sample A-Nb500 presents isotropic morphology and reveals the presence of abundant nanoparticles agglomerate; this agglomeration infers that the calcination temperature used is still low. Figure 3(b) shows a typical SEM image of Nb_2O_5 nanostructured. It can be seen that the nanoparticles surface is well defined. Since these samples were not calculated it was difficult to count the mean particle size since. Figure 3(c) shows a uniform distribution of the nanoparticles; this indicates that the thermal treatment at 750 *°C* for 2*h* promotes the growth of the particles. These results suggest that the increase in the synthesis temperature allowed significant morphological modifications.

Journal of Physics: Conference Series **687** (2016) 012084 doi:10.1088/1742-6596/687/1/012084

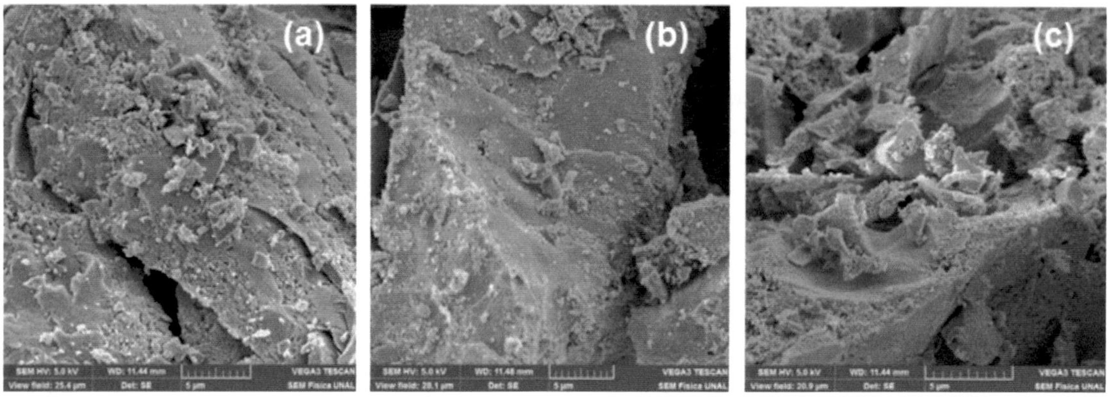

Figure 3. Imagery obtained by SEM for A samples heated at (a) 500, (b) 650 and (c) 750 °C.

4. Conclusions

It was verified that the synthesis method is suitable to obtain the desired oxide phase. In the four routes, the Pechini method helps significantly to improve the structural properties studied in this work since the T-phase Nb_2O_5 was accepted, in which an orthorhombic structure was confirmed by XRD. The thermal conditions that best favoured the growing of the particles were obtained to thermal treatment at 750 °C for 2 hours. According to the results presented, it can be noted that the best synthesis conditions for the obtaining of the T-phase Nb_2O_5 corresponded to the samples that employed an aqueous solution of HNO_3.

References

[1] Pechini MP 1967 *Method of preparing lead and alkaline earth titanates and niobates and coating method using the same to form a capacitor* (US: Patent No.3.3306.97)

[2] Vargas M A, Franco Y, Ochoa Y, Ortegón Y and Rodriguez J E 2011 *Bol Soc Esp Ceram Vidr* **50** 267

[3] Sreethawong T, Ngamsinlapasathian S, Lim S H and Yoshikawa S 2013 *Chemical Engineering Journal* **215** 322

[4] Lin H Y, Yang H C, Wang W L 2011 *Catalysis Today* **174** 106

[5] Zhao Y, Zhou X, Ye L and Edman Tsang S C 2012 *Nano Reviews* **3** 17631

[6] Prado A G S, Bolzon L B, Pedroso C P, Moura A O and Costa L L 2008 *Applied Catalysis B Environmental* **82** 219

[7] Lopesa O F, Paris E C and Ribeiro C 2014 *Applied Catalysis B Environmental* **144** 800

Energy transfer in two-wave mixing quasi-degenerated in photorefractive crystals

B Barrera[1], N A Arias Hernández[1], M L Molina Prado[1] and M Tebaldi[2]
[1] Universidad de Pamplona, Pamplona, Colombia.
[2] Universidad Nacional de La Plata, La Plata, Argentina.

E-mail: molinaprado@gmail.com

Abstract. Wave equations describing the quasi-degenerated two-beam coupling in the photorefractive crystals for high frequency gratings have been solved. In the case of the quasi-degenerate two-beam coupling, equations depend upon the coupling coefficient, the response time of the medium, the input beams frequencies, the absorption coefficient and the input intensity ratio. The response time of the medium is function of the diffusion field, the drift field, the saturation field and the concentration ratio, i.e., the ratio between N_A density of acceptor impurities and N_D density of donor impurities. The effect of these parameters on the gain has been studied in details.

1. Introduction

Photorefractive crystals make possible an energy transfer from the reference or pump beam into the signal beam. The coupling of interacting beams in dynamic hologram recording in photorefractive media has been extensively studied due to their potential applications in signal processing, optical communications, optical computing, real-time holography, image enhancement and holographic memories. The conventional theoretical description of the waves coupling is known as the coupled wave's theory.

The nonuniform pattern received by the crystal results in a refractive index change due to the photorefractive and linear electro-optic effect. This index grating can be considered a volume dynamic grating. In experiments using non stationary conditions of recording, the energy transfer between the two beams takes place due to a phase mismatch between the incident light intensity grating in the crystal and the photoinduced index grating produced inside the crystal. In our proposal, wave equations describing the quasi-degenerated two-beam coupling in the photorefractive crystals have been analysed. The beams intensities dependence on the material response, the diffusion field, the saturation field, the drift field, the characteristic time of the medium and the concentration ratio are discussed.

2. Theoretical discussion

When two laser beams with different frequencies are incident in a photorefractive medium, a non-stationary interference fringe pattern is generated. Let us define w_1 and w_2 as the frequency of the high (reference) and low (signal) intensities beams, respectively. The electric field of each beam can be written as:

$$E_j = A_j \mathrm{e}^{i(w_j t - \bar{k}_j \cdot \vec{r})} \quad j = 1,2 \tag{1}$$

Where A_1 , A_2 are the amplitudes and $\vec{k}_1, \vec{k}_2$ are the wave vectors. The resulting intensity is:

$$I = A_1 A_1^* + A_2 A_2^* + A_1 A_2^* e^{-i(\Omega t - \vec{k}_1 \cdot \vec{r})} + A_2 A_1^* e^{i(\Omega t - \vec{k}_2 \cdot \vec{r})} \tag{2}$$

Where $\Omega = w_2 - w_1$; $\vec{K} = \vec{k}_2 - \vec{k}_1$ and * indicate the complex conjugate. The intensity distribution given by Equation (2) represents a fringe pattern moving with a speed: $v = \frac{\Omega}{k} = \frac{\Omega\Lambda}{2\pi}$ where Λ is the fringe pattern period. The refractive index including the fundamental component of the intensity-induced grating can be written [1] as,

$$n = n_0 + \left[\frac{n_1}{2} e^{i\varphi} \frac{A_1^* A_2}{I_0} e^{i(\Omega t - \vec{k} \cdot \vec{r})} + c.c \right] \tag{3}$$

Where $I_0 = I_1 + I_2 = |A_1|^2 + |A_2|^2$, n_0 is the nonpertubated index of the material, ϕ indicates the shift of the photoinduced index grating with respect to the recording interference pattern and the values of ϕ is [2, 3]:

$$\phi = \phi_0 + tan^{-1}(\Omega\tau)$$

Where ϕ_0 is a constant phase shift related to the non-local response of the crystal under the interference fringe illumination and n_1 is

$$n_1 = \frac{2\Delta n_s}{\left(1 + \Omega^2 \tau^2\right)^{1/2}} \tag{4}$$

Where Δn_s is the saturation value of the photo-induced index change. Δn_s and ϕ_0 depend not only on the grating spacing and its direction but it depend on the material properties. In photorefractive medium when only the diffusion transport mechanism is consider (i.e., no external static electric field), $\varphi_0 = \frac{\pi}{2}$ [4]. On the basis of band transport model in which the materials rate equations are solved for moving grating under the assumption $I_1 << I_0$, the response time τ is given by the relation [6, 7]:

$$\tau = t_0 \left[\frac{E_d + E_\mu}{E_d + E_q} \right]; \qquad t_0 = \frac{N_A}{N_D s I_0} = \frac{1}{r s I_0}; \qquad E_d = \frac{k K_B T}{q}; \qquad E_q = \frac{q N_A}{\varepsilon_e k}; \qquad E_\mu = \frac{\gamma_r N_A}{\mu_e k} \tag{5}$$

Where t_0 is the characteristic time constant of the medium, E_d is the diffusion field, E_q is the saturation field and E_μ is the characteristic pseudo field, N_A is the density of acceptor impurities, N_D is the density of donor impurities, s is the cross-section of photo-excitation, K_B is the Boltzmann constant, T is the temperature, q is the electronic charge, γ_r is the recombination constant, ε_e is the effective dielectric constant and μ_e is the effective mobility.

The presence of the phase shift between the interference pattern and the photoinduced volume index grating allows the possibility of the non-reciprocal steady-state transfer of energy between the reference and signal beams. The coupled wave equations for the case quasi-degenerated are given by:

$$\frac{dI_1}{dz} = -\left[\frac{\gamma_0}{1 + \Omega^2 \tau^2} \right] \frac{I_1 I_2}{I_0} - I_1 \alpha \tag{7}$$

$$\frac{dI_2}{dz} = \left[\frac{\gamma_0}{1+\Omega^2\tau^2}\right]\frac{I_1 I_2}{I_0} - I_2\alpha \tag{8}$$

Where γ_0 is coupling constant in the degenerated case. Equations (7) and (8) can be integrated to yield:

$$I_1(z) = I_1(z)\frac{1+m^{-1}}{1+m^{-1}\exp\left[\gamma_0 L\left(1+\Omega^2\left(\frac{1}{rsI_0}\left(\frac{E_d+E_\mu}{E_d+E_q}\right)\right)^2\right)^{-1}\right]}\exp(-\alpha L)$$

$$I_2(z) = I_2(z)\frac{1+m}{1+m\exp\left[-\gamma_0 L\left(1+\Omega^2\left(\frac{1}{rsI_0}\left(\frac{E_d+E_\mu}{E_d+E_q}\right)\right)^2\right)^{-1}\right]}\exp(-\alpha L) \tag{9}$$

3. Results

Equation (9) is the beams intensities analytical expression when transfer of energy in two-wave mixing for the quasi-degenerated case is produced. It is evident that the intensity of beam I_2 increases with increasing crystal thickness reaches a maximum and then decreases exponentially due to the material absorption. Note that, for values of r higher than 500, the intensity I_2 is independent of concentration ratio r. Moreover for larger value of m, the maximum of intensity reaches at higher values of crystal thickness Figure 1(a–c).

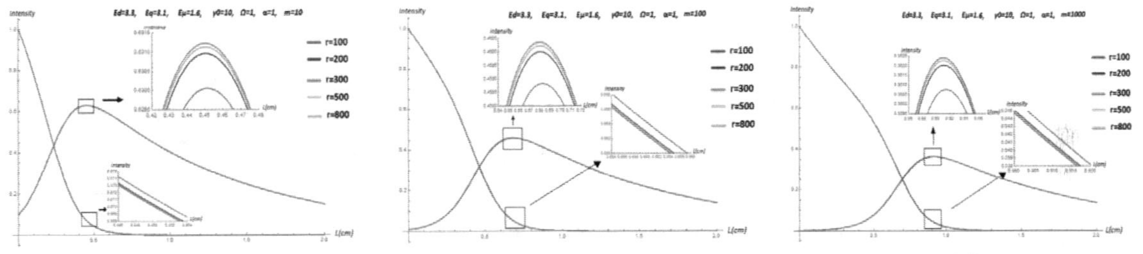

(a) (b) (c)

Figure 1. Dependence of the beams intensities on the crystal thickness and concentration ratio $r=N_D/N_A$ in quasi-degenerated two wave mixing (grating spatial frequency 2000lines/mm, Ed=3.3kV/cm; Eq=3.1kV/cm; Eμ=1.6kV/cm; γ_0=10; Ω=1 at various values of (a) m=10, (b) m=100, (c) m=1000.

The intensity of the beams I_1 and I_2 in term of the crystal thickness in a quasi-degenerated two wave coupling for a 2000lines/mm recording grating and different values of energy coupling coefficient γ_0=5,10,15,20,25cm^{-1} is show in Figure 2.

The results presented in Figure 3 confirm that the optimum transfer energy occurs at the same interaction length (L=0.68cm) for the different values of Ω=0.1, 0.3, 0.6, 0.8, 1.0Hz. For values of Ω lower than 0.3, the maximum intensity I_2 is independent of oscillation frequency shift.

The intensities of the reference and signal beams with respect to the concentration r for BSO and $BaTiO_3$ crystals are plotted in Figure 4(a) y 4(b) for m=100. Note that the beam I_1 decreases rapidly for lower values of r for BSO compared with $BaTiO_3$ while that the beam I_2 increases with increasing concentration ratio and intensity maximum is higher for BSO at lower values of concentration ratio than the $BaTiO_3$ material it due to the values of dielectric constant.

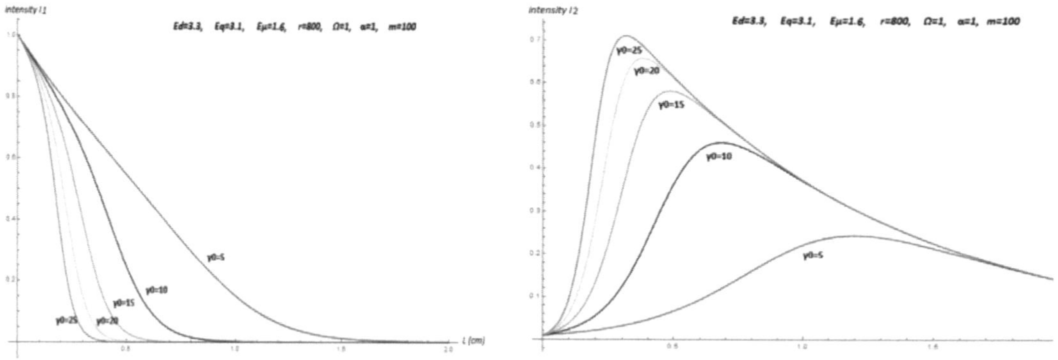

(a) (b)

Figure 2. Intensity of the beams in two wave coupling quasi-degenerated case, (a) I_1 (b) I_2, in terms of the crystal thickness corresponding to a 2000 lines/mm recording grating and different values of energy coupling coefficient γ_0=5, 10, 15, 20, 25cm^{-1} (Ed=3.3kV/cm; Eq=3.1kV/cm; Eμ=1.6kV/cm; m=100; r=800; Ω=1).

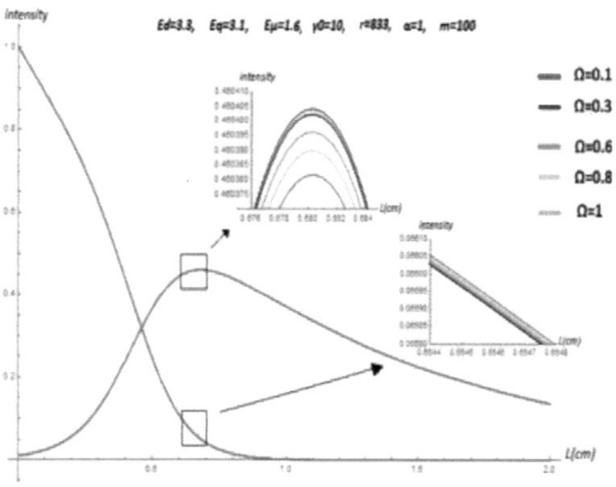

Figure 3. Dependence of the beams intensities in term of the crystal thickness for different values of Ω (Ed=3.3kV/cm; Eq=3.1kV/cm; Eμ=1.6kV/cm; m=100; γ_0=10; r=833).

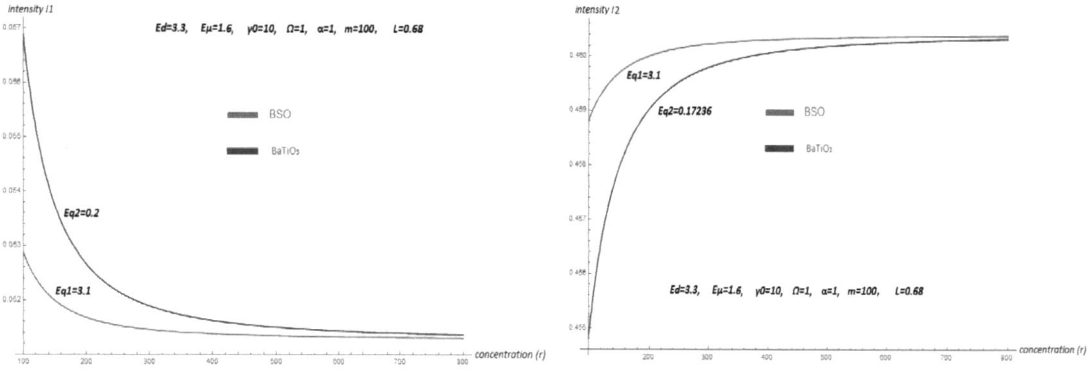

(a) (b)

Figure 4. Intensities of the reference and signal beams (a) I_1 and (b) I_2 with respect to the concentration ratio r=N_D/N_A for BSO and $BaTiO_3$ materials in two beam coupling quasi-degenerated case (2000 lines/mm recording grating, Ed=3.3kV/cm; Eμ=1.6 kV/cm; m=100; γ_0=10; Ω=1; L=0.68).

Journal of Physics: Conference Series **687** (2016) 012085 doi:10.1088/1742-6596/687/1/012085

4. Conclusions

It is clear that the intensity of the two beams (reference and signal) inside of the crystal thickness not only depend on the coupling coefficient γ_0, the crystal thickness L, the modulation ratio m, the absorption coefficient and the oscillation frequency shift Ω. Also, the mentioned intensities change with the density of acceptor impurities N_A and the density of donor impurities ND ($r=N_D/N_A$). The transfer energy between the two beams increases with crystal thickness. The maximum intensity I_2, increase when the concentration ratio increases. Indeed, it is observed higher values for BSO crystal at lower values of concentration ratio than $BaTiO_3$ material.

Acknowledgments

This research was performed under the grants: UNIPAMPLONA, PR130-00-033 (GA160-BP-II-2013-2.1.2.2.1) and CONICET PIP No. 0549/12 (Argentina), Facultad de Ingeniería, Universidad Nacional de La Plata No. 11/I168 (Argentina). Martha Lucía Molina acknowledges financial support under CONICET and TWAS-UNESCO Associateship Scheme at Centers of Excellence in the South, and Universidad de Pamplona (Pamplona, Colombia).

References

[1] Fischer B, Cronin-Golomb M, White J O and Yariv A 1981 *Opt Lett* **6** 519
[2] Huignard J P and Marrackchi A 1981 *Opt Commun* **38** 249
[3] Mosquera L, de Oliveira I, Frejich J, Hernandes A C, Lanfredi S and Carvalho J F 2001 *J Appl Phys* **90** 2635
[4] P Yeh 1993 *Introduction to Photorefractive Nonlinear Optics* (New York: Wiley)
[5] Kukhtarev NV, Markov V B, Odulov S G, Soskin M S and Vinetskii V L 1979 *Ferroelectrics* **22** 961
[6] Kwak C H, Park S Y, Jeong J S, Suh H H and Lee E H 1994 *Opt Commun* **105** 353
[7] Kwak C H, Park S Y and Lee E H 1995 *Opt Commun* **115** 315
[8] Maurya M K, Yadav T K and Yadav R A 2010 *Opt Laser Technol* **42** 465
[9] Yadav T K, Maurya M K and Yadav R A 2012 *Optik* **123** 1329

Structure, mechanical and thermal behaviour of mixtures of polyester resin and dental ceramic waste

G Peña Rodríguez[1], L Martínez Maldonado[1] and H J Dulce Moreno[1]
[1] Universidad Francisco de Paula Santander, San José de Cúcuta, Colombia.

E-mail: gpenaro@ufps.edu.co , lore_3612@hotmail.com

Abstract. The tensile strength and bending strength, structure and thermal behaviour of mixtures of polyester resin (P-2000) and powders (ASTM sieve 200, <75μm) of dental ceramic wastes (dentals impressions, alginate and gypsum) was reported. The samples consisted of mixtures with percentage weights of 50-50%, 60-40%, 70-30%, 80-20%, 90-10%, where the resin was the majority phase, the Mekc (4% wt) was used as catalyst. The structure was studied using SEM and XRD, the thermal behaviour using DSC, TGA and DMA, while the mechanical strength was tested using standards ASTM D790 and D638. Irregular morphology and presence of small agglomerations was observed, with particle sizes between 29.63 and 38.67μm, the presence of different phases of calcium sulphate was found, and that to the increasing the concentration of the powder, the materials becomes more crystalline, increasing its density. An average service temperature of 69.15±4.60°C was found. Vickers hardness values are reported in ranges from 18.65 to 27.96. Considering the elastic modules was established that the materials become more rigid by having more powder concentration.

1. Introduction

In the manufacture of dental prosthesis, dentists use alginate and gypsum for dental impressions of patients, which once used is discarded following the protocols manual integrated waste management in each country. In Colombia, there is no knowledge of the recycling of these dental impressions, while in countries such as Greece, Brazil and Jordan among others, there are studies of the composition of dental solid waste, with the aim of recycling potential [1-3].

Thermosetting resin products are some of the most adaptive materials in the marketplace and hundreds of industries, for its easy handling and shaping. In recent years it has been used as filling material recycled silicon oxide powders, aluminium oxide, rice husk, wood powders, among others, in order to find new materials for technological applications [4-6].

Therefore, the structural, thermal and mechanical characterization, of mixtures of polyester resins and waste of alginate powders and gypsum used in dentistry are reported.

2. Materials y methods

The pre-accelerated unsaturated polyester resin (P-2000) was supplied by Industries of Resins S.A.S (Medellin Colombia), which is commonly used in non-aggressive environments, pink colour, with approximately 65% of polyester. For curing, Mekp catalyst was used in concentration of 4% by weight, being gelation time at room temperature of 20 minutes, with a maximum exothermic temperature of approximately 165°C.

After obtaining dental impressions discarded of gypsum and alginate (see Figure 1), were subjected to autoclaving at pressure of 15 pounds, temperature of 121±2°C for one hour, then manually grinding was performed, followed by trituration using hammer mill then sieved on a 200 ASTM sieve (≤75 microns). With the above powders, a mixture was prepared with 50% by weight of gypsum and 50% alginate, which was used as filler in mixtures with polyester resin, the following mixtures defined percentage by weight: 50-50, 60-40, 70-30, 80-20 and 90 -10, where the majority phase was the resin and the minority the powder mixture of gypsum and alginate. There was established a time of 10 minutes of constant mixing without the catalyst and then a time frame of 2 minutes of constant mixing with the catalyst. The moulding process was then strained in an aluminium mould with the dimensions according to the standards used.

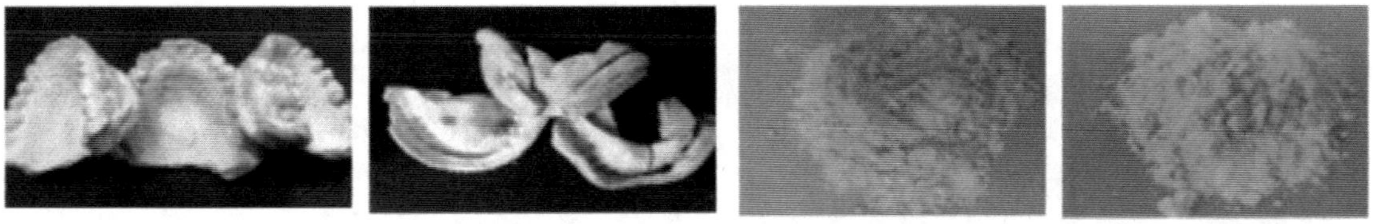

(a) (b) (c) (d)

Figure 1. a) Dental gypsum impressions, (b) Dental alginate impressions, (c) Powders gypsum, (d) Powders alginate. Fount: authors.

The morphology and structure of the samples were studied using the scanning electron microscope (SEM) FEI Quanta 650 FEG, and the X-ray diffractometer (XRD) Bruker D8 Discover with DaVinc geometry. Thermal behaviour is performed using thermogravimetric balance Discovery series TA Instruments under nitrogen purge gas flow of 50ml/min, temperature range 30°C to 1000°C and heating rate 10°C/min, while the differential scanning calorimetry (DSC) was performed according to ASTM E2160 for which was used a DSC of the same series and marks in a temperature range of 25°C to 500°C with 10°C/min temperature ramp. For the dynamic mechanical analysis there was used a DMA Q800 TA Instruments equipment, with temperature ranges between -145°C and 600°C and ramps up to 20°C/min. The effective thermal conductivity (K) at room temperature was determined using the KD2 Pro® system, and the SH-1 sensor, system operating with the physical principle of linear transient heat flow.

3. Analysis and discussion of results

From analysis using SEM images, for alginate, an irregular morphology was found with the presence of diatomic particles of circular shape with diameters between 15.4±0.8 and 17.06±0.76 microns, which are found in Chrysophyta and correspond to algae, Bacillariophycea class, which are part of inorganic composition of alginate [8], while gypsum powder, consists of very fine particles of different geometries. Microanalysis, showed that sodium alginate is kind, moreover, gypsum main elements were found as calcium and sulphur, results that match those reported in references [9-10].

In Figure 2 the SEM micrograph at 500X for 70-30% mixture is presented, the morphology of alginate and gypsum particles is irregular with average sizes between 11.5 and 74.83μm, immersed in the resin matrix. Microanalysis by EDS-SEM, shows that as the concentration of filler in the sample increases, the weight percentage of carbon and oxygen decrease, while the sodium, aluminium, silicon and calcium increase.

Qualitative analysis using XRD (see Figure 3), evidence the presence of hydrated calcium sulfate, Cristobalite low, Anhydrita, Quartz, Chantalita, Basinita and Dolomite. Also, it is observed that as concentration of the filler (powdered gypsum and alginate) to the sample increases, the crystallinity increases, which is evidenced by the decrease of the curvature at the base of the diffraction patterns [11].

The decrease in curves thermogravimetry (TG), allow to infer a sample decomposition in a simple process, *ie*, the material undergoes mass loss, where the first drop in the curve corresponds to evaporation catalyst [12], a proportional relationship between the amount of resin present in the composition and weight loss is also shown, since at these temperatures the matrix phase is tending to undergo changes [13]. It was found that the glass transition temperature (Tg) increases with increasing concentration of filler [14], reporting an average value of 69.15±4,596°C Tg, very close to the value reported in the literature [15], the lack of significant variability of Tg, is based on that it is a function of the degree of cure. In Figure 4, the behaviour of the heat capacity of the sample 70-30wt%, wherein the observed endothermic processes to 108.30°C and 364.50°C respectively is presented. With regard to dynamic mechanical analysis (DMA) increased elastic modulus with increasing concentration of filler, from 1917MPa for the sample with 0% filler, to 4845MPa for the sample with 50% wt of plaster dust is evident and alginate, showing a higher thermo-mechanical added to the filler material [16].

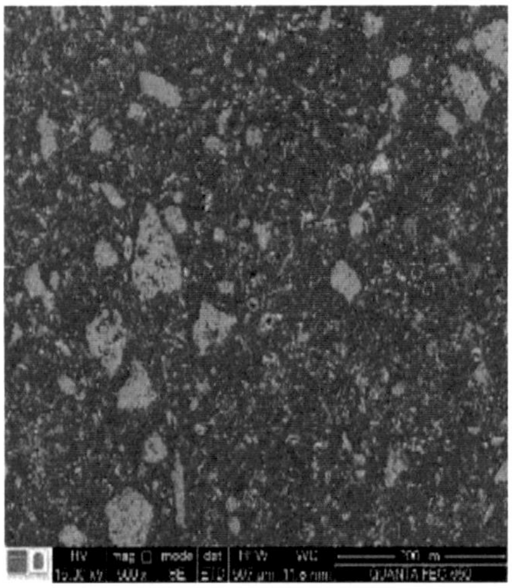

Figure 2. SEM micrograph at 500X using the sample 70-30% wt.

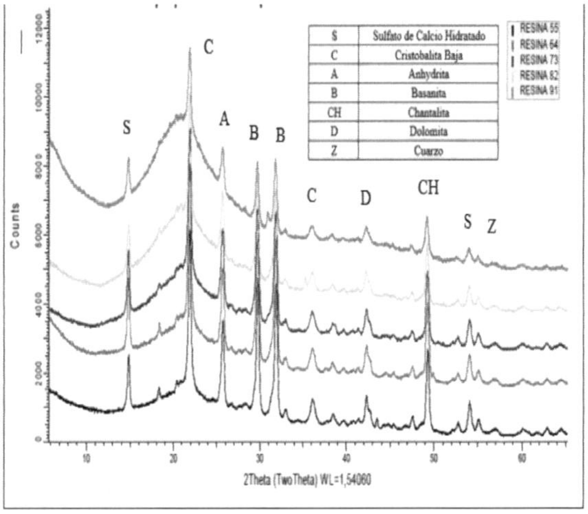

Figure 3. Patterns XRD of different mixtures.

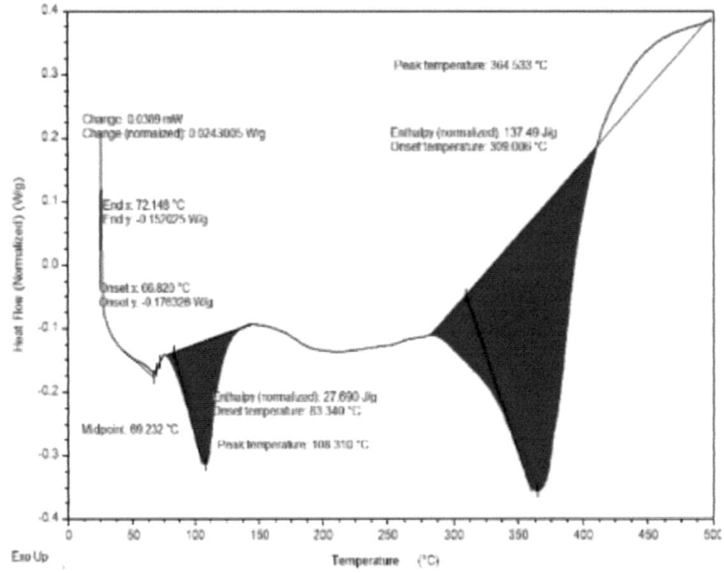

Figure 4. DSC graph corresponding to 70-30% mix.

Table 1. Thermal conductivity, hardness, elastic modulus and maximum stress of the samples.

Mixture	*K* (W/m°C)	Hardness Vickers	Elastic Modulus (MPa)	Maximum Stress (MPa)
100	0.1635 ± 0.001	22.35 ± 1.29	1534.22	110.000
90-10	0.1887 ± 0.004	18.65 ± 1.84	1751.74	63.893
80-20	0.2132 ± 0.009	21.21 ± 0.78	2168.18	57.494
70-30	0.2506 ± 0.016	21.15 ± 1.62	2193.71	47.952
60-40	0.2810 ± 0.006	26.26 ± 1.33	2818.14	45.912
50-50	0.3105 ± 0.004	27.96 ± 0.98	2595.96	42.802

From the results of Table 1, an increase in the effective thermal conductivity (K), the hardness and elastic modulus in tension is seen, with increasing proportion of powders, while the maximum stress in bending decreases with increasing material filler, which can be attributed to the decrease in the proportion of resin in the material, since it is responsible for granting flexibility compound.

4. Conclusions

It was developed and characterized a material based on polyester resin, which was used as material powder filling dental waste of dental impressions, which are added in weight percentage of 50, 40, 30, 20 and 10% to resin matrix. The morphology and structure of the sample reported a heterogeneous size distribution of particles with average sizes ranging between 11.50 and 74.83 microns, with the presence of hydrated calcium sulphate, Cristobalite low Anhydrita, Quartz, Chantalita and basanita, the crystallinity increases with increasing filler concentration. The thermal behaviour was found that the average temperature was 69.15±4.60°C, also the thermal conductivity, the hardness and elastic modulus in tension is increased with increasing concentration of filler, while the maximum stress in bending decreases. This work contributes to the mitigation of environmental impact by recycling dental solid waste, to be used as fillers in composite materials based on polyester resins.

References

[1] C D Vieira, M A Roque, N A Menezes, M A Alvarez, S G dos Santos, R M Fonseca, M X Silva and L M Farias 2009 *Waste Management* **29** 1388
[2] N Washizawa, H Narusawa, Y Tamaki and T Miyazaki 2012 *Dental Materials Journal* **31-4** 629
[3] D P Komilis, E A Voudrias, S Anthoulaskis and N Losifidis 2009 *Waste Management* **29** 1208
[4] Torre C, Sosa G, Rodríguez R, Robles F 2009 *Revista de Ingeniería* **13** 65
[5] Elizetxea C, Gurmendi U, Pijoan T 2010 *Revista de metalurgia* **46** 138
[6] S H Mohammand 2013 *Journal od Engineering and development* **17-6** 201
[7] A JadahFarhan and N Jerjak 2014 *Int J of Application or Innovation in Engineering Management* **3-1** 68
[8] Lemes C, Borges R, Souza L, Correr L, Soares C, Coelho S 2010 *Mat Res* **13-2** 3
[9] Bermudez C, Florez I, Cardona L 2009 *Manual de procedimientos para el montaje y análisis de modelos articulados* (Medellín: Universidad CES)
[10] Li-Hong H, Vuuren L, Planitz N, Swain M 2010 *Dental Materials Journal* **29-4** 433
[11] Klug H, Lee Alexander 1974 X-Ray *diffraction procedures for polycrystalline and amorphous materials 2nd ed* (United States: Wiley)
[12] Gheno G, Ganzerla R, Bortoluzzi M & Paganica R 2015 *Progress in Organic Coatings* **78** 239
[13] Laoubi K, Hamadi Z, Ahmed Benyahia A, Serier A & Azari Z 2014 *Composites Part B Engine* **56** 520
[14] Gharbi A, Hassen R B & Boufi S 2014 *Industrial Crops and Products* **62** 491
[15] Morote-Martínez V, Torregrosa-Coque R & Martín-Martínez J M 2011 *International Journal of Adhesion and Adhesives* **31-3** 154
[16] Jacquelot E, Galy J, Gérard J F, Roche A, Chevet E, Fouissac E & Verchère D 2009 *Progress in Organic Coatings* **66-1** 86

Structural and magnetic characterization of the new $GdMn_{1-x}Fe_xO_3$ perovskite material

J A Cardona Vasquez[1,2], D A Landínez Téllez[1], C A Collazos[2] and J Roa Rojas[1]
[1] Universidad Nacional de Colombia, Bogotá, Colombia.
[2] Universidad Manuela Beltrán, Bogotá, Colombia.

E-mail: joacardonava@unal.edu.co

Abstract. In this paper we presents the synthesis process of the $GdMn_{1-x}Fe_xO_3$ perovskite material by conventional solid state reaction method. Crystalline phase evolution during the synthesis was studied by X-ray Diffraction (XRD) in powder of the materials, observing that the chemical reaction of the precursor oxides was significant above 1000°C. Rietveld refinement of DRX patterns shows a perovskite structure with octahedral distortions (space group Pbnm, # 62) for studied values of x (0, 0.1 and 0.2). The degree of substitution generates an increasing tendency on lattice parameters a and c, while for b is decreasing just as for the volume of the unit cell. The effect of the change in the lattice parameters directly affects the octahedral distortions, ie, with increasing degree of substitution (increased parameter c) octahedra tend to arrange one above the other aligned with the c axis. Magnetization measurements as a function of temperature were performed above room temperature between 300K and 860K with an applied field of 20Oe and below room temperature in Field Cooling (FC) and Zero Field Cooling modes (ZFC) between 4.2K and 300K with an applied field of 200Oe. Magnetic behavior above room temperature is paramagnetic for used values of x, on the other hand at low temperatures (T<30K) magnetic phase transitions associated to the apparition of an antiferromagnetic phase are observed. In addition for x=0.1 the derivative of magnetization shows a peak around 31K, associated to the ferrimagnetic transition for this material. Curie-Weiss fit reveals the antiferromagnetic (ferrimagnetic) behavior of the materials, also shows that the configurations with x=0 and x=0.2 have an effective magnetic moment very similar to the reported value of undoped material, while for x=0.1 a higher value is observed confirming the ferrimagnetic behavior of this configuration.

1. Introduction

Perovskites are ceramic materials that combine metal with non-metallic elements. The perovskite structure is represented as ABX_3, there are variety of behaviours and properties associated with the materials with structure perovskite due to the relative ease of the election of the cations A and B, which may be occupied by different elements. Commonly the A position is occupied by alkali, alkaline earth or rare earth ions; and the B position usually is occupied by transition metals, giving a large number of possible combinations associated with different properties [1]. An example of the variety of properties are the rare earth manganites.

In recent years, rare earth manganites have been the subject of intense research due to its attractive properties [2-6], mainly associated with the interaction between the 4f orbitals of the rare earth ions with 3d orbitals of the manganese ions, these interactions generate interesting electrical and magnetic behaviours and transport properties [3]. Some of the best known properties in this type of material are

Journal of Physics: Conference Series **687** (2016) 012087 doi:10.1088/1742-6596/687/1/012087

the antiferromagnetism, ferroelectricity (spontaneous and/or induced by magnetic fields), multiferroicity, colossal magnetoresistance and exchange interactions between the magnetic moments of manganese Mn^{3+}, as well as some of the ions rare earth R^{3+} [5]. The properties mentioned above depend of the temperature range in which to study, and the ion R.

The aim of this paper is to report the synthesis process of the $GdMn_{1-x}Fe_xO_3$ system (x=0; 0.1 and 0.2) and study the effect of Fe substitution in the structural and magnetic properties in order to determinate if the multiferroic behavior can be improved by this substitution.

2. Experimental details

$GdMn_{1-x}Fe_xO_3$ system was synthesized by the conventional solid state reaction method, from high purity oxides commercially available. The mass required was obtained by stoichiometrically calculations. Precursor oxides were dried at 160°C for 1 hour and the calculated masses are weighted. Oxides were manually mixed in an agate mortar (around 2h), heat treatments at 700, 800 and 900°C were performed on the resultant powders. Subsequently, the materials were pressed at 10ton/cm^2, and subjected to heat treatment at 1000, 1100 and 1200°C. Between each heat treatment the powders were grinded and analysed by XRD.

XRD measurements were carried out at room temperature using X'pert PRO PANalytical powder X-ray diffractometer with copper anode (λ=1.54056Å) in the 2θ range between 20° and 70° with a step of 0.02° and 2 seconds of time step. Magnetic characterization was performed through curves of magnetization as a function of temperature, measured in a Quantum Design MPMS in a temperature range from 4K to 300K, and 200 Oe applied field.

3. Results and discussion

According to the literature the $GdMnO_3$ has an orthorhombic perovskite structure with octahedral distortion [4,6]. Evolution of the crystal structure during the heat treatments was studied by XRD. Figure 1 shows the study of structural evolution by XRD to different heat treatment temperatures of the $GdMn_{0.9}Fe_{0.1}O_3$ compound, the other two materials show the same behaviour. In Figure 1 can be seen that below 1000°C the formation of the orthorhombic phase of $GdMn_{1-x}Fe_xO_3$ is minimal, ie, until this temperature XRD patterns mostly agree with those of precursors. Above this temperature the formation of the orthorhombic phase associated with the material without substitution is observed. At 1200°C the XRD patterns did not show significant changes compared with the results obtained at 1100°C. Comparison of the diffraction peaks for different x values show small shifts in the 2θ positions as well as changes in the intensity ratio as a result of the inclusion of Fe in the crystal structure of the materials. Identification of the symmetry and space group was performed using the PANalitical X'Pert HighScore Plus software. The Rietveld refinement of crystal structures was performed using the GSAS + EXPGUI software [8,9]. The Rietveld refinement of the $GdMn_{1-x}Fe_xO_3$ confirms that as expected, Fe^{3+} ions take the same atomic positions that Mn^{3+} ions.

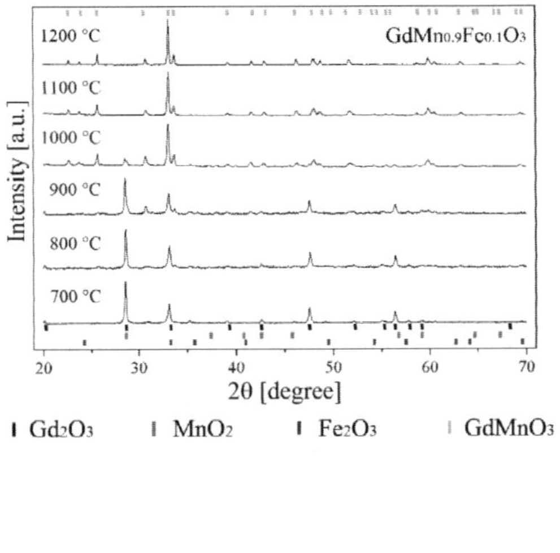

Figure 1: Study of structural evolution by XRD to different heat treatment temperatures of the $GdMn_{0.9}Fe_{0.1}O_3$ compound. At the bottom are shown the 2θ positions of reported peaks for precursors (Gd_2O_3, MnO_2 and Fe_2O_3) and on top of the peak positions of $GdMnO_3$.

Where is more visible the influence of the percentage of substitution is on the lattice parameters of material which can be seen with increasing of substitution degree the parameters *a* and *c* grow, while the parameter *b* and the unit cell volume decrease as shown in Table 1. More significant changes are observed in the parameters *b* and *c* (0.5% and 0.3% for 0.1 of substitution respectively). Figure 2 shows the distorted perovskite structure $GdMn_{1-x}Fe_xO_3$ where distortions MnO_3 octahedra are observed, besides in Figure 2 can be seen the results of Rietveld refinement for three configurations. One effect of changes in the lattice parameters directly affects the distortion of the structure, ie with increasing degree of substitution (increase of parameter *c*), octahedra tend to arrange one above the other aligned with the *c* axis.

Table 1: Refined lattice parameter and cell volume of $GdMn_{1-x}Fe_xO_3$ compounds.

Compound	*a* [Å±0.0001]	*b* [Å±0.0001]	*c* [Å±0.0001]	*Volume* [$Å^3$±0.01]
$GdMnO_3$	5.3179	5.8345	7.4440	230.97
$GdMn_{0.9}Fe_{0.1}O_3$	5.3191	5.8082	7.4694	230.76
$GdMn_{0.8}Fe_{0.2}O_3$	5.3203	5.7765	7.4945	230.33

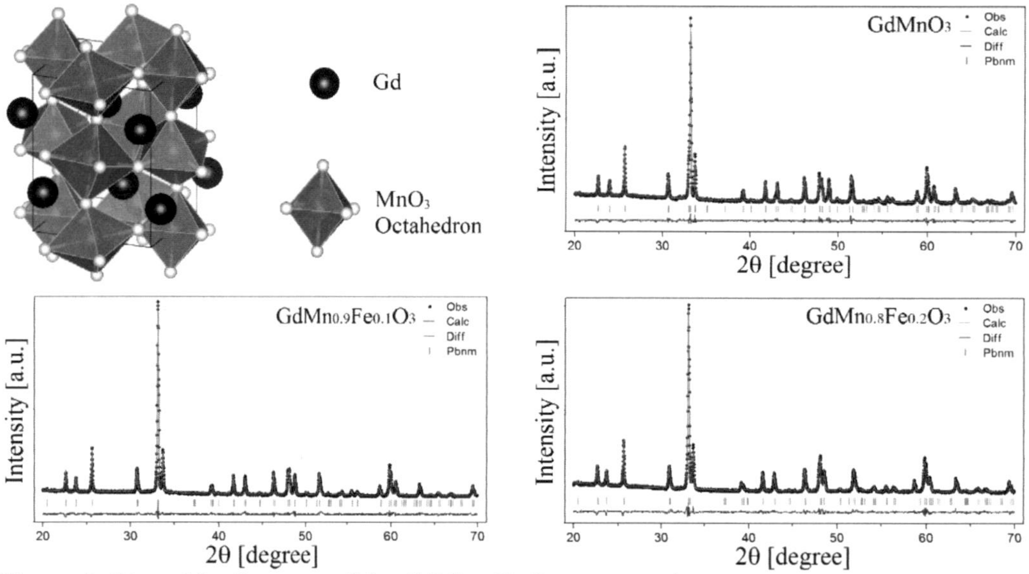

Figure 2: Rietveld refinement of the $GdMn_{1-x}Fe_xO_3$ compounds.

Literature indicates that the undoped material $GdMnO_3$ has A type antiferromagnetism [6]. Similarly three phase transitions are reported for this material [9], the first associated with the transition to the A type antiferromagnetic is reported close to 23K, the second associated with the orientation of the spins of Mn^{3+} ions around 14K and finally the transition due to the orientation of the spins of Gd^{3+} at 5K [9]. Figure 3 (right) shows the ZFC magnetization curves as a function of temperature of the three materials at the same scale, the inset shows the derivative of these curves, where are observed phase transitions around the reported temperatures for the ordering of Mn^{3+} and Gd^{3+}; besides is indicated other temperature (30.7K) corresponding to an anomaly only noted in the material with x=0.1. This anomaly may be associated with ferrimagnetic character related with the imbalance created by the inclusion of Fe^{3+} in the structure. In Figure 3 (left) are shown the graphs of derivative of the inverse of magnetic susceptibility versus temperature, in these graphs can be seen the transitions Para-Antiferromagnetic states (~21.8K - 25.2K) and the transition observed at 30.7K again.

Table 2 shows the results of Curie-Weiss fit of magnetic susceptibility, where the magnetic parameters of the materials were calculated. Effective magnetic moment (μ_{eff}) obtained for $GdMnO_3$

($9.18\mu_B$) agrees with reported value for this material ($9.2\mu_B$) [9]. Figure 3 shows strong similarity between the materials with x=0 and x=0.2, in Table 2 is observed the same similarity in μ_{eff}. For x=0.1 a higher value is observed in μ_{eff}, which reinforces the hypothesis of an imbalance in the antiferromagnetic arrangement generated by the distribution of iron in the crystal structure of $GdMn_{0.9}Fe_{0.1}O_3$.

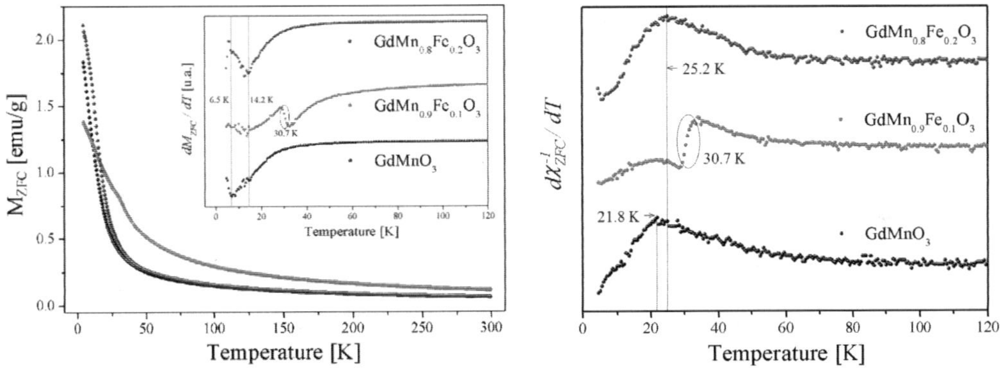

Figure 3: Magnetic behaviour and phase transitions of $GdMn_{1-x}Fe_xO_3$ compounds.

Tabla 2: Curie-Weiss fit of $GdMn_{1-x}Fe_xO_3$ compounds.

Material	C [KJm/TAkg] $\pm$0.000002	θ_P [K] $\pm$0.5	μ_{eff} [μ_B] $\pm$0.08
$GdMnO_3$	0.000494	-34.4	9.18
$GdMn_{0.9}Fe_{0.1}O_3$	0.000897	-22.1	12.34
$GdMn_{0.8}Fe_{0.2}O_3$	0.000491	-27.4	9.14

4. Conclusions

We have synthesized the system $GdMn_{1-x}Fe_xO_3$ by the solid state reaction method. Rietveld refinement shows an orthorhombic phase (Pbnm #62) for all values of x. We observed that the chemical reaction of the precursor oxides occurs for temperatures above 1000°C, obtaining phases with high purity and crystallinity at 1200°C. Study of the magnetization as a function of temperature showed a paramagnetic behavior at room temperature with phase transitions to the antiferromagnetic below 25K for all configurations. We assume that the inclusion of Fe in the crystal structure induces a ferrimagnetic state for x=0.1 which can be seen to 30.7K.

Acknowledgements

Authors acknowledge to direction of research from Bogota (DIB) at the National University of Colombia and direction of laboratories, University Manuela Beltran.

References

[1] J A Cardona-Vasquez, D A Landínez-Téllez, J Roa-Rojas 2014 *Physica B* **455** 39
[2] Xiang Li, Chengliang Lu, Jiyan Dai, et al 2015 *Scient Rep* (DOI: 10.1038/srep07019)
[3] Shuai Dong and Jun-Ming Liu 2012 *Modern Physics Letters B* **26-9** 1230004
[4] T Kimura, T Goto, H Shintani, K Ishizaka, T Arima & Y Tokura 2003 *Nature* **426** 55
[5] P P Rout, S K Pradhan, S K Das, B K Roul 2012 *Physica B* **407** 2072
[6] T Goto, T Kimura, G Lawes, A P Ramirez and Y Tokura 2004 *Phys Rev Let* **92-25** 257201
[7] A C Larson and R B Von Dreele 2004 *General structure analysis system (GSAS)* (USA: Los Alamos National Laboratory Report - LAUR) p 86
[8] B H Toby 2001 *J Appl Cryst* **34** 210
[9] W S Ferreira, J Agostinho Moreira, A Almeida, et al 2009 *Phys Rev B* **79** 054303

Hardness and compression resistance of natural rubber and synthetic rubber mixtures

J M Arguello[1] and A Santos[1]
[1] Universidad Pontificia Bolivariana Seccional Bucaramanga, Floridablanca, Colombia.

E-mail: juan.arguello@upb.edu.co

Abstract: This project aims to mechanically characterize through compression resistance and shore hardness tests, the mixture of hevea brasiliensis natural rubber with butadiene synthetic rubber (BR), styrene-butadiene rubber (SBR) and ethylene-propylene-diene monomer rubber (EPDM). For each of the studied mixtures were performed 10 tests, each of which increased by 10% the content of synthetic rubber in the mixture; each test consisted of carrying out five tests of compression resistance and five tests of shore hardness. The specimens were vulcanized on a temperature of 160°C, during an approximate time of 15 minutes, and the equipment used in the performance of the mechanical tests were a Shimadzu universal machine and a digital durometer. The results show that the A shore hardness increases directly proportional, with a linear trend, with the content of synthetic BR, SBR or EPDM rubber present in the mixture, being the EPDM the most influential. With respect to the compression resistance is observed that the content of BR or SBR increase this property directly proportional through a linear trend; while the EPDM content also increases but with a polynomial trend.

1. Introduction

Polymers have had a quick expansion in recent years, which has happened at the expense of traditional materials in already established applications, also the emergence of new applications and markets. This boom is due to the great versatility of properties that present these materials, among which highlights its low weight, corrosion resistance, ease of processing and economy, among others.

The natural and synthetic polymers known as elastomers have as main characteristic its ability to present a fully recoverable and unlimited instant elasticity to high deformations; due to this characteristic are widely used in the manufacture of elastic bands, tires, seals, hoses, etc.

Due to the large amount of elastomers available in the market and to the variation in mechanical properties that each of these materials presents, this project aims to mechanically characterize, through compression resistance and shore hardness tests, the mixture of hevea brasiliensis with BR, SBR and EPDM; to determine how vary these properties as a function of the amount of synthetic rubber present in the sample.

2. Conceptual framework

2.1. Elastomers

Elastomers are materials which characteristic property is the deployment of a large amount of deformation, generally greater than 200%, when a force is applied to them [1]. These materials have high molecular weights and flexibility. Rubbers are classified according to their origin in natural and synthetic. Natural rubber is obtained from a white liquid called latex, which is found in many plants, being the most representative the tree of the specie Hevea Brasiliensis; synthetic rubbers are prepared from saturated hydrocarbons through chemical reactions and there is a wide variety of these rubbers, some of which have better mechanical properties than natural rubber.

The tire manufacturing industry is the largest consumer of natural and synthetic rubber [2-4] being the most commonly used natural rubber, styrene butadiene rubber, butadiene rubber and the rubber ethylene propylene diene monomer [5,6].

2.2. Vulcanization
The rubbers are polymers derived from dienes and as a result they possess double bonds in the main chains. The existence of these is used to insert in a controlled manner the clips between the chains through a process known as vulcanization. Through this process, the polymer that has very poor properties is transformed into a tough, tenacious and elastic rubber. The degree of vulcanization affects significantly these properties. The vulcanization, which uses sulphur atoms, is a common method of forming these cross-links; this process was accidentally invented by Charles Goodyear [7,8].

2.3. Mechanical Properties of rubbers
The stress-deformation curve for an elastomer shows that virtually all the curve represents elastic deformation; therefore, elastomers show a nonlinear elastic behaviour. At the beginning the elasticity modulus (E) is reduced due to the unwinding of the chains. However, once extended the chains, any additional elastic deformation happens thanks to the stretch of the links, which generates a high E.

The number of cross links determines the elasticity of the rubber or the quantity of Sulphur added to the material. Low Sulphur additions leave a soft and flexible rubber; while percentage increase, this restrict the unrolling of the chains and the rubber is made harder, more rigid and fragile. Many more efficient vulcanization systems that do not have sulphur have been developed in recent years [9].

A rubber hardness is an indication of its rigidity against moderate stress, as those that often has to bear in service. The most popular method of rubbers hardness is the shore A in which is performed the measurement with an instrument called a durometer, based on the penetration of a frusto-conical tip against a calibrated metal spring reaction.

3. Methodology

3.1. Used materials
The raw material used for the preparation of the different specimens were: natural rubber Hevea Brasiliensis, BR, SBR and EPDM. The accelerators added during the mixing process were the 2-mercapto benzothiazole (MBT), which helps the vulcanized process and curing of the rubber; and Thiuram, which is a secondary accelerator that is used to give consistency to the mixture in the vulcanization and supports the MBT.

Other used materials were sulphur, since being mixed with the raw rubber improve its hardness and resistance and is compatible with the selected accelerators, has low cost and present a high efficiency in the vulcanization process; and the black smoke, which has carbon particles that are mixed with the rubber and other components that allow a better mixture and composition in the vulcanization.

3.2. Preparation of the mixtures
For the manufacture of the specimens was used the same amount of additives in the mixing of all the specimens in order to not add additional variables that prevent analyse the relationship between the percentage of used rubber and the mechanical properties of hardness and compression resistance. In Table 1 are related the percentages of used materials in each of the tests performed, and were defined according to the permissible ranges found in the bibliographic information collected, and that correspond to: accelerators (0.2-0.8%), sulphur (0.5-3%) and black smoke (20-40%). For each specimen a total weight of 800 grams of rubber was taken as a reference.

Table 1. Amount of each material used in the rubber mixtures.

Test	Natural Rubber		Synthetic Rubber		Black Smoke		Sulphur		MBT		Thiuram	
	%	Weight (g)	%	Weight (g)	%	Weight (g)	%	Weight (g)	%	Weight (g)	%	Weight (g)
1	0	0	100	800	20	160	1	8	0.75	6	0.25	2
2	10	80	90	720	20	160	1	8	0.75	6	0.25	2
3	20	160	80	640	20	160	1	8	0.75	6	0.25	2
4	30	240	70	560	20	160	1	8	0.75	6	0.25	2
5	40	320	60	480	20	160	1	8	0.75	6	0.25	2
6	50	400	50	400	20	160	1	8	0.75	6	0.25	2
7	60	480	40	320	20	160	1	8	0.75	6	0.25	2
8	70	560	30	240	20	160	1	8	0.75	6	0.25	2
9	80	640	20	160	20	160	1	8	0.75	6	0.25	2
10	90	720	10	80	20	160	1	8	0.75	6	0.25	2

The time used to homogenize the mixture was between 50 and 60min. The mixing process was carried out in the following way: initially are laminated independently for 2min the synthetic rubber and natural rubber for 7min; subsequently two rubbers are mixed by an approximate time of 8min. Then is added the MBT accelerator and laminated the mixture for 5min; and for last is added the Thiuram and mixed for 5min; and finally is added the black smoke and laminated the mixture for 10min, time in which you get a homogeneous mixture and compact mixture, as you can see in Figure 1.

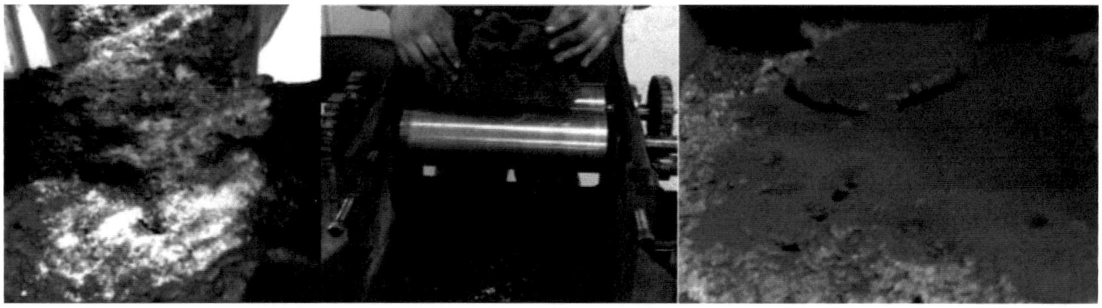

Figure 1. Lamination process of components that conforms the rubber mixtures.

3.3. Vulcanization process of the specimens

For the process of vulcanization was used a total weight of the mixture between 650 and 700 grams, in order to avoid excess material in moulds; the temperature and pressure used were 160°C and 90kPa, respectively. The processing time was 15min.

3.4. Tests performed

Specimens were subjected to hardness shore A, and to compression resistance of 25% deformation; the equipment used was a digital durometer and a Shimadzu universal machine. In total 8 hardness tests and 14 tests of compression for each test were performed.

4. Obtained results

The results of the average values of hardness and resistance to compression, for each of the mixtures of rubbers are listed in Figures 2 and 3 respectively.

Figures 2 and 3 allow to observe that both the Shore A hardness, as the resistance to compression (25%) increase when the natural rubber Hevea Brasiliensis is mixed with synthetic rubbers, being the most influential the EPDM. The tendency of these properties is directly proportional to the content of synthetic rubber in the mixture and they usually have a linear behaviour.

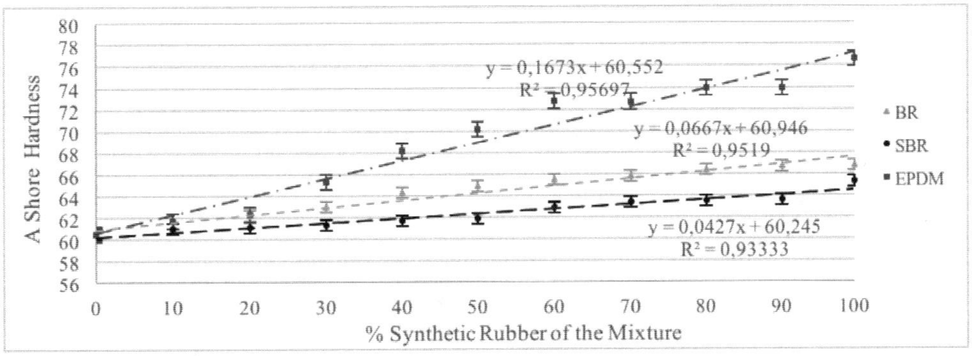

Figure 2. Variation of Shore A Hardness in Hevea Brasiliensis natural rubber and Synthetic rubber.

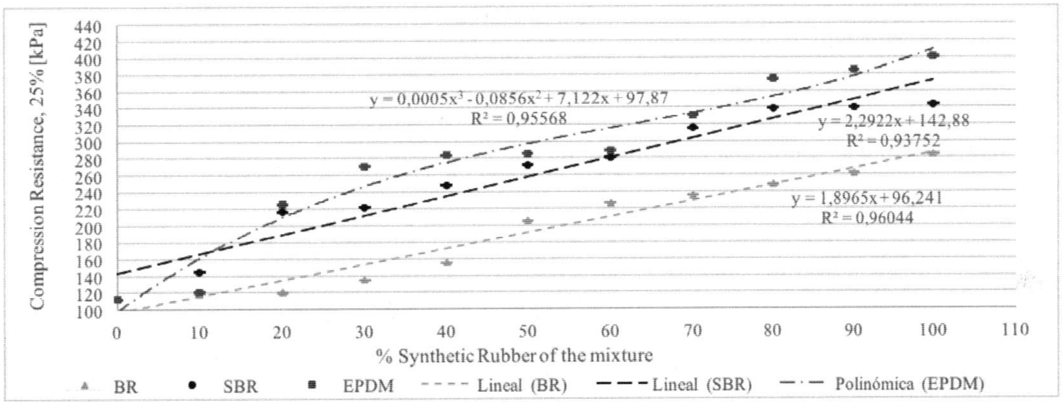

Figure 3. Variation of Compression Resistance at 25% in Hevea Brasiliensis natural Rubber and Synthetic rubber.

5. Conclusions

Shore A hardness in the mixtures present a directly proportional behaviour with a linear tendency, in function of the content of BR, SBR and EPDM present, being the EPDM the more influential.

The compression resistance at 25% is increased in a directly proportional way through a linear tendency with the content of synthetic rubber BR and SBR; while with the content of EPDM also is increased but with a polynomial tendency of third order.

References

[1] D R Askeland and P P Phulé 2004 *Ciencia e ingeniería de los materiales 4th ed* (México: Thompson)
[2] T M Letcher and D A Vallero 2011 *Waste: A handbook for management 1st ed* (Boston: Academic Press) chapter 21 pp 297–320
[3] J Karger-Kocsis, L Mészáros and T Bárány 2013 *Journal of Materials Science* **48-1** 1
[4] P J H Van Beukering and M A Janssen 2001 *Resources Conservation and Recycling* **33-4** 235
[5] K Fukumori, M Matsushita, H Okamoto, N Sato, Y Suzuki and K Takeuchi 2002 *JSAE Review* **23-2** 259
[6] H Nabil, H Ismail and A R Azura 2013 *Materials and Design* **50** 27
[7] V V Rajan, W K Dierkes, R Joseph and J W M Noordermeer 2006 *Progress in Polymer Science* **31-9** 811
[8] M Akiba and A S Hashim 1997 *Progress in Polymer Science* **22-3** 475
[9] C A Harper 2007 *Handbook of materials products design 3rd ed* (Unite States: Mc Graw Hill)

Journal of Physics: Conference Series **687** (2016) 012089 doi:10.1088/1742-6596/687/1/012089

Production a characterization of new system superconductor $TR_3Ba_8Cu_{11}O_\delta$

S Sandoval Gutiérrez[1], I Supelano G[1], J Roa-Rojas[2] and C A Parra Vargas[1]
[1] Universidad Pedagógica y Tecnológica de Colombia, Tunja, Boyacá, Colombia
[2] Universidad Nacional de Colombia, Bogotá D.C., Colombia.

E-mail: carlos.parra@uptc.edu.co

Abstract. The production and structural & magnetic characterization of new superconductor system TR:3811 (TR=Sm, Y, Yb) is presented in this work. The samples were produced by the solid state reaction method. The analysis of the measures of the diffraction X-ray, were made with the Rietveld refinement technique, that allowed to infer that the system presents a Pmm2 type structure with the following lattice parameters (a=3.88Å, b=3.82Å, c=42.82Å). The magnetization measurements were made by using the VSM-VERSALAB equipment with a magnetic field variation from a 100Oe to 7000Oe, and a temperature from 50K to 350K, showing the superconducting transition of the order of 90K.

1. Introduction

Since high critical temperature HTSC's superconductors were discovered, a great job has been devoted to raise that temperature of transition between the normal and superconducting state. These HTSC's may present a simple or complex perovskite structure ABOx, according to the amount of strange earth atoms that are arranged in the stoichiometry. One of the earliest and most famous was the compound $YBa_2Cu_3O_7$ (Y123) that exceeded the 90K for Tc [1,9]. At the same time, trying to understand the behaviour of HTSC's, a lot of compounds have been used with different stoichiometric weights and structurally varying the primitive cell of the perovskite. The Y358 occurred subsequently obtaining a Tc of around 94K, with a structure type Pmm2 [6,11]. The amount of oxygen in the superconductor plays a very important role in the creation of Cooper pairs and the anisotropy in the structure increases the Tc. In perovskites type HTcS, the number Barium and rare earth must match the number of copper atoms, in this way, for the TR123 would have $TR_{x-1}Ba_xCu_{x+1}O_y$, for TR237 is $TR_{x-1}Ba_xCu_{2x+1}O_y$ and so on until getting the formula of TR3811, $TR_{x-5}Ba_xCu_{x+3}O_y$ [2]. The amount of oxygen must balance the system according to the electronegativity, in the case of TR3811 there are 24 oxygens. With the record of previous studies to the TR3811 structure, two important situations have been observed in the HTSC's, one of them is the relation of the oxygen with the Cooper pairs. Clearly CuO chains present in the CuO_2 provide the Cooper pairs in the superconductivity [15,19]. Furthermore, the more anisotropic the sample is, the more CuO_2 planes in the system, and because of this the aforementioned result. This work shows the superconductor system behaviour TR(3811) with the analysis of magnetization measurements.

2. Experimental

The superconducting samples $TR_3Ba_8Cu_{11}O_{24}$ (TR=Y, Yb, Sm), were produced by the solid-state reaction method using commercial oxide precursors with purities of 99.99%. Homogenates compounds were brought to a temperature of 800°C of calcination in a normal atmosphere for 24 hours, and then was compacted under a pressure of 5ton/m_2. The sintering step was performed at 860°C for 48 hours. Oxygenation was performed between 860°C and 550°C, for 24 hours. The samples were analysed with X-ray diffraction (XRD) in an Analytical Expert equipment with Cu-Kα radiation and λ=1,54064Å; with values of 2θ between 20 and 90 degrees were made. These diffraction patterns were refined by the Rietveld method using GSAS+EXPGUI code. Magnetization measurements depending on the temperature H(T) for each of the samples of TR:3811 were made in the VSM-VersaLab equipment, with the routine ZFC-ZFC (Zero Field Cooled- Field Cooled) between 50K and 350K and with magnetic fields from 100Oe to 7000Oe.

3. Results and discutions

The theoretical structure for compound $TR_3Ba_8Cu_{11}O_y$ (P_{mm2}) was shown in Figure 1, with a perovskite type structural arrangement. The difractogramas shows peaks of intensity at 33° and 47°, typical of families with these tetragonal structures, as can be seen in the Y237 and Y358 reported by another authors [2,7,13]. Experimental results the XRD on the samples of Sm3811, Yb3811 and Y3811 are shown in Figure 2 in a first view, the samples are not in pure crystalline phase, the peaks at 23°, 25°, 28°, 38°, confirm the phases, TR237 and TR358, a result that might interfere with the final temperature of the superconducting state. The Table 1 shows the lattice settings for $Tr_3Ba_8Cu_{11}O_{24}$ (TR=Sm, Yb, Y), in addition to the values of critical temperature.

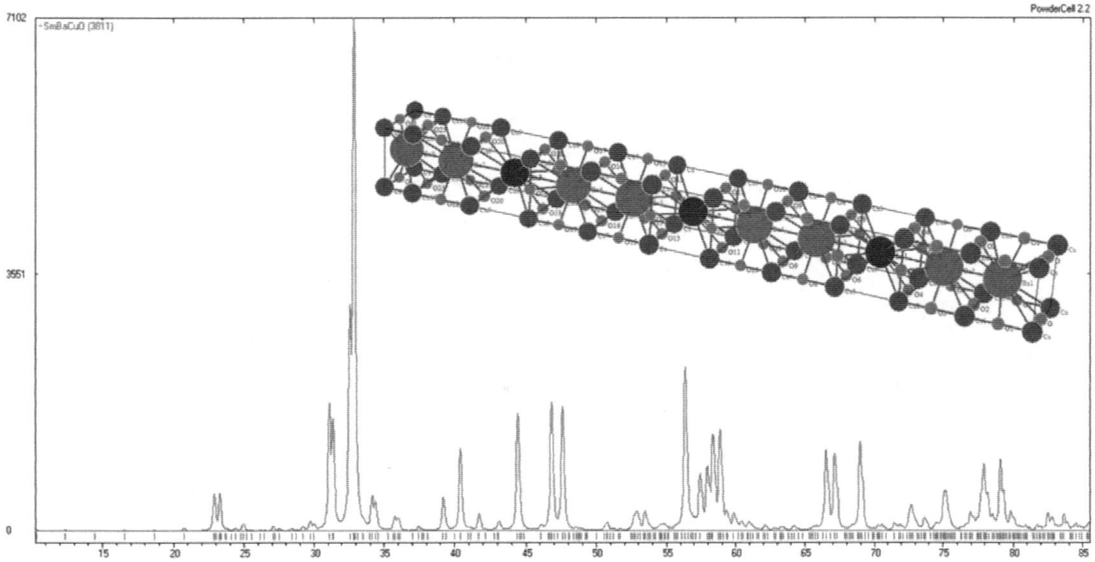

Figure 1. Theoretical diffractogram for the superconductor system Sm3811 by using PCW23.

Table 1. Lattice parameters and superconducting critical parameters For samples TR3811 (TR=Sm, Yb, Y).

Sample	a(Å)	b(Å)	c(Å)	Tc(K)	T_{irr}(K)
$Sm_3Ba_5Cu_8O_\delta$	3,89	3,821	42,83	69	61
$Yb_3Ba_5Cu_8O_\delta$	3.88	3,82	42,824	89	77
$Y_3Ba_5Cu_8O_\delta$	3,88	3,82	42,82	69	71

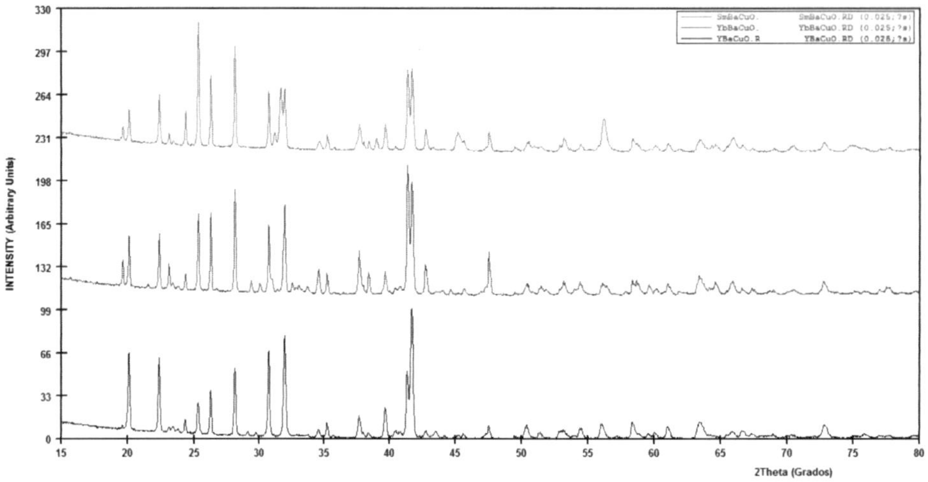

Figure 2. Structural experimental results showing TR3811 (Sm, Yb, Y).

In addition, Figures 3 to 5 shows the ZFC-FC magnetization of samples in functions of the temperature. The diamagnetic contribution observed, is characteristic of the superconducting state below Tc. The value of critical temperature, Tc, was determined by the intersection between the linearly extrapolated magnetizations in the normal and superconducting phases and the irreversible temperature, T_{irr}, was determined in the crossing between ZFC and FC curve, the values were obtained with a magnetic field of 500Oe (Table 1). Figures 6 to 8 shows the hysteresis loop at 50K for the three samples, evidencing the characteristic behaviour of superconducting materials of high critical temperatures.

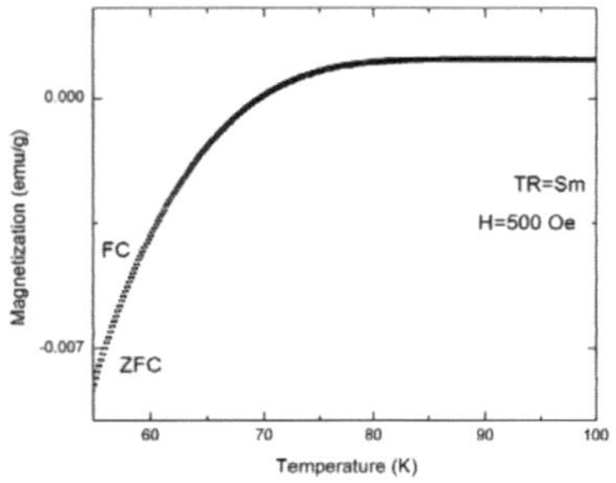

Figure 3. Magnetization ZFC-FC curve for TR=Sm

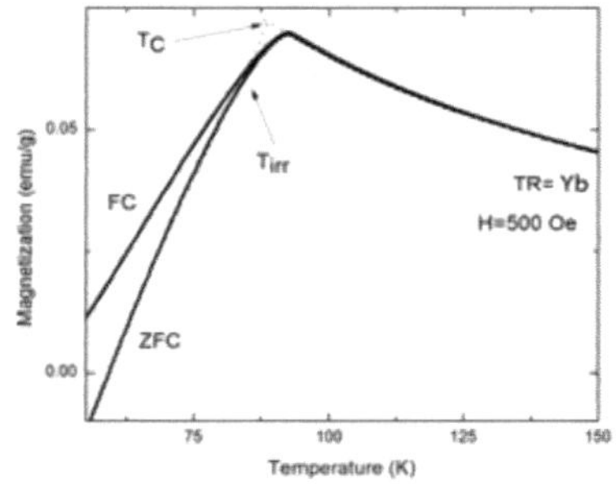

Figure 4. Magnetization ZFC-FC curve for TR=Yb

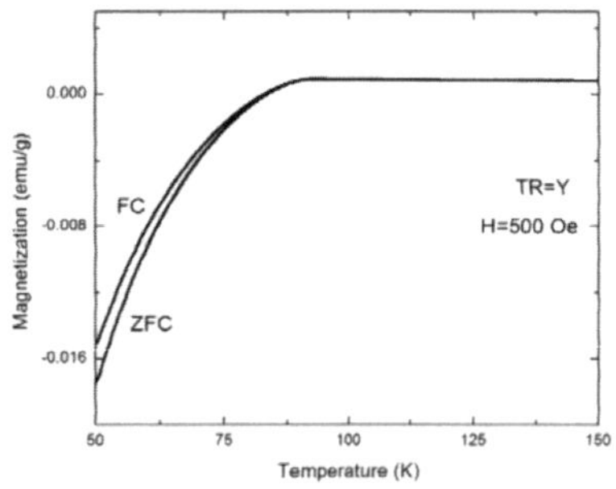

Figure 5. Magnetization ZFC-FC curve for TR=Y

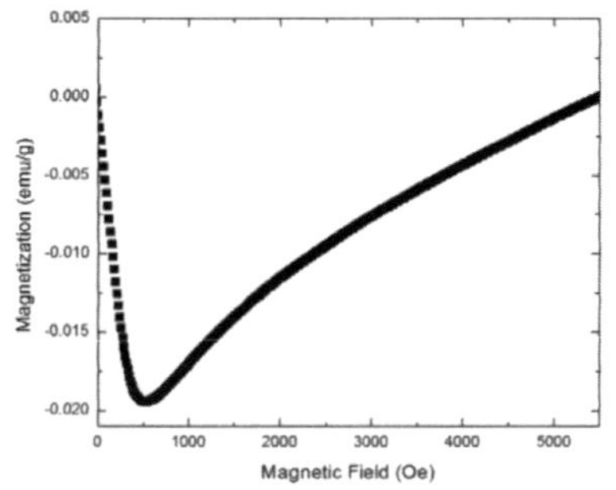

Figure 6. Hysteresis loop for TR=Y

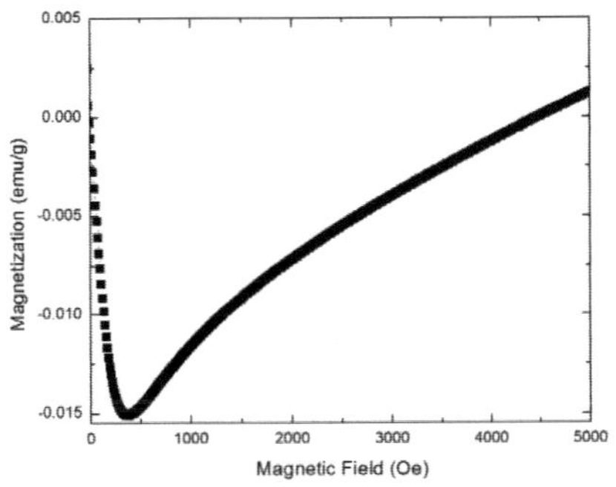

Figure 7. Hysteresis loop for TR=Sm

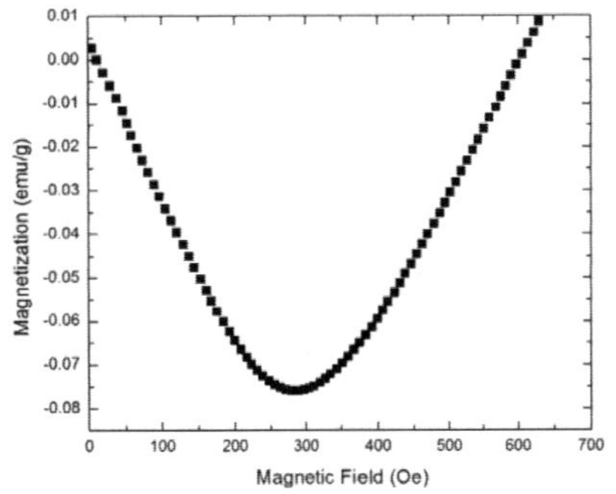

Figure 8. Hysteresis loop for TR=Yb.

The Table 1 present a summary of the structural and superconducting parameters found. The Yb3811 sample shows a temperature critical higher compared to the other samples.

4. Conclusions

The crystal structure has no pure phase. XRD reading shows diffraction peaks related to the structures Y237 and Y358. This mixture of phases does not contribute to a higher Tc, although it exposes the high anisotropy of the sample, as well as the increased of CuO_2 planes. In the magnetization results no higher temperatures of transition, than the ones reported in the sample Y358 by Aliabadi [2].

References

[1] Michel Cyrot, Davor Pavuna 1992 *Introduction to superconductivity and high Tc materials* (Singapore: World Scientific Publishing Co)
[2] Aliabdi A, Akhavan F, et all 2009 *Physica C* **469** 2010
[3] Marsh P, Fleming R M, Mandich M L, DeSAntolo A M, Kwo J, Hong M, Martinez Miranda L J 1988 *Nature* **334** 660
[4] Karpinski J, Rusiecki S, Bucher B, Kaldis E, Jilek E 1989 *Physica C* **161** 618
[5] Genoud J Y, Graf T, Triscone G, Junod A, Muller J 1992 *Physica C* **192** 137
[6] Wu K, Ashburn J R, Torng C J, Hor P H, Meng R L, Gao L, Huang Z J, Wang Y Q, Chu C W 1987 *Phys Rev Lett* **58** 908
[7] Kruaehong T 2014 *International Journal Of Physics Sciences* **9-16** 360
[8] Charles P Poole, Horacio A Farach and Richard J Creswick 1995 *Superconductivity* (Amsterdam: Academic Press)
[9] A Tavana and M Akhavan 2010 *Eur Phys J B* **73** 79
[10] Burns, Gerald 1992 *High temperatura superconductivity an introduction* (San Diego: Academic Press)
[11] D Martinez Buitrago, N C Reyes Suarez, J P Peña, O Ortiz Diaz, J Otalora, C A Parra Vargas 2011 *J Supercond Nov Magn* (DOI 10.1007/s10948-012-1456-x)
[12] Tinkham M I 1996 *Introduction to superconductivity* (New York: Mac-GrawHill)
[13] A Schmid 1969 *Phys Rev* **180** 527
[14] F M Barros, V N Vieira, F W Fabris, M P Cantao, A R Jurelo, P Pureur, J Schaf 2004 *Physcica C* **408-410** 632
[15] D V Livanov, A A Varlamov, M Putti, M R Cimberle and C Ferdeghini 2000 *Phys J B* **18** 401
[16] C A Parra Vargas, D A Landinez Tellez, J Roa-Rojas 2007 *Physica B* **398** 301
[17] C A Parra Vargas, J Roa-Rojas, D A Landinez Tellez 2006 *Revista Colombiana de Física* **38** 2
[18] C A Parra Vargas, J L Pimente Jr, P Pureur, D A Landinez Tellez, J Roa-Rojas 2009 *Physica B* **404** 2766
[19] Z Hao and J Clem 1991 *Phys Rev Lett* **67** 2371
[20] S Sujinnapramc, P Udomsamuthiruna, T Kruaehonga, T Nilkamjona and S Ratreng 2011 *Bull Mat Sci* **34-5** 1053
[21] P Udomsamuthirun, T Kruaehong, T Nilkamjon, S Ratreng 2010 *J Supercond Nov Magn* **23** 1377

Experimental set up of a magnetoelectric measuring system operating at different temperatures

K Gil[1], J Gil[1], B Cruz[1], A Ramírez[1], M Medina[1] and J Torres[2]
[1] Universidad Tecnológica de Pereira, Pereira, Colombia.
[2] Universidad de Caldas, Manizales, Colombia.

E-mail: kthegilmonsalve@gmail.com

Abstract. The magnetoelectric effect is the phenomenon whereby through a magnetic stimulation can be produced an electrical response or vice versa. We implement a magnetoelectric voltage measuring device through the dynamic method for a different range of temperatures. The system was split into an electric set and an instrumentation and control set. Design and element selection criteria that the experimenter must take into account are presented, with special emphasis in the design of the sample holder, which is the fundamental component that differentiates the system operating at high temperature and the one operating at room temperature. The experimental equipment consists of an electromagnet with DC magnetic flux density (B) in a range of (0.0 to 1.6) KOe, a Helmholtz coil which operates with a sinusoidal B between (0.0 and 0.016) KOe and a PT100 temperature sensor. A tubular heating resistance, a Checkman temperature control and an SSR 40A were used for controlling the temperature. As an application of the system, the transverse and longitudinal magnetoelectric coefficient was measured for a thin film of $BiFeO_3$ at room temperature and 307K. It was observed that the behaviour of the longitudinal and transverse magnetoelectric coefficient matches the reported value and decreased with increasing temperature.

1. Introduction

The magnetoelectric effect is the phenomenon whereby a magnetic field (H) produces an electrical polarization, or an electric field (E) produces a magnetization. This effect has drawn attention from the scientific community because of their potential applications in the new generation of magnetic field detectors, sensors, transducers and improvements in magnetic storage [1]. The experimental methods for measuring the magnetoelectric coefficient (α_{ME}) are the static, quasi-static, dynamic and pulsed dynamic methods [2]. For the dynamic method, the sample is subjected to the action of a superimposed AC field in a variable DC magnetic field, which generates at the ends of the material a voltage response (ME signal) which permits to obtain, based on Equation 1, the α_{ME} value indirectly [3]. This method has had a great reception since it reduces the problems of charge accumulation at the edge of the sample. The magnetoelectric coefficient expression is given by:

$$\alpha_{ME} = \frac{V}{h_0 d} \tag{1}$$

Where h_0 is the amplitude of the applied AC magnetic field, d is the sample thickness and V is the voltage measured at the ends of the test specimen [2].

For the design and implementation of a α_{ME} coefficient measurement system through the dynamic method, two effects can be take into account, depending on the material to be studied. Single phase materials have a weak response and require low temperatures measurements while composites can be measured at room temperature (RT) and high temperature.

The aim of this work is to perform a comparative analysis of a magnetoelectric system implemented by the dynamic method, which operates at different temperatures. The voltage signal generated across the ends of a sample with magnetoelectric response, particularly using a $BiFeO_3$ thin film, is studied.

2. Materials and methods

2.1. Design criteria for the DC magnetic field generator

- Depending on the maximum density of the magnetic flux generated by a DC electromagnet, set the current range in which to operate the DC source that feeds it.
- Define the range of the DC magnetic flux density (B_{DC}) with which the sample is stimulated.
- Set the size of the AC coil, which will define the separation between the electromagnet's cores.

2.2. Design Criteria for the AC magnetic field generator

- Set the maximum value of the AC magnetic flow density (B_{AC}), which depends on the maximum current to be generated and the geometrical characteristics of the AC Helmholtz coil.
- Define the operation range of the AC source that depends on the desired maximum B_{AC}, set the type of wave to be generated, ensuring high stability and accuracy, also establish the operating frequency range.
- Set the size of the thin film to be studied, which will determine the radius of the AC coil.

2.3. Design criteria for the instrumentation and control

This set should treat the signal from the DC source and condition the ME signal and the Hall signal, the design criteria are:

- For the control of the DC power source, it must be guaranteed: linearity, stability and flexibility in managing the sampling time.
- Ensure synchronicity between the B_{DC} growth and the displayed voltage signal generated by the sample.
- Depending on the instrumental characteristics of the Hall sensor, determine the maximum value of B_{DC} in order to measure the DC field.

2.4. Magnetoelectric measuring system operating at room and high temperature

For implementing a measurement system operating at different temperatures, using the dynamic method, the system was divided into two functional units, an electrical assembly and instrumentation and control set. The first functional unit contains the generators B_{DC} and B_{AC}; while the second regulates the electrical signals from the electric set, the EM signal and the Hall signal. The curve is displayed via a user interface developed in LabVIEW.

2.4.1. Sample holder design criteria at low and high temperature

- Depending on the size of the sample, define the size of the sample holder and the AC coil.
- Set the temperature range to which the sample will be studied, that defines the thermocouple and resistance heating type that allows varying the temperature.
- To control the temperature, it should be ensured stability and flexibility.

3. Results and discussion

3.1. Magnetoelectric measuring system operating at room temperature

The result of applying the design criteria previously mentioned was studied by setting a device that measures the transverse and longitudinal magnetoelectric coupling in a material. Since the spacing between the cores of the electromagnet is limited by the size of the AC coil (that was previously designed and built); the system operates with a B_{AC} between (0.0 and 0.016) KOe and a B_{DC} in the range of (0.0 to 1.6) KOe. Magnetoelectric coefficient curves as function of the B_{DC} are presented on a user interface developed in LabVIEW. The coefficient was obtained indirectly from the voltage response of the material, which is amplified and limited in band, to improve the signal-to-noise ratio. Figures 1 and 2 exhibit the experimental setup and the sample holder employed, respectively.

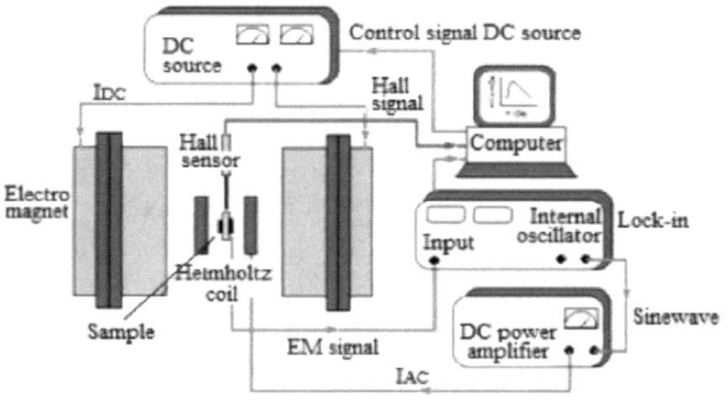

Figure 1. Experimental setup of a ME measurement system operating at room and high temperature.

3.2. Magnetoelectric measuring system operating at high temperature

This system differs from the one that operates at room temperature in the sample holder and the temperature control system. This control was implemented so the user can operate at the desired temperature. Figure 3 presents the design of the sample holder, which has three sections. The innermost containing the sample and the sensor that monitors the material temperature during the α_{ME} measurement and the outer section which has the heat resistance that controls the temperature of the sample. If the ME system operates at low temperatures, the middle section of the sample holder contains the cooling fluid that allows to diminish the temperature.

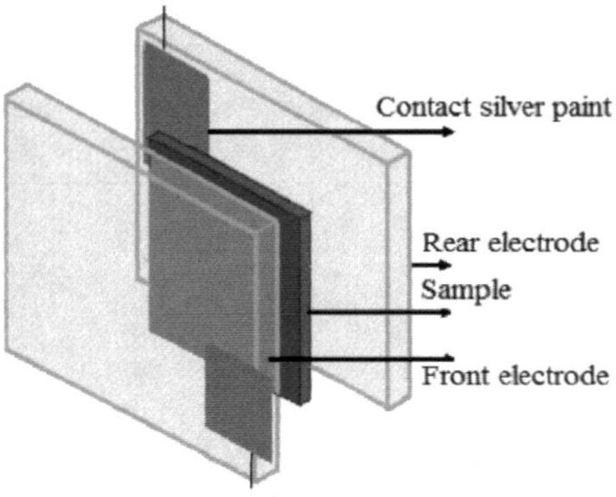

Figure 2. Sample holder used when the system is operating at RT. Adapted from [4].

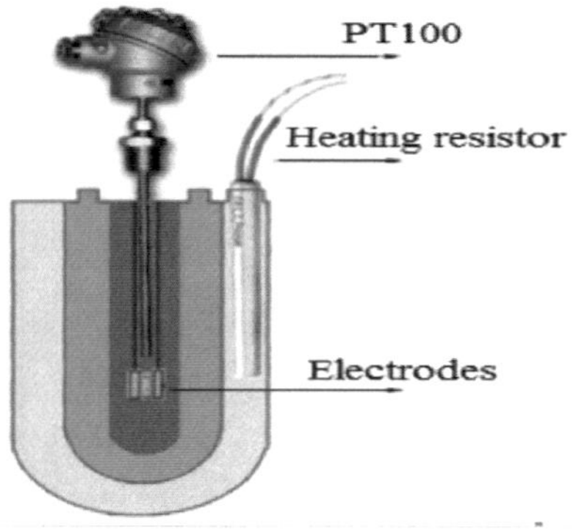

Figure 3. Sample holder implemented when the system is operating at lower and higher temperatures than RT.

For the temperature control a tubular heating resistance, a Checkman temperature control, an SSR 40A and a PT100 for measuring temperature, were used. The designed control is limited by the type of temperature sensor that is connected. In the case of PT100, the working range is from 83K to 874K, which gives flexibility to the equipment to perform magnetoelectric characterizations both at low and high temperature.

3.3. Test run

Figure 4 shows the magnetoelectric response of a thin film of $BiFeO_3$ at 300K (RT) and 307K for a frequency of 200Hz. By comparing the response of the material (except the behaviour of the transverse test at RT), it can be noticed a rapid increase of α_{ME} for DC fields below 0.25KOe and a subsequent saturation can be attributed to the alignment of the electric dipoles induced in the direction of the applied field. At room temperature, saturation values of α_{ME} are 37.5V/cmOe and 109V/cmOe, respectively, in transverse and longitudinal configuration. This behaviour and values are consistent with those reported [5,6]. On the other hand, the response coefficient for both configurations decreases with increasing temperature. We have not come across similar data from the published literature; however, it is possible to infer that there is temperature dependence with the magnetoelectric response in $BiFeO_3$ according to [6].

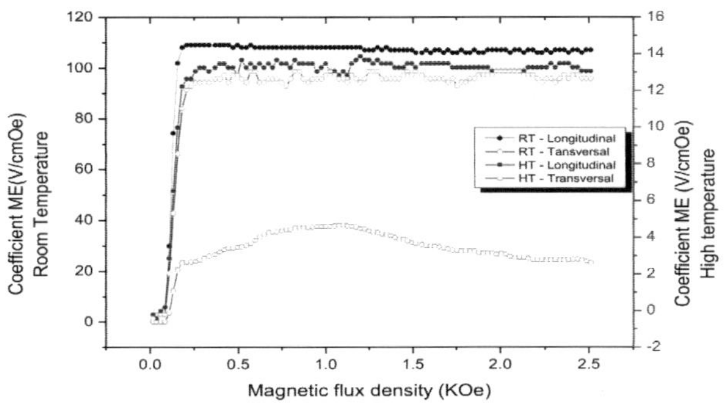

Figure 4. Magnetoelectric coefficient variation depending on the DC magnetic flux density at room temperature and 307K in a sample of $BiFeO_3$ for a frequency of 200Hz.

4. Conclusions

According to the design criteria that were considered, we have implemented a system to measure the magnetoelectric coefficient (α_{ME}) using the dynamic method which operates at different range of temperatures. In order to test the equipment, the behaviour of the longitudinal and transversal magnetolectric coefficient in a thin film of $BiFeO_3$ was observed at high and room temperature. It was noticed that the α_{ME} significantly decreased with the increase of temperature, showing a possible loss of the magnetoelectric characteristic in the specimen.

Acknowledgments

The authors want to thank Colciencias public call for proposals "Jóvenes Investigadores e Innovadores (2014)" and the branch of Electronics and Telecommunications of TecnoParque (Pereira) for their help on the assembly of the AC source.

References

[1] Bueno D, Grossinger R and Corral V 2006 *J Appl Phys* **8-D** 908
[2] Fuentes L, Matutes J and Fuentes M 2011 *Handbook of magnetic materials* vol 19 (Amsterdam: Elsevier B V) p 130
[3] Bhattacharya G, Mazumder S and Baisnab B 2004 *IJEMS* **11** 185
[4] Hoffman T 2012 *Domain patterns and dynamics in the magnetoelectric switching os spin-spiral multiferroics* (Germany: University of Bonn)
[5] Ordoñez J E, Nevaez J, Gomez M E and Lopera W 2012 *Rev Mex Fıs* **58** 97
[6] Naik V and Mahendiran R 2009 *Elsevier Solid State Commun* **49** 754

Structural study of yttrium substituted $BiFeO_3$

J A Mejía Gómez[1], C Canaria[2], R Ochoa Burgos[1,2] C A Ortiz[2], G I Supelano[2] and C A Parra Vargas[2]
[1] Universidad Antonio Nariño, Boyacá Colombia.
[2] Universidad Pedagógica y Tecnológica de Colombia, Tunja, Boyacá, Colombia.

E-mail: juliethmejia@uan.edu.co

Abstract. Yttrium-substituted $Bi_{1-x}Y_xFeO_3$ (x=0, 0.03, 0.07, 0.15, 0.2 and 0.5) samples were prepared by solid state reaction technique. Morphological analysis was obtained by Scanning Electron Microscopy (SEM) technique indicating mainly granular behaviour. In addition, the substitution of yttrium promotes smaller particle size of $BiFeO_3$. The obtained samples were also studied by X-ray diffraction (XRD). The crystal structure and the lattice parameters were confirmed by XRD. Rietveld refinement of experimental X-ray diffraction patterns showed that substituted $BiFeO_3$ compounds crystallize in a R3c type structure and the lattice parameters decrease as Y concentration increases.

1. Introduction

$BiFeO_3$ (BFO) is a multiferroic material with special structural and magnetic properties (magnetoelectric coupling at room temperature) [1-2]. This perovskite has a rhombohedral R3c crystallographic structure with larger Bi^{3+} ions occupying the cube-octahedral positions and the medium sized Fe^{3+} ions in octahedral coordination (see Figure 1). The lattice parameters are a=5.57874Å and b=13.8688Å [3].

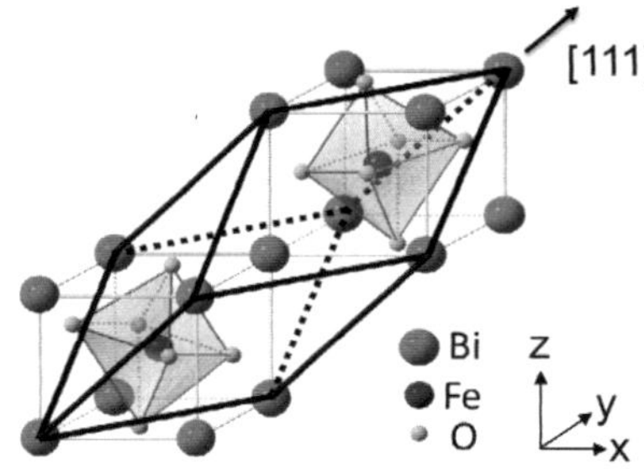

Figure 1. Crystal structure of $BiFeO_3$.

BFO can be obtained by different methods of synthesis [4-6]. One of them is solid state reaction route, where the resulting final product (presence of impurities, changes in the structure) depends on the temperature conditions [6].

Rare-earth elements can be used to substitute Bi ions in the structure of BFO, which can affect the electrical, magnetic and specifically, structural properties [7-13] (by comparison, all the references here mentioned, used solid state route as synthesis method). Different Dy concentrations have been used to investigate the substitution of this element in BFO. Zhang et al. [7] synthesized $Bi_{1-x}Dy_xFeO_3$ (x=0-0.3).

They found that the presence of dysprosium prevents the appearance of impurities. The size of the grains are smaller as Dy concentration increases. These results are also obtained for other authors independent of the sintering conditions of solid state route [8-9]. The structural, optical, magnetic and dielectric properties of $Bi_{1-x}Eu_xFeO_3$ samples (x=0, 0.03, 0.05, 0.07, 0.1, 0.12 and 0.15) were investigated [10]. It was established that for x>0.03 impurities such as $Bi_2Fe_4O_9$ are eliminated. However, for x>0.15 distortion of the rhombohedral structure is evidenced. Small size and also better densification of the grains are obtained. Samples of BFO doped with yttrium were obtained for different few authors [11-13]. The investigations of these researchers show similar results as obtained for the others rare earth elements (presence of impurities, structure distortion and small grain size). But, low concentrations have been used (up to 0.2).

In the present work, $Bi_{1-x}Y_xFeO_3$ (x=0, 0.03, 0.07, 0.15, 0.2 and 0.5) ceramics were synthesized by the solid state route. The purpose is to make a rigorous analysis of the structural properties using a broad range of concentrations. The results are based on X-ray Diffraction (XRD) and Scanning Electron Microscopy (SEM).

2. Experimental

Samples of $Bi_{1-x}Y_xFeO_3$ (x=0, 0.03, 0.07, 0.15, 0.2, and 0.5) were prepared by using high purity Fe_2O_3, Bi_2O_3 and Y_2O_3 powders in a conventional solid state reaction technique. The ground oxide mixtures were pressed into pellets. Then, the samples were presintered at 750°C for 6h in an alumina crucible. The resulting mixtures were ground, pressed and subjected to a second heat treatment at 750°C followed by quenching at room temperature (the temperature and time were kept constant for all x values). Phase analysis was done by X-ray diffraction. The XRD patterns were recorded using an X´Pert PRO PANalitycal diffractometer with Co-K radiation (λ=1.7902Å). GSAS software was used for Rietveld refinements. Morphological study of the samples was performed by Scanning Electron Microscopy (JEOL JSM 6490-LV).

3. Results and discussion

XRD patterns of the $Bi_{1-x}Y_xFeO_3$ samples are shown in Figure 2. It was found that the main phase in all the samples is rhombohedral structure $BiFeO_3$. Few other peaks in the XRD patterns have been observed for secondary phases. For x≤0.2, traces of $Bi_2Fe_4O_9$ are found, which increase as Y concentration increases. For x=0.2, an Yttrium rich component starts to appear, becoming dominant for higher Y concentrations.

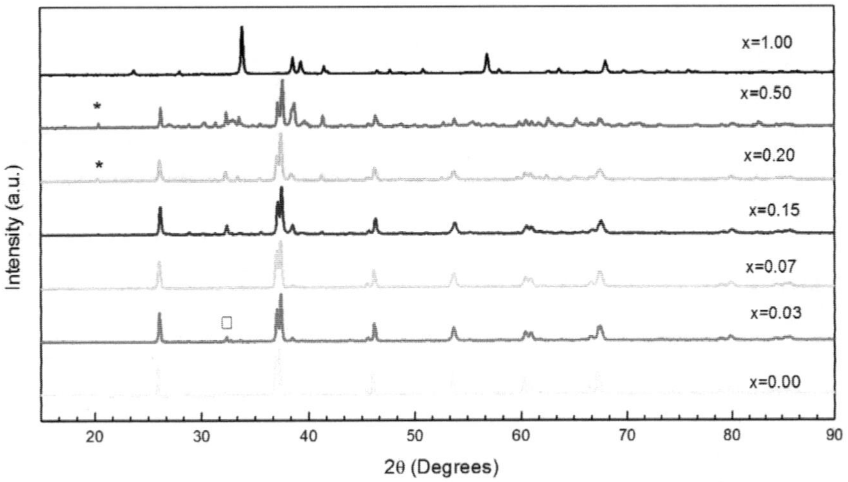

Figure 2. XRD patterns obtained for $Bi_{1-x}Y_xFeO_3$ compounds. The symbol ( ) represents the $Bi_2Fe_4O_9$ phase and (*) $Y_3Fe_5O_{12}$.

The coexistence of orthorhombic phase of Pn21a/Pnma/Pbnm space group along with R3c space group is ruled out. The addition of some rare-earth elements in the BFO synthesis can cause a compositional phase transition from rhombohedral to other type of structure [10,14].

The lattice parameters for $Bi_{1-x}Y_xFeO_3$ samples were obtained from Rietveld analysis. They are summarized in Table 1. It shows that *a* and *c* values become smaller with increasing yttrium concentration. Similar values for x = 0.15 and 0.2 were obtained as those reported by Singh et al. [12] (independent of the sintering conditions during the synthesis in both investigations). This fact evidences that the substitution of Yttrium in the BFO structure is not affected for the synthesis procedure in solid state reaction method. The decrease in the lattice parameters values can be attributed to the substitution of Y^{3+} (0.9Å) for Bi^{3+} (1.03Å) in A-sites of BFO structure [12]. Other earth rare elements substituting Bi^{3+} ions can cause the same effect [7-10,14].

The average crystallite size of the samples was calculated using Scherrer's equation ($D=K\lambda/\beta_{1/2}\cos\theta_{hkl}$), where K (shapefactor) was set to 0.9; λ=1.7902Å, $\beta_{1/2}$ is the full width at half maxima (FWHM) and θ_{hkl} is the Bragg angle. These values are displayed in Table 1. The results show that the average crystallite size values decrease as Y substitution increases. The reduction in crystallite size creates high lattice strain [8].

Table 1. Comparison of lattice parameters *a* and *c*, the grain size (GS) obtained by SEM and crystallite size (D) by Scherrer´s formula. All the values are for the phase $BiFeO_3$.

x	*a* (Å)	*c* (Å)	GS (μm)	D (nm)
0	5.5792	13.8723	6.42	95.9(5)
0.03	5.5757	13.8542	5.66	77.3(1)
0.07	5.5723	13.8471	5.10	56.4(9)
0.15	5.5713	13.8431	4.01	61.2(7)
0.20	5.5637	13.8325	-----	52.0(6)
0.50	5.5576	13.8310	-----	42.4(6)

SEM micrographs of some $Bi_{1-x}Y_xFeO_3$ samples are displayed in Figure 3. For x=0-0.07, the grains have a rounded shape and the average grain sizes decrease (GS, summarized in Table 1) with increasing Y doping level. The micrographs for x=0.15-0.5 show high interconnection between the grains, which makes difficult to calculate the average grain size. However, visual inspection evidences smaller size compared with lower yttrium concentrations. Doping can suppress oxygen vacancy concentration which lead to smaller grain sizes, which is common in the substitution of rare earth elements into BFO structure [7-13]. The difference between the values for crystallite and grain size can be explained for the formation of polycrystalline grains.

It can also be observed that there are secondary phases (which agrees with the XRD measurements) in a form of agglomerates.

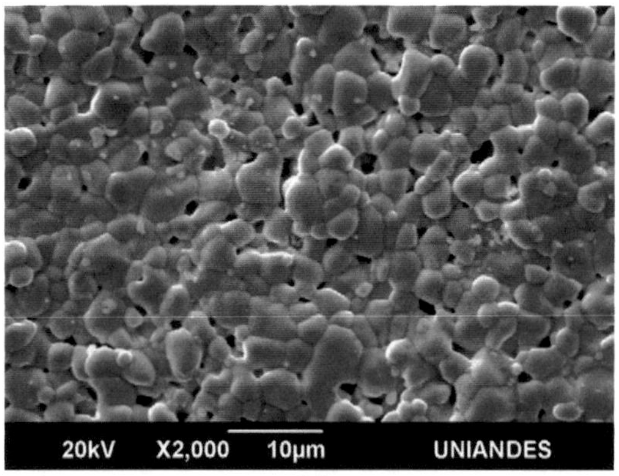

(a)

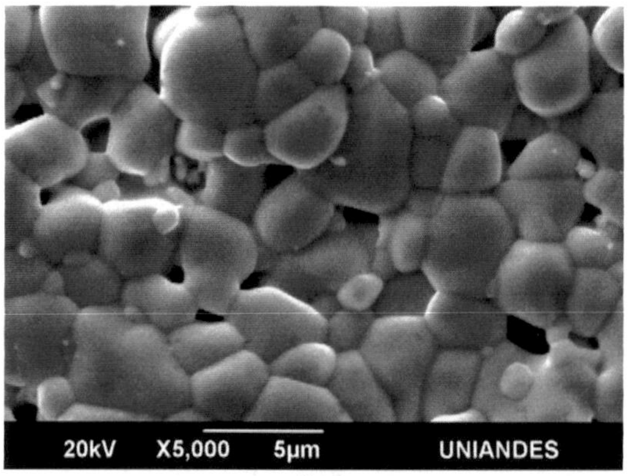

(b)

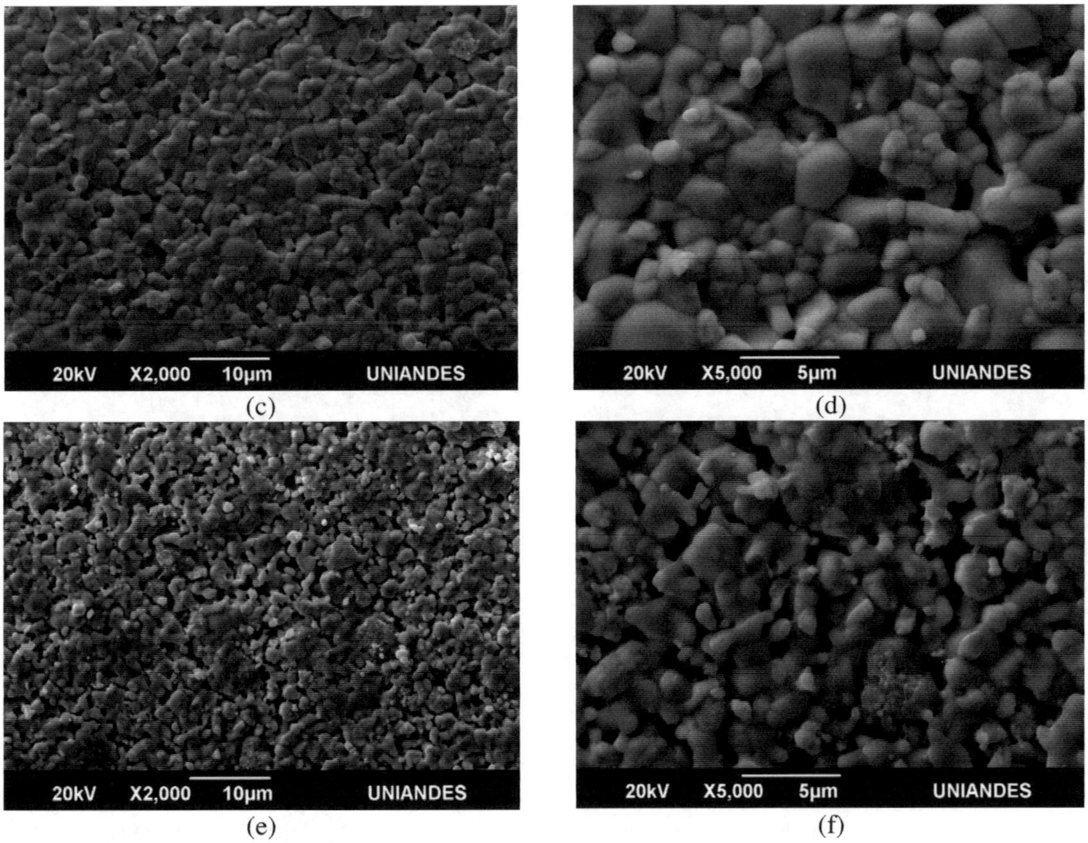

Figure 3. SEM micrographs of some $Bi_{1-x}Y_xFeO_3$ samples. (a) and (b) x=0, (c) and (d) x=0.03, (e) and (f) x=0.2.

4. Conclusions

Samples of BFO were prepared by conventional solid-state reaction technique. The results obtained from XRD are consistent with the changes in the structural properties by Y doping. SEM images indicates the reduction of the grain sizes of BFO by the replacement of Y in A-sites.

References

[1] Fischer P, Polomska M, Sosnowska I and Szymanski M 1980 *J Phys C Solid State* **13** 1931
[2] Moreau J M, Michel C, Gerson R and James W J 1971 *J Phys Chem Solids* **32** 1315
[3] Kubel F and Schmid H 1990 *Acta Cryst B* **46** 698
[4] Fruth V, Mitoseriu L, Berger D, Ianculescu A, Matei C, Preda S and Zaharescu M 2007 *Prog Solid State Chem* **35** 193
[5] Maurya D, Thota H, Nalwa K S and Garg A 2009 *J Alloys Compd* **477** 780
[6] Jha P K, Jha P A, Srivastava G, Jha A K, Kotnala R K and Dwivedi R K 2014 *J Magn Magn Mater* **349** 95
[7] Zhang S, Luo W, Wang D and Ma Y 2009 *Mater Lett* **63-21** 1820
[8] Pattanayak S, Choudhary R N P, Das P R and Shannigrahi S R 2014 *Ceram Int* **40-6** 7983
[9] Xu J, Ye G and Zeng M 2014 *J Alloy Compd* **587** 308
[10] Sati P C, Kumar M, Chhoker S and Jewariya M 2015 *Ceram Int* **41-2** 2389
[11] Rao T D, Karthik T and Asthana S 2013 *J Rare Earth* **31** 370
[12] Singh V, Sharma S, Jha P K, Kumar M and Dwivedi R K 2014 *Ceram Int* **40-1** 1971
[13] James V, Prabhakar Rao P, Sameera S and Divya S 2014 *Ceram Int* **40** 2229
[14] Pandit P, Satapathy S and Gupta P K 2011 *Physica B* **406-13** 2669

Prediction of the quality of coke obtained from vacuum residues by using spectroscopy infrared FTIR-ART

A Y León[1], N A Rodríguez[1], E Mejía[1] and R Cabanzo[1]
[1] Universidad Industrial de Santander (UIS), Bucaramanga, Colombia.

E-mail: yovani_96@hotmail.com

Abstract. According to the trend of the heavy crudes and high demand of fuels, it is projected a considerable increase in the production of vacuum residues. With the purpose of taking advantage of these loads, the refineries have been improving conversion processes for the production of better quality distillates. However, as increasing the severity conditions and the species content of resins and asphaltenes high concentrations of coke are obtained. To provide an insight into the quality and cokes properties, in this study fifty (50) coke samples obtained from vacuum residues processed under conditions of thermal cracking and hydroconversion were selected. Each coke was analysed in detail with properties such as fixed carbon, volatile material, ash, and calorific value. Subsequently, a characterization methodology was developed to predict the properties of cokes, by using partial least squares regression, and infrared spectroscopy (FTIR-ATR) in the spectral range from 4000 to 500cm^{-1}. The models obtained by chemometrics allowed to predict the quality of the coke produced from vacuum residues with reliable responses in short periods of time.

1. Introduction

The increase in the production of heavy oil, the demand for clean fuels and the increase in the production of vacuum residues require the implementation of new studies and tools to improve technology conventional processes. Within the refining processes of vacuum residues, conversion technologies play an important role in the treatment of hydrocarbons 525°C, for obtaining products of low molecular weight and low content of contaminants such as metals and heteroatoms [1, 2]. In the hydroconversion processes and thermal cracking simultaneously large amounts of coke are generated, which contain high metal content, heteroatoms and aromatic structures polycondensates [3]. Reza et al. [3], Martinez at al. [4], and Hauser et al. [5], performed previous research related to the conversion of heavy crudes and vacuum residues under different conditions of severity. The results indicated that yields of liquids, gases and coke are clearly affected by temperature and catalyst. Meanwhile, hydrocarbons of large molecular size are converted to smaller fractions oriented towards light products with boiling points below 525°C, and a high relation hydrogen/carbon, H/C. The solids are mainly formed by the thermal cracking of aromatic polycondensates. Coke is an economic byproduct of refining crude oil with a high calorific value, but its physicochemical properties and its composition depend on the nature of vacuum residue and the process conditions. For this reason, it is important to perform an extensive characterization of the coke to determine its quality. Currently, methodologies have been developed that provide characterization information from the quality of products, in which monitoring and process control is common based on information provided by analytical techniques. Several authors have used the technique of chemometrics from partial least squares regression and spectroscopy techniques to predict physicochemical properties of heavy crude oil fractions and vacuum residues. For example, Falla et al.

[6] and Satya et al. [7] developed methodologies for estimating of distillation curves, SARA composition, CCR Conradson carbon residue, density and molecular weight for crudes and its heavier fractions using infrared spectroscopy technique. On the other hand, Alciaturi et al. [8], developed correlations to predict the properties in mineral coals as the percentages of moisture, volatile matter, fixed carbon, total sulphur, calorific value and C/H, from infrared spectroscopy. In this study, four models were developed to determine the properties as ash, fixed carbon, volatile material and calorific value of coke produced from vacuum residue in conversion conditions, by using partial least squares regression, and infrared spectroscopy (FTIR-ATR).

2. Experimental methodology

For the development of this work 50 coke samples obtained in conversion processes of vacuum residues of crude oil mixtures were selected. The characterization of samples was performed in order to determine its quality from analysis such as volatile matter (D 3175), ash (D 3174) and fixed carbon. The calorific value was determined using the norm ASTM D 5865. The characterization methodology petroleum coke was developed using multivariate analysis with principal components (PCA) and spectroscopy in the region of the middle infrared (FTIR-ATR). The measurements of spectra were performed in a Nicolet FTIR Thermo Scientific, which has a diamond ATR cell with incidence angle of 45°. Measurements were made in the region between 500 and 4000cm^{-1} with a resolution of 4cm^{-1} and 32 scans (scans). The air was taken as a reference spectrum. Prediction models of quality properties of coke were performed by principal component analysis (PCA), using the software The Unscrambler X version 10 (CAMO) and information of mid-infrared spectroscopy FTIR-ATR. Principal component analyses allow to determine the discrepant samples in the data set of calibration. For each one of the studied properties different spectral regions included between 650-1650cm^{-1} and 2850-3000cm^{-1} were selected. The root-mean-square error of cross-validation (RMSEP) and variance were used to estimate the optimum number of latent variables for each model.

3. Results and discussion

Table 1 shows the range of analysed properties of cokes selected in this study.

Table 1. Characterization of cokes obtained from vacuum residues.

Property	Min	Max
Volatile matter, % wt	9.21	22.68
Fixed carbon, % wt	73.94	85.64
Ash, %wt	0.44	1.09
Calorific value, KJ/g	30.53	42.71

It is evident from data that the cokes obtained from the vacuum residues differ significantly in their properties. The volatile matter content, fixed carbon and ashes vary in a factor of 2.46, 1.16 and 2.48, respectively. Meanwhile the calorific value has a variation factor of 1.4. Figure 1 shows the FTIR-ATR spectra for five (5) cokes obtained from vacuum residues in the process of thermal cracking and hydroconversion under different operating conditions. The results show that cokes differ in its vibration characteristics across the spectrum, which indicates that these samples have significant variations in chemical composition. In the spectrum are identified vibrations of symmetric stretch and asymmetric for aliphatic bonds CH2 and CH3 between 2920-2850cm^{-1}, and deformation vibrations of the methyl and methylene group δCH3 and δCH3-CH2 near 1375cm^{-1} and 1460cm^{-1} were identified, respectively.

The wavelengths of deformation vibration of methyl groups are identified between 720-727cm^{-1}. Also, wavelengths were identified in the spectrum for polyaromatic rings between 870 and 740cm^{-1} are identified. The assignment of stretching near the wavelength at 1600cm^{-1} is attributed to double bond C=C aromatic. The displacement of the intensities in the infrared spectra for samples of coke is due to changes caused by the cracking reactions that experience the class compounds of the vacuum residues, particularly the condensation of aromatic rings in the coking reactions. Therefore, the treatment of principal component models (PCA) was performed with the spectral information obtained in the range of wavelengths from 1700-650 and 3000-2800cm^{-1}. Table 2 shows the principal components required to explain the properties of the petroleum cokes. Root Mean Square Error of Cross Validation (RMSEP) and variance were used to estimate the optimal number of principal components.

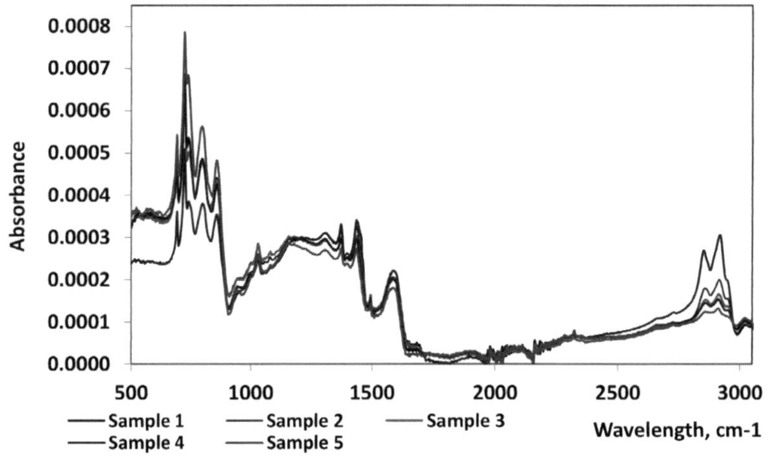

Figure 1. FTIR-ATR spectrum of cokes obtained from vacuum residues.

Table 2. Principal component analysis.

Property	PC	RMSEP	R^2
Volatile matter, VM, % wt	3	1.006	0.977
Fixed carbon, FC, % wt	7	0.245	0.988
Ash, %wt	6	0.022	0.976
Calorific value, Cv, KJ/g	6	0.209	0.985

The models show deviations between 0.2-1 and correlation coefficients higher than 97%, indicating high reliability of the developed models. Figure 2 shows that the results of prediction of fixed carbon and calorific value obtained from FTIR-ATR spectroscopy have small differences, with respect to the data obtained with the standard methodologies.

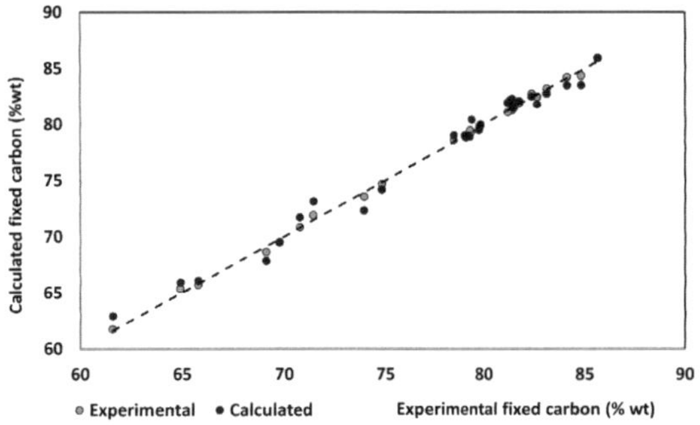

(a)

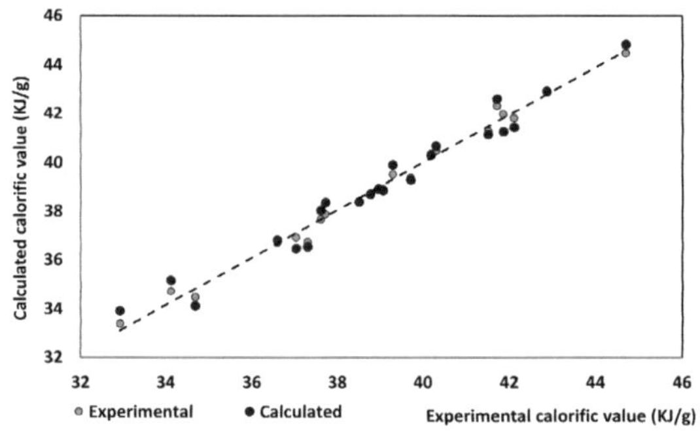

(b)

Figure 2. Calibration curve for the properties of petroleum coke: (a) volatile matter and (b) calorific value.

To determine the reliability of the models, it was necessary to calculate the overall error (%δo) in the prediction models, using the expression %δo=(RMSEP/MPE)*100. Where the % average error (MPE) was determined from all samples. Figure 3 shows the results obtained for the overall error.

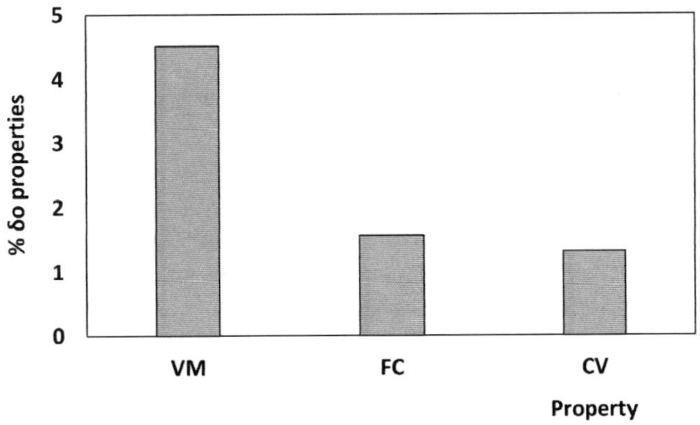

Figure 3. Overall percentage of error (%δo) for the properties of petroleum cokes.

The developed models have overall error below to 5%, so they are favourable and can be used with confidence in predicting properties of petroleum cokes such as volatile, fixed carbon and calorific value.

4. Conclusions

The technique of infrared spectroscopy FTIR-ATR allows to identify regions with greater variance, which are in the ranges of 3000-2600cm^{-1} and 1700-650 cm^{-1}. In these spectral regions were obtained the predictive models of properties such as volatile material, fixed carbon and calorific with satisfactory results, because the chemical structure of samples of petroleum coke are related with the aromatic and aliphatic functional groups.

The proposed methodology for characterizing from information FTIR-ATR spectroscopy and chemometrics technique using PCA can predict properties such as fixed carbon, volatile matter, ash and calorific value of cokes produced from vacuum residues in short periods of time.

Acknowledgements

The authors would like to thank at the Universidad Industrial de Santander, laboratory spectroscopy, fluids laboratory of metallurgical engineering, for its scientific support and human for development of this work.

References

[1] Stratiev D, Dinkov R, Petkov K and Stanulov K 2010 *Petroleum & Coal* **52-1** 35
[2] Castañeda L C, Muñoz J A D and Ancheyta J 2012 *Fuel* **100** 110
[3] Reza A S, Mohammad G and Ramin K 2011 *Fuel Processing Technology* **92** 2226
[4] Martinez G H, Ortiz H M, Sanchez F M, Ramírez J and Cuevas R C and Ancheyta J J 2014 *Catalysis Today* **220–222** 295
[5] Hauser A, AlHumaidan F, Al-Rabiah H and Absi Halabi M 2014 *Energy Fuels* **28** 4321
[6] Falla F S, Larini C, Le Roux G A, Quina F H, Moro L F and Nascimento C A 2006 *Journal of Petroleum Science and Engineering* **51** 127
[7] Satya S, Roehner M R, Milind D D and Hanson V F 2007 *Energy and Fuels* **21** 998
[8] Alciaturi E, Escobar M and De la Cruz C 2003 *Rev Tec Ing Univ Zulia* **26-3** 197

Prediction and extension of curves of distillation of vacuum residue using probability functions

A Y León[1], P A Riaño[1] and D Laverde[1]
[1] Universidad industrial de Santander (UIS), Bucaramanga, Colombia.

E-mail: yovani_96@hotmail.com

Abstract. The use of the probability functions for the prediction of crude distillation curves has been implemented in different characterization studies for refining processes. The study of four functions of probability (Weibull extreme, Weibull, Kumaraswamy and Riazi), was analyzed in this work for the fitting of curves of distillation of vacuum residue. After analysing the experimental data was selected the Weibull extreme function as the best prediction function, the fitting capability of the best function was validated considering as criterions of estimation the AIC (Akaike Information Criterion), BIC (Bayesian information Criterion), and correlation coefficient R2. To cover a wide range of composition were selected fifty-five (55) vacuum residue derived from different hydrocarbon mixture. The parameters of the probability function Weibull Extreme were adjusted from simple measure properties such as Conradson Carbon Residue (CCR), and compositional analysis SARA (saturates, aromatics, resins and asphaltenes). The proposed method is an appropriate tool to describe the tendency of distillation curves and offers a practical approach in terms of classification of vacuum residues.

1. Introduction

Global trends show an increase in the production of heavy crude oil, with which high yields are projected in the production of atmospheric residues and vacuum residues. In order to take advantage of the products obtained from the new diets refining on fuel demand, it has been showing great interest in the study of the characterization of the vacuum residues, to expand their knowledge about the behaviour new refining schemes. The vacuum residues have atmospheric points higher than 525 ° C and operational are classified as the non-distillable fraction of petroleum vacuum distillation conditions [1, 2]. These charges are considered as a complex mixture of many components and their processing is related to the chemical structure of their species [3, 4].

The characterization of the crudes and vacuum residues plays an important role on predicting yields products obtained in refining processes. One method of characterization is the distillation curve TBP (True Boiling Point); it provides information on the composition of raw materials and performance. For heavier hydrocarbon fractions TBP curve is obtained from the ASTM D1160 and ASTM D7169. However, the vacuum residues with high concentrations of asphaltenes and resins have the limitation on the distillate yield, because their end boiling point is about 50% of the distillate volume. For reasons of operation it is important extend the curves of distillation of the vacuum residues, from the available data. To this end, it is possible to use probability functions that help to predict such points from the known properties.

Several authors have used probability functions to classify and study the trend of the crude distillation curves of different nature. For example, Behrenbruch et al. [5], they developed a methodology for ranking of the crude, based on the form of the distillation curve parameters using the probability function Gamma. Sanchez et al. [6] determined that the adjustment of the curves of distillation for oil fractions, from probability functions with four parameters, shown best fit compared with experimental data. Haitham et al [7] estimated the prediction of the curves of distillation of distillates obtained in the hydrotreating of vacuum gas oil, using the probability function of Riazi and adjusting the parameters with the density of the products.

The main objective of this work is to develop a methodology for predicting and extension distillation curves TBP of the vacuum residues, using probability functions. For this purpose, four probability functions were evaluated in order to find the function with greater adjustment capacity for data TBP of the vacuum residues. For the evaluation of these functions the statistical criteria AIC (Akaike Information Criterion) and BIC (Bayesian Information Criterion) were used.

2. Experimental methodology

For the development of this work, fifty-five (55) vacuum residues were selected. The characterization was carried out using the ASTM methods for the following properties: Conradson Carbon (CCR) D-4530, D-7169 simulated distillation and compositional analysis SARA D-2007.

2.1. 2.1 Evaluation of probability functions

50 samples were selected to evaluate the adjustment of the probability functions Weibull extreme, Weibull, Kumaraswamy and Riazi. Subsequently, 5 samples for validation of the selected probability function were taken. For this purpose, temperatures were normalized so that all temperature values were between values of zero and one, using the following equation:

$$\theta_i = \frac{T_i - T_0}{T_l - T_0} \tag{1}$$

Where θ_i is the dimensionless temperature, T_i is the average temperature current boil, and T_0 and T_l are reference temperatures, that for the development of this study were 150 and 750, respectively. The parameters of probability functions were calculated with the optimization of the residual sum of squares (RSS), from the difference between the experimental weight fractions and calculated for each point of temperature. The predictive ability of the probability functions; it was evaluated from the statistical criteria AIC (Akaike Information Criterion) and BIC (Bayesian Information Criterion) [8]. The parameters AIC and BIC are indicators that take into account the complexity of a model and the good fit of the data with a minimal number of parameters. These statistical criteria are determined by the following expression:

$$AIC = 2k + nln\left(\frac{RSS}{n}\right) \tag{2}$$

$$BIC = kln(n) + nln\left(\frac{RSS}{n}\right) \tag{3}$$

Where k represent the number of parameters, n the number of observations and RSS the residual sum of squares. To evaluate the extent of the curves of distillation, from the functions of probability were selected the points T_5, T_{10}, T_{20}, T_{30}, T_{40} y T_{50}. To adjust the maximum point of distillation for the extension of the distillation curves was chosen as a reference the percentage of total distillate maltenes (saturates aromatics and resins). For this purpose, the TBP curves of vacuum residues were compared with the TBP curves reconstructed from the data TBP fractions of saturates, aromatics and resins.

To extend the study of predicting the TBP curve was divided into three study areas. The first region corresponds to the region between the T_5 and T_{50}. The region 2 corresponds to the region between the T_{50} and the final boiling point FBP of the vacuum residues validation. The last region is the extension of the distillation curve from the final boiling point FBP to the point where the maximum proportion corresponding to maltenes distillable fraction is obtained.

3. Results and discussion

The Table 1 shows that data from the selected vacuum residues differ significantly in their properties. The content of saturates, aromatics, resins and asphaltenes vary by a factor of 4, 1.5, 2.5 y 6, respectively.

Table 1. Characterization of vacuum residues.

Property	Method	Min.	Max.
CCR (%wt)	ASTM D-4530	15	30
Saturated (%wt)	ASTM D-2007	6	23
Aromatic (%wt)	ASTM D-2007	36	55
Resins (%wt)	ASTM D-2007	15	37
Asphaltenes (%wt)	ASTM D-2007	5	30
T50, °C	ASTM D- 7169	638	747

3.1. Selecting of the probability function

The Figure 1 shows the results of AIC and BIC for different randomly selected samples and evaluated with the four (4) probability functions. Considering the results of the statistical criteria AIC and BIC it is noted that the probability function Weibull Extreme presents the best adjustability for the data of TBP curves of vacuum residues. The analysis of the remaining 46 samples presented a similar behaviour.

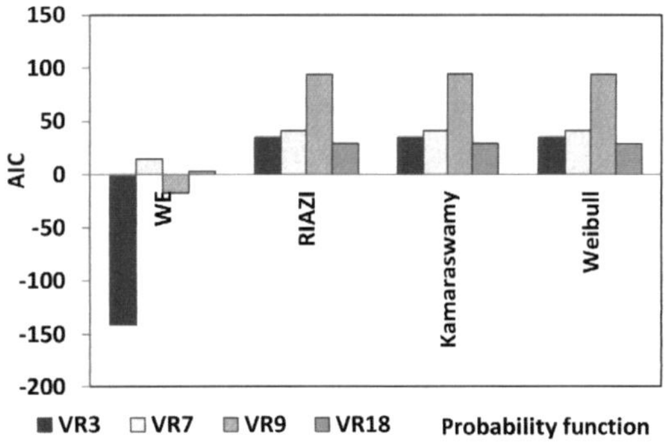

(a)

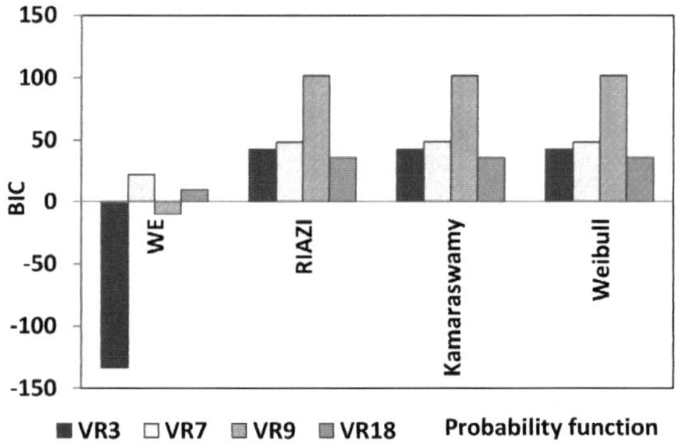

(b)

Figure 1. Statistical parameters: (a) AIC and (b) BIC.

3.2. Extension of distillation curves

In Figure 2 an insignificant difference was observed in Zone 1 between the temperatures T5 to T50 experimental and calculated with the probability function Extreme Weibull (WE), for the vacuum residue 2. In general, it was determined that the average deviation for validation vacuum residues in zone 1 was 1.23°C. Meanwhile, the extension of the distillation curve in zone 2 with the probability function, from the T50 until the end boiling point corresponding to 61% in weight of distilled of the vacuum residue 2, has an average deviation of 1.15°C. The results indicate that the probability function evaluated has a satisfactory performance in the prediction and extension of distillation curves until the end boiling point FBP of the vacuum residues.

Figure 2 shows that the boiling temperature for validation vacuum residue is available to 61% by weight of distillate, so it is necessary a greater extent considering as reference the percentage of total maltenes distillates. The TBP curve of the vacuum residues depends of the nature of his components, for this reason the TBP curve of validation samples was reconstructed from the TBP curve of fractions SAR, for the purpose of comparing their behaviour with extended curve with the proposed methodology. The results indicate that the boiling points of the curve extended with the probability function Extreme Weibull are close to the values obtained with the distillation curve recombined.

Therefore, at the time of the extension through the proposed methodology, this have a behaviour similar to the nature of the sample trends.

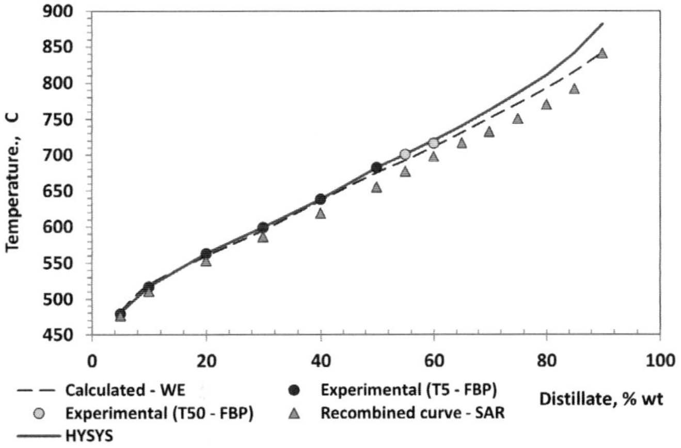

Figure 2. Extension of the distillation curve of vacuum residue 2.

Comparing the curve extension with the proposed methodology and the extended curve with the program ASPEN HYSYS V8.0, it is observed that the prediction curve program ASPEN HYSYS has a greater deviation from the end boiling point of the total percentage of distillate curve reconstructed from the SAR fractions.

4. Conclusion

The probability function with greater ability to fit the data of the curve of distillation of the vacuum residues was the Weibull Extreme, and as second option is the three parameters Riazi.

The recombined curve of the vacuum residue from total percentage of the fraction of maltenes is a criterion of reference to evaluate the methodology of extension from the probability functions.

Acknowledgements

Special acknowledgments for the Industrial University of Santander (UIS) Chemical engineering faculty, for financial support, scientific and human development of this work.

References

[1] Celio Pasquini, Aerenton Ferreira Bueno 2007 *Fuel* **86** 1927
[2] Shaoping Li, Qi Zhang, Yao Yao, Xin Sun, Qiang Fan and Jianliang Chen 2007 *Energy & Fuels* **21** 1145
[3] Stratiev D, Dinkov R, Petkov K, Stanulov K 2010 *Petroleum & Coal* **52-1** 35
[4] Mohan S Rana, Vicente Sámano, Jorge Ancheyta, J A I Diaz 2007 *Fuel* **86** 1216
[5] Sanchez S, Ancheyta J and McCaffrey W 2007 *Energy & Fuels* **21-5**
[6] Peter Behrenbruch P and Dedigama T 2007 *Journal of Petroleum Science and Engineering* **57** 166
[7] Haitham L, Dduha C, Riazi M, Ayman A and Hamad A 2011 *Fuel* **90** 719

Reactivity study on thermal cracking of vacuum residues

A Y León[1], S D Díaz[1], R C Rodríguez[1] and D Laverde[1]
[1] Universidad industrial de Santander (UIS), Bucaramanga, Colombia.

E-mail: yovani_96@hotmail.com

Abstract. This study focused on the process reactivity of thermal cracking of vacuum residues from crude oils mixtures. The thermal cracking experiments were carried out under a nitrogen atmosphere at 120psi between 430 to 500°C for 20 minutes. Temperature conditions were established considering the maximum fractional conversion reported in tests of thermogravimetry performed in the temperature range of 25 to 600°C, with a constant heating rate of 5°C/min and a nitrogen flow rate of 50ml/min. The obtained products were separated in to gases, distillates and coke. The results indicate that the behaviour of thermal reactivity over the chemical composition is most prominent for the vacuum residues with higher content of asphaltenes, aromatics, and resins. Finally some correlations were obtained in order to predict the weight percentage of products from its physical and chemical properties such as CCR, SARA (saturates, aromatics, resins, asphaltenes) and density. The results provide new knowledge of the effect of temperature and the properties of vacuum residues in thermal conversion processes.

1. Introduction

The increase of heavy oil in the diet of the refineries has been paying great attention on the oil industry due to the projected high increases in the production of vacuum residues [1]. These loads are hard to process because of their high content of asphaltenes, heteroatoms and metals. Vacuum residues have elevated equivalent atmospheric points and are classified as the non-distillated fraction in vacuum distillation [2]. To take advantage of the heavier fractions of oil, refineries have been applying and improving their thermic conversion processes to treat the low quality loads and generate lighter products with commercial value [3].

Thermic conversion processes depend of operating conditions as temperature, reaction time and charge nature [4]. Francoise et al. [5] and Jasvinder et al. [6] achieved to establish that the convertible fraction of the heavier fractions of vacuum residues type C14+ and asphaltenes toward distillates, is due to the release of molecules lower in molecular size, but their thermal reactivity increases in structures with a greater degree of aromaticity and content of side chains. Moreover, Marsh et al. [7] and Van Heek et al. [8], proved that the yields of distillate fraction decreases with increasing temperature, resulting in products with orders of yield as ACPM>Kerosene>Nafta>Gases.

In studies of thermal conversion reactivity, the analysis by thermogravimetry has played an important role. Thermogravimetry is a technique that measures the mass changes in function of time and temperature. Also it is widely used in thermal analysis to determine reactivity zones and kinetic parameters that characterize the heavy crude oil, vacuum residues and its derivate [9,10].

The objective of this study consists of determining the reactivity on thermal cracking of vacuum residues in a batch reactor. The research was developed in two stages: the first one consisted in a thermogravimetric analysis (TGA) in order to find the higher reactivity zone. Secondly, thermal cracking tests were made in a temperature range of 430–500°C and a 120psi pressure. Additionally, predictive correlations of the yield to coke, distillates and gases, were established from physicochemical properties of vacuum residues. Furthermore the purpose of this work is to provide new knowledge related to the effect of the temperature and nature of vacuum residues in thermal conversion processes.

2. Methodology

For this work were selected six (6) vacuum residues from crude oil mixtures. Also the characterization was realized using the ASTM methods for carbon conradson (CCR) D-4530, simulated distillation D-7169 and compositional analysis SARA D-2007.

2.1. Thermogravimetric analysis

(TA Instruments TGA/2050, Thermogravimetric Analizer). In trials was analysed approximate amounts of 10mg. The thermal decomposition was performed by increasing the temperature up to 600 with flow 50ml/min of nitrogen at a heating rate of 5/min.

2.2. Thermal cracking tests

The samples of vacuum residues are added to the batch reactor; it is pressurized with nitrogen to guarantee inert atmosphere. Later the reactor is submitted to the tests cracking within a temperature range of 430–500°C. Finally, it is cooled with water and the products are recovered in order of realize mass balances.

3. Results discussion

Table 1 shows the used ASTM methods for the characterization of the two vacuum residues which content of asphaltenes is appreciably different from one another.

Table 1. Vacuum residues characterization.

Property	Method	FV1	FV2
Density gr/ml, 15°C	ASTMD-4052/70	1.0	1.1
CCR (%wt)	ASTM D-4530	17	35
Saturates (%wt)	ASTM D-2007	16	13
Aromatics (%wt)	ASTM D-2007	42	41
Resins (%wt)	ASTM D-2007	33	20
Asphaltenes (%wt)	ASTM D-2007	9	26
T_{50}, °C	ASTM D- 7169	640	707

3.1. Thermogravimetric analysis

Figure 1 shows the differential curve of the mass derivate respect to the temperature. This curve allows demonstrating the temperature where there is greater loss of mass, approximately to 430°C. The results of the two (2) vacuum residues selected indicate that the higher reactivity zone is in a range of temperatures between 400 to 500°C. Therefore these results allowed to establish the range of temperature for the study of thermal treatment in a reactor batch which was determinate to be between 430–500°C.

3.2. Temperature effect

Figure 2 shows the coke formation, distillates and gases yields for two vacuum residues respectively. It indicates the direct relation between the reaction temperatures and coke yield for both vacuum residues. However, the coke formation depends on its nature.

The asphaltenes content is an important indicator of coke. This allows clarify that the FV 1 shows lower percentages of coke with respect to FV 2. Meanwhile, distillate yield decreases with increasing temperature due to its transformation to gases and coke. The tendency of gas formation is a result of cracking of the molecules of less to more complexity of SARA fractions.

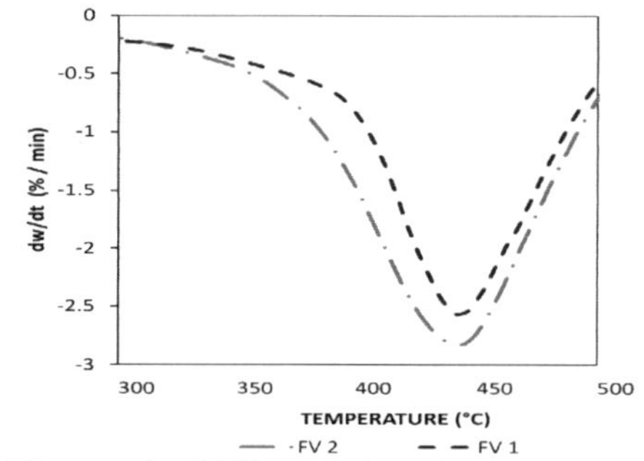

Figure 1. Differential weight loss for vacuum residues at 5°C/min.

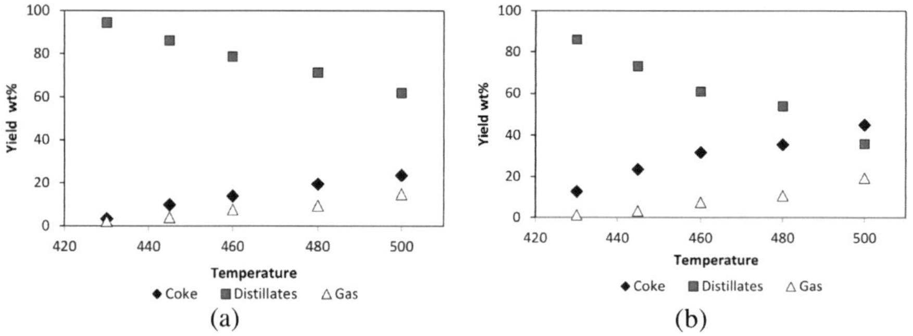

Figure 2. Product yields. (a) FV 1 and (b) FV 2.

3.3. Products formation predictions

The correlations for predicting coke yield, distillates and gases were made from the (6) vacuum residues using multivariate analysis with Stapgraphic Centurion XVI. Therefore, normalized correlations for prediction of performance of products were made by the following general expression:

$$\%\mathrm{Coke} = \mathrm{m} * \mathrm{T} + \mathrm{a} \tag{1}$$

$$a = 523.5 - 0.96 * \mathrm{Tbp50} - 8.2 * \mathrm{CCR} + 3.5 * \mathrm{SAT} + 10.5 * \mathrm{ASP} \tag{2}$$

$$m = -1.2 + 0.002 * \mathrm{Tbp50} + 0.02 * \mathrm{CCR} - 0.008 * \mathrm{SAT} - 0.02 * \mathrm{ASP} \tag{3}$$

$$\%Distillates = b * T^2 + c * T + d \tag{4}$$

$$b = 0.04 + 0{,}004 * \mathrm{API} + 0{,}0023 * \mathrm{CCR} - 0{,}0002 * \mathrm{TB50} + 0{,}{,}0004 * \mathrm{RES} \tag{5}$$

$$c = -35.8 - 2.16 * \mathrm{CCR} - 0.38 * \mathrm{RES} - 3.77 * \mathrm{API} + 0{,}18 * \mathrm{TB50} \tag{6}$$

$$d = 8008.3 + 855.4 * \mathrm{API} + 492 * \mathrm{CCR} - 41.2 * \mathrm{TB50} + 89.3 * \mathrm{RES} \tag{7}$$

$$\%\mathrm{Gases} = \mathrm{e} * T^2 + \mathrm{f} * \mathrm{T} + \mathrm{g} \tag{8}$$

$$\mathrm{e} = -0.05 + 0.00011 * \mathrm{Tb50} + 0.00001 * \mathrm{API} - 0.00027 * \mathrm{RES} - 0.00051 * \mathrm{ASP} \tag{9}$$

$$f = 48.6 - 0.144 * Tb50 - 0.096 * API + 0.016 * RES + 0.27 * ASP \qquad (10)$$

$$g = -11067.2 + 21.9 * Tb50 + 5.03 * API - 61.8 * RES - 106.3 * ASP \qquad (11)$$

Figure 3 shows coke yield and distillates experimental and calculated. The predictive correlations for coke, distillates and gas production, showed standard deviations between 1,13 y 2,95wt%. The results are satisfactory and can be applied to predict the coke yield, distillates and gas thermal cracking processes.

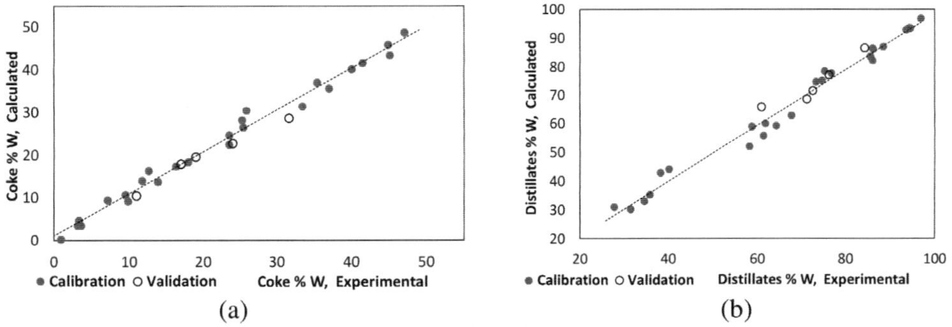

Figure 3. Product yields. (a) % coke y (b) % distillates.

4. Conclusions

Thermogravimetric analysis is an important tool to determine the operational ranges in the thermal cracking treatment of vacuum residues; therefore favourable results can be obtained from small sample amounts under short periods of time.

The increase in reaction temperature favours the coke and gases production. Compared to the distillates production where it yields decreases with the increase in temperature, as a result from its transformation into gases and coke.

The predictive correlations for coke, distillates and gas production in thermal treatment of vacuum residues are satisfactory, taking into account that the average standard deviation is less than 3%.

Acknowledgments

Special acknowledgments for the Industrial University of Santander (UIS) Chemical engineering faculty, for financial support, scientific and human development of this work, and Corrosion Research Group (GIC), for its attention and cooperation during the experiments.

References

[1] Stratiev D, Dinkov R, Petkov K, Stanulov K 2010 *Petroleum & Coal* **52** 35
[2] Hedrick B W, Seibert K D and Crewe C A 2006 *A new approach to heavy oil and bitumen upgrading* (Saudi Arabia: 16th Annual Symposium on Catalysis in Petroleum Refining and Petrochemicals (http://www.uop.com/?document=a-new-approach-to-heavy-oil-and-bitumen-upgrading&download=1)
[3] Sanfilippo D 2009 *Bottom of the barrel technology within refining extracting additional value from oil feedstocks using EST technology* (Italy: WTG Webinar Connected from San Donato Milanese)
[4] Reza Asgharzadeh Shishavan, Mohammad Ghashghaee, Ramin Karimzadeh 2011 *Fuel Processing Technology* **92** 2226
[5] Francoise Behar, Francois Lorant, Laurent Mazeas 2008 *Organic Geochemistry* **39** 764
[6] Jasvinder Singh, M M Kumar, Alok K Saxena, Surendra Kumar 2005 *Chemical Engineering Journal* **108-3** 239
[7] Marsh H 1991 *Carbon* **29-6** 703
[8] Van Heek K H 2000 *Fuel* **79-1** 1
[9] AijunGuo, Xuejun Zhang, Zongxian Wang 2008 *Fuel Processing Technology* **89** 643
[10] Alvarez E, et al 2011 *Fuel* **90** 3602

Journal of Physics: Conference Series **687** (2016) 012095 doi:10.1088/1742-6596/687/1/012095

Treatment of vacuum residues in hydroconversion conditions

A Y León[1], D L Mendoza[1], J O Espinosa[1] and D Laverde[1]
[1] Universidad industrial de Santander (UIS), Bucaramanga, Colombia.

E-mail: yovani_96@hotmail.com

Abstract. In this paper the use of a liquid homogeneous catalyst has been studied in reactivity vacuum residues by hydroconversion under different conditions. To cover a wide range of compositions, six (6) vacuum residues were selected from crude mixtures. Hydroconversion test were performed in batch reactor with hydrogen atmosphere at about 2000psi in a temperature range between 430 and 480°C. The results allowed to establish that the reactivity hydroconversion conditions about coke formation is higher in vacuum residues with higher content of resins and asphaltenes. The reaction conditions promote distillate formation, however, with increasing stringency conditions, the distillate yield decreases due to distillate transformation into temperature range 430 and 460°C compared to the tests performed without catalyst demonstrating that the use of homogeneous catalyst is an alternative to treating vacuum residues and results are satisfactory in the conversion processes. Finally, predictive expressions have been developed in the formation of products depending on the conditions of temperature and physicochemical properties of processed vacuum residue.

1. Introduction

Oil is known as one of the main non-renewable energy resources in the world, is processed for the purpose of generating fuel and value aggregate products. With increased heavy oil production and high demand for fuel, the refineries have been forced to rethink their processes with new diets what it's projected a high increase in the production of vacuum residues with numbers that can range between 35 and 50vol% with respect to oil processed [1]. Oil is considered a complex fluid formed of hundreds of thousands of molecules, with significant differences in their molecular structures and its more complex molecules are concentrated at the vacuum residues [2, 3]. With the high availability of these residues, it has been promoted to the search for processes that allow its transformation as thermal carbon rejection processes and hydroconversion [4, 5]. In hydroconversion processes the hydrocarbon is contacted with hydrogen in the presence of a catalyst under temperature conditions in the range between 300 and 450°C, and the hydrogen pressure between 800 and 2900psi [6]. In studies oriented to catalytic conversion, the reactivity was evaluated in processes that include the use of middle distillates and streams from heavier fractions. For example, Ortega et al. [7], Fortain et al. [8] and Samina et al. [9] were able to establish that the increased of conversion yields of vacuum residues and heavy crudes is mainly due to increase in temperature. However, with the presence of catalysts were achieved conversions up to 70% of heavy hydrocarbons, with reductions in sulphur and nitrogen content above 60 and 55%, respectively. The objective of this study is to evaluate the effect of a homogeneous catalyst under different temperature conditions in the vacuum residues treatment.

Journal of Physics: Conference Series **687** (2016) 012095 doi:10.1088/1742-6596/687/1/012095

2. Experimental

For the development of this work were selected six (6) vacuum residues from crude mixtures. The characterization was carried out using the ASTM methods for the following properties: Conradson Carbon (CCR) D-4530, simulated distillation D-7169 and compositional analysis SARA D-2007.

2.1. Treatment of vacuum residues by hydroconversion

An amount of vacuum residue and liquid catalyst NM is added to the batch reactor. The react is pressurized with hydrogen to effect the optimal reaction. Then the batch reactor is subjected to severe conditions up to the reaction temperature for 20 minutes. After the reaction, the reactor is cooled in ice water in order to stop the reaction. With the objective to vary the conversion and to determine the performance of the reaction, experiments were performed in the temperature range of 430 to 480°C. Finally, the obtained products are recovered, in order to perform the respective balance mass.

3. Results and discussion

Table 1 shows the characterization of vacuum residues selected for this study.

It is evident from the data selected vacuum funds differ significantly in their properties. The content of saturates, aromatics, resins and asphaltenes vary by a factor of 4, 1.6, 1.9 and 6 respectively.

Table 1. Characterization of vacuum residues.

Property	Method	Min	Max
Density gr/ml	ASTMD-4052/70	1.01	1.06
CCR (%wt)	ASTM D-4530	17.1	34.6
Saturates (%wt)	ASTM D-2007	4	16
Aromatics (%wt)	ASTM D-2007	34	56
Resins (%wt)	ASTM D-2007	20	37
Asphaltenes (%wt)	ASTM D-2007	5	30
T50, °C	ASTM D- 7169	638	747

3.1. Effect of temperature

Figure 1 shows the variation in the reaction temperature significantly modifies the performance of reaction products. As the reaction temperature increases coke formation has an increase which depends on the nature of the sample. Coke formation can be attributed largely to the asphaltenes content. For this reason, the vacuum residue VI with a percentage of asphaltenes of 30% by weight and the vacuum residue I with a percentage of asphaltenes of 5% by weight showed the highest and lowest coke formation during the entire range temperature.

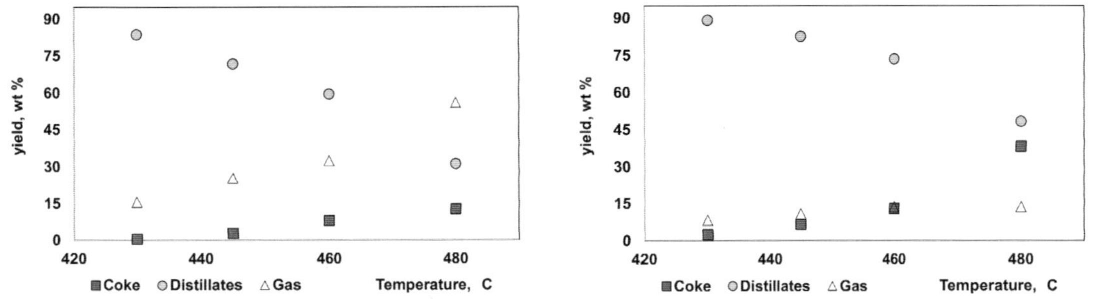

(a) (b)

Figure 1. Hidroconversion experimental results: (a) Vacuum residue I and (b) Vacuum residue VI.

Meanwhile, the results show a decrease in the percentage of distillates as the temperature increases mainly due to cracking of the newly formed structures which contribute to the formation of gases and low coke yields. The above allows clarify the formation of gases where the percentage of these increases when increase the reaction temperature.

3.2. Effect of catalyst

To evaluate the effectiveness of the catalyst with the temperature on the inhibition of coke formation, cracking tests were conducted to vacuum residue VI without catalyst. Figure 2 shows that the production of coke for the hydroconversion process with catalyst is lower with respect to the tests without catalyst for vacuum residues I and VI in the range of temperatures between 430°C and 480°C. The results show a consistent effectiveness in inhibiting catalyst coke formation from 79 to 63% by weight and 89 to 35% by weight in the temperature range from 430 to 460°C, for vacuum residues I and VI respectively. However, in conditions above of 460°C the coke conversions are similar with and without addition of catalyst, due to problems of diffusion of hydrogen. The results corroborated that at high temperatures the dominant process is the thermal cracking. Therefore, although the catalyst plays an important role in hydrogenation processes, it has little effect on the control of cracking mechanisms.

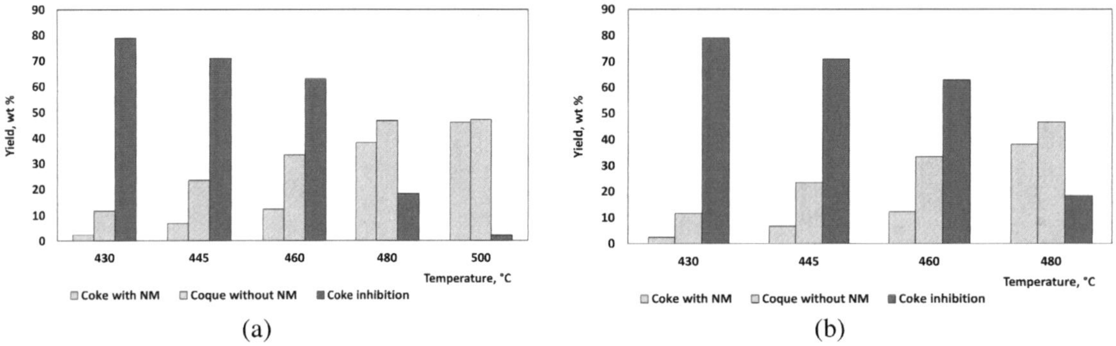

Figure 2. Coke yield with and without catalyst: (a) Vacuum residue I y (b) Vacuum residue VI.

3.3. Prediction of product formation

The development of predictive correlations for production of coke, distillates and gas from six (6) vacuum residues was performed by multivariate analysis using Stapgraphic Centurion XVI. As mentioned above, the temperature has a predominant effect on the formation of the products obtained in the process of hydroconversion of vacuum residues. Therefore, normalized correlations for prediction of performance of products were made by the following general expression:

$$\%Products_i = \alpha_i + \beta_i T \tag{1}$$

Where, i=% Coke, % Distillates y % Gas.

% Coke is:

$$\alpha = 1781.9 - 1451.2Density + 1.5Saturates - 0.6T50 \tag{2}$$

$$\beta = -4.1 + 3.33Density - 0.003Saturates + 0.001T50 \tag{3}$$

% Distillates:

$$\alpha = 3604.2 + 38.3CCR + 1.5Aromatics - 6.2T50 \tag{4}$$

$$\beta = -8.1 - 0.08CCR - 0.003Aromatics + 0.001T50 \tag{5}$$

% Gas:

$$\alpha = -5359.8 + 27.9Saturates + 16.2Aromatics + 6.2T50 \quad (6)$$

$$\beta = 12.3 - 0.06Saturates - 0.033Aromatics - 0.013T50 \quad (7)$$

Figure 3 shows the percentage values of experimental and calculated for distillates and coke. Correlations for predicting the percentage of coke, gases and distillates showed standard deviations between 1.7 and 2.6% by weight. The results indicate that the correlations obtained are appropriate for predicting the performance of the products obtained from vacuum residue at hydroconversion conditions.

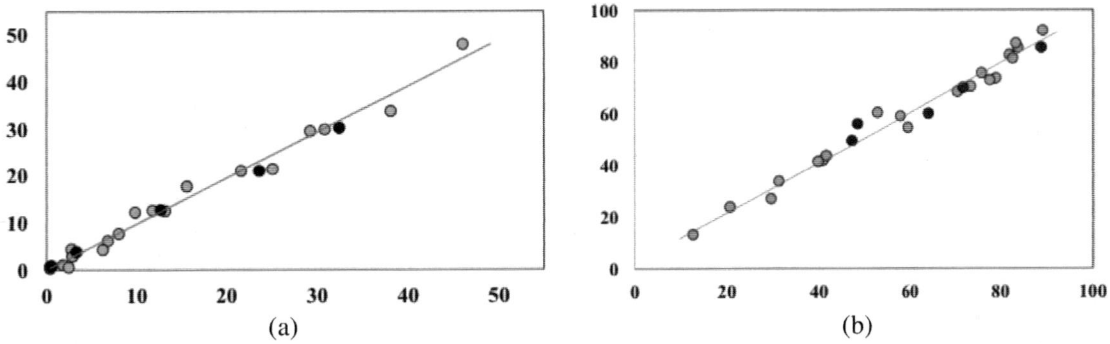

Figure 3. Yields of products: (a) experimental coke vs % coke calculated y (b) experimental distillates vs % distillates calculated.

4. Conclusions

Increasing the reaction temperature favours the production of coke and gases. The increase in coke formation at temperatures above 480 ° C is due to the decrease of hydrogen diffusion. In the case of gases and distillates it is observed at temperatures above 460 ° C the distillate yield decreases and promotes the production of gases.

The inhibition of coke formation produced by the catalyst in the temperature range from 430 to 460°C for vacuum residues I and VI decreases from 79 to 63% by weight and 89 to 35% by weight. However, for temperatures above 460°C to coke conversions tend to be similar with and without addition of catalyst. However, trends in inhibiting coke formation depends on the nature and percentage of compositional fractions SARA of vacuum residues.

Acknowledgments

Special acknowledgments for the Universidad Industrial de Santander (UIS) Chemical engineering faculty, for financial support, scientific and human development of this work, and Corrosion Research Group (GIC), for its attention and cooperation during the experiments.

References

[1] Stratiev D, Dinkov R, Petkov K, Stanulov K 2010 *Petroleum& Coal* **52** 35
[2] Wiehe I and Liang K 1996 *Ruid Phase Equilibria* **117** 201
[3] Cindy-Xing Yin, Jeffrey M and Murray R 2009 *Energy & Fuels* **23** 2600
[4] Mohammad F and Saeed A 2006 *Fuel Processing Technology* **87** 573
[5] Bridjanian H and Khadem S 2011 *Petroleum & Coal* **53** 13
[6] Mohan S, Vicente S, Jorge A and Diaz J 2007 *Fuel* **86** 1216
[7] Celio P and Aerenton F 2007 *Fuel* **86** 1927
[8] Shaoping Li, Qi Zhang, Yao Yao, Xin Sun, Qiang Fan and Jianliang Chen 2007 *Energy & Fuels* **21** 1145
[9] Kejing Li, Babajide A, Andrew M, Yuan Yang, Michael B, Robert J, John R, Dorgan M 2013 *Fuel* **103** 843

Structural magnetic and Mössbauer study of ^{57}Fe doped $La_{2/3}Ca_{1/3}Mn_{1-x}Fe_xO_3$ ($0 \leq x \leq 0.1$) manganites

L C Sánchez[1], **D Durango**[1] **and O Arnache**[2]
[1] Universidad de Córdoba, Montería, Colombia.
[2] Universidad de Antioquia, Medellín, Colombia.

E-mail: lsanchezpacheco@gmail.com

Abstract. We have studied the effect of ^{57}Fe substitution on structural and magnetic properties of $La_{2/3}Ca_{1/3}Mn_{1-x}Fe_xO_3$ (LCMFO; x=0, 5 and 10%) manganites. The powders were prepared by the solid state reaction method, and the final products were characterized by X-Ray Diffraction (XRD), Mössbauer Spectrometry and Magnetic Measurements. Only one single phase corresponding to the manganite phase was detected without the presence of impurities or iron phases. The hyperfine parameters of Mössbauer spectra (MS), taken at different temperatures for the doped samples, reveal the only presence of Fe^{3+} ions in octahedral coordination. At low temperatures, the MS spectra were fitted with a model of three components associated to three magnetic phases of iron (ferromagnetic and paramagnetic states). These ordered and not ordered magnetic phases are related to the effects of grain size distribution in LCMFO samples. Finally, significant changes were observed in the saturation magnetization, Curie temperature and coercivity field with increasing of iron doping level.

1. Introduction

Colossal magnetoresistance (CMR) in manganites has generated a rise on studying these materials because of both interesting fundamental properties and potential device applicability. In particular, the manganite $La_{1-x}A_xMnO_3$ (A=Ca, Sr, Ba, Pb) presents a remarkably complicated phase diagram as the relative concentration of Mn^{3+} and Mn^{4+} changes. Ferromagnetic order is due to double exchange (DE), via Mn^{3+}–O^{2-}–Mn^{4+}, whereby ferromagnetic (FM) interactions between localized spins are mediated by hopping of itinerant spins [1]. In order to study the influence of Fe doping on the electrical, structural and magnetic properties of manganites, early studies have shown the direct replacement of Mn^{3+} by Fe^{3+}, considering their similar ionic sizes. Mössbauer spectroscopy (MS) is an important technique to provide information about the nature of the local induced magnetic effects, small changes in the local electronic and vibrational state of ^{57}Fe via hyperfine interactions. In this context, we thought it is interesting and useful to show the results of the effects of varying the content of ^{57}Fe on the structural and magnetic properties in $La_{2/3}Ca_{1/3}Mn_{1-x}Fe_xO_3$ ($0 \leq x \leq 0.10$) manganite powder samples, prepared by solid state reaction method.

2. Experimental

Powder samples of $La_{2/3}Ca_{1/3}Mn_{1-x}Fe_xO_3$ with x=0, 0.05 and 0.10 nominal composition were prepared by the solid state reaction method with high purity (>99.9%) precursors of La_2O_3 (Lanthanum oxide), MnO_2 (Manganese dioxide), $CaCO_3$ (Calcium carbonate), and α-Fe_2O_3 (Hematite) 95.06% enriched with ^{57}Fe. The powders of these oxides were mixed and heated in several stages with temperatures among 900°C and 1300°C to obtain $La_{2/3}Ca_{1/3}MnO_3$ (LCMO), $La_{2/3}Ca_{1/3}Mn_{0.95}Fe_{0.05}O_3$ (LCMFO-5%), and $La_{2/3}Ca_{1/3}Mn_{0.90}Fe_{0.10}O_3$ (LCMFO-10%) powders. The samples were analysed by XRD (x-ray diffraction) using a Bruker Advance 8.0 with a Cu (Kα) source, in Bragg-Brentano geometry. The data were collected in the range 20–80° in 2θ with steps of 0.013°, and the patterns were fitted by employing the Rietveld refinement analysis implemented in the FULLPROF-SUITE program. The absorption ^{57}Fe Mössbauer spectra were recorded in a conventional Mössbauer spectrometer with a ^{57}Co (Rh) source, while the absorbers were collected through a cryostat at variable temperatures range from 7 to 300K. These spectra were analysed using the NORMOS program to obtain the hyperfine parameters. Magnetic measurements, zero field cooling (ZFC) and field cooling (FC) curves, and hysteresis loops (at 7, 50, 100, 200 and 300K) were performed via vibrating-sample magnetometer technique (VSM) with the physical properties measurement system (PPMS, Quantum Design). The ordering temperature (T_C) was determined from dM/dT derivative of the ZFC curves. The difference between FC and ZFC curves was used to estimate the irreversibility temperature (T_I).

3. Results and discussion

The structural parameters of all samples were obtained by fitting with the Rietveld refinement method the XRD patterns (see Figure 1), assuming the perovskite structure (orthorhombic Pnma-(62) space group). The obtained lattice parameters and cell volumes are shown in Table 1. Peaks of other phases such as secondary oxides or impurities were not detected. Both the lattice parameters and the cell volumes slightly increase with the Fe doping. These results indicate that due to both Fe^{3+} and Mn^{3+} have almost identical ionic radii, thus, the crystalline structure is not modified by Fe doping in our concentrations. Similar results were reported by Cai [2] and Ogale [3] et al. Additionally we estimated the average crystallite size about of 305, 328, and 340nm, for LCMO, LCMFO-5% and LCMFO-10%, respectively. For this, the patterns were analysed by using MAUD program, which combines the Rietveld method and a Fourier transform analysis.

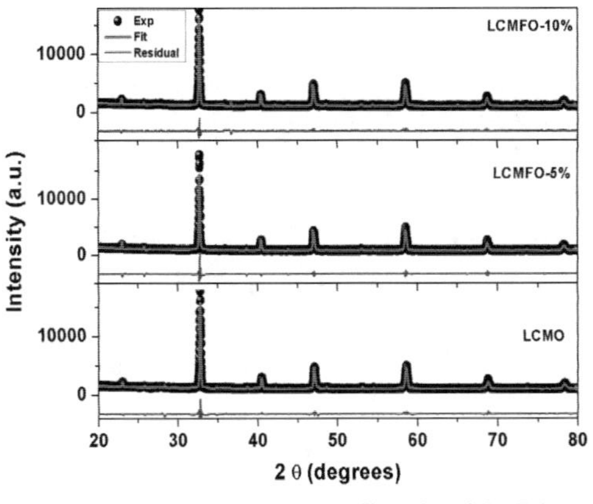

Figure 1. XRD patterns fitted with Rietveld refinement. Continuous lines represent the fit.

Table 1. Crystallographic and magnetic parameters of LCMO, LCMFO-5% and LCMFO-10% samples. a, b, and c are the cell parameters, V the cell volume, T_C and T_I are the Curie and irreversibility temperature, respectively.

Sample	a (Å)	b (Å)	c (Å)	V (Å)^3	T_C(K)	T_I(K)
LCMO	5.449	7.695	5.464	229.1	271	260
LCMFO 5%	5.460	7.706	5.466	230.0	179	175
LCMFO 10%	5.461	7.707	5.467	230.1	160	136

The temperature evolution of Mössbauer spectra (MS) for the LCMFO-5% and LCMFO-10% samples recorded at temperatures between 7 and 300K are shown in Figure 2 and Figure 3, respectively. At 300K, the spectra were fitted considering a paramagnetic doublet (D2) [4]. The values of isomer shift are typical of Fe^{3+} ions in octahedral coordination, while the small quadrupole splittings of the samples indicate a slight distortion of the octahedron. As temperature is further reduced, we detected a six-line spectrum (S1) with hyperfine parameters that corroborate the presence of Fe^{3+} ions. For this fitting model it was assumed Lorentzian line shape with the same line width, isomer shift and quadrupole interaction, based on a discrete distribution model used of Simopoulos et al [5]. The obtained values of quadrupole splitting might be explained by a possible structural transformation of the perovskite matrix (Mn site), in according to reports made by other authors [6,7]. In the temperatures intermediate range between 150 and 100K, we found the same behaviour of the distributed magnetic sextet with asymmetric line broadenings; it was attributed to early manifestation of magnetic ordering with decreasing temperature. At low temperatures, the fits were obtained by introducing an additional crystalline sextet (S2), with hyperfine parameters associated again to the presence of Fe^{3+}. At 7K the MS displays a well-defined crystalline component, a broad distributed magnetic sextet, which corresponds to smaller effective hyperfine magnetic fields, and a paramagnetic doublet. This validates the presence of three different iron sites leading to a moderate distribution in the hyperfine magnetic field (see Figure 4). Figure 5 shows M(T) in both ZFC and FC modes for applied external magnetic field of 500 Oe. The obtained values of T_C were 271, 179 and 160K, for LCMO, LCMFO-5% and LCMFO-10%, respectively. We found that paramagnetic contribution remains in the spectra below T_C and decreases with the diminution of the temperature. The presence of paramagnetic fractions PM in our Mössbauer spectra could be associated to a dynamic phase separation, i.e. coexistence of small ferromagnetic FM domains (clusters) undergoing superparamagnetic like relaxation and small PM insulating clusters in the vicinity of T_C.

From the estimated average crystallite size for our samples (around 300nm), the temperature evolution of the MS cannot be attributed to classical superparamagnetism [7]. In the doped samples, the main effect of the substitution of a Fe by a Mn site is the reduction of both T_C and saturation magnetization M_S regarding to the undoped sample. The M_S estimated at 7K of the hysteresis curves (not shown) were 88.3, 76.8 and 57.4emu/g for the LCMO, LCMFO-5% and LCMFO-10% samples, respectively. Furthermore, a broader ferromagnetic-paramagnetic FM-PM transition in the M(T) curves (see Figure 5) with the Fe concentration were observed. Following the results of X. Wang et al. [8] we could attribute this broadening to the effect of antiferromagnetic (AF) couplings between Fe and its neighbors, which causes a weaker magnetic interaction between Mn-Mn, in according to our Mössbauer and magnetic measurements.

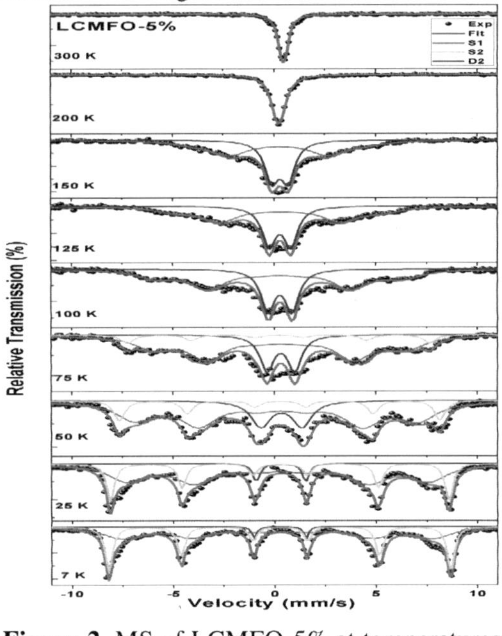

Figure 2. MS of LCMFO-5% at temperatures from 7 to 300K. Closed circles are experimental data, colour lines are different components (S1, S2 and D2) and red line is the fit.

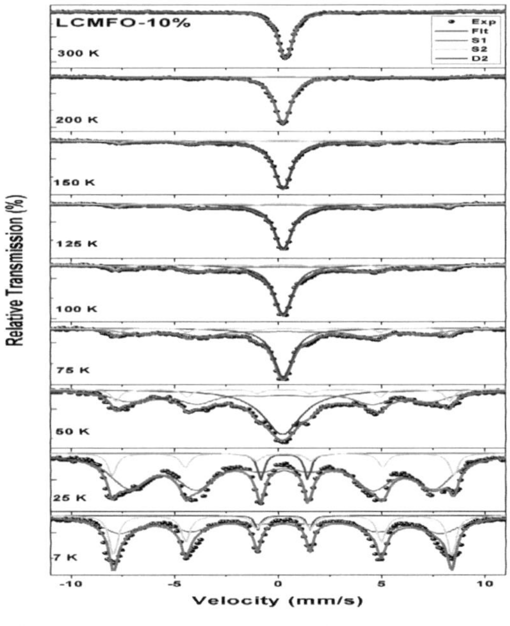

Figure 3. MS of LCMFO-10% at temperatures from 7 to 300K. Closed circles are experimental data, colour lines are different components (S1, S2 and D2) and red line is the fit.

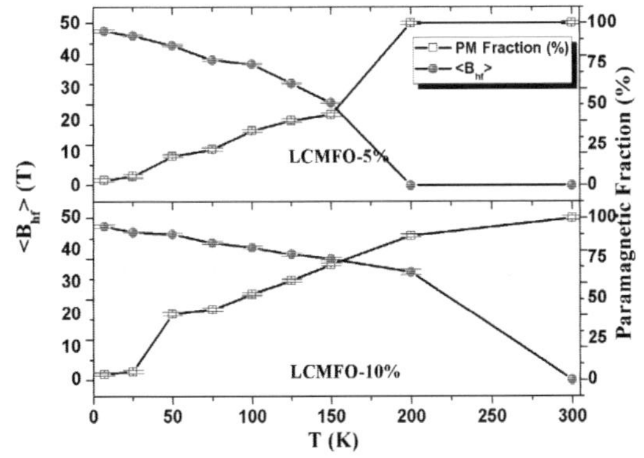

Figure 4. Average hyperfine magnetic field $<B_{hf}>$ and paramagnetic fraction versus temperature for LCMFO-5% and LCMFO-10% samples.

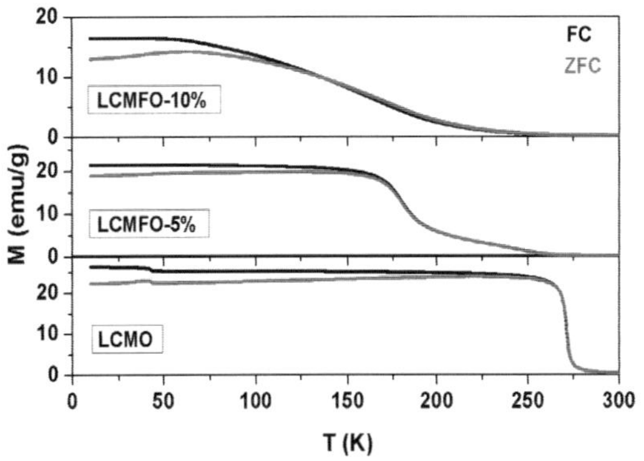

Figure 5. Magnetization M(T) versus temperature (ZFC and FC modes) in a 500 Oe field for LCMO, LCMFO-5% and LCMFO-10% samples.

4. Conclusions

Fe^{3+} ions not only introduce an important antiferromagnetic superexchange interaction, but also lead to a substantial decrease in the double exchange mechanism. The possible explanation for the temperature evolution of the Mössbauer spectra and the persistent paramagnetic contribution in our system is the presence of ferromagnetic clusters with size distribution. These clusters with small sizes display paramagnetic behaviour, caused by the fast fluctuation of the magnetic moment at a certain temperature and as the temperature decreases, the fluctuation decreases and the cluster displays magnetic hyperfine splitting. Finally, the distortion of the octahedron according to the Fe by Mn substitution could result in a tilting of the magnetic moments; this may explain the decrease in the Curie temperature, consistent with Mössbauer and magnetic measurements.

References

[1] Zener C 1951 *Phys Rev* **81** 440
[2] Cai J W, Wang C, Shen B G, Zhao J G and Zhan W S 1997 *Appl Phys Lett B* **71** 1727
[3] Ogale S B, R Shreekala, Bathe R, Date S K, Patil S I, Hannoyer B, Patil F, et al 1998 *Phys Rev B* **57** 7841
[4] Lampen-Kelley P, Kamzin A S, Romachevsky K E, Hue D T M, Chinh H D, Skiranth H and Phan M H 2015 *J Alloys Comp* **636** 323
[5] Simopoulos A, Pissas M, Kallias G, Devlin E, Moutis N, Panagiotopoulos I, Niarchos D, Christides C and Sonntag R 1999 *Phys Rev B* **59** 1263
[6] Hannoyer B, Marest G, Greneche J M, Ravi Bathe, Patil S I and Ogale S B 2000 *Phys Rev B* **61-14** 9613
[7] Pissas M, Kallias G, Devlin E, Simopoulos A and Niarchos D 1997 *J Appl Phys* **81-8** 5770
[8] Wang X, Xu Z and Ma R 1999 *Sci China Ser A* **42-7** 750

Spectroscopy characterization and quantum yield determination of quantum dots

S N Contreras Ortiz[1], E Mejía Ospino[1] and R Cabanzo[1]
[1] Universidad Industrial de Santander, Bucaramanga, Colombia.

E-mail: silviancontreras@gmail.com , rcabanzo@uis.edu.co

Abstract. In this paper we show the characterization of two kinds of quantum dots: hydrophilic and hydrophobic, with core and core/shell respectively, using spectroscopy techniques such as UV-Vis, fluorescence and Raman. We determined the quantum yield in the quantum dots using the quinine sulphate as standard. This salt is commonly used because of its quantum yield (56%) and stability. For the CdTe excitation, we used a wavelength of 549nm and for the CdSe/ZnS excitation a wavelength of 527nm. The results show that CdSe/ZnS (49%) has better fluorescence, better quantum dots, and confirm the fluorescence result. The quantum dots have shown a good fluorescence performance, so this property will be used to replace dyes, with the advantage that quantum dots are less toxic than some dyes like the rhodamine. In addition, in this work we show different techniques to find the quantum dots emission: fluorescence spectrum, synchronous spectrum and Raman spectrum.

1. Introduction

In the last 20 years, quantum dots (QDs) have been widely used as luminescent materials. Among their advantages are that they are versatile, and have the tight fluorescence bands, high quantum yield, heat stability and light stability [1]. QDs are formed by hundred to thousands atoms, and their size can be between 1nm and 20nm [2,3]. In this work we used QDs with size between 2nm and 7nm. By changing its size, the fluorescence wavelength can be in the visible spectrum or NIR [4].

The optical properties of QDs can be modelled as a particle in a box model. Energy in the excited state decreases when the particle´s size increases [4,5]. The energy in the excited state depends on the QDs material [6]. We characterized CdTe and CdSe/ZnS with spectroscopy techniques like UV-Vis, fluorescence and Raman [7,8]. The efficiency in the fluorescence mechanism is measured by the quantum yield. It is defined as the relation between the number of emitted photons and the number of absorbed photons [9–11].

2. Experimental procedure

2.1. Preparation of samples

We prepared the samples using 2.5mg of CdTe in water and 2.5mg of CdSe/ZnS in n-hexane. The first samples had a concentration of 100ppm, so we did two dilutions: 40ppm and 20ppm.

2.2. UV-Vis and fluorescence characterization

We used the UV-Vis spectrophotometer Shimadzu Corporation 2401PC with photomultiplier detector R-928. To determine the maximum absorption for each sample, we realised a full scanning from 800nm to 300nm, with monochromator from 190 to 900nm±0,2nm, and a step beam of 0.1nm. We used quartz cell with 10mm optical path.

For the fluorescence spectrum, we used spectrofluorimeter Photon Technology International (PTI) QM-40, xenon lamps and photomultiplier detector 814. We added the samples in the Starna quartz cell with 10mm optical path and 3.5mL nominal volume.

For the spectrum determination, we used the same parameters: slit width in excitation and emission spectrum of 0.25nm, step size of 1nm and integration time of one second.

2.3. Raman characterization

We used the spectrometer LabRam HR Evolution, Horiba Scientific. We used two kinds of laser: blue (473 nm) and red (785 nm). The detector was a charge coupled device with wavelength in UV-Vis and NIR.

3. Results and discussion

3.1. Preparation of quinine sulphate standard

We prepared a $1.25x10^{-4}$M solution of quinine sulphate standard (QS) with sulphuric acid 0,1N as solvent. UV-Vis absorption spectra was obtained in a range from 800nm to 300nm in the UV-Vis spectrophotometer. The highest absorbance was in 347nm.

3.1.1. Fluorescence determination to QS. In order to obtain the fluorescence spectra, we used the wavelength of highest absorbance in UV-VIS (347nm). From this information, we used the sweep range of fluorescence from 353nm to 630nm to determine the highest emission wavelength. It was in 445nm. In order to obtain better results, we used different wavelengths for emission and excitation, and obtained the synchronic spectrum. With the synchronic spectrum, we could see the highest fluorescence with different excitation wavelength and it was in 446nm.

3.2. CdTe characterization

A similar approach was used for the CdTe characterization. We determined the highest absorbance in UV-Vis. In this case, the highest intensity was in 505nm when we did a sweep range in 800-300nm. With this result we determined the fluorescence changing the emission wavelength between 510-620nm. The highest value was in 544nm. Then, we changed the emission and the excitation wavelength and obtained the synchronic spectrum (Figure 1(a)). The highest value was in 547nm. We used Raman spectroscopy because it is a fast and sensitive method to determine changes in atoms vibration [12]. We used red laser light (785nm). The highest peaks were in $83.4cm^{-1}$ and $155cm^{-1}$. They represent tellurium vibrations. In $893cm^{-1}$ there was a peak with lower amplitude due to Cd-Te vibrations (Figure 1(b)).

3.3. CdSe/ZnS

Using a similar approach, we used the UV-Vis spectroscopy to determine the highest absorption. In this case it was in 511nm. With this sample, we had to thin more, because with the sweep range in UV-Vis, the sample grew up more than the allowable range. Therefore, this sample was diluted in the concentrations: 40ppm and 20ppm.

Then, by changing the excitation and the emission wavelength, we got the synchronic spectrum with the highest data in 531nm (Figure 2(a)). For the Raman characterization, we found two special peaks: one in $143cm^{-1}$ owing to Cd-Se interactions, and the other in $202.4cm^{-1}$ corresponding to the shell interactions (Zn-S) (Figure 2(b)).

3.4. Quantum yield determination

For the quantum yield (QY) we used the equation below. Where F and F_{std} are the fluorescence areas in the sample and the standard; A and A_{Std} are the absorbance in the sample and the standard and; n and n_{std} are the refraction index [13].

Journal of Physics: Conference Series **687** (2016) 012097 doi:10.1088/1742-6596/687/1/012097

We used the QS because it is very common to determine the QY for good stability. QY was 0.546 corresponding to 54.6% [14]. In the table below, we can see the data for QY determination.

$$\Phi_F = \Phi_{F(Std)} \cdot \frac{F \cdot A_{Std} \cdot n^2}{F_{Std} \cdot A \cdot n_{Std}^2} \quad (1)$$

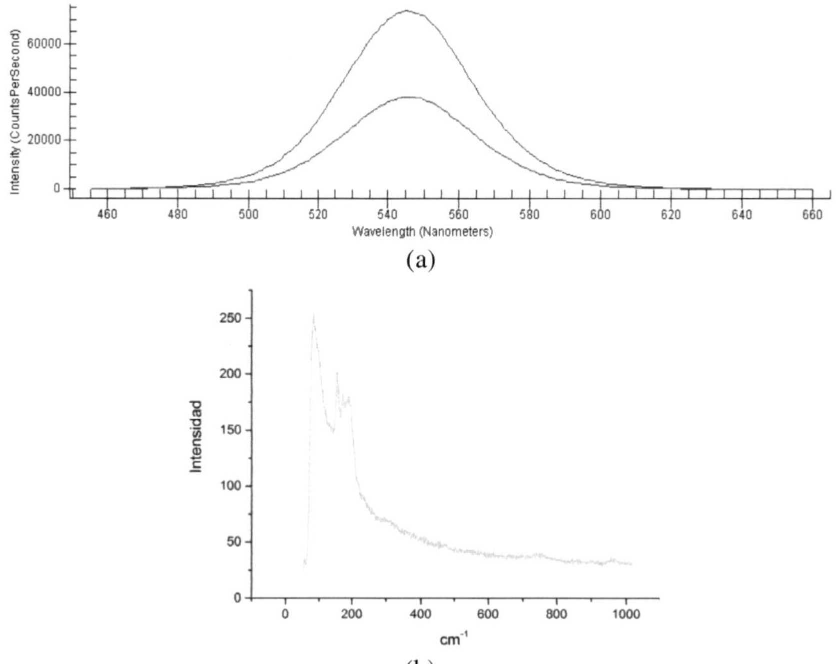

Figure 1. (a) Synchronic spectrum, the green line the CdTe in 100ppm and the violet line is in 40ppm. (b) Raman spectrum with different vibrational modes.

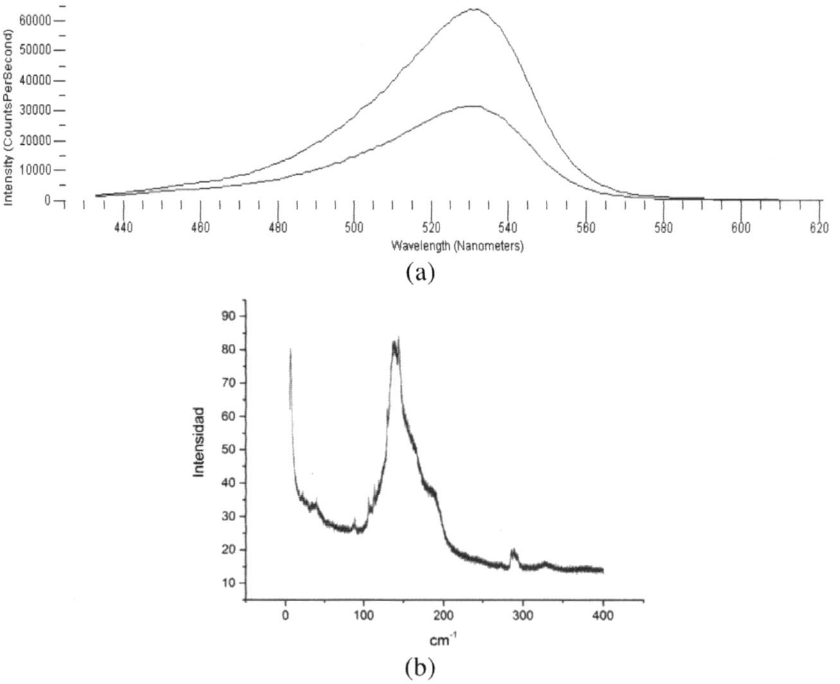

Figure 2. (a) Synchronic spectrum, the purple line the CdSe/ZnS in 20ppm and the green line is in 40ppm. (b) Raman spectrum with different vibrational modes.

Table 1. Data of different wavelength for each sample and standard, and the quantum yield in each case.

Sample		Solvent	Refractive index	λ absortion	λ emission	λ excitation	λ synchronoun	Quantum yield
Quinine sulfate		H_2SO_4	1.63	347nm	445nm	359nm	446nm	0.546
CdTe	100ppm	Water	1.33	506nm	544nm	339nm	545nm	0.465
CdTe	40ppm	Water	1.33	505nm	545nm	326nm	546nm	0.465
CdSe/ZnS	40ppm	n-hexane	1.38	511nm	527nm	339nm	530nm	0.497
CdSe/ZnS	20ppm	n-hexane	1.38	511nm	525nm	340nm	531nm	0.497

4. Conclusion

The CdSe/ZnS has a better quantum yield (49.7%) than the CdTe (46.5%). It is because the CdSe has a shell that gives it more stability and fluorescence. We could see that the QDs are very fluorescent and have a good quantum yield. This is important because in the future, we can use the QDs as substitution to dyes.

From the conducted experiment, we can conclude that the CdSe/ZnS is a better marker than the CdTe. However, in biological applications it is better to use CdTe, as it is hydrophilic and has less toxicity.

Acknowledgments

The authors gratefully acknowledge the support of this research by LEAM and the Universidad Industrial de Santander.

References

[1] Thomsen C and Hens Z 2013 *Chem Mater* **25-12** 2388
[2] Liu B R, Winiarz J G, Moon J-S, Lo S-Y, Huang Y-W, Aronstam R S and Lee H-J 2013 *Colloids Surfaces B Biointerfaces* **111** 162
[3] Alivisatos A P 1996 *Science* **271** 933
[4] Lakowicz J R 2006 *Principles of fluorescence spectroscopy 3rd ed* (Baltimore: Springer)
[5] Mattoussi H, Palui G and Na H Bin 2012 *Adv Drug Deliv Rev* **64** 138
[6] Laverdant J, Marcillac W D de, Barthou C, Chinh V D, Schwob C, Coolen L, Benalloul P, Nga P T and Maître A 2011 *Materials* **4** 1182
[7] Ibnaouf K H, Prasad S, Hamdan A, Alsalhi M, Aldwayyan A S, Zaman M B and Masilamani V 2014 *Spectrochim Acta - Part A Mol Biomol Spectrosc* **121** 339
[8] Scheerschmidt K, Conrad D, Kirmse H, Schneider R and Neumann W 2000 *Ultramicroscopy* **81** 289
[9] Crosby G A and Demas J N 1971 *J Phys Chem* **75** 991
[10] Xiao Z, Liu D, Tang Z, Li H and Yuan M 2015 *Mater Lett* **148** 126
[11] El-Nahass M M, Youssef G M and Noby S Z 2014 *J Alloys Compd* **604** 253
[12] Schreder B, Schmidt T, Ptatschek V, Spanhel L, Materny A and Kiefer W 2000 *J Cryst Growth* **214** 782
[13] Masilela N and Nyokong T 2011 *J Photochem Photobiol A Chem* **223** 124
[14] J C Scaiano 1989 *Handbook of organic photochemistry* (Florida: CRC Press)

The relationship between ferroelectric domain pattern and properties of PLT21 ceramics

F A Londoño[1], J A Eiras[2] and D Garcia[2]

1 Universidad de Antioquia, Medellín, Colombia.
2 Universidade Federal de São Carlos, São Carlos, Brasil.

E-mail: fernandoa.londono@udea.edu.co

Abstract. The light passing through a ferroelectric crystal tends to split into several separate beams due to refraction and reflection at the domain walls. This interaction of light with ferroelectric domains becomes one of the most important features of modern optoelectronic and nonlinear optics materials. Recently, lanthanum modified lead titanate, $Pb_{(1-x)}La_xTiO_3$, (PLT) has become popular because it possesses interesting properties such as a lower Curie temperature, a lower coercive field, and smaller remanent polarizations than PZT and has great potential for nonlinear optical and electro optical applications. In this work, we propose a simple model taking into account the optical transmission and rearrangement of the ferroelectrics domains structure on the PLT21 ceramic in temperature function. The variation of dielectrical and optical measurements in function of temperature were observed and correlated with the optical visualization of ferroelectric domains.

1. Introduction

Lead titane $PbTiO_3$ (PT), a ferroelectric material, is known for its interesting properties; high Curie temperature, pyroelectric coefficient, and spontaneous polarization, and low dielectric constant. These properties make it suitable for numerous applications: ultrason transducers thermistors, optical electronic devices etc [1]. At room temperature, PT has a tetragonal perovskite structure, and combined with other oxides it forms materials such as (Pb,La)TiO_3, (PLT) [1,2]. $Pb_{(1-x)}La_xTiO_3$ is a ferroelectric system with excellent dielectric and piezoelectric properties [3]. However, this material can also be obtained as transparent ferroelectric ceramics for relatively high La content, $0.18 \leq x \leq 0.22$ [4,5]. Transparent ferroelectric ceramics (TFC) have recently acquired a high interest due to their potential application as electro optics devices [6]. The degree of transparency of TFC decrease due to extrinsic defect (as example, pores and secondary phases at the grain boundaries) and intrinsic defect (as example, ferroelectric domains pattern) [7]. Therefore, the main objective of this work was to correlate the optical and electrical response with domain pattern behavior as a function of the temperature for the ceramics $Pb_{(1-x)}La_xTiO_3$ with x=0.21 (PLT21), (PLT21 has major transmittance that the other compositions [4]) prepared through the solid-state reaction method.

2. Experiments

The raw materials PbO purity (99.3%), TiO_2 (99.9+%), and La_2O_3 (99.9%), were weighed according to the formula $Pb_{(1-x)}La_xTiO_3$, with x=0.21. The primary oxides were ball-milled

Journal of Physics: Conference Series **687** (2016) 012098 doi:10.1088/1742-6596/687/1/012098

in distilled water, with zirconia as the grinding medium, for 3h. The resulting slurries were then dried at 120°C and later calcined at 850°C for 3h in air. The calcined powders were ball-milled in distilled water for 3h and dried again. Binder was added to the powders, which were uniaxially cold-pressed into disks (10mm in diameter and 10mm thick), labeled according to nominal composition, PLT21 (x=0.21). Subsequently, isostatic pressure was applied to reduce the density gradients. After burning out the binder, the disks were hot-pressing at 1220°C, for 3h, under a uniaxial pressure of $\sim$ 6.0 MPa, in O_2 atmosphere (30kPa).

The purity of the phases and the structural parameters were determined by X-ray diffraction (XRD) using a Rigaku diffractometer, with CuK radiation. The lattice parameters were calculated by least-square fitting from the positions of the diffraction peaks. The ideal density was calculated from the unit cell volume found by XRD analysis and the molecular weight. For dielectric measurement, gold electrodes were deposited on both faces of the disk samples (5 mm diameter and $\sim$ 1.0mm thick). The relative permittivity was measured by impedance analysis (HP4194A). The transmittance was measured in a spectrophotometer (Micronal-B582), at wavelengths ranging from 200 to 1000nm for optically polished samples, 600m thick [8]. The visualization of the domains was carried out with a microscope Nikon, Eclipse ME600, interfaced to a control system of temperature Linkam (room temperature to 200°C). The microscopic patterns are studied using transmittance light. Magnifications in the 100X to 200X range were realized for domain visualization in ceramics with thickness 0.1mm width 4mm and length 5mm.

3. Results

The results revealed, homogeneous microstructure, grain size of $(10.3\pm0.8)\mu m$, high optical transmittance and high relative density (97.8%) [4, 7]. In the Figure 1(a) can be seen DRX patterns of PLT21 ceramic. It is observed and calculated the absence of second phases with perovskite structure and tetragonality factor, c/a=1.013. The anisotropy as measured by the tetragonality of the unit cell, c/a can be considered interest for electrical properties studies [9].

The temperature profile of real relative permittivity under frequency of electric field 1kHz for PLT21 is shown in Figure 1(b). The value of relative permittivity ϵ' is increased with increasing temperature and reach a maximum value at 129° C, then it is decreased with increasing temperature. High dielectric permittivity and only transition can be seen. This measurement is typical of material ferroelectrics normal with phase transition (ferro-para transition) [9, 10].

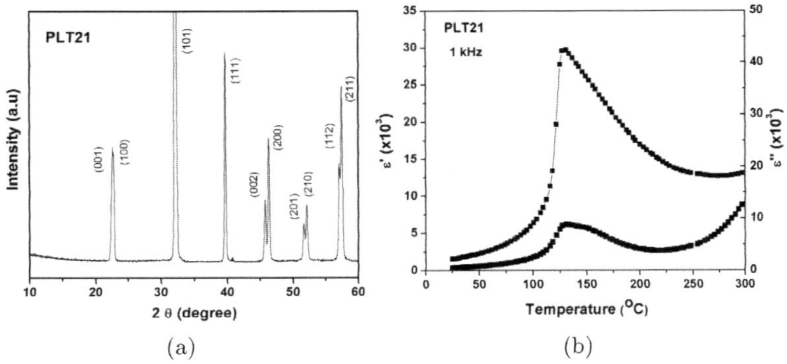

Figure 1. (a) XRD patterns of PLT21 ceramics to room temperature, (b) temperature dependence of ϵ' and ϵ', measured at 1kHz for PLT21 ceramic.

Figure 2 shows optical micrograph to different temperatures of PLT21 ceramic. Also in these figure is inserted the dielectrical measurements in function of temperatures. In the first micrograph can be observed clearly many 180° domains under polarized light to 27°C, however, the domains disappear with the increase of temperature. In the same way, for temperatures higher than the the phase transition temperature (129°C) it is can be seen as the ferroelectric domain disappears completely, however, the domains reappear when the temperature decrease. This observation can be used for the relationship between ferroelectric domain pattern and dielectrical properties of PLT21 ceramics, especially in determining the phase transition temperature of normal ferroelectrics materials.

The most important result of this observation is that we can easily and quickly distinguish ferroelectric domains of PLT21 ceramics.

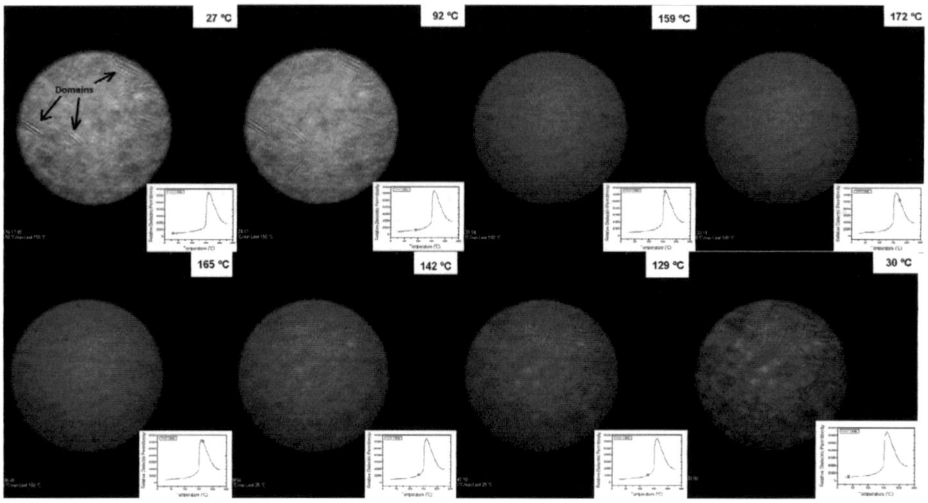

Figure 2. Relationship between ferroelectric domain pattern and dielectrical properties of PLT21 ceramics.

4. Conclusions

The assembly described in this paper operates as a light beam deflector. The deflection is accomplished by total internal reflection of the beam at the boundaries of a 180° domain wall in a ceramic and can be used for the determination of phase transition temperature in normal ferroelectrics materials.

This work is the first in the optical visualization of ferroelectric domains of the system PLT21.

5. Acknowledgements

This research was partially supported by Colombian Agencies: CODI-Universidad de Antioquia (Estrategia de Sostenibilidad 2014-2015 de la Universidad de Antioquia and project: "Síntesis y caracterización de cerámicas transparentes multifuncionales para aplicaciones en dispositivos electro-ópticos y para la construcción de matrices para láseres de estado sólido"), FaCEN-Universidad de Antioquia and to CAPES, FAPESP and CNPq for the financial support.

References

[1] K Limame and M Kellati and S Sayouri 2005 *Moroccan Journal of Condensed Matter* **6-1** 44
[2] D Garcia, F A Londono and J Eiras 2010 *Revista Latinoamericana de Metalurgia y Materiales* **31-1** 52
[3] Vasco E, Vzquez L and Zaldo C 1999 *Applied Physics A* **69-1** S827
[4] D Garcia, F A Londono and J Eiras 2012 *Boletn de la Sociedad Espaola de Cermica y Vidrio* **51-6** 353
[5] Yamamoto T, Igarashi, Hideji, Okazaki, Kiyoshi 1983 *Journal of the American Ceramic Society* **66-5** 363
[6] A Sternberg 1989 *Ferroelectrics* **91-1** 53
[7] D Garcia, F A Londono and J Eiras 2012 *Optical Materials* **34** 1310
[8] J Eiras, F Milton, F Londono and D Garcia 2012 *Optics and Photonics Journal* **2-3** 157
[9] G Shirane and S Hoshino 1952 *Journal of the Physical Society of Japan* **6-4** 265
[10] Vijendra A Chaudhari and Govind K Bichile 2013 *Smart Materials Research* **2013-147524** 1 (doi:10.1155/2013/147524)

Influence of the pH of the synthesis using sol-gel method on the structural and optical properties of TiO_2

J Jaramillo[1], B A Garzón[1] and L Tirado Mejía[1]
[1] Universidad del Quindío, Armenia, Quindío, Colombia.

E-mail: jjaramillo@uniquindio.edu.co

Abstract. The photocatalysis process using semiconductor materials, in particular TiO_2, is one of the most attractive treatment for polluted waters decontamination because of its advantages over other oxidation processes [1-6]. In this study the effect on the physical properties of TiO_2 due to the pH used during the manufacturing of the semiconductor is studied. Different samples were synthesized using ammonium hydroxide (NH_4OH) and nitric acid (HNO_3) as catalysts to provide basic and acid pH environments, respectively. Changes in composition, structure and morphology of the samples were studied and its dependence with the pH of the synthesis is discussed. Results indicate that the base catalysis favours the formation of anatase TiO_2 crystalline phase with crystallite size ~26nm obtained by Rietveld refinement; the spherical particles formed agglomerates of ~100nm; the average pore size is in the range of mesopores and the surface area increases with the amount of NH_4OH added in the process. On the other hand, with acid catalysis, a mixture of two crystalline phases, anatase and rutile, was obtained with crystallite sizes around 26 and 49nm, respectively. The grain size is several orders of magnitude higher than those obtained by basic catalysis. The photocatalytic activity was measured using methylene blue solutions to determine their degradation with radiation. Greater efficiency was observed in the photocatalysts synthesized with NH_4OH.

1. Introduction

The sol-gel technique is one of the most common methods used in research laboratories to grow TiO_2 [7-9]. The size and the entanglement of particles are strongly dependent of the acidity or basicity of the system (pH dependent). Due to the small particle size, the Van der Waals interaction is significant; this interaction increases exponentially as the particle size decreases. These facts favor the growth of clusters. In order to conserve the sample in a powder morphology with a high surface area, it is necessary to stabilize the gel. For low pH values a positive charged transition state is generated and groups –OR from the alkoxide are able to stabilize the solution better than the –OH groups. The low-hydrolyzed species will exhibit a high condensation velocity favoring the formation of long low-branched chains and big size particles. Instead, if the catalyst is basic, the transition state charge is negative, the –OH group stabilizes the solution and the high-hydrolyzed species will condense faster, favoring the formation of highly branched chains and small sized particles [10-11]. In this work we present the influence of the synthesis-pH on the physical properties of TiO_2 semiconductor grown by sol-gel method. The photocatalytic behaviour of both type of samples was compared measuring the degradation of methylene blue (MB) solutions exposed to solar radiation.

2. Experimental procedures

The TiO_2 photocatalyst was synthesized by the sol-gel method using alkoxide tetraisopropyl orthotitanate ($Ti(OC_3H_7)_4$) 98%, and as catalyst HNO_3 70% for the acid catalysis, and NH_4OH 28-30% for the basic catalysis. Because of the higher photocatalytic activity of basic samples in a MB solution exposed to solar radiation, in this work samples with different NH_4OH concentration are studied.

2.1. Sample preparation

Under N_2 flow, 3mL of alkoxide were added to 15mL of isopropanol. Solution was stirred for 20min. Then, a mixture of water 3mL, isopropanol 3mL and catalyst was added to the solution, and stirred for 20min. The resulting preparation was aged for 48h. After it, sample was dried at 100°C for 24h, calcinated at 500°C for 2h and finally, macerated. In this work, the chosen sample grown under acid catalysis with 0.3mL of HNO_3 70% is named A-TiO_2. Samples grown under basic catalysis with different NH_4OH volumes (vmL) are named B-TiO_2(v). The (v) values are: 0.1, 0.2, 0.3, 0.4, 0.5, 0.6, 0.7, 0.8, 0.9, 1.0, 1.5; 2; 2.5.

2.2. Photocatalyst characterization

The determination of the functional groups was achieved with a FTIR Shimadzu Prestige 21 using the middle infrared range. The presence of the TiO_2 crystalline phases was determined by X-ray diffraction analysis. Samples were measured in an X-ray diffractometer Bruker D8 Advance, with Cu anode, over a 2Θ range of 20°-80° using a 0.02° step. Crystallite size and phase percentage were calculated by Rietveld refinement using TOPAS P3. Morphological and chemical analysis were obtained by SEM and EDS using a JEOL JSM-6490LV at 25keV, and TEM images were obtained in a JEOL-1200EX with 80keV and an amplification of 200kX.

The surface area was measured in an ASAP 2020Micrometrics by the BET model measuring the adsorbed nitrogen at 77K previous degasification of samples at 180°C during 1h. Measures were made on the range from 0.01 to 0.99 for the relative pressure (P/P_0).

2.3. Photocatalytic activity characterization

Samples were prepared adding to the photocatalyst (50mg) 30mL of MB solution at 20ppm. All samples were exposed to solar radiation during 5h from 11 am to 4pm, for 11 cycles, under natural conditions in our geographic region. The calculated mean intensity was 27000 luxes. The evaluation of the photocatalytic activity was obtained by the measurement of MB concentration in the solution after light exposure using UV-Vis absorption technique. The absorbance value at 664nm is the standard used to measure the colorant degradation percent $[((C_0-C_t)/C_0)*100\%]$.

One of the main goals in the decontamination processes is to use the same photocatalyst sample several times. In this work 11 times was repeated the discoloration process of a MB solution with the same photocatalyst. After each cycle, solutions were dried to recover the photocatalyst and a new solution of MB was added. In the eleventh cycle (last cycle), 1mL aliquot was taken each 30min during the first 2.5h. Each aliquot was diluted in 10mL of water and filtered with a 0.22μm Fisherbrand filter to avoid the absorption due to the presence of solid particles in the solution. Then, absorbance at 664nm was measured.

3. Results and discussion

3.1. Characteristics of A-TiO_2 and B-TiO_2 samples

FTIR measurements are presented in Figure 1 for B-$TiO_2$0.3 and A-TiO_2 samples synthesized in this work, and for commercial TiO_2. All spectra exhibit the 700 and 500cm^{-1} bands associated to the Ti-O bond vibrations. Synthesized samples present bands in 3500cm^{-1} and 1630cm^{-1} produced by O-H bonds, attributed to the presence of H_2O probably adsorbed in the TiO_2 lattice or to residuary titanium hydroxide groups. The line shape of the B-sample is near to the commercial TiO_2 spectrum. In the X-ray diffractograms (Figure 2) it is observed the presence of crystalline phases, anatase (A) and rutile (R) in the A-TiO_2 sample, and just one, anatase, in the B-$TiO_2$0.3 sample.

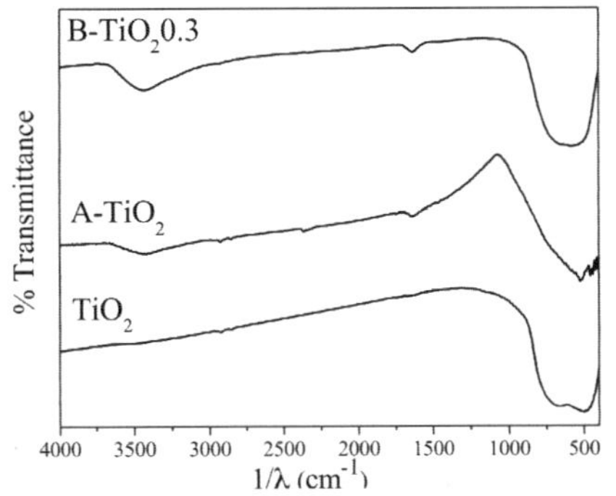

Figure 1. FTIR spectra in the MIR range of TiO_2 photocatalyst obtained by basic catalysis and acid catalysis, and commercial TiO_2.

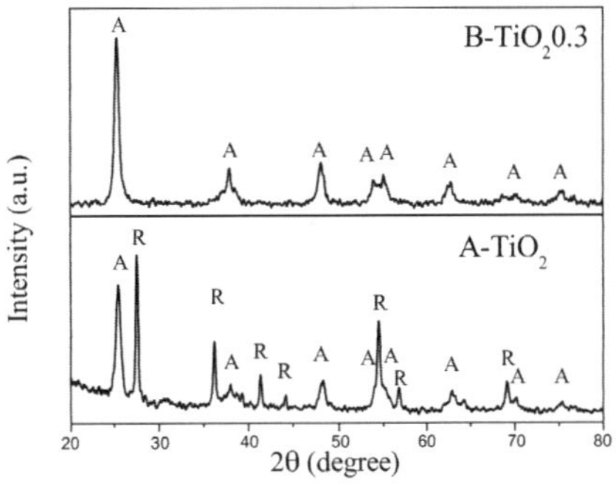

Figure 2. X-ray diffractograms of B-$TiO_2$0.3 and A-TiO_2. The diffraction peaks of the anatase phase are labeled with A and the rutile one, with R.

The images obtained by SEM (Figure 3) evidenced a great difference between the morphology of samples synthesized using basic catalyst and those obtained using acid catalyst. For the first ones the grains are well defined, spherical and, as the catalyst concentration increases the size of grains decreases. The average grain size value is ~0.55μm for B-$TiO_2$0.1 and ~0.20μm for B-$TiO_2$2.5. For samples synthesized using acid catalyst, grain size are ~2400μm and have no regular forms, nor regular sizes.

(a) (b) (c) (d) (e)

Figure 3. SEM micrographs of (a) B-$TiO_2$0.1, (b) B-$TiO_2$2.5, and A-TiO_2 (c), and TEM images of B-$TiO_2$2.5 (d) and A-TiO_2 (e).

TEM images (Figure 3) exhibit spherical particles for B-samples, homogeneous in their form, with diameters ~9nm, and they are incorporated in clusters of ~200nm diameter. A-sample also exhibit spherical particles but much bigger than B-samples, with diameters about 98nm. These morphologic characteristics are responsible for the high surface area per weight unit: less grain size, higher surface area, higher contact surface with the dye, higher degradation activity. A-TiO_2 sample have morphologic characteristics quite different from the B-TiO_2 samples; A-samples have big grains formed by two different crystalline structures. By Reitveld refinement of the diffraction patterns it was obtained a crystallite size for rutile about 49.1nm and for the anatase phase, 26.3nm.

In agreement with morphologic analysis, the porosimetry results indicated that B-samples have a higher surface area (74.79m^2/g) than A-sample (40.94m^2/g), unless the porous size is in the mesoporous range for both.

The chemical composition of samples was obtained by EDS analysis. In these spectra it was evidenced the presence of Ti, O, and C, probably as a residuary element of titanium alkoxide used in the synthesis process.

3.2. Photocatalytic activity and stability
The photocatalytic activity of samples was determined through the evolution of the degradation percent on MB solution after radiation exposure. As is shown in Figure 4, the degradation rate decreases with time. This behaviour is explained considering two mechanisms in the first stages of the process: degradation due to the photocatalytic activity and adsorption of the dye by the porous microstructures.

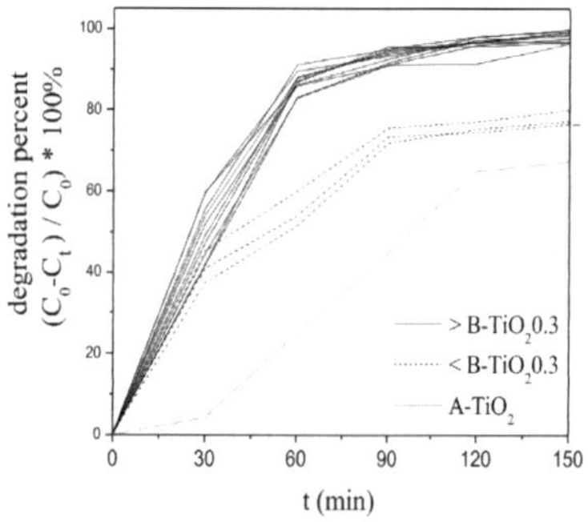

Figure 4. Photocatalytic activity of samples synthesized with acid catalysis (A-TiO_2) and samples synthesized with basic catalysis with different NH_4OH concentration (B samples).

As the process continues, the porous could be saturated but the photocatalyst remains active. In Figure 5, for 150min it is evidenced a difference between three types of samples: sample synthesized with acid catalysis, samples synthesized with basic catalysis with NH_4OH concentration 0.3 or below, and samples synthesized with basic catalysis with NH_4OH concentration higher than 0.3. The A-TiO_2 sample exhibited a low efficiency degradation percent (67%). The B samples synthesized with low NH_4OH concentration (≤0.3) presented a higher efficiency (78±2%) than A sample. And the efficiency degradation percent for samples with higher NH_4OH concentration (B-$TiO_2$0.4) was 96% or higher. The degradation percent obtained for B-$TiO_2$0.6 sample was 99.9%.

4. Conclusions
TiO_2 photocatalyst was obtained by the Sol-Gel method. In this work we use two different catalyst in the synthesis process: HNO_3 (acid) and NH_4OH (basic). The TiO_2 synthesized with the acid catalyst exhibit two crystalline phases, rutile and anatase. The average grain size is 2400μm. The morphologic characteristics of these samples are in agreement with the low surface area obtained by BET analysis and its low photocatalytic activity. Instead, TiO_2 samples synthesized using a basic catalyst exhibit just one crystallographic phase, anatase. The grain size are not higher than 0.7μm, value that became lower as the base concentration increases. These small sizes are related to a high surface area, a high adsorption area and a high efficiency photocatalytic activity. The stability of samples as well as its high efficient photocatalytic activity is demonstrated by the results obtained after eleven cycles.

References
[1] Wang P, Xie T, Wang D and Dong S 2010 *Journal of Colloid and Interface Science* **350** 417
[2] Han H and Bai R 2010 *Separation and Purification Technology* **73** 142
[3] Tryba B 2008 *Journal of Hazardous Materials* **151** 623
[4] Lu S Y, Wu D, Wang Q L, Yan J, Buekens A and Cen KF 2011 *Chemosphere* **82** 1215
[5] Matilsinen A and Sillanpää 2010 *Chemosphere* **80** 351
[6] Magalhães F, Moura F, Lago R 2011 *Desalination* **276** 266
[7] Yao N and Yeung K L 2011 *Chemical Engineering Journal* **167** 13
[8] Chong M N, Jin B, Chow C and Saint C 2010 *Water Research* **44** 2997
[9] Huang D G, Liao S J, Liu J M, Dang Z and Petrik L 2006 *Journal of Photochemistry and Photobiology A Chemistry* **184** 282
[10] Brinker C J and Scherrer G W 1990 *Sol-gel science: The physics and chemistry of sol-gel processing vol 1* (San Diego: Elsevier) p 112
[11] Malato S, Fernández P, Maldonado M, Blanco J and Gernjak W 2009 *Catalysis Today* **147** 1

Measurement of nonlinear refractive index of organic materials by z-scan

M Fontalvo[1], A Garcia[1], S Valbuena[2] and F Racedo[1]
[1] Universidad del Atlántico, Barranquilla, Colombia.
[2] Universidad de la Costa (CUC), Barranquilla, Colombia.

Email: fran@mail.uniatlantico.edu.co

Abstract. The nonlinear effects characterization by using the Z-Scan transmission technique in many materials has generated great in forest according to the technological necessities. The majority part of the nonlinear effects can be described by the classic electromagnetic theory, with the electrical susceptibility in the constitutive equation that relates the electrical polarization with the electrical field. In this work the sign and refractive index magnitude and the nonlinear absorption coefficient of the following organic substance were determined: methylene-blue, rodamine LD, vegetable powder and gentian violet a hundred percent pure dissolved in isopropyl alcohol, a laser Nd: YAG was used as a source excitation. The bunch of laser was focused with a lens of ten centimeters of focal length; by using a displacement system the sweeping of twenty centimeters was realized. The following results of the normalized curves of the transmittance in function of the z position were obtained applying the *Sheik- Bahae* theory: The nonlinear refractive index of the dye shows an increase in function of its concentration and the power of exciting of the laser with negative nonlinear sign in the majority of the sample.

1. Introduction

With the creation of the laser in 1960 by Theodore H. Maiman [1,2] was possible to produce luminous beams with high intensities and when they interact with the matter, they presented behaviours that were not described by the theories before known, Fraken and his collaborators observed that a crystal of quartz emits light with the double of the frequency of the incident light that is transmitted by a laser of ruby creating in this way one of the areas of the physics, The nonlinear optic. Giving beginning to the study of the phenomenon that happen as consequence to the modification of the optical properties of a material when is exposed to high densities of energy.

2. Z–Scan

Z–Scan technique proposed by Sheik Bahae [3,4,5] and his collaborators is based on the spatial distortion of one beam due to the self-modulation of phase of the sample. It is one of the most used techniques by scientists who study the phenomenon related with the nonlinear optic in the determination of the positive no linearity (self–focalization) and negative (self-defocusing) in organic samples such as the dyes (methylene-blue, vegetable powder, rodamine LD and gentian-violet) in its pure state and dissolved in isopropyl alcohol.

In the basic configuration of the Z–Scan technique (Figure 1). The nonlinear refractive index n_2 of a sample is measured through the simple relation between the variation of the transmittance of the potency observed in a distant field and the distortion of phase induced acquired when it passes through the sample. The potency transmitted through the aperture with radius r is captured when one Gaussian beam crosses a sample that moves along its propagation axis, the sample moves around the focal position to receive different intensities generating in the material one of the nonlinear response known as effect Kerr.

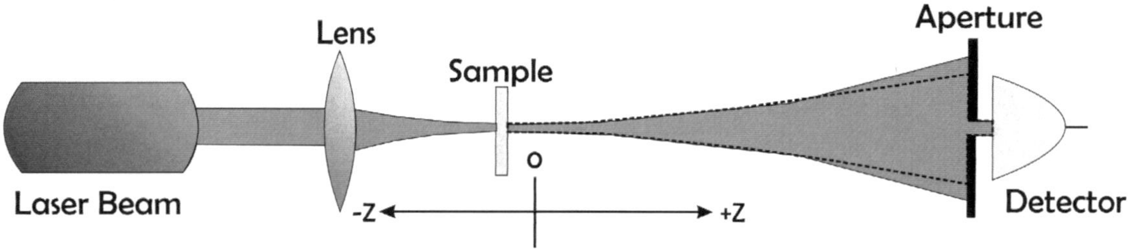

Figure 1. Schematic diagram of experimental arrangement for the Z-scan technique.

The linear effects are eliminated when the potency transmitted to a sample in a position Z is divided by the potency transmitted when the simple is located far away of the focus.

$$T(z) = \frac{P(z)}{P(\infty)} \quad (1)$$

When the beam passes through the nonlinear sample suffers a change in its phase, this difference can be determined through the equation:

$$\Delta\Phi_o = kn_2I_oL_{eff} \quad (2)$$

Where I_o is the initial intensity of the laser beam. L_{eff} is the thickness of the sample, S is the diameter of the aperture, P is the potency of the incident beam and $k$is vector of wave.

$$\Delta T_{pv} = 0.406(1 - S)^{0.25}|\Delta\Phi_o| \quad (3)$$

ΔT_{pv} is the difference between the transmittance pick-valley or valley pick and this variation amplitude of transmittance pick-valley is proportional to the distortion of phase [6,7,8].

$$\Delta T_{pv} = 0.406|\Delta\Phi_o| \quad (4)$$

Starting from the equation (3) it can be determined and if $S << 1$ is easy to show that the nonlinear refractive index (n_2) is given by the following expression [9,10]:

$$\Delta Z_{pv} = 1.7z_o \quad (5)$$

$$z_o = \frac{k\omega_o^2}{2} \quad (6)$$

3. Experimental arrangement

The experimental arrangement is similar to the (Figure 1) A laser of Nd: YAG of emission continuous with a longitude of wave of $532nm$, a stable potency of $50mW$ and a $L_{eff} = 1mm$ was used. The minimum radius of the laser $\omega_o = 63.04\mu m$ and $I_o = 800.973\ W/cm^2$ was determined.

The beam was focalized with a lens of 10cm of focal distance, the wide area detector was put 20cm from the position of the lens on the axis of sweeping Z and previously an iris of $0.01mm$ of diameter was placed in order to capture only the intensity around the optical axis getting in that way a better sensibility in the measurements. Cells with 1mm of thickness were used to measure the nonlinear properties of the dyes: methylene blue, vegetable powder, gentian violet and rodamine LD. Several concentrations dissolved in isopropyl alcohol were prepared, and the nonlinear refractive index was determined for each one of them.

4. Results and experimental analysis

The study in four different sample of dyes was done. Methylene-blue, vegetable powder, gentian-violet and rodamine LD. During the measuring, a potency of excitation of the laser beam of $50mw$ was kept and sweepings of $20cm$ with rising of $0.21mm$ in a bank of line displacement were carried out, the sweepings were performed of +z to –z.

The Figure 2 correspondent respectively to the normalized curves of transmittance for: methylene-blue, vegetable powder, gentian-violet and rodamine LD.

Using the equations (5) and (6) can be obtained the value of w_0 in an experimental way.

$$\omega_o^2 = \frac{2\Delta Z_{pv}}{1.7k} \tag{6}$$

These values were replaced in the equation (1) and solving, in this way we obtained:

$$n_2 = \frac{\Delta T_{pv}\Delta Z_{pv}\lambda^2}{2.7608\pi P L_{eff}} \tag{7}$$

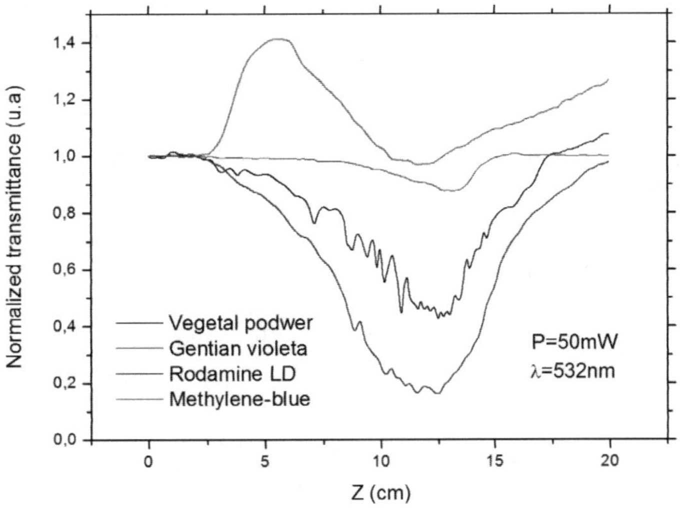

Figure 2. shows normalized curves of the transmittance in function of distance Z(cm) for the dyes.

The methylene-blue by using a concentration of 0.02 dissolved in $12ml$ of isopropyl alcohol with a value of the nonlinear refraction index (n_2) of 6.43084×10^{-8} and negative sign of (n_2). For gentian-violet was used a concentration of 0.0002 dissolved in $106ml$ of isopropyl alcohol with a refraction index of 5.7014×10^{-7} and positive sign, in the rodamine LD was used 0.0058 dissolved in 18ml of isopropyl alcohol, the refraction index was calculated 5.9495×10^{-7} and the negative sign and finally the refraction index for 0.1308 vegetable -powder was calculated 2.99869×10^{-7}.

We studied the concentration of 0.03 M.B with potency's variations of laser, and we obtained this results: In $30mW$ we obtained a refraction´s index 1.54×10^{-11}, we increased the potency´s laser in 40 and $50mW$, obtaining this results respectly 8.79261×10^{-8} and 8.84293×10^{-8}. This results shows a increase of nonlinear refractive index. as shown in Figure 3 and in Table 1

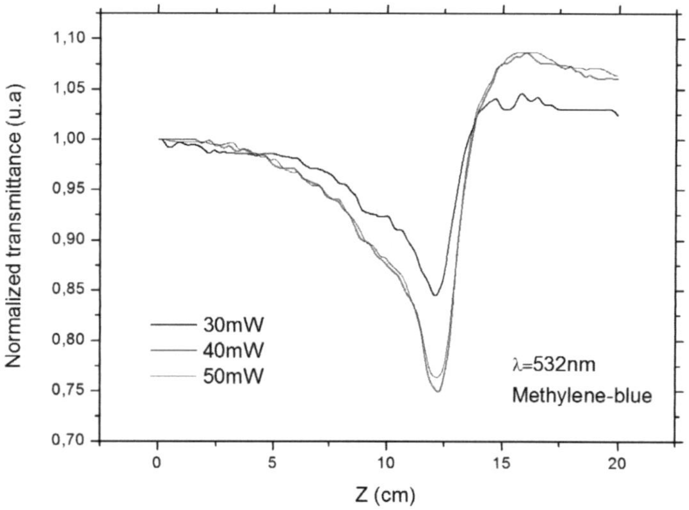

Figure 3. Variation of nonlinear refractive index with the incident beam intensity for the methylene-blue.

Table 1. The measured values of nonlinear refractive index.

Dye	ΔT_{pv}	ΔZ_{pv}(cm)	n_2(cm^2/W)
Blue of methylene	0.179839	5.46274	$6{,}43084\text{x}10^{-8}$
Vegetable powder	0.665161	6.88702	$2{,}99869\text{x}10^{-7}$
Rodamine LD	0.837419	10.8534	$5{,}9495\text{x}10^{-7}$
Violet gentian	1.41396	6.15986	$5{,}7014\text{x}10^{-7}$

5. Conclusion

The nonlinear properties of the dye were characterized through the technique Z–Scan for $\lambda = 532nm$. The normalized curve of transmittance for the nonlinear refractive index, magnitude and sign for different concentrations of the dye dissolved in ethyl alcohol were obtained showing negative nonlinear refractive index or positive according the cases. The obtained results demonstrate a variation in the nonlinear optical properties of the dyes.

References

[1] R Boyd 2007 *Nonlinear optics 3rd ed* (New York: Academic Press)

[2] M Sheik-Bahae, A A Said, T Wei, D J Hagan, E W V Stryland 1990 *IEEE J Quantum Elect* **26** 760769

[3] A García, S Valbuena, R sarmiento, F Racedo 2015 *Óptica Pura y Aplicada* **48-1** 55

[4] New Geoffrey 2011 *Introduction to nonlinear optics* (New York: Cambridge University Press)

[5] Piovesan E, De Bonl L, Mendoca C R 2006 *Efeitos ópticos lineares e não lineares em derivados de perilenos - workshop da pós-graduação do IFSC* (São Carlos: IFSC)

[6] Gomez S L, Cuppo F L S and Figueiredo A M 2003 *Brazilian Journal of Physics* **33-4** 813

[7] Linares R y Nicho Diaz M A 2010 *Sociedad Mexicana de Ciencias y Tecnología de superficies y Materiales, Superficie y Vacío* **23-S** 85

[8] Tingchao He Yongguang Cheng, Yabing Du, Yujun Mo 2007 *Optics Communications* **275-1** 240

[9] Qusay Mohammed Ali and P K Palanisamy 2005 *Optik International Journal for Light and Electron Optics* **116-11** 515

[10] A Anwaar AL- Dergazly and Aqueel Fadhel 2010 *Advanced research in physics and engineering nanotecnology Proccedings of the 2nd WSEAS International Conference on Nanotechnology* (UK: WSEAS Press) pp 144-149

Considerations on electrical impedance measurements of electrolyte solutions in a four-electrode cell

C V Chaparro[1], L V Herrera[1], A M Meléndez[1] and D A Miranda[1]
[1] Universidad Industrial de Santander, Bucaramanga, Colombia.

E-mail: dalemir@uis.edu.co

Abstract. A tetrapolar probe to measure the electrical properties of electrolyte solutions was implemented with gold electrodes according to the van der Pauw method. Electrical impedance spectroscopy (EIS) measurements of different concentrations of phosphate buffer saline (PBS) solution and an oral mucosal tissue sample dispersed in PBS were performed in the galvanostatic mode using a four-electrode cell (tetrapolar probe). Taking advantage of using a potentiostat/galvanostat for carrying out the electrical measurements, a simple and rapid method using a three-electrode electrochemical cell is described for: a) cleaning of electrodes, b) verification of surface chemical state of electrode material and c) choice of current supplied to electrodes for EIS measurements. Results of this research shown a depolarization effect due to the addition of oral mucosa tissue cells into the PBS solution.

1. Introduction

Electrical impedance spectroscopy (EIS) is a powerful method to study electrical properties of electrolytic solutions and biological tissues. Electrical properties of biological materials can be studied by two-, three- and four-electrode cells [1]. Conductivity measurements with two and three electrodes are used, however they depend on the accuracy of standards [2,3], and also they exhibit electrode polarization impedance effects. Four-electrode arrays minimize the influence of electrode polarization at low frequencies [3]. A four-electrode cell based on var der Pauw method, originally used to measure the electrical resistivity of semiconductor materials, allows obtaining an averaged reading of the impedance measurements of electrolyte solutions [4]. Furthermore, EIS is a non-invasive characterization technique that can be used to study the frequency response of the electrical impedance of biological tissues [5,6].

The electrical measurements for determining van der Pauw DC resistivity in a four-electrode cell setup require a current source and a voltmeter or electrometer. In AC measurements an impedance analyser is required. Nowadays, a growing number of laboratories possess a potentiostat/galvanostat, which is an instrument designed for a wide range of electrochemical applications: cleaning of metallic surfaces, characterization and evaluation of materials or liquid solutions, as well as determination of their thermodynamic, kinetic, electronic or electrical properties. An impedance analyser is frequently integrated to potentiostat/galvanostat, thereby expanding the possibilities of this instrument through measurement. Taking advantage of aforementioned capabilities, herein is proposed a methodology for prepare the measuring electrodes and how to set the measurement conditions.

2. Experimental

2.1. Preparation of solutions and tissue sample

Both solutions, piranha H_2SO_4/H_2O_2 (7:3) and 0.5M H_2SO_4, were prepared for chemical and electrochemical cleaning of gold electrodes. Electrolyte solution of phosphate buffered saline (PBS) was prepared with a final concentration of 0.680M NaCl, 0.0134M KCl, and 0.0405 Na_2HPO_4 adjusted at pH 7.4 with HCl (hereinafter referred to as "1 time solution"), and it was diluted to the appropriate concentration. All solutions were prepared in deionized water from a Milli-Q ultrapure water system with a resistance of 18.2MΩ. All chemical reagents used were analytical grade. Mucosa tissue cell samples were collected by rolling the cotton swab firmly on the inside of the cheek. The swab was then placed in a 100μL volume of 1 time PBS solution and the sample was then dislodged by shaking. The final volume of the aliquot was 500μL by adding 1 time PBS solution.

2.2. Elaboration of four-electrode cell

Four Au wires (18 kilates, 1mm of diameter and 15mm in long) were welded each one in an angle of 90° to a silver wire with the same dimensions (Figure 1). Electrodes were chemically cleaned by dipping into a piranha solution for 5 minutes, followed by rinsing with copious amounts of deionized water. Each electrode was electrochemically cleaned by cyclic voltammetry (see below). After that, electrodes were rinsed with deionized water. The cell was elaborated in Teflon as is shown in Figure 1.

Figure 1. Gold electrode and cylindrical four-electrode cell. Front view (top) and side view (bottom) of the cell.

2.3. Electrochemical and electrical measurements

Cyclic voltammetry measurements were performed in a conventional three-electrode cell at room temperature. Dissolved oxygen was removed from all solutions by bubbling nitrogen for 15min. The working electrode was each one of prepared gold electrodes. The counter electrode was a high purity graphite rod (99.9995%, 5mm in diameter) and an electrode Ag/AgCl/3.0M KCl placed in a Luggin capillary was used as reference. The galvanostatic mode (current constant) impedance measurements were conducted in the four-electrode cell shown in Figure 1, the way to connect electrodes to potentiostat is indicated below. EIS measurements were carried out at $1x10^{-7}$A between 10Hz and 0.1MHz. A volume of 500μL of each solution was used for the PBS impedance measurements. All electrochemical and electrical measurements were performed with a potentiostat/galvanostat Autolab PGSTAT302N and NOVA 1.10.4 software.

3. Results and discussion

In spite of chemical cleaning of gold electrodes with piranha solution, cyclic voltammetry measurements showed a different electrochemical behaviour among electrodes. Indeed, different electrochemical reactivity evidence differences at the surface of each electrode, which could lead to thermoelectric effect during the impedance measurements. In order to avoid possible interference, oxidation-reduction cycling in 0.5M H_2SO_4 was done for each electrode by scanning the potential between -0.1 and 1.6V at a scan rate of $0.1Vs^{-1}$, the cycling was stopped until to obtain a reproducible and characteristic electrochemical behaviour (data not shown). According to anodic behaviour of gold in PBS solutions, it can be oxidized if the current supplied is large enough, hence, current must not exceeded about $1x10^{-7}A$ to ensure that impedance measurements are performed under linear conditions.

Impedance measurements with a set of four electrodes can minimize the errors resulting from polarization effects [2]. However, it is important to properly connect the terminals of the potentiostat to each of the four electrodes of the cell. Improper electrode connection gives incorrect results. In a four-electrode configuration, current is delivered through two drive electrodes and voltage is measured via two sensing electrode. When a potentiostat/galvanostat is used, current should be supplied through working and counter electrodes, and thereby voltage is measured at reference electrode and working electrode sense, as illustrated in Figure 2. Thus, the conductivity of electrolyte solution is obtained by measuring the impedance and the geometry of the cell, resistivity values were calculated from Equation (1) [7].

$$\varrho(\omega)=Z(\omega)\,\pi h/\ln 2 \qquad (1)$$

Where $\varrho(\omega)$ is the electrical resistivity (ω is the frequency), $Z(\omega)$ is the impedance of the measured electrolyte solution, and h is the height of the liquid filling the cylindrical cell, to be 10mm.

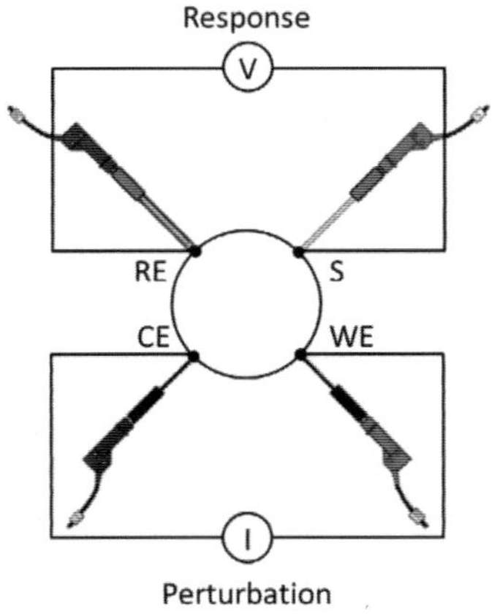

Figure 2. Schematic view of a tretapolar setup (van der Pauw geometry) for resistivity measurements of electrolyte solutions whether a potentiostat/galvanostat is used. RE, reference electrode; S, working sense; CE, counter electrode; WE, working electrode. Bare Au wires are used as electrodes (see Figure 1).

Figure 3 shows the resistivity values for 1.0, 0.4, 0.2 and 0.1 times PBS over the frequency range of 10 to 105Hz. All curves show a consistent decrease in resistivity with the increase in PBS concentration. With the exception of 1 time PBS solution, the resistivity/impedance of the others solutions is constant over the entire range of frequencies measured. In the case of 1 time PBS solution, the increase of resistivity at low-frequency values could be due to the polarization effect, which occurs when there is an accumulation of ionic charge at the interface between the electrodes and electrolyte solution [8]. The calculated standard deviation over the frequency range where resistivity is approximately constant is lower than 1.03%. In spite of polarization effect at low frequencies, the high frequency data show that the conductivity (it is the reciprocal of resistivity) of PBS solutions varies linearly with concentration as would be expected for a stock solution and its successive dilution (Figure 4).

EIS measurements of an oral mucosal tissue sample dispersed in 1 time PBS solution was performed. The magnitude of the impedance obtained for a volume of 500μL of oral tissue-PBS mixture is greater than oral tissue-free PBS solution (Figure 3). Indeed, oral mucosal tissue cells are not a simple electrolyte and ion-exchange process are needed to stabilize their membranes [8]. Surprisingly, addition of oral mucosal tissue sample to PBS depolarizes the electrodes, and hence the impedance measurement becomes constant at low frequencies. The impedance of the system gives insight into the adhesive behaviour of the cells on the electrodes.

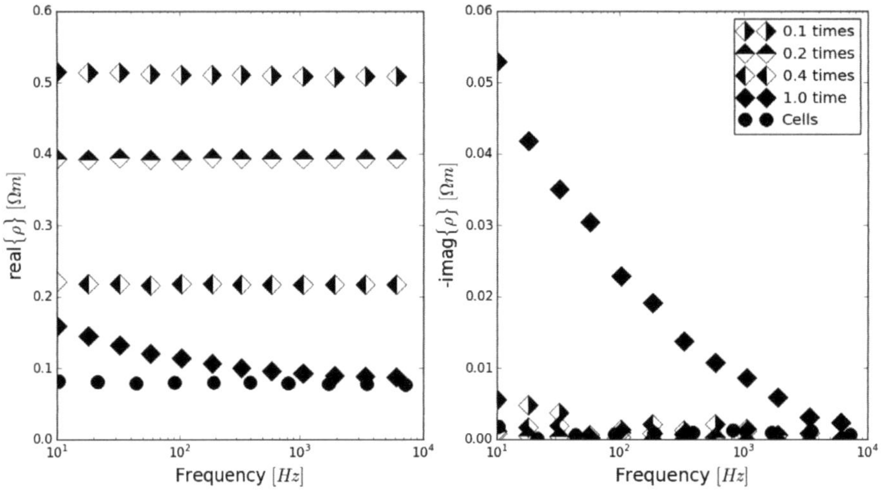

Figure 3. Electrical resistivity, ϱ, measurements for PBS solutions of different concentrations and mucosal tissue cell sample dispersed in 1 time PBS solution.

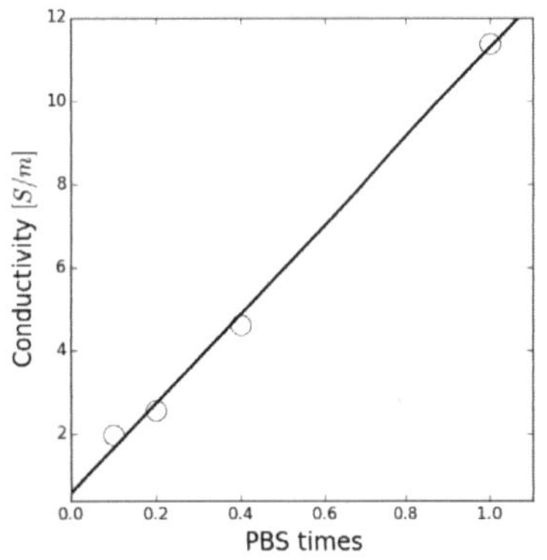

Figure 4. Electrical conductivity of PBS solutions at different concentrations. PBS times equal to 1.0 mean the PBS without dilutions and e.g. 0.2 means five dilutions of PBS.

4. Conclusions

Potentiostat/galvanostat offers a host of features and benefits to establish appropriate conditions for carried out conductivity measurements in a four-electrode cell. EIS measurements can provide useful information about resistivity and conductivity of electrolytic solutions; furthermore it can apply to study ions moving through the cell membranes.

Acknowledgements

This research was supported by COLCIENCIAS and the Universidad Industrial de Santander with the program "Jóvenes Investigadores" in the call 617-2013 and VIE-5742 Grant.

References

[1] Grimnes S and Martinsen Ø G 2015 *Bioimpedance and biolectricity basics 3rd ed* (London: Academic press) pp 163-191
[2] Moroń Z, Rucki Z and Szczepanik Z 1997 *IEEE Trans Instr Meas* **46** 1268
[3] Moroń Z 2003 *Measurements* **33** 282
[4] Moroń Z and Grysinski T 2009 *XIX IMEKO world congress* (Portugal: Fund Appl Metrology) p 1640
[5] Bera T K 2013 *J Medical engineering vol 2014* (Korea: Hindawi Publishing Corporation) pp 2-3
[6] Miranda D, Corzo S, González-Correa C 2013 *J Phys Conf Ser* **434** 012056
[7] Zhang B, Lin Z, Zhang X, Yu X, Wei J and Wang X 2014 *Meas Sci Technol* **25** 055005
[8] Prodan C and Bot C 2009 *J Phys D Appl Phys* **42** 175505

Magnetic effect in viscosity of magnetorheological fluids

H A Fonseca[1], E Gonzalez[1], J Restrepo[2], C A Parra[1] and C Ortiz[1]
[1] Universidad Pedagógica y Tecnológica de Colombia, Tunja, Boyacá, Colombia.
[2] Universidad de Antioquia, Medellín, Antioquia, Colombia.

E-mail: hugoalexander.fonseca@uptc.edu.co

Abstract. In this work the study of viscosity is presented for a magnetorheological fluid made from iron oxides micrometre, under an external magnetic field. The material was characterized by magnetic loops in a vibrating sample magnetometer and its crystal structure by X-ray diffraction. The results show that saturation magnetization and coercive field have dependence with the powder size. The material has different crystal structure which lattice parameters were determined by Rietveld refinement. The viscosity of the magnetorheological fluid was measured by a viscometer with rotational symmetry with and without external field. This result evidence a dependency on the size, percentage iron oxide and the applied magnetic field, it is due to the hydrodynamic volume of iron oxide interacts with the external magnetic field, increasing the flow resistance.

1. Introduction

The magnetorheological fluids are non-Newtonian fluids consisting of magnetic particles, example magnetite, dispersed in a liquid carrier that can change its rheological properties under application of an external magnetic field. Moreover, these fluids are characterized by an apparent viscosity, which decreases as spindle speed increase. Some of its applications are in suspension systems, magnetic separation industry and biomedicine [1-3].

In the preparation this class of fluids, particles of size micrometric or nanometre coated with a surfactant. It is necessary to generate a hydrodynamic volume that increase surface interaction between the magnetic particles and the carrier liquid and thus to reduce effects of aggregation and interaction in the presence of external magnetic field. These particles can be achieved chemically (cooprecipitation) or mechanical (milling) processes.

2. Experimental

Iron oxide powders were characterized by vibrating sample magnetometer to have an appreciation of behaviour magnetic and X-ray diffraction to quantify your composition and to know the crystalline structure. The magnetorheological fluid was prepared with micrometre-sized powders (T_5=15μm y T_2=40μm), oleic acid was used as surfactant, and the oxide/oleic acid ratio was kept at 1/1.5 in volume and SAE engine oil as carrier liquid, the samples are labelled T_2(15%), T_2(20%) and T_5(20%), the percentage value corresponds to the amount of oxide in the ferrofluid. The samples were heat treated (45°C up to 60°C in oleic acid and up to 50°C in engine oil SAE) in a mechanical shaker at cycles of 0.5, 1, 1.5, 2 hours to 43rpm. This process an activation chemical surface and suspension stability of magnetite in the liquid carrier is present due to the surfactant. The viscosity measurement was performed for various spindle speed without magnetic field and to different values of magnetic field with constant shearing rate at room temperature.

3. Result and discussion

In order to detect and quantify the phases, we refined the X-ray pattern of the sample by the Rietveld method. The experimental pattern as shown in Figure 1 was fitted using two phases, 89% magnetite, and 11% hematite, with cubic (Fd3m) and hexagonal (R3c) structures, respectively. With a refined adjustment of χ^2=1.905.

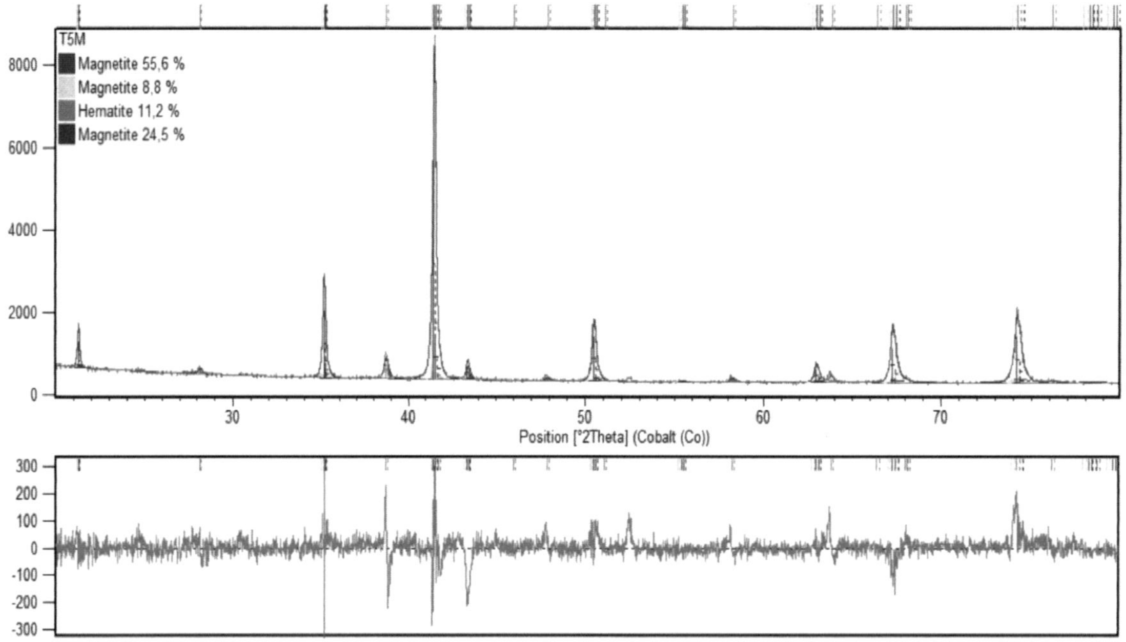

Figure 1. X-ray diffraction pattern and Rietveld fit of iron oxide sample.

In the Rietveld refinement it was necessary to introduce three cubic structures magnetite and a hematite hexagonal structure, which lattice parameters, are listed in Table 1.

Table 1. Refined lattice parameter of phases presents in the iron oxide: three of magnetite and one of hematite sample.

	Structural parameters		
Phase	a	b	c
	8.3893(4)	8.3893(4)	8.3893(4)
Magnetite	8.420(3)	8.420(3)	8.420(3)
	8.373(3)	8.373(3)	8.373(3)
Hematite	5.026(1)	5.026(1)	13.735(3)

The hysteresis loops show that the magnetic parameters such as saturation magnetization, coercive force and remanent magnetization are influenced by the magnetic domains state of the samples, which in turn it is a function of grain size [4,5]. That is, as it is increase the size of the oxide, saturation magnetization lowers (Figure 2) which values are within the range of a magnetically soft material as shown in Table 2, which is an important factor in the development of magnetorheological fluids. The magnetic behaviour of the oxide is due to the contribution of the crystal structures present in 89% magnetite and hematite 11%. Hematite presence generates slight changes in the magnetic properties of the sample compared to a pure magnetite [4,5].

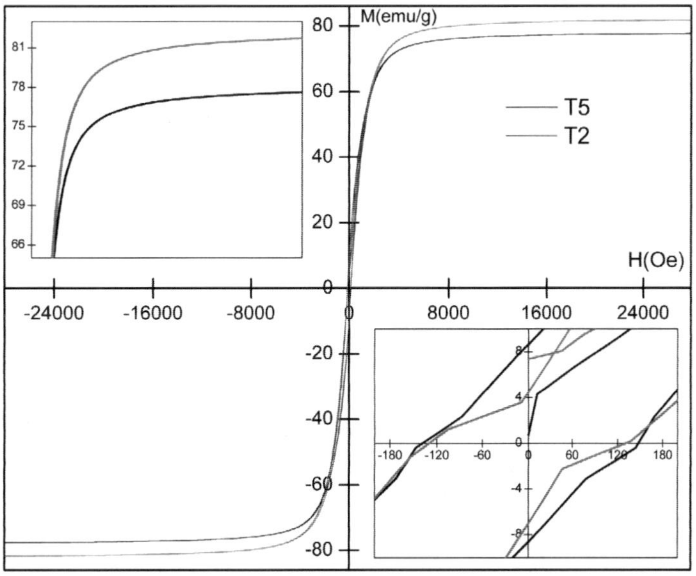

Figure 2. Iron oxide Hysteresis loops.

Table 2. Characteristic magnetic parameters: saturation magnetization Ms, remanent magnetization Mr, coercive field Hc.

	T_2	T_5
Ms (emu/g)	81.59	77.53
Mr (emu/g)	7.02	8.67
Hc (Oe)	132.2	144.62

The behaviour viscosity with spindle speed without applying field is shown in Figure 3(a). The viscosity versus spindle speed evidences a non-Newtonian behaviour. The values increase when the size of the oxide decreases, indicating better stability of oxide-oleic acid-oil [6,7]. The hydrodynamic volume is uniform and stable with the oxide size decrease, increasing flow resistance [8].

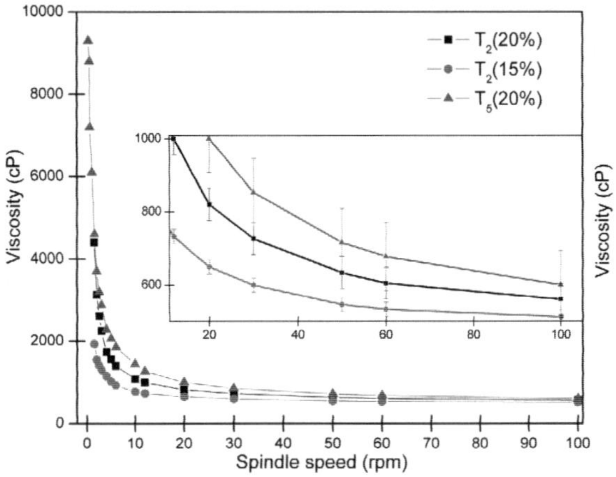

Figure 3(a). Viscosity behaviour varying the spindle speed with absence of magnetic field.

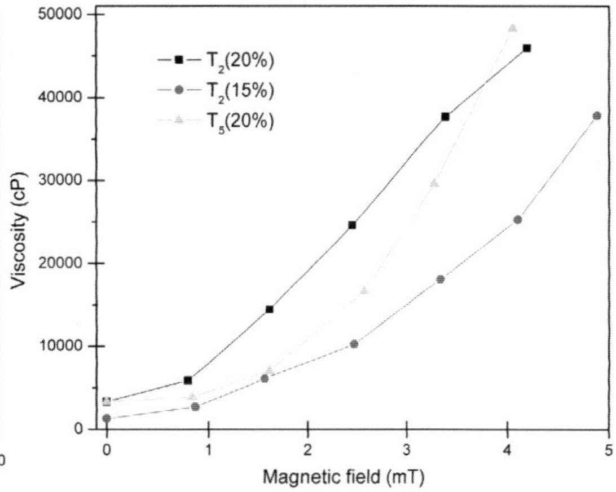

Figure 3(b). Viscosity behaviour varying the magnetic field with spindle speed to 2rpm.

In the Figure 3(b) show the viscosity in function of field applied for the spindle speed in this case was 2rpm. The change appears in several orders of magnitude in viscosity, due to micrometre oxide tends to align the applied field, increasing the torque in the spindle which registers the viscosity of the magnetorheological fluid [8] and it is caused by the formation of a robust chain structure.

The behaviour observed in Figure 3(a) can be characterized by a Bingham model with a yield stress, meaning that the suspensions acts as a solid-liked material when exposed to an external shear stress [9,10]. Regarding the oxide volume concentration, the viscosity increases due to the increases the hydrodynamic volume net giving the material a better response to the magnetic field. In the case of size of oxide, the viscosity does not depend on size in the presence of a magnetic field because the powder rotates seeking their preferred orientation. Magnetite, which contributes more in the magnetization, is the one that gives the orientation of the oxide in the fluid due to the external field.

4. Conclusions

The magnetorheological fluids prepared have a dependency on spindle speed acting as a plastic fluid conforms to the model Bingham. A clear dependence of the magnetorheological fluid with the magnetic field applied it was found that, because the micrometre oxide easily responds generating a preferred orientation to the applied field, thus increasing the resistance to fluid flow. Oleic acid surfactant satisfactorily fulfils the function of preventing the aggregation of magnetite due to magnetic interactions, thus behaving as a paramagnetic fluid. The crystal structures give an unusual magnetic response of the sample although their values are within those reported in the literature for a pure magnetite.

Acknowledgments

The authors thank the Centro de Gestión de Investigación y Extensión (CEDEC) of the Faculty of Engineering, call 001 for research groups.

References

[1] Scherer C and Figueiredo Neto A M 2005 *Brazilian J Phys* **35-3a** 718
[2] Yavuz C T, Arjun Prakash, Mayo J T and Colvin V L 2009 *J Chem Eng Sci* **64** 2510
[3] Shokrollahi H 2013 *Mater Sci Eng* **C33** 4485
[4] Christopher P Hunt, Bruce M Moskowitz and Subir K Banerjee 1995 *Magnetic properties of rocks and minerals in rock physics & phase relations: A handbook of physical constants* (Washington: Wiley & Sons)
[5] Özden Özdemir, Dunlop D J and Moskowitz B M 2002 *Earth Planet Sci Lett* **194** 343
[6] Guardia P, et al 2007 *J Mag Magn Mater* **316** e756
[7] López López M T, de Vicente J, González Caballero F and Durán J D G 2005 *Colloids Surf A Physicochem Eng Aspects* **264** 75
[8] Korobko E V, Novikova Z A and Zhurauski M A 2015 *J Phys Conf Ser* **602** 1
[9] Sedlacík M, et al 2010 *Smart Mater Struct* **19** 1
[10] Brookfield Engieering Labs 2014 *More solutions to sticky problems: A guide to getting more from your brookfield viscometer & rheometer* (http://www.viscometers.org/PDF/Downloads/More%20Solutions.pdf)

Measurement of the thermal expansion coefficient of Guadua angustifolia-Kunth using the photoacoustic technique

A Bedoya[1], F Gordillo Delgado[1] and E Marín[2]
[1] Universidad del Quindío, Armenia, Quindío, Colombia.
[2] Instituto Politécnico Nacional, Centro de Investigación en Ciencia Aplicada y Tecnología Avanzada, México D.F., México.

E-mail: adrian_filipo@hotmail.com

Abstract. In this paper, the Linear Thermal Expansion Coefficient of *Guadua angustifolia*–Kunth samples was measured using the Photoacoustic (PA) technique in a heat transmission configuration and considering the thermoelastic bending as a PA signal generation mechanism in addition to the thermodiffusion ones. The obtained value of $(27\pm7)\times10^{-6}K^{-1}$ is a reasonable value compared with that reported for similar materials such as wood.

1. Introduction

Guadua angustifolia–Kunth (GAK) has high population density in Asia and South America [1]. The mechanical and thermal characterization of this plant and by-products have been reported in several works [2,3,4], proving scientifically the reason of why this material is widely used in the construction industry [5] since ancient times, being comparable in several features with concrete and mild steel [6]. Additionally, this natural material is a renewable and sustainable resource and this has application for replacing the wood and for manufacturing Natural Fibres Reinforced Composites (NFRC) [7]. In this paper the measurement of the linear thermal expansion coefficient, α_T, of GAK is reported. Measurements were made using the photoacoustic (PA) technique [8] in heat transmission mode and using a model for data processing that takes into account the thermoelastic bending signal generation mechanism in addition to the heat diffusion process. The thermal diffusivity, α, is reported too.

2. Materials and methods

In this study, Macana biotype GAK was taken from a harvest located in a municipality of La Tebaida, in the Quindío, Colombia, at a height of 1480 meters above the sea level. The middle part of the stem was chosen and divided through its internodals. These sections were dried using a dehumidifier system with controlled temperature (313K) until a moisture content about 15% was reached. For measurement of α and α_T, thin sheets were extracted through a longitudinal cut along the fibre direction in an intermediate region.

The PA technique was used in a heat transmission configuration, in which the intensity modulated laser beam impinges and is absorbed on the sample's surface opposite to that facing the PA air chamber. The generated PA signal was detected with a microphone connected to the air chamber by a small diameter duct and was measured with a Lock-In amplifier synchronized at the light modulation frequency, f.

In order to measure the linear thermal expansion coefficient measurements were performed at frequencies at which the sample is thermally thick (tk), so that the thermoelastic bending (T) signal generation mechanism (dependent on the α_T value) becomes important, being added to the thermodiffusion (D) mechanism (dependent on α). To eliminate the frequency response of the detector, it was necessary to normalize the PA signal from the sample ($S_{D,tk} + S_{T,tk}$) with that of a reference sample, which in our case is in the thermally thin (*tn*) regime ($S_{D,tn} + S_{T,tn}$). The ratio of the amplitudes of both signals is given by [9]:

$$S = \frac{|S_{D.tk}+S_{T,tk}|}{|S_{D.tn}+S_{T,tn}|} = \sqrt{f}\sqrt{\frac{\left(ge^{-\sqrt{\frac{f}{f_c}}}\right)^2 + \left(hC^{1/2}\right)^2 + 2ghC^{1/2}e^{-\sqrt{\frac{f}{f_c}}}\cos\left(-\pi-\sqrt{\frac{f}{f_c}}-\varphi\right)}{\left(1+m^2\sqrt{f}^6+2m\sqrt{f}^3\cos\left(-{}^{7\pi}/_{4}\right)\right)}} \tag{1}$$

Where g, h, m are constants, $C = \left(1-\frac{1}{x}\right)^2 + \left(\frac{1}{x}\right)^2$, $\varphi = \arctan\left(\frac{1}{x-1}\right)$, $x = \sqrt{\frac{f}{f_c}}$ and f_c is the cutoff frequency of the material, which is defined as:

$$f_c = \frac{\alpha}{\pi {l_s}^2} \tag{2}$$

The methodology consists in measurement of S as a function of f. Then a least squares fit of the experimental data using Equation (1) is performed, leaving g, h, m and f_c as fit parameters. From f_c the thermal diffusivity can be determined using Equation (2) if the sample's thickness, l_s, is well known.

On the other hand, α_T can be obtained using the following expression [9]:

$$\alpha_{Te} = \frac{4R_C^2 l_S^2 \alpha_g{}^{1/2}}{3R^4 \alpha_s{}^{1/2} T_0}\frac{h}{g} \tag{3}$$

In which this parameter was redefined as α_{Te}. This experimental value will agree with the expected α_T only under conditions in which the sample is in the ideal thermally thick regime, i.e. when $f_c \to 0$ [9].

Measurements were performed in seven samples of different thicknesses. From the slope of the plot of f_c as a function of l_s^{-2} the thermal diffusivity was determined using Equation (2), while α_T is calculated from the α_{Te} versus f_c graph as the α_{Te} for $f_c \to 0$.

3. Results

The measured values of thicknesses, f_c, α_s and α_{Te} of the samples are presented in Table 1.

Figure 1(a) shows the dependence of f_c on l_s^{-2} from which a value $\alpha = (0{,}11 \pm 0{,}01) \times 10^{-6} m^2 s^{-1}$ was obtained following the methodology described in previous section. This value agrees well with that reported previously [10] for this kind of material and can be considered as a mean value of those obtained for the individual samples (Table 1).

Figure 1(b) shows α_{Te} as a function of f_c. A decreasing behaviour is observed, which can be fitted using the Equation 4.

$$\alpha_{Te} = a_1 e^{\frac{-f_c}{t_1}} + a_2 \tag{4}$$

Where a_1, t_1 and a_2 are fit parameters, for which the values $(25 \pm 7) \times 10^{-6} K^{-1}$, $(2{,}0 \pm 0{,}2) \times 10^{-6} K^{-1}$ and $(0{,}12 \pm 0{,}03)$ Hz were obtained respectively. Note that when $f_c \to 0$, the linear thermal expansion coefficient becomes $\alpha_{Te} = \alpha_T = a_1 + a_2 = (27 \pm 7) \times 10^{-6} K^{-1}$. To the best of our knowledge, there is not a previous report on this parameter. However, the order of magnitude is preserved when the obtained value is compared with that corresponding to some types of wood. For example, the α_T value of wood varies between about $2 \times 10^{-6} K^{-1}$ and $30 \times 10^{-6} K^{-1}$ [11] in the transversal direction, and for composites reinforced with bamboo fibres α_T is larger than $40\times10^{-6}K^{-1}$ [12,13], a value which increases with respect to those corresponding to its separate components (fibre-matrix). Therefore, we believe that the linear thermal expansion coefficient found in this study for GAK is a reasonable value.

Table 1. Results of thermal diffusivity and experimental thermal expansion coefficient of *Guadua angustifolia*-Kunth.

Thickness (μm)	fc (Hz)	α_S (m^2s^{-1})x10^{-6}	α_{Te} (°K^{-1})x10^{-6}
212±8	0.737±0.007	0.104±0.008	2.4±0.4
241±9	0.684±0.001	0.130±0.020	1.7±0.3
254±6	0.510±0.030	0.103±0.005	2.9±0.6
289±9	0.367±0.006	0.963±0.006	3.1±0.9
357±7	0.282±0.002	0.110±0.030	4.5±0.9
423±6	0.216±0.002	0.122±0.003	6.3±0.8
508±9	0.1587±0.0005	0.129±0.005	8.9±04

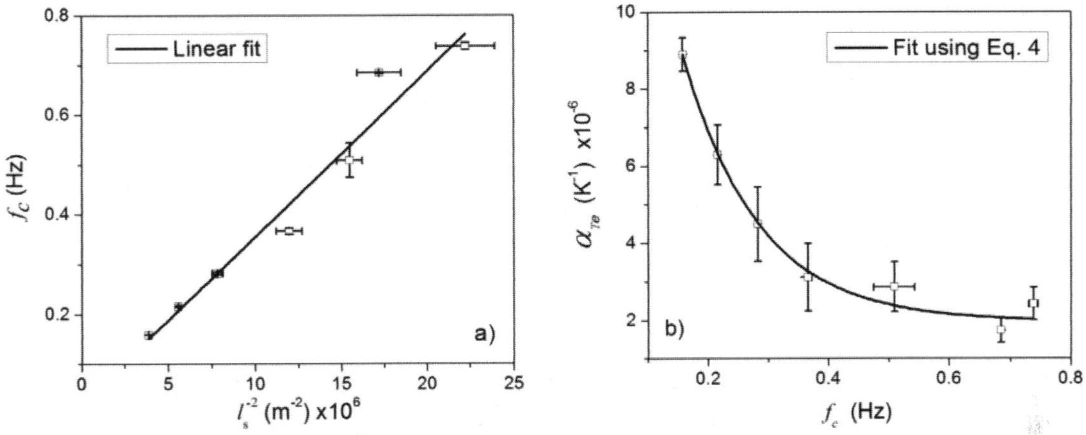

(a) (b)

Figure 1. (a) f_c as a function of l_s^{-2}. The solid line is the result of the best least squares linear fit (b) α_{Te} as a function of f_c and best least squares fit using Equation (4).

4. Conclusions

The linear thermal expansion coefficient of GAK was obtained using the PA technique. The obtained value is comparable with that of similar materials such as wood. This technique has advantages respecting common used methods for measurement of this parameter, because it makes use of samples with small dimensions without any preparation, and without the need of impose large temperature gradients that can affect the samples properties during the measurement. The methodology presented allowed also the measurement of the thermal diffusivity of GAK in agreement with literature reported values.

Acknowledgements

The authors thank to the University of Quindío and COLCIENCIAS by support through the young researcher program. E Marín thanks the support through research grants from SIP-IPN (1638, 20140006 and 20150390), CONACyT and COFFAA-IPN.

References

[1] Lo Y T, Cui H, Tang W C P and Leung H C 2008 *Const Build Mat* **22-7** 1532
[2] Sharma B, Gatóo A and Ramage M H 2015 *Const Build Mat* **83** 95
[3] Nian-Ping L, Ji-bo L, Lin S, Li W and Shan Z 2012 *J Cent South Univ* **19** 600

[4] Xu Q, Harries K, Li X, Liu Q and Gottron J 2014 *Eng Struct* **81** 230
[5] Li Y, Shan W, Shen H, Zhang Z and Liu J 2015 *Thin Walled Struc* **89** 17
[6] Bhardwaj S, Sharma R and Kumar R 2014 J *Eng Comp & Appl Sciences* **3-10** 14
[7] Abdul Khalil H P S, Bhat I U H, Jawaid M, Zaidon A, Hermawan D and Hadi Y S 2012 *Material and Design* **42** 353
[8] Almond D P and Patel P M 1996 *Photothermal science and techniques 1st ed* (London: Chapman and Hall) chapter 4 pp 63-77
[9] Bedoya A, Marín E, Mansanares A M, Zambrano-Arjona M A, Riech I, Calderón A 2015 *Thermochimica Acta* **614** 52
[10] Gordillo-Delgado F, Marín E, Cortés-Hernández D M 2011 *Research Journal of Chemical Sciences* **1-9** 17
[11] Weatherwax R C and Stamm A J 1956 *The coefficients of thermal expansion of wood and wood products* (USA: Forest products laboratory, department agriculture)
[12] Hebel D E, Heisel F and Javadian A 2013 *Proceedings of the Annual International Conference on Architecture and Civil Engineering* 94 (DOI: 10.5176/2301-394X_ACE13.26)
[13] Huang R, Zhang Y, Xu X, Zhou D and Wu Q 2012 *BioResources* **7-4** 4563

Synthesis, structural and morphological characterization of the perovskite $LaYbO_3$

A Cabrera Ramírez[1], F E Charry Pastrana, J Roa Rojas[1], D A Landinez Tellez[1] and F Fajardo
[1] Universidad Nacional de Colombia, Bogotá, Colombia.

E-mail: lacabrerar@unal.edu.co

Abstract. In this work we report synthesis of the $LaYbO_3$ ceramics material by the solid state reaction technique and its structural and morphological characterization from X-ray diffraction and Scanning Electron Microscopy experiments. Rietveld refinement of the diffraction patterns reveals that this material crystallizes in an orthorhombic perovskite, *Pnma* space group (#62) with lattice parameters a=6.0233Å, b=8.2080Å and c=5.7203Å. Systematic monitoring of the synthesis process is carried out through analysis of results of X-ray diffraction. The surface morphology of the samples was qualitatively analysed as a function of the sintering process.

1. Introduction

In recent years perovskite ceramics have been extensively studied due their physical and chemical properties. These mixed oxides have the ABO_3 stoichiometry, where A is an alkaline-earth or rare-earth element, B represents a transition metal or rare-earth element, and O is the oxygen, and exhibit exotic properties that change when making modifications to the structure and composition of the material [1]. Our particular interest to study the $LaYbO_3$ ceramic material is raised in the controversial literature, which report that this material presents and orthorhombic highly distorted structure [2] associated to the symmetry groups Pna2 and Pnma. These conclusions were established from X-ray diffraction (XRD) and neutron diffraction experiments and reported in different works, but there is a discussion still about which of these two groups corresponds to the spatial structure of the $LaYbO_3$ perovskite [3]. This study is also motivated because of the potential applications of the simple lanthanide perovskites, such as its implementation in communications technology to manufacture filters and resonators [4]. In this paper the synthesis process, and the structural and morphological properties of the perovskite $LaYbO_3$ synthesized by the method of solid state reaction, are analysed from information obtained from the of X-ray diffraction technique, Rietveld refinement of the experimental data and Scanning Electron Microscopy (SEM) studies.

2. Experimental

One sample of $LaYbO_3$ material was synthesized by the standard solid state reaction method from the precursor oxides Yb_2O_3 and La_2O_3 (99.99% purity), which were first dehydrated and mixed according with the stoichiometry of the nominal composition $LaYbO_3$. Mixture was submitted to a maceration process in an Agatha mortar for 3,5 hours and successively calcined at 1000, 1100 and 1190°C for 24 hours with intermediate macerations and XRD measurements in a X-Pert Pro MPD Panalytical equipment ($\lambda_{Cu\text{-}K\alpha}$=1.5405Å) up to obtain a single crystallographic phase. A second sample was directly calcined at 1190 °C in order to compare the reflection peaks of the XRD patterns. XRD diffractograms were obtained in the Bragg-Brentano configuration with 2θ from 10° to 80°. Rietveld refinement of the experimental data was performed by using XpertHighScore Plus program and the theoretical model was developed through the Structure Prediction Diagnostic Software SPuDs [5]. The morphological study was carried out from SEM images by the utilization a VEGA3 TESCAN.

3. Results and discussion

Figure 1 shows the diffraction patterns for the synthesis process of the $LaYbO_3$ perovskite from the precursor powders of La_2O_3 (Figure 1(a)) and Yb_2O_3 (Figure 1(b)) to form the mixture of Figure 1(c). Then, a thermal sequence of sintering, at 1000°C (Figure 1(d)), 1100°C (Figure 1(e)) and 1190°C (Figure 1(f)), was developed in order to obtain the single crystallographic phase of the $LaYbO_3$ material. Figure 1(g) corresponds to the sample which was directly calcined at 1190°C without a sequence of sintering process.

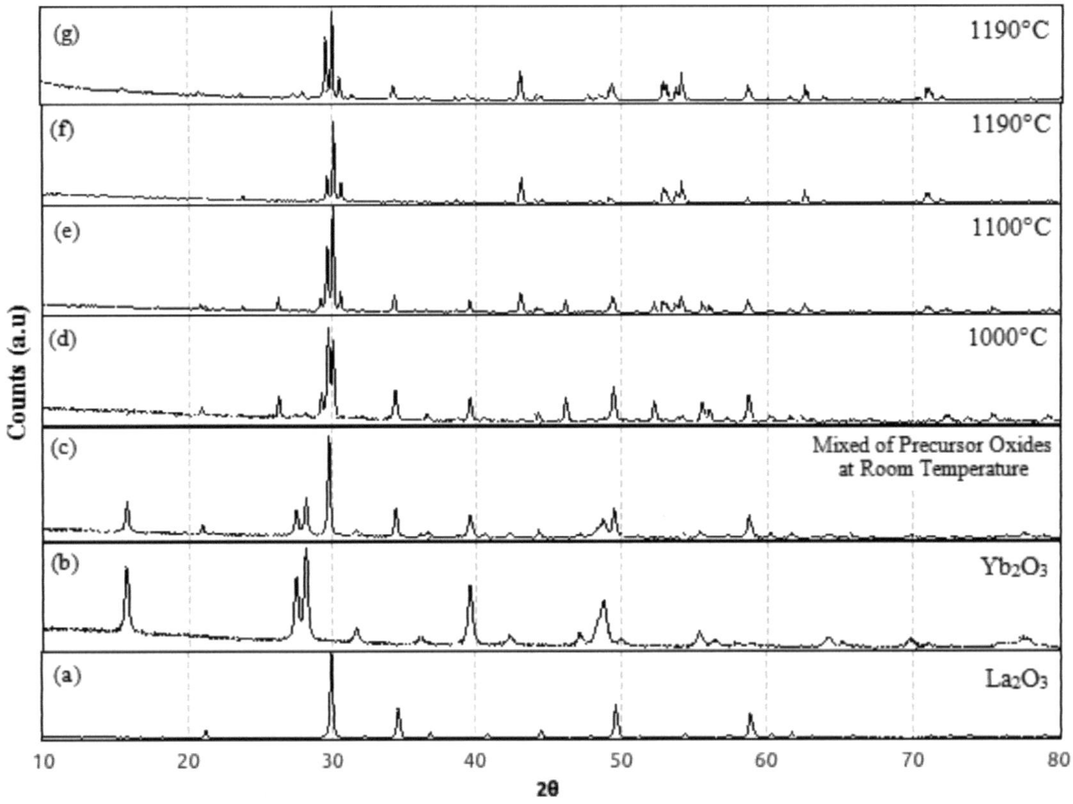

Figure 1. Diffraction patterns for (a) La_2O_3 (b) Yb_2O_3 (c) mixture of precursor oxides at room temperature, and successive sintering process at temperatures (d) 1000°C (e) 1100°C (f) 1190°C and (g) sample directly calcined at 1190°C.

It is observed in Figure 1 that in the thermal treatment, reflection peaks of the precursor oxides disappear progressively and characteristic peaks of the $LaYbO_3$ rise to form a single crystallographic structure. Particularly, Figures 1(a) and 1(b) evidence peaks with high intensity in 2θ=29.94°, 49.65° and 58.90° for the La_2O_3 oxide and in 2θ =15.78°, 27.43°, 28.11°, 39.63° and 48.83° for the Yb_2O_3 oxide. These reflection peaks are clearly observed in the mixture of precursor powders at room temperature (Figure 1(c)). After the calcined and sintering processes at 1000°C and 1100°C, peaks of the precursor oxides appears but new reflections are evident for 29.42° at 1000°C which is dislocated up to 30.03°, and other close to 40° at 1100°C, both due to the Yb_2O_3 oxide. On the other hand, an observed peak in 58° at 1100°C appears from the La_2O_3 precursor. However, other peaks that are not present in the precursor oxides appear for the thermal treatment between 1000°C and 1100°C which are related to the transition to form a new crystallographic phase. These are located in 2θ=26.16°, 29.19°, 46.18° and 52.23°. At last, for T=1190°C, all peaks of the precursor powders disappear and the $LaYbO_3$ perovskite is obtained with the structure previously reported [3]. When the sample is obtained from a single thermal treatment at T=1190°C, the structure is obtained but a difference in the intensities of the diffraction peaks is observed when the diffractogram of Figure 1(g) is compared with the Figure 1(f).

The analysis of the diffraction patterns was performed through the Rietveld refinement, which permits to minimize the difference between the experimental data and a theoretical model calculated from the atomic positions and lattice constants on the Pnma space group, from the analysis of the Global Instability Index given by the SPuDs program [5]. These values are shown in Table 1.

Table 1. Atomic parameters for $LaYbO_3$ used in the theoretical model.

Atom	Site	x	y	z
La	4c	0.6014	0.2500	0.5369
Yb	4b	0.5000	0.0000	0.0000
O	4c	-0.0334	0.2500	0.3613
O	8d	0.3101	0.0694	0.6806

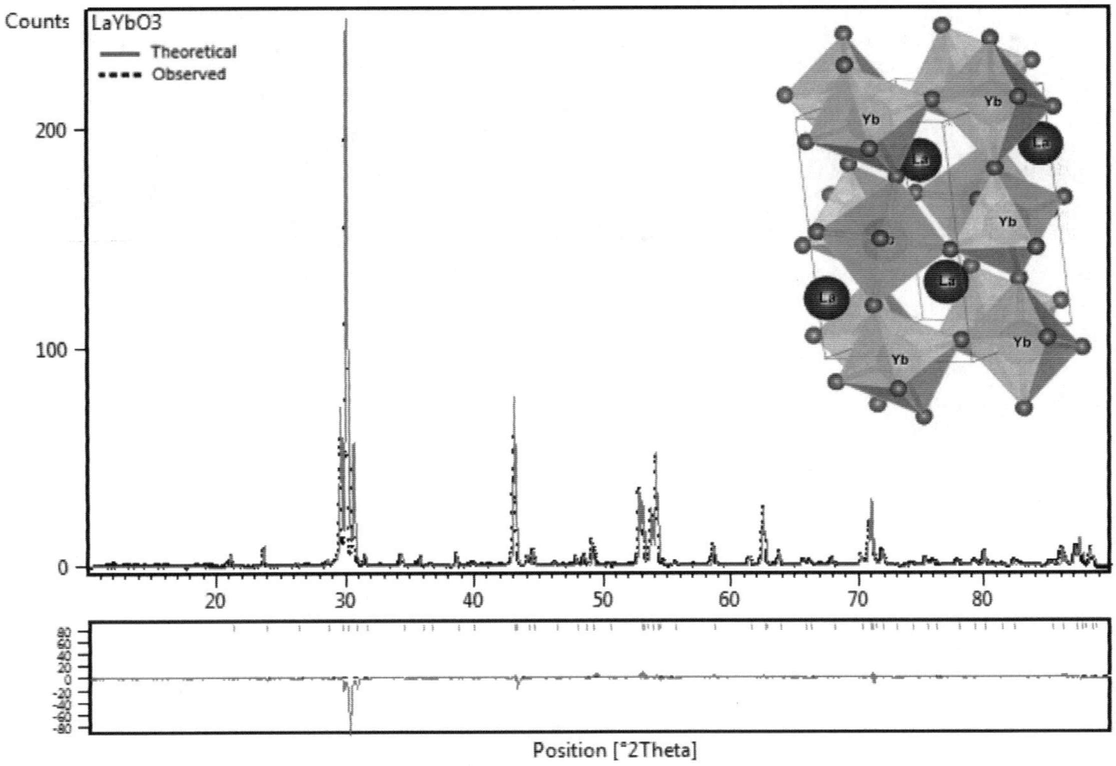

Figure 2. Rietveld refinement of the experimental diffraction data for $LaYbO_3$.

Figure 2 exemplifies the experimental and the simulated patterns in the refinement of the XRD data. From the XpertHighScore software it was obtained that this material crystallizes in an orthorhombic structure, internal angles $\alpha=\beta=\gamma=90°$ and lattice parameters $a \neq b \neq c$, with a difference of 2% when compared with other reports [6].

Table 2. Refined parameters for $LaYbO_3$.

	Rietveld Refinement	Database	Relative error
a(Å)	6.0233	6.0100	0.22%
b(Å)	8.2080	8.3900	2.17%
c(Å)	5.7203	5.8100	1.50%

Morphological studies of the sample across the synthesis process were carried out through SEM images. As observed in Figure 3 for treatment temperatures of 1000°C, 1100°C and 1190°C, there is a gradual variation of the grain size, with a systematical growth with the increasing of the sintering temperature. This occurs because the high temperature gives rise to an effective nucleation of grains and diffusion between them, by chemical bonds, to increase the grain size, diminishing their porosity [7]. When the crystalline phase is reached at T=1190°C, the mean grain size is 2μm.

Figure 2. Surface microstructure of the $LaYbO_3$ perovskite observed from SEM images.

4. Conclusions

The ceramic material $LaYbO_3$ was synthesized by the solid state reaction method. XRD patterns reveals that at T=1190°C this material adopts an orthorhombic perovskite structure, space group Pnma, with lattice parameters a=6.0233Å, b=8.2080Å and c=5.7203Å. Surface morphology of samples by means SEM images permitted to determine the mean grain size of the grains to be 2μm.

References

[1] D A Landínez Téllez, D Llamosa P, C E Deluque Toro, Arles V Gil Rebaza, J Roa-Rojas 2013 *J Mol Struct* **1034** 233
[2] Moreira R L, Feteria A, Dias A 2005 *J Phys Condens Matter* **17** 2775
[3] Feteira A, Gillie L, Elsebrock R, Sinclair D 2007 *J Am Ceram Soc* **90** 1475
[4] Chen X, Bai S, Li M, Zhang W 2013 *J Eur Ceram Soc* **33** 3001
[5] M W Lufaso, P M Woodward 2001 *Acta Crystallogr B* **57** 725
[6] Mueller-Buschbaum H, Teske C 1968 *Inorg Nucl Chem Lett* **4** 151
[7] J Cuervo, J J Olaya, E Vera, D A Landínez Téllez, J Roa-Rojas 2012 *Physica B* **407** 3089

Influence of thermal spraying parameters on the corrosion resistance of aluminium oxide coatings deposited on steel 1020

Y Salas[1], E Vera[2], M Moreno[2] and Y Pineda[2]
[1] Universidad Antonio Nariño, Bogotá, Colombia.
[2] Universidad Pedagógica y Tecnológica de Colombia (UPTC), Tunja, Colombia.

E-mail: yecsalas@uan.edu.co

Abstract. Parameters required for the preparation of coatings of aluminium oxide deposited on AISI 1020 steels were determined according to their thickness and type of flame to differentiate their behaviour against corrosion. Commercial powders were used by the method of thermal spraying deposition. The coatings were analysed by OM (optical microscopy), the thickness was measured by means of a coating thickness gauge and electrochemical techniques variables measured was the Linear Polarization Resistance (LPR) and approximation Tafel potentiodynamic curves. The corrosion current for steel 1020 with Na_2SO_4 electrolyte of 3.5% is of the order of hundreds of A/cm^2 and coated steel given in the order of A/cm^2, which leads to think that the projection produces coatings uniform low closed porosity, although techniques DC indicate a significant porosity as is observable current response to the potentiodynamic curve. The observed thicknesses fall into the hundreds of microns and little uniformity was noted in this coatings. The coatings deposited by oxidizing flame was better performance in corrosion than the coating deposited by neutral flame.

1. Introduction

An Alumina coatings improve the tribological properties of the substrates where it is deposited and increasing the corrosion resistance of metallic materials that are subject to aggressive environments [1].

Thermal spraying is an effective method to obtain coatings of aluminium oxide as it is conducted at temperatures which allow melt the alumina in powder form, producing coatings with good adhesion and uniform porosity that allows its use in applications of thermal barrier with a very good mechanical performance and outstanding corrosion resistance [2].

Alumina based coatings have been described as uneven in their thickness and with remarkable levels of porosity, as well as excellent mechanical performance but with a bond to the anchoring layer (nickel-based) that can present problems [2].

The purpose of this work is to characterize the corrosion behaviour of Al_2O_3 coatings on 1020 steel using potentiodynamic curves and linear polarization resistance measures, which give us evidence of coating performance in aggressive media, in addition to observing the effect of thickness on electrochemical measurements.

2. Methodology

1020 steel coupons, 1inch in diameter and 1cm high, were used. This was surface prepared with sandblasting, so as to have a surface free of impurities and adequate roughness allowing better adherence.

Preheating the substrate to a temperature of 130°C was made, thereby eliminating moisture and improve adhesion. A first thermal spray coating is made, Ultrabond 50000, a Ni-based alloy [3], over which a layer of Al_2O_3 is deposited using two different flame conditions (neutral and oxidizing flame).

To the coated specimens were performed thickness measurements and electrochemical tests in order to find correlation between thicknesses (given by the number of layers) and type of flame (see Table 1).

2.1. Experimental design

The average coating thickness was calculated over multiples measurements using a magnetic thickness gauge following SSPC PA_2, dry thickness magnetic gages.

Table 1. The flame was characterized by means of gases fluxes and the rate of powder.

Flame	Neutral	Oxidizing
Powder Flux	40 g/min	40 g/min
Argon Pressure	20 psi	12 psi
Argon Flux	15 L/min	8 L/min
O_2 Pressure	60 psi	100 psi
O_2 Flux	6 L/min	11 L/min
C_2H_2 Pressure	5 psi	5 psi
C_2H_2 Flux	10 L/min	10 L/min

The experimental matrix considered two factors (Table 2), type of flame, with two levels, neutral and oxidizing flame; and thickness average, with three levels, 2 layers, 4 layers and 6 layers (determinates for the number of pass of the deposition mechanism).

Table 2. Experimental Matrix.

Anchoring mechanism	Flame	Layers	Average thickness (m)
	Neutral	2	46,32
	Neutral	4	106,53
UltraBond50000	Neutral	6	134,33
100 m aprox.	Oxidizing	2	46,32
	Oxidizing	4	106,53
	Oxidizing	6	134,33

2.2. Materials and methods

For the electrochemical test was used Na_2SO_4 electrolyte of 3.5%, a scan rate of 2mV/s, an exposed area of 0,93cm^2 with a sweep from -500mV to +500mV with respect to Ecorr. The steel was previously characterized by means of arc spectroscopy and other experimental details are summarized immediately.

2.2.1. Thermal spray process

In this process the powder on oxyfuel flame (oxy-acetylene), where it is melted and led by a jet of air and the flame to the workpiece, is fed. The particle velocity is relatively low (<100m/s) and the adhesion force of the deposits is generally less than faster processes. The porosity may be high and the cohesive force is generally lower. The deposition rate is 0.5 to 9kg/hour while materials with low melting point can be sprayed at higher deposition rates. The substrate temperature can be quite high [4].

2.2.2. Experimental equipment

2.2.2.1. Thermal spray equipment. The equipment, located in the laboratories of Metallurgy Engineering (UPTC, Tunja, Colombia), consists of a closed chamber, a gun CastoDyn DS 8000^{TM} with a programmable mechanism through LabView^{TM} that automates the process of spraying.

Journal of Physics: Conference Series **687** (2016) 012105 doi:10.1088/1742-6596/687/1/012105

2.2.2.2. Electrochemical equipment. The Gamry G-750 Potentiostan-Galvanostats, an electrode Ag/AgCl and a flat electrochemical cell was used for generate the potentiodynamics and LPR curves and is located in the Materials Lab of the Universidad Antonio Nariño, Tunja, Colombia.

2.2.2.3. Magnetic thickness gauge. Al_2O_3 coating Thickness can be measured on magnetic 1020 steel surfaces using a digital Elcometer^{TM} coating thickness gauge. The principle of electromagnetic induction is used for non-magnetic coatings, like Al_2O_3, on magnetic substrates such as steel + Ultrabond 50000.

3. Results

The obtained results were resume in the next potentiodynamics graphics and in the Table 3. In the potentiodynamics graphics, the x axis is the base 10 logarithm of current (Amperes) and the y axis is the Potential E in Volts (see Figure 1).

3.1. Potentiodynamics curves

The potentiodynamics curves have different behaviour depending on the type of flame deposition. Likewise it shows its variation depending on the number of deposited layers.

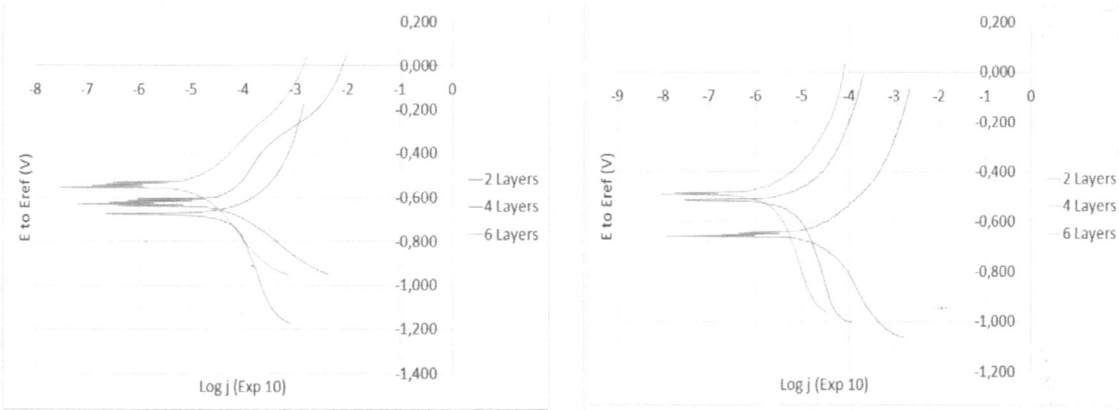

(a) (b)

Figure 1. Curves for Neutral (a) and Oxidizing flame (b) for the different number of layers.

3.2. Resume for experimental data obtained

The main response variables for each experiment were the corrosion and current potential, anodic and cathodic pending by means of Tafel approximation and LPR, obtained by linear regression.

Table 3. Experimental data from the electrochemical tests.

Flame	Layers	Ecorr (mV)	Icorr (A/cm^2)	a (mV/dec)	c (mV/dec)	CR (mm/y)	LPR (*cm^2)
Neutral	2	-666,61	69,85	184,5	-612,0	0,8132	177,28
Neutral	4	-622,48	51,48	336,2	-185,0	0,5993	229,89
Neutral	6	-536,48	15,34	237,4	-302,5	0,1785	345,49
Oxidizing	2	-627,98	29,00	168,2	-326,8	0,3376	286,19
Oxidizing	4	-539,12	11,42	317,4	-651,7	0,1330	360,92
Oxidizing	6	-496,62	3,25	251,5	-572,7	0,0378	440,97

4. Discussion

Comparisons for each factor are made in the graphs below. The scale for each plot is particular, but it indicates the change in the response variable for each of the considered factors.

In the Figure 2 it is compared the corrosion current for the two types of flame. It shows that as the number of layers decreases, current value is increased.

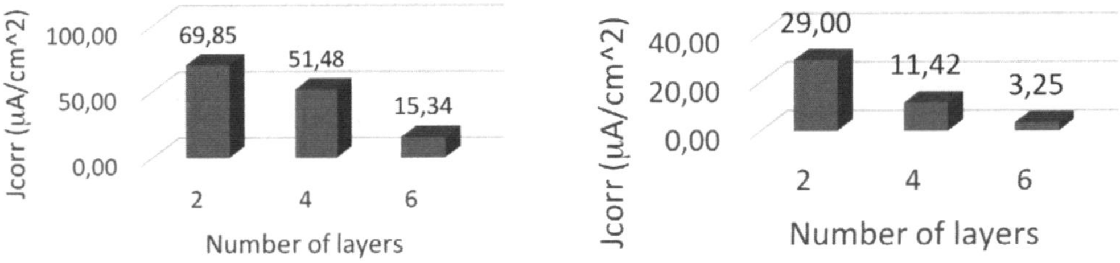

Figure 2. j_{corr} comparison for neutral and oxidizing flame.

The best performance of coatings applied with oxidizing flame is evident if is considered a constant thickness. It follows that the oxidizing flames generate less porous coatings respect to neutral flame, which contributes to better sealing and therefore the barrier effect is better, reducing the redox processes of the steel substrate (see Figure 3).

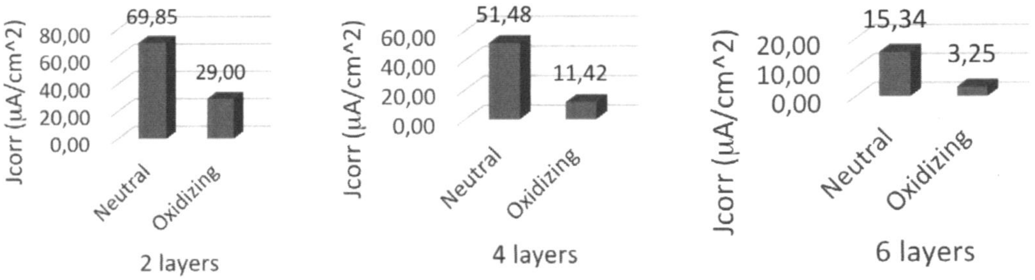

Figure 3. j_{corr} comparison for number of layers deposited.

5. Conclusions

Given that the thermal spraying is a method that generates coatings whose purpose is to reduce wear [2], it is also true that they have remarkable porosity and corrosion performance is improved if the number of layers is increased as it improves the sealing of the film. The porosity thus decreases.

The anchoring layer based on nickel and deposited with oxidizing flame [2], has better compatibility with oxidizing flame deposition. This indicates that the additional oxygen in the deposition inhibits chemically reacts with the electrolyte.

By increasing the number of layers, therefore the thickness, the corrosion potential tends to passive states, that is, its value decreases in magnitude in both types of flame.

The thickness has an ohmic behaviour when considering that as the number of layers increases, the corrosion current decreases and the lineal polarization resistance increases.

References

[1] Pawlowski L 2008 *The science and engineering of thermal spray coatings* (New York: John Wiley & Sons)
[2] Ferrer M, et al 2014 *Revista Colombiana de Materiales* **5** 317
[3] Castolin Eutectic 2013 *Data sheet ultrabond 50000* (Frankfurt: Castolin Eutectic Manuals)
[4] Davis J *Handbook of thermal spray technology* (Quincy: Davis & Associates)

Evaluation of the structural steel corrosion behaviour, covered with epoxy-type paints, by means of electrochemical DC techniques

Y Salas[1], L Guerrero[2], R Martinez[3], T Chicino[3] and C Devia[3]
[1] Universidad Antonio Nariño, Bogotá, Colombia.
[2] Universidad Pedagógica y Tecnológica de Colombia, Tunja, Colombia.
[3] Universidad Antonio Nariño, Tunja, Colombia.

E-mail: yecsalas@uan.edu.co

Abstract. In this work we have studied the behaviour of the electrochemical corrosion of structural steel AISI SAE 1007 with epoxy coatings, using epoxy-type paints, through techniques such as DC resistance Polarization and Potentio-dynamic tests. In order to determine potential and corrosion rates of these coatings, have been correlated this results with different used electrolytes. For this, coatings were characterized by thickness measurement and continuity measurements. The coatings showed a slight degradation in the testing time, due to defects present in their structure, and the attack by the electrolyte; however, epoxy coating system tends to react with the electrolytes based on their chemical composition.

1. Introduction

Epoxy coatings on steel have opened a number of potential uses such as waterproofing, sealing, bonding, damping, reinforcing and protecting load-bearing structures. The slow degradation and low relative cost of these coatings make them ideal for these purposes [1].

The coating system has properties such as: Low permeability, flexibility, heat resistance, abrasion resistance, resistance to acids and alkalis and other properties that make them viable for coating metal surfaces that want protection against corrosion [2]. Given these properties it is very helpful to use the epoxy system which is the most prominent industrial maintenance for their excellent stiffness, adhesion, alkali resistance, and humidity, temporary or permanent immersion to mineral acids, solvents and moderately good resistance at high temperatures (80 to 90°C) [3].

The purpose of this work is to characterize the corrosion behaviour of epoxy coatings on 1007 steel, using only D.C. electrochemical test (instead of the usual alternating current A.C. method: Electrochemical Impedance Spectroscopy EIS) [4,5], in function of the media (acid, neutral and basic), which give us evidence of coating performance in this different media.

Due to the thickness of the paint the current response was expected it was very small, but the potentiostat used had a very good definition that reached the order of 10^{-12} Amperes (pA), a fact which was verified in preliminary tests that yielded significant and visible data.

2. Methodology

Structural Steel AISI SAE 1007 specimens were used, supplied by the company SIKA Colombia and previously characterized by spectroscopy arc, with dimensions of 20cm long and 10cm wide and 1.035mm thick, with the respective type epoxy coating systems: A Self-priming Grey epoxy FZ HS, A epoxyphenolic S400 Ivory, Sika Epoxy 90HS Series A 200 and A 300 series 100HS Epoxy Sika, whose information was compiled manual metal coatings Sika. Electrolytic solutions NaCl, NaOH and H_2SO_4 with concentrations of 3.5, 3.75 and 5% respectively was used, as suggested by the NACE, Standard-TM0169-2000 [6]. These percentages of concentration are determinate to work the electrochemical tests simulate harsh environments, taking into account two aspects, the first is that the pH of the solution does not exceed the reference electrode (Ag/AgCl) so as to avoid this saturation, and the second is the ohmic overpotential having electrolyte solutions especially NaOH and H_2SO_4.

2.1. *Experimental design*

An experimental matrix design features two factors: Epoxy system with four levels determined by the paint supplied; and type of electrolyte, with three levels acid (H_2SO_4), neutral (NaCl) and basic (NaOH) (See Table 1).

Table 1. Experimental matrix.

Epoxy System	Electrolyte (Media)	Variables Response
	Acid: H_2SO_4 5%	
A autoimprimante epóxico gris HS FZ F0898	Neutral: NaCl 3.5%	
	Basic: NaOH 3.75%	Thickness (mm)
	Acid: H_2SO_4 5%	Continuity
A epoxifenolico S400 marfil F0909-14	Neutral: NaCl 3.5%	Ecorr (mV)
	Basic: NaOH 3.75%	Icorr (μA)
	Acid: H_2SO_4 5%	βa (mV/dec)
A sika epoxi 90HS serie 200 F0829	Neutral: NaCl 3.5%	βc (mV/dec)
	Basic: NaOH 3.75%	CR (mm/yr)
	Acid: H_2SO_4 5%	LPR (Ω*cm^2)
A sika epoxi 100HS serie 300 F0994	Neutral: NaCl 3.5%	
	Basic: NaOH 3.75%	

2.2. *Materials and methods*

For the electrochemical test was used the electrolytes mentioned in the experimental matrix, a scan rate of 2mV/s, an exposed area of 9,62cm^2 with a sweep from -250mV to +250mV with respect to Ecorr [7]. The SAE 1007 steel was previously characterized by means of arc spectroscopy and painted in the sika's plant in Tocancipa, Colombia. The reference electrode used was Ag/AgCl.

2.2.1 Electrochemical methods. Potentio-dynamic curves performed were analysed by means of Tafel approximation and in the LPR test values of linear polarization resistance were deducted.

2.2.2. Continuity tester. For continuous measurements epoxy coatings systems of all samples the same procedure was performed considering the suggestions described ASTM D 5162-A [8], both for the preparation of the specimens, as for the measurement and use Meter Compact equipment continuity PCWI™.

2.2.3. Electrochemical equipment. The Gamry G-750 Potentiostan-Galvanostats, an electrode Ag/AgCl and a flat electrochemical cell was used for generate the potentio-dynamics and LPR curves and is located in the Materials Lab of the Universidad Antonio Nariño, Tunja, Colombia.

2.2.4. Magnetic thickness gauge. Epoxy coating thickness can be measured on magnetic 1020 steel surfaces using a digital Elcometer™ coating thickness gauge. The principle of electromagnetic induction is used for non-magnetic coatings, on magnetic substrates such as SAE 1007 steel.

3. Results

The obtained results are reported in the Figure 1 and in the Table 2. In the potentio-dynamics graphics (Figure 1), the x axis is the base 10 logarithm of current (Amperes) and the y axis is the potential E in volts respect to E reference.

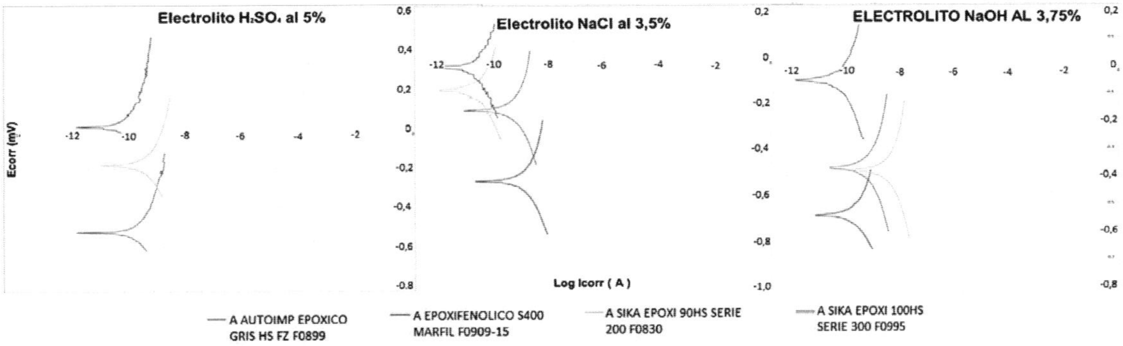

Figure 1. Curves for the three types of electrolyte with the four coating systems.

Table 2. Resume for experimental data obtained.

Epoxic system	NaCl 3,5% electrolyte			NaOH 3,75% electrolyte			H_2SO_4 5% electrolyte		
	CR mm/y	LPR Ω*cm^2	Thick mm	CR mm/y	LPR Ω*cm^2	Thick mm	CR mm/y	LPR Ω*cm^2	Thick mm
A autoimprim epoxic gris HS FZ	8,41E-07	1.92E+09	0.34	8,86E-07	9.62E+08	0.45	1,22E-07	1.92E+09	0.31
A epoxifenolico S400 marfil	2,05E-06	5.77E+08	0.61	1,62E-07	1.92E+09	0.37	-	-	-
A sika epoxi 90HS serie 200	3,26E-08	9.62E+09	0,3	5,53E-06	2.89E+08	0,3	3,84E-07	9.62E+08	0,3
A sika epoxi 100HS serie 300	3,25E-08	9.62E+09	0,95	7,61E-08	4.81E+09	0,57	1,78E-07	3.85E+09	0,43

4. Discussion

In the Figure 2 it is compared the corrosion rate for the four types of paint systems. It is noteworthy that the system epoxyphenolic saturate the equipment when subjected to H_2SO_4 electrolyte. The epoxy 90HS system present a notable corrosion rate for a basic media, the same way that the epoxyphenolic with neutral medium. The epoxy 100HS system had the best performance to maintain their corrosion rates around 10^{-7}mm/y for all electrolytes.

For all the systems the polarization resistance is practically the same values remaining at around tenths of Giga Ohms per square centimeter.

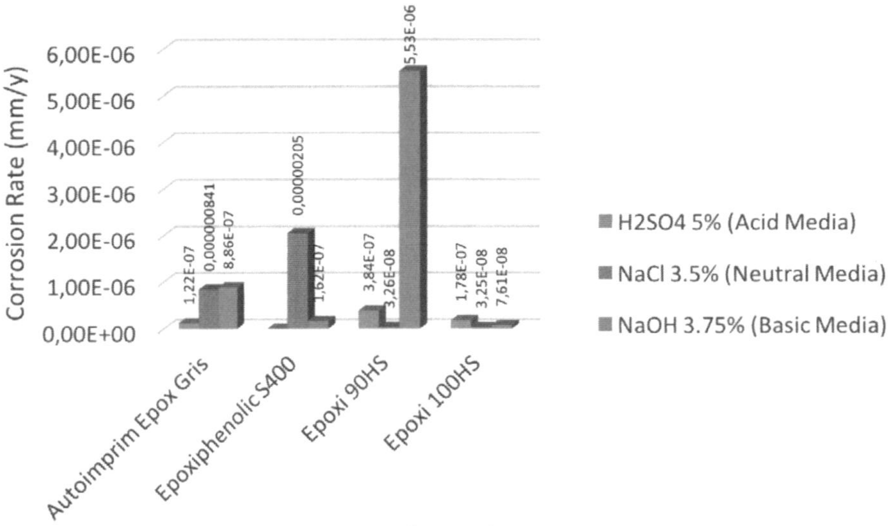

Figure 2. Corrosion Rate comparison for the four paint systems.

5. Conclusions

The Electrochemical Techniques D.C. are a very good method to determine the quality of epoxy coatings systems and application media, based on their respective corrosion and graphical analysis.

The coating system A epoxyphenolic S400 Ivory did not react to the acid environment by having polyamide in their chemical composition.

Based on the results obtained in the study and evaluation of epoxy systems with respect to pH range, i.e. the means in which we evaluated, we can say that more stable system and better performance in the three media both rich chlorides such as alkali and acid was sika epoxy system A 100HS Series 300, which does not mean that the other systems are not as efficient as this, since the lineal polarization resistance as was observed were in the order of G Ω*cm^2.

Coating systems possessing polyaminoamides have better performance than polifenilamida systems present in their chemical structures.

References

[1] Carbonelli C 2010 *Pinturas y recubrimientos introduccion a su tecnologia* (Madrid: Diaz Santos)
[2] Gupta A and Dev A 2008 *J Plastic Industry* **32** 1
[3] Lee H and Neville K 1967 *Industrial & Engineering Chemistry* **59-9** 16
[4] Haynes G S 1985 *Laboratory corrosion tests and standards a symposium by ASTM committee G-1 on corrosion of metals* (Bal Harbour: ASTM International)
[5] Diaz F 2001 *Estudio de la corrosion atmosferica del cinc y el acero galvanizado* (Santa Cruz de Tenerife: Universidad la Laguna)
[6] Guzman A and Ocampo C L 2011 *Evaluacion de la resistencia a la corrosion del sistema primer epoxico rico en Zinc/acabado polisiloxano por medio de espectroscopia de impedancia electroquimica* (Medellin: Universidad Nacional de Colombia)
[7] ASTM-G5 1994 *Standard reference test method for making potentiostatic and potentiodynamics anodic polarization measurements* (West Conshohocken: ASTM International)
[8] ASTM-G5 2008 *Standard practice for discontinuity (Holiday) testing of nonconductive protective coat* (West Conshohocken: ASTM International)

Processing of spent Ni-MH batteries for the recovery of cobalt, nickel and rare earth elements bearing materials by means of a chemical and electrochemical sequential process

P Delvasto[1], R Orta Rodríguez[2] and S Blanco[3]
[1] Universidad Industrial de Santander, Bucaramanga, Colombia.
[2] Universidad Simón Bolívar, Caracas, Venezuela
[3] Fundación Instituto de Ingeniería para la Investigación y Desarrollo (FIIID), Caracas, Venezuela.

E-mail: delvasto@uis.edu.co

Abstract. Rechargeable Ni-MH batteries contain strategic metal values which are worth to be recovered. In the present work, a preliminary sequential chemical and electrochemical procedure is proposed, in order to reclaim materials bearing Ni, Co and rare earth elements (REE) from Ni-MH spent batteries. Initially, spent batteries are disassembled to separate the electrode materials (anode and cathode), which are then leached with an aqueous solution of 5w% sulphuric acid. The metal content of this solution is checked by atomic absorption spectrometry techniques. The obtained solution is pH-adjusted (with NaOH), until pH is between 4.0 and 4.3; then, it is heated up to 70°C to precipitate a rare earth elements sulphate (Nd, La, Pr, Ce), as determined by means of x-ray fluorescence techniques. The solids-free solution is then electrolyzed, in order to recover a Ni-Co alloy. The electrolysis conditions were established through a cyclic voltammetry technique.

1. Introduction

Once household batteries are depleted, they become, by far, the domestic solid waste that bears the most heavy metals on a mass basis. Since heavy metals are deleterious for the environment and human health, it is clear that spent batteries are becoming nowadays a real environmental problem, particularly in emerging economies such as the Latin American region [1,2]. Despite developed countries already have environmentally sound technologies for the recycling of household spent batteries, that is not the case elsewhere. Recycling facilities in Europe and Japan take advantage of pyrometallurgical processes and are capable of processing spent batteries in large amounts, as part of the burden in the production of ferroalloys and other specialty metal products. Weak collection systems, inappropriate waste management and low consumption rates make this kind of solutions inadequate for most Latin American countries. In the search for small-scale local solutions for the problem of spent batteries recycling, research is being conducted on hydrometallurgical processes, because they seem to adapt better to the particular economic conditions in this part of the world [1-5]. There are several types of batteries available in the market. Among them, nickel metal hydride (Ni-MH) rechargeable batteries highlight, because of their interesting chemical composition. Spent Ni-MH batteries contain nickel (in metal, oxide and oxyhydroxide form) and rare earth elements (REE) metal alloys, both comprising the electrode (anode and cathode) active materials [1]. High prices, relative scarcity and trade restrictions have made nickel and rare earth elements become strategic metals, so they are worth to be reclaimed from secondary resources such as spent batteries. It is with this background in mind that the present work will explore the viability of using a sequence of chemical and electrochemical processes to recover rare earth chemical compounds and Ni-Co alloys from the solutions obtained after leaching the active electrode materials of spent Ni-MH batteries with sulphuric acid.

2. Experimental procedure

Spent Ni-MH batteries (different brands and sizes) were obtained from a community collection system in Caracas, Venezuela [2]. The batteries were disassembled manually, in order to extract the anode and cathode powders. The chemical characteristics of these powders are given elsewhere [1]. Five grams of the powders were leached in 100ml of 5% sulphuric acid solution in shake flasks (250rpm) at room temperature (25°C). The leaching time was 1 hour. After leaching the flask contents are filtered and a dark green pregnant solution is obtained for further processing. The chemical composition of this dark green solution, determined by atomic absorption spectroscopy (AAS), was as follows: Ni 0.37, Co 0.045, Mn 0.001, Fe 0.003, La 0.02 and Zn 0.02 (all values given in Mole/Litre). In order to precipitate the REE contained in 100ml of the pregnant solution, NaOH is added, as needed, until a pH around 4.0-4.3 is reached. Then, the solution temperature is increased to 70°C for 5 minutes. This procedure grants the full precipitation of a white compound comprising the REE metals and only traces of Ni and Co. The white precipitate was filtered, dried and submitted for chemical analysis by means of x ray fluorescence (XRF). To study the electrochemical behaviour of the Ni-Co resulting solution (depleted in rare earth elements), a cyclic voltammetry test was performed [6]. It was carried out using a potentiostat Autolab Pgstat 302N. A three electrode cell type was employed, with a Pt wire as auxiliary electrode, a gold disk as working electrode and an Ag/AgCl electrode as reference. The result was compared with the electrochemical response of reagent grade $NiSO_4$ (0.2M) and $CoSO_4$ (0.02M) standard solutions in H_2SO_4. The aim of the test was to stablish whether or not Ni and Co could be co-deposited from such a solution and foresee the possibility of using the solution as electrolyte for the production of electroplated Ni-Co alloys.

3. Results and discussion

The dark-green solution obtained after leaching the Ni-MH electrode powders contains mainly Ni, Co and rare earth metals (only La is shown due to analysis limitations). There are some other ions present, in lesser proportion, such as Fe, Mn, Zn, Na and K, the last two arising from the electrolyte of the battery. The solubility of the REE is lower around pH 4 and temperatures above 25°C. As a result, these conditions were chosen to extract REE from the pregnant solution. Once the precipitation is completed, a white salt is obtained. The chemical composition of this precipitate is shown in Table 1.

Table 1. Semi quantitative[a] FRX chemical composition of the REE bearing white precipitate obtained from the pregnant solution at pH 4 and 70°C.

Element	Concentration (mass %)	Element	Concentration (mass %)
S	18.90	Nd	8.89
La	42.64	K	5.02
Ce	17.72	Ca	0.19
Pr	6.57	Ni	0.07

[a] Oxygen was not quantified in the assay. Results are normalized.

As it can be observed, presumably a mixed REE sulphate is obtained as solid product. The REE are concentrated in this product in such a manner that it is richer than most commercial REE mineral concentrates. Another interesting point is that no cobalt and very little Ni are found in the product, leaving the solution enriched in both metallic elements. In order to determine if Ni and Co can be co-deposited directly from this remaining solution, a cyclic voltammetry was performed, as explained in the experimental part. The result is shown in Figure 1.

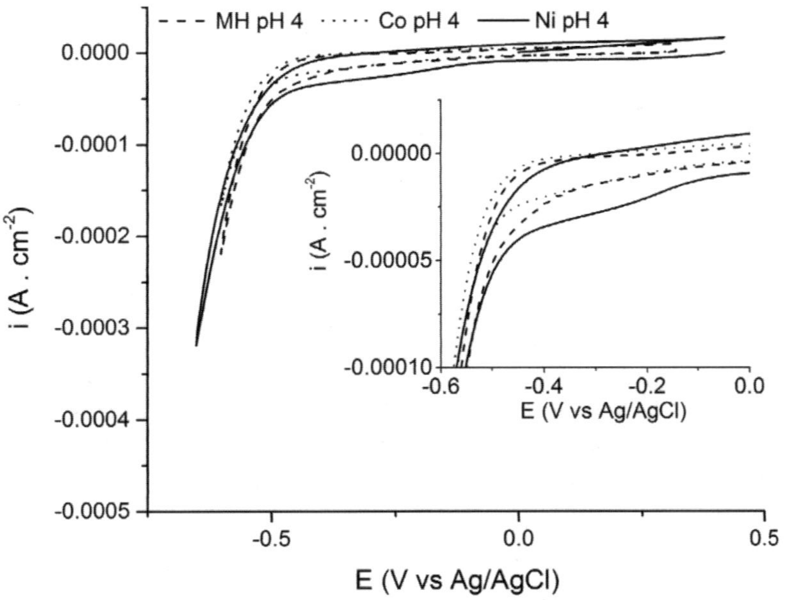

Figure 1. Cyclic voltamperograms on gold electrodes of the Ni-MH battery leaching solution (MH), Ni standard solution (Ni) and Co standard solution (Co). Experimental conditions: pH 4, temperature 22°C, scan rate 50mV/s. Embedded: detail at lower currents.

In this Figure, particularly for Ni, a negative current is observed corresponding to the reduction of hydrogen (hydrogen evolution). This reduction process occurs according to the mechanism proposed by Volver-Heyrovsky-Tafel, given by reactions (1) to (4):

$$2H^{+} + 2e^{-} \rightarrow H_2 \quad (1)$$

$$M + H_3O^{+} + e^{-} \leftrightarrow MH_{ads} + H_2O \quad (2)$$

$$MH_{ads} + H_3O^{+} \leftrightarrow M + H_2 + H_2O \quad (3)$$

$$MH_{ads} + MH_{ads} \leftrightarrow H_2 + 2M \quad (4)$$

The rate of reaction (1) is affected by the surface of the material where it occurs. The electrodeposition of Co is given by reaction (5), while Ni deposition is given by reaction (6), as follows:

$$Co^{+2} + 2e^{-} \rightarrow Co \quad (5)$$

$$Ni^{+2} + 2e^{-} \rightarrow Ni \quad (6)$$

It can be seen in Figure 1 that the hydrogen reduction starts (or becomes significantly visible) at potentials below -0.5V vs Ag/AgCl. This opens a window of potentials between -0.3 and -0.5V vs Ag/AgCl for the electrodeposition of Co from the solution (Co in Figure 1). When the same analysis is performed for the $NiSO_4$ synthetic electrolyte, it is observed that once a layer of Ni is created on the surface of the gold electrode, a lower potential will be required for the hydrogen reduction reaction. This is due to the catalytic effect of Ni on reaction (1). This catalytic effect shortens the potential range on which Ni can be reduced and, thus, be deposited on the electrode surface without hydrogen evolution [6]. In addition, the current for the reduction of Ni and Co from the solution obtained after leaching of the Ni-MH battery (MH in Figure 1), between -0.3 and 0.5V vs Ag/AgCl is lower than the current obtained when Ni is reduced from the synthetic Ni solution (Ni in Figure 1). This variation is a consequence of the presence of several other ions in the leachate solution, as shown in the experimental procedure. This result indicates the feasibility of obtaining Ni-Co materials or Ni-Co coatings from the electrolysis of the Ni-MH leaching solutions [6], once they are depleted on REE.

4. Final considerations

The results indicate that a hydrometallurgical procedure, based on the use of sulphuric acid as leaching agent, can be applied to obtain pregnant solutions containing Ni, Co and REE. Form such a solution, selected pH and temperature conditions permit the precipitation of a REE mixed sulphate, with little or no Ni and Co present. As it was shown by the cyclic voltammetry test, once the solution has been depleted in REE, Ni and Co can be co-deposited by electrolysis. Simple and small-scale solutions for the recycling of batteries, such as the one depicted in this work, open the possibility of establishing modular recycling facilities in undeveloped countries from the Latin American region.

References

[1] Zambrano A, Romero C, Moccia A, Orta R, López J and Delvasto P 2015 *Producción + Limpia* **10-1** 51
[2] Moccia A, Romero C, Orta R and Delvasto P. 2015 *Av Quím* **10** 11
[3] Pietrelli L, Bellomo B, Fontana D and Montereali M R 2002 *Hydrometallurgy* **66** 135
[4] Zhang S, Huang X and Wang D 2015 *Rare Metal Mat Eng* **44** 73
[5] Dixini, P, Celante V, Lelis M and Freitas M *J Power Sources* **260** 163
[6] Chi B, Li J, Yang X, Gong Y and Wang N 2005 *Int J Hydrogen Energ* **30** 29

Journal of Physics: Conference Series **687** (2016) 012108 doi:10.1088/1742-6596/687/1/012108

Geometric, electronic and intrinsic chemical reactivity properties of mono- and bi-substituted quinoline derivatives for the ground state in gas phase

O L Neira Bueno[1], L Hincapié H[1] and C García Madrid[1]
[1] Universidad Popular del Cesar, Valledupar, Colombia.

E-mail: oscarneira@unicesar.edu.co

Abstract. The study of geometric, electronic properties and intrinsic chemical reactivity is presented for the case of Quinoline and three-derived molecules (4-Amino-Quinoline, 3-Phenyl-Quinoline, 4-Amino-3-phenylquinoline). The study was carried for the ground state in gas phase in the context of the functional theory density using B3LYP/6 31+G (d) model. The purpose of the study is aimed for identifying a compound derived from quinoline, on based to mono- or bi-substitution, using the amino fragment and the phenyl group.

1. Introduction

Organic compounds (OC) with π-conjugated bonds chains occupy an important place in different fields of science [1-12]. Its potential use depends on its optical, optoelectronic and photostability properties among others. All these properties are closely related to the electronic properties and intrinsic chemical reactivity of molecular structures that constitute each material [1-4], OCs play an important role to develop materials of interest as organic electronic materials (semiconductors), organic light emitting diodes [1], field effect transistors [3], solar cells [4], active media [9]. It is known that with an adequate alteration in the molecular structure of organic compounds it is possible to modify its properties [1-12]. On the other hand, the role of the quinoline [12] structure as the basic unit in the family of organic compounds that have ideal properties of absorption and fluorescence has allowed for laser effect in the UV and violet regions [11]. In work [12] found that derivate compounds obtained from the replacement of one of the atoms bonded to a carbon atom into quinoline has good optical properties (See Figure. 1).

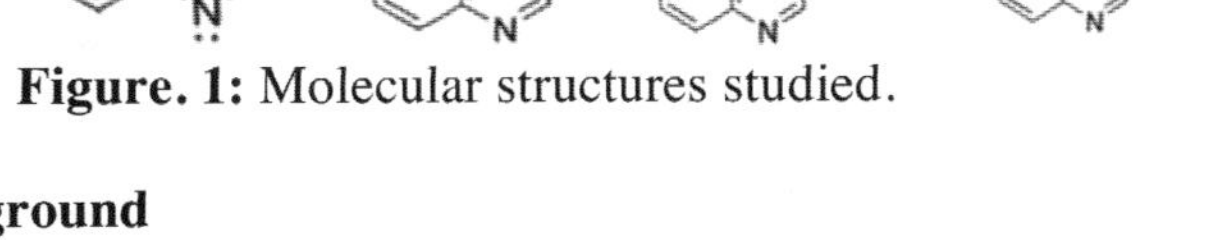

Figure. 1: Molecular structures studied.

2. Theoretical background

The prediction of the reactivity and the selectivity of a chemical process is critical into many organic, inorganic, and biochemical reactions. Thus different theories have been proposed to explain chemical reactivity and/or selectivity. Each theory has proposed a set of global and/or local indexes aiming to describe the chemical processes. Global indexes such as electronegativity (μ) [1] and hardness (η) [2]

have been introduced to characterize the reactivity of a compound. Local indexes, like Fukui's frontier molecular orbital [6-10] or local softness [6] have been introduced to explain the region and/or stereo selectivity of a reaction. In Density Functional Theory (DFT), the total energy E of a system (the ground-state energy of an atom or a molecule) is expressed as a functional of the electronic density $\rho(r)$, and $v(r)$, in the form [1-6].

$$E[\rho(r)] = F[\rho(r)] + \int_r v(r)\rho(r)d\tau \tag{1}$$

Where $v(r)$, is the external potential acting on the interacting system (at minimum, for a molecular system, the electron-nuclei interaction) and $F[\rho]$ is the universal Hohenberg–Kohn functional: $F[\rho] = T[\rho] + V_{ee}(\rho)$, $T[\rho]$ represents the electronic kinetic energy functional (T_s or T_{sk} is the Kohn–Sham kinetic energy). This energy corresponds to the sum of the nuclear interaction energy and the electronic energy of the molecule, which was obtained for the fixed spatial configuration of the nuclei to ensure the minimum energy of the molecular system by approximation of Born-Oppenheimer.

$$E[\rho] = T_s[\rho] + \int \upsilon_{ext}(r)\rho(r)d\tau + V_H[\rho] + E_{xc}[\rho] \tag{2}$$

DFT provides the theoretical framework for rationalizing molecular reactivity in term of the response of the system toward variation of N and $v(r)$. The response to changes in the number of electrons, when the external potential remains constant, is measured at first order by the chemical potential [3,4],

$$\mu = \left(\frac{\partial E}{\partial N}\right)_{v(r)} \tag{3}$$

and at second order by the hardness [3,4],

$$\eta = \left(\frac{\partial^2 E}{\partial N^2}\right)_{v(r)} \tag{4}$$

Definitions of μ and η were given by Parr and Pearson [6-9]. Using a two-point finite difference approximation for chemical potential and a three-point finite difference approximation for the hardness, it leads to the following working definitions of these quantities:

$$\mu \cong \frac{1}{2}[E(N+1) - E(N-1)] \tag{5}$$

$$\eta \cong \frac{1}{2}[E(N+1) + E(N-1) - 2E(N)] \tag{6}$$

Where $E(N-1) = IP$ and $E(N+1) = EA$ are the first vertical ionization potential and electron affinity of the neutral molecule, respectively, while $E(N)$ is the energy of the neutral state system. Using the Koopmans' theorem, ($IP = -\varepsilon_H$ and $EA = \varepsilon_L$) allows one to write μ and η in terms of the energy of frontier HOMO (ε_H) and LUMO (ε_L) molecular orbitals, is obtained finally:

$$\mu \quad \frac{1}{2}(\varepsilon_L + \varepsilon_H) \text{ , } \eta \quad \frac{1}{2}(\varepsilon_L - \varepsilon_H) \tag{7}$$

3. Methodology and computational design

All the calculations obtained have been carried out with the package of programs Gaussian 09 [13], within the framework of density functional theory (DFT), using the three parameter Becke parametrization [6] for the Lee et al [7] exchange-correlation (B3LYP) and 6-31+G(d) basis set.

4. Analysis of obtained results and discussion

The comparison is made for each descriptor of each molecule derived with the value of the parent molecule descriptor. The influence of the substituent in the values of similar chemical bonds between identical atoms within each molecular structure atoms are presented in Table 1 and Table 2 and the values of energy, dipolar moment, reactivity descriptors in Table 3.

Table 1. length values of similar chemical bonds between identical atoms within each molecular structure.

Atoms	Q	4AQ	3FQ	4A3FQ
C1-C2	1.378	1.379	1.379	1.381
C2-C3	1.421	1.422	1.422	1.421
C3-C4	1.433	1.435	1.432	1.430
C4-C5	1.421	1.421	1.420	1.421
C5-C6	1.379	1.378	1.379	1.379
C6-C1	1.419	1.419	1.419	1.415
C1-H2	1.087	1.087	1.087	1.087
C2-H3	1.088	1.085	1.088	1.087
C3-C7	1.420	1.427	1.417	1.437
C4-N1	1.367	1.366	1.367	1.367
C5-H4	1.086	1.086	1.086	1.086
C6-H5	1.087	1.087	1.087	1.087
C7-C9	1.376	1.381	1.383	1.399
C8-C9	1.420	1.417	1.430	1.420
C8-N1	1.318	1.318	1.316	1.317
C8-H6	1.090	1.090	1.089	1.090

Table 2. Descriptive statistics approach based on dispersion and central tendency for length values of similar chemical bonds.

Atoms	Q	Media	Mediane	Rango	Varianza	D. Estandar	Coef. Var
C1-C2	1,378	1,379	1,379	0,002	0,0000013	0,0012	0,00084
C2-C3	1,421	1,422	1,422	0,001	0,0000003	0,0006	0,00041
C3-C4	1,433	1,433	1,432	0,005	0,0000063	0,0025	0,00176
C4-C5	1,421	1,421	1,421	0,001	0,0000003	0,0006	0,00041
C5-C6	1,379	1,379	1,379	0,001	0,0000003	0,0006	0,00042
C6-C1	1,419	1,418	1,419	0,004	0,0000053	0,0023	0,00163
C1-H2	1,087	1,087	1,087	0	0,0000000	0,0000	0,00000
C2-H3	1,088	1,087	1,087	0,003	0,0000023	0,0015	0,00141
C3-C7	1,420	1,425	1,427	0,02	0,0001000	0,0100	0,00701
C4-N1	1,367	1,367	1,367	0,001	0,0000003	0,0006	0,00042
C5-H4	1,086	1,086	1,086	0	0,0000000	0,0000	0,00000
C6-H5	1,087	1,087	1,087	0	0,0000000	0,0000	0,00000
C7-C9	1,376	1,385	1,383	0,018	0,0000973	0,0099	0,00713
C8-C9	1,420	1,422	1,420	0,013	0,0000463	0,0068	0,00479
C8-N1	1,318	1,317	1,317	0,002	0,0000010	0,0010	0,00076
C8-H6	1,090	1,090	1,090	0,001	0,0000003	0,0006	0,00053

The most significant changes happen for the angles between identical chemical bonds at the C8-C9-N1, C3-C4-C7 and C1-C6-H5and C7-C9 bonds. The statistical descriptors of central tendency and dispersion, which are shown in the Tables 1 and 2, it can be inferred that values of different chemical bond lengths are very close to the values of the corresponding bonds in the parent molecule, which shows that despite the changes that causes the fragments substituents in such bonds, from the geometrical point of view, an additional element emerges to affirm that new derivative structures belong to a family of chemical compounds, appearance which is ratified by the range values (10^{-2} to 10^{-3}), variance (10^{-4} to 10^{-5}), the mean square deviation ((10^{-2} to 10^{-3}).

Table 3. Global molecular descriptors of the Quinoline and its three derived molecules for the ground state in gas phase.

Parameter	Q	4AQ	3FQ	4A3FQ
Molecular descriptors	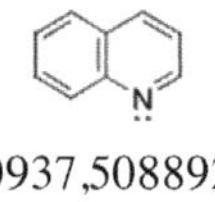	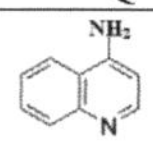	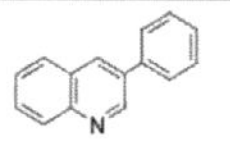	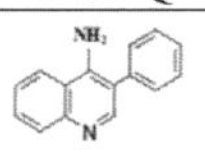
Energy (eV)	-10937,508892614	-12443,637280553	-17225,098294597	-18731,511286025
Dipolar moment (D)	2,1864	2,6419	2,2203	3,6206
HOMO (eV)	-6,574546354	-6,428149022	-6,345154252	-5,882560452
LUMO (eV)	-1,750509362	-1,684657774	-1,835136816	-1,506695218
Gap (eV)	4,824036992	4,743491248	4,510017436	4,375865234
Chemical potential (eV)	-4,162527858	-4,056403398	-4,090145534	-3,694627835
Hardness	2,412018496	2,371745624	2,255008718	2,187932617
Electronegative	4,162527858	4,056403398	4,090145534	3,694627835

For total energy value, the effect of substituents caused in 4-Amino-Quinoline a decrease of 13.7%, in 3-phenyl-Quinoline of 57.4%, in 4-amino.3-phenyl-Quinoline of 71.2%. i.e. from the energy point of view, they would be more stable. The effects of the respective substituents caused in the dipole moments are in 4-Amino-Quinoline a decrease of 20,8%, in 3-phenyl-Quinoline of 1,5%, in 4-amino.3-phenyl-Quinoline of 65,6%. i.e. from the dipolar moment point of view, It is suggesting that new structures should provide better absorption and fluorescence properties in relation to the quinolone. The values obtained for HOMO energies, allow us to infer that effects of the respective substituents caused an increase of 2,2% in the value of HOMO in 4-Amino-Quinoline, 3,5% in 3-phenyl-Quinoline and 10,5% in 4-amino-3-phenyl-Quinoline i.e. from this point of view, It is suggesting that new structures can more easily donate an electron under similar conditions in a electrophilic solvent or a photon of lower energy is required to ionize the molecules, i.e. ionization potentials of derivate molecules are lower compared to that of the parent molecule. Also new molecules have less ability to nucleophilic attack. The values obtained for LUMO energy, allow us to infer that effects of the respective substituents caused an increase of 3,7% in the value of LUMO in 4AQ, decrease 4,6% in 3FQ and increase 13,9% in 4A3FQ, i.e. from this point of view, it is suggesting that is more easy of an attack by nucleophiles for in 3FQ.

Table No 4. Theoretical prediction of the position of an absorption spectrum band.

Name	ΔE_z (eV)	$\lambda_{H\rightarrow L}$ (nm)	$\Delta^{Q\text{-}XQ}\ \lambda_{H\rightarrow L}$ (nm)
Q		4,824	257,01533
2AQ	4,743	261,458	4,44267
3FQ	4,510	274,90952	17,89419
4A3FQ	4,375	283,32768	26,31235

In Table 4 are presented the values of the wavelengths of the photons that should influence on molecular structures occurring the transition of the electron that is occupying the HOMO to LUMO. The values obtained suggest that the absorption bands of these compounds to the Quinoline, part of it should be located about 257nm, for 4-AQ at 261.6nm, for 3-phenyl-Quinoleina at 274.9 and finally for 4A3FQ in 283,32768nm.

5. Conclusions

It was possible to characterize four organic compounds, which constitute Quinoline derivatives, which by their intrinsic properties, chemical reactivity electronic augur possess better optical properties than the parent compound.

References

[1] D Tomkute-Luksiene, J Keruckas, T Malinauskas, J Simokaitiene, V Getautis 213 *Dyes and Pigm* **96** 278
[2] Fuli Zhang, Lian Duan, Juan Qiao, Guifang Dong, Liduo Wang, Yong Qiu 2012 *Organic Electro* **13** 1277
[3] Hungshin Fua, Yi-Ming Chengb, Pi-Tai Choua and Yun Chib 2011 *Materials Today* **14** 476
[4] Frederik C Krebs 2009 *Solar Energy Materials & Solar Cells* **93** 394
[5] Matthew Wright, Ashraf Uddin 2012 *Solar Energy Materials & Solar Cells* **107** 87
[6] Chengteh Lee, Weitao Yang and Robert G Parr 1988 *Phys Rev B* **37-2** 785
[7] D Becke 1988 Phys Rev A 38-6 3098
[8] Parr R G, Yang W 1989 *Density-functional theory of atoms and molecules* (New York: Oxford University Press)
[9] A V Kukhta, I N Kukhta, N A Kukhta, O L Neyra and E Meza 2008 *Journal of Physics B Atomic Molecular and Optical Physics* **41-20** 205701
[10] R G Parr and W Yang 1995 *Annual Review of Physical Chemistry* **46-1** 701
[11] W Kohn and L J Sham 1965 *Phys Rev A* **140-4** 1133
[12] Pereira Hugo 2015 *Influencia de sustituyentes donantes y aceptores de electrones en las propiedades estructurales, de absorción óptica y fluorecencia en el compuesto quinoleína: Un enfoque desde la teoría del funcional de la densidad (DFT)* (Colombia: Universidad Popular del Cesar)
[13] M J Frisch, G W Trucks, H B Schlegel, G E Scuseria, M A Robb, J R Cheeseman, G Scalmani, et al 2009 *Gaussian 09 revision D.01* (Wallingford: Gaussian Inc)

AC magnetic susceptibility and influence of heat treatment on obtaining the nanocrystalline structure for the amorphous alloy of $Fe_{37}Co_{35}Nb_6B_{11}Si_{10}Cu_1$ composition

A Orozco[1], Z Caamaño[1] and A Rosales[2]
[1] Universidad del Atlántico, Barranquilla, Colombia.
[2] Universidad Nacional de Colombia sede Manizales, Manizales, Colombia.

E-mail: amikomodo@hotmail.com

Abstract. This work presents a study of AC magnetic susceptibility and the influence of heat treatment on obtaining the nanocrystalline structure for the amorphous alloy of $Fe_{37}Co_{35}Nb_6B_{11}Si_{10}Cu_1$ composition. AC magnetic susceptibility at different frequencies and sample lengths, using a variable magnetic field was measured. The influence of heat treatment on obtaining the sample nanocrystalline structure was evidenced by means of XRD technique. The crystallite size, the crystalline volume fraction and the distance between grains for each heat treatment temperature were calculated using the XRD peaks refinement software GSAS. Samples were obtained in the amorphous state by melt spinning technique and all measurements by AC magnetic susceptibility and XRD were performed at room temperature.

1. Introduction

During the last 15 years, the importance of soft magnetic FeCo based alloys has increased due to their excellent soft magnetic properties, which allowed the development of improved technological applications such as magnets and sensors for high temperatures [1-3]. Nanocrystalline magnetic alloys show superior soft magnetic properties when their amorphous phase is submitted to a controlled heat treatment [1-4]. In Colombia, over the last decade, there has been a growing interest in research on Fe based soft magnetic materials, mainly because these alloys exhibit interesting magnetic phenomena [5-7]. However, AC magnetic susceptibility of FeCo-based soft magnetic alloys has not been studied yet. In this work, AC magnetic susceptibility, as a function of frequency and length of $Fe_{37}Co_{35}Nb_6B_{11}Si_{10}Cu_1$ amorphous alloy, has been studied. The sample was prepared by the melt spinning technique and then was partially crystallized by heat treatments from room temperature up to different temperatures, in which crystallization process occur. The evolution of the crystallization process with the heat treatment temperatures increase was evidenced by XRD technique. From the XRD diffraction maxima refinement, the crystallite size, the crystalline volume fraction and the distance between grains [1] were obtained according to the heat treatment temperature. The evaluation of these parameters is crucial for the achievement of excellent soft magnetic properties.

2. Experimental procedure

Amorphous ribbons of $Fe_{37}Co_{35}Nb_6B_{11}Si_{10}Cu_1$ composition were prepared by the melt spinning technique. AC magnetic susceptibility measurements was performed on an AC susceptomer at the Magnetism and Advanced Materials Laboratory from Universidad Nacional de Colombia, sede

Manizales, using a frequency range between 10Hz and 5kHz and sample lengths between 0.4 and 1cm, with a constant amplitude of about 1Vrms. The magnetic field was varied from -74 to 74Oe. To obtain the nanocrystalline structure, amorphous samples were heat treated from room temperatures up to 550, 600, 650, 700 and 750°C in a GSL-1500X (MTI) vacuum furnace. DRX analysis was performed on a PANalytical X'Pert difractometer from the DRX Laboratory of Universidad del Valle. Measurements were made at room temperature using Cu radiation, over 30-90° 2θ and 0.54°/2.22s step. The crystallite size and the crystalline volume fraction were estimated taking into account the XRD peak refinement software GSAS (General Structure Analysis System) and the distance between adjacent grains was calculated using the method of the literature [1].

3. Results and discussion

3.1. AC magnetic susceptibility measurements

Figure 1 shows real (a) and imaginary (b) components, χ'_{AC} and χ''_{AC}, of the AC magnetic susceptibility as a function of the applied magnetic field, for the $Fe_{37}Co_{35}Nb_6B_{11}Si_{10}Cu_1$ alloy at a frequency of 40Hz at room temperature. A peak at zero magnetic field, corresponding to the maximum value of the AC magnetic susceptibility, is observed. This behaviour could be explained in terms of the domain walls movement, as it was indicated by O. Moscoso [7]. By the other side, the narrower peak observed in the χ''_{AC} imaginary part, is attributed to the minimum energy loss in magnetization processes of soft magnetic materials.

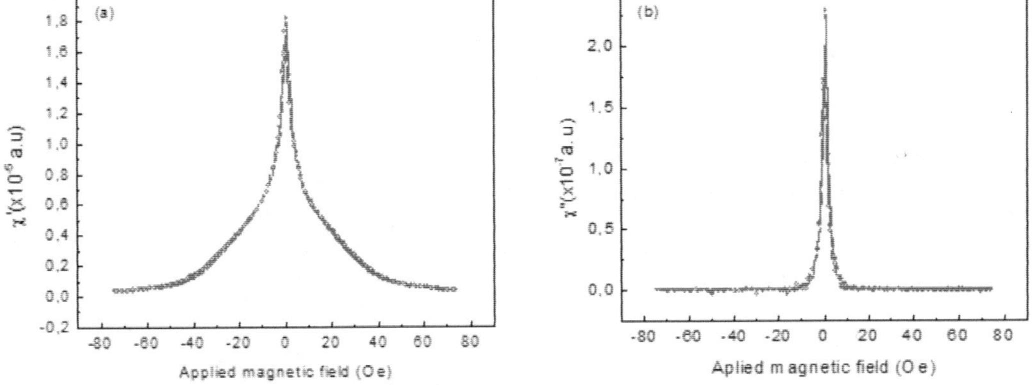

Figure 1. Real AC susceptibility (a) and imaginary AC susceptibility (b) Vs. Applied magnetic field at 40 Hz.

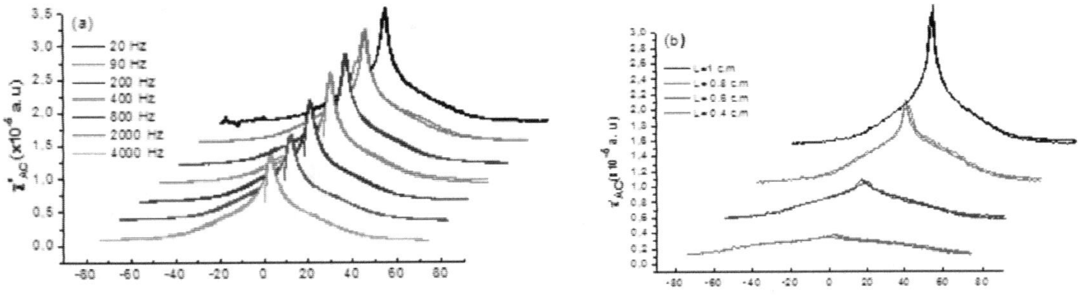

Figure 2. (a) Real part of AC magnetic susceptibility vs. external magnetic field at different frequencies and (b) at different lengths of the amorphous alloy of $Fe_{37}Co_{35}Nb_6B_{11}Si_{10}Cu_1$ composition.

The AC magnetic susceptibility dependence on frequency and length of the sample is shown in Figure 2. It can be seen that the real AC magnetic susceptibility tends to widen as the frequency increases, while the AC magnetic susceptibility response enhances as the sample length increases; indicating the strong dependence of AC magnetic susceptibility on frequency and sample length. This dependence could be due to a possible magnetic hardening of the sample when the frequency is increased and the length is decreased, because the magnetic domain configuration is strongly influenced by the demagnetizing field, as it was explained by L. Goncalves et al. [8].

3.2. Structural characterization by XRD

In Figure 3, it is shown the XRD pattern obtained for the as-quenched sample and heat treated at different temperatures. It is observed that the sample is still amorphous at 550°C and from 600°C to 750°C, the sample is partially crystalized. This is evident by the appearance of the α-FeCo and $(Fe,Co)_3Si$ primary crystalline phases and the $(Fe,Co)_2B$ secondary phase. These crystalline phases were also reported for alloys of similar compositions in [1, 9].

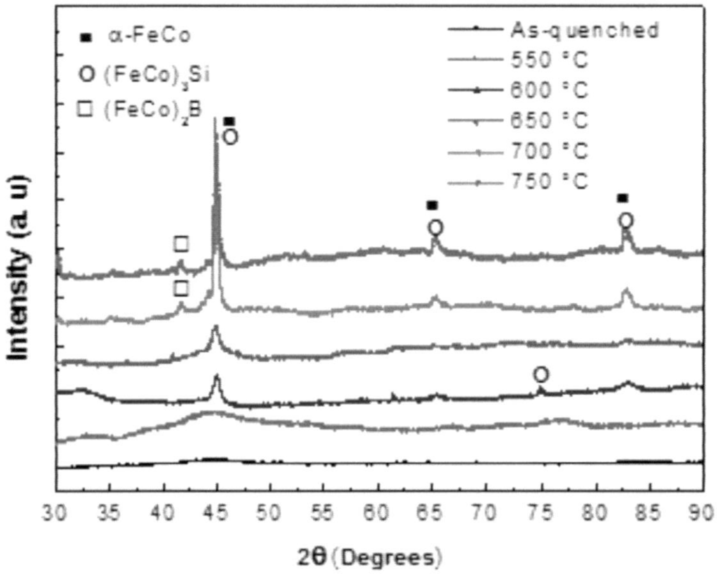

Figure 3. XRD pattern of the sample $Fe_{37}Co_{35}Nb_6B_{11}Si_{10}Cu_1$ as-quenched and heat treated at different temperatures.

Table 1. Crystallite size (*D*), crystalline volume fraction of α-FeCo phase ($V_{Cr\text{-}\alpha FeCo}$), distance between grains (*Λ*) for the sample heat treated at different temperatures.

Heat treatment temperature (°C)	D (nm)	$V_{Cr\text{-}\alpha FeCo}$ (%)	Λ (nm)
550	-	-	-
600	9.9	3.5	20.7
650	27.4	4.1	56.7
700	25.6	11.4	27.5
750	21.1	24.9	13.0

From the principal diffraction maxima, the α-FeCo phase crystallite size (*D*) and the crystalline volume fraction ($V_{Cr\text{-}\alpha FeCo}$) as well as the distance between adjacent grains (*Λ*) were calculated, as shown in Table 1. It can be noticed that $V_{Cr\text{-}\alpha FeCo}$ increases with the heat treatment temperature increase, indicating that the sample evolves from an amorphous state to a crystalline state. While, D increases from 9.9nm (600°C) up to 27.4nm (650°C), and tends to decrease slightly at higher heat treatment temperatures. Similar behaviour is observed for Λ, where at 650°C a maximum value is obtained. The variation in the obtained values D and Λ can be due to the formation of the α-FeCo nanocrystaline phase. Besides, the ordered $(Fe, Co)_3Si$ phase and the $(Fe, Co)_2B$ hard magnetic phase cause the magnetic decoupling between two adjacent grains at the heat treatment temperature of 650°C, as was reported in [10].

4. Conclusions

AC magnetic susceptibility depends strongly on the frequency and length of the sample $Fe_{37}Co_{35}Nb_6B_{11}Si_{10}Cu_1$ and its response is closely related to a minimum energy loss, which is characteristic for this type of alloys.

A nanocrystalline structure with the following heat treatment of the amorphous sample in the range 600–750°C was obtained. The α-FeCo and $(Fe, Co)_{3Si}$ primary crystalline phases and the $(Fe,Co)_2B$ secondary phase could be identified.

The partial crystallization of the sample was also confirmed by the increase of the crystalline volume fraction with the heat treatment temperature. Although, the crystallite size as well as the distance between two adjacent grains decreased, as a consequence of the existing crystalline phases.

Acknowledgments

To the group of Physics of Materials II of the Universidad Autónoma de Barcelona (Bellaterra, Spain), for the collaboration in the preparation of the samples by the melt spinning technique.

References

[1] Jia Wang, Zhi Wang, Yun-yun Jia, Rui-min Shi, Zhuan-ping Wen 2013 *J Magn Magn Mater* **328** 62
[2] Kong F, Chang C, Inoue A, Shalaan E, Al-Marzouki F 2014 *J Alloys Compd* **615** 163
[3] Blázquez J, Ipus J, Conde C, Cabrera D, Franco V, Conde A 2014 *J Alloys Compd* **615** 5213
[4] Lashgari H, Chu D, Shishu Xie, Huand Sun, Ferry M, Li S 2014 *J Non-Cryst Solids* **391** 61
[5] Caamaño Z, Costa L and Rosales-Rivera A 2012 *Rev Mex Fís S* **58-2** 35
[6] Valencia V 2004 *Study of Giant magnetoimpedance phenomenon GMI in soft magnetic ribbons of alloys* $Fe_{73.5-x}Cr_xNb_3Cu_1Si_{13.5}B_9$ *with* $0 \leq x \leq 10$ (Manizales: Universidad Nacional de Colombia)
[7] Moscoso O 2008 *Influence of boron content in the magnetic, thermal and structural properties of Fe based amorphous alloys* (Manizales: Universidad Nacional de Colombia)
[8] Goncalves L, Soares J, Machado F, De Acebedo W 2006 *Physica B* **384** 152
[9] Ye-Mei Han, Zhi Wang, Xiang-hui Che, Xue-gang Chen, Wen-run Li, et al 2009 *Mat Sci Eng B* **156** 57
[10] Orozco A 2013 *AC magnetic susceptibility, magnetization and influence of heat treatment on obtaining the nanocrystalline structure for the amorphous alloy of composition* $Fe_{37}Co_{35}Nb_6B_{11}Si_{10}Cu_1$ (Barranquilla: Universidad del Atlántico)

Changes on electrical and structural properties of polyaniline and polypyrrol by mullite doping

C P González[1], A M Montaño[1,2], S E Estrada[1] and C A Ortiz[3]
[1] Universidad Pontificia Bolivariana, Floridablanca, Colombia.
[2] Universidad Industrial de Santander, Bucaramanga, Colombia.
[3] Universidad Pedagógica y Tecnológica de Colombia, Tunja, Colombia.

E-mail: claudia.gonzalez@upb.edu.co

Abstract. In this work, the effect of mullite dopant (mull) on the structural and electrical properties of two conductive polymers: polyaniline (PAni) and polypyrrole (PPy), which have great interest in science and materials engineering, has been evaluated. Doped polymers were synthesized with different contents of mineral and at two polymerization times (t_P) by an in situ chemical synthesis in acidic aqueous solution and using Ammonium Persulfate (APS) as the reaction initiator. Structural characterization was performed by X-Ray Diffraction (XRD) and Fourier Transform Infrared Spectroscopy (FTIR). The conductivity values (σ_{ac}) were estimated from Electrochemical Impedance Spectroscopy (EIS). It was found that, at concentrations of work, were obtained higher values σ_{ac} of conductivity at the highest concentration of the mineral, however, in some compounds the conductivity decreased. The variations in conductivity were attributed to PC-dopant interactions.

1. Introduction

Conductive polymers (PCs) are used as conductive glues, non-static textiles, organic compounds sensors [1,2], capacitors, and electrodes for rechargeable batteries [3] between many others. Recently, many studies have been focus on to optimize the structure and to improve electrical properties of PCs, which can be done by adding doping agents. It has been reported that before doping, PCs shows very low electrical conductivities (values $\approx 10^{-10}$S/cm), after doping this values can rise even in twelve magnitude orders ($\approx 10^{2}$S/cm) depending on polymeric matrix and dopant agent type [3-6].

The fundamental process of doping is a charge-transfer reaction between the organic polymer and a dopant. The dopants for conducting or semiconducting PANI include several types of molecules as acids, salts or mineral phases. In this study mullite mineral (mull) is used as dopant agent. Some outstanding properties of mullite are low thermal expansion, high temperature strength and good chemical stability. Besides mullite has been used recently as corrosion resistive coatings even at high temperatures [7,8].

The goal of this work is to present synthesize, structural and electrical characterization of doped PCs with a mineral phase. Studied PCs were polyaniline as emeraldine salt (PAni-ES) and protonated polypirrol (PPy). These polymers were obtained by chemical polymerizacion in acidic medium. Both of these polymers were doped with mullite mineral ($3Al_2O_3.2SiO_2$) at three concentrations: 20, 40 and 60%. Doped PCs were analysed by XRD, FTIR and EIS.

2. Experimental

PAni/mull and PPy/mull compounds were synthesised in oxidative polymerization in acidic medium with dopant agent in situ. Three quantities of mull were added to obtain mixtures of 20, 40 and 60% weight and reaction was performed at two polymerization times (t_P) of 8 and 14 hours. General procedure was: 0.01mol of monomer (Ani or Py) were added to 100mL of HCl 2M. Precipitates formed were colected by filtration and washes with distilled water and the proper solvent (acetone for PAni and methanol for PPy) to discoloration. Later mull was added with continuous and vigorous stirring keeping temperature under 4°C. Then 0.01mol (2,28g) ammonium persulfate was slowly added (PSA) [$(NH_4)S_2O_8$], as oxidant agent. Each mixture was allowed to react for 8 and 14h with continuous stirring. Doped PCs were dried at room temperature in a desiccator and then structurally and electrically studied.

3. Result and discussion

3.1. Structural characterization

3.1.1. XRD. XRD for PAni/mull and PPy/mull are shown in Figure 1(a) and Figure 1(b) respectively, with 20, 40 and 60% of dopant and t_P of 8h. It is seem PAni characteristic signal at 25.35° and at 25.47° for PPy besides of mull typical peaks. XRD for both PCs at t_P of 14h has the same pattern (not shown here).

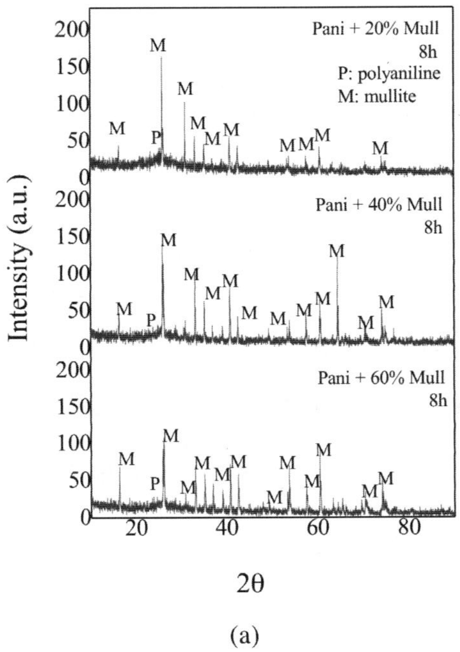

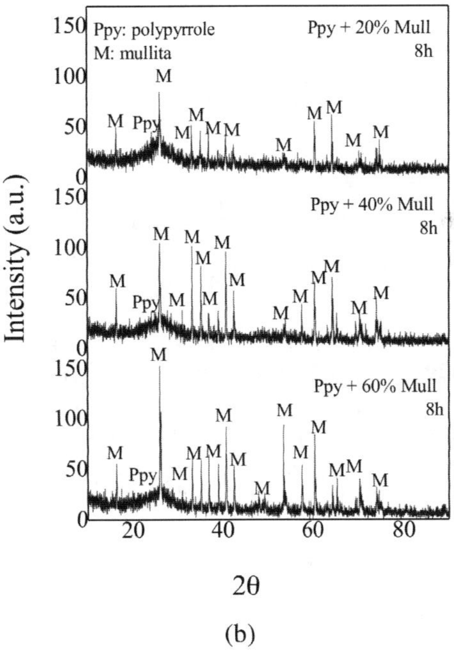

Figure 1. Ray diffraction patterns x of (a) PAni/mull and (b) Ppy/mull at t_p of 8h.

Diffraction profile for doped PCs indicate that mineral phase is already incorporated into polymeric matrix: there is a principal peak, of low intensity, for each PC obtained ($2\theta \approx 25.35°$ for PAni and $2\theta \approx 25.47°$ for PPy); intense signals for mull also they appear. The most intense peak for mull is near to PCs peak at 26°. Mineral insertion to polymer chains causes and increase in crystallinity of PAni and PPy.

3.1.2. FTIR. FTIR spectra for PAni/mull and PPy/mull are shown in Figure 2(a) and Figure 2(b) respectively. All bands for PC and some key Mull bands prove the successful doping. In every PAni/mull FTIR spectra (Figure 2(a)), all PAni bands appear. Further it is observed a band next to 704cm^{-1} related to stretching vibrations of Al–O bond in tetrahedral coordination and to flexion modes of Al–O–Al mull bond. Bands close to 584cm^{-1} and 497cm^{-1} are due to stretching vibrations of Al–O bond in octahedral structural units. In FTIR spectra for PPy/mull (Figure 2(b)) bands corresponding to PPy it seems and there are two bands at 550cm^{-1} and 426cm^{-1} associated with flexion modes of Al–O-Al bond in mull.

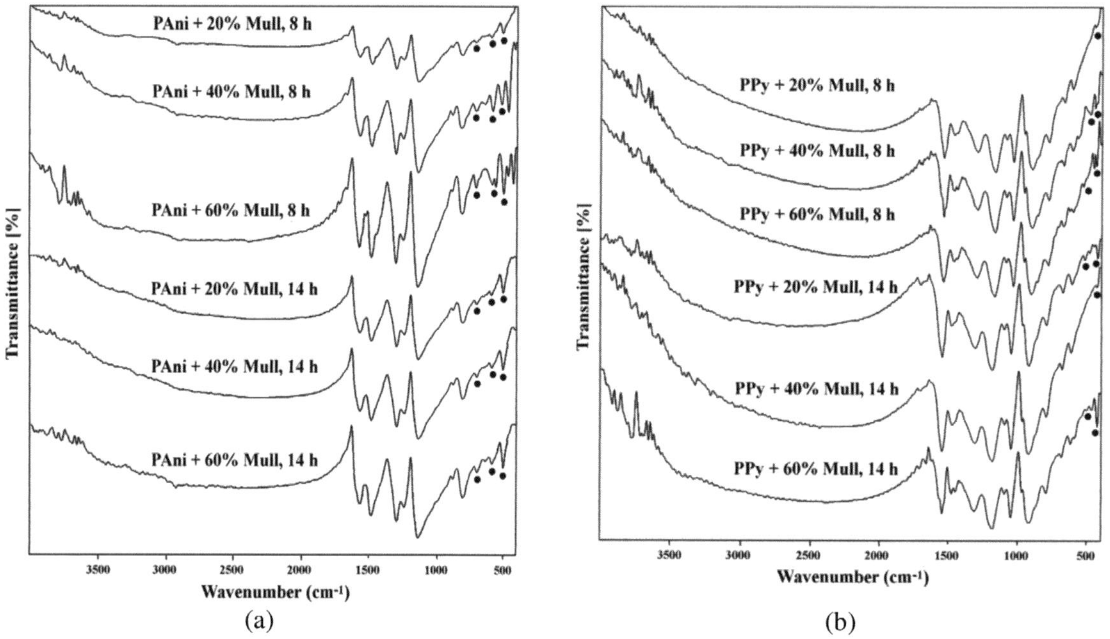

Figure 2. FTIR spectra of (a) Pani/mull and (b) Ppy/mull composites.

FITR spectra for doped PCs (Figure 2) shows that there are changes in intensity and number of bands that appear because of the mull adding. Those bands become more intense for more mineral concentration in most of the cases. Intensity alterations are presents no just in doping agent bands but in those for PC. This behaviour proves that the doping process involves a doping agent interaction with some specific places of polymeric chain, which carries out a structural modification. If it is true that this interaction it is not well known at molecular level, nevertheless it is know that this interaction is associated to changes in several properties as, for example, conductivity increase or decrease, depending on the electrical character of doping agent.

3.2. *Electrical characterization (EIS)*

With the purpose of to appreciate the changes in conductivity for each PC/doping agent system, caused by the synthesis variables (mull concentration and polymerization time), results of EIS are shown in bar graphs. In Figure 3 variation of σ_{ac} as a function of mull content for both t_P for PAni/mull is plotted.

The 8h PCs sample with lowest mull content (20%), has a lower conductivity than that exhibited by non-doped PAni (pure emeraldine) synthesized at this t_P. On the other hand in PCs with 40% and 60% mull content σ_{ac} values increase according to increment of mineral doping. To 14h PCs every single sample with some mull content has higher σ_{ac} values that for non-doped PAni. The higest σ_{ac} value for PAni/mull it is exhibit by 60% mull PC at 8h.

In Figure 4 shift of σ_{ac} as a function of mull content for both t_P for PPy/mull is plotted. For 8h PCs σ_{ac} values are always higher than those of non-doped PPy. The highest conductivity it is shown for 40 and 60% mull concentration. For 14h simple with 20% mull, presents a σ_{ac} much less tan σ_{ac} for non-doped PPy. The conductivity value for compounds 40 and 60% mull increase with increasing mineral quantity, but values were lower than those obtained for the same concentrations synthesized to 8h.

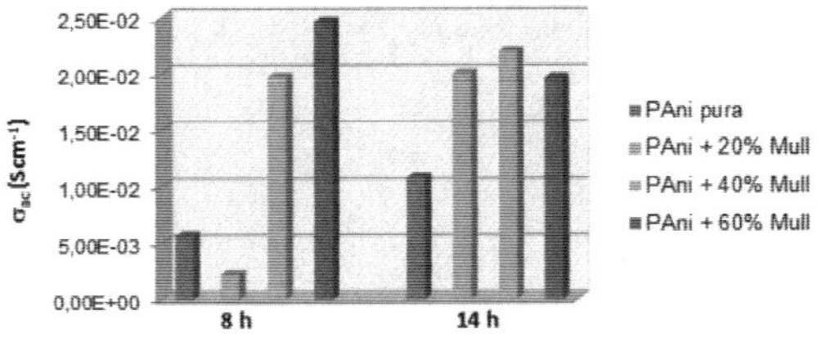

Figure 3. σ_{ac} versus composites PAni/mull t_P=8 and 14h and f=1x10^5Hz.

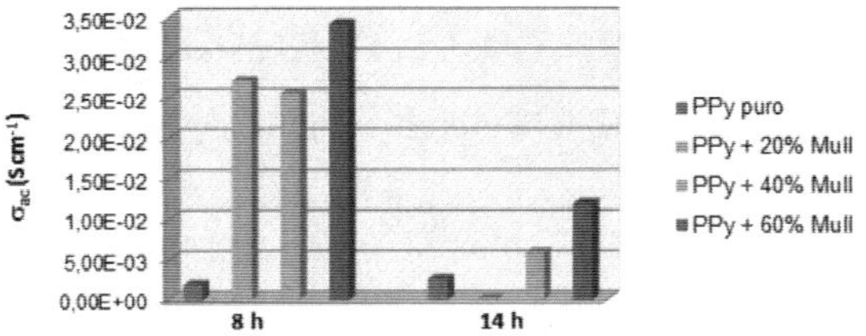

Figure 4. σ_{ac} versus % mullite composites Ppy/mull t_P=8 and 14h and f=1x10^5Hz.

4. Conclusions

From analysis of the diffraction patterns and FTIR spectra it was concluded that the addition of mullite to PCs involves a dispersion in the polymer dopant, which involves an increase in the crystalline behaviour of PCs. Similarly, there is an interaction between PC and dopant, possibly associated with PCs conductivity changes.

We found that in most PC-doped conductivity values increased relative to the values obtained for pure mullita and PCs. The highest value of conductivity for PAni/mull system was obtained for 60% mullita concentration synthesized during 8h. For the PPy /mull system is noted that for the compounds synthesized 8h, the conductivity values are always higher than that obtained for pure PPy, being the highest value for the sample with 60% of mullita. By comparing the two systems studied it was found that the interaction between the PAni and mullita and is less strong than between PPy and mullita, and so the increase in conductivity is lower. Therefore, the effect of doping on the electrical conductivity of the PCs depends on the nature of the dopant, the polymer and the interaction between them under certain conditions. Doping agents act as a charge carries which is seems more efficient on PPy. Conductivity in PPy specifically is due to *p*-type (bipolar) conduction, in the inter-chain hoping of electrons and the motion of anions or cations within the material [9]. It is important also to keep present that the efficacy of doping is strongly dependent on external factors as pH and temperature. So, the values found in this study must be expanded to the same PC-doping composites but obtained and tested at different conditions.

References

[1] Bhadra D, Khastgi D, Singha N and Hee Lee J 2009 *Prog Polym Sci* **34** 783
[2] Pirsa S and Alizadeh N 2010 *Sens Actuators B* **147** 461
[3] Armelin E, Meneguzzi A, Ferreira C and Aléman C 2009 *Surf Coat Technol* **203** 3763
[4] Akasaki I, Sota S, Sakai H, Tanaka T, Koike M and Amano H 1996 *Electron Lett* **32** 1105
[5] O'Leary S K, Foutz B E, Shur M S, Bhapkar U V and Eastman L F 1998 *J Appl Phys* **83** 826
[6] Narayan H, Alemu H, Somerset V, Iwuoha E, Hernández M, Hernández J, Montaño A and Henao J 2009 *Fly ash composites with polyaniline synthesis, characterization and conductivity measurements* (New York: Nova Science Publishers) chapter 5 p 111
[7] Tushar Kulkarni, Wang H Z, Basu, S N and Sarin V K 2011 *Surf Coat Technol* **205** 3313
[8] Huang Jian-Feng, Zeng Xie-Rong, Li He-Jun, Xiong Xin-Bo and Huang Min 2003 *Carbon* **41** 2825
[9] Richard Balint, Cassidy N J, and Cartnell, S H 2014 *Acta Biomat* **10** 2341

Surface characterization of stainless HP-40 steel using laser induced μ-breakdown spectroscopy (μ -LIBS)

M Pinto[1], X Calderón[1], E Mejía Ospino[1], R Cabanzo[1] and Juan C Poveda[1]
[1] Universidad Industrial de Santander, Bucaramanga, Colombia.

E-mail: monica.pinto@correo.uis.edu.co

Abstract. In the present study, optical microscopy in stereoscopic mode coupled to laser-induced μ-breakdown spectroscopy (μ-LIBS) was applied for analysing HP-40 steel samples. microLIBS (μ-LIBS) is a new growing area that employs low energy laser pulses for the generation of plasma emission, which allow the realization of localized microanalysis [1]. This new LIBS instrument was used for the surface characterization of the steel samples in the spectral range from 356 to 401nm. Elements such as Cr, Ni, Fe, Nb, Pb, Mo, C, Mn and Si in the steel samples were investigated. The results allowed the construction of elemental distribution profiles of the samples. Complementarily the HP-40 steel samples were superficially characterized by Scanning Electron Microscope (SEM).

1. Introduction

Reformer tubes are commonly used in furnaces to produce hydrogen in the refining and petrochemical. Their most critical components are the radiant tubes, where extreme pressure and temperature conditions required HP grade steels [2-4], owing to their high mechanical and corrosion resistances at high temperatures. They are designed for a nominal life of 100,000h in service conditions, typically 980°C with an internal pressure of 10 to 40 bars. Nevertheless, the HP grade alloys undergo damage with time, caused by the extreme work conditions. With reference to the foregoing arises the fact analyse microstructural changes that occur in steels using laser-induced breakdown spectroscopy (LIBS). Laser-induced breakdown spectroscopy is an analytical technique where atoms and ions are mainly formed in their excited states as product of interaction between a tightly focused laser beam and the material sample. The interaction among matter and high-density photons creates a plasma plume, which evolves with time and may eventually acquire thermodynamic equilibrium. One of the main features of this technique is that it does not need any sample preparation, unlike conventional spectroscopic analytical techniques [5]. The present paper report the results of the characterization of the HP-40 steel coming from the reformer furnaces pipes out of service by the presence of cracks and structure damages. Implemented laser induced breakdown spectroscopy coupled with optical microscopy that generates a microLIBS device (μ-LIBS). This device allowed carry out a localized microanalysis of surfaces of HP 40 steels. The HP-40 steel samples were superficially characterized by Scanning Electron Microscope (SEM).

2. Experimental description

HP-40 steels samples from tubes of reformer furnaces removed from the industry with 65000-66000 hours. Table 1 shows the elemental composition reported in accordance with global standards of steel. μ-LIBS device was implemented as shown in Figure 1. Nd: YAG laser (Quanta-Ray INDI) with a fundamental wavelength of 532nm, pulse width of 7ns, repetition frequency of 10Hz, is employed to generate the plasma. The laser intensity is reduced using an attenuator in a range from 50% to 60%. The size laser beam spot is reduced through an aluminium sheet with a hole of approximately 3.18mm in diameter and reflected by a triangular prism to the side entrance of the microscope SM-LUX-POL (Leitz Wetzlar, Germany), allowing to focus the laser beam perpendicularly into the area of interest, using an Olympus 10x objective. Finally, we obtain a laser beam with an intensity of 2.2mJ per pulse. The microscope was used in stereoscopic mode (i.e. with an external light source). A Shamrock spectrometer 500i (Andor Technology), coupled to a detector ICCD (intensified charge-coupled device) iStart 720 Gen II (Andor Technology) of 1024x256 pixels, with detectors of 26x26μm was used. The experimental operations were controlled by PC and carried out in air at atmospheric pressure.

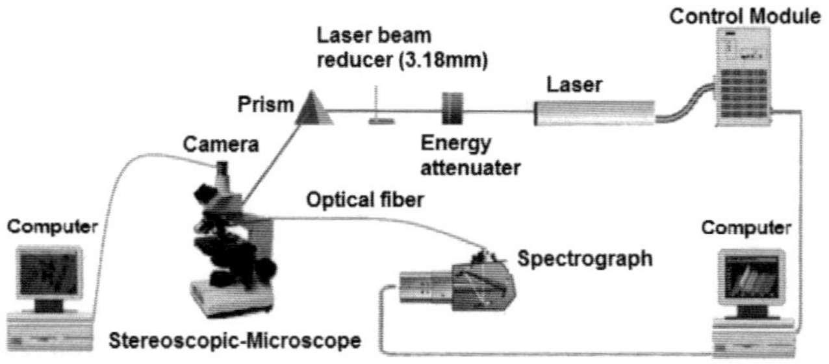

Figure 1. Experimental assembly μ-LIBS.

In addition, we used a camera Guppy GF046C (Allied) on the top of the microscope to localize the microanalysis region. The elemental distribution profiles were obtained over the surface in the range 356-401nm, using a grid of 1200lines/mm. The acquisition time of signal was 1μs with a delay of 1μs after laser shooting the surface. With these times and an accumulation of 5 pulses was possible to spectrally solve, the emission lines of the elements present in the μ-plasma.

The Scanning Electron Microscopy micrographs of the HP-40 steel were obtained using a Digital Scanning Electron Microscope (FEI Quanta 650 FEG SEM), under the following analytical conditions: magnification=800-1024x, WD=9.2-10.0mm, HV=20kV, signal=Z Cont detector=ETD.

Table 1. Elementary composition of HP-40 steel [6].

Elementary composition (%)								
Cr	Ni	Si	C	Mn	S	Pb	Nb	Mo
23.5	34.0	1.5	0.37	Máx. 1.25	Máx. 0.003	Máx. 0.01	0.7-1.5	Máx. 0.5
26.5	37.0	2.0	0.45					

3. Results and discussion

3.1. Qualitative identification

The main goal of LIBS is to realize a chemical analysis at the atomic level; a key step is the appropriate identification of each emission line of a particular element in a neutral or ionized state. If the sample composition is identified approximately, the set-up can be adjusted to know the optimal spectral range where there are emission lines of the elements in analysis or to discard emission lines of elements outside the sample. Figure 2(a) shows spectrum of atomic emission in a spectral range from 356 to 401nm. Most of the lines in this region belong to transitions in neutral state (Cr, Pb, Ni, Fe, Mo and Mn), also emission lines of ionized state were identified (Si, C). All transitions are listed in the National Institute of Standards (NIST) database [7]. Laser-induced breakdown spectroscopy (LIBS) coupled with a microscope was utilized to find and focus the laser beam directly on precipitations observed in the samples (removed of service). Figure 2(b) shows spectrum of atomic emission in a spectral range from 356 to 401nm, the black line shows the spectrum of a zone of the sample without precipitations and the blue line is the spectrum with the laser beam focus in a precipitation. It is notable in the blue spectrum a rise in signal intensities, which suggests an increase in concentration of the elements in this area.

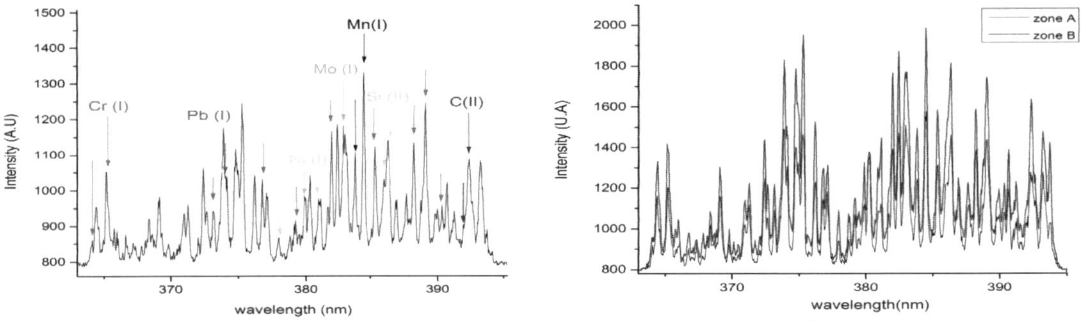

(a) (b)

Figure 2. LIBS Spectra of HP-40 steels (spectral range 356-401nm) (a) the arrows of colours indicate the location of atomic emission lines of elements in the steel. (b) Spectrum produced focusing the pulse laser in a zone affected by precipitation.

3.2. Elemental profiling

With μ-LIBS has been possible a micro-spatial analysis of steels. The distribution of selected trace element (Cr, Pb, Ni, Fe, Mo, Mn, Si, and C) was measured on $1cm^2$ area approximately. Observed revealed similar accumulation for all the tracked elements across the scanned lines, we detected homogeneity in elemental distribution inside the sample of steel without any type of work. The opposite showed the samples extracted from the pipe (exposed to high pressures and temperatures), an elemental inhomogeneity can be clearly distinguished. The variations in composition of Pb, Ni, Fe, Mo, Mn, Si, and C were not very drastic compared with the changes in the composition of Cr. Figure 3(a) depicts chrome distribution with the more intense atomic emission (381.95nm). The Chromo content increases toward the precipitations, the results are confirmed with the observations in SEM.

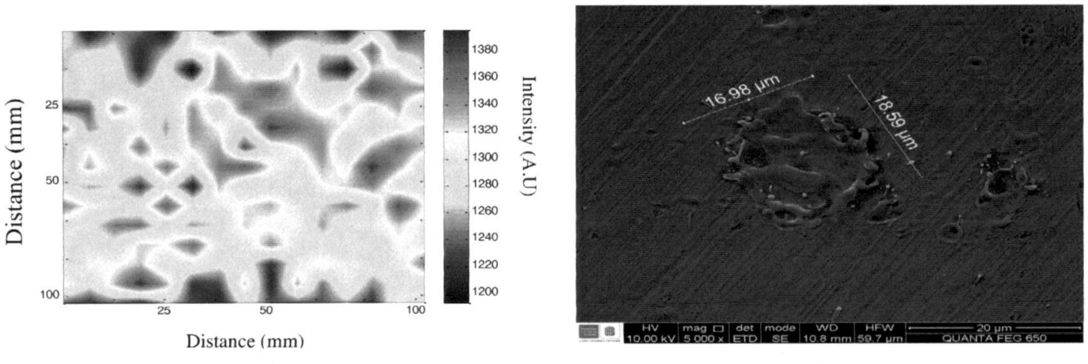

(a) (b)

Figure 3. (a) Distribution graph of the chrome (381.95nm) (b) Surface image of the crater of ablation produced by a pulse laser 7ns (FWHM) and 2.2mJ of energy on a sample of steel HP-40.

3.3. Superficial damage

The superficial damage of the analysed samples depends principally on the laser wavelength, the duration of the pulse and the positioning system, which will limit the superficial area of the crater produced by the ablation laser. LIBS is a technique considered practically not destructive, since the craters produced by the impacts of the pulses laser on flat surfaces of steel produce approximately craters of 200μm of diameter and a depth of penetration less than 1μm [1]. The area of approach can be diminished by the suitable optics. Spectral information of the steels was obtained producing craters of minor area, diminishing the superficial damage in the analysed sample. The use of Objective Lenses of microscope, an attenuator of energy and a laser beam reducer (3.18mm of diameter), it generated diameters of superficial craters about 20μm (Figure 3(b)). With the parameters and the suitable optic, it was achieved to obtain spectra of the zones affected by the precipitations.

3.4. Scanning electron microscopy

Figure 4 shows the SEM image of a precipitation in the HP-40 steel and the spectra overall composition of the inclusion and surface texture. Detail of analysis show that the concentration of Cr in the two spectra is (i) 25.84%, (ii) 62.12%; Fe is (i) 51.62%, (ii) 22.25%; Ni is (i) 20.56%, (ii) 3.46% and Si is (i) 0.48%, (ii) 1.21%. The detailed analysis per SEM of the precipitation showed the variation in chemical compositions in the steel HP-40, supporting the findings obtained by LIBS.

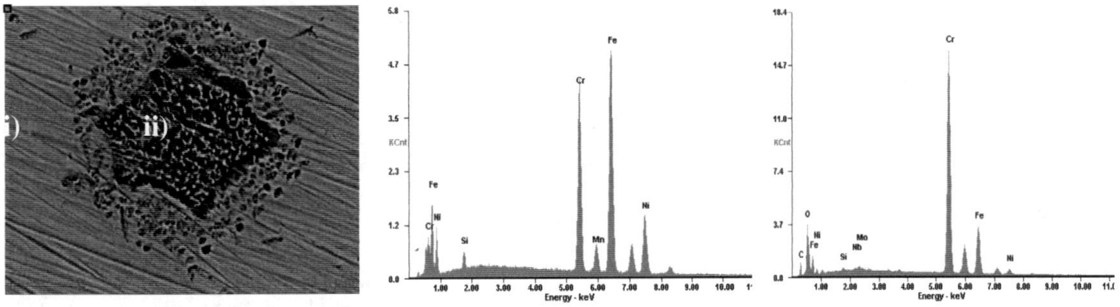

(a) (b)

Figure 4. (a) SEM elemental analysis, (b) Surface spectrum, (c) Precipitation spectrum.

4. Conclusions

This work demonstrates of LIBS capabilities for mapping in surfaces, mainly by optimization of LIBS ablation crater diameter. LIBS proved to be suitable for the fast analyses superficial. The measured Cr, Pb, Ni, Fe, Mo, Mn, Si, and C profiles across the sample showed fluctuations and the agglomeration of the elements in certain areas. We have shown that LIBS with microscope coupled can be successfully applied as direct or complementary techniques in spatially resolved microchemical analysis of steel samples. The sensibility, the selectivity and the versatility of this analytic technique make it ideal for the surface characterization. The results indicate that the laser focal conditions and the energy of the laser pulse are important parameters that determine the lateral resolution. Exist a direct correlation between the sizes of the crater of ablation with the energy of the incident beam; craters of different sizes for the same optical conditions can be generated.

Acknowledgments

The authors acknowledge the financial support of the Research VIE and Laboratorio de Microscopía of the Universidad Industrial de Santander, Parque Tecnológico de Guatiguará.

References

[1] J P Singh and N S Thakur 2007 *Laser-induced breakdown spectroscopy 1st ed* (Amsterdam: Elsevier) chapter 8 p 173
[2] Alvino A, Lega D, Giacobbe F, Mazzocchi V and Rinaldi A 2010 *Eng Fail Ana* **17** 1526
[3] Almeida L H, Ribeiro A F, May L I 2003 *Materials Characterization* **49** 219
[4] Silveira T L and May L I 2006 *The Arabian Journal for Science and Engineering* **31** 99
[5] Contreras W A, Cabanzo R and Mejía E 2011 *Journal of Physics Conference Series* **274** 1
[6] Bringas J E 2004 *Handbook of comparative world steel standards 3rd ed* (West Conshohocken: ASTM International)
[7] NIST 2015 *NIST Atomic Spectra Database* (http://www.nist.gov/pml/data/asd.cfm)

Adsorption of vapreotide on gold colloids studied by surface enhanced Raman spectroscopy

J A Gómez[1], R Cabanzo[1] and E Mejia Ospino[1]
[1] Universidad Industrial de Santander, Bucaramanga, Colombia.

E-mail: emejia@uis.edu.co

Abstract. Surface Enhanced Raman Spectroscopy (SERS) has been used to investigate the somatostatin (SST) analogue Vapreotide (VAP) in gold colloids. The optimum conditions to detect SERS signals of VAP have been studied. The observed SERS bands correspond to different vibrational modes of the peptide; being the most dominant SERS signals the ones derived from the aromatic amino acids Tryptophan (Trp), Phenylalanine (Phe) and Tyrosine (Tyr). Changes in enhancement and wavenumber of the proper bands upon adsorption on gold colloid are consistent with VAP adsorption, primarily through Tryptophan residues.

1. Introduction

Vapreotide (VAP) is a synthetic Somatostatin (SST) analogue that comprises eight amino acids (D-Phe-Cys-Tyr-D-Trp-Lys-Val-Cys-Trp-NH_2) and bears an internal disulphide bridge in its backbone (Cys^2-Cys^7) (see Figure 1).

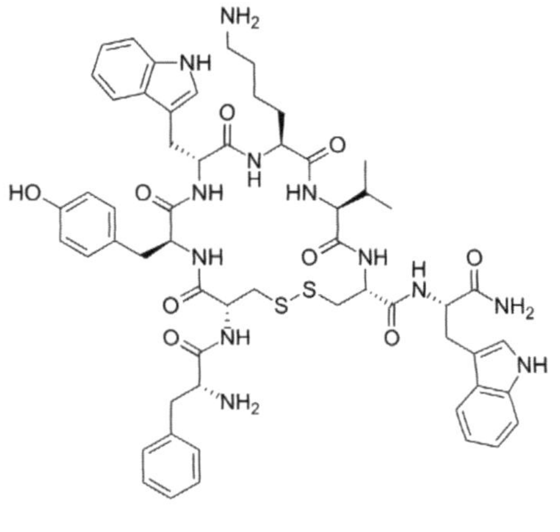

Figure 1. Molecular Structure of VAP.

SST is a tetradecapeptide which acts as a physiological inhibitor of growth hormone secretion. It also inhibits the secretion of a wide variety of stimulatory gastrointestinal hormones and decreases gastrointestinal motility and blood flow [1]. SST binds to specific receptors (sst) of which five principal subtypes have been identified: sst1, sst2, sst3, sst4 and sst5. SST receptors are well known to be overexpressed in a broad range of tumour cells. However, SST has a short half-life in vivo due to its rapid degradation by various peptidases [2].

In recent years, a variety of SST analogues have been prepared and studied because of their potential in diagnosis and clinical treatment: octreotide, lanreotide, vapreotide, somatuline, etc. [3]. Some of these analogues have longer plasma half-lives than SST and, therefore, their therapeutic effects are extended throughout the time. These peptides can be used in the treatment of some tumours (acromegaly, thyrotropin secretory and non-secretory pituitary adenomas and carcinoid tumours). VAP, specifically, is a widely studied SST analogue with anti-neoplastic properties. It has a higher binding affinity to SST receptor subtype 2 (sst2) than native SST. Additionally, it is up to 500 times more potent than octreotide and somatuline in inhibiting the synthesis/release of hormones from SST-expressing pituitary and malignant neuroendocrine cells [2].

The main motivation to use VAP arises from its biological function as synthetic SST analogue which specifically targets SST receptors. Despite the biological importance of the aforementioned peptide analogue, limited information regarding its structure or the manner in which it interacts with its receptors can be found in the literature. An improved understanding of the adsorption mode of this peptide on a specially prepared gold colloidal surface will enable future predictions about the interaction between a peptide (with a known amino acid sequence and structure) and a given gold surface.

2. Experimental

Gold nanoparticles were prepared by chemical reduction of chloroauric acid with sodium citrate as reducing agent [4]. In brief, 50μL of $HAuCl_4{\cdot}3H_2O$ (2wt%) was heated to boiling and 500μL of $Na_3CH_5O_7{\cdot}2H_2O$ (1wt%) solution were added under stirring. After addition of sodium citrate solution, stirring continued until solution turned brilliant red. Gold colloid was aged 24 hours prior to use. This solution was stored at 4°C for further use.

Raman and SERS measurements were registered with a LabRam HR high resolution Raman spectrometer (Horiba Jobin-Yvon) equipped with a microscope and a cooled (-70°C) CCD camera and using a 600grooves/nm holographic grating. The signal was calibrated by using the 520cm^{-1} line of a Si wafer and a 50X objective. As an excitation source, a 785nm irradiation from a diode laser with a power of 25mW was used. Spectral scanning conditions were chosen to avoid sample degradation. No spectral changes that could be associated with peptide decomposition were observed during these measurements.

3. Results and discussion

Figure 2 presents the Raman and SERS spectra of VAP in the range of 400-1800cm^{-1}. The spectral analyses were carried out on the basis of early Raman and SERS band assignments of the amino acid residues in the peptide and some VAP analogues or shorter peptides with similar sequences [5-12].

It is evident that the Raman signals are mostly associated with the characteristic vibrational modes of Phe, Trp and Tyr residues present in the peptide amino acid sequence. This was not surprising in light of previous studies that have shown that modes due to aromatic amino acid vibrations dominate the Raman spectra [5].

3.1. Raman spectra of VAP

The most intense band at 1002cm^{-1} corresponds to the aromatic ring breathing mode in Phe. At the higher frequency side of this band, around 1030, 1181, 1204 and 1604 cm^{-1}, several characteristic Phe bands appear. These bands are due to in-plane CH bending of the aromatic ring, the combination of C-H in-plane bending mode with a stretching vibration of the phenyl ring [6], the stretching of the aromatic carbons in Phe and the ring stretching mode of benzene ring in Phe [7], respectively. On the other hand, in the lower frequency region of these spectra, only one Phe mode appears at 620cm^{-1}, and it is due to the in-plane ring deformation vibrations.

For VAP, the Phe modes are accompanied by the Raman markers of Trp and Tyr residues. Specifically, these bands are observed at around 757, 877, 1009, 1338, 1359, 1549 and 1577cm^{-1} for Trp and 641, 827, 851, 1181, 1204, 1604 and 1617cm^{-1} for Tyr. These bands are related to the vibrations of the indole ring breathing, indole ring H-scissoring and a N-H deformation, indole ring breathing, Trp Fermi doublet (which arises from a Fermi resonance between a fundamental mode at $134cm^{-1}$ and a combination of two out-of-plane modes involving the benzene and pyrrole rings), indole ring vibration mainly contributed from the C=C stretch [8] and benzene and pyrrole ring skeletal stretching [9], respectively for Trp. Likewise, for Tyr the bands come from aromatic ring deformation, Tyr Fermi doublet (rise to a pair of relatively strong vibrations, assigned to the ring breathing mode and the first overtone of an out-of-plane ring deformation of the phenolic ring, respectively [10,11], C-H in-plane bending, symmetric stretching of the carbons in the para-substituted aromatic ring, ring stretching mode of benzene ring and C=C stretching mode.

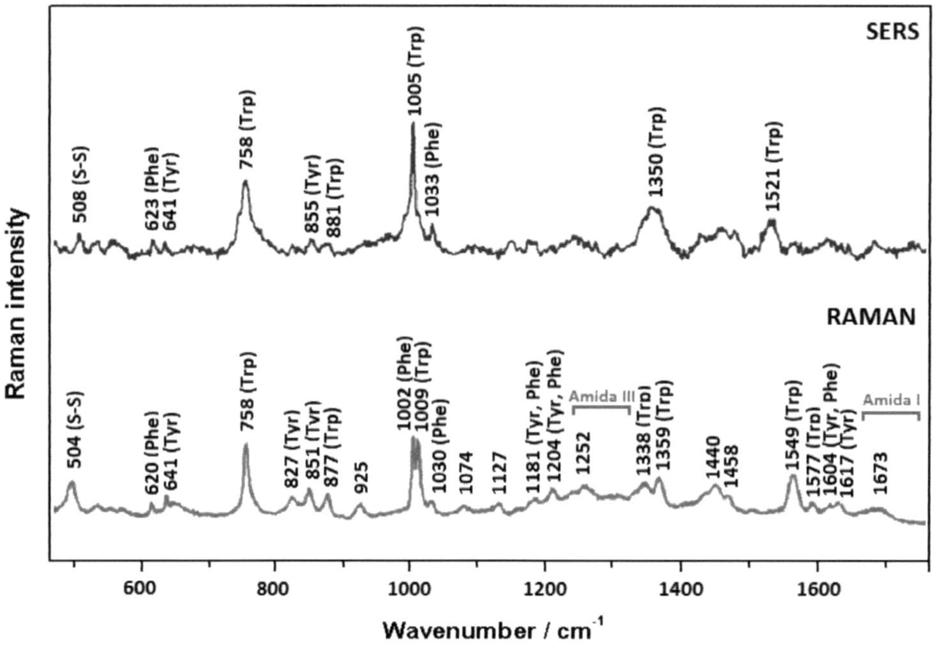

Figure 2. Raman (in red) and SERS (in blue) spectra of VAP. Measurement conditions: sample concentration, 20mM and $7.5x10^{-6}M$, respectively. Excitation wavelength 785nm, power at laser output, 25mW.

There are three other noteworthy spectral characteristics. The first one is in the region from 500 to $600cm^{-1}$, where a band at $504cm^{-1}$ is observed, assigned to the S-S stretching vibrations. The second one belongs to the vibrational modes corresponding the set of signals Amide III observed between $1310\text{-}1210cm^{-1}$, that mostly come from N-H deformation modes due to the amide backbone of the peptide and contain information about the secondary structure of the biomolecules. And, finally, a band at $1673cm^{-1}$ in the region of the so-called amide I, corresponding to complex vibrations with contributions COO- asymmetric stretching modes (80%), CN (10%) NH stretching and bending (10%) [12].

3.2. SERS spectra of VAP

In order to examine the interaction of the gold nanoparticles with the peptide, an analysis of the changes in the intensity, width and shift in the bands between corresponding Raman and SERS spectra was carried out. Conclusions about the adsorption process of VAP on gold surface are drawn below.

The SERS spectrum of VAP is shown in Figure 2(in blue) compared to that of an aqueous solution (Figure 2(in red)). With no doubt, the medium-strong intensity bands due to the indole ring system of

Trp are observed at around 758, 881, 1005, 1338-1359 and 1521cm^{-1} in the SERS spectra of VAP. Particularly important are the Raman bands at 1009 and 1549cm^{-1} due to the out-of-phase benzene and pyrrole ring breathing and the pyrrole ring stretching vibration ν(C=C), respectively. These two bands appear at 1005 and 1521cm^{-1} in the VAP SERS spectrum (Figure 2, in blue). The first-mentioned spectral feature (at 1005cm^{-1}) increases in intensity and slightly shifts (Δ=-4cm^{-1}) from its position in the Raman spectrum, while the second one (at 1521cm^{-1}) shifts by 28cm^{-1} to lower wavenumbers. Additionally, the band width of these bands increases. Likewise, bands at 1138, 1359cm^{-1} are broader in the SERS spectrum and look like one single band. Hence, this could indicate that the Trp strongly binds to the gold nanoparticles.

On the other hand, the absence or very weak intensity of bands due to the Tyr or Phe vibrations in the SERS spectrum suggests that these side chain groups either lie parallel to the gold nanoparticle or are sufficiently far from the surface. Finally, among a variety of interactions in peptides and proteins, disulfide bridges are very important for the protein folding. The Raman and SERS spectra yield excellent information about disulfide bond conformations, based on the frequency of the S-S stretching vibration. The appearance at 508cm^{-1} a band assigned to the S-S mode suggests that the disulfide bond interacts with the metal surface without bridge cleavage.

4. Conclusions

The peptide vapreotide, a somatostatin analogue, has been vibrationally studied by Surface Enhanced Raman Scattering. Optimum conditions for the detection of a strong Raman signal have been found. SERS spectrum of VAP adsorbed on colloidal gold surface differs considerably from its Raman spectrum and exhibits features that can be used to identify the specific residues in the adsorbed peptide. The VAP SERS spectrum is dominated by signals coming from the tryptophan amino acid. Changes in the SERS spectra of VAP adsorbed on the gold surface showed that adsorption appeared to be stablished via the Trp side chain.

Acknowledgements

This work has been supported by the Vicerrectoria de Investigación y Extensión (VIE) of Universidad Industrial de Santander, Project No. 5197. Authors would like to acknowledge to Laboratorio de Espectroscopia (LE) for the generous collaboration. J.G. Thanks to Departamento Administrativo de Ciencia, Tecnologia e Innovación, COLCIENCIAS for a doctoral scholarship (Francisco Jose de Caldas grants, No. 511, 2010).

References

[1] A Rothen-Weinhold, K Besseghir, Y De Zelicourt and R Gurny 1998 *J Control Release* **52** 205
[2] K J O'Byrne, N Dobbs, D J Propper, J P Braybrooke, M I Koukourakis, K Mitchell, J Woodhull, D C Talbot, A V Schally and A L Harris 1979 *Br J Cancer* **79-9–10** 1413
[3] S Banerjee, W J Wiggins, J L Geoghegan, C T Anthony, E A Woltering and D S Masterson 2013 *Org Biomol Chem* **11-37** 6307
[4] G Frens 1973 *Nat Phys Sci* **241-105** 20
[5] D Sobolewski, E Proniewicz, D Skołuba, A Prahl, Y Ozaki, Y Kim and L M Proniewicz 2013 *J Raman Spectrosc* **44-2** 212
[6] E Podstawka, Y Ozaki and L M Proniewicz 2004 *Appl Spectrosc* **58-5** 570
[7] E Podstawka, E Sikorska, L M Proniewicz and Bernard Lammek 2006 *Biopolymers* **83-3** 193
[8] H Takeuchi 2003 *Biopolym Biospectroscopy Sect* **72-5** 305
[9] A Combs, K McCann, D Autrey, J Laane, S A Overman and G J Thomas 2005 *J Mol Struct* **735–736** 271
[10] G Zhu, X Zhu, Q Fan and X Wan 2011 *Spectrochim Acta Part A Mol Biomol Spectrosc* **78-3** 1187
[11] J Laane 2009 *Frontiers of molecular spectroscopy* (Amsterdam: Elsevier) p 565
[12] E Podstawka, Y Ozaki and L M Pronicwicz 2008 *Langmuir* **24-19** 10807

Journal of Physics: Conference Series **687** (2016) 012113 doi:10.1088/1742-6596/687/1/012113

Physico-chemical characterization of products from vacuum oil under delayed coking process by infrared spectroscopy and chemometrics methods

L V Meléndez[1], R Cabanzo[1], E Mejía-Ospino[1] and A Guzmán[2]
[1] Universidad Industrial de Santander (UIS), Bucaramanga, Colombia.
[2] Instituto Colombiano del Petróleo (ICP), Ecopetrol S.A., Piedecuesta, Santander, Colombia.

E-mail: emejia@uis.edu.co, vivilesme@hotmail.com

Abstract. Eight vacuum residues and their delayed coking liquids products from Colombian crude were study by infrared spectroscopy with attenuated total reflectance (FTIR-ATR) and principal component analysis (PCA). For the samples the structural parameters of aromaticity factor (fa), alifaticity (A2500-3100cm^{-1}), aromatic condensation degree (GCA), length of aliphatic chains (LCA) and aliphatic chain length associated with aromatic (LACAR) were determined through the development of a methodology, which includes the previous processing of spectroscopy data, identifying the regions in the IR spectra of greatest variance using PCA and molecules patterns. The parameters were compared with the results obtained from proton magnetic resonance (^{1}H-NMR) and ^{13}C-NMR. The results showed the influence and correlation of structural parameters with some physicochemical properties such as API gravity, weight percent sulphur (% S) and Conradson carbon content (% CCR).

1. Introduction

The delayed coking is a chemical conversion process for exposure of samples at high temperatures. This is normally applied to heavy crudes and vacuum residues (VR) for light products such as gasoline and gas oils, also is obtained as residue carbon materials concentrated called coke with impurities presents in the initial charge as sulphur, nitrogen, vanadium, among others [1]. Vacuum residues (VR) are a heavy fraction of the process of vacuum distillation of oil, which until recently, was consider as a by-product of low commercial value because it concentrates the most complex molecules and many of the impurities present in the crude. these impurities include up to 6% w of sulphur, 0.1-2% nitrogen, 0.005-1.5% oxygen, forming functional groups such as carboxylic acids, aldehydes, ketones, thiophenes, pyrrole rings nitrogen bases such as pyridine, trace metals such as nickel, copper and vanadium, present as oxides, salts or also occupying the centre of organometallic porphyrin type structures [2]. Nevertheless, because oil is a nonrenewable energy resource and demand is growing, has generated the need to modify and implement new refining processes and characterization of this material.

Therefore many qualitative and quantitative researches approach have been conducted in order to characterize VR, in which; the nuclear magnetic resonance spectroscopy (NMR) of 1H and 13C is a technique that has been widely used since 1958 [3]. This technique can provide information at the molecular level, as a percentage of aromatic carbons, n-paraffinic naphthenic, average length of the side alkyl chains, the degree of substitutions and condensations, in addition to providing the percentage of compounds aromatic, aliphatic and naphthenic, etc. [4]. However, in recent studies have found that through spectroscopic data from infrared spectroscopy and chemometrics methods as principal components analysis (PCA) can calculated structural parameters related to physicochemical properties of samples. The aim of this paper is to demonstrate the capability of FTIR-ATR for use as a complementary technique or as an alternative method to the characterization of complex samples.

2. Experimental description

The properties (API gravity, percentage S and percentage CCR) of the VR and liquid products from the delayed coking process (10-20 psi and 490-510°C) were obtained according to standard tests [5-7]. FTIR-ATR analysis was carried out in a Nicolet is 50 equipped with an ATR diamond cell with a simple reflexion. The angle of incidence of the IR beam was 45°. Each spectrum was acquired with 32 scans, $4cm^{-1}$ resolution and wave number from 4000 to $400cm^{-1}$.

^1H and ^{13}C NMR spectra were recorded on a Bruker Avance III NMR spectrometer equipped with a 5mm and 10mm probe resonating at the frequency of 400.16 and 100.62 MHz, for ^1H and ^{13}C respectively. The conventional ^1H spectra were carried out using 4% w/v sample solutions in $CDCl_3$ (99.8% Merck) with zg30 pulse sequence, eight number of scans, 10s relaxation delay. The ^{13}C NMR spectra of the sample were recorded using 20% (w/v) in CDC13 solutions, with Cr $(acac)_3$ 0.05M as paramagnetic relaxant. Total 1024 numbers of scans were collected using a 20s relaxation delay. All the ^{13}C spectra were obtained with zgig30 bruker pulse sequence. Before analysis, the spectra obtained were corrected for phase and baseline and then each of them were separated into different regions which correspond to different types of protons and carbons according to their position in the molecule. Later, each spectrum was integrated thrice and averaged within the indicated regions.

The patterns molecules to find the most important signals in the infrared spectra are shower in Figure 1. This was selected according with the typical molecules structures present in VR [8].

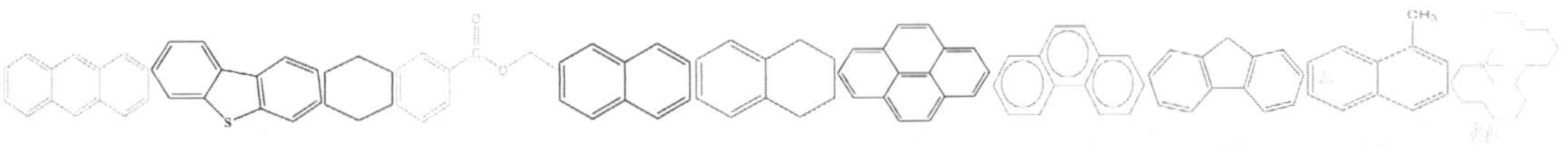

(B) (K) (C) (D) (J) (H) (I) (E) (F) (G) (A)

Figure 1. Patterns molecules. (B) Anthracene, (K) dibenzothiophene, (C) cyclohexane, (D) benzoate ethyl, (J) naphthalene, (H) tetrahydronaphthalene, (I) pyrene, (E) phenanthrene, (F) fluorine, (G) methyl naphthalene and (A) hexadecylamine.

3. Results and discussion

3.1. Data processing

The processing and chemometric analysis were performed with the software Unscrambler 10.3. The best methods were determined calculating the relative standard deviation (RSD) for signals representative signals of the spectra (see Figure 2).

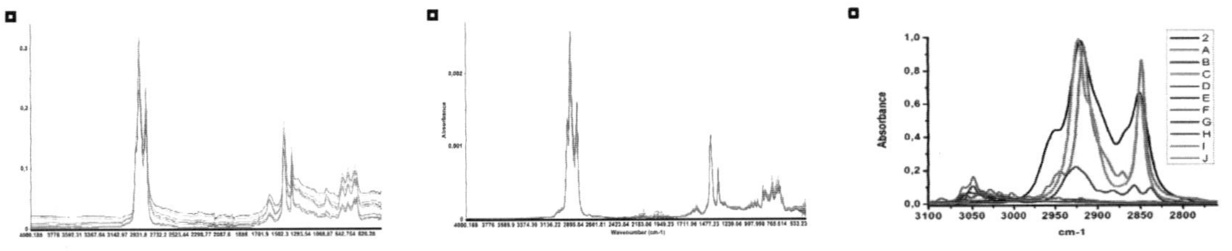

(a) (b) (c)

Figure 2. (a) Original spectra. (b) Area normalization spectra. (c) Identification of signals with patterns molecules (aliphatic chains and mono and polyaromatics rings)

3.2. FTIR-ATR spectrum analysis

In general for crudes and fractions, the most intense bands in the spectra appear near 3000cm^{-1}: C–H stretch in CH_2 at 2918cm^{-1} and 2853cm^{-1}. There are also intense bands at 1495cm^{-1} and 1377cm^{-1}, corresponding to twisting scissors (symmetric and asymmetric) of the CH_2 group. Other bands are found at low frequencies although of more complex assignment such as the deformation bands of aromatic C–H in the plane and outside of it.

3.3. Principal component analysis

Principal component analysis (PCA) is a well-known chemometric technique in multivariate data analysis for size reduction, displaying most of the original variability of data in a small space. PCA is commonly used for making a new dimension of data set by reducing the total number of variables into a small number, which are linear combinations of the original variables, orthogonal and contain the maximum variance within them [9]. As shown in Figure 3(a) the graph of scores allows correlation between neighbouring patches in the PC1-PC2 plane, in the Figure 3(b) loadings plot shows the X-variables in the PC1 that best describe each group, for distillates around 720cm^{-1} ((CH2)n, n>4), finally the explained variance chart shows that the first three PCs contain 100% of the variance of the data (see Figure 3(c)).

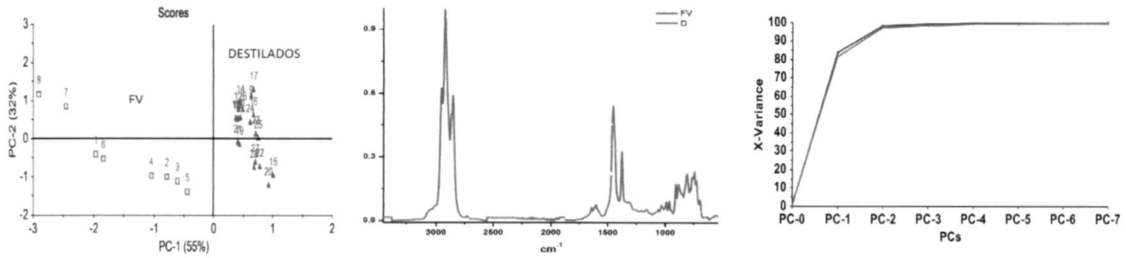

(a) (b) (c)

Figure 3. PCA plots (a) Scores. (b) Loadings in spectra. (c) Explained variance.

3.4. Structural parameters

The variables used to determine the structural parameters were found from the PCA analysis of the IR data and the signals identified with patterns molecules for deconvolution; taking into account a previous work done in our research group [8].

$$LCA = \frac{\Sigma A_{2913-2929} + \Sigma A_{2850-2857}}{\Sigma A_{2860-2873} + \Sigma A_{2943-2960}} \tag{1}$$

$$GCA = \frac{A_{835-899cm^{-1}}}{A_{733-775cm^{-1}}} \tag{2}$$

$$fa = \frac{A_{900-700cm^{-1}}}{A_{3000-2800cm^{-1}}} \tag{3}$$

$$LACAR = \frac{A_{704-730cm^{-1}}}{A_{732-920}cm^{-1}} \tag{4}$$

$$Alifaticity = A_{2800-3000cm^{-1}} \tag{5}$$

For all samples were calculated the found structural parameters (see Table 1). The results show that the structural parameters (1), (2) and (4) are higher for VRs than products. This is because the delayed coking process produces the bond breaking and therefore a shortening of aliphatic chains occurs, and also the accumulation of the polycondensates rings in the residual coke decreases GCA in distillates products. For the same reasons the alifaticity is higher for products (light content is greater). LCA, *fa* and LACAR parameters are commonly calculated by NMR, so the comparison of these was realized by IR and NMR spectroscopy. In the Figure 4 the black line represents the parameters calculated by NMR spectroscopy and the red bars the parameters calculated by FTIR-ATR. Since the data obtained by both techniques are different, trend of parameters is evaluated, not its numerical value. In general the trend is the same for both techniques.

Table 1. Structural parameters of VRs and products.

	Vacuum Residues					Products				
	LCA	LACAR	GCA	Alifaticity	*fa*	LCA	LACAR	GCA	Alifaticity	*fa*
Minimum	1,994	0,123	0,588	77,241	0,184	1,668	0,109	0,474	74,808	0,214
Máximum	30,729	0,227	12,246	853,487	0,319	21,177	0,190	0,734	818,553	0,407
Average	23,265	0,183	2,216	701,386	0,238	16,058	0,148	0,559	755,222	0,269

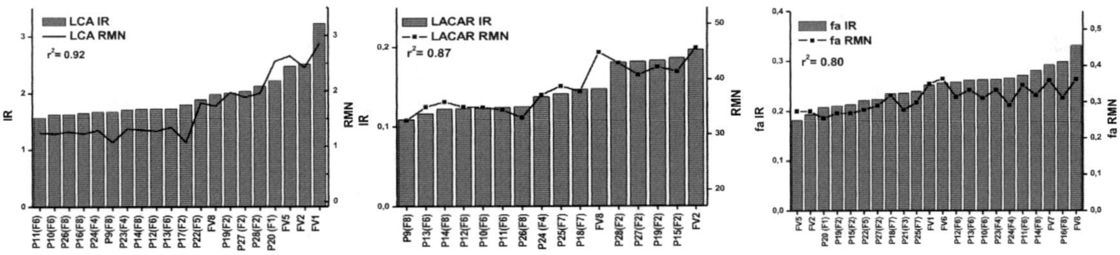

Figure 4. Structural parameters calculated by NMR and IR-ATR data.

3.5. Correlation between structural parameters and properties

To correlate the physicochemical properties of the samples with the calculated parameters, a multiple lineal regression (MLR) was performed (see Table 2).

The F ratio, MS_{model}/MS_{error}, was positive and greater than 66.6 for all models indicating that much of the total variance is being described by the fitted model. This is confirmed by the R^2 value which was close to 1.

Table 2. Results of MLR models.

Properties	F ratio	Residual Average	R^2
PI	139,8	± 1,3	0.99
%w S	156,5	± 0,2	0.98
%w CC	66,6	± 0,6	0.98

4. Conclusions

This work demonstrates through determining structural parameters is possible to identify and correlate the molecular changes of VRs when they are submitted to the Delayed Coking process with physicochemical properties.

Acknowledgments

This work was financially supported by COLCIENCIAS under the project No FP44842-039-2015. Also, the authors thank to the Universidad Industrial de Santander (UIS) and the Instituto Colombiano de Petroleo (ICP) for the support through technological cooperation agreement ICP-UIS 003-2007.

Journal of Physics: Conference Series **687** (2016) 012113 doi:10.1088/1742-6596/687/1/012113

References

[1] Wang B, et al 2012 *Research on Chemical Intermediate* **38** 2295
[2] Boduszynski M 2002 *Petroleum Chemistry* **47** 329
[3] Rodriguez J, Tierney J W & Wender I 1994 *Spectroscopy Fuel* **73** 1863
[4] Altgelt K & Boduszynski M 1992 *Energy and Fuels* **6** 68
[5] ASTM D4294 2010 *Standard test methodfor sulfurin petroleum and petroleum products by energy dispersive X-ray fluorescence spectrometry* (West Conshohocken: ASTM international)
[6] ASTM D4530 2006 *Standard test method for determination of carbon residue "micro method"* (West Conshohocken: ASTM international)
[7] ASTM D4052 2011 *Standard test method for density, relative density and API gravity of liquids by digital density meter* (West Conshohocken: ASTM international)
[8] Orrego J and et al 2012 *Energy and Fuels* **26** 586

Effects of substitutions of C atoms by Al and N in the w-AlN compound

J F Murillo[1], C Ortega[1] and M J Espitia[2]
[1] Universidad de Córdoba, Montería, Colombia.
[2] Universidad Distrital Francisco José de Caldas, Bogotá, Colombia.

E-mail: juanfmg1973@gmail.com

Abstract. We present ab-initio calculations in the pseudopotential approximation for the carbon atoms substitutions by aluminium and nitrogen sites at the AlN compound in wurtzite phase. Structural parameters for the AlN with and without carbon were optimized. Subsequently, the electronic and magnetic properties are determined by the density of states (DOS). Also it was determined that Carbon substitutions (AlN:C) are quite stable. This substitution exhibits magnetic properties, indicating that these compounds are good candidates for possible application in diluted magnetic semiconductors, spin injectors, and other spintronics applications.

1. Introduction

The aluminium nitride (AlN), which crystallizes in the wurtzite structure, is a semiconductor material with a wide band gap of approximately 6.2eV, not only is a suitable material for the design Light-Emitting Diodes (LEDs) and Laser Diodes (LDs) [1], also has been used in electronics and optoelectronics [2]. The AlN present stability to high temperature, a considerable thermal conductivity, low thermal expansion and high resistance to gases and chemicals [3]. Recently, AlN has received extensive attention because of its possible use as a diluted magnetic semiconductor (DMS) with potential application in the field of spintronics. For these applications, ferromagnetism at room temperature is a requirement. In recent years, high-temperature ferromagnetism has been reported by many researchers in several types of transition metal (TM)-doped semiconducting oxides and nitrides [4-6]. However, it was found that the magnetic TM dopants in TM-doped DMS segregate to form ferromagnetic clusters, precipitate, or secondary phases [7-9]. This represents a big obstacle for practical applications of diluted magnetic semiconductors. To avoid this problem, many researchers have focused on investigating the effect of nonmagnetic ion doping in semiconductors in order to obtain high-temperature ferromagnetic semiconductors. Experimental results have proved that is possible to obtain high-temperature ferromagnetism in nonmagnetic anion, for example nonmagnetic C- and N-doped ZnO, TiO_2 and C-doped GaN [10-12]. In this paper, we present first-principles total energy calculations of the structural, electronic, and magnetic properties of $AlC_{0.0625}N_{0.9375}$ and $Al_{0.9375}C_{0.0625}N$ compounds. First, we calculated the structural parameters of pure AlN and the two ternary compounds considered in this study. Finally, we made a detailed of the density of states. We found that the ternary compounds exhibit magnetic properties, and therefore they can be used in dilute magnetic semiconductors and spintronics.

2. Computational method

The calculations were performed within the DFT framework using the Quantum ESPRESSO package [13]. The correlation and exchange effects of the electrons were treated using the generalized gradient approximation (GGA) of Perdew, Burke, and Ernzerhof (PBE) [14]. Electron–ion interactions were treated with the pseudopotential method [15,16]. The electron wave functions were expanded into plane waves with a kinetic-energy cutoff of 40 Ry. For the charge density, a kinetic energy cutoff of 400Ry was used. A 6x6x4 Monkhorst-Pack mesh [17] was used to generate the k-points in the unit cell. The calculations were performed taking into account the spin polarization. To calculate the structural, electronic, and magnetic properties of pure AlN, a 32-atom 2ax2bx2c wurtzite supercell was considered. The concentrations $AlC_{0.0625}N_{0.9375}$ (C by N) and $Al_{0.9375}C_{0.0625}N$ (C by Al) were obtained by substituting one N-atom and one Al-atom, respectively.

3. Results and discussion

3.1. Structural parameters

To determine the structural properties in the ground state, such as the lattice constant (a_0), the bulk modulus (B_0), the c/a ratio, and the total energy (E_0) of $AlC_{0.0625}N_{0.9375}$ and $Al_{0.9375}C_{0.0625}N$ concentrations in the wurtzite structure, the total energy was calculated as a function of the volume, and the results were fitted to the Murnaghan equation of state [18]. Figure 1 shows the energy-volume curves for $AlC_{0.0625}N_{0.9375}$ (C by N) and $Al_{0.9375}$ $C_{0.0625}N$ (C by Al).

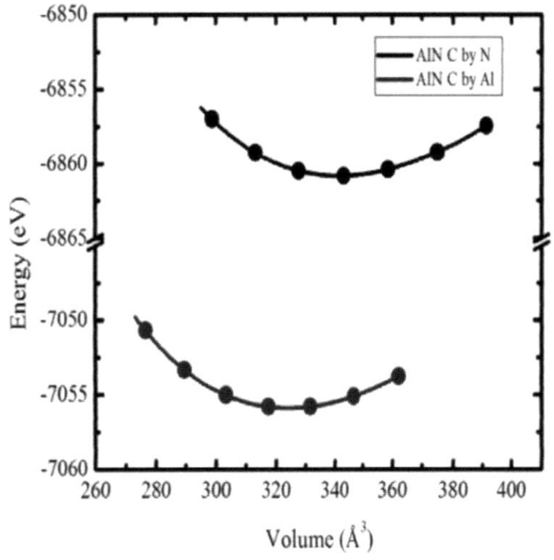

Figure 1. Total energy as a function of volume for AlN:C. Up C by N and bottom C by Al.

The lattice constant, the c/a value, the bulk modulus (B_0), the total energy (E_0), and the magnetic moment (μ_β) per cell are shown in Table 1.

Table 1. Lattice constant, c/a ratio, bulk modulus, total energy, and magnetic moment per cell of pure AlN, $AlC_{0.0625}N_{0.9375,}$ and $_{Al0.9375}C_{0.0625}N$.

Structure	a_0 (Å)	c/a	V_0 (Å^3/atom)	B_0 (GPa)	E_0 (eV/atom)	$\mu(\mu_\beta$/cell)
AlN	3.121	1.602	10.551	192.93	-221.66	0.0
	3.110[a]	1.601[a]	10.432[c]	192[b]	-	
$AlC_{0.0625}N_{0.9375}$	3.129	1.609	10.679	184.71	-214.40	~ 2.0
$Al_{0.9375}$ $C_{0.125}N$	3.084	1.594	10.135	182.88	-220.49	~ 0.4

[a] Experimental Reference [19], [b] Theoretical Reference [20], [c] Experimental Reference [21].

Figure 1 show that $Al_{0.9375}C_{0.0625}N$ compound is energetically most favourable than the $AlC_{0.0625}N_{0.9375}$ compound; additionally, the lattice constant *a* and c/a ratio calculated for the pure AlN accords well with values reported experimentally [19], since it differs by less than one percent. The values of the bulk modules of the pure AlN, $AlC_{0.0625}N_{0.9375}$ and $Al_{0.9375}C_{0.0625}N$ concentrations are higher, which confirms that they are quite rigid, making them good candidates for possible application in devices operated at high temperature and high power, as well as in hard coatings.

As we can see in the Table 1, when one N atom in the 2ax2bx2c supercell is substituted with a C atom, the lattice constant in the $AlC_{0.0625}N_{0.9375}$ compound changes only slightly with respect to pure AlN. This small change in the parameters may be because the radius of the C atom (0.914Å) is very close to the atomic radius of N (0.92Å). Moreover, when one Al-atom is substituted by a C atom, the lattice constant in the $Al_{0.9375}C_{0.0625}N$ compound decreases because the radius of the C atom (0.914Å) is most smaller than the atomic radius of Al (1.43Å). The spin polarization calculations for the $AlC_{0.0625}N_{0.9375}$ and $Al_{0.9375}C_{0.0625}N$ compounds result in a magnetic moment of 2.0 and $0.4\mu_\beta$ per supercell, respectively. Similar magnetic behaviour has been found in C-doped GaN [12] and ZnO [22].

3.2. Electronic properties

Figure 2 shows the total and partial density of states for pure AlN and $AlC_{0.0625}N_{0.9375}$ (C by N) compound.

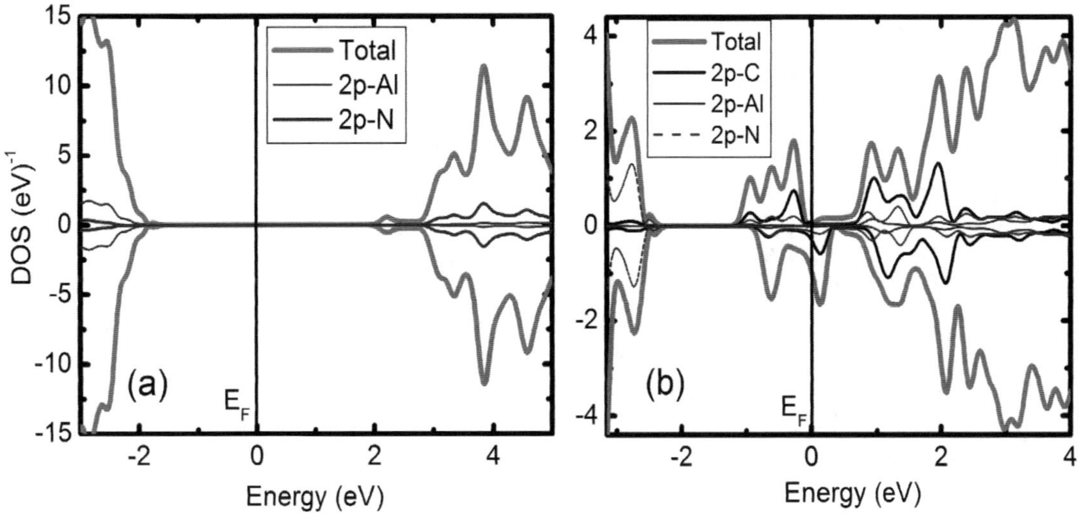

Figure 2. Total and partial density of states of (a) AlN, (b) $AlC_{0.0625}N_{0.9375}$

Figure 2(a) show the density of state pure AlN. This confirms the semiconductor behaviour. Figure 2(b) show the density state of the ternary $AlC_{0.0625}N_{0.9375}$ compound. We can observe that the majority spin (up) states preserve a band gap, but in the minority spin (down) states there is a penetration towards the prohibited energy zone of the state C-p in greater proportion and states 2p-N and 2p-Al in lesser proportion. Therefore, due to the introduction of a C atom into the structure of AlN, it loses its semiconductor nature. The allowed ternary compound exhibit half-metallic behaviour, determined by such states. We can see in Figure 2(b) that in the majority spin, the partial density of states belonging to the C-2p orbital exhibits a high peak near the Fermi level, about -0.4eV for $AlC_{0.0625}N_{0.9375}$ indicating localized states in that region. In the valence band around the Fermi level, the C-2p states overlap with another two peaks of smaller amplitude belonging N-2p and Al-2p. Therefore, there is strong hybridization and polarization between the C-2p, N-2p, and Al-2p orbitals. Additionally, the TDOS confirms the presence of some unoccupied bands above the Fermi level, because there is no

Journal of Physics: Conference Series **687** (2016) 012114 doi:10.1088/1742-6596/687/1/012114

contribution of spin up and the minority spin exhibits a strong hybridization and polarization between the C-2p, N-2p, and Al-2p states, resulting in a magnetic moment of $2.0\mu_\beta$ the main contribution to the magnetic moment comes from the carbon atom. A similar result was obtained by Lin Yu et al. [12] in their first-principles study of the effects of carbon on the ferromagnetism in doped GaN.

4. Conclusions

First-principles total energy calculations to determine the structural, electronic, and magnetic properties of the $AlC_{0.0625}N_{0.9375}$ and $Al_{0.9375}C_{0.0625}N$ compounds were carried out. The calculated values of the bulk modules were quite high; therefore, the ternary compounds are quite rigid, which makes them attractive for potential applications at high temperatures and for hard coatings. Also, we found that the compounds exhibit magnetic properties with magnetic moments $2\mu_\beta$ and $0.4\mu_\beta$ per supercell, respectively. These properties show that these compounds are good candidates for possible application in diluted magnetic semiconductors, spin injectors, and other spintronics applications.

References

[1] Taniyasu Y and Kasu M 2008 *Diam Relat Mater* **17** 1273
[2] Venkataraj S, Severin D, Drese R, Koerfer F and Wutting M 2006 *Thin Solid Films* **18** 235
[3] Beheshtian J, Baei M T, Bagheri Z and Peyghan A A 2012 *Microelectr J* **43** 452
[4] Gonzalez N, Majewski J A and Dietl T 2011 *Phys Rev B* **83** 184417
[5] Vargas-Hernández C, Espitia-Rico M, Bez Cruz R 2015 *Computational Condensed Matter* **4** 1
[6] Frazier R M, Thaler G T, Leifer J Y, Hite J K, Gila B P, et al 2005 *Appl Phys Lett* **86** 052101
[7] Park J H, Kim M G, Jang H M, Ryu S, Kim Y M 2004 *Appl Phys Lett* **84** 1338
[8] Kaspar T C, Droubay T, Heald S M, Engelhard M H, Nachimuthu P, et al 2008 *Phys Rev B* **77** 201303
[9] Zhou S, Potzger K, Von-Borany J, Grotzschel R, Skorupa W, Helm M, et al 2008 *Phys Rev B* **77** 035209
[10] Herng T S, Lau S P, Wei C S, Wang L, Zhao B C, Tanemura M, et al 2009 *Appl Phys Lett* **95** 133103
[11] Cruz M M, Da-Silva R C, Franco N, Godinho M 2009 *Condens Matter* **21** 206002
[12] Yu L, Wang Z Y, Guo M, Liu D, Dai Y, Huang B 2010 *Chem Phys Lett* **487** 251
[13] Giannozzi P, Baroni S, Bonin N 2009 *J Phys Condens Matter* **21** 395502
[14] Perdew J, Burke K, Ernzerhof M 1996 *Physical Review Letter* **77** 3865
[15] Vanderbilt D 1990 *Phys Rev B* **41** 7892
[16] Laasonen K, Pasquarello A, Car R, Lee C, Vanderbilt D 1990 *Phys Rev B* **47** 10142
[17] Monkhorst H, Pack J 1990 *Phys Rev B* **13** 5188
[18] Murnaghan F D 1944 *Proceedings of the National Academy Science USA* **30** 244
[19] Schulz H and Thieman K H 1977 *Solid State Commun* **23** 815
[20] Peng F, Chen D, Fu H and Cheng X 2008 *Physica B* **403** 4259
[21] Strite S, Morkoc H and Vacuum J 1992 *Sci Technol B* **10** 1237
[22] Peng H W, Xiang H J, Wei S H, Li SS, Xia J B, Li J B 2009 *Phys Rev Lett* **102** 017201